GENETICS

GENETICS
Second Edition

URSULA GOODENOUGH

Harvard University

1978   SAUNDERS COLLEGE

Philadelphia

Saunders College
West Washington Square
Philadelphia, PA 19105

cover photo: Histone-depleted metaphase chromosome from a human HeLa cell. A scaffold or core, having the shape characteristic of a metaphase chromosome, is surrounded by a halo of DNA. The halo consists of many loops of DNA, each anchored in the scaffold at its base; most of the DNA exists in loops at least 10–30 μm long. (From J. R. Paulson and U. K. Laemmli, Cell **12:**817–828, 1977, copyright © M.I.T.)

Acquiring Editor: Kendall Getman
Managing Editor: Jean Samson
Senior Project Editor: Peggy Middendorf
Production Manager: Robert Ballinger
Design Supervisor: Renée Davis
Text Design: Nancy Axelrod   **Cover Design:** Sheila Granda

Library of Congress Cataloging in Publication Data

Goodenough, Ursula.
  Genetics.

  Includes bibliographies.
  1. Genetics.   I. Title.
QH430.G66   1978   575.1   77-26245
ISBN 0-03-019716-3

Printed in the United States of America
  0 1   071   9 8 7 6 5 4

PREFACE

Children are like pancakes, they say: you should throw away the first one. First children have their redeeming features, but the aphorism certainly applies to textbooks. The combination of rapid changes in the field of genetics and numerous suggestions from users of the first edition of this text has resulted in my writing an almost completely new book. Retained is a molecular/chromosomal approach to the subject and the use of human examples whenever feasible; retained also is an extensive use of illustrative materials. Major changes include the following:

1. Mitosis, meiosis, and Mendelism now follow the introductory chapter.
2. Topics such as recombination mechanisms, histocompatability genetics, and somatic cell genetics have been expanded considerably, and such new topics as environmental mutagenesis, recombinant DNA, and restriction mapping are presented in detail.
3. Strictly biochemical topics are found in Chapter 7, "Molecular Biology of Chromosomes," and the focus on genetics is sharpened throughout the text.
4. The two population genetics chapters, while still only an introduction to a very large field, have been expanded.
5. Chapter sections are numbered and extensively cross-referenced. Many new problems have been added, and "worked out" problems have been included for such topics as three-factor crosses and complementation tests.

There appear to be two extremes in textbook writing: some texts present little more than would be covered in a semester's worth of lectures, while others border on the encyclopedic. I have attempted to chart a middle course. Many more topics are included than would ordinarily be covered in a one-semester genetics course, allowing both instructors and students to pick and choose. Moreover, the topics are generally considered in more depth than would be possible in 30–40 hours of lecture. At the same time, I have tried to select my examples carefully and to focus on their important features. My goal has been to document and "flesh out" key genetic concepts in a clear and readable fashion; the reader who then desires additional facts on a particular subject can consult the detailed reference list at the end of each chapter.

Individual revised chapters were reviewed and carefully criticized by Drs. John Drake (National Institute of Environmental Health Sciences, Triangle Park, North Carolina), Nancy Martin (University of Minnesota), Howard Schneiderman (University of California, Irvine), and Peter Cherbas, David Dressler, Sarah Elgin, Argiris Efstratiadis, and Jan Pero (Harvard University); the entire book was reviewed by Julian Adams (University of

Michigan), Richard Siegel (UCLA), Edward Simon (Purdue University), and S. R. Snow (University of California, Davis). Extensive appraisals of the first edition were also received from many sources, particularly helpful being those of Sally Allen (University of Michigan), John Brumbaugh (University of Nebraska, Lincoln), A. Gib De Busk (Florida State University), Søren Nørby (University of Copenhagen), and Michael Freeling, James Fristrom, Leonard Kelly, and Philip Spieth (University of California, Berkeley). Errors that persist are, of course, of my own making. The skilled combination of patience and persistence practiced by Ken Getman and Peggy Middendorf at Holt, Rinehart and Winston cannot be overpraised, nor can I adequately express my gratitude to members of my laboratory, and to Jason and Mathea, for bearing with all too many months of "the book." The retirement from textbook writing of Paul Levine, my former co-author, did not allow him much escape from its pervasive tedium, since he has remained my husband, scientific colleague, and best friend.

A professional story writer remarked in a recent interview that yes, writing did give him pleasure, perhaps two minutes per day. Having just emerged from six months of galleys and page proof, the pleasures of researching and writing this text seem remote. When, however, it was a matter of reading, thinking, and writing about the structure and transmission of genes, there were many very pleasurable hours, for which I thank many hundreds of inspired geneticists.

March 1978   Ursula Goodenough

PREFACE TO THE FIRST EDITION

This book is intended to accompany a one-semester course in genetics for college or medical students who have a knowledge of introductory biology and chemistry. A certain amount of biochemistry is included in the text, but biochemical concepts and techniques are explained at the time they are introduced; no prior knowledge is assumed.

As anyone who has organized a course in genetics is aware, the field has come to play a central role in virtually all biological disciplines. To give full treatment to the many and diverse applications of genetics would result in a diffuse and thereby useless text, but to ignore these applications would produce a limited and sterile view of genetics. We have therefore made the necessary compromises. We give major attention and emphasis to the principles and methodology that are unique to genetics, but we also indicate how these genetic principles relate to such active research areas as biochemistry and molecular biology, embryology, and evolutionary biology.

The teaching of genetics has traditionally followed an historical approach: the experiments that have led to our present understanding of genetic systems are presented in their chronological order. This approach has an immediate strength, for the development of scientific ideas is classically illustrated by the history of genetics and the student can often come away with a deep appreciation of how scientific discoveries, combined with careful experimental observation, can generate a complex science within a relatively short period of time. We do not avoid presenting genetic ideas in their historical context when this seems the clearest way to develop a particular concept, but neither do we adhere to a "Mendel first, phages later" formula. It is our experience that students enter a genetics course with a knowledge, however superficial, of DNA, RNA, and protein synthesis, and they are eager to learn how such subjects relate to genes. To delay discussion of these subjects becomes both confusing and somewhat artificial, whereas to master these subjects first and then apply them to the more classical genetic observations becomes fascinating.

We have therefore organized the material according to key topics in genetics. The sequence of topics progresses from more simple to more complex orders of genetic organization. Thus we first focus on the molecular properties of the genetic material: its ability to replicate, to recombine, to mutate, and to dictate RNA and protein synthesis. We then discuss the ordering of genes within chromosomes and the capacity of genes to segregate and assort. Finally, we discuss functional interactions between genes, genetic regulation, and genes in populations. Realizing that a different sequence of chapters may well suit individual instructors, we have tried to write each chapter with enough internal consistency and cross-referencing so that a different ordering of chapters can still be meaningful. We have also made our index as comprehensive as possible. We often found it necessary to deal separately with viruses, bacteria, and nucleate organisms, since each has evolved genetic systems that are often quite distinct. We try, however, to relate these systems to one another whenever possible so that the student can come away with a grasp of the common denominators of genetics rather than segregated information classed, for example, as "molecular genetics" or "animal genetics."

In writing this text we sought to present genetics in a readable fashion, avoiding the telescopic "dictionary style" of textbook writing. We also sought to convey that genetics is a way of thinking as well as a collection of important facts, and we have avoided detailed summaries of procedures and data in the belief that such catalogues offer a poor substitute for the reading of original research papers. It is our hope that the text will provide a sufficiently up-to-date and integrated view of genetics that students can explore the original literature in a meaningful way. To this end we have included an extensive bibliography with each chapter and have made special note of those papers that describe experiments cited in the chapter. Instructors may well choose to assign particular research papers and to focus lectures or class discussions directly on the papers themselves.

Each chapter concludes with questions and problems. For many of these we are indebted to the past and present instructors of Natural Sciences 5 and Biology 14 and 140 at Harvard University. To the best of our knowledge, all of the problems are original with us or with these instructors; if, however, any were in fact culled from published sources, we express our gratitude to their originators.

Several genetics textbooks have included photographs of distinguished geneticists, and we have continued this enjoyable tradition. In making our choice, we focused on those who had not happened to appear in other collections. The choices remained most difficult, however, and numerous persons of high distinction remain for future textbooks to include.

While we are responsible for any errors in this text, a number of persons read all or portions of various drafts of the manuscript and offered invaluable comments and criticisms. These include Drs. Jonathan Beckwith, Peter Carlson, John Drake, David Dressler, Maurice Fox, Martin Gorovsky, Guido Guidotti, Joel Huberman, Roger Milkman, Janice Pero, John Preer, Herbert Riley, Robert Stellwagen, Andrew Travers, Thomas Wegman, J. A. Weir, and Ms. Nancy Hinckley. We are also grateful to the many persons who sent us photographs and micrographs. We are indebted to the late Sir Ronald Fisher, Dr. Frank Yates, F.R.S., Rothamsted, and to Messrs. Oliver & Boyd Ltd., Edinburgh, for permission to reprint Table III from their book, *Statistical Tables for Biological, Agricultural, and Medical Research.*

The preparation of the manuscript and the final publication processes were aided enormously by the following persons: Ms. Susan F. Klinger who presided over endless drafts and bibliographies and almost never lost heart;

Mr. Donald Schumacher and Ms. Lyn Peters of Holt, Rinehart and Winston who guided us through the early stages of the project; the artists of Eric Hieber Associates who transformed our scribblings into intelligent and pleasing drawings; and Ms. Dorothy Crane of Holt, Rinehart and Winston whose expert editing and unfailing sense of humor permitted the final six months of the project to be almost pleasurable.

Ursula Goodenough
Robert Paul Levine

August 1973

CONTENTS

Chapter 20 Population Genetics II: Genetic Polymorphism, Species Formation, and Molecular Evolution 788

GENETICS

CHAPTER
1

DNA (and RNA) as the Genetic Material

INTRODUCTION

Today, when the terms **gene** or **genetics** are mentioned, most biologists immediately think of DNA. DNA, or deoxyribonucleic acid, is well known as the chemical bearer of genetic information; RNA (ribonucleic acid) serves this function in certain viruses.

In the history of genetics as a science DNA became the center of attention only relatively recently. Focus first centered on **heredity,** on the patterns of inheritance of a given trait (blue eyes, red flower color, short tail) from parent to offspring. It was postulated that these inherited traits were somehow dictated by genes and that genes were linearly aligned along the chromosomes of higher animals and plants. "Maps" of gene order on chromosomes were constructed, and many of the details of gene transmission from generation to generation were worked out well before much was known about what a gene is and how it acts.

As the science of genetics developed, increased attention was given to how genes function and more experimental use was made of microorganisms, notably bacteria and bacterial viruses. During this period it was proposed, with good evidence, that the function of most genes is to specify the formation of proteins. When it was eventually established that most genes are borne within molecules of DNA, primary attention was given to the chemical nature of the gene itself.

In beginning our text with DNA and RNA and in developing a molecular picture of genes and gene function at the same time as we establish patterns of heredity we are, in one sense, violating the sequence set by scientific history. In another sense, of course, we are more closely following evolutionary history, since genes almost certainly developed their fundamental properties well before the hereditary patterns exhibited by modern organisms were established.

THE REQUIREMENTS TO BE MET BY GENETIC MATERIAL

Certain requirements must be met by any molecules if they are to qualify as the substances that transmit genetic information. These requirements extend directly from what is known about the continuity of species and the process of evolutionary change.

1. The genetic material must contain biologically useful information that is maintained in a stable form.
2. The genetic information must be reproduced and transmitted faithfully from cell to cell or from generation to generation.
3. The genetic material must be able to express itself so that other biological molecules, and ultimately cells and organisms, will be produced

and maintained. Implicit in this requirement is that some mechanism be available for decoding, or translating, the information contained in the genetic material into its "productive" form. A narrow, but important, distinction is thus made between a molecule that can generate only its own kind and a molecule that can also generate new kinds of molecules. A salt crystal can "seed" a salt solution so that new salt crystals are formed, but this is the extent of its influence over its surroundings.

4. Genetic material must be capable of variation. This requirement is somewhat contradictory to the first requirement which demanded stability of the genetic material. There is, in fact, no a *priori* reason why genetic material should have built-in provisions for change; one could certainly design a hypothetical genetic system in which information would be rigidly conserved from one generation to another. The dominant theme in the history of life is, however, organic evolution, and this demands that genetic material be capable of change, if only infrequently.

Table 1-1 Definitions from Organic Chemistry

	Benzene ring, with = indicating double bonds where two carbon atoms share four electrons between them. The common abbreviated version of a benzene ring is shown at right.
$-CH_3$	Methyl group
$-OH$	Hydroxyl group
$=O$	Keto group
$-\overset{H}{C}=O$	Aldehyde group
$-\overset{\overset{O}{\parallel}}{C}-OH$	Carboxyl group, characteristically acidic ($-COO^-$)
$-NH_2$	Amino group, characteristically basic ($-NH_3^+$)
Covalent bond	A strong bond formed when two atoms share a pair of electrons between them.
Hydrogen bond	A weak attractive force between an electronegative (electron-seeking) atom (usually N or O) and a hydrogen atom covalently linked to a second electronegative atom (usually O—H or N—H).
Hydrolysis	Breaking a large molecule into two or more smaller molecules by adding water.

Two sources of change have been recognized in present-day genetic systems: **mutation** and **recombination.** A mutation changes the nature of the information transmitted from parent to offspring, and thus represents a relatively drastic way of bringing about variation. If the change is deleterious (and it usually is), the offspring may be greatly handicapped and may die soon after conception, or else it may introduce a deleterious gene into the population. Recombination is a more moderate way of producing variation. It occurs during the course of some sort of sexual process, and it involves the precise shuffling of parental genetic information such that new combinations of genes are produced. These are then inherited by the offspring.

With these four requirements in mind we can study the physical and chemical properties of DNA and RNA, putting the molecular facts into a genetic context. Table 1-1 reviews some key definitions from organic chemistry that are relevant to the next few sections of this chapter.

THE STRUCTURE OF DNA AND RNA

1.1   THE POLYNUCLEOTIDE

DNA and RNA are composed of two different classes of nitrogen-containing bases, the purines and the pyrimidines. The two most commonly occurring purines in DNA are **adenine** and **guanine,** and the common pyrimidines are **cytosine** and **thymine.** Their structures are shown in Figure 1-1. Thymine is not found in most species of RNA; instead, one finds the pyrimidine **uracil,** which is also shown in Figure 1-1. Modified forms of these bases (5-methyl cytosine, for example) are occasionally found as well, particularly in certain specialized forms of RNA described in Chapter 8.

The purines and the pyrimidines can be seen in Figure 1-1 to contain several conjugated double bonds. Molecules containing such bonds are potentially able to exist in a number of different chemical forms, for their hydrogen atoms have a certain freedom. In such a molecule a hydrogen atom can, for example, move away from an amino group ($-NH_2$), leaving an imino group ($-NH$) and a net negative charge that is absorbed by the conjugated ring system of the molecule. Such chemical fluctuations are called **tautomeric shifts** and the different molecular structures that result are called **tautomers.** It turns out that under physiological conditions, the purines and pyrimidines exist almost invariably in the forms that have been drawn in Figure 1-1; the other tautomeric forms of these bases rarely occur. In other words, even though the bases possess potentially unstable bonds, they remain chemically stable, in one tautomeric form, most of the time. This stability is, of course, an important genetic attribute.

Purines and pyrimidines can form chemical linkages with pentose (5-

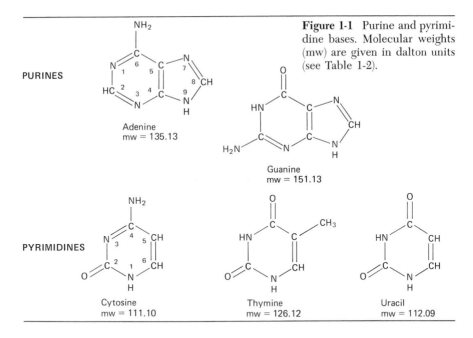

Figure 1-1 Purine and pyrimidine bases. Molecular weights (mw) are given in dalton units (see Table 1-2).

carbon) sugars. The carbon atoms on the sugars are designated 1', 2', 3', 4', and 5', and it is the 1' carbon of the sugar that becomes bonded to the nitrogen atom in position 1 of a pyrimidine or the nitrogen atom in position 9 of a purine. The resulting molecules are known as **nucleosides** (Figure 1-2) and they can serve as elementary precursors for DNA or RNA synthesis. DNA precursors contain the pentose **deoxyribose;** in RNA the sugar is **ribose** and not deoxyribose, the difference being that ribose contains an additional single oxygen atom at position 2' (Figure 1-2).

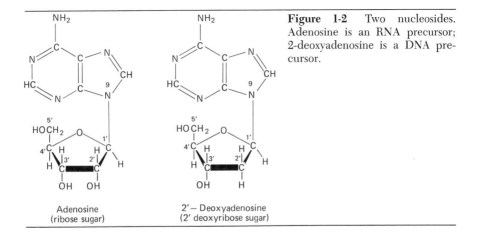

Figure 1-2 Two nucleosides. Adenosine is an RNA precursor; 2-deoxyadenosine is a DNA precursor.

Before a nucleoside can become part of a DNA or RNA molecule it must become complexed with a phosphate group to form a **nucleotide** (Figure 1-3) or, more specifically, a deoxyribonucleotide or ribonucleotide. Nucleotides with a single phosphate group are known as nucleoside monophosphates, an example being adenine ribonucleoside monophosphate or AMP. Nucleotides can also possess two or three phosphate groups (for example, ADP or ATP). It is the nucleoside triphosphates that serve directly as precursors for DNA and RNA synthesis, as we shall see in Chapter 5.

DNA and RNA are simply long polymers of nucleotides, called **polynucleotides.** Only one phosphate group of each precursor nucleotide triphosphate is included in the polymer. This phosphate group, which is bound to the 5′-carbon of the pentose sugar on one nucleotide (Figure 1-3), also becomes chemically bound to the 3′-carbon of the sugar of a second nucleotide, and so on (Figure 1-4), so that a long series of **5′-3′ phosphate linkages** holds the nucleotides together along the length of the polymer. The phosphate bonds are known as covalent ester bonds and are extremely strong. The phosphate residues (PO_4^{--}) along the chain are acidic, leading to the term **nucleic acid,** but are commonly neutralized so that the molecule is more appropriately considered a salt.

Once a sugar-phosphate "backbone" has been formed, the position of the purine and pyrimidine bases in a nucleic acid is quite rigidly fixed. Each base will be stacked on top of the next, like a stack of pennies, with one base 3.4 Å away from its neighbor (see Table 1-2 for a list of physical units). Since the purines and pyrimidines are essentially flat, two-dimensional structures (Figure 1-1), there is no stereochemical interference between the bases in a stack. Indeed, weak bonds form between the stacked bases and add to the stability of the molecule.

The structure of a polynucleotide chain, as drawn in Figure 1-4, can be abbreviated as in Figure 1-5. Adjacent base-sugar complexes are represented

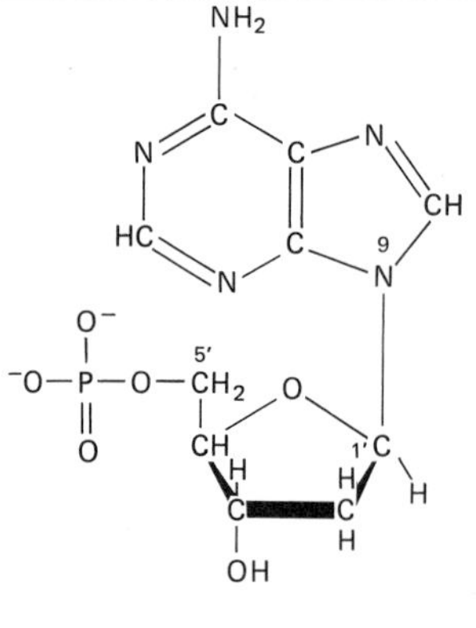

Figure 1-3   A nucleotide, deoxyadenosine 5′-monophosphate, also a DNA precursor.

Deoxyadenosine 5′—monophosphate
(deoxyadenylic acid; dAMP)

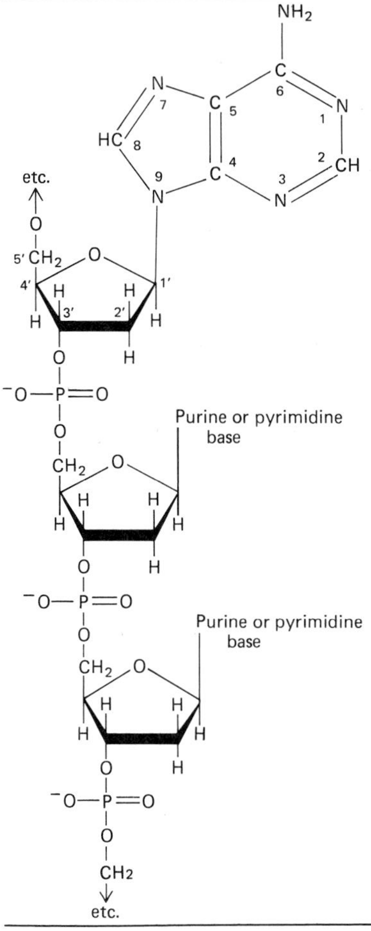

Figure 1-4 A polynucleotide chain, or a single strand of DNA.

Table 1-2 Units of Length and Mass

Length

1 angstrom (Å) = 10^{-8} cm ≃ diameter of a hydrogen nucleus
1 nanometer (nm) = 10 Å
1 micron (μ) = 1000 nm
1 millimeter (mm) = 1000 μ
1 meter (m) = 1000 mm

Mass

1 dalton = mass of 1 hydrogen atom = 3.32×10^{-24} g

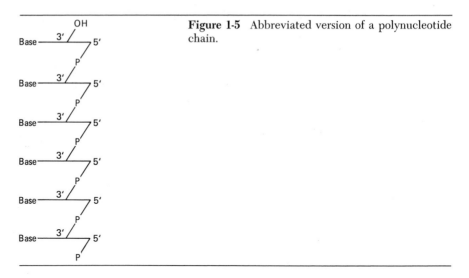

Figure 1-5 Abbreviated version of a polynucleotide chain.

by the parallel lines that are oriented at right angles to the long axis of the molecule. The covalent 5'-3' phosphate linkages are represented by the sloping lines that extend from median (3') positions to terminal (5') positions on adjacent sugar molecules. This schematic rendering of the polynucleotide emphasizes an important fact about these molecules: they are **structurally polarized,** which means that a 3'-hydroxyl (3'-OH) end and a 5'-phosphate (5'-P) end can be identified for any polynucleotide chain. In the illustration (Figure 1-5) the 3'-OH end is at the top, but clearly the same molecule could be redrawn so that the 5'-P end would be at the top.

Single-stranded polynucleotide chains have been adopted as the genetic material in certain viruses. Five kinds of virus are presently known to contain single-stranded DNA chromosomes, and most RNA-containing viruses, including the familiar polio and influenza viruses, possess single-stranded RNA. Single-stranded forms of DNA and RNA should not be confused with the more familiar **double helix** forms of DNA (and RNA) which are described in the next section.

1.2 THE DOUBLE HELIX

In 1953, J. D. Watson and F. H. C. Crick proposed that most DNA is found in a double helix having very specific properties. Their hypothesis was based on a number of important pieces of information that had been established by the work of others.

1. The primary structure of a single polynucleotide chain, diagrammed in Figure 1-4, was known at that time. It was also suspected that most naturally occurring DNAs did not exist as single chains; instead, two or

more chains appeared to interact with one another in some way. The resulting macromolecules were known to be long, thin, and rigid, such that they formed highly viscous solutions in water.

2. E. Chargaff had established that when DNA from any particular species is subjected to chemical hydrolysis so as to release its component purines and pyrimidines, the total amount of adenine released is always equal to the total amount of thymine (A = T), as seen in the third and fourth columns of Table 1-3. Similarly, the total amount of guanine always equals that of cytosine (G = C), as shown in the fifth and sixth columns of Table 1-3. **Chargaff's Rule,** in other words, states that, in natural DNAs, the base ratio A/T is always close to unity and the G/C ratio is always close to unity.

3. R. Franklin, working at the same time as Watson and Crick, was studying the X-ray diffraction patterns produced by isolated DNA fibers. These patterns revealed that the fibers contained highly ordered, helical molecules. Knowing that DNA was constructed from polynucleotide chains, she realized that two or more of these chains must be coiled around one another, in a helical fashion, to produce the secondary structure of a DNA macromolecule.

Watson and Crick, in attempting to visualize such a macromolecule, started with the simplest assumption; namely, that two polynucleotide

Table 1-3 Molar Properties of Bases (as Moles of Nitrogenous Constituents per 100 g-Atoms Phosphate in Hydrolysate) in DNAs from Various Sources

Organism	Tissue	Adenine	Thymine	Guanine	Cytosine	$\dfrac{A + T}{G + C}$
Escherichia coli (K-12)	–	26.0	23.9	24.9	25.2	1.00
Diplococcus pneumoniae	–	29.8	31.6	20.5	18.0	1.59
Mycobacterium tuberculosis	–	15.1	14.6	34.9	35.4	0.42
Yeast	–	31.3	32.9	18.7	17.1	1.79
Paracentrotus lividus (sea urchin)	Sperm	32.8	32.1	17.7	18.4	1.85
Herring	Sperm	27.8	27.5	22.2	22.6	1.23
Rat	Bone marrow	28.6	28.4	21.4	21.5	1.33
Human	Thymus	30.9	29.4	19.9	19.8	1.52
Human	Liver	30.3	30.3	19.5	19.9	1.53
Human	Sperm	30.7	31.2	19.3	18.8	1.62

From E. Chargaff and J. Davidson, Eds., *The Nucleic Acids.* New York: Academic, 1955, pp. 356–359.

chains form a double-helical or **duplex** molecule. Working with "ball and stick" type models, they experimented with the idea of placing the sugar-phosphate backbones on the outside and pointing the bases toward the inside of the helix. By placing the two polynucleotide chains in phase with each other—an arrangement suggested by the regular X-ray diffraction patterns—opposing bases on the two adjacent chains tended to line up with one another. Here Watson and Crick made an exciting observation.

They realized that a stable helical structure could form *only* if adenine lined up opposite thymine or if guanine lined up opposite cytosine. In the former case the molecular structures of the adenine and thymine were such that two hydrogen bonds could readily form between them (Figure 1-6). In the latter case they found that cytosine could form three hydrogen bonds with guanine (Figure 1-6). These bonds, it was reasoned, would greatly stabilize a large helical molecule and would help to explain the physical properties of naturally occurring DNA.

By assuming a duplex molecule containing only A—T and G—C base pairs, Watson and Crick further found that they could construct a model

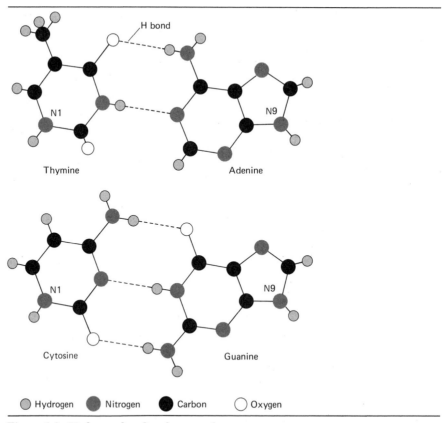

Figure 1-6 Hydrogen bonding between bases.

with a uniform diameter corresponding to about 20 Å. Purine-purine base pairs, they discovered, were too large to fit within such a helix, whereas pyrimidine-pyrimidine pairs were so far apart that hydrogen bonds could not form between them at all. A distortion of the helical structure also occurred when an attempt was made to pair A with C or G with T.

In short, when all possible base-pair combinations were considered, it became clear that A—T and G—C pairs were the only pairs that could form stable, hydrogen-bonded entities having the correct molecular dimensions to fit inside a polynucleotide double helix with a regular diameter of 20 Å. For this reason A and T were said to be **complementary** to each other, and G and C were said to be complementary to each other.

This conclusion related immediately to Chargaff's rule, which, as summarized earlier, states that in the DNA of any species, the A/T and G/C ratios are invariably close to unity (Table 1-3). No such consistent relationship had been found for any other sets of bases. Thus the Watson-Crick model was supported at its inception by some careful experimental evidence.

The full structure for the DNA molecule proposed by Watson and Crick is shown in Figure 1-7, and a more detailed molecular model appears in Figure 1-8. Having such a model in hand, it is possible to count 10 bases in each full turn of a helical polynucleotide strand. It will be recalled that in a polynucleotide chain the distance between bases in the pennylike stack was

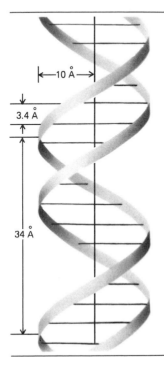

Figure 1-7 The Watson-Crick model of a DNA double helix. (After J. D. Watson and F. Crick, *Cold Spring Harbor Symposia on Quantitative Biology* **18**:123, 1953.)

10 Å

3.4 Å

34 Å

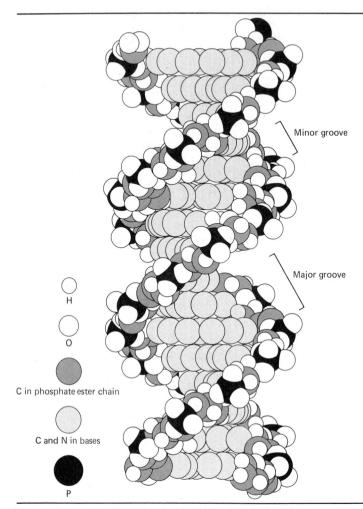

Minor groove

Major groove

○ H

○ O

● C in phosphate ester chain

○ C and N in bases

● P

Figure 1-8 Molecular model of DNA double helix. (Reprinted from M. Feughelman et al., *Nature* **175:**834, 1955, courtesy of M. H. F. Wilkins.)

3.4 Å, and it follows that in the double helix a strand makes one complete turn every 34 Å. The actual numbers are less important to the geneticist than the concept that the DNA duplex is an extremely ordered structure composed of relatively stable organic molecules that pair in a precise manner.

It is important to look at the complementary base pairs in Figure 1-6 and note that they are drawn in a specific way, with the relevant hydrogen-bonding groups precisely aligned. In the DNA duplex such an alignment of the bases can occur *only* when the two polynucleotide strands are oriented in opposite directions. As pointed out earlier and shown in Figure 1-5, a polynucleotide possesses a 3'-OH and a 5'-P end. The Watson-Crick model requires that one strand be oriented in a 3′ ⟶ 5′ fashion in the helix, the other in a 5′ ⟶ 3′ fashion, as shown schematically in Figure 1-9. When an attempt is made to construct a double helix with two similarly directed strands, the bases do not align with each other to allow the formation of the requisite hydrogen bonds.

The Watson-Crick hypothesis was advanced for DNA, but double helical RNA molecules have since been described for certain RNA-containing viruses. Moreover, most single-stranded RNA molecules probably contain at least some regions in which the strand folds back on itself to form short double-helical segments. When RNA exists as a double helix and serves as the genetic material of viruses, most of the statements we have made about the secondary structure of DNA are probably applicable, except that adenine forms hydrogen bonds with uracil and not with thymine.

We now leave the physical and chemical properties of helical nucleic

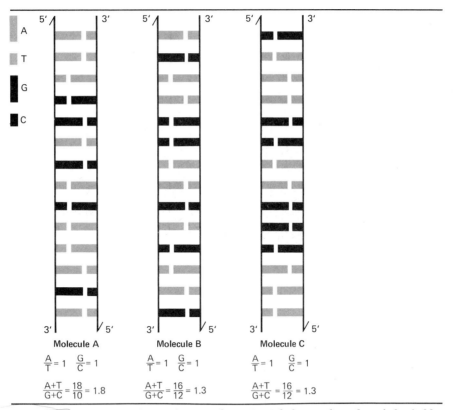

Molecule A	Molecule B	Molecule C
$\dfrac{A}{T}=1 \quad \dfrac{G}{C}=1$	$\dfrac{A}{T}=1 \quad \dfrac{G}{C}=1$	$\dfrac{A}{T}=1 \quad \dfrac{G}{C}=1$
$\dfrac{A+T}{G+C}=\dfrac{18}{10}=1.8$	$\dfrac{A+T}{G+C}=\dfrac{16}{12}=1.3$	$\dfrac{A+T}{G+C}=\dfrac{16}{12}=1.3$

Figure 1-9 "Ladder" scheme showing three DNA helices. The sides of the ladder represent the sugar-phosphate "backbone," and the rungs are formed by the base pairs. Arrowheads indicate a 5′-phosphate. All three molecules carry distinct nucleotide *sequences*, but molecules B and C happen to have the same overall nucleotide *composition*.

acids and go on to consider some of their genetic properties, focusing particularly on DNA. In so doing, however, it seems appropriate to quote from the concluding paragraph of a recent review article by J. Josse and J. Eigner.

What has perhaps emerged most clearly from all the work which has been reported here is the enduring validity of the double-helical model of DNA structure proposed by Watson and Crick. Down to almost all of its finest details, this model has been convincingly confirmed from virtually every physical approach with which DNA can be examined.

RELATING DNA STRUCTURE TO ITS GENETIC REQUIREMENTS

1.3 DNA AS A CODED MOLECULE

Genetic material must carry information—the first requirement on our list. Once the structure for DNA had been proposed, the general mechanism for its ability to encode genetic information became apparent, namely, that the **sequence** of bases along a polynucleotide chain contains this information. Thus one chain might read AAAATT. . . , another GAAATT. . . , another GTAATT. . . , and so on through an endless series of possibilities, and we can simply imagine, as did a number of geneticists in the 1950s, that one such nucleotide sequence specifies one sort of genetic information and another sequence codes for another sort.

The fact that most DNA exists in duplex form does not at all contradict this hypothesis. We simply extend the concept and propose that the information in a DNA duplex is encoded in the sequence of nucleotides along its **two** polynucleotide chains. This has been diagrammed in Figure 1-9, in which three hypothetical DNA duplexes are shown, each with a different sequence of nucleotides. The bases in such double helical molecules, as we saw with single polynucleotide chains, are stacked, with one flat surface on top of the next, like pennies (Figure 1-8). Accordingly, no stereochemical limitations are placed on the **order** of bases within the helix, as contrasted with the strict limitation placed on the **pairing** of bases within the helix. This means that any nucleotide sequence is chemically possible in the helix, as long as the two strands of the helix are oriented in opposite directions and the bases in one strand are complementary to those in the other.

If the genetic code is indeed inherent in the nucleotide sequence of a DNA molecule, we might predict that different species of organisms, having different genetic information, will possess DNAs with different nucleotide sequences, whereas all DNA from the same species will be identical. Biochemical techniques in the 1950s did not permit determination of the base sequences in DNA, but the prediction was certainly supported by Chargaff's experiments (Table 1-3). We have already noted that Chargaff's data established the universality of the A = T and G = C rule. In addition, his data reveal that there is not necessarily any similarity between the **total nucleotide composition** of DNA isolated from one type of organism, and the DNA isolated from another type of organism. Expressing total nucleotide composition as the ratio A + T/G + C, we find (Table 1-3, last column) a ratio of 1.00 for the colon bacterium *Escherichia coli,* 1.59 for the bacterium *Diplococcus pneumoniae,* and so on, a result that is readily explained by assuming that the various species possess DNA with different nucleotide sequences (Figure 1-9). When, on the other hand, we compare different cell types from the *same* species of organism—from the thymus, liver, and sperm cells of humans, for example—all are found to have nearly identical A + T/G + C ratios (Table 1-3), as if all possessed the same genetic information.

Nucleotide sequences and nucleotide compositions are two quite distinct properties of a DNA molecule, as we demonstrate by example in Figure 1-9. We therefore stress that the Chargaff data do not *prove* that DNA contains genetic information. They served, however, to provide important support for the concept of DNA as the genetic material at the time this concept was being formulated. Proof that DNA is a coded molecule is forthcoming in succeeding chapters of this book.

1.4   DNA REPLICATION

The double-helical model for the structure of DNA immediately suggested a mechanism for faithful DNA replication, the second on our list of requirements for the genetic material. Watson and Crick saw that once a particular nucleotide sequence was established on one strand of the helix, the sequence on the other strand had to be complementary and could be predicted. For example, knowing that one strand reads 5′. . . ACGAT . . . 3′, the other strand, by the rules of base pairing, must read 3′. . . TGCTA. . . 5′.

Watson and Crick therefore proposed that if two strands in a given helix were to separate by an unwinding process and so become exposed to a solution containing nucleotides, each strand might serve as a **template** for the laying down of new polynucleotide strands. The sequences of bases in the new daughter strands would, by the rules of base pairing, complement the sequences in the parental strands, as shown in Figure 1-10. At the end of this round of replication two new helices would be formed, each containing one parent and one daughter strand, and both new helices would be exact replicas of the original helix.

This proposal is a particularly ingenious feature of the Watson-Crick model. Moreover, it appears to be correct: predictions can be made about DNA and RNA replication on the basis of the Watson-Crick proposal, and many of them have been verified experimentally, as we describe in Chapter 5.

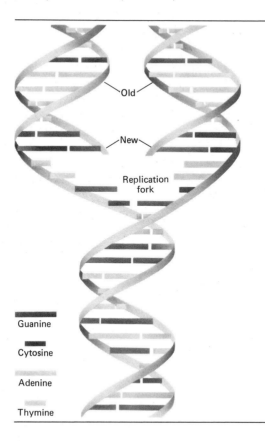

Figure 1-10 The replication of DNA according to Watson and Crick. Purines are drawn as long rectangles and pyrimidines as short rectangles; the new strands are drawn in gray. (After J. D. Watson and F. H. Crick, *Nature* **171:**737–738, 1953.)

Old

New

Replication fork

Guanine

Cytosine

Adenine

Thymine

1.5 DNA EXPRESSION

The third requirement we imposed on the genetic material was that its encoded information be somehow "expressible," that is, translatable into such processes as growth and development. At the time the Watson-Crick model was proposed it was not clear how a code composed of a sequence of bases could be biologically effective, but the concept soon evolved that the sequence of nucleotides in a DNA molecule might be **translated** into the sequence of amino acids in a polypeptide, and that the group of nucleotides which specified such a single polypeptide could be considered a **gene.** It also became clear that genes did not ordinarily participate directly in the synthesis of polypeptides; instead, their nucleotide sequences were first **transcribed** into complementary nucleotide sequences in the RNA molecules known as **messenger RNA.** These messenger RNA molecules then went on to engage in polypeptide synthesis. The transcription and translation of genes are discussed in detail in Chapters 8 and 9.

1.6 DNA VARIATION

Our fourth requirement of the genetic material was that it be subject to occasional variation by mutation and recombination. The process of mutation could be readily visualized with the Watson-Crick model, for any alteration in the sequence of its nucleotides, produced either physically or chemically, would bring about a change in the information contained in the DNA molecule and would thereby change, in some aspect, the construction of the cell. The physical and chemical basis for mutation, the subject of Chapter 6, could in fact be meaningfully explored only after a structure for the genetic material had been proposed.

On the other hand, the Watson-Crick model did not, in itself, immediately suggest a mechanism for recombination. As we shall see in Chapter 13, the physical basis for recombination remains only partially understood.

EXPERIMENTS INDICATING DNA AND RNA AS THE GENETIC MATERIAL

Watson and Crick did not choose DNA as the subject for their model-building simply out of hand. A paper describing the nucleic acids was published as early as 1874 by F. Miescher, and by 1953 it was clear that proteins, long favored as the genetic material, were poor candidates for a number of reasons. R. Shoenheimer had noted, in 1938, that the DNA of a cell was unusually stable, in contrast to the rapidly turning-over proteins, and A. Mirsky and H. Ris had found, in 1949, that whereas all cells of an organism contained similar amounts of DNA, different cell types contained quite different amounts and kinds of protein. Its stability and constancy therefore favored DNA as the genetic material.

A number of experiments, moreover, had indicated that DNA alone could transmit genetic information from one generation to the next. Several of these experiments are described in the following paragraphs, including some that were performed after 1953. In the course of this presentation certain organisms and laboratory techniques are discussed rather sketchily and only as they relate to the experimental results. In later chapters the organisms and the techniques are both covered in much greater detail. At this point the experiments serve as an introduction to the kinds of approaches used by the molecular geneticist.

1.7 TRANSFORMATION EXPERIMENTS

One series of experiments indicating DNA as the genetic material utilized the bacterium *Diplococcus pneumoniae,* or pneumococcus. To understand the experiments, several facts are pertinent.

1. Pneumococcus can exist in two forms: smooth (S) cells are surrounded by a smooth polysaccharide capsule and are **virulent,** or infectious, while rough (R) cells lack the capsule and are **nonvirulent.**

2. Several S strains have been isolated (Types IIS and IIIS, for example), each having a polysaccharide coat with a slightly different chemical composition. The various S and the R coat types are inherited from generation to generation and are thus said to be genetically determined.

3. A change from S ⟶ R will occur by mutation in about one S cell in 10 million. The R colony that grows from this mutant will, in turn, sometimes give rise to an S cell by a second reverse mutation, R ⟶ S. The resultant S colony is found to possess the same capsule type as the S strain from which the R mutant was originally derived; for example, Type IIS ⟶ R ⟶ Type IIS, not IIIS.

The early experiments with S and R pneumococcus were performed by F. Griffith. Mice were injected with a small number of living (but nonvirulent) R pneumococci which had originally arisen, by mutation, from the Type IIS strain. At the same time the mice were injected with a large number of heat-killed (and therefore no longer virulent) Type IIIS bacteria. Surprisingly, many of the mice developed pneumonia, and blood samples of the diseased mice revealed the presence of large numbers of living, virulent, smooth pneumococci. Moreover, these were all of Type IIIS, meaning that they could not have arisen by reverse mutation of the living injected R cells. The conclusion, therefore, was that some of the dead Type IIIS bacteria had **transformed** the living R bacteria into smooth, virulent Type IIIS cells during their coexistence within the mouse (Figure 1-11). It could be further shown

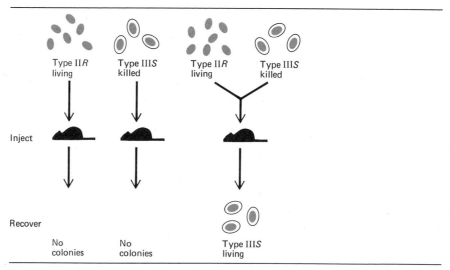

Figure 1-11 *In vivo* pneumococcus transformation experiment with mice. (Redrawn from Sutton, *Genes, Enzymes, and Inherited Diseases.* New York: Holt, Rinehart and Winston, 1961.)

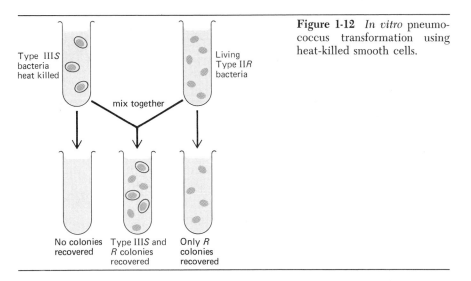

Figure 1-12 *In vitro* pneumococcus transformation using heat-killed smooth cells.

that the newly formed Type IIIS cells reproduced true to type for many generations, indicating that the transformation had directly affected their genetic material.

As other investigators repeated these experiments it was found that the mouse could be eliminated: heat-killed S cells and live R cells could be grown together in a test tube, and virulent S cells with the same capsule type as the heat-killed S cells were recovered (Figure 1-12). In a further experiment it was found that even intact, heat-killed S cells were not required: an extract of heat-killed S cells could transform the R cells *in vitro*. The remaining task, therefore, was to determine which chemical substance in the extract was responsible for this transformation process.

The substance, or **transforming principle,** as it was then called, proved to be DNA. In 1944 O. T. Avery, C. M. MacLeod, and M. McCarty incubated R cells in the presence of a highly purified DNA fraction obtained from Type IIIS bacteria, and some virulent Type IIIS bacteria were recovered from the test tube. If the Type IIIS DNA fraction was first treated with **deoxyribonuclease,** an enzyme that destroys DNA, no transformation of R cells occurred in the subsequent experiment.

At the time these experiments were published they were interpreted by some to mean that the transforming principle, or DNA, was capable of inducing a type-specific, "directed" mutation in the genetic material of the R cells such that S cells were produced. The mechanism by which such a directed mutation might be induced was obscure, but at that time it was certainly no more obscure than the correct interpretation of the experiments, namely, that the purified, S-type DNA is capable of entering the R cells and recombining with the R-cell DNA to produce a small number of living, smooth (III S) recombinants. The details of this bacterial transformation as it is now understood are discussed in Chapters 11 and 13.

1.8   HERSHEY-CHASE EXPERIMENTS

In 1952 A. D. Hershey and M. Chase published experiments that indicated DNA as the genetic material in a more direct manner. Their experiments were concerned with the bacteriophage T2, a virus that can only replicate in the bacterium *Escherichia coli.* The structure of T2 is diagrammed in Figure 1-13; it has a hexagonal head containing DNA, a tail, and tail fibers. The first step in the T2 phage infection of *E. coli* was already known at that time to be the attachment, or **adsorption,** of a phage, via its tail fibers, to the outside of the host cell. Phage material then entered the bacterium in some way and multiplied at the expense of the bacterium until the bacterium finally burst open, or **lysed,** liberating perhaps a hundred new phage progeny.

T2 phages were known to be composed of approximately equal amounts of DNA and protein. Since DNA contains phosphorus but no sulfur, and most proteins contain no phosphorus but (usually) some sulfur, the two can be distinguished by the use of radioactive isotopes of phosphorus and sulfur. Hershey and Chase therefore grew *E. coli* in a medium containing either the radioactive isotope of phosphorus (^{32}P) or that of sulfur (^{35}S) and then allowed T2 phages to infect and multiply within these labeled host cells (Figure 1-14). The progeny phages that appeared after lysis of the bacteria were collected and found to be similarly labeled. In this way Hershey and Chase acquired two stocks of radioactive T2, one containing ^{32}P-labeled DNA and the other, ^{35}S-labeled protein.

They next took suspensions of the labeled phages and subjected them to osmotic shock that burst the phages. When the ^{32}P-labeled phages were so treated, almost all the radioactivity was found in solution, whereas after lysis of the ^{35}S-labeled phages, almost all the label remained in particulate form.

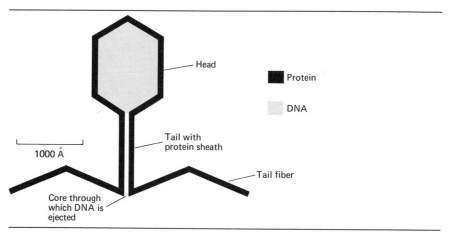

Figure 1-13 Simplified diagram of a T2 bacteriophage. (Redrawn from Sutton, *Genes, Enzymes, and Inherited Diseases.* New York: Holt, Rinehart, and Winston, 1961.)

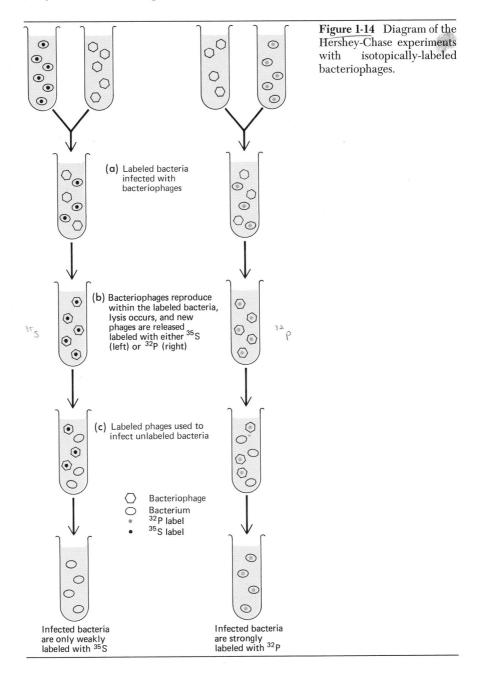

Figure 1-14 Diagram of the Hershey-Chase experiments with isotopically-labeled bacteriophages.

(a) Labeled bacteria infected with bacteriophages

(b) Bacteriophages reproduce within the labeled bacteria, lysis occurs, and new phages are released labeled with either ^{35}S (left) or ^{32}P (right)

(c) Labeled phages used to infect unlabeled bacteria

⬡ Bacteriophage
◯ Bacterium
⊙ ^{32}P label
● ^{35}S label

Infected bacteria are only weakly labeled with ^{35}S

Infected bacteria are strongly labeled with ^{32}P

Electron microscopy of these particles revealed empty-looking phage "ghosts": only the outer walls of the phages were present in the preparation. Thus it was established that a phage is constructed of an outer protein coat surrounding an inner mass of DNA, as diagrammed in Figure 1-13, and that the DNA and the protein could be separated experimentally.

Labeled phages were then used to infect unlabeled *E. coli* cells (Figure 1-14c) and a significant result was obtained. When the infection was brought about by the ^{32}P-labeled phages, almost all the radioactivity was subsequently found *within* the host bacteria; moreover, following bacterial lysis, some of the label appeared in the progeny phages. On the other hand, when ^{35}S-labeled phages were used, very little label appeared within the host bacteria or in the progeny phages; almost all of it remained on the *outside* of the bacterium, adsorbed to the bacterial cell wall. Thus it was demonstrated that the phage DNA and protein become separated during the infection process: the DNA enters the host cell and brings about phage replication; the protein coat appears to function primarily in the external adsorption process.

The Hershey-Chase experiment does not, in fact, offer unambiguous proof that DNA is the phage genetic material, for about 20 percent of the ^{35}S label enters the host along with the DNA, and it could certainly be argued that this small amount of protein carries the genetic information. The Watson-Crick model was published in the following year, however, and the era of DNA-oriented research was launched.

A "clean" Hershey-Chase type experiment proved not to be possible so long as the infection of intact bacteria by bacteriophages was a part of the experimental procedure, for the injection of a small amount of protein is an obligate feature of the natural phage infection process. If, however, bacteria are first stripped of their cell walls to form **protoplasts,** an intact phage is no longer required for infection; purified phage DNA can be introduced into the protoplasts and infectious phage progeny will still appear. Thus DNA alone clearly contains all the information necessary to construct virulent T2 phages.

1.9 EXPERIMENTS WITH RNA VIRUSES

The demonstration that RNA is the genetic material in RNA-containing viruses came in 1956, when A. Gierer and G. Schramm showed that tobacco plants could be inoculated with purified RNA from the tobacco mosaic virus (TMV) and TMV-like lesions could later be identified on the tobacco leaves. A different approach was taken by H. Fraenkel-Conrat and B. Singer in experiments published in 1957. They first found that they could separate the RNA from the protein of TMV viruses, much as Hershey and Chase had done with the DNA and the protein of T2. They then developed techniques for forming "reconstituted" viruses containing the protein from one mutant strain of TMV and the RNA from another, or vice-versa. Such hybrid viruses were allowed to infect tobacco leaves, and the progeny were examined. In all cases the progeny were of the parental RNA type and *not* the parental protein type. Such an outcome is reminiscent of the pneumococcus transformation results, in which the parental DNA type determined the progeny type.

In a final refinement of this kind of experiment N. Pace and S. Spiegelman in 1966 purified RNA of two quite distinct base compositions from two different mutant strains of the RNA phage Qβ. The isolated RNAs were then incubated separately in the presence of an *E. coli* cell extract containing an enzyme capable of RNA replication. The new RNA synthesized was in each case identical in base composition to the particular phage RNA introduced into the *in vitro* system, thus indicating that the phage RNA can serve as a template for its self-replication.

References†

The Structure of DNA and RNA as the Genetic Material

*Chargaff, E. "Structure and function of nucleic acids as cell constituents," *Fed. Proc,* **10**:654–659 (1951).

Davidson, J. N. *The Biochemistry of the Nucleic Acids,* 7th ed. London: Chapman and Hall, 1972.

Felsenfeld, G., and H. T. Miles. "The physical and chemical properties of nucleic acids," *Ann. Rev. Biochem.* **36**:407–448 (1967).

*Franklin, R. E., and R. Gosling. "Molecular configuration of sodium thymonucleate," *Nature* **171**:740–741 (1953).

Josse, J., and J. Eigner. "Physical properties of deoxyribonucleic acid," *Ann. Rev. Biochem.* **35**:789–834 (1966).

Lomant, A. J., and J. R. Fresco. 1975. "Structural and energetic consequences of noncomplementary base oppositions in nucleic acid helices," *Prog. Nucl. Acid Res. Mol. Biol.* **15**:185–218.

*Watson, J. D., and F. H. C. Crick. "A structure for desoxyribose nucleic acids, *Nature* **171**:737–738 (1953). [Reprinted in J. A. Peters, *Classical Papers in Genetics.* Englewood Cliffs, N.J.: Prentice-Hall, 1959.]

*Watson, J. D., and F. H. C. Crick. "The structure of DNA," *Cold Spring Harbor Symp. Quant. Biol.* **18**:123–131 (1953). [Reprinted in G. L. Zubay, *Papers in Biochemical Genetics.* New York: Holt, Rinehart and Winston, 1968.]

*Wilkins, M. H. F., A. R. Stokes, and H. R. Wilson. "Molecular structure of desoxypentose nucleic acids," *Nature* **171**:738–740 (1953).

Experiments Indicating DNA and RNA as the Genetic Material

*Avery, O. T., C. M. MacLeod, and M. McCarty. "Studies on the chemical nature of the substance inducing transformation of pneumococcal types. Induction of transformation by a desoxyribonucleic acid fraction isolated from pneumococcus Type III," *J. Exp. Med.* **79**:137–158 (1944).

†The selection of references stresses recent review articles, an excellent source of numerous research articles.
*Denotes articles described specifically in the chapter.

*Fraenkel-Conrat, H., and B. Singer. "Virus reconstitution: combination of protein and nucleic acid from different strains," *Biochim. Biophys. Acta* **24**:540–548 (1957).

*Gierer, A., and G. Schramm. "Infectivity of ribonucleic acid from tobacco mosaic virus," *Nature* **177**:702–703 (1956). [Reprinted in G. L. Zubay, *Papers in Biochemical Genetics.* New York: Holt, Rinehart and Winston, 1968.]

Goulian, M., and A. Kornberg. "Enzymatic synthesis of DNA, XXIII. Synthesis of circular replicative form of phage ϕX174 DNA," *Proc. Natl. Acad. Sci. U.S.* **58**:1723–1730 (1967).

*Griffith, F. "Significance of pneumococcal types," *J. Hygiene* **27**:113–159 (1928).

*Guthrie, G. D., and R. L. Sinsheimer. "Infection of protoplasts of *Escherichia coli* by subviral particles of bacteriophage ϕX174, *J. Mol. Biol.* **2**:297–305 (1960). [Reprinted in G. L. Zubay, *Papers in Biochemical Genetics.* New York: Holt, Rinehart and Winston, 1968.]

*Hershey, A. D., and M. Chase. "Independent functions of viral protein and nucleic acid in growth of bacteriophage," *J. Gen. Physiol.* **36**:39–56 (1952). [Reprinted in G. S. Stent, *Papers on Bacterial Viruses,* 2nd ed. Boston: Little, Brown, 1965.]

*Mirsky, A. E., and H. Ris. "Variable and constant components of chromosomes," *Nature* **163**:666–667 (1949).

*Pace, N. R., and S. Spiegelman. "*In vitro* synthesis of an infectious mutant RNA with a normal RNA replicase," *Science* **153**:64–67 (1966). [Reprinted in G. L. Zubay, *Papers in Biochemical Genetics.* New York: Holt, Rinehart and Winston, 1968.]

(a) (b)

(a) Rosalind Franklin performed early X-ray diffraction studies of DNA. (b) Erwin Chargaff (Columbia University) performed early biochemical studies on the base composition of DNA.

Questions and Problems

1. Distinguish between a purine, pyrimidine, nucleoside, nucleotide, polynucleotide, and double helix. How do these structures differ in DNA and RNA?

2. Diagram the pairing between adenine and uracil. Is there any difference between the way in which adenine pairs with uracil or thymine?

3. If one strand of a helix reads 5'-AGCAGCA-3', what would be the structure of the complementary strand in a DNA helix? An RNA helix?

4. Diagram two DNA duplexes, each having 15 nucleotide pairs and an A + T/G + C ratio of 2.0, but each having different nucleotide sequences.

5. Diagram a hypothetical duplex with five base pairs to represent a portion of the gene specifying the Type IIS pneumococcus capsule. Now make an arbitrary change in the sequence to represent the following mutational event: Type IIS → R. What change would be required for the event R → Type IIS? Would you expect both events to occur with equal frequency?

6. Why does the Watson-Crick model require that a DNA helix must unwind before it is replicated?

7. Outline experiments using radioactive isotopes in which you could show that transformation occurs when S-type pneumococcal DNA, and not protein, enters R cells.

8. At a certain time after the addition of ^{3}H-thymidine to a growing culture of *E. coli*, all the DNA is found to possess one radioactive strand and one nonradioactive strand. Is this result demanded by the Watson-Crick hypothesis for the replication of DNA? Explain.

9. A DNA duplex possesses major and minor "grooves" (Figure 1-8). Are any of the atoms in the purines and pyrimidines exposed in these grooves? Are the hydrogen bond-forming atoms exposed? (This question can most readily be answered if a space-filling model of a double helix is available.)

10. Which of the following base-composition ratios confirms the Watson-Crick DNA model? Which contradicts it? Explain for each.

(a) $\dfrac{G + C}{A + T} = 1$ (c) $\dfrac{\text{pyrimidines}}{\text{purines}} = 1$ (e) $\dfrac{GA}{C} = T$

(b) $\dfrac{G + A}{C + T} = 1$ (d) $\dfrac{G + C + A}{T} = 1$

CHAPTER
2

Mitosis and Chromosome Structure in Eukaryotes

INTRODUCTION

This chapter focuses on the structure and organization of **chromosomes,** the physical bearers of genetic information in nucleate (as well as non-nucleate) organisms. If we think of all the genetic information possessed by an individual as its **genome** then, for nucleate organisms, the genome is subdivided into a collection of chromosomes, each carrying a distinct subset of genes and, therefore, distinct nucleotide sequences. It is clearly essential to the survival of the species that when one cell gives rise to two daughter cells, each daughter receives a complete chromosome collection. Such equitable chromosome distributions are effected by **mitosis,** a process also described in this chapter. Finally, the chapter reviews features of nucleate cells and organisms that are directly relevant to genetic analysis.

THE EUKARYOTIC CELL AND CELL CYCLE

In all organisms except bacteria, blue-green algae, and viruses, the chromosomes of each cell are ordinarily confined within a membrane-limited region called the **nucleus.** Such organisms are referred to as **nucleate** or **eukaryotic** (a *karyon* is a nucleus in Greek, and the prefix *eu-* denotes "true"), in contrast to the **prokaryotic** bacteria and blue-green algae. Of the vast array of single-celled eukaryotic organisms in existence, those most extensively used at present in genetic research include the yeast *Saccharomyces cerevisiae;* the cellular slime mold *Dictyostelium discoideum;* the alga *Chlamydomonas reinhardi,* and the ciliate *Paramecium aurelia.* Certain multicellular fungi such as *Neurospora crassa* and *Aspergillus nidulans* also enjoy the attention of geneticists. All such creatures are often referred to as "lower" eukaryotes to distinguish them from the "higher" eukaryotes; of the latter, the fruit fly (*Drosophila melanogaster*), corn (*Zea mays*), the mouse (*Mus musculus*), and the human being (*Homo sapiens*) are perhaps the most extensively studied from a genetic point of view.

In the sections that follow, the structure of a eukaryotic cell is first examined with emphasis on its nucleus. There follows a description of the eukaryotic cell cycle and its culminating event, the apportionment of chromosomes from parent to daughter cells, known as mitosis.

2.1 EUKARYOTIC CELL STRUCTURE

A eukaryotic nucleus as seen with the electron microscope is illustrated in Figure 2-1. This nucleus is not engaging in mitosis and is thus called an **interphase** nucleus. The chromosomes in an interphase nucleus cannot be distinguished as individual entities but instead appear as an amorphous

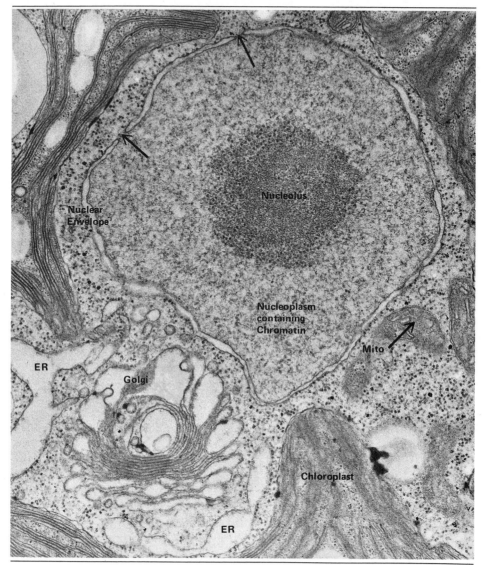

Figure 2-1 Structure of a eukaryotic cell, *Chlamydomonas reinhardi*. Arrows point to nuclear pores.

network; for this reason, interphase chromosomal material is commonly referred to as **chromatin** (Figure 2-1). The most prominent landmark of an interphase nucleus is its **nucleolus** (Figure 2-1)—a large, deeply staining spherical body that contains RNA and protein and represents the site of synthesis and storage of the cell's cytoplasmic **ribosomes**. These ribosomes are the structures that are involved in protein synthesis, and detailed information on the nucleolus and ribosomes will be found in Chapters 8 and 9.

The chromatin in an interphase nucleus is surrounded by a membrane that folds back on itself to form an **envelope.** The envelope appears in Figure 2-1 as two dense lines—the inner and outer membranes—that enclose a narrow channel or cisterna. At intervals the envelope is perforated by **pores,** each perhaps 400 Å in diameter, which are usually covered by a septum of a moderately dense and still unidentified material (Figure 2-1). At other intervals the envelope extends out into the cytoplasm as a network of channels and large cisternae known as the **endoplasmic reticulum (ER).** The cell's ribosomes are often bound to the outer surface of the ER, forming what is known as **"rough" ER** (Figure 2-1).

The non-nuclear portion of a eukaryotic cell is called the **cytoplasm.** The cytoplasm contains soluble enzymes, free ribosomes, and additional systems of membranes which serve to divide the cell into a number of compartments called **organelles.** One of these organelles is the endoplasmic reticulum, and this is often associated with the **Golgi apparatus** (Figure 2-1), a relationship important for the synthesis of many proteins that are secreted from the cell. Other organelles include mitochondria and chloroplasts (Figure 2-1). The former, found in all eukaryotic cells, are devoted primarily to respiration; the latter, found only in eukaryotic plant cells, are devoted primarily to photosynthesis. Chloroplasts and mitochondria possess chromosomes of their own, and the genetics of these chromosomes has become of major interest in recent years, as will become apparent in Chapter 14.

2.2 THE TYPICAL EUKARYOTIC CELL CYCLE

In simple eukaryotes such as yeast, the DNA synthesis leading to chromosome replication takes place throughout the interphase and ceases only during the brief period of mitotic nuclear and cell division. In higher plants and animals, however, DNA replication occurs only during a discrete interval of the interphase known as the **S period** (S standing for synthesis). Before and after the S period, cells engage in growth and metabolic activity but not in chromosome replication and are said to be in the G_1 or G_2 (G standing for gap) periods of interphase. The G_2 period is ordinarily followed by mitosis **(M),** and the sequence $G_1 \longrightarrow S \longrightarrow G_2 \longrightarrow M$, followed by another G_1, is known as the **cell cycle.** When nutrients become scarce the cells shift into a **stationary** or G_0 phase in which cellular metabolism essentially shifts into a holding pattern until nutrients are replenished.

As shown in Figure 2-2, different eukaryotic cells vary in the length of time they take to complete an entire cell cycle; they also differ in the relative proportions of time allotted to each of the four stages of the cycle. In general, the length of an S phase is a constant property of any one cell type, but the length of its G_1 phase may vary considerably with environmental fluctuations, such as nutrient supply. It is also apparent in Figure 2-2 that the amount of time allotted to the actual mitotic segregation of daughter chromosomes is usually relatively brief. Even here, however,

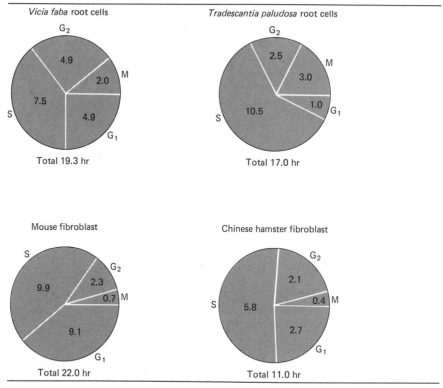

Figure 2-2 The cell cycle in various cell types. (From B. Kihlman, T. Eriksson, and G. Odmark, *Hereditas* **55**:386, 1966.)

exceptions are known: in the grasshopper neuroblast, for example, mitosis may take as long as eight hours after an interphase of only 30 minutes.

From a geneticists point of view, the two key events in the eukaryotic cell cycle are the DNA/chromosome replication which occurs during the S phase and the precise allotment of these replicated chromosomes to daughter cells which occurs at mitosis.

2.3   DNA REPLICATION IN THE CELL CYCLE

Each chromosome in a eukaryotic cell is believed to contain a single, enormous DNA helix which is associated with a variety of proteins, a statement amplified in later sections of this chapter. During the S phase, this helix must be copied so that it will yield 2 daughter helices. The molecular details of this process are considered in Chapter 5, where we discuss bacterial cells; here we convey the essential events of DNA replication in the context of the eukaryotic cell cycle.

DNA is replicated by the action of a variety of **enzymes** (the properties of enzymes are reviewed in Box 2.1 should they be unfamiliar). The most critical enzyme in the process—**DNA-dependent DNA polymerase** (or, more simply, **DNA polymerase**)—copies each strand of a DNA helix into a complementary strand, so that two daughter helices are generated from an original parent helix (see Figure 1-10). Various studies have indicated that in human cells, DNA is replicated at an average rate of about 0.5μ/min. Since an average human chromosome contains about $30,000\mu$ of DNA, this means that if replication were to initiate at one end of a chromosome and proceed sequentially to the other end, replication would take 1000 hours, a prediction clearly in conflict with the typical S-phase duration of 6–8 hours. J. Huberman and A. Riggs discovered a solution to this paradox when they exposed cultured human cells in S-phase to thymidine labeled with the radioactive isotope tritium (^{3}H) (recall that thymine is a DNA-specific base) for brief periods; they then isolated large DNA fragments from these cells, spread the fragments on a flat surface, and subjected the fragment preparation to **autoradiography** (see Box 2.2). They found that a given DNA fiber exhibits multiple sections that are heavily labeled with tritium (Figure 2-3), demonstrating that each chromosome is "worked on" by replication enzymes at many different places at once. In other words, each chromosome contains a number of **replication units** or **replicons,** each with an average length of about $30\ \mu$, with the $30\ \mu$ daughter strands that are formed alone in each unit eventually linking up to form long, continuous daughter strands.

**Box 2.1
PROPERTIES
OF ENZYMES
RELEVANT TO
MOLECULAR
GENETICS**

An enzyme is a protein that speeds up the rate of a chemical reaction in a cell. Any biochemical reaction—the synthesis of a large sugar molecule from two smaller molecules, for example—is theoretically capable of occurring at intracellular temperatures and pH in the absence of an enzyme, but it will occur only infrequently. This is because a certain amount of energy, called the **activation energy,** must be supplied to the two molecules so that they can form the proper chemical bonds; the molecules must also be oriented properly with respect to each other so that the bond-forming atoms can react. An enzyme has the effect of allowing molecules to interact more often and more readily so that the activation

energy for the reaction is lowered. Enzymes therefore permit biochemical reactions to take place at reasonable rates under intracellular conditions and are said to **catalyze** such reactions. An enzyme is not "used up" as it catalyzes a reaction; instead, it is free to interact with fresh substrates and to catalyze repeated reactions.

Enzymes catalyze biochemical reactions by interacting with the reacting molecules, known as **substrates.** For many enzymes this interaction is highly specific, and a given enzyme will interact only with particular molecules or, on occasion, with slightly modified versions of these molecules. The basis of this specificity is thought to reside in the three-dimensional conformation of the enzyme itself. The polypeptide chain or chains that interact to form an enzyme become oriented to create what is called an **active site.** This site is so constructed that it has a high specific affinity for the substrate molecule(s). The upper diagram below illustrates, quite schematically, the relation between substrates and active sites in the enzyme glucokinase.

Certain enzymes are specific for particular classes of covalent bonds, rather than for particular substrate molecules, and it is these enzymes that are of special interest to the geneticist. Some of them act to catalyze the *breakdown* of a particular bond: the **deoxyribonucleases,** for example, catalyze the hydrolysis of the phosphodiester bonds that link nucleotides in a DNA molecule. Other enzymes are commonly involved in the *formation* of particular bonds, an appropriate example being the **DNA polymerases** which promote the formation of DNA polymers. The lower diagram below illustrates, again schematically, the mode of action of a polymerase enzyme. (Diagrams from J. D. Watson, *Molecular Biology of the Gene.*, Copyright © 1965 by J. D. Watson. W. A. Benjamin, Inc., Menlo Park, Calif.)

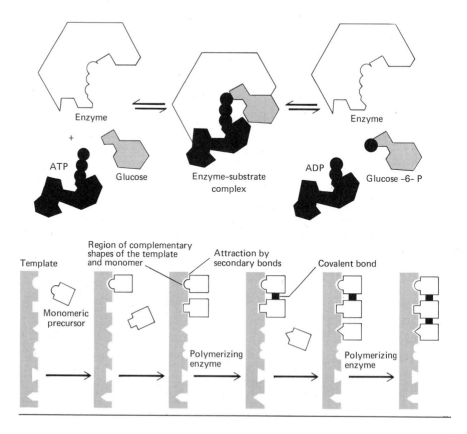

Box 2.2
AUTO-RADIOGRAPHY

The sub-atomic particles (for example, β-particles) emitted from a radioisotope will expose the silver grains in a photographic emulsion. This fact can be used to localize the position of molecules carrying such isotopes as ^{14}C, ^3H, or ^{125}I. The sample to be studied may be overlaid by a photographic film, or a slurry of photographic emulsion can be layered over the sample and allowed to dry. In either case the sample is stored in the dark after preparation, to allow for radioactive decay; the film or the emulsion-covered sample is then developed in much the same way as a photographic film. The blackened grains in the emulsion localize the position of the isotope with great precision.

In many respects, replication units behave as if they were independent of one another: in Figure 2-3, a short exposure to ^3H-thymidine labels certain segments whereas adjacent segments remain unlabeled. Since all the DNA in the nucleus must be copied in advance of mitosis if daughter nuclei are to receive complete sets of genetic information, some mechanism must exist which recognizes when all the replication units have completed their synthetic activities and triggers the onset of the G_2 and M phases of the cell cycle. The nature of this mechanism is totally unknown.

2.4 CHROMOSOME REPLICATION IN THE CELL CYCLE

As we noted earlier, the genome of a eukaryotic cell is typically found not in one chromosome but in several chromosomes. As an example we can cite the cellular slime mold *Dictyostelium discoideum*. In the G_1 phase, each nucleus of a *Dictyostelium* cell contains 7 chromosomes which collectively contain all the information (that is, all the DNA sequences) required to make a *Dictyostelium* cell. During the S phase, all of the chromosomes are replicated, so that each G_2 nucleus has twice the number of *Dictyostelium* chromosomes; for the G_2 nucleus of *Dictyostelium,* this number is 14. Finally, during mitosis, one copy of each chromosome goes to each of the two daughter cells so that each daughter nucleus enters its own G_1 phase with the fundamental chromosome number of 7 again. These relationships are diagrammed in Figure 2-4a.

If we focus next on one particular chromosome of the 7 in *Dictyostelium* (Figure 2-4b), the net result of the S phase is that two identical copies of its information are present in the nucleus, where only one copy was present before. To avoid confusion these identical copies are best called **sister chromatids,** with the term chromosome being reserved to specify an unreplicated structure. (In practice, "chromosome" is frequently used as a catchall term to mean both replicated and unreplicated forms of genetic material; in this text we distinguish between chromatids and chromosomes when to do so increases the clarity of an explanation.) Sister chromatids, like unreplicated chromosomes, cannot be recognized as such in an inter-

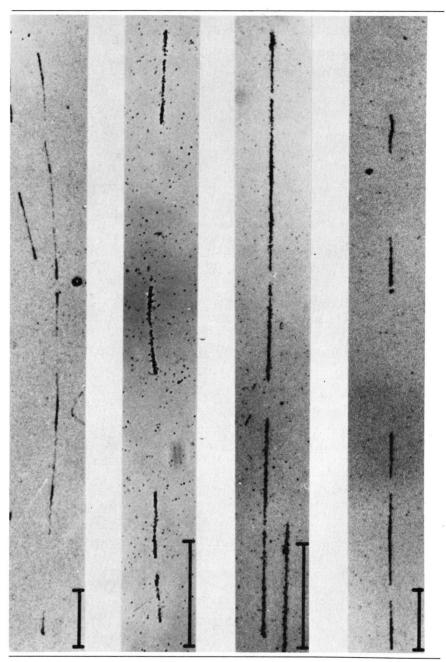

Figure 2-3 Autoradiograms of HeLa and Chinese Hamster DNA that has been pulse-labeled with ³H-thymidine during replication. Scale indicates 50 μ. (Used with permission from J. A. Huberman and A. Tsai, *J. Mol. Biol.* **75**:5–12, 1973. Copyright by Academic Press, Inc., (London) Ltd.)

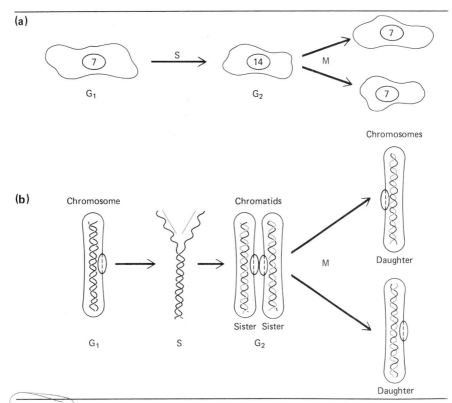

Figure 2-4 The mitotic replication and segregation of chromosomes. The replication and allotment of the entire chromosome set of *Dictyostelium discoideum* is shown in (a); a single chromosome is shown in (b). The centromere is represented by an oval structure in (b).

phase (that is, in the G_1, S, or G_2 phase) nucleus. It is only when the nucleus is preparing to divide, either by mitosis or by meiosis (see Chapter 3), that individual chromatids undergo a dramatic condensation process and become distinguishable through the microscope as small, deeply staining bodies (the original structures for which the term chromosome was coined). Each sister chromatid pair is seen at this time to be connected at a region known as a **centromere** (Figure 2-4b). During mitosis, the centromere region will effectively split in two, with one sister chromatid and its centromere-half going to one daughter nucleus and the other chromatid with its centromere-half to the second daughter nucleus. The separated chromatids are now called **daughter chromosomes,** as diagrammed in Figure 2-4b.

2.5 MITOSIS

Strictly speaking, mitosis is unrelated to DNA and chromosome replication, since the latter events occur during the S period of interphase (Figure 2-4b).

Mitosis is, however, an essential adjunct to chromosome duplication since it ensures that each daughter cell gets one sister of each chromatid pair and therefore a complete set of chromosomes. The four mitotic stages—*prophase, metaphase, anaphase,* and *telophase*—are in fact arbitrarily defined, since mitosis takes place as a continuous process. Figure 2-5 diagrams the sequence of events during mitosis as it appears under the light microscope, while Figure 2-6 is a series of photographs of mitosis in the peony.

Prophase   Mitotic prophase is heralded by the onset of chromosome coiling (Figure 2-5b), a phenomenon considered later in this chapter. As coiling and condensation progress (Figure 2-5c and d), the two sister chromatids of each chromosome can often be distinguished. The other notable morphological feature of prophase is that the nucleolus becomes undetectable with the light microscope. With the electron microscope it is apparent that the component particles of the nucleolus disperse throughout the nucleus during mitotic prophase, but the reason for this dispersal is not known.

Metaphase   The onset of metaphase is marked in most cells by the breakdown of the nuclear envelope, releasing the chromosomes and nucleoplasm to fill the greater part of the cell (Figure 2-5e). In such lower eukaryotes as protozoa and fungi, however, the nuclear envelope and nuclear integrity characteristically remain intact throughout mitosis. A more diagnostic feature of metaphase is therefore the presence of a complete **spindle.** Under a polarizing light microscope the spindle appears as a refractile zone (Figure 2-7) from which cytoplasmic components of the cell are excluded. Under the electron microscope each spindle is seen to be composed of numerous long fibers, each approximately 250 Å in diameter, known as **microtubules** (Figure 2-8, M).

In most animal cells the spindle exhibits two **poles** located on opposite sides of the cell, each pole being marked by the presence of one or two small complex structures known as **centrioles** (Figure 2-8, C). Plant cells do not ordinarily possess centrioles and their spindles, unlike those of animal cells, are not tapered at the poles; otherwise the plant and animal spindle structures are similar.

Spindle microtubules originate (or terminate, depending on one's point of view) at the poles. In animal cells they typically appear to insert into some dense material of unknown composition that surrounds the centriole. Some spindle microtubules, known as continuous microtubules, extend from one pole to the other. Others, called chromosomal microtubules, after beginning at one or the other of the two poles, extend only to the chromosomes, where they insert into the centromere regions of the sister chromatid pairs.

The site of microtubule insertion in the centromere region is called the **kinetochore.** The kinetochore may exhibit no ultrastructure which differ-

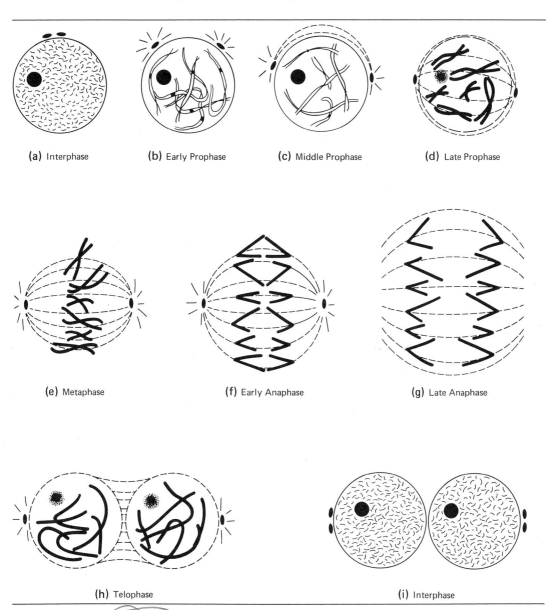

(a) Interphase

(b) Early Prophase

(c) Middle Prophase

(d) Late Prophase

(e) Metaphase

(f) Early Anaphase

(g) Late Anaphase

(h) Telophase

(i) Interphase

Figure 2-5 Mitosis in an animal cell.

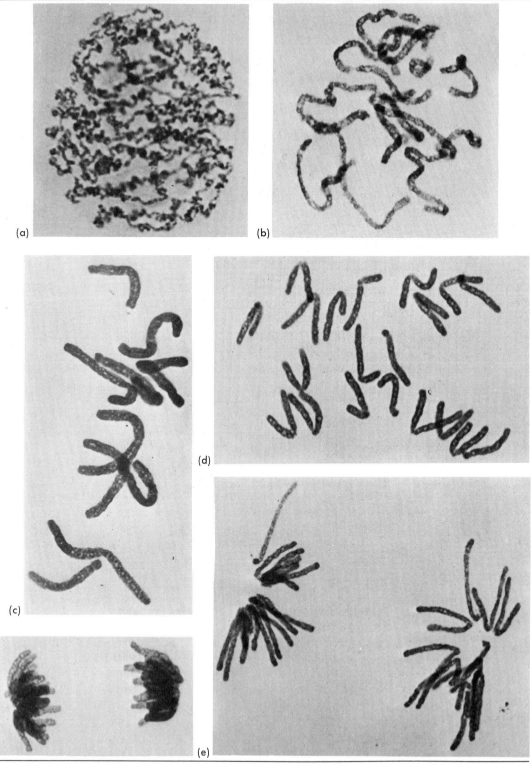

Figure 2-6 Mitosis in the peony. (a) Early prophase, (b) late prophase, (c) metaphase, (d) early anaphase, (e) late anaphase, (f) telophase. (Courtesy of M. Walters and S. Brown.)

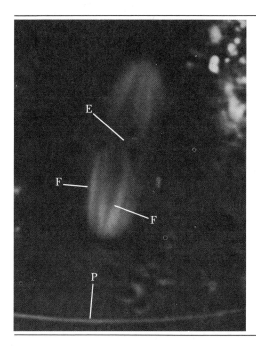

Figure 2-7 Polarizing micrograph of the spindle in a living egg from the worm *Chaetopterus*. Metaphase chromosomes (not visible) lie along the equator (E). The fibrous (F) nature of the spindle is indicated. P, plasma membrane of cell. (Courtesy of Dr. Shinya Inoué, University of Pennsylvania, from *Chromosoma* **5**:491, 1952; Berlin-Göttingen-Heidelberg: Springer.)

entiates it from the surrounding chromosome, but in a number of organisms it appears as a platelike, often banded, structure (Figure 2-8, K). In favorable material a kinetochore at metaphase can be seen to exhibit two faces, one pointing toward one pole and the other toward the other pole, and comparable numbers of chromosomal microtubules attach themselves to either face (Figure 2-8).

With the insertion of the chromosomal microtubules into the kinetochores at metaphase, the sister chromatid pairs become aligned in a plane that usually passes approximately through the cell midline (the **metaphase plate**). Since the sister chromatid pairs are held together only at their common centromere regions, the arms of each pair are free to dangle (Figure 2-5e).

Anaphase Mitotic anaphase begins with the division of the centromeres or, more precisely, of the kinetochores. The sister chromatids of a given pair now separate, and each moves, kinetochore first, toward the spindle pole nearest to it (Figure 2-5f). Once the chromatids have separated they are considered as **daughter chromosomes** (Figure 2-4b). The difference between sister chromatids and daughter chromosomes is thus an operational one; physically they are equivalent structures except that sister chromatids share a centromere, while daughter chromosomes do not. Obviously, a daughter chromosome is synonymous with a chromosome and the term is encountered only when mitotic events are being discussed.

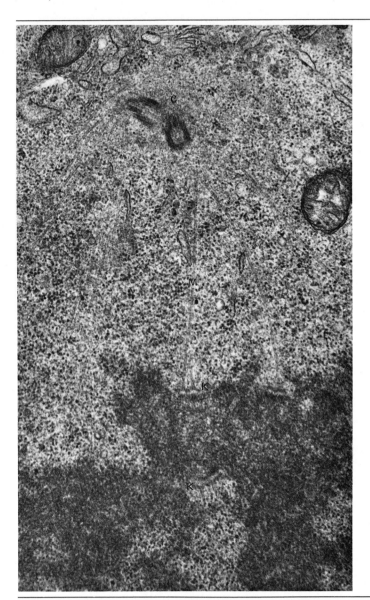

Figure 2-8 Electron micrograph of a cat kidney tubule cell in metaphase. A pair of centrioles (C) lies at one pole of the spindle, with spindle microtubules (M) radiating toward the chromosomes. The two kinetochore faces (K) of a sister chromatid pair are indicated, and chromosomal microtubules insert into each face from opposite poles. (From P. Jokelainen, *J. Ultrastruct, Res.* **19:**19, 1967.)

The alignment of chromosomes along the metaphase plate and their migration to the poles at anaphase constitutes one of the most dramatic activities exhibited by eukaryotic cells, and much research is currently focused on how such movements are generated. There is general agreement that the spindle microtubules are involved, since the drug **colchicine** has been shown to interfere specifically with spindle microtubule formation, and also to block chromosomal movements. It has been proposed that the contractile proteins actin and myosin may also participate in mitotic movements, but it is not yet clear how these various filamentous macromolecules might interact with one another to generate both the medial movements of metaphase and the poleward movements of anaphase.

Telophase and Cytokinesis Once poleward migration is complete (Figure 2-5g), a nuclear envelope forms around each set of daughter chromosomes, a nucleolus reforms, the spindle microtubules disappear, the chromosomes uncoil, and each daughter nucleus gradually assumes an interphase morphology. These events characterize telophase, the final mitotic stage (Figure 2-5h). Under most circumstances telophase is followed by cell division, or **cytokinesis,** and the nuclei return to interphase (Figure 2-5i).

At the conclusion of mitosis, therefore, both daughter nuclei contain an identical chromosome complement, the result of the equal separation of the identical sister chromatids at anaphase. The orderly alignment of the chromatids at the metaphase plate and the division of each centromere into two halves that move to opposite poles ensure that the genetic material is transmitted exactly from parent to daughter cells.

THE KARYOTYPE

2.6 GENERAL FEATURES OF KARYOTYPES

When a human metaphase cell is photographed and examined, a total of 46 sister-chromatid pairs is encountered (Figure 2-9a). Further examination reveals that each of these pairs is of different size from the others and/or differs in the position of its centromere. Some are **metacentric** (M in Figure 2-9a), with the centromere positioned medially so that the four chromatid arms are about equal in length; others are **acrocentric** (A in Figure 2-9a), with two of the arms much longer than the other two; the rest are intermediate between these two extremes.

A complete description of all the chromatid pairs possessed by a given cell type constitutes its **karyotype,** and a karyotype is usually prepared by cutting out individual chromatid pairs from a photograph like Figure 2-9a and arranging them in series according to size. When this is done for humans, it becomes apparent that each type of chromatid pair is in fact represented twice: there are two representatives of the longest metacentric

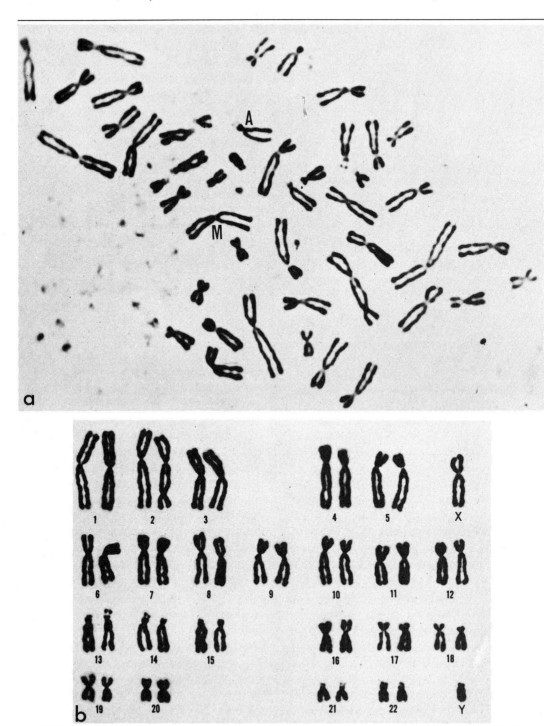

Figure 2-9 (a) Mitotic chromosomes of a human male. (b) Same mitotic chromosomes arranged in homologous pairs and numbered. (Courtesy of M. W. Shaw).

type, two representatives of the shortest metacentric type, and so on (Figure 2-9b). Therefore, while a human metaphase cell indeed possesses 46 chromatid pairs, it actually possesses only 23 *different kinds* of chromatid pairs.

In Figure 2-9b, each of the largest metacentric homologues is called chromosome 1 and each of the smallest is called chromosome 22. Chromosomes 1 to 22 are all classified as **autosomes** to contrast them with the **sex chromosomes.** The sex chromosomes are considered as the 23rd chromosomes, but because they differ in size they are placed in different positions in the display: the **X chromosome** is included in the same grouping as the middle-sized autosomes, whereas the tiny **Y chromosome,** which determines maleness in humans, is present at the end of the series (Figure 2-9b).

A karyotype such as Figure 2-9b, with each chromosome type represented twice, is diagnostic of a **diploid** nucleus, and can be contrasted with a **haploid** nucleus such as is found in egg and sperm cells, where only one representative of each chromosome type is present. The haploid chromosome complement contains, collectively, all the information in the genome, and is thereby comparable to the 7 chromosomes present in the haploid nucleus of a *Dictyostelium* cell. Each set of haploid chromosomes, in other words, will be expected to contain a minimum of one copy of every kind of gene possessed by the organism (the sex chromosomes being an exception to this rule). The number of chromosomes in a haploid complement is commonly symbolized as n, so that $n = 23$ for humans and $n = 7$ for *Dictyostelium.* Representative haploid chromosome numbers for other animals and plants are found in Table 2-1.

The origin of a human diploid nucleus is, of course, the fusion of a haploid egg nucleus with a haploid sperm nucleus to form a **zygote.** Because of its mode of formation, the zygote and all its descendant diploid cells possess two complete sets of chromosomes, one representing the maternal haploid set and the other, the paternal haploid set. These two sets will not, under ordinary circumstances, carry identical nucleotide sequences, any more than will two parents ordinarily carry identical genetic information. The chromosome sets usually carry the same kinds of genes, however, and are usually similar in morphology. For this reason they are termed **homologues,** and a diploid cell is spoken of as possessing pairs of homologous chromosomes. Sex chromosomes present the most common exception to this generalization: they may be homologous over certain regions but, as we saw for humans (Figure 2-9), they often differ in size and in the kind and number of genes they carry.

The display of a full chromosome complement as in Figure 2-9b has come to be called a karyotype, even though the term also has the more general meaning of a description of the complement in any of its guises. Karyotypes are useful in that they permit the rapid recognition of an aberration in chromosome number or morphology, as we shall see in Chapter 6. They are also of interest in establishing evolutionary relationships

Table 2-1 Haploid Chromosome Numbers

Common and Scientific Names	Chromosomes	Common and Scientific Names	Chromosomes
Human, *Homo sapiens*	23	Frog, *Rana pipiens*	13
Rhesus monkey, *Macaca mulata*	21	Toad, *Bufo americanus*	11
Horse, *Equus caballus*	32	Toad, *Xenopus laevis*	18
Pig, *Sus scrofa*	19	Goldfish, *Carassius auratus*	50
Cattle, *Bos taurus*	30	Housefly, *Musca domestica*	6
Sheep, *Ovis aries*	27	Fruit fly, *Drosophila melanogaster*	4
Cat, *Felis catus*	19	Garden pea, *Pisum sativum*	7
Dog, *Canis familiaris*	39	Bean, *Phaseolus vulgaris*	11
Rat, *Rattus norvegicus*	21	Tobacco, *Nicotiana tabacum*	24
Mouse, *Mus musculus*	20	Corn, *Zea mays*	10
Chicken, *Gallus domesticus*	ca. 39	Yeast, *Saccharomyces cerevisiae*	17
Slime mold, *Dictyostelium discoideum*	7		
Ciliate, *Paramecium aurelia*	30–63	Bread mold, *Neurospora crassa*	7
Ciliate, *Tetrahymena pyriformis*	5	*Chlamydomonas reinhardi*	16

In part from P. L. Altman and D. S. Dittmer, Eds., *Biology Data Book,* Vol. 1, 2nd ed. Bethesda, Md.: Federation of American Societies for Experimental Biology, 1972.

between different species. The mouse karyotype, for example, bears little resemblance to the human; it is made up of fewer chromosomes, all of which are small and acrocentric. The various primate karyotypes, in contrast, are all similar to one another (see Figure 20-4).

The allotment of the genetic material to more than one major chromosome is a feature that distinguishes eukaryotic organisms from bacteria and viruses (Chapter 5). The multichromosome state appears to be a more manageable way to package and distribute the enormous amount of DNA present in the eukaryotic nucleus; this, however, is not its only attribute. As we shall see in the next chapter, the processes of meiosis and sexual fusion in organisms with multiple chromosomes result in the creation of numerous new combinations of genes. Thus the multichromosomed species possess great potential for diversity, and hence for evolutionary flexibility.

2.7 KARYOTYPES USING BANDED CHROMOSOMES

It should be apparent with close study of Figure 2-9 that it is virtually impossible to distinguish many of the sister chromatid pairs from one another. For this reason, the construction of karyotypes from such conventionally stained preparations often proves to be both tedious and frustrat-

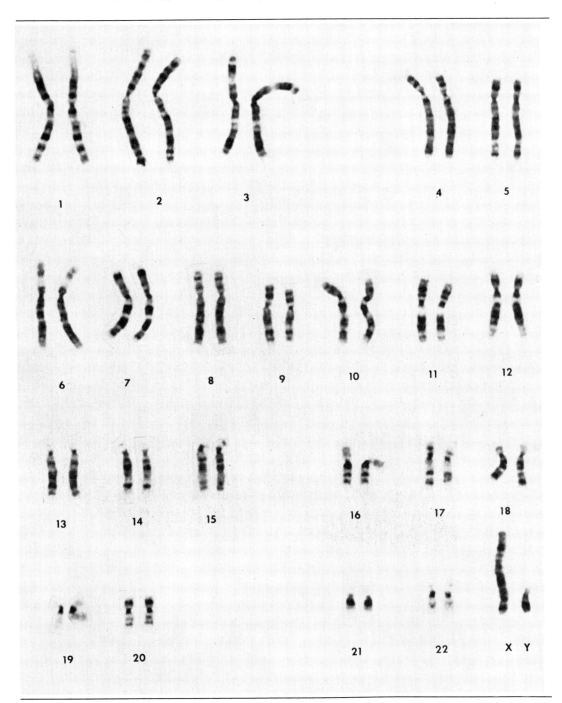

Figure 2.10a G-banding pattern of metaphase chromosomes from a normal human male. (From J. J. Yunis and M. E. Chandler, *Am. J. Pathol.* **88:**466, 1977.)

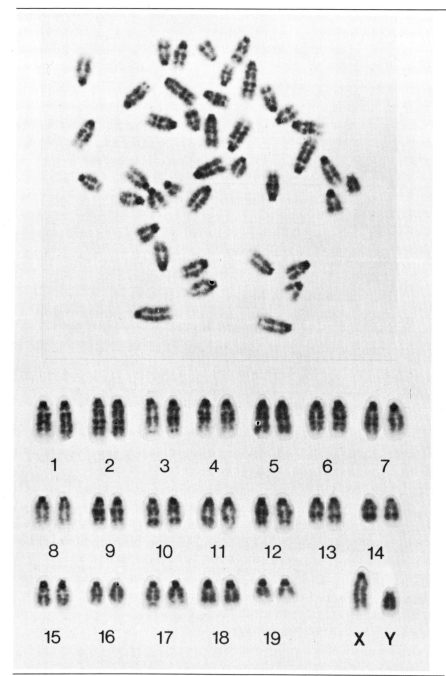

Figure 2.10b G-banding pattern of metaphase chromosomes from the mouse. (From R. A. Buckland, et al., *Exptl. Cell Res.* **69:**232, 1971.)

ing. Fortunately, staining procedures have been developed in the past 10 years which reveal discrete bands along the lengths of sister chromatids. These readily permit the recognition of individual chromatid pairs and the matching of homologues having similar banding patterns. A banded human karyotype is shown in Figure 2-10a, and its enhanced information over Figure 2-9 should be obvious. Figure 2-10b shows, for comparison, the shape and banding patterns in a mouse karyotype.

Of the many staining protocols that have been devised in various laboratories to produce banded chromosomes, the two most generally used are the Acid-Saline-Giemsa (ASG) technique, which reveals **G-bands** (for Giemsa stain), and the Quinacrine mustard technique, which produces fluorescent **Q bands;** technical details for both are described in Box 2.3. G- and Q-bands have the same locations and are presumed to reveal the same underlying chromosome structures. While the molecular basis for the staining reactions remains a matter of speculation, the techniques themselves have revolutionized the discipline known as cytogenetics.

Box 2.3
STAINING AND
BANDING
CHROMOSOMES

Feulgen Staining Cells are subject to a mild hydrolysis in 1 N HCl at 60°, usually for about 10 minutes. This treatment produces a free aldehyde group in deoxyribose molecules. The aldehyde will then react with a chemical known as Schiff's reagent (basic fuchsin bleached with sulfurous acid) to give a deep pink color. The ribose of RNA will not form an aldehyde under these conditions, and the reaction is thus specific for DNA.

Q Banding The Q bands (from *quinacrine*) are the fluorescent bands observed after quinacrine mustard staining and observation with ultraviolet light. The distal ends of each chromatid are not stained by this technique. The Y chromosome becomes brightly fluorescent, both in the interphase nucleus and in metaphase.

R Banding The R bands (from *reverse*) are those located in the zones that do not fluoresce with quinacrine mustard, that is, they are between the Q bands. They can be visualized as green, brightly fluorescent bands with acridine orange staining.

G Banding The G bands (from *Giemsa*) have the same location as the Q bands and do not require fluorescent microscopy. Many techniques are available, each involving some pretreatment of the chromosomes. In the acid-saline-Giemsa (ASG) technique, for example, cells are incubated in citric acid and NaCl for 1 hr at 60°C and are then treated with the Giemsa stain. Proteolytic enzyme treatment also reveals these bands.

C Banding The C bands correspond to *constitutive heterochromatin* (see Section 2.13).

At a conference in Paris in 1971, the bands visualized in human mid-metaphase karyotypes by such techniques were assigned a nomenclature in which the letters p and q represent, respectively, the short and long "arms" of a metaphase chromatid; these arms are then subdivided by numbers. It has recently been found that in late prophase, when mitotic chromosome condensation is less advanced, many more bands can be visualized; these proceed to coalesce with one another to produce the major bands seen in metaphase, as illustrated in Figure 2-11. Thus, in late-prophase karyotypes, it is presently possible to identify 1256 bands per human haploid chromosome complement. In Figure 2-12, the Paris-Conference numbers are given for the left chromatid of each pair and the late-prophase numbers are found in the right-hand chromatids. The utility of such band numbers will become evident in Chapters 6 and 12.

ATYPICAL EUKARYOTIC CELL CYCLES

2.8   CELL CYCLE OF CILIATED PROTOZOA

Ciliated protozoa such as *Tetrahymena* and *Paramecium* are atypical in that each cell contains two types of nuclei: a **macronucleus** which controls the growth and development of the organism, and one or two **micronuclei** which function only at the time of sexual reproduction. Each micronucleus

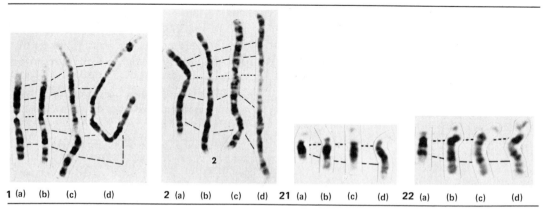

1 (a)   (b)   (c)   (d)   **2** (a)   (b)   (c)   (d)   **21** (a)   (b)   (c)   (d)   **22** (a)   (b)   (c)   (d)

Figure 2-11   Human G-banded chromosomes 1, 2, 21, and 22 at mid-metaphase (a), early metaphase (b), early prometaphase (c) and late prophase (d), showing the progressive coalescence of the multiple fine bands of late prophase into the thicker and fewer dark bands of metaphase. (From J. J. Yunis, *Science* **191:**1268, 1976.)

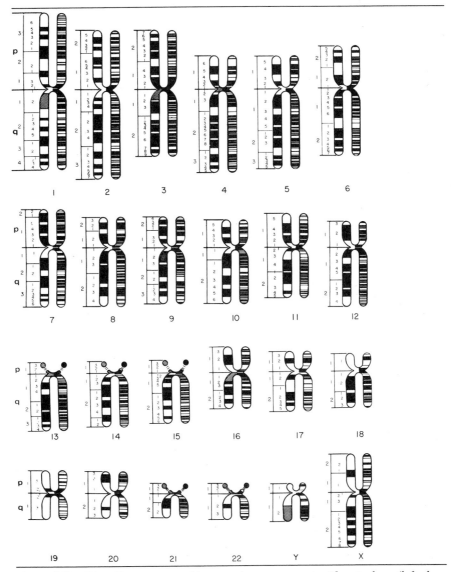

Figure 2-12 G-banding patterns of human chromosomes at mid-metaphase (left chromatid of each chromosome) and at late prophase (right arm). Numbering system for the mid-metaphase bands is that established at the Paris Conference, 1971, with p and q being the short and long arms, respectively. The uppermost band of the first chromosome is designated 1p36, the next as 1p35, and so on.

is diploid, whereas the macronucleus is said to be **hyperpolyploid,** having anywhere from 45 (*Tetrahymena*) to 800 (*Paramecium*) times the amount of DNA in the haploid complement. A macronucleus originates from a micronucleus during the sexual act of conjugation (Section 3.14), proceeds to amplify its chromosome complement, and then transmits this enormous quantity of DNA to daughter macronuclei during asexual reproduction. Since no conventional mitotic configurations are observed during macronuclear division, the division process is said to be amitotic. It is not yet known whether a precise allotment of "genome equivalents" is distributed to each daughter during amitosis or whether the process is more random; nor indeed is it clear whether in *Paramecium,* for example, the entire genome is represented 800 times per macronucleus or whether certain chromosomes are represented more abundantly than others.

2.9   CELL CYCLE OF DIPTERAN LARVA CELLS

When a eukaryotic microorganism such as yeast is starved of nutrients, it stops dividing, enters a stationary G_0 phase, and eventually dies. Several cell types in dipteran larva, the best known being the salivary gland cells of the fruit fly *Drosophila,* also stop mitotic divisions after perhaps 18 hours of larval development, but chromosomal DNA replication and cell growth continue apace. As a result, huge cells are formed, and each comes to contain, by a $2 \longrightarrow 4 \longrightarrow 8 \longrightarrow 16$ doubling series, as much as 1024 times the haploid amount of DNA. The numerous copies of each chromosome type remain associated with one another, moreover, so that each of the 8 chromosomes in the nucleus becomes greatly amplified in size. Finally, homologous chromosomes tend to associate with one another, a phenomenon known as **somatic pairing,** so that when the nuclei are broken open and their contents examined with the light microscope, it appears as though 4 enormous chromosomes are present (Figure 2-13).

In selected cells of the *Drosophila* larva, therefore, an **endopolyploidization** of the diploid genome is said to occur, and the multistranded chromosomes that result from this process are termed **polytene chromosomes.** The endopolyploid cells are "terminal" in the sense that they never again divide and are eventually discarded during pupation. They remain, however, very much alive in the sense that they respond to environmental stimuli and produce specific classes of proteins (Section 8.20).

For the geneticist, polytene chromosomes are perhaps most fascinating in that they exhibit a series of **bands** of varying widths, each band flanked by lightly staining **interbands,** also of varying widths (Figure 2-13). The bands serve the same "landmark" function as the bands in the ASG-stained chromosomes (Figure 2-10) permitting a ready identification of each of the 4 polytene chromosomes. Even more important, the chromosomes are so

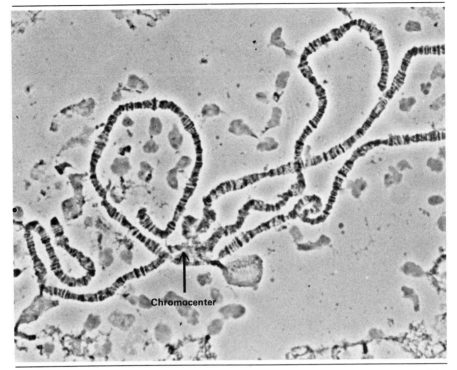

Figure 2-13 Salivary gland chromosomes from *Drosophila melanogaster* as visualized by phase contrast microscopy. (From L. Silver and S. Elgin, *Proc. Natl. Acad. Sci. U.S.* **73:**423, 1976.)

large and the bands so numerous that it becomes possible to identify very narrow intervals along the length of each chromosome. In 1938, C. Bridges reported that the 4 polytene chromosomes of this organism, when widely stretched, reveal approximately 5000–6000 bands; more recent studies estimate 6900 bands. The mode of polytene band formation is discussed later in the chapter.

STRUCTURAL PROPERTIES OF EUKARYOTIC CHROMOSOMES

The structural complexity of a human mitotic chromosome becomes apparent when it is examined at high magnifications (Figure 2-8): a veritable thicket of fibers is seen. Considerable research is presently focused on describing the chemical and physical properties of eukaryotic chromatin in its interphase state and the mechanism whereby it becomes packaged into discrete mitotic chromosomes.

2.10   CHROMOSOMAL DNA

Table 2-2 lists the total amount of DNA found per haploid nucleus or cell in a variety of prokaryotic and eukaryotic organisms. This amount of DNA is often refered to as the **C value** (C standing for the DNA concentration per nucleus), and it is clear from Table 2-2 that C values range over several orders of magnitude, a fact considered again in Section 7.6. If the C value for a given species is divided by its haploid chromosome number (Table 2-1), a value for the average amount of DNA per chromosome can be calculated; for example, the average human chromosome contains about 6×10^{10} daltons of DNA, which corresponds to approximately 3 cm of duplex DNA (see Table 2-3). It is presently impossible to isolate molecules of this size without breaking them into pieces. In an organism such as yeast with a smaller C value (9×10^9 daltons; Table 2-2) and probably 17 chromosomes, however, the average DNA content per chromosome is only 5×10^8 daltons, and single DNA molecules have been isolated from yeast nuclei with molecular weights in the 3- to 6×10^8 dalton range. Chromosome-sized DNA molecules have also been detected in *Drosophila* nuclear extracts. It is presently believed, therefore, that the DNA in each eukaryotic chromosome exists as a single, continuous molecule having the form of an enormously long double helix; the electron micrograph illustrating the cover of this text displays such DNA in a particularly dramatic fashion.

2.11   CHROMOSOMAL PROTEIN

The chromosomal DNA seen on the cover of the text could be visualized only after extracting proteins from the chromosomes. Native chromosomal DNA is invariably associated with two major classes of protein—the histones and the nonhistones—to form a **nucleoprotein fiber.**

Histones   Histones are basic proteins, meaning that they carry a net positive charge at physiological pH. This charge is carried by $-NH_3^+$ groups of the amino acids lysine and arginine (Figure 2-14) which represent a significant proportion (20 to 30 mole-percent) of the total amino acids in each histone molecule. The lysines and arginines, moreover, tend to be clustered towards one end of each histone so that one end of the protein has a very high density of positive charge. DNA, as you will recall from Chapter 1, carries a high density of exposed, negatively charged $PO_4^=$ groups along its "backbone." These negative charges are believed to interact with the positively charged ends of the histones to produce an intimate complex, often called a **nucleohistone.**

Table 2-2   DNA Content per Nucleus of Various Organisms

Organism	Daltons	Nucleotide Pairs
Viruses		
Bacterial		
ϕX174	1.6×10^6	5500[b]
λ	30×10^6	45×10^3
T5	85×10^6	130×10^3
T2	130×10^6	200×10^3
Animal		
Shope papilloma	5×10^6	7.5×10^3
Adenovirus 12	14×10^6	20×10^3
Fowlpox	230×10^6	350×10^3
Bacteria[a]		
Mycoplasma gallisepticum	0.2×10^9	0.3×10^6
Hemophilus influenzae	0.7×10^9	1×10^6
Escherichia coli	2.6×10^9	4×10^6
Pseudomonas sp.	2.4×10^9	4×10^6
Bacillus subtilis	2×10^9	3×10^6
Fungi		
Saccharomyces cerevisiae	9×10^9	14×10^6
Invertebrates		
Tube sponge	0.06×10^{12}	0.1×10^9
Jellyfish[a]	0.2×10^{12}	0.3×10^9
Sea urchin[a]	0.5×10^{12}	0.8×10^9
Nereid worm[a]	1×10^{12}	1.4×10^9
Snail, *Tectorius muricatus*[a]	4×10^{12}	6.3×10^9
Cliff crab	1×10^{12}	1.4×10^9
Drosophila melanogaster	0.12×10^{12}	0.2×10^9
Vertebrates		
Lung fish	60×10^{12}	94×10^9
Frog	28×10^{12}	45×10^9
Salamander (Amphiuma)	120×10^{12}	180×10^9
Shark, *Carcharias obscurus*	3.4×10^{12}	5.2×10^9
Carp	2×10^{12}	3.3×10^9
Chicken	1.2×10^{12}	2.1×10^9
Mouse	3.0×10^{12}	4.7×10^9
Human	3.6×10^{12}	5.6×10^9
Plants		
Arabidopsis thaliana	2.6×10^{12}	4×10^9
Zea mays	20×10^{12}	30×10^9
Tradescantia paludosa	40×10^{12}	60×10^9

[a] Haploid cells. Other numbers are for diploid cells.
[b] Nucleotides; chromosome is single-stranded.

From B. J. McCarthy, in *Handbook of Cytology* (A. Lima-de-Faria, Ed.). Amsterdam: North-Holland, 1969, Table 1.

Table 2-3   Rule-of-Thumb Conversion Factors for Duplex DNA

Average molecular weight of 1 nucleotide pair $\simeq$ 660 daltons.
1 pg (10^{-12} g) duplex DNA $\simeq$ 6×10^{11} daltons $\simeq$ 10^9 nucleotide pairs.
1 μ duplex DNA $\simeq$ 2×10^6 daltons $\simeq$ 3×10^3 nucleotide pairs.
1 structural gene $\simeq$ 10^3 nucleotide pairs

Five major types of histone are found in most cells, each type differing in its relative content of arginine and lysine (Table 2-4). Within each major type of histone a number of subtypes are found, all similar to one another. The five types appear to be quite constant from one cell to another and from one tissue to another. In other words, there is generally no evidence that specific types of histones are found only in specific cell types. The most prominent exception to this generalization is provided by sperm cells. The DNA of these cells frequently loses most of its histone and becomes associated with other kinds of basic proteins.

Histones and DNA are present in mammalian chromatin in approximately equal proportions on a weight-per-weight basis. This proportion is maintained throughout the mammalian cell cycle via a strict "coupling" between histone synthesis and DNA replication: histones are synthesized only during the S phase, and an experimentally induced interruption of DNA replication is rapidly followed by a cessation of histone biosynthesis. Similarly, a disruption of histone synthesis results in a block in DNA replication.

Nucleosomes   When chromatin is gently released from a nucleus onto an electron microscope grid, beaded structures can be visualized (Figure 2-15); these may or may not appear connected by thin threads or "linkers," depending on the method of preparation. Each bead is called a **nucleosome** or **nu body,** and each is about 110 Å in diameter.

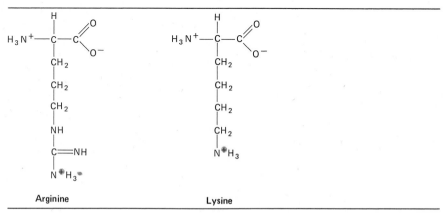

Arginine   Lysine

Figure 2-14   The structures of the basic amino acids, arginine and lysine.

Table 2-4 Characterization of the Histones From Calf Thymus

Class	Fraction	Lys/Arg Ratio	Total Amino Acids	Molecular Weight
Very lysine rich	H1	22.0	~215	~21,500
Lysine rich	H2A	1.17	129	14,004
	H2B	2.50	125	13,774
Arginine rich	H3	0.72	135	15,324
	H4	0.79	102	11,282

From S. C. R. Elgin and H. Weintraub, *Ann. Rev. Biochem.* **44**:725, 1975.

Single nucleosomes prove to be readily isolated by treating chromatin with the enzyme staphylococcal nuclease, and their conformation and composition have been intensively studied. Figure 2-16 presents a working model of nucleosome structure. The helix is seen to enter and leave the nucleosome from the same side, forming two adjacent loops consisting of

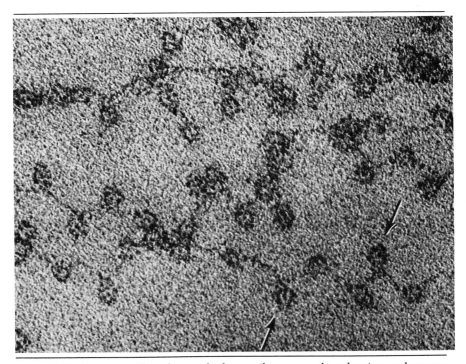

Figure 2-15 Chromatin fibers from a chicken erythrocyte nucleus showing nucleosomes (arrows) (~100 Å) and connecting strands (~140 Å in length). The strands become evident when the nucleosomes "unravel" during certain isolation procedures. (Courtesy D. E. Olins and A. L. Olins.)

Figure 2-16   A working model of the topological organization of a nucleosome, with plastic tubing representing the two turns of DNA and plasticine representing the histones. Two each of the four histone species are present. H4 (north-south) and H3 (east-west) molecules bridge the nucleosome, making contacts in two regions with the DNA; H2A and H2B have only one (identified) region of contact each. A rotational axis of symmetry exists along the north-south axis. (Courtesy of John C. Wooley.)

140 base pairs of DNA. Eight histone molecules fill the central portion of the nucleosome and make contacts with the helix at specific sites. The histones in the core have a very regular composition, namely, 2 molecules of H2A, 2 of H2B, 2 of H3, and 2 of H4.

What do nucleosomes represent? They are much too small to represent genes (as we shall see, an average gene is believed to comprise about 1000 nucleotide pairs), and do not appear to contain specific nucleotide sequences. Nucleosome formation is therefore presently regarded as a "packaging device" which serves to compact long, slinky DNA molecules into a more managable form. As we consider in future chapters how eukaryotic DNA is replicated, recombined, mutated, and transcribed, these molecular activities will ultimately have to be visualized as being acted out not on a DNA duplex but on DNA organized into nucleosomes.

Nonhistones   The nonhistone chromosomal proteins are defined as those proteins other than the histones that are isolated together with DNA in purified chromatin. Since these proteins are often more loosely associated with DNA than are the histones, and since most estimates indicate that the nonhistone class includes at least 20 major proteins and perhaps hundreds of minor proteins, these proteins are understandably difficult to isolate and study. Interest in these proteins is, nonetheless, extremely high, since a number of reports claim that particular nonhistone proteins are either confined to particular types of cells or else are far more abundant in certain types of tissues than in others. Such tissue specificity suggests, of course, a role for the nonhistone proteins in determining which sectors of the chromosomes are to express their genetic information, a concept explored more fully in Section 18.7.

The nonhistones also appear to play a structural role in chromosome organization. If chromatin is so treated that all its histone is removed then, as illustrated on the cover of this text, DNA unravels from individual chromosomes and a proteinaceous **scaffold,** shaped like a chromosome, remains. Presumably this scaffold is involved in the mitotic coiling process.

2.12 CHROMOMERES

When chromosomes are in the process of condensing during mitotic or meiotic (Chapter 3) prophase, a beaded substructure is evident along their lengths in favorable light-microscope preparations (Figure 2-6b). Each of these beads is much larger than a nucleosome, and each is referred to as a **chromomere.** An increasing number of cytogeneticists believe that chromomeres may correspond to genes or small groups of genes, an idea first proposed by B. McClintock in 1931. The most convincing evidence for this thesis, to be considered at length in later chapters, comes from the salivary-gland polytene chromosomes of *Drosophila* (Figure 2-13); in these, each band behaves, in genetic and cytogenetic studies, as though it were one or a small number of genes, and each band appears to originate as a side-to-side alignment of chromomeres, as diagrammed in Figure 2-17.

Chromomeres are also believed to be ultimately related to the bands detected in mitotic chromosomes. As we described earlier in this chapter, the thick G-bands of metaphase chromosomes derive from the progressive coalescence of many fine bands seen during late prophase (Figure 2-11). Furthermore, when early or mid-prophase nuclei are examined, some 2000 to 3000 dark and light units are detectable by the same G-banding techniques; these are believed to be chromomere equivalents, and they apparently coalesce to give the late-prophase bands. Since estimates as to the number of human genes are of the order of 30,000–50,000 (Section 18.9), each of these several thousand early-prophase units may well represent a small number of genes.

One cannot at present offer any explanation as to why or how it is that, when chromosomes coil for mitosis or meiosis, they display their constitu-

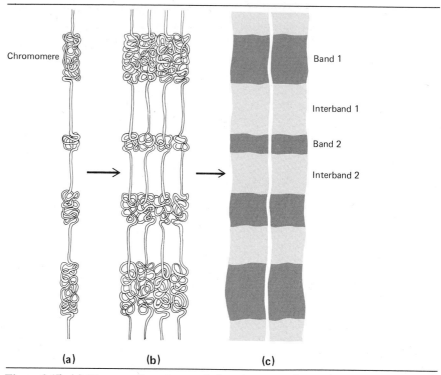

Figure 2-17 (a) Diagram of an elemental chromosome within a giant polytene chromosomes showing a folded (or coiled) configuration in the chromomeres and an extended configuration in the intervals between chromomeres. (b) Lateral alignment of four such elemental chromosomes showing the amplification of individual chromomeres into bands. (c) Somatic pairing between two polytene homologues to form what is apparently a single giant structure, equivalent to the chromosome arms seen in Figure 2-13. (After E. J. DuPraw and P. M. M. Rae, *Nature* **212**:598, 1966.)

ent genes as periodic regions of aggregated chromatin. Also unknown at present is whether a nonvisible counterpart to the chromomere exists in the interphase nucleus—that is, whether individual interphase genes are structurally demarcated in some fashion.

2.13 CONSTITUTIVE HETEROCHROMATIN

An important exception to the rule, stated earlier, that chromatin condensation occurs only in the mitotic phase of the cell cycle is given by a particular class of chromatin known as **constitutive heterochromatin.** Throughout the interphase, constitutive heterochromatin remains condensed, and therefore deeply staining, in contrast to the faintly staining **euchromatin** that fills the rest of the nucleus. Blocks of heterochromatin in the interphase nucleus are known as **chromocenters,** and these often fuse to form a single large composite chromocenter. At early prophase, constitutive heterochromatin is considerably more condensed than the remaining chromatin, while at late prophase it is identified as those regions where sister chromatids appear joined together.

In metaphase karyotypes, regions of constitutive heterochromatin are best identified by yet another banding technique, which produces the so-called **C-bands** (see Box 2.3), C standing for constitutive heterochromatin. These "bands," which are perhaps more properly described as zones, are always localized to particular chromosome sites: in the mouse, for example, C-bands are restricted to the centromere regions of every chromosome, while in *Drosophila* the entire Y chromosome, perhaps 1/3 to 1/2 of the X chromosome, and 1/4 of each of the large metacentrics (chromosomes 2 and 3) are stained by the C-banding procedure and remain condensed throughout interphase.

Constitutive heterochromatin differs from euchromatin not only in its condensation patterns but also in its replication patterns during the cell cycle. In both *Drosophila* and humans, for example, replication of constitutive heterochromatin (visualized by autoradiography of ^{3}H-thymidine-pulsed cells) is restricted to a brief period late in the S phase. Thus, even though it is located in a number of different chromosomes, the whole class of constitutive heterochromatin receives the same kind of differential treatment from the replication enzymes. During the "atypical" cell cycle that leads to polytene-chromosome formation in *Drosophila,* moreover, the **late-replicating** behavior of constitutive heterochromatin gives way to an **under-replicating** behavior: those chromosome segments known to contain constitutive heterochromatin are represented, at most, only a few times in polytene chromosomes, while the rest of the genome is copied as many as 10 times.

Finally, constitutive heterochromatin differs from euchromatin in its genetic content. Genetic analyses, presented in later chapters, indicate that it contains very few if any functional genes. Moreover, the DNA sequences of constitutive heterochromatin are different from those of euchromatin (Section 7.13), and heterochromatin can affect the expression of genes in contiguous, euchromatic portions of a chromosome (Section 18.18).

2.14 CHROMOSOME CONDENSATION

As noted earlier, the chromatin in an interphase nucleus can be said to have as its "primary structure" the nucleosome configuration. Metaphase chromatin is found to be organized in the same nucleosome units as interphase chromatin; therefore, it is clear that the condensation of chromatin that occurs during mitosis (and meiosis), and in constitutive heterochromatin as well, must be the result of secondary and tertiary kinds of interactions within a nucleosome "string."

Such interactions must indeed be remarkable. If we think of an average-sized human chromosome with perhaps 7 cm of DNA and assume that it is packaged into nucleosome units, the resultant chromatin fiber will measure perhaps 1 cm in length and 100 Å in width. This must somehow condense into a single sister chromatid that is perhaps 0.5μ wide and 10μ long. Such a feat is comparable to packaging an extremely thin strand of material 35 meters long into a cylinder 8 mm long and 1 mm in diameter.

As we have seen (Figure 2-6c), the early-prophase manifestations of the coiling process are the chromomeres, which then condense upon themselves to form the thick metaphase chromatids. Therefore, it seems likely that condensation begins at numerous sites along the length of each chromatin fiber, causing short segments of the fiber to condense about their own axes to produce the chromomeres. No molecular basis can yet be given for these events. Experiments have been performed in which cell cycles have been synchronized so that many nuclei are entering prophase at the same time, and it is reported that H1 histone molecules are specifically phosphorylated just at the time that chromosome condensation becomes evident with the phase microscope. Other studies, cited earlier, have suggested a role for nonhistone proteins in chromosome condensation. Whatever its nature, there is no doubt that the coiling process occurs in a highly complex fashion.

References

The Organization of Eukaryotic Cells

Cavalier-Smith, T. "The origin of nuclei and of eukaryotic cells," Nature **256**:463–468 (1975).
DuPraw, E. J. DNA and Chromosomes. New York: Holt, Rinehart and Winston, 1970.
Keller, J. M., and D. E. Riley. "Nuclear ghosts: a nonmembranous structural component of mammalian cell nuclei," Science **193**:399–401 (1976).
Novikoff, A. B., and E. Holtzman. Cells and Organelles, 2nd ed. New York: Holt, Rinehart, and Winston, 1976.

The Cell Cycle and Mitosis

Blumenthal, A. B., H. J. Kriegstein, and D. S. Hogness. "The units of DNA replication in Drosophila melanogaster chromosomes," Cold Spring Harbor Symp. Quant. Biol. **38**:205–223 (1974).
Darlington, C. P., and K. R. Lewis. Chromosomes Today, Vol. II. Edinburgh: Oliver & Boyd, 1969.
Dutrillaux, B., J. Couturier, C. Richer, and E. Viegas-Péquinot. "Sequence of DNA replication in 277 R- and Q-bands of human chromosomes using a BrdU treatment," Chromosoma **58**:51–61 (1976).
Hartwell, L. H., J. Culotti, J. R. Pringle, and B. J. Reid. "Genetic control of the cell division cycle in yeast," Science **183**:46–51 (1974).
*Hohmann, P., R. A. Tobey, and L. R. Gurley. "Phosphorylation of distinct regions of f1 histone. Relationship to the cell cycle," J. Biol. Chem. **251**:3685–3692 (1976).

*Denotes articles described specifically in the chapter.

Housman, D., and J. A. Huberman. "Changes in the rate of DNA replication fork movement during S phase in mammalian cells," *J. Mol. Biol.* **94:**173–181 (1975).

*Huberman, J. A., and A. D. Riggs. "On the mechanism of DNA replication in mammalian chromosomes," *J. Mol. Biol.* **32:**327–341 (1968).

*Inglis, R. J., T. A. Langan, H. R. Matthews, D. G. Hardie, and E. M. Bradbury. "Advance of mitosis by histone phosphokinase," *Exp. Cell Res.* **97:**418–425 (1976).

*Kavenoff, R., and B. H. Zimm. "Chromosome-sized DNA molecules from *Drosophila,*" *Chromosoma* **41:**1–27 (1973).

Kubai, D. F. "The evolution of the mitotic spindle," *Int. Rev. Cytol.* **43:**167–228 (1975).

Mazia, D. "Mitosis and the physiology of cell divisions." In *The Cell,* Vol. 3, *Meiosis and Mitosis,* p. 77–412 (J. Brachet and A. E. Mirsky, Eds.). New York: Academic, 1961.

Monesi, V. "DNA, RNA, and protein synthesis during the mitotic cell cycle." In *Handbook of Cytology* (A. Lima-de-Faria, Ed.). Amsterdam: North Holland, 1969.

Naha, P. M., A. L. Meyer, and K. Hewitt. "Mapping of the G_1 phase of a mammalian cell cycle," *Nature* **258:**49–53 (1975).

Rattner, J. B., and M. W. Berns. "Centriole behavior in early mitosis of kangaroo rat cells," *Chromosoma* **54:**387–395 (1976).

Roos, U.-P. "Light and electron microscopy of kangaroo rat cells in mitosis. III. Patterns of chromosome behavior during prometaphase," *Chromosoma* **54:**363–385 (1976).

Karyotyping

Comings, D. E., B. W. Kovacs, B. E. Avelino, and D. G. Harris. "Mechanisms of chromosome banding. V. Quinacrine banding," *Chromosoma* **50:**111–145 (1975).

Comings, D. E. "Mechanisms of chromosome banding. VIII. Hoechst 33258-DNA interaction," *Chromosoma* **52:**229–243 (1975).

Darlington, C. D., and K. R. Lewis, Eds. *Chromosomes Today,* Vol. I. New York: Plenum, 1966. (Numerous articles concerned with karyotypes.)

Dutrillaux, B., and J. Lejeune. "New techniques in the study of human chromosomes: methods and applications," *Adv. Human Gen.* **5:**119–156 (1975).

Gatti, M., S. Pimpinelli, and G. Santini. "Characterization of *Drosophila* heterochromatin. I. Staining and decondensation with Hoechst 33258 and quinacrine," *Chromosoma* **57:**351–375 (1976).

Hamerton, J. L. *Human Cytogenetics.* New York: Academic, 1971 (2 volumes).

Hsu, T. C. "Longitudinal differentiation of chromosomes," *Ann. Rev. Genet.* **7:**153–176 (1973).

Jalal, S. M., R. W. Clark, T. C. Hsu, and S. Pathak. "Cytological differentiation of constitutive heterochromatin," *Chromosoma* **48:**391–403 (1974).

Lefevre, G. Jr. "A photographic representation and interpretation of the polytene chromosomes of *Drosophila melanogaster* salivary glands. In *The Genetics and Biology of Drosophila v. Ia.* (M. Ashburner and E. Novitski, Eds.). London, Academic Press, 1976, pp. 32–66.

Ohno, S. *Sex Chromosomes and Sex-linked Genes.* New York: Springer, 1967.

*Paris Conference (1971). *Standardization in Human Cytogenetics.* New York: National Foundation-March of Dimes, 1972.

Schnedl, W. "Banding patterns in chromosomes," *Int. Rev. Cytol.,* **Suppl. 4:**237–272 (1974).

Sharma, A. K., and A. Sharma. *Chromosome Techniques, Theory and Practice,* 2nd edition. Baltimore: University Park Press, 1972.

*Sumner, A. T., H. J. Evans, and R. A. Buckland. "New technique for distinguishing between human chromosomes," *Nature New Biol.* **232:**31–32 (1971).

*Yunis, J. J. "High resolution of human chromosomes," *Science* **191:**1268–1270 (1976).

Chromosomal Proteins

Bustin, M., D. Goldblatt, and R. Sperling. "Chromatin structure visualization by immuno-electron microscopy," *Cell* **7:**297–304 (1976).

Elgin, S. C. R., and H. Weintraub. "Chromosomal proteins and chromatin structure," *Ann. Rev. Biochem.* **44**:725–774 (1975).

Hnilica, L. S. *The Structure and Biological Function of Histones.* New York: CRC Press, 1972.

*Phillips, D. M. P., Ed. *Histones and Nucleohistones.* London: Plenum, 1971.

Silver, L. M., and S. C. R. Elgin. A method for determination of the *in situ* distribution of chromosomal proteins. *PNAS* **73**:423–427 (1976).

Chromatin Substructure and Nucleosomes

Camerini-Otero, R. D., B. Sollner-Webb, and G. Felsenfeld. "The organization of histones and DNA in chromatin: evidence for an arginine-rich histone kernel," *Cell* **8**:333–347 (1976).

Crick, F. H. C. and A. Klug. "Kinky helix," *Nature* **255**:530–533 (1975).

Crick, F. H. C. "Linking numbers and nucleosomes," *Proc. Natl. Acad. Sci.* **73**:2639–2642 (1976).

*Kornberg, R. D. "Chromatin structure: a repeating unit of histones and DNA," *Science* **184**:868–871 (1974).

Martinson, H. G., M. D. Shetlan, and B. J. McCarthy. "Histone-histone interactions with chromatin. Cross-linking studies using ultraviolet light," *Biochem.* **15**:2002–2007 (1976).

*Noll, M., and R. D. Kornberg. "Action of micrococcal nuclease on chromatin and the location of histone H1," *J. Mol. Biol.* (1976).

*Olins, A. L., R. D. Carlson, and D. E. Olins. "Visualization of chromatin substructure: ν^{28} bodies," *J. Cell Biol.* **64**:528–537 (1975).

*Oudet, P., M. Gross-Bellard, and P. Chambon. "Electron microscopic and biochemical evidence that chromatin structure is a repeating unit," *Cell* **4**:281–300 (1975).

*Paulson, J. R., and U. K. Laemmli. "The structure of histone-depleted metaphase chromosomes," *Cell* **12**:817–828 (1977).

Sollner-Webb, B., R. D. Camerini-Otero, and G. Felsenfeld. "Chromatin structure as probed by nucleases and proteases: Evidence for the central role of histones H3 and H4," *Cell* **9**:179–193 (1976).

*Wooley, J. C., and J. P. Langmore. "Scanning transmission electron microscopy studies on chromatin architecture." In: *Molecular Human Cytogenetics,* R. S. Sparkes, D. Comings, and C. F. Fox, Eds. New York: Academic, 1977.

(a)   (b)

(a) Arthur Riggs (City of Hope Medical Center in Duarte, California) and Joel Huberman (Roswell Park Memorial Institute, Buffalo, N.Y.) provided early evidence of discontinuous DNA replication in mammalian cells. (b) Ada and David Olins (Oak Ridge National Laboratory) were among the first to describe and characterize the nucleosome structure of eukaryotic chromatin.

Questions and Problems

1. Distinguish between the following: kinetochore, centriole, centromere, chromocenter, colchicine, cytokinesis.

2. Cells entering the S phase of their cycle were exposed to ^{3}H-thymidine. When these cells reached mitotic metaphase, some were fixed and subjected to autoradiography. Would you expect all the sister chromatids to be labeled, half the sister chromatids to be labeled, or some other labeling pattern? Explain with diagrams such as Figure 2-4b, where the gray DNA strands can be considered ^{3}H-labeled and the black strands unlabeled.

3. The unfixed daughter cells resulting from the mitosis described in problem 2 were allowed to undergo a second cell cycle, this time without any ^{3}H-label present. When the daughter cells reached mitotic metaphase, some were fixed and subjected to autoradiography. Would you expect all the sister chromatids to be labeled, half the sister chromatids to be labeled, or some other labeling pattern? Explain with diagrams such as Figure 2-4b.

4. When polytene cells are exposed to ^{3}H-thymidine pulses during an endo-duplication cycle and autoradiographs are made, certain bands are radioactively labeled and others are not. By following the labeling pattern of known marker bands in many different samples, it can be demonstrated that certain bands start replication before others, and that each requires distinctive lengths of time to complete replication. How might you relate these observations to the autoradiograms shown in Figure 2-3?

5. An average polytene chromosome band is about 30 μ in length. In what way is this fact relevant to your answer to problem 4?

6. If we let X be the minimum amount of genetic material that carries all the information of a species, then in an organism with a diploid chromosome number of 2, how much DNA (X, 2X, 4X, etc.) is found in the following: an egg nucleus; a sister chromatid; a daughter nucleus following mitosis; a homologue following mitosis; a nucleus at the onset of mitotic coiling?

7. Given the DNA-content-per-nucleus values provided in Table 2-2, calculate C values for the 8 vertebrates listed. Does C value increase with evolutionary advance? Explain.

8. (a) Using the conversion factors provided in Table 2-3 and the human C value calculated in problem 7, how many genes would be present in the human genome if all its DNA contained genetic information?
(b) Various estimates set the number of genes in the human genome at about 50,000 (Chapter 18). What percent of human DNA would, in such a case, be devoted to carrying genes, and what percent would be involved in other functions? Can you speculate as to what some of these other functions might be?

9. Distinguish between two sister chromatids and two homologous chromosomes; distinguish between a diploid and a haploid nucleus.

10. A karyotype of a mitotic nucleus from a female cat shows 76 sister chromatids. What is the diploid and haploid chromosome number of the cat? How many homologous chromosome pairs are present?

11. Related organisms frequently have similar chromosome numbers. Document this statement with several examples from Table 2-1, and note as well some exceptions evident in the Table.

12. Comparing Figures 2-10a and 2-10b, do you think it would be easier to construct a mouse or a human karyotype? Explain.

13. Speculate as to why histone phosphorylation (addition of PO_4-groups) might alter the structural organization of chromatin.

14. Relate the following to one another in terms of size and proposed function: chromomere, gene, metaphase chromosome bands, nucleosome, polytene chromosome bands.

15. The DNA of a eukaryotic chromosome was until recently thought to be surrounded by a layer of histone. How does the model in Figure 2-16 alter such a concept?

CHAPTER
3

The Meiotic Transmission
of Chromosomes and the
Principles of
Mendelian Genetics

INTRODUCTION

The transmission of chromosomes from a parent cell to its daughter cells *via* mitosis (Chapter 2) is an asexual process: one parent cell can give rise, through successive mitotic divisions, to a **clone** of cells that are genetically identical. Such a **vegetative** or **somatic** pattern of cell proliferation is responsible for the growth of multicellular organisms and for the self-propagation of unicellular eukaryotes. Most eukaryotes, however, possess an alternate, **meiotic** mode of chromosome transmission; this mode is inevitably coupled with a **sexual** phase in their life cycle wherein the genes from two different parents come to reside in a single cell. Specifically, meiosis represents the avenue by which haploid **gametes** (ova, sperm, and pollen) arise from diploid cells in the germinal tissues of higher plants and animals; in lower eukaryotes, it is also the avenue by which haploid vegetative cells are generated from diploid zygotes.

This chapter presents a detailed description of meiosis. The life cycles of organisms of genetic interest are then described with emphasis on the alternation of mitotic and meiotic modes of chromosome transmission in each case. Finally, the segregation of allelic genes and the independent assortment of unlinked genes are described in the context of meiosis. These last two phenomena are the essential features of what has come to be called Mendelian genetics, and their application to gene inheritance is examined in detail in Chapter 4.

MEIOSIS

3.1   KEY FEATURES OF MEIOSIS

Like mitosis, meiosis does not involve chromosome replication *per se:* the meiotic equivalent of the S phase (Section 2.3) occurs well before meiosis begins, so that a diploid cell enters both meiosis and mitosis with a 4C amount of DNA. An important difference between meiosis and mitosis, however, is that meiotic DNA replication is followed by two nuclear divisions in succession rather than one. It should be obvious that if a diploid cell (2C) replicates its DNA (4C) and then undergoes two meiotic divisions, the first division will produce two cells with a 2C amount of DNA while the second division will result in four cells that are each haploid or C in the amount of DNA they contain. In other words, the net effect of meiosis is to **reduce a cell's chromosome number by half,** usually from an initial 2n to a final n. This is a key feature of the meiotic process.

As noted in Section 2.6, diploid cells contain two complete sets of chromosomes, one derived originally from a paternal gamete and the other from a maternal gamete. Since meiosis gives rise to such gametes, it follows that the $2n \longrightarrow n$ meiotic reduction must occur in such a way that **each haploid product of meiosis is allotted one complete set of chromosomes** containing all the genetic information pertaining to that species. This complete allotment, then, constitutes the second key feature of the meiotic process.

A third key feature of meiosis is that it is the stage in eukaryotic development in which most **new gene combinations are generated.** These gene combinations come about in two ways. First, the maternally- and paternally-derived homologous chromosomes that coexist in a $2n$ organism are distributed among the organism's haploid meiotic products in numerous combinations. Thus if we think of a diploid cell as having three sets of homologous chromosomes, 1*M* and 1*P*, 2*M* and 2*P*, and 3*M* and 3*P*, where *M* stands for maternally-derived and *P* stands for paternally-derived, the haploid products of the cell will include 1*M*2*M*3*M* cells, 1*M*2*M*3*P* cells, 1*M*2*P*3*P* cells, and so on (in this case there are eight possible combinations in all). Second, maternally- and paternally-derived homologous chromosomes frequently take part in genetic exchange during meiotic prophase. When this occurs, a typical haploid product will contain some maternally-derived, some paternally-derived, and some recombinant chromosomes as members of its complete chromosome set, the recombinant chromosomes containing information derived from both maternal and paternal chromosomes. This cell, with its unique combination of genes, will fuse with a second haploid cell, also carrying a unique gene combination, to produce a $2n$ diploid zygote. The genetic makeup of this zygote will clearly be quite different from the makeup of either of its diploid parents.

In the description of meiosis that follows, the mechanical basis for the above three key features of meiosis will become apparent.

3.2 THE STAGES OF MEIOSIS

Compared with mitosis, meiosis is a lengthy process, the complete cycle usually taking days or weeks rather than hours. Its first stage, prophase I, is particularly complex, and in animals it frequently takes at least four to five days to complete (in contrast to the typical one-half to one hour occupied by mitotic prophase). Prophase I is therefore commonly broken down into five substages: leptonema, zygonema, pachynema, diplonema, and diakinesis (the adjectives corresponding to the first four stages are leptotene, zygotene, pachytene, and diplotene and are often used as nouns). Each of these substages is defined arbitrarily and all, of course, flow from one to the next. Prophase I is followed by metaphase I, anaphase I, and telophase I, and these are followed by prophase II, metaphase II, anaphase II, and telophase II. A short interphase usually separates the first from the second meiosis in plants. Figure 3-1 shows micrographs of meiosis in the grasshopper; Figure 3-2 diagrams the various meiotic stages that have been defined, beginning with a premeiotic interphase cell (Figure 3-2a).

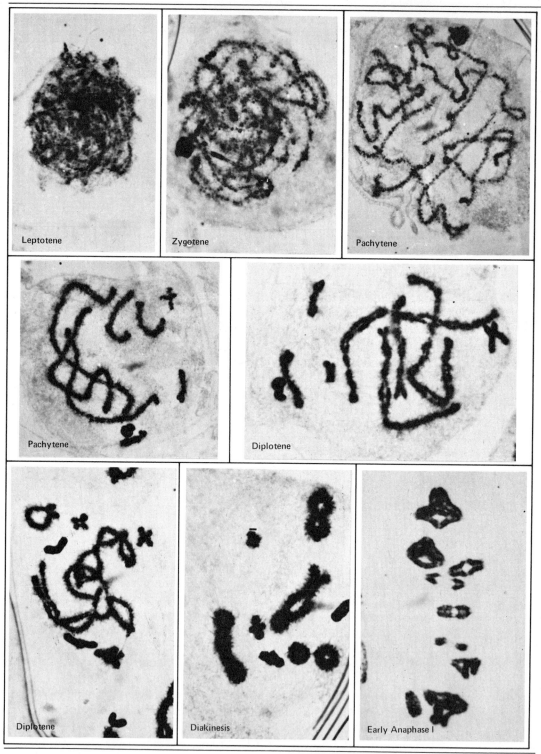

Figure 3-1 Meiosis in the grasshopper. (Courtesy James L. Walters, University of California, Santa Barbara.)

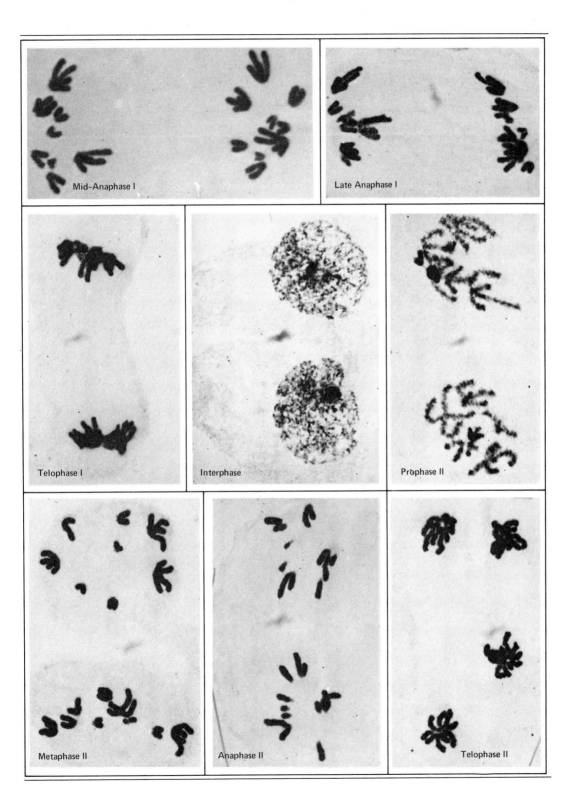

Mid-Anaphase I

Late Anaphase I

Telophase I

Interphase

Prophase II

Metaphase II

Anaphase II

Telophase II

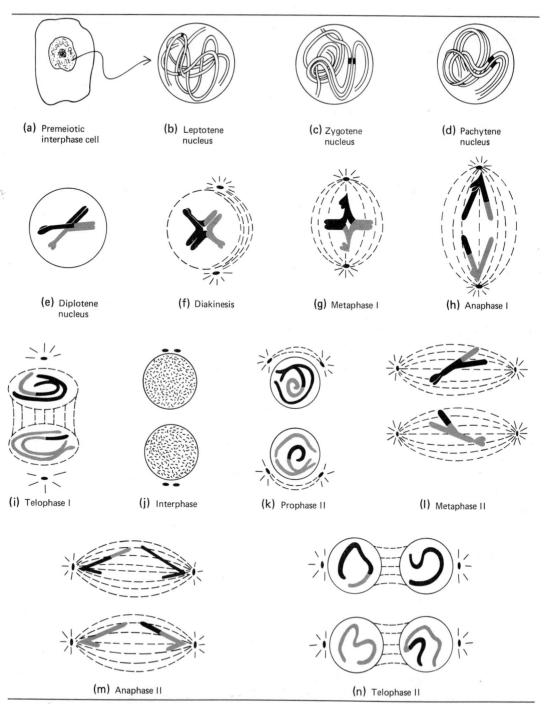

(a) Premeiotic interphase cell

(b) Leptotene nucleus

(c) Zygotene nucleus

(d) Pachytene nucleus

(e) Diplotene nucleus

(f) Diakinesis

(g) Metaphase I

(h) Anaphase I

(i) Telophase I

(j) Interphase

(k) Prophase II

(l) Metaphase II

(m) Anaphase II

(n) Telophase II

$(b \rightarrow f) =$ prophase I

Figure 3-2 Diagram of the stages of meiosis in an animal with one pair of chromosomes (n = 1). (After McLeish, J., and Snoad, B., *Looking at Chromosomes*, New York: St. Martin's Press, Macmillan & Co., Ltd., 1972.)

3.3 LEPTONEMA (Figure 3-2b)

Leptonema marks the end of premeiotic interphase. During the S period of this interphase, chromosome replication has occurred, a replication that to all appearances seems identical to the replication that precedes mitosis. Indeed, certain kinds of cells that will ordinarily undergo meiosis can, after an S phase, be so manipulated experimentally that they proceed instead through mitosis. Once a cell enters the leptotene stage, however, it is apparently committed to a meiotic course.

Leptonema means "slender thread," and this stage is heralded by the presence of threadlike chromosomes in their initial phase of meiotic coiling. Each thread in fact represents a pair of sister chromatids which are identical (as in mitosis) and which are held together by a common centromere. The leptotene threads may also exhibit the chromomeric periodicity that was noted in Section 2.12 for the early prophase chromatids of the mitotic nucleus.

With the electron microscope it becomes apparent that a band of material called a **lateral component** or **lateral element** is laid down between or alongside each pair of sister chromatids during leptonema. Such a lateral component is perhaps 500 Å wide and appears to be a complex of RNA and protein (ribonucleoprotein) associated with DNA. Lateral component synthesis is apparently a joint activity of both chromatids in each chromatid pair, and DNA from both sister chromatids is believed to be included in the lateral component complex.

In addition to chromosome condensation and lateral element synthesis, the leptonema marks the onset of an association between the ends (**telomeres**) of the chromosomes and a particular sector of the nuclear envelope. This sector of the envelope often lacks nuclear pores and may exhibit a thickened region termed an **attachment site.** This differentiated sector of nuclear membrane is believed to serve as a focal site where homologous chromosomes begin to align with one another.

3.4 ZYGONEMA (Figure 3-2c)

Zygonema (from the Greek *zygon,* a yoke) is defined as the stage in which homologous sets of sister chromatids complete their side-to-side alignment, an association called **synapsis** (from the Greek *syn-,* together + *apsis,* a joining). Synapsis appears to begin in the telomere-envelope attachment zone described above and is widely assumed to initiate with short-lived physical contacts between homologous lateral elements. Such contacts have not yet been documented by electron microscopy, however. Instead, aligned lateral elements are observed to move to within 200–300 nm of one another and then participate in the formation of a **synaptonemal complex.** The complex consists of the two lateral elements and a **central region,** approximately 100 nm wide in the mature state, which is bisected by a narrow band, the 20 nm-wide **central component.** A mature synaptonemal complex is illustrated in Figure 3-3. It is not yet clear whether the material of the central region is assembled as a cooperative effort of the synapsing, homologous chromosomes or whether prefabricated central region portions are inserted between two correctly spaced lateral components; both modes of assembly have been reported. It is, however, generally agreed that the central components, like the lateral elements, contain both RNA and protein and that, if DNA is present at all, it is present in very small amounts.

The establishment of a synaptonemal complex is a crucial genetic event because these complexes mediate the meiotic exchange of genetic infor-

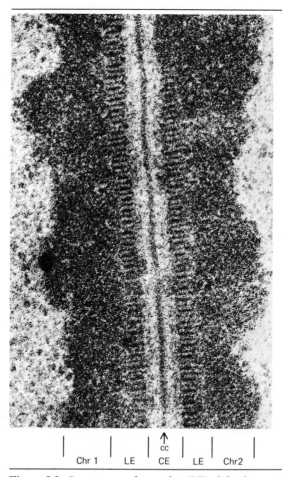

| Chr 1 | LE | CE | LE | Chr2 |

Figure 3-3 Synaptonemal complex (SC) of the fungus *Neotiella* separating two homologous chromatid pairs (Chr 1 and Chr 2). The lateral elements (LE) are banded, which is not the case for many organisms. CC, central component; CE, central element. (Courtesy M. Westergaard and D. von Wettstein.)

mation, known as **crossing over** or **recombination,** between homologous chromosomes. Chromosomes that are to participate in recombination must be sufficiently alike (homologous) to line up with one another, gene for gene. The chromosomes then exchange pieces of their DNA in such a way that information is neither gained nor lost from either chromosome. Specifically, if the two chromosomes undergoing pairing are designated A B C D and a b c d, then one recombination event might yield A B c d and a b C D chromosomes, another might yield A b c d and a B C D chromosomes, and so on. In the case of the synaptonemal complex, then, its formation must involve both mutual recognition between sister-chromatid pairs and their precise, point-for-point alignment, so that crossing over does not produce chromosomes that transmit too little, or too much, of the requisite genetic information of the species.

As will become clear in later chapters, precise recognition, alignment, and crossing over can occur between chromosomes that are not conjoined by synaptonemal complexes. Why, then, are these complexes ubiquitous in all cells that undergo meiotic synapsis and crossing over, and what role(s) do they perform? One plausible, although not yet proved, explanation is that lateral components serve as scaffolds into which key "recognition sequences" of DNA are carefully threaded during leptonema. At zygonema, such sequences could participate in recognition and alignment, eliminating the staggering problem of how two homologous chromatids could possibly synapse along their entire lengths when each can contain up to half a meter of DNA. Once recognition takes place, central elements would be added between the homologues to stabilize the (often prolonged) synaptic state that develops.

Synaptonemal complexes can only be observed with the electron microscope. With the light microscope, however, a zygotene nucleus can be distinguished in two ways: in many organisms the association of telomeres with the nuclear envelope causes a "bouquet" arrangement of the chromosome strands, and in most organisms, zygotene chromosome strands appear thicker than they do in leptonema. Although this increase in thickness comes partly from continued chromosome condensation, it is primarily the result of sister chromatids at this stage becoming intimately intermeshed with one another so that it becomes impossible to distinguish them as individual entities. For this reason, each synapsed set of homologues, which in fact includes four chromatids, appears to be composed of only two chromosomes (Figure 3-3). Each synapsed homologue set is therefore referred to as a **bivalent.**

3.5 PACHYNEMA (Figure 3-2d)

Pachynema means "thick strand" and conveys the continued shortening and thickening of the bivalents that occurs during this stage of meiosis. It is generally agreed that synaptonemal complex formation and the con-comitant synapsis of homologues is complete by the onset of pachynema, and that the actual physical exchanges that result in chromosomal crossing over occur during the pachytene stage. Again, detailed discussion of these events at the molecular level will be presented in Chapter 13. The completion of synapsis is accompanied by the dispersal of the "zygotene bouquet" arrangement of chromosome fibers.

3.6   DIPLONEMA (Figure 3-2e)

The pachynema-diplonema transition is heralded by an apparent repulsion between all 4 chromatids in a bivalent, the result being that the "split" between sister-chromatid pairs becomes evident and a separation also appears between the homologues in each set. With the electron microscope it is seen that most of the synaptonemal complex material is shed during diplonema, and repulsion is presumed to be the consequence of this shedding.

Two kinds of constraints appear to prevent the repulsing chromosomes from separating completely. First, the sister chromatids continue to be held together by centromeres. Second, nonsister chromatids in a bivalent are usually held together at one or several positions along their lengths by regions of apparent contact called **chiasmata** (singular, **chiasma**). Chiasma means a cross, and chromatids connected by a single chiasma typically assume a crosslike configuration when viewed under the light microscope (Figure 3-4a). With the electron microscope chiasmata appear to be comprised of short lengths of synaptonemal complex that have not been shed and continue to hold nonsister chromatids together at comparable positions along their lengths.

Two, three, or all four chromatids in a bivalent may participate in chiasma formation, and a given chromatid may form more than one chiasma. If, for example, we designate the four as A, A', B, and B', then chromatid A' may form a chiasma with chromatid B at one position, while at another position it may exhibit a chiasma with chromatid B or, alternatively, with chromatid B'. These relationships are illustrated in Figure 3-4b.

Chiasmata occur frequently. In a normal meiosis there is at least one chiasma per bivalent, and in a typical human oocyte containing 23 bivalents, perhaps 52 chiasmata are observed. Chiasmata are almost certainly related to crossing over (See Chapter 13).

In most organisms the diplotene stage is quickly followed by the remaining stages of meiosis, but in the oocytes of many animals the diplotene stage is extremely prolonged. The human female fetus, for example, possesses approximately 3,400,000 oocytes in each ovary, and these go through the first stages of meiosis during the fourth to seventh months of fetal life. The oocytes then remain within the ovary in a diplotene stage (sometimes

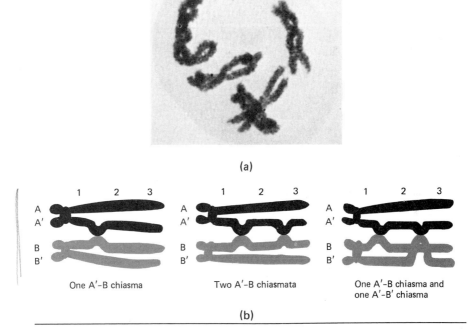

(a)

(b)

Figure 3-4 (a) Chiasmata visualized during diakinesis in the lilly (*Trillium erectum*). (Courtesy A. H. Sparrow and R. F. Smith, Brookhaven National Laboratory.) (b) Several chiasma interrelationships among homologous chromatids A-A′ and B-B′.

called a **dictyotene** stage) which may last as long as 50 years! At puberty and in the presence of follicle-stimulating hormone (FSH) and leuteinizing hormone (LH) one oocyte per menstrual cycle is ovulated and goes on to complete meiosis in the fallopian tubes if fertilization occurs.

3.7 DIAKINESIS (Figure 3-2f)

Throughout prophase I the chromosomes have continued to coil, and by diakinesis they appear to be maximally condensed. Chiasmata now appear in electron micrographs to be free of any synaptonemal complex material, instead consisting of uninterrupted spans of chromatin.

During diakinesis chiasmata frequently go through a process known as **terminalization,** in which they appear to move down the chromatids until they reach the ends of the bivalents. The bivalents are thus characteristically interconnected by one or two terminal (or partly terminal) chiasmata as they enter metaphase I. Both the mechanism and function of terminalization are obscure.

3.8   METAPHASE I (Figure 3-2g)

Metaphase of the first meiotic division is characterized by spindle formation, as in mitotic metaphase, but the two processes are otherwise different. Each bivalent includes two distinct centromeres, each of which holds two chromatids together. The two centromeres dispose themselves on either side of a plane that is analogous to a metaphase plate.

For meioses involving more than one bivalent (that is, $n > 1$), a key feature of metaphase I is readily visualized by recalling that each bivalent possesses a maternally derived centromere and a paternally derived centromere. For bivalent #1, the maternal centromere may lie above the metaphase plate and the paternal below. Looking now at bivalent #2, its maternal centromere may also lie above, and its paternal centromere below, the metaphase plate. It is equally probable, however, that the reverse alignment will occur, with the paternal centromere lying above and the maternal centromere lying below the plate. In other words, the maternal-paternal arrangement of a given set of homologous centromeres with respect to the metaphase plate is *totally independent* of the arrangements assumed by all other sets, and all possible combinations of arrangements thus occur with equal frequency when large numbers of metaphase I cells are considered. This fact has important genetic consequences, as specified shortly.

3.9   ANAPHASE I AND TELOPHASE I (Figure 3-2h and i)

Bivalents typically continue to be held together by (terminal) chiasmata until anaphase I, at which point these connections are severed and the two centromeres of the bivalent move to opposite poles of the cell. In contrast to anaphase in mitosis, anaphase I of meiosis involves only a *separation* of independent centromeres. **No centromeric divisions occur in anaphase I.**

The nuclear envelope typically disperses during metaphase I and anaphase I. Depending on the organism, the envelope may then reorganize around the two separated sets of homologous chromosomes during telophase I or the chromosomes may enter directly into the second meiotic division. Similarly, a furrow may or may not divide the cell into two daughter cells.

The germ cells of most female animals take part in **asymmetric** meiotic divisions, meaning that the metaphase I plate lies near the cell surface rather than at the cell equator, and, at anaphase I, one set of homologues moves into a tiny bleb of cytoplasm while the other remains behind in the enormous egg cell, known at this stage as an **oocyte.** At telophase I the bleb is separated from the main cell by a furrow, and the resultant tiny cell is called the **first polar body** (Figure 3-5).

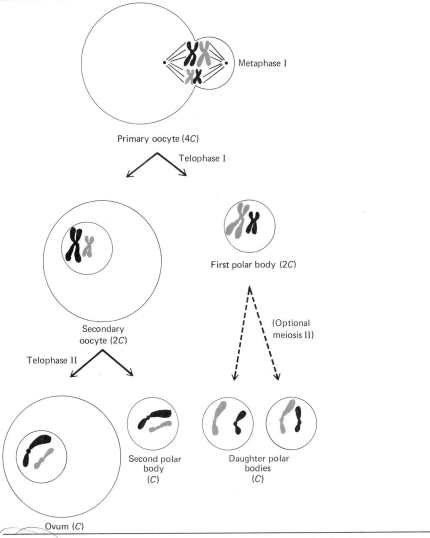

Primary oocyte (4*C*)

Metaphase I

Telophase I

First polar body (2*C*)

Secondary
oocyte (2*C*)

(Optional
meiosis II)

Telophase II

Second polar
body
(*C*)

Daughter polar
bodies
(*C*)

Ovum (*C*)

Figure 3-5 Oogenesis in an animal showing the formation of polar bodies.

3.10 INTERPHASE (Figure 3-2j)

The interphase that separates the two meiotic divisions may be brief. Its most notable feature is that it includes **no replication of chromosomal DNA,** and thus it is unlike a mitotic interphase or the interphase preceding meiosis I.

3.11   PROPHASE, METAPHASE, ANAPHASE, AND TELOPHASE II (Figure 3-2k and n)

Cells that bypass a telophase I and enter meiosis II directly do not experience a true prophase II, but otherwise the second meiotic division is morphologically indistinguishable from a mitotic division. Centromeres connecting pairs of chromatids move to a metaphase plate, divide into two halves (for the first time in the meiotic process), and move to opposite poles at anaphase. At the completion of telophase II, therefore, four haploid cells, or a **tetrad,** have derived from each original diploid cell and each haploid cell now returns to an interphase state.

In animal oocytes anaphase II and telophase II are again asymmetric and a **second polar body** is formed (Figure 3-5). The first polar body may also undergo a meiosis II so that four haploid products result—three polar bodies and the **ovum**; alternatively, the first polar body may not divide, in which case there will be only three meiotic products, one diploid and two haploid. In either case the haploid ovum is the only viable gamete, in contrast to male animal spermatogenesis in which all four meiotic products are viable.

LIFE CYCLES OF SEXUALLY REPRODUCING ORGANISMS

3.12   ANIMAL LIFE CYCLES

In animals such as *Drosophila* and humans, the sexual phase of reproduction is experienced exclusively by cells of the **germ line.** These cells become differentiated from the remaining **somatic line** cells early in embryogenesis, and they alone have the potential to undergo meiosis and become gametes. This potential is realized only after the somatic portion of the organism has reached maturity via numerous mitoses. The maturation process may involve several immature stages, as in *Drosophila* (Figure 3-6), or it may occur in the original organism, as in humans.

Gonad differentiation and activity proceeds quite differently for male and female higher animals. In human females, for example, the two original germ-line cells of each ovarian primordium undergo approximately 21 mitoses to generate the 3.4×10^6 ($\simeq 2^{21}$) oocytes present in the fetal ovary. As noted earlier, these then enter meiotic prophase, in which they await a one-per-month ovulation during the post-pubescent years. In human males the original germ-line cells also divide by mitosis to populate the fetal testes. At puberty, however, these give rise to cells called **stem spermatogonia** which divide by mitosis, on an average of once every 16 days, to produce spermatogonia. The spermatogonia, in turn, undergo meiosis to generate the millions of spermatozoa produced daily. It can be calculated that 380 rounds of mitosis have occurred in the testis of a 28-year-old man

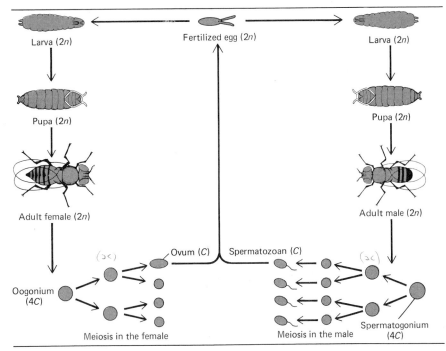

Figure 3-6 Life cycle of *Drosophila*.

and 540 rounds in a 35-year-old man. There thus exists at least a ten-fold difference in the number of mitoses undergone by the human male and by the human female germ lines.

3.13 HIGHER PLANT LIFE CYCLES

An immediate distinction between an animal and a higher plant is that the latter does not possess a germ line. Instead, among those 2n vegetative cells that make up the flowering or **sporophyte** portion of the plant, some are induced to differentiate into **megasporophyte** (♀) and **microsporophyte** (♂) cells that then undergo meiosis. Meiosis produces haploid (n) **spores** (megaspores and microspores) which divide mitotically, sometimes many times, to produce what is called the **gametophyte.** In mosses the haploid gametophyte is large and leafy while the diploid sporophyte is inconspicuous. In the ferns, the sporophyte is far more prominent, but the gametophyte is still multicellular and green. The gametophyte is most reduced in the Angiosperms. As shown in Figure 3-7 for corn, one of the megaspore nuclei undergoes three mitotic divisions to produce an embryo sac containing eight haploid nuclei; one of these then serves as the egg nucleus. The microspore comes to contain three haploid nuclei, two of which serve

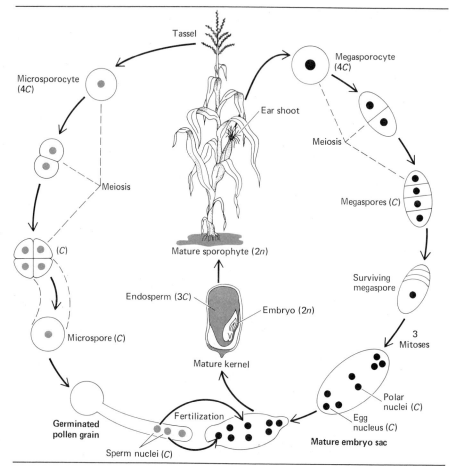

Figure 3-7 Life cycle of corn (*Zea mays*). (From *General Genetics*, Second Edition, by Adrian M. Srb, Ray D. Owen, and Robert S. Edgar. W. H. Freeman and Company. Copyright © 1965.)

as sperm nuclei in the pollen grain. At the time of fertilization one of the sperm nuclei fuses with the egg nucleus to form a 2n zygote. The second sperm nucleus fuses with two other nuclei remaining in the embryo sac (the **polar nuclei**) to form the triploid (3n) **endosperm,** which helps to nourish the developing embryo. The remaining haploid nuclei play accessory roles in fertilization and seed production.

In comparing higher plants and animals, therefore, it is important to realize that the haploid phase of the organism's life cycle is often far more conspicuous and long-lived in plants than it is in animals. The haploid or gametophyte phase in plants is often a valuable one for genetic studies, as shown in Section 4.4.

3.14 LOWER EUKARYOTE LIFE CYCLES

As noted in Chapter 2, there is no sharp distinction between higher and lower eukaryotes in a taxonomic sense, but geneticists tend to regard the fungi, the algae, and the protozoa as lower versions of eukaryotic cells. The following sections describe the life cycles of four such organisms that have become particularly important in genetic research, namely, *Chlamydomonas, Neurospora, Saccharomyces,* and *Paramecium.*

Chlamydomonas (Figure 3-8) Vegetative cells of *Chlamydomonas reinhardi* are haploid and each is a free-swimming, flagellated green organism. When the growth medium becomes depleted of nitrogen, these cells undergo certain morphological changes that permit them to mate. No separate germ line is involved in this gametogenic process.

Despite the absence of a separate germ-line, there are two sexes of *C. reinhardi,* one called mating type *plus,* the other *minus.* Gametes of the two sexes appear to be identical under the light microscope, and thus *C. reinhardi* is said to be **isogamous.** Similarly sexed gametes ignore one another, however, whereas *plus* and *minus* gametes undergo rapid pairing and fusion. The two gamete nuclei then fuse, and a diploid zygote results. After several days of maturation this zygote undergoes meiosis to yield four haploid products. Each of the meiotic products then divides mitotically to produce four clones, with each clone comprising genotypically identical haploid vegetative cells. Under the appropriate conditions any one of these haploid vegetative cells can become differentiated into a gamete and the life cycle of *C. reinhardi* can be repeated.

Neurospora (Figure 3-9) Like *C. reinhardi,* the bread mold *Neurospora crassa* also exists in the form of two mating types—in this case, *A* and *a.* Cells of both types can reproduce vegetatively by mitosis, with the daughter cells remaining attached to one another such that long filaments, called **hyphae,** are formed. At the tips of the hyphae, cells called **conidia** will pinch off. Under certain conditions these pinched-off conidia will give rise to new haploid vegetative hyphae. Under other conditions the same conidia may instead take part in sexual reproduction.

During the sexual reproduction process a conidium from mating type *A* finds its way to a large specialized hypha, a **protoperithecium,** which develops within a mating-type *a* organism; conversely, a type *a* conidium can find its way to an *A* protoperithecium. In either case the two cell types will fuse so that two nuclei will be present in a single cell. Since the protoperithecium donates most of the cytoplasm to this cell, it is considered the female element in *Neurospora* sexuality; the conidium is considered the male element.

The two nuclei do not immediately fuse in the newly formed cell. Instead, each undergoes several mitotic divisions so that a multinucleate

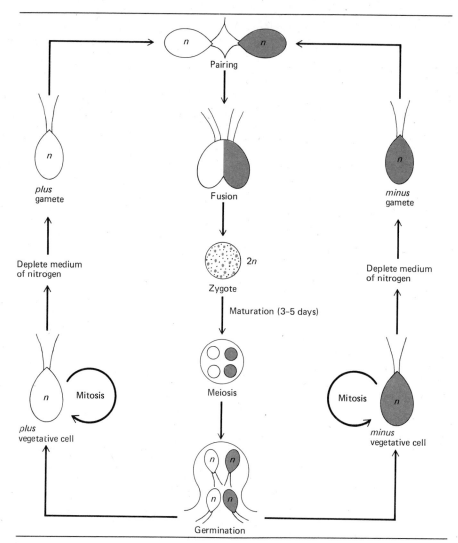

Figure 3-8   Life cycle of *Chlamydomonas reinhardi* as it occurs in the laboratory.

cell is formed. Individual nuclei of opposite mating types then fuse to form diploid nuclei, and each of these diploid nuclei becomes walled off in a cell having a thick wall known as an **ascus** (plural, **asci**). Meiosis now occurs to yield four haploid nuclei. These in turn divide mitotically so that each ascus contains eight haploid nuclei, each meiotic product being represented twice. Walls now form to create eight uninucleate cells within the original ascus. Each of these cells is known as an **ascospore.** The ascospores are released when the ascus ruptures, and each can give rise to vegetative hyphae.

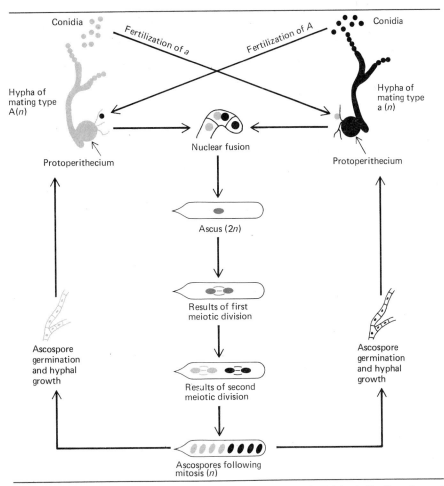

Conidia

Fertilization of *a*   Fertilization of A

Conidia

Hypha of
mating type
A(*n*)

Hypha of
mating type
a (*n*)

Protoperithecium

Nuclear fusion

Protoperithecium

Ascus (2*n*)

Results of first
meiotic division

Ascospore
germination
and hyphal
growth

Results of second
meiotic division

Ascospore
germination
and hyphal
growth

Ascospores following
mitosis (*n*)

Figure 3-9 Life cycle of *Neurospora*. (From R. Wagner and H. Mitchell, *Genetics and Metabolism*. New York: Wiley, 1955.)

Yeast (Figure 3-10) The life cycle of the single-celled fungus *Saccharomyces cerevisiae,* or baker's yeast, is less intricate than that of *Neurospora,* and it differs from both *Neurospora* and *Chlamydomonas* in that individual vegetative cells can be haploid or diploid. Thus vegetative haploid cells of opposite mating type (a and α) will fuse to form vegetative diploid cells that continue to propagate by mitosis. The cytokinesis that follows a haploid or diploid yeast mitosis is usually unequal; that is, a smaller cell (the "daughter") appears to bud from a larger cell (the "mother"); the two, however, are identical in their nuclear composition.

A vegetative diploid cell of *S. cerevisiae* may also be induced, by nitrogen starvation, to undergo **sporulation,** a complex process that endows the

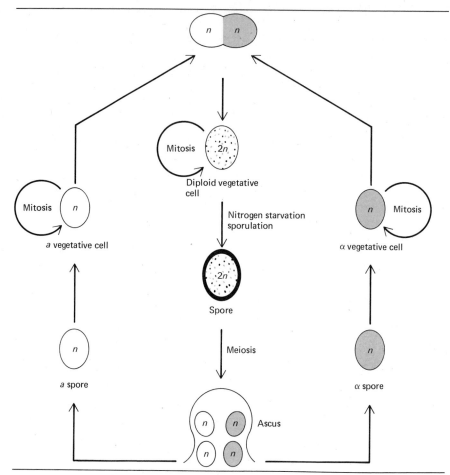

Figure 3-10   Life cycle of yeast (*Saccharomyces cerevisiae*).

diploid cell with the ability to undergo meiosis. The four resulting meiotic products are contained within a mother cell wall, also called an ascus. Haploid ascospores are then released, each divides to produce a clone of vegetative haploid cells, and the reproductive cycle of the organism can be repeated.

Paramecium (Figures 3-11 and 3-12).   The life cycles of ciliated protozoa such as *Tetrahymena* and *Paramecium* include some of the most elaborate nuclear behaviors known in eukaryotes. A somatic *Paramecium aurelia* cell, as noted in Section 2.8, contains a hyperploid macronucleus and two diploid micronuclei. Sexuality is induced by nutrient starvation, and begins with a **conjugation** process between cells of opposite mating type (Figure 3-11a). Conjugal contact triggers macronuclear breakdown (Figure 3-11a)

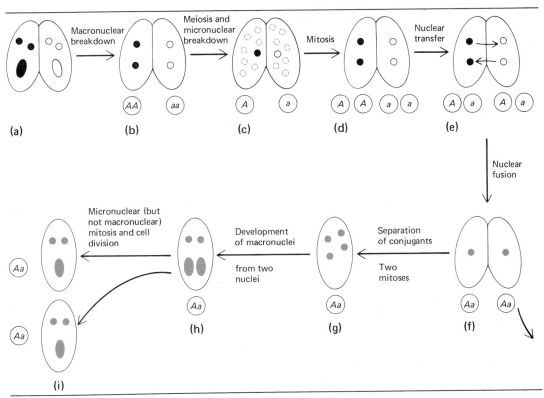

Figure 3-11 Life cycle of *Paramecium aurelia.*

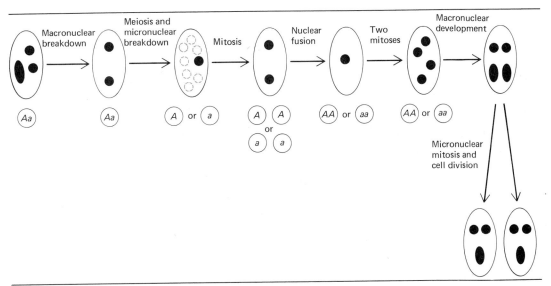

Figure 3-12 Autogamy in *Paramecium aurelia.*

and meiosis in both micronuclei (Figure 3-11b); all but one of the 8 meiotic products then disintegrates in each conjugant (Figure 3-11c). The remaining haploid nucleus in each cell undergoes a single mitotic division to give two haploid nuclei (Figure 3-11d). Each of the conjugating cells then donates one of these haploid nuclei to its mate via a bridge that develops to connect them (Figure 3-11e). The two haploid nuclei in each cell fuse to form new diploid micronuclei (Figure 3-llf) and the conjugants separate. Following separation, each *Paramecium* experiences two mitotic nuclear divisions (Figure 3-11g) followed by macronuclear development (Figure 3-11h). A final round of mitosis is followed by one cell division (Figure 3-11i).

In addition to the sexual events shown in Figure 3-11, *P. aurelia* can also undergo self-fertilization, a process known as **autogamy,** in which meiosis, nuclear breakdown, and a single mitotic division occur as before, but in which the two haploid nuclei in a single cell proceed to fuse together. These events are illustrated in Figure 3-12.

RELATING MEIOSIS TO MENDEL'S LAWS

3.15   GENETIC TERMINOLOGY

Before we relate the meiotic transmission of chromosomes to the Mendelian transmission of genes, it is essential to define some terms.

You should by now feel comfortable with the concept that a diploid cell contains two homologues of each chromosome type possessed by the species (Figure 2-9), one of these homologues deriving from one gamete (for example, a *plus* mating-type cell or an egg), and the other homologue deriving from the other gamete (for example, a *minus* cell or a sperm) during a sexual process. Each homologue in a pair contains the same *kinds* of genes and in the same order, a statement documented in later chapters. Thus each X chromosome in a *Drosophila* female will carry a gene determining body color at one end, a gene determining bristle length at the other end, and so on. The physical location of a particular gene along a chromosome is called a **locus,** so we can restate our case by saying that each *Drosophila* X chromosome possesses a body-color gene at one terminal locus and a bristle gene at the other terminal locus.

However, while the genes found at a particular locus on homologous chromosomes are of the same general kind, they need not be identical. For *Drosophila,* the gene dictating a *gray* body color is said to be the **wild-type** gene, meaning that most flies found "in the wild" have gray bodies, whereas a **mutant** form of this gene, which confers *yellow* body color, is also known. The wild-type gene is usually symbolized as + and the mutant gene by a descriptive and italicized abbreviation (in this case, *y* for *yellow*). The + and *y* genes are said to be **alleles** (from the Greek *allelon,* of one another), indicating that they represent two alternative genes that occupy the same chromosomal locus.

Returning to a female *Drosophila*, it is clear that with respect to this body-color locus, three possibilities exist: 1) She may have received a *y* allele from her mother and a *y* allele from her father, in which case her genetic endowment at that locus is expressed in genetic shorthand as *y/y*, the slash denoting that she is diploid; 2) she may have received a wild-type allele from each parent and thus be +/+; and finally, 3) she may have received one allele of each kind and thus be +/*y*. An organism with two identical alleles at a given locus (for example, a +/+ or a *y/y* fruit fly) is said to be **homozygous,** while an organism such as a +/*y* fly is said to be **heterozygous** (*hetero* meaning different and *homo* the same).

Other systems of genetic notation should also be set forth here. In some cases geneticists may refer to the above pair of body-color alleles as y^+ and *y* or as y^+ and y^-. In other cases one allele may be designated by a capital and another by a lowercase letter, as in *Y* and *y*, the capital letter here indicating the wild-type gene and the lower case letter the mutant gene. However, all such notation systems summarize the genetic makeup or **genotype** of an organism, a term that is traditionally contrasted with phenotype.

The term **phenotype** derives from the Greek root *phain,* to appear, and connotes the particular expression of a gene—yellow body, short tail, amino-acid-requiring—each particular feature being known as a **trait**. Phenotype is also used as a collective term to refer to the sum of the traits that characterize an organism: thus the standard phenotype of humans might be said to include 10 fingers, a characteristic growth rate, and several vitamin requirements. The genotype of any individual organism is relatively stable throughout its lifetime, whereas its phenotype will vary, depending on its state of development, environmental influences, and so on. The genotype of a female mammal, for example, includes a number of genes that do not find expression in the phenotype until sexual maturity. The phenotype, in short, includes all the products or manifestations of the organism's genes—the amino acid sequence of its proteins, its enzyme activities, its appearance, and even its behavior—while the genotype is synonymous with the collection of genes in the organism's chromosome(s).

In a haploid cell, with only one copy of each gene per nucleus, the presence of a given phenotypic trait can at once be attributed to the presence of a particular gene. Thus if a *C. reinhardi* cell behaves in mating as a *plus* gamete, then it must have the mating-type *plus* (mt^+) genotype. Such conclusions cannot immediately be drawn for diploid cells, however, because of the existence of **dominant** and **recessive** genes. A diploid cell possesses two alleles at every chromosomal locus. If one of these alleles turns out to specify a nonfunctional protein while the other specifies a normal protein, then enough normal protein may be produced to endow the cell with a normal phenotype, and the normal allele is said to be dominant to the abnormal allele. Many other kinds of gene-gene interactions can lead to dominant/recessive relationships between alleles, as described in later chapters. For present purposes, the important feature of a recessive gene is that it will contribute to the phenotype only if it is present in homozygous form. Thus for the recessive *bn* (*brown* eyes) gene and its + allele in *Drosophila,* a fly will develop a *brown-eyed* phenotype only if it has the homozygous (*bn/bn*) genotype; a +/+ homozygote and a +/*bn* heterozygote will both develop wild-type (reddish) eyes, the + allele in the heterozygote dictating the synthesis of sufficient amounts of red pigment that a normal eye color is generated.

3.16   MENDEL'S FIRST LAW: THE SEGREGATION OF ALLELES

In Figure 3-2 the homologous chromosomes and their centromeres in each bivalent are colored gray or black. As we noted earlier, these colors can be considered as denoting parental origin; for example, the gray chromosomes might be maternal and the black paternal in origin. As stressed earlier, the gray and black (or maternal and paternal) members of a given bivalent align themselves above and below the metaphase I plate and segregate at anaphase I.

Now consider that one of the maternal chromosomes carries the gene *A* and that the homologous paternal chromosome carries the *a* allele of this gene. When such an *A/a* heterozygote undergoes meiosis, the segregation of homologous chromosomes dictates that the haploid cells that are produced will contain either *A* or *a*, but never both; this is illustrated in Figure 3-13a. This phenomenon is called the **Principle of Segregation** and is often referred to as **Mendel's First Law** after its enunciator, Gregor Mendel, who undertook a series of pioneering genetic experiments with the garden pea in the mid-nineteenth century (see Gregor Johann Mendel, Box 3.1).

Up to now we have been considering each chromosome as a discrete physical unit, so that all the genes located in a given chromosome should move together as a unit throughout the stages of meiotic (or mitotic) division. The genes of any one chromosome are said to be **linked,** and it should be obvious that just as alleles segregate, so should all of the linked genes on a given chromosome segregate from all of the linked genes on its homologous chromosome at anaphase I.

An exception to the generalization that alleles segregate at anaphase I arises in the event of chiasma formation and genetic exchange (crossing over) between homologues during prophase I. Returning to the example of the maternal chromosome carrying gene *A* and the homologous paternal chromosome carrying its allele *a*, consider a single genetic exchange that occurs anywhere between the centromeres and the *A* and *a* loci, as diagrammed in Figure 3-13b. The maternal centromere now holds together two sister chromatids that are no longer identical: one chromatid carries the original sequence of genes, including the gene *A,* but the other carries a new segment, including the paternally-derived gene *a.* Similarly, the paternal centromere holds together two nonidentical chromatids, including gene *A,* transmitted to it during exchange, and the original gene *a,* so that the

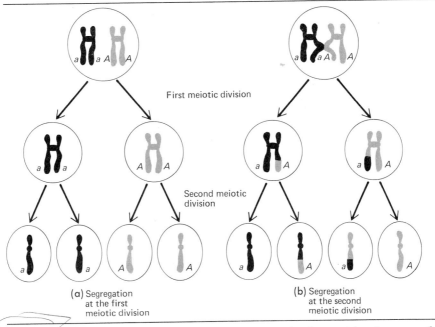

First meiotic division

Second meiotic division

(a) Segregation at the first meiotic division

(b) Segregation at the second meiotic division

Figure 3-13 The segregation of alleles during meiosis, in the absence (a) and presence (b) of crossing over. Maternally derived chromosomes are shown in gray and paternally derived chromosomes in black.

Box 3.1
GREGOR
JOHANN
MENDEL

Mendel was born in 1822 in Heinzendorf, which is now in Czechoslovakia. He was of peasant stock but came to study for the priesthood at the Augustinian monastery at Brünn. He failed his first examination for a teaching certificate in natural science, but then went on to study science at the University of Vienna before returning to Brünn. It was in the garden at the monastery that he grew his peas, and in 1856 he began his genetic experiments. These studies resulted in a paper that was published in 1866 in the Proceedings of the Brünn Society for Natural History. The paper was read by few and apparently understood by no one until 1900, long after Mendel's death. Meanwhile Mendel tried unsuccessfully to repeat his observations with another plant species, *Hieracium*, which, it turns out, forms seeds without a true meiosis. Mendel knew nothing about meiosis, and the failure of these experiments must have represented a great disappointment to him. He also studied several other species of plants, kept bees and mice, and was interested in meteorology. We have, in short, a picture of a man of many and considerable talents whose major work was not appreciated and who undoubtedly came to regard himself as a scientific failure. By 1871, when he assumed administrative duties as abbot of the Brünn monastery, most of his productive scientific work seems to have ceased. He died in 1884.

paternal centromere also holds together two nonidentical chromatids. At anaphase I maternal and paternal centromeres segregate as usual, but the genes *A* and *a* are now held together by common centromeres and so move together. They do not segregate until anaphase II, at which time the centromeres divide and the chromatids separate (Figure 3-13b).

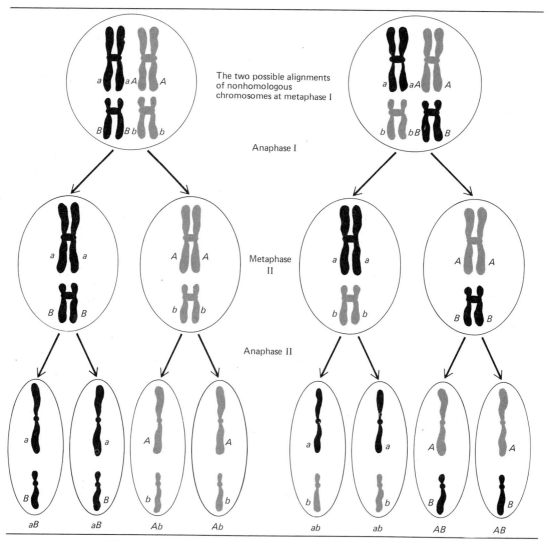

Figure 3-14   The independent assortment of nonhomologous chromosomes during meiosis, with maternal chromosomes in gray and paternal in black. At metaphase I, the two sets of nonhomologues may happen to line up on the metaphase plate such that paternally derived chromosomes are on one side and maternally derived on the other, as shown in the cell on the left. Alternatively, the chromosomes may assume a staggered arrangement, as shown in the cell on the right. The genetic makeup of the meiotic products from each of these two arrangements is shown.

Segregation of alleles therefore can occur at either of the meiotic divisions. Segregation at anaphase I (**first-division segregation**) occurs in the absence of any crossover, whereas segregation at anaphase II (**second-division segregation**) occurs if a crossover has occurred at any point between the locus of the two alleles in question and the locus of the centromere (Figure 3-13b).

3.17 MENDEL'S SECOND LAW: INDEPENDENT ASSORTMENT OF ALLELES ON NONHOMOLOGUES

The Principle of Segregation applies to homologous chromosomes. Consider now the behavior of nonhomologous chromosomes during meiosis. We noted in Section 3.8 that the different pairs of homologous chromosomes arrange themselves on the metaphase I equator in an independent manner and remain independent throughout meiosis. As a consequence, genes that are located on nonhomologous chromosomes (in other words, genes that are not linked) undergo an **independent assortment** during meiosis.

In our example of a cell with one homologous chromosome pair carrying the alleles A and a, a second homologous pair in the same cell might carry a different pair of alleles B and b. Since the two homologues assort independently, probability dictates that one-quarter of all the haploid cells resulting from meiosis will contain the genes A and B, one-quarter a and b, one-quarter a and B, and one-quarter A and b. This outcome, diagrammed in Figure 3-14, is known as **The Principle of Independent Assortment** and represents the **Second Law** to emerge from Mendel's studies.

References

Baker, B. S., and J. C. Hall. "Meiotic mutants: Genetic control of meiotic recombination and chromosome segregation." In *Genetics and Biology of Drosophila,* pp. 352–435 (M. Ashburner and E. Novitski, Eds.) London: Academic Press, 1976.

Elliot, A. M., Ed. *Biology of Tetrahymema.* Stroudsburg: Dowden, Hutchinson & Ross, Inc., 1973.

Gillies, C. B. "Synaptonemal complex and chromosome structure," *Ann. Rev. Genetics* **9**:91–109 (1975).

Grell, R. F. "Distributive pairing." In: *The Genetics and Biology of* Drosophila. pp. 435–486 (M. Ashburner and E. Novitski, Eds.) London: Academic Press, 1976.

Heywood, P., and P. T. Magee. "Meiosis in protists. Some structural and physiological aspects of meiosis in algae, fungi, and protozoa," *Bact. Rev.* **40**:190–240 (1976).

John, B. "Myths and mechanisms of meiosis," *Chromosoma* **54**:295–325 (1976).

Luciani, J. M., M. Marazzani, and A. Stahl. "Identification of pachytene bivalents in human male meiosis using G-banding technique," *Chromosoma* **52**:275–282 (1975).

*Mendel, G. "Experiments in plant hybridization." [Reprinted in J. A. Peters, Ed., *Classical Papers in Genetics*. Englewood Cliffs, N.J.: Prentice-Hall, 1959].

*Denotes articles described specifically in the chapter.

Rhoades, M. M. "Meiosis." In *The Cell,* Vol. 3, *Meiosis and Mitosis,* p. 1–75 (J. Brachet and A. E. Mirsky, Eds.) New York: Academic Press, 1961.

Roth R. "Temperature-sensitive yeast mutants defective in meiotic recombination and replication," *Genetics* **88:**675–686 (1976).

Sonneborn, T. M. *Paramecium aurelia.* In *Handbook of Genetics 2,* pp. 469–594 (R. C. King, Ed.) N.Y. Plenum, 1974.

Westergaard, M., and D. von Wettstein. "The synaptonemal complex," *Ann. Rev. Genetics* **6:**74–110 (1972).

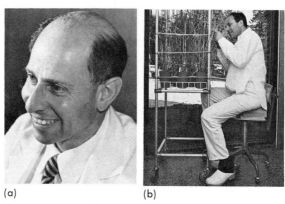

(a)   (b)

(a) Curt Stern (University of California, Berkeley) was among the first to establish the relationship between chiasmata and genetic exchange. (b) Diter von Wettstein (Carlsberg Laboratory Copenhagen) is a plant cytogeneticist who, with M. Westergaard, has analyzed the ultrastructure of meiosis.

Questions and Problems

1. Examine Figure 3-1 and determine the haploid chromosome number of the grasshopper. Does terminalization of chiasmata occur in this organism?

2. In an organism with a haploid chromosome number of 7, how many sister chromatids are present in its mitotic metaphase nucleus? In its meiotic metaphase I nucleus? In its meiotic metaphase II nucleus?

3. During which stages (or prophase I substages) of meiosis do you expect to find the following (more than one stage may obtain): sister chromatids; synaptonemal complexes; daughter chromosomes; bivalents; chiasmata; spindle microtubules; DNA replication?

4. In an organism with a haploid chromosome number of 3, write out the eight possible combinations of maternal and paternal homologues in its gametes (for example, $1M2M3P$). For each case, diagram the alignments of the nonhomologues with respect to the metaphase 1 plate.

5. For Figure 3-2, imagine that $n = 2$ and that the second bivalent does not undergo crossing over. Show *all possible* metaphase I alignments and telophase II tetrads that could result.

6. Cite the essential differences between mitosis and meiosis.

7. Diagram first- and second-division segregation when a maternal chromosome carries genes *F* and *g* and its paternal homologue carries alleles *f* and *G*.

8. The X and Y sex chromosomes behave like homologues during meiosis. What type(s) of gametes are produced by a male (XY) and a female (XX) with respect to X chromosomes? Does a father transmit copies of his X chromosome to his sons, daughters, or both? Explain.

9. For each of the three following statements, indicate which (if any) are true for mitosis, for the first meiotic division, and for the second meiotic division. (a) The segregation of centromeres is reductional (that is, centromeres of different parentage always separate). (b) The reciprocal products of a single crossover always segregate (that is, become separated). (c) Chromosomes generally duplicate between this division and the previous one.

10. Explain the following statement: Autogamy in *Paramecium aurelia* results in diploid organisms homozygous at all loci.

11. For Figure 3-4b, imagine that the paternal and maternal homologues are heterozygous at the loci marked 1, 2, and 3. (a) Assign hypothetical genotypes to the four original sister chromatids at loci 1-3. (b) Assuming each chiasma generates recombinant chromatids, diagram the genotypes associated with centromeres A, A′, B, and B′ at the onset of metaphase I for each of the three bivalents shown in Figure 3-4b.

12. Cite 6 human traits that you would classify as wild type (that is, highly invariant) and 6 traits that commonly appear in a number of variant forms in the human population.

13. How many bivalents would you expect to find in a human female pachytene nucleus?

14. A pair of alleles is always observed to segregate at the first meiotic division. How might this observation be explained?

15. Life cycles generally represent variations on a central theme, namely, the alternation of meiotic and mitotic divisions between fertilizations (F) or sexual generations. Which of the patterns given below best applies to the following: *Drosophila*, maize, *Chlamydomonas*, *Neurospora*, yeast, and *Paramecium*?

(a) F $\longrightarrow$ meiosis $\longrightarrow$ mitosis $\longrightarrow$ F
(b) F $\longrightarrow$ mitosis $\longrightarrow$ meiosis $\longrightarrow$ F
(c) F $\longrightarrow$ mitosis $\longrightarrow$ meiosis $\longrightarrow$ mitosis $\longrightarrow$ F

CHAPTER
4

Mendelian Inheritance
of Genes Carried by
Autosomes and
Sex Chromosomes

INTRODUCTION

The preceding chapter presented the details of meiosis and showed how the behavior of chromosomes at meiosis predicted that alleles should segregate from one another and that unlinked genes should assort independently. In this chapter the same phenomena are approached in a different way: phenotypic traits are first identified and sexual crosses are then performed to determine whether the genes specifying these traits are allelic and whether they are linked or assort independently. The first portion of the chapter considers the general practice of such **Mendelian genetics,** after which the Mendelian inheritance of the sex chromosomes is considered.

GENETIC DEMONSTRATION OF ALLELIC SEGREGATION

Mendel's First Law states that when two allelic genes are present in a cell that undergoes meiosis, two of the haploid meiotic products will possess copies of one gene and the other two will possess copies of the other gene (Figure 3-13). In other words, the alleles will segregate and will appear in a 2:2 ratio (or, by simplification, a 1:1 ratio) among the meiotic products. In the following sections we present various genetic demonstrations of this principle.

4.1 SEGREGATION IN TETRAD ANALYSIS

The genetic demonstration of the equivalence of chromosomal and allelic transmission is most vividly given by **tetrad analysis.** Asci of yeast or *Neurospora* or the mature zygotes of *Chlamydomonas* can be manipulated individually to liberate their four (or eight) haploid products, and these products can be separated. With *Chlamydomonas,* for example, perhaps 10 zygotes are aligned to one side in a petri dish that contains agar and a growth medium, as diagrammed in Figure 4-1a. After meiosis has been completed, each zygote wall bursts open, and the four individual meiotic products of each zygote can be manipulated into a line, as shown in Figure 4-1b. Each of these haploid cells then reproduces itself mitotically to form a clone of genetically identical haploid cells (Figure 4-1c).

If necessary, one or more **replica plates** can be made from this original or master plate so that the growth properties of the meiotic products can be tested on a number of different media. The principles of replica plating are described and illustrated in Figure 4-2.

Let us follow a cross in which wild-type gametes that do not require arginine are mixed with gametes that derive from an arginine-requiring

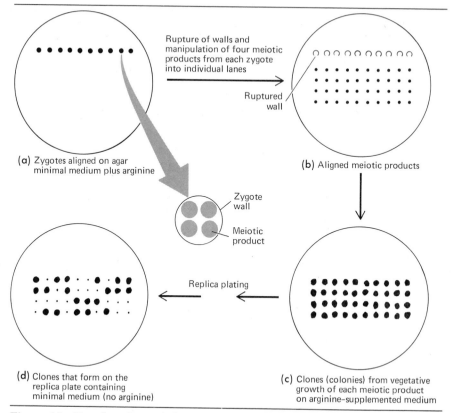

(a) Zygotes aligned on agar minimal medium plus arginine

Rupture of walls and manipulation of four meiotic products from each zygote into individual lanes

Ruptured wall

(b) Aligned meiotic products

Zygote wall

Meiotic product

Replica plating

(d) Clones that form on the replica plate containing minimal medium (no arginine)

(c) Clones (colonies) from vegetative growth of each meiotic product on arginine–supplemented medium

Figure 4-1   Tetrad analysis of the cross *arg* × + in *Chlamydomonas*. (From B. A. Kihlman, *Actions of Chemicals on Dividing Cells*, p. 60, © 1966. Adapted by permission of Prentice-Hall, Inc., Englewood Cliffs, New Jersey.)

strain. The cross is thus written + X *arg*. The growth medium of the master plate is a minimal medium supplemented with arginine so that all the meiotic products are able to survive (Figure 4-1b). After colonies have formed, replicas are made from the master plate to plates containing minimal medium and, as a control, to plates of arginine-supplemented minimal medium. After about 72 hours colonies are visible on both kinds of medium. On the arginine-supplemented plate each of the original tetrads is again represented by four colonies. In contrast, on the minimal-medium plates colonies develop only from two of the four meiotic products of each tetrad (Figure 4-1d). Microscopic examination of these plates will show that cells from the other two meiotic products were indeed transferred during replica plating but that these cells failed to grow on minimal medium. Each tetrad, therefore, consists of *equal numbers of two kinds of cells:* half of these cells require arginine for their growth and thus carry the *arg* gene, while half do not, and thus carry its + allele. The outcome predicted by a knowledge of meiosis is thus confirmed by genetic testing.

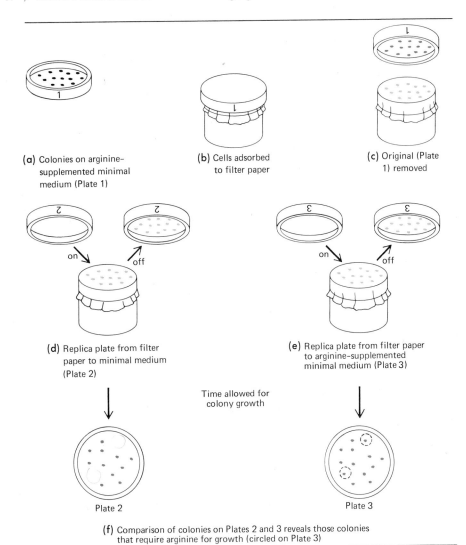

(a) Colonies on arginine-supplemented minimal medium (Plate 1)

(b) Cells adsorbed to filter paper

(c) Original (Plate 1) removed

(d) Replica plate from filter paper to minimal medium (Plate 2)

(e) Replica plate from filter paper to arginine-supplemented minimal medium (Plate 3)

Time allowed for colony growth

Plate 2

Plate 3

(f) Comparison of colonies on Plates 2 and 3 reveals those colonies that require arginine for growth (circled on Plate 3)

Figure 4-2 Replica-plating technique, as devised by J. and E. Lederberg, used to screen for arginine-requiring mutant strains of a microorganism. (a) Cells are plated onto a solidified agar medium containing nutrients and each forms a colony. (b) An adsorbent material such as filter paper is brought into contact with the agar surface such that some cells adhere to, and are picked up by, the paper (c). The paper is now pressed onto fresh media and a few cells are transfered (d and e); in the experiment illustrated, plate 2 lacks arginine whereas plate 3 contains arginine. Comparison of these plates reveals cells that require arginine for growth.

4.2   SEGREGATION IN RANDOM MEIOTIC PRODUCTS

Genetic analysis of higher eukaryotic animals is based on two assumptions. First, it is assumed that among the four products of a female meiosis any one may happen to become the egg cell with equal probability, the other three becoming polar bodies (Figure 3-5). Second, it is assumed that any of the meiotic products of spermatogenesis has an equal probability of fertilizing the ovum in a given cross. In other words, it is assumed that *all* the meiotic products of the male and the female are equally likely to participate in the successful formation of a zygote.

Granted these assumptions (we shall note later the unusual circumstances in which they do not hold), then, if we look at a sizable number of progeny arising from a sizable number of zygotes, these progeny should derive from a *random sampling* of all meiotic products. It can thus be expected that the patterns of gene distribution found in the offspring of mating will reflect the patterns of gene transmission followed during meiosis.

We can now examine a cross between a stock of true-breeding wild-type (and therefore $+/+$) female flies of *D. melanogaster* and a stock of true-breeding vestigial (vg/vg) male flies, *vestigial* being a recessive trait wherein the wings are small and rumpled. Meiosis in the females will produce eggs that carry only a copy of the $+$ allele, while the males will produce sperm that carry only the *vg* allele. Fertilization will thus produce zygotes with a $+/vg$ genotype, and, because of dominance, all of the resulting flies will be of the wild type. These results are shown in Figures 4-3a and b.

How is it demonstrated that these flies of the **F₁** or **first filial** generation are indeed heterozygous? The most direct answer is given by what is known as a **test cross,** and it is here that we are in a position to demonstrate the segregation of alleles in higher organisms. In a test cross (Figure 4-3c and d) the F_1 flies are crossed to true-breeding *vestigial* flies, which we know to be homozygous for *vg* and to produce only *vg*-carrying gametes. Since the *vg* gene is recessive, the *vestigial* parents can be thought of as contributing neutral gametes to the cross against which the genetic makeup of the F_1 flies can be tested. The results of the test are shown in Figure 4-3c: among 984 progeny approximately half are phenotypically of the wild type and half of the *vestigial type*. Therefore, as diagrammed in Figure 4-3d, half the gametes produced by the F_1 flies must carry the $+$ gene and half the *vg* gene. Since we assume that a random sampling of gametes participated in formation of the 984 progeny scored, we can conclude that the F_1 flies had a $+/vg$ genotype, and that the two alleles segregated in a 1:1 ratio during meiosis. Figure 4-4 diagrams the meiotic events that can be inferred from the test cross we have just analyzed.

What occurs if the $+/vg$ F₁ flies are not test-crossed but are instead **inbred** (allowed to breed with one another)? Each fly, whether female or

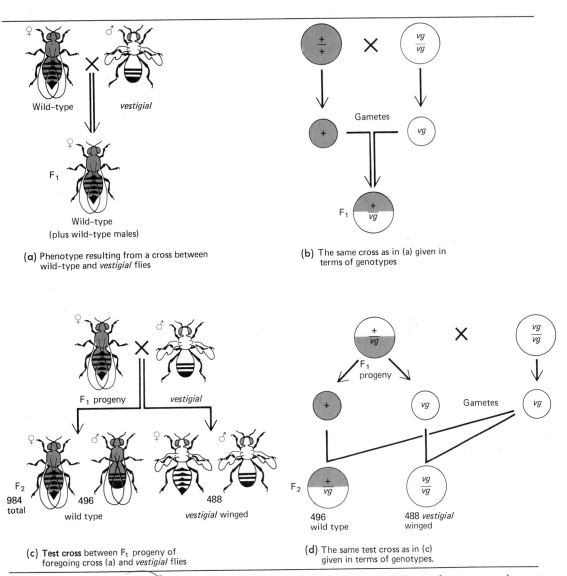

(a) Phenotype resulting from a cross between wild-type and *vestigial* flies

(b) The same cross as in (a) given in terms of genotypes

(c) **Test cross** between F₁ progeny of foregoing cross (a) and *vestigial* flies

(d) The same test cross as in (c) given in terms of genotypes.

Figure 4-3 A cross between wild-type and *vestigial* flies of *D. melanogaster* and a test cross involving the F₁ progeny.

male, should produce equal numbers of +- and vg-carrying gametes by the Principle of Segregation. Granted the assumption that each type of sperm has an equal probability of combining with each type of egg, there are four possible unions: + with +, + with vg, vg with +, and vg with vg (Figure 4-5). We can thus predict that in the **F₂ (second filial)** generation of flies one-quarter should be +/+, one-half +/vg, and one-quarter vg/vg, or that

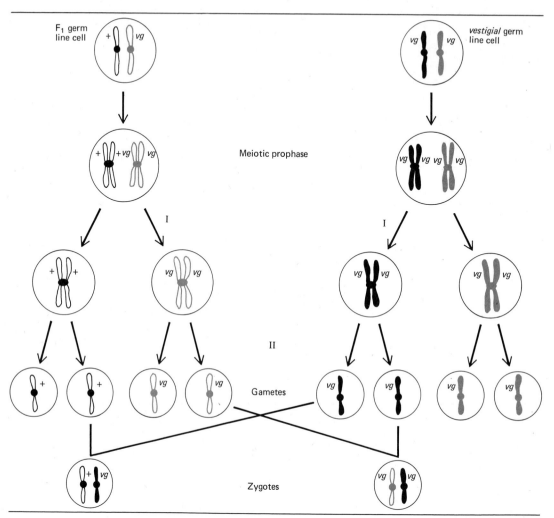

Figure 4-4 The test cross of Figure 4-3 interpreted in terms of meiosis. The homologues in each parent are colored differently so they can be readily followed. I and II indicate the two meiotic divisions.

the **ratio of genotypes** should be 1:2:1 (Figure 4-5). Since the + gene is dominant to the vg gene, however, the **ratio of phenotypes** should be three wild type to one *vestigial*. Actual results of inbreeding F$_1$ +/vg heterozygotes are given in Figure 4-5 along with the numerical data, and it is clear that the data agree with the predicted phenotypic ratio of 3 wild:1 *vestigial*.

One additional prediction can be made and verified with the flies in this inbred cross. The F$_2$ flies with a wild phenotype should be either homozygous or heterozygous for the + allele, and we would predict, from the expected genotypic ratio, that the ratio of +/+ to +/vg flies should be 1:2.

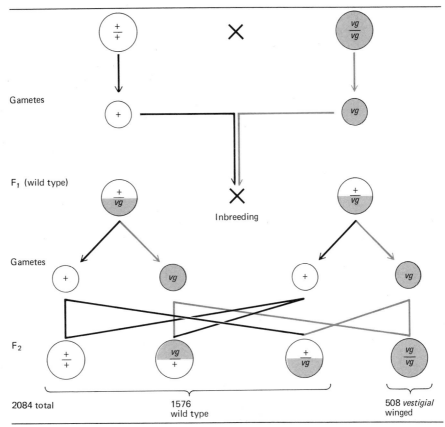

Figure 4-5 Results of inbreeding a heterozygous F_1 population in *Drosophila*, showing a 3:1 ratio of phenotypes in the F_2 and a 1:2:1 ratio of genotypes.

This can be demonstrated by test-crossing individual F_2 wild-type flies to *vg*/*vg* flies.

4.3 SEGREGATION IN HUMAN PEDIGREES

The Principle of Segregation is well exhibited in the inheritance of human traits. A classic example is the case of the skeletal abnormality called **brachydactyly,** in which the fingers are very short. A pedigree for brachydactyly, shown in Figure 4-6, represents one of the first studies of single-gene inheritance in humans. In a pedigree, females are represented by circles and males by squares; when the sex of an individual is unknown, a diamond-shaped symbol is used. A black symbol represents an individual who expresses the trait. When it is relevant to indicate the phenotypes of two spouses, their marriage is symbolized as ⚬–◻ and their children as ⚬⟂◻.

When information on a spouse is irrelevant or unavailable, a person's children are indicated as ♀ or ♂.

The pedigree in Figure 4-6 shows that an initial marriage (generation I) between an affected female and a normal male produced eight children (generation II), four of whom were brachydactylous and four of whom were normal. This ratio is reminiscent of a test cross between heterozygous and homozygous recessive individuals, in which half the progeny have the phenotype of one parent and half have the phenotype of the other because of gene segregation. Thus we can adopt the hypothesis that the gene "B," conferring the brachydactylous trait, is dominant to its + allele, and that the original mother was a B + heterozygote. This hypothesis is confirmed by subsequent generations in the pedigree. Whenever brachydactylous individuals have children (we can assume that their spouses are normal since brachydactyly is a rare affliction), approximately half their offspring are brachydactylous and half are normal, generation after generation.

We can now ask what the outcome would be if two heterozygous brachydactylous individuals from this lineage were to marry and have children. Statistically we would expect three of every four of their children to be afflicted. Moreover, we would predict that one of every three brachydactylous individuals would be a B/B homozygote (you should convince yourself of these ratios by making diagrams such as those in Figure 4-5). Such a B/B homozygous individual would, of course, produce only B

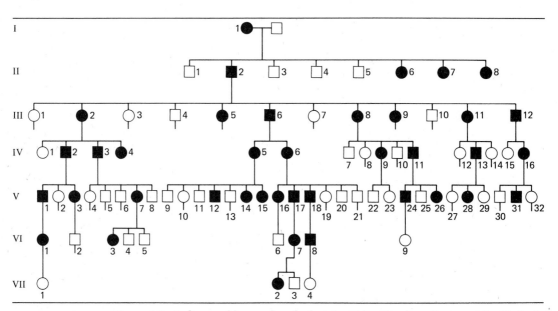

Figure 4-6 Pedigree of human brachydactyly. (After Farabee, Papers of the Peabody Museum, Harvard University, Vol. 3, 1905, updated by V. A. McKusick, *Human Genetics*, 2nd ed., © 1972. Adapted by permission of Prentice-Hall, Inc., Englewood Cliffs, N.J.)

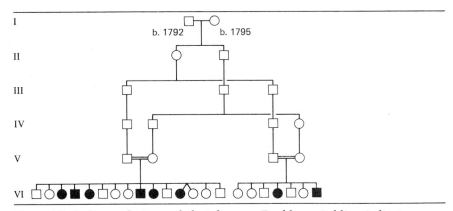

Figure 4-7 Pedigree of microcephaly in humans. Double marital lines indicate consanguinity. (From V. A. McKusick, *Human Genetics*, 2nd ed., © 1969. Reprinted by permission of Prentice-Hall, Inc., Englewood Cliffs, New Jersey.)

gametes, and thus all of her or his children would be brachydactylous, regardless of the genotype of her or his spouse.

As a second example of the analysis of human pedigrees, Figure 4-7 shows a pedigree for the trait of **microcephaly,** in which the affected homozygous individual has a small head and is mentally retarded. It is evident that the pedigree for this trait differs strikingly from the pedigree for brachydactyly (Figure 4-6): the only individuals exhibiting the microcephaly trait derive from consanguineous marriages (symbolized by double marital lines). In such cases we can suspect strongly that a rare recessive rather than a dominant gene is responsible for the trait. Homozygotes for this gene should be exceptionally rare and should arise most commonly in consanguineous marriages in which two individuals have a statistically better chance of being heterozygous for the same pair of alleles. It is, in fact, this increased chance of expressing rare recessive (and deleterious) genes by homozygosity that argues against consanguineous marriages.

4.4 SEGREGATION ANALYZED IN POLLEN

Genetic analysis of higher plants follows, in general, the same lines that we have outlined for *Drosophila* and for humans. In addition, however, the gametophyte phase of higher plants can sometimes be utilized for genetic analysis. This is the case with pollen. Pollen sometimes travels long distances before it makes contact with the pistil of a flower, and consequently it often has a stable and relatively long-lived existence. Since each pollen grain typically contains two identical haploid nuclei, the phenotype of the grain will reflect its genotype. The segregation of alleles can thus be demonstrated directly in pollen.

As an example, we can consider the genes that determine the carbohy-drate composition of maize pollen. The gene *Wx* dictates the synthesis of starch (amylose), and starch-containing pollen grains turn blue when treated with iodine. Pollen carrying the mutant allele *wx*, however, cannot synthesize amylose and these grains turn a reddish-brown in iodine. In one experiment in which 6919 grains from a heterozygous *Wx/wx* plant were examined, 3437 grains were found to stain blue and 3482 to stain red, giving a 1:1 ratio.

4.5 THE STAGE OF MEIOSIS AT WHICH SEGREGATION OCCURS

We noted in Section 3.16 (Figure 3-13) that there are two stages of meiosis at which alleles can segregate: namely, anaphase I and anaphase II, de-pending on whether or not a crossover has separated the alleles during meiotic prophase I. We can now present genetic evidence for the occurence of such first-division and second-division segregations using tetrad analysis of *Neurospora*.

In *Neurospora* (Figure 3-9) the ascus is sufficiently narrow that neither nuclei nor ascospores are normally able to move about within it. Conse-quently the alignment of the eight ascospores in a mature ascus reflects the positions of the various spindles that mediated the two meiotic and the one (postmeiotic) mitotic division within the ascus. Thus, if we follow the fate of the four chromatids making up any one bivalent in prophase I, as we have done in Figure 4-8, the first pair of ascospores would derive from one chromatid and the second pair from its sister, whereas the third and fourth pairs would derive from the two sister chromatids of the homologous chromosome set. For this reason *Neurospora* is said to produce **ordered tetrads.**

When a cross is made between wild-type *Neurospora*, which produces pink conidia, and a mutant strain called *albino* (*al*), which has white conidia, tetrad analysis reveals that there is indeed a one-to-one segregation of wild type and *albino* alleles. When, however, the ascospores are removed from the ascus in order, two types of ascus are found. In one type the order of genotypes (and phenotypes) within the ascus is +, +, *al*, *al*; in the other it is +, *al*, +, *al*. Segregation in both cases is one-to-one, but the sequence of wild-type and mutant products within the ascus has changed.

These two patterns can be understood by examining the events that take place during meiosis (Figure 4-8). The sequence +, +, *al*, *al* (Figure 4-8a) results from the segregation of alleles at the first meiotic division or, more precisely, from segregation of the homologous centromeres at anaphase I. The sequence +, *al*, +, *al* results whenever a crossover occurs between two of the homologous chromatids during prophase I, provided that the ex-change takes place anywhere *between the position of the centromere and the locus of al and its + allele.* Because of this crossover, the + and *al*

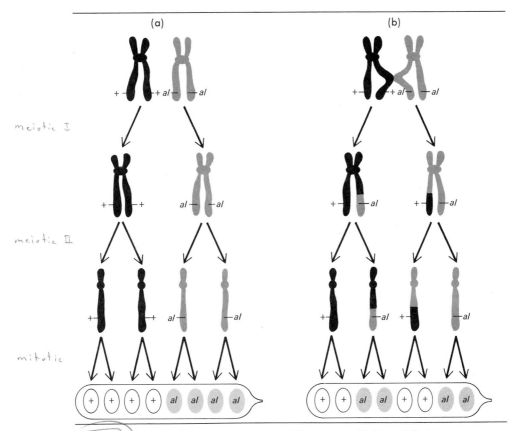

Figure 4-8 (a) First-division segregation in *Neurospora*. (b) Second-division segregation in *Neurospora*.

alleles are held together by a common centromere and are unable to segregate at anaphase I, as diagrammed in Figure 4-8b. It is only at anaphase II, when the centromeres of sister chromatids divide and the sister chromatids separate, that the two alleles can reside in separate nuclei.

4.6 FACTORS THAT BIAS NORMAL SEGREGATION RATIOS

The preceding sections have amply demonstrated that for a diploid *A/a* heterozygote the *A* and *a* alleles should be transmitted meiotically in a 1:1 ratio. Any marked or consistent discrepancy from this ratio—for example, a 2:1 ratio in favor of the *a* allele—could have one of three general causes which we shall explore in turn.

First, zygotes carrying the *A* allele could suffer developmental difficulties and thus only infrequently mature into viable progeny. This is the most frequently encountered factor responsible for distorted segregation ratios and is generally referred to as a **marker effect:** the allele selected as a gene "marker" for a particular cross proves to adversely influence the mortality of the offspring, and this is reflected in the genetic data.

Second, a consistent abnormality in meiosis could preferentially exclude the A allele from the gametes. This situation is particularly applicable to female flower parts and female animals in which only one of the four meiotic products becomes a functional egg. A well studied case is known in maize. When maize is heterozygous for a heterochromatic "knob" on an abnormal chromosome 10 (K10/k10), as many as 70 percent of the functional megaspores are found to carry the knob. It has been proposed that the knobbed region has **neocentromere** activity, meaning that it associates with spindle fibers prematurely and thus experiences a precocious migration to one pole at both meiosis I and II. For this reason the knobbed chromosome is thought to have a better chance of reaching the basal megaspore, the cell destined to become the functional egg (Figure 3-7), than the knobless homologue.

The third potential cause of biased segregation ratios is that an a-carrying sperm experiences markedly greater success than its A-counterpart in reaching an egg and/or fertilizing it. This applies particularly to species in which the male gametes take the active role in finding and fertilizing the egg. In the case of pollen, for example, the genotype of the gamete is known to be expressed during the complex process of pollen tube formation, nuclear migration, and the like. Thus pollen grains carrying gene a might well direct a faster rate of pollen tube growth than grains of genotype A, and an excess of a-carrying embryos might thereby appear among the offspring in a given season.

In the case of Drosophila, on the other hand, there is strong evidence that the genotype of the sperm is not expressed at all during transmission and fertilization. Appropriate genetic crosses can be made to yield sperm that carry only the Y chromosome and the tiny chromosome IV. Such sperm can fertilize eggs normally, even though most of their genes are missing (the resulting zygote is, of course, inviable). It is believed, therefore, that when one type of animal sperm appears to function better than another the difference must be conferred on the sperm during early stages of sperm production, perhaps even when the precursor of the sperm cell is still heterozygous. As a specific example of this, we can examine the **segregation distorter** locus in D. melanogaster, found in or near the constitutive heterochromatin in the right arm of chromosome 2. In males heterozygous for SD and its + allele, virtually all of the progeny inherit SD, even though + organisms are not inviable, and the spermatogenesis and transfer of +-carrying sperm to the storage organ of the female fly appear to occur normally. It seems that a gene product is dictated by the SD locus during the early stages of meiosis, when the spermatocytes are still +/SD heterozygotes. This product apparently has some adverse effect on the fertilizing competence of the gametes that carry the + allele, but is not deleterious when included in the gametes carrying *SD*. The true cause of sperm dysfunction at the time of fertilization is unknown in the case of *SD,* and in similar cases as well, but two possibilities can be visualized. The sperm carrying the dysfunctional allele might experience some mechanical difficulty and so be unable to penetrate the egg or migrate to the female pronucleus. Alternatively, the egg might somehow "recognize" that the sperm is in some way unacceptable and actively prevent sperm penetration or the formation of a zygotic nucleus.

INDEPENDENT ASSORTMENT

Returning to the general rule that any given pair of alleles segregates in a one-to-one fashion, we must consider next what happens when we deal with the segregation of two or more pairs of alleles, each residing in *different sets* of homologous chromosomes. Keep clearly in mind that the examples we are giving do *not* concern two or more pairs of alleles that reside in the same chromosome, since the behavior of such linked genes is considered in Chapter 12.

As stressed in Chapter 3, any two pairs of nonhomologous chromosomes are not only independent of each other in the way they become aligned at the metaphase plate during the first meiotic division, but they also remain independent throughout meiosis. Thus, if the homologous chromosomes of one pair carry alleles *A* and *a* and those of the other pair carry *B* and *b*, then meiosis should yield an equal number of the combinations *AB, Ab, aB,* and *ab* (Figure 3-14). In other words, as predicted by Mendel's Second Law, two pairs of unlinked genes should assort independently of each other.

4.7 INDEPENDENT ASSORTMENT IN TETRAD ANALYSIS

This prediction can be tested by tetrad analysis. A cross can be made between an *arginine*-requiring strain (*arg*) and an *acetate*-requiring strain (*ac*) of *Chlamydomonas*. The meiotic products are separated and allowed to form colonies on a medium containing both arginine and sodium acetate. The colonies are then replica-plated to four different kinds of media: minimal medium, minimal medium plus arginine, minimal medium plus sodium acetate, and minimal medium plus arginine and sodium acetate.

The cross of the two mutant strains is symbolized as *arg* + × + *ac;* remember that the *arg* and *ac* loci are in different chromosomes. Assuming independent assortment, four kinds of meiotic products, and therefore four kinds of haploid vegetative cells, should result from the cross: *arg* +, + *ac*, + +, and *arg ac*. These four can be distinguished from one another by the media on which they will or will not grow (Table 4-1).

Table 4-1 Types of Progeny Produced in the Cross of *arg* + X + *ac* in *Chlamydomonas* and the Media on Which They Will Grow

Type of Progeny	Minimal	Minimal with Arginine	Minimal with Acetate	Minimal with Arginine and Acetate
arg ac	−	−	−	+
+ *ac*	−	−	+	+
arg +	−	+	−	+
+ +	+	+	+	+

Growth is indicated by a plus sign, absence of growth by a minus sign.

Tetrad analysis of the cross gives the results shown in Table 4-2. Three classes of tetrads are found. The first consists of colonies derived from two sorts of meiotic products, *arg* + and + *ac*. Since two of the products have the genotype (and phenotype) of one parent and two have the genotype (and phenotype) of the other, this class of tetrad is called a **parental ditype, or PD.** Note that each pair of alleles has segregated in a one-to-one fashion.

The second class of tetrad also shows a one-to-one segregation for each pair of alleles, but it consists of two new genotype combinations: two are *arg ac* and two are + +. This tetrad, with two types of product each genotypically different from either of the original parents, is called a **nonparental ditype, or NPD.**

Four different genotypes are found in the third class of tetrad. Two are the parental combinations (*arg* + and + *ac*) and two are the nonparental types (*arg ac* and + +). A tetrad of this sort is called a **tetratype, or T.** Again each pair of alleles segregates one-to-one.

The origin of the PD and NPD tetrads relates directly to the independent alignment of the chromosomes at metaphase I, as illustrated in Figure 4-9a. If the occurrence of the two alternative alignments is equally probable, the ratio of PD:NPD tetrads in any cross involving unlinked genes should be equal to one. The data given in Table 4-2 show that there are 71 PD and 69

Table 4-2 The Three Classes of Tetrads Produced in the Cross *arg* + X + *ac* in *Chlamydomonas*

	PD	Tetrad Class NPD	T	
	arg +	*arg ac*	*arg* +	
	arg +	*arg ac*	+ +	
	+ *ac*	+ +	*arg ac*	
	+ *ac*	+ +	+ *ac*	
Number	71	69	95	Total = 235

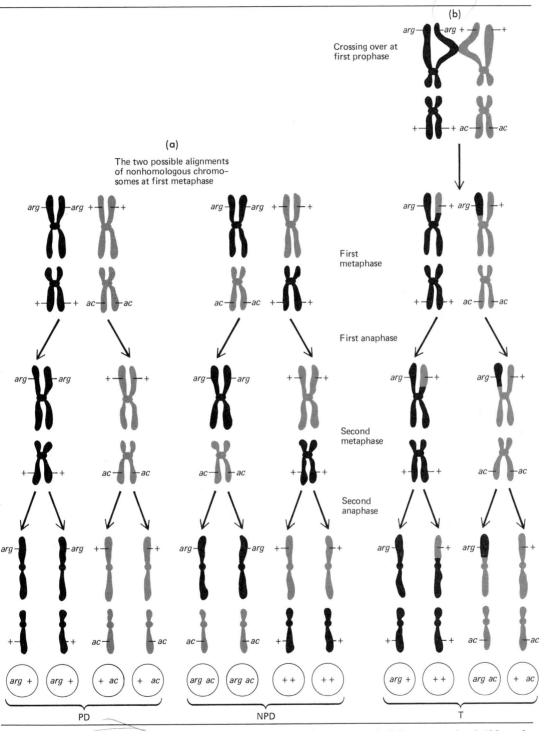

Figure 4-9 (a) Independent assortment of two pairs of alleles in tetrads of *Chlamydomonas* yielding PD and NPD tetrads. (b) Independent assortment of two pairs of alleles in tetrads of *Chlamydomonas* where a single exchange has occurred, yielding a T tetrad.

NPD tetrads, confirming the prediction. *When a cross yields roughly equal numbers of PD and NPD tetrads, the pairs of alleles being followed can be assumed to assort independently of one another.* In other words, it can be concluded that each set of alleles resides in a different set of homologous chromosomes, the assumption we in fact made about the *arg* and *ac* loci at the outset.

Consider next the tetratype tetrads. As we noted earlier, segregation can occur at the first or second division of meiosis. If a crossover occurs between either of the allelic loci and their respective centromeres, so that one pair of alleles segregates at the second meiotic division, the tetrad formed will be a tetratype. This is shown in Figure 4-9b, in which second-division segregation has occurred for *arg* and its + allele, and first-division segregation for *ac* and its + allele. (If second-division segregation occurs for both pairs of alleles, the resulting tetrads will be either PD or NPD, an outcome you should work out for yourself by following the pattern in Figure 4-9.)

The occurrence of second-division segregation in no way violates the rule of the independent assortment of nonlinked genes. Tetratype tetrads consist of one of each of the four possible genotypes. Thus the ratio of the genotypes emerging from a cross, as seen in Table 4-2, remains 1:1:1:1 for *arg* +, + *ac*, + +, and *arg ac*, regardless of the number of tetratypes present. This ratio, along with the fact that equal numbers of PD and NPD tetrads are present, demonstrates that each of the alleles of a given pair can *assort independently with respect to any member of another allelic pair in a nonhomologous chromosome.*

4-8   ANALYSIS OF RANDOM MEIOTIC PRODUCTS

In the preceding paragraph, when we determined the genotypic ratio from the data in Table 4-2 without concern for tetrad class, we were essentially analyzing a random sample of meiotic products. This is the approach that must be taken to demonstrate independent assortment in diploids.

A cross of female *Drosophila* homozygous for *vestigial* to males homozygous for the body-color mutation *ebony* (e) gives an F_1 that is wild in phenotype. We have already seen that *vg* is recessive, and the occurrence of a wild-type F_1 in the present cross indicates that e is recessive as well. The F_1, therefore, is heterozygous for both *vg* and e and has the genotype $\frac{+ \quad +}{vg \quad e}$. These F_1 flies should produce equal numbers of four types of gametes, *vg* +, + e, + + and *vg e*, if the two pairs of alleles indeed assort independently.

To check this prediction a test cross can be made in which the F_1 flies are crossed to flies homozygous for both *vg* and e. When this is done, four different kinds of flies are obtained in essentially equal numbers, as shown in Figure 4-10. Thus a random sampling of gametes gives a ratio of flies that is 1 *vg* + :1 + e:1 + + :1 *vg* e. There is, of course, no direct way to

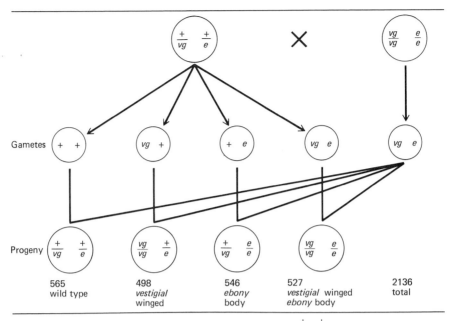

Figure 4-10 Independent assortment in a test cross of $\dfrac{+}{vg}\ \dfrac{+}{e}$ *Drosophila*.

determine what proportion of the gametes has arisen by second-division segregation of one or the other pair of alleles, but we know from tetrad analysis that the occurrence of second-division segregation does not alter the 1:1 ratio of parental (in this case, *vestigial* and *ebony*) to nonparental (wild type and *vestigial-ebony*) genotypes.

There is another, though more indirect, way to demonstrate independent assortment in a diploid: the F_1 heterozygotes can be inbred rather than test-crossed. Four different phenotypes are found among the progeny F_2 flies, as in the test cross, but this time they appear in quite different proportions; for example, when the $F_1\ \dfrac{+}{vg}\ \dfrac{+}{e}$ are crossed among themselves, we obtain the F_2 data shown in Table 4-3. The ratio of phenotypes is approxi-

Table 4-3 The Phenotypes Obtained by Inbreeding *Drosophila* of the Genotype $\dfrac{+}{vg}\ \dfrac{+}{e}$

Phenotype	Number
wild type	2834
vestigial	920
ebony	951
vestigial ebony	287
Total	4992

Table 4-4 The Genotypes Obtained by Inbreeding *Drosophila* of the Genotype $\dfrac{+\ +}{vg\ e}$

Eggs	Sperm			
	$+\ +$	$+\ e$	$vg\ +$	$vg\ e$
$+\ +$	$\dfrac{+\ +}{+\ +}$	$\dfrac{+\ e}{+\ +}$	$\dfrac{vg\ +}{+\ +}$	$\dfrac{vg\ e}{+\ +}$
$+\ e$	$\dfrac{+\ +}{+\ e}$	$\dfrac{+\ e}{+\ e}$	$\dfrac{vg\ +}{+\ e}$	$\dfrac{vg\ e}{+\ e}$
$vg\ +$	$\dfrac{+\ +}{vg\ +}$	$\dfrac{+\ e}{vg\ +}$	$\dfrac{vg\ +}{vg\ +}$	$\dfrac{vg\ e}{vg\ +}$
$vg\ e$	$\dfrac{+\ +}{vg\ e}$	$\dfrac{+\ e}{vg\ e}$	$\dfrac{vg\ +}{vg\ e}$	$\dfrac{vg\ e}{vg\ e}$

phenotypes

mately 9/16:3/16:3/16:1/16 or 9:3:3:1; the wild-type flies are about nine times more frequent than the *vestigial-ebony* flies.

The nine different genotypes that account for the four phenotypes (*vestigial, ebony,* wild type, and *vestigial-ebony*) that appear in the F_2 flies are listed in Table 4-4. This table is constructed to demonstrate that the nine genotypes in fact derive from all possible combinations of the four sorts of gametes produced, in equal numbers, by each of the heterozygous parents. To prove that the nine genotypes in fact exist among the F_2 one must resort to test crosses between individual F_2 flies and $\dfrac{vg\ e}{vg\ e}$ flies. Table 4-4 predicts that if single F_2 flies with a wild-type phenotype are isolated and subjected to such test crosses, 1/9 would prove to be homozygous for both wild-type alleles, 2/9 would be heterozygous for *vg*, 2/9 would be heterozygous for *e*, and 4/9 would be heterozygous for both *vg* and *e*. Similar predictions could be made regarding the genotypes of the F_2 *vestigial* and *ebony* flies; you can work these out yourself.

4.9 THE CHI-SQUARE TEST

The foregoing paragraphs have offered certain predictions that are based on the hypothesis of independent assortment. We now wish to learn whether the data obtained from either test crosses or F_1 inbreeding are indeed in agreement with the predictions and thus with the hypothesis; for example, we can ask whether the F_2 data in Table 4-3, obtained by inbreeding $\dfrac{+\ +}{vg\ e}$ flies, fit a ratio of 9:3:3:1. We can determine the "goodness of fit" of the data to the predicted ratio by a simple statistical test known as the chi-square (χ^2) test.

The χ^2 test involves, first, determining a predicted ratio and, second, establishing how closely the observed data fit this ratio. The χ^2 test is made by ascertaining the probability that the deviation of the observed ratio from the predicted ratio is due to chance and not to some other factor such as experimental conditions, biased sampling, or even the wrong hypothesis. The usual statistical procedure establishes an arbitrary criterion for what degree of deviation is considered a significant deviation from that which would be expected from chance alone. If the *probability* (P) of obtaining the observed ratio is equal to or less than five in 100 ($P = 0.05$), the deviation between the expected and the observed ratio is considered significant and not simply attributable to chance. If the probability is one in 100 or less ($P = 0.01$), the deviation is highly significant, and some non-chance factor is almost certainly operating. When the P value is greater than 0.05, the deviation is not considered statistically significant and can be expected on the basis of chance alone.

A χ^2 test for the data given in Table 4-3 is worked out as follows:

1. *The hypothesis:* That the observed phenotypic ratio is in accord with a predicted phenotypic ratio of 9/16 wild-type:3/16 *vestigial*:3/16 *ebony*:1/16 *vestigial-ebony.*
2. *Expected or predicted ratio:* By dividing the total number of 4992 by 16 the expected number of *vestigial-ebony* flies is determined to be 312. The expected number of wild-type flies is therefore 9 X 312, or 2808, and the expected number of *vestigial* and *ebony* flies is accordingly 3 X 312, or 936, each.
3. *Deviation between observed and expected ratios and the calculation of* χ^2:

	wild type	vestigial	ebony	vestigial ebony	total
Observed (*O*)	2834	920	951	287	4992
Expected (*E*)	2808	936	936	312	4992
Deviation (*D* = *O*—*E*)	26	16	15	25	
D^2	676	256	225	625	
D^2/E	0.24	0.27	0.24	2.00	
	$\Sigma(D^2/E) = \chi^2 = 2.75$				

4. *Calculating degrees of freedom:* To determine a P value from a χ^2 value it is essential to establish the number of *degrees of freedom*. The degrees of freedom (*df*) in an analysis involving *n* classes is usually equal to *n* − 1. What this means is that if a given total number of individuals (4992 in the foregoing example) is divided into *n* classes

Table 4-5 Table of χ^2 (Chi-Square)

df	P = 0.99	0.98	0.95	0.90	0.80	0.70	0.5	0.30	0.20	0.10	0.05	0.02	0.01
1	0.000157	0.00628	0.00393	0.0158	0.0642	0.148	0.455	1.074	1.642	2.706	3.841	5.412	6.635
2	0.0201	0.0404	0.103	0.211	0.446	0.713	1.386	2.408	3.219	4.605	5.991	7.824	9.210
3	0.115	0.185	0.352	0.584	1.005	1.424	2.366	3.665	4.642	6.251	7.816	9.837	11.345
4	0.297	0.429	0.711	1.064	1.649	2.195	3.357	4.878	5.989	7.779	9.488	11.668	13.277
5	0.554	0.752	1.145	1.610	2.343	3.000	4.351	6.064	7.289	9.236	11.070	13.388	15.086
6	0.872	1.134	1.635	2.204	3.070	3.828	5.348	7.321	8.558	10.645	12.592	15.033	16.812
7	1.239	1.564	2.167	2.833	3.822	4.671	6.346	8.383	9.803	12.017	14.067	16.622	18.475
8	1.646	2.032	2.733	3.490	4.594	5.527	7.344	9.524	11.030	13.362	15.507	18.168	20.090
9	2.088	2.532	3.325	4.168	5.380	6.393	8.343	10.656	12.242	14.684	16.919	19.679	21.666
10	2.558	3.059	3.940	4.865	6.179	7.267	9.342	11.781	13.442	15.987	18.307	21.161	23.209

Abridged from Table II of Fisher & Yates: *Statistical Tables for Biological, Agricultural and Medical Research* (1953); published by Longman and by permission of the authors and publishers.

(four phenotypic classes in the foregoing example) then, once we have calculated the numbers expected in three of these classes, the fourth number is set. It follows that there are only three degrees of freedom in the analysis. Specifically, when we add the expected values of three of the genotypic classes (2808 + 936 + 936) and get 4680, the expected value for the fourth class *must* equal 312 if the total is to be 4992.

5. *The P value:* Tables of χ^2 values have been prepared from which it is possible to determine the corresponding *P* values, given a particular number of degrees of freedom. A part of such a table is given in Table 4-5, and it is seen that when $\chi^2 = 2.75$ and there are three degrees of freedom, the *P* value is greater than 0.30 and less than 0.50 ($P = 0.50 - 0.30$). Thus from 30 to 50 times out of 100 we could expect chance deviations of the magnitude observed. Since the value of *P* is clearly greater than 0.05, the observed results for a 9:3:3:1 ratio are in good agreement with those to be expected for the independent assortment of two pairs of alleles.

4.10 PHENOTYPIC RATIOS

The standard way of determining the genotypic ratios that emerge in a cross is to construct a "checkerboard" like that in Table 4-4, with all possible eggs on one ordinate and all possible sperm on the other, and then to make all possible combinations of these gametes. There is, however, a simpler method for determining phenotypic ratios. When doubly heterozygous individuals such as $\dfrac{+ \ +}{vg \ e}$ are inbred and we follow *vg* and its + allele alone, we expect the genes to segregate in a 1:1 ratio and that the phenotypic ratio will be 3 wild type:1 *vestigial*. The same is true for e and its + allele. Therefore, assuming that the two gene pairs assort independently and that all genotypes have equal viability, the probability of their appearance will be as follows:

3 out of 4 F_2 flies should have wild-type wings and
1 out of 4 F_2 flies should have *vestigial* wings;
3 out of 4 F_2 flies should have wild-type body color and
1 out of 4 F_2 flies should have *ebony* body color.

A rule established by probability theory states that the probability that two independent events will occur simultaneously is the product of their separate probabilities. Thus the probability that various combinations will occur is as follows:

wild-type wing and wild-type body color should be $3/4 \times 3/4 = 9/16$;
wild-type wing and *ebony* body color should be $3/4 \times 1/4 = 3/16$;

vestigial wing and wild-type body color should be 1/4 X 3/4 = 3/16; *vestigial* wing and *ebony* body color should be 1/4 X 1/4 = 1/16.

We can also use this method to determine expected phenotypic ratios when inbreeding individuals are heterozygous for *three* unlinked pairs of alleles; for example, true-breeding pea plants that have *yellow* and *round* seeds and *red* flowers can be crossed with true-breeding plants that have *green* and *wrinkled* seeds and *white* flowers, the corresponding traits being allelic. The F_1 all have *yellow, round* seeds and *red* flowers, which indicates that these are the dominant traits. When these plants are self-fertilized, then, following our previous reasoning, three out of four should have *yellow* seeds, while one out of four should have green seeds; three out of four should have *round* seeds; and so on. By multiplying these probabilities we find that the combinations should occur with the following frequency:

round-yellow-red:	3/4 X 3/4 X 3/4 = 27/64
round-yellow-white:	3/4 X 3/4 X 1/4 = 9/64
round-green-red:	3/4 X 1/4 X 3/4 = 9/64
wrinkled-yellow-red:	1/4 X 3/4 X 3/4 = 9/64
round-green-white:	3/4 X 1/4 X 1/4 = 3/64
wrinkled-yellow-white:	1/4 X 3/4 X 1/4 = 3/64
wrinkled-green-red:	1/4 X 1/4 X 3/4 = 3/64
wrinkled-green-white:	1/4 X 1/4 X 1/4 = 1/64

Summing these classes, we see that for every 64 individuals in the F_2, 27 will be expected to exhibit all three dominant traits, 27 will show two dominant traits, 9 will show one dominant trait, and 1 will exhibit all three recessive traits. The only individuals in the F_2 whose genotype is known by inspection are the *wrinkled-green-white* plants: these plants must be homozygous recessive, a state usually symbolized by plant geneticists as *rr yy cc,* where the c locus controls pigment formation in the flower. In addition to the *wrinkled-green-white,* seven other phenotypic classes also emerge, and 26 different genotypes are represented within them. The genotypes can be derived by a checkerboard analysis, assuming that equal numbers of *RYC, RYc, RyC, Ryc, rYC, rYc, ryC,* and *ryc* are present in both eggs and pollen. The resulting genotypes are shown in Table 4-6. Clearly, the genetic analysis becomes increasingly tedious as more independently-assorting alleles are followed in a cross.

Following these lines of reasoning, a general relationship can be derived for the number of gene combinations expected when diploids that are heterozygous for numerous unlinked genes are crossed. This is shown in Table 4-7. The major limit placed on such calculations is the number of chromosome pairs (*n*) possessed by a given species, so that *n* cannot be greater than 4 for *D. melanogaster* nor greater than 7 for the pea. Nonetheless, it is clear that the independent assortment of genes can yield enormous

Table 4-6 The Theoretical Number of Individuals, with Their Genotypes and Breeding Behavior, Expected in F_2 from a Cross of a Round, Yellow-Seeded, Red-Flowered Variety of Pea with a Wrinkled, Green-Seeded, White-Flowered One

Number of Individuals	Genotype Class	Phenotype Class	Ratio of Phenotypes	Breeding Behavior When Self-Fertilized
1	RR YY CC			Breeds true
2	Rr YY CC			Segregates round-wrinkled, 3:1
2	RR Yy CC			Segregates yellow-green, 3:1
2	RR YY Cc	Round		Segregates red-white, 3:1
4	Rr Yy CC	Yellow Red	27	Segregates round-wrinkled, yellow-green, 9:3:3:1
4	Rr YY Cc			Segregates round-wrinkled, red-white, 9:3:3:1
4	RR Yy Cc			Segregates yellow-green, red-white, 9:3:3:1
8	Rr Yy Cc			Segregates round-wrinkled, yellow-green, red-white, 27:9:9:9:3:3:3:1
1	RR YY cc	Round		Breeds true
2	RR Yy cc	Yellow	9	Segregates yellow-green, 3:1
2	Rr YY cc	White		Segregates round-wrinkled, 3:1
4	Rr Yy cc			Segregates round-wrinkled, yellow-green, 9:3:3:1
1	RR yy CC	Round		Breeds true
2	RR yy Cc	Green	9	Segregates red-white, 3:1
2	Rr yy CC	Red		Segregates round-wrinkled, 3:1
4	Rr yy Cc			Segregates round-wrinkled, red-white, 9:3:3:1
1	rr YY CC	Wrinkled		Breeds true
2	rr Yy CC	Yellow	9	Segregates yellow-green, 3:1
2	rr YY Cc	Red		Segregates red-white, 3:1
4	rr Yy Cc			Segregates yellow-green, red-white, 9:3:3:1
1	rr yy CC	Wrinkled		Breeds true
2	rr yy Cc	Green Red	3	Segregates red-white 3:1
1	rr YY cc	Wrinkled		Breeds true
2	rr Yy cc	Yellow White	3	Segregates yellow-green, 3:1
1	RR yy cc	Round		Breeds true
2	Rr yy cc	Green White	3	Segregates round-wrinkled, 3:1
$\frac{1}{64}$	rr yy cc	Wrinkled Green White	1	Breeds true

From E. W. Sinnott, L. C. Dunn, and T. Dobzhansky, *Principles of Genetics*. New York: McGraw-Hill, 1958, Table 6.1.

Table 4-7 The Relation Between the Number of Genes Involved in a Cross and the Number of Phenotypic and Genotypic Classes in F_2

Number of Genes Involved in the Cross	Number of Visibly Different F_2 Classes of Individuals if Dominance is Complete	Number of Different Kinds of Gametes Formed by the F_1 Hybrid	Number of Genotypically Different Combinations	Number of Possible Combinations of F_1 Gametes
1	2	2	3	4
2	4	4	9	16
3	8	8	27	64
4	16	16	81	256
n	2^n	2^n	3^n	4^n

From E. W. Sinnott, L. C. Dunn, and T. Dobzhansky, *Principles of Genetics,* New York: McGraw-Hill, 1958, Table 6.2.

genetic variability in sexually reproducing organisms and that this variability is in no way dependent on crossing over between homologous chromosomes. On the assumption that the bringing together of new gene combinations is of selective advantage to a species, the organization of genes into multiple linkage groups and their random assortment during meiosis is indeed beneficial to eukaryotes.

SEX-LINKED INHERITANCE

Perhaps the most unambiguous examples of the segregation of alleles and the independent assortment of unlinked genes come by following genes that are located in the sex chromosomes. This section first examines the sex chromosomes themselves and then explores the patterns by which these chromosomes are transmitted from parent to offspring.

4.11 THE SEX CHROMOSOMES

Many higher plants and certain animals are hermaphroditic, meaning that any one organism can give rise to both male and female gametes; in such cases the organism in question must carry the genetic information required for both avenues of sexual differentiation. When the two sexes occur as separate organisms, some of the relevant genetic information can be separated as well. The actual number of genes controlling all aspects of sexual differentiation is unknown for any eukaryote, but in most diploid eukaryotes, key genes have come to be restricted to one pair of chromosomes, the **sex chromosomes,** usually referred to as the **X** and the **Y** chromosomes.

It is generally believed that the sex chromosomes were once ordinary homologous chromosomes occurring in hermaphroditic organisms and that they carried a number of ordinary genes in addition to certain sex-determining genes. Subsequent evolution led to the modification of one homologue. In many cases a physical modification of one of these once-homologous chromosomes can be visualized with the light microscope: one of the two sex chromosomes has become much smaller than the other (for example, the human Y chromosome, Figure 2-9) and is often highly heterochromatic. In some cases, including that of *D. melanogaster,* the heterochromatic Y chromosome is longer than its X homologue and may have a different shape. It is important to bear in mind that the postulated transformation affected one homologue only; modern X chromosomes continue to carry a full complement of ordinary, sex-unrelated genes.

In *Drosophila,* males and females are most readily differentiated by the appearance of their abdominal segments: the female abdomen has a tapered end and a different pattern of stripes than the male abdomen, as seen in Figure 3-6. The sex of a fly is determined, in part, by the number of X chromosomes which that individual possesses. Normally a female fly will have two X chromosomes, whereas a male will have only one X chromosome plus one copy of the heterochromatic Y chromosome. Females are therefore typed as XX and males as XY. An occasional abnormal fertilization will give rise to a fly with one X but no Y chromosome, and such an XO individual, although sterile, is phenotypically a male. Flies with two X chromosomes and a Y chromosome have also been obtained, and they appear as normal females. The Y chromosome in *Drosophila* therefore plays no apparent role in the determination of sex: in addition to an apparent contribution made by autosomal genes, femaleness results when two X chromosomes are present and maleness results when one X is present. Indeed, except for the gene(s) somehow involved in producing fertile sperm, the Y chromosome in *Drosophila* appears to carry very few active genes at all.

In mammals normal females are also XX and males XY. In contrast to *Drosophila,* however, it is the presence of the small, heterochromatic Y chromosome that determines maleness and its absence that determines femaleness. In humans, mechanical errors in the separation of homologues during meiosis, described in Chapter 6, can produce exceptional XO individuals that have an X but no Y chromosome; these are found to be sterile females exhibiting **Turner Syndrome** (Table 4-8); XXY exceptions, on the other hand, are sterile males exhibiting **Klinefelter Syndrome** (Table 4-8). Similarly, XXX humans are phenotypically female and XXXY are phenotypically male. It is not yet known how many male-determining genes reside in the Y chromosome, but present evidence suggests that there may be only a few, perhaps distributed along the length of the chromosome in multiple copies. An extra dosage of the Y chromosome, as in XYY persons, produces normally fertile males.

Table 4-8 Sex Chromosome Abnormalities in Humans

Sex Chromosome Constitution	Syndrome	Frequency	Sex Phenotype	Fertility	Characteristic Features (Some Are Not Invariable)
XO	Turner	One in 3500 "female" births	Female	−	Short stature, webbed neck, low-set ears, broad chest, wide-spaced nipples and underdeveloped breasts, small uterus, abortive ovary development
XXY	Klinefelter	One in 500 "male" births	Male	−	Long legs, small testes, sparse body hair, femalelike breast development
XYY		At least one in 2000 male births	Male	+	Unusually tall
XXX		One in 1400 "female" births	Female	Reduced	Mentally retarded

V. A. McKusick, *Human Genetics*, 2nd ed., © 1972. Adapted by permission of Prentice-Hall, Inc., Englewood Cliffs, N.J.

A third kind of sex chromosome organization is found in birds (and also in butterflies): the male carries the two large chromosomes, called Z, and the female carries one Z and one smaller chromosome called W. Males thus are ZZ and females ZW.

4.12 PATTERNS OF SEX-LINKED GENE TRANSMISSION

In animals with X and Y chromosomes the female is termed the **homogametic** and the male the **heterogametic** sex, meaning that as far as sex chromosomes are concerned females produce only X-carrying eggs, whereas males produce both X- and Y-carrying sperm. Because offspring that are XX are female and those that are XY are male, the inheritance of the X chromosome will follow a specific pattern (Figure 4.11): **a male will transmit his X chromosome only to his female offspring, whereas a female will transmit her X chromosomes to both her male and her female offspring.** It follows that a male will always inherit his X chromosome from his mother, since his father must have contributed the Y. Moreover, a male can transmit the information in his X chromosome to his grandchildren only through his daughters.

Any genetic traits that are transmitted by way of this specific pattern are said to be **sex-linked.** As an example we can follow the inheritance of the *white* phenotype in *Drosophila;* affected flies have no eye pigments and carry a mutation at the *w* locus. When a cross is made between wild-type females and *w* males, all of the F_1 flies are wild-type, indicating that *w* is recessive to +. When the F_1 flies are inbred, three-quarters of the resulting flies possess the wild phenotype and one-quarter the *white* phenotype. This

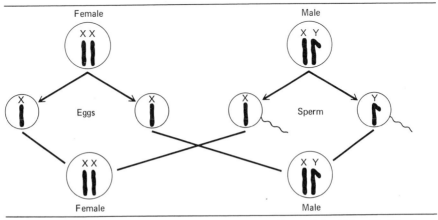

Figure 4-11 Inheritance of X and Y chromosomes in Drosophila. (After E. Altenburg, *Genetics.* New York: Holt, Rinehart and Winston, 1957.)

result appears to be a straightforward example of the segregation of a pair of alleles, w and $+$. When, however, the F_2 flies are classified for both eye color and sex, it is found that all of the females are wild type, whereas half of the males are wild type and the other half have *white* eyes (Figure 4-12a).

We can now examine the reciprocal cross: *white* females crossed to wild-type males (Figure 4-12b). In this case the F_1 is no longer all of the wild type; instead, we find wild-type females and *white* males. Moreover, when the F_1 is inbred, half the progeny are wild in phenotype and half have *white* eyes. Equal numbers of males and females are represented in both phenotypic classes (Figure 4-12b).

Whenever reciprocal crosses give markedly different F_1 and F_2 phenotypic ratios, the trait being studied is likely to be sex-linked. Figures 4-13a and b illustrate the transmission of w and its wild-type allele by assuming that w is located in the X chromosome, and in both reciprocal crosses this assumption readily accounts for the observed phenotypic ratios.

By looking more closely at Figure 4-13 it becomes clear that the presence

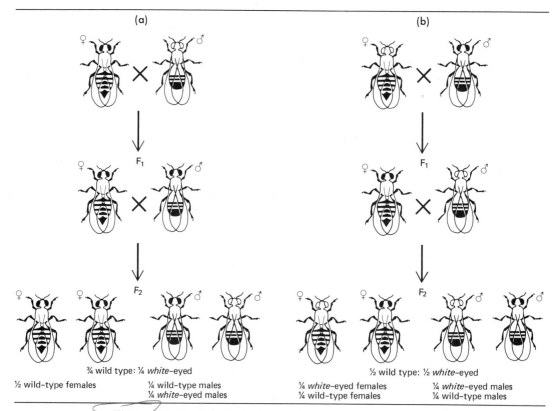

(a)	(b)
¾ wild type: ¼ *white*-eyed	½ wild type: ½ *white*-eyed
½ wild-type females	¼ *white*-eyed females
¼ wild-type males	¼ wild-type females
¼ *white*-eyed males	¼ *white*-eyed males
	¼ wild-type males

Figure 4-12 Sex-linked inheritance of *white* eye color in *Drosophila* in terms of phenotypes. Reciprocal crosses are shown in (a) and (b).

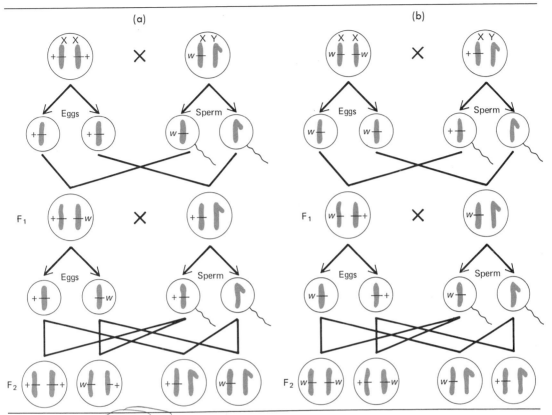

Figure 4-13 Inheritance of *white* eye color in terms of genotypes and the XY chromosomes. (a) Crosses correspond to Figure 4-12a. (b) Crosses correspond to Figure 4-12b.

of the Y chromosome is without effect on the eye-color phenotype. Thus in cross *a*, in which all of the original female parents produce +-bearing eggs, all of the F_1 males have normal eye color. In cross *b*, in which the female parents produce *w*-bearing gametes, all the F_1 males have *white* eye color. The Y does not appear to carry any gene that modifies the expression of either the dominant + or recessive *w* allele. It therefore behaves as a genetically inert chromosome, which is what we would expect since, as noted in Section 2.13, constitutive heterochromatin carries very few genes. For this reason it is inappropriate to regard *Drosophila* males as being either homozygous or heterozygous for most of the genes on their X chromosome. Instead the male flies are referred to as being **hemizygous,** meaning that they possess only one set of these genes. This set they inherit exclusively from their mothers, and their Y chromosomes are "silent" as far as the expression of the sex-linked genes is concerned.

Figures 4-13a and b show the predicted genotypes of F_1 and F_2 flies in the reciprocal crosses we have been considering. We expect, for example, that one-half of the F_2 females in both crosses should be heterozygous for w. This prediction can be verified by crossing individual F_2 females to w males; you can work this out yourself with diagrams. (Is it in fact necessary to use w males in such test crosses or would males of wild phenotype serve the same purpose?)

Another example of sex-linked inheritance in *Drosophila* is given by the gene mutation *Bar* (*B*). In females homozygous for *B* the eyes are highly reduced in size as compared with those of wild-type flies. A *Bar*-eyed male has a phenotype similar to that of a *B/B* female. When salivary gland chromosome preparations are made of *Bar*-eyed individuals and their X chromosomes are examined, a segment known as 16A, which contains several bands, is found to be present in double dose compared with the amount of this segment found in a wild-type X chromosome, as shown in Figure 4-14. In females heterozygous for *Bar* (*B/+*) an intermediate phenotype is observed: the eyes are neither normal nor decidedly bar-shaped. Thus *Bar* is dominant to its wild-type allele. Nonetheless, its inheritance is typically sex-linked. As shown in Figure 4-15, the genotype of a male with respect to *Bar* is dependent on the genotype of his mother, so that male offspring of a *B/B* female are *Bar*-eyed, whereas those of a *B/+* female are both wild type and *Bar* in equal numbers. We also see that the X chromosome contributed by a male is transmitted only to his female offspring and that it does not appear in male progeny until the following generation.

Sex-linked genes not only segregate but also assort independently from genes located on autosomes. A problem that illustrates this fact is included at the end of this chapter.

4.13   SEX-LINKED INHERITANCE IN BIRDS

In *Drosophila* and in mammals the heterogametic sex is the male, whereas, as noted earlier, the female is the heterogametic (ZW) sex in birds and butterflies. Sex-linked inheritance in birds, however, follows the pattern to be expected from this reversal of chromosomal determinants, as shown in Figure 4-16 for the dominant gene *B* (*Barred* feathers) in chickens.

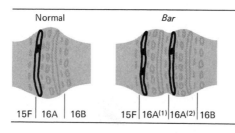

Normal	Bar
15F │ 16A │ 16B	15F │ 16A(1) │ 16A(2) │ 16B

Figure 4-14 Normal and *Bar* polytene X chromosomes of *Drosophila*, showing duplicated 16A segment in the *Bar* chromosome.

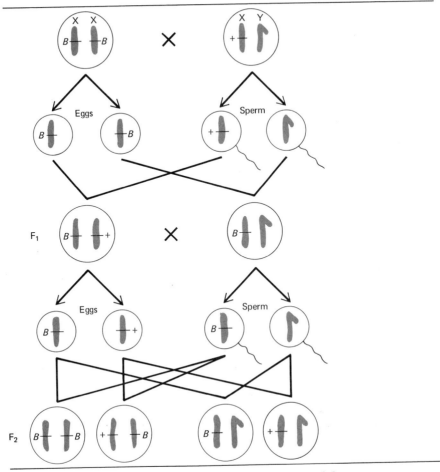

Figure 4-15 The sex-linked inheritance of *Bar* eye in *Drosophila*.

4.14 SEX-LINKED INHERITANCE IN HUMANS

We saw earlier, in the case of brachydactyly and microcephaly, that human traits are inherited in a Mendelian fashion. The construction of the brachydactyly pedigree (Figure 4-6) was straightforward in that brachydactyly is caused by a dominant gene mutation and thus all carriers express the mutant phenotype. In the case of the gene for a rare autosomal recessive trait, on the other hand, its presence is usually masked by a dominant allele, and it is only in the relatively rare cases in which two heterozygous carriers have children (as in consanguineous marriages) that the gene is permitted expression in homozygous individuals. Moreover, the standard method of ascertaining the genotypes of heterozygotes, the test cross, cannot be

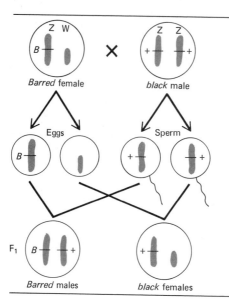

Figure 4-16 Sex-linked inheritance in chickens (*Barred* vs. *black* feathers). (After E. Altenberg, *Genetics*. New York: Holt, Rinehart and Winston, 1951.)

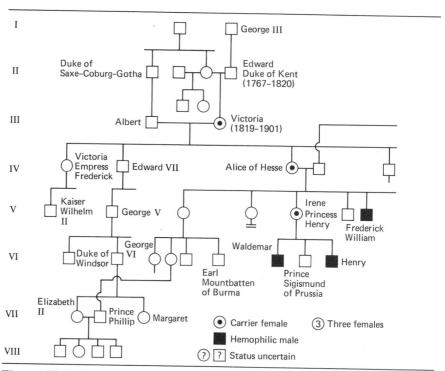

Figure 4-17 Pedigree of the descendants of Queen Victoria showing carriers and afflicted males possessing the X-linked gene conferring the disease hemophilia. (From V. A. McKusick, *Human Genetics*, 2nd ed., © 1972. Adapted by permission of Prentice-Hall, Inc., Englewood Cliffs, N.J.)

feasibly applied to humans. There is thus only limited pedigree data on the inheritance of recessive human autosomal genes (see, for example, Figure 4-7). In contrast, there is a wealth of familial data on the inheritance of recessive sex-linked genes, since a woman effectively subjects her two X chromosomes to a test cross every time she bears sons. Her mate contributes only the silent Y chromosome to their male offspring, and thus the phenotypes of her sons will directly reflect the genetic constitution of one or the other of her two X chromosomes. Studies of sex-linked inheritance in humans have amply demonstrated that patterns of gene transmission in humans follow those first recognized in flies.

Certainly the most famous case of human sex-linked inheritance is that of hemophilia A. Individuals suffering from this disorder cannot synthesize a normal blood protein called antihemophilic globulin (AHG), a substance required for the formation of thromboplastin and thus for normal blood clotting. The AHG gene is located in the X chromosome, as demonstrated by the inheritance of a mutant AHG gene in the royal families of Europe during the last century (Figure 4-17). Queen Victoria was probably the original carrier of the mutant gene, since the disease was unknown in

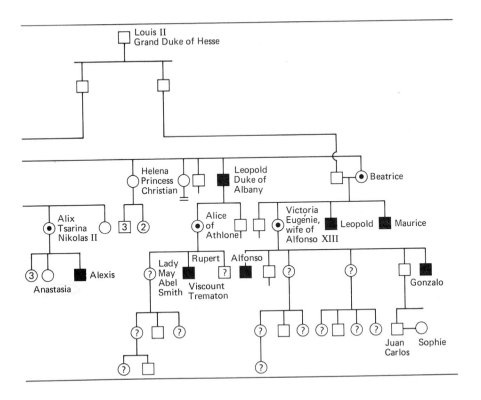

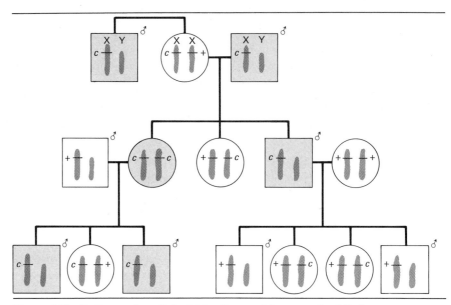

Figure 4-18 Pedigree of sex-linked inheritance of human color blindness.

preceding generations of highly inbred royalty, and thus the gene probably arose as a new mutation in either the egg or the sperm from which she ultimately developed. One of her nine children, a son, was hemophilic, or a "bleeder," and two daughters proved to be carriers. One daughter, Alice, passed the gene on to her daughter, Alix, who became Empress Alexandra of Russia and bore a hemophilic son, the Tsarevich Alexis.

A frequently occurring form of red-green color blindness is also a sex-linked recessive characteristic in humans. A pedigree is given in Figure 4-18. In the first generation a normal woman, who had a color-blind brother, married a color-blind man. Of their three children, one male and one female were color-blind and one female was normal. The fact that a color-blind daughter was born to this couple at once indicates that the mother must have been a heterozygous carrier, as shown in Figure 4-18. The son then married a woman with normal color vision and all their children were likewise normal (although the daughters were carriers). The son's color-blind sister (who must be homozygous for the gene) married a man with normal color vision; all of their sons were color-blind, whereas their one daughter had normal vision but was a carrier.

Table 4-9 presents a list of the known X-borne mutations in humans. It is of interest in this regard that at least two of the genes that are sex-linked in humans—the gene for AHG and the structural gene for a subunit of the enzyme glucose-6-phosphate dehydrogenase—are also sex-linked in a number of other placental mammals, including the mouse, whose karyotype is strikingly different from that of the human (Figure 2-10b). This and other

Table 4-9   Human Sex-Linked Traits

Addison's disease (adrenal insufficiency) (one form)
Agammaglobulinemia (immunoglobulin insufficiency)
Albinism (various forms)
Albright's osteodystrophy
Aldrich syndrome (immunological disorder)
Amelogenesis imperfecta (tooth malstructure)
Anemia, hypochromic
Angiokeratoma (Fabry's disease) (kidney disfunction)
Cerebellar ataxia
Cerebral sclerosis, Scholz type
Charcot-Marie-Tooth peroneal atrophy
Choroideremia (progressive blindness)
Cleft palate
Color blindness (complete and several partial types)
Deafness
Diabetes insipidus (several types)
Dyskeratosis
Ectodermal dysplasia
Ehlers-Danlos syndrome (bruising tendency)
Endocardial fibroelastosis
Faciogenital dysplasia
Fibrin-stabilizing factor deficiency
Focal dermal hypoplasia
Glucose-6-phosphate dehydrogenase deficiency
Glycogen storage disease
Granulomatous disease
Hemophilia A
Hemophilia B (Christmas disease)
Hydrocephalus
Hypomagnesemic tetany
Hypophosphatemia (vitamin D-resistant rickets)
Hypoxanthine guanine phosphoribosyl transferase deficiency
Ichthyosis (epidermal scaling)
Incontinentia pigmentii
Iris, hypoplasia of
Keratosis follicularis spinulosa
Lowe's oculocerebrorenal syndrome
Macular dystrophy
Megalocornea
Menkes syndrome (kinky hair disease)
Mental deficiency
Microphthalmia
Mucopolysaccharidosis II (Hunter syndrome)
Muscular dystrophy, progressive Becker type (onset between age 30–40)
Muscular dystrophy, Duchenne type (onset in childhood)
Muscular dystrophy, tardive, Dreifuss type (not lethal)
Night blindness
Norrie's disease (blindness)

Table 4-9
(*continued*)
Nystagmus (one type)
Ophthalmoplegia (myopia)
Oral-facial-digital syndrome
Parkinsonism (one type)
Pelizaeus-Merzbacher disease (cerebral sclerosis)
Phosphoglycerate kinase deficiency
Pituitary dwarfism
Pseudohermaphroditism, male
Reticuloendotheliosis
Retinitis pigmentosa (blindness)
Retinoschisis
Spastic paraplegia
Spinal and bulbar muscular atrophy
Spinal ataxia
Spondylo-epiphyseal dysplasia, late
Testicular feminization syndrome
Thrombocytopenia
Thyroxine-binding globulin reduction
Van den Bosch syndrome (mental deficiency)
XG blood group system
XM system (macroglobulin serum protein)

From V. A. McKusick, *Mendelian Inheritance in Man*, 3rd ed., Baltimore: Johns Hopkins Press, 1971.

lines of evidence suggest that once a certain chromosome in some ancestral mammalian population became entrusted with the all-important role of sex determination, very little gene rearrangement within this chromosome was tolerated by successive evolutionary lines, even though extensive divergence occurred for most other chromosomes. In other words, the X chromosome is probably a very ancient chromosome.

4.15 SEX LINKAGE AND THE CHROMOSOME THEORY

The development in the early part of this century of the **chromosomal theory of inheritance,** which proposes that genes reside in chromosomes, made elegant use of sex linkage. L. V. Morgan, for example, studied exceptional strains of *Drosophila* in which the females transmit sex-linked traits to their daughters rather than to their sons. Specifically, when the females are wild type and are crossed to *Bar*-eyed males, the F_1 consists solely of wild-type females and *Bar*-eyed males, a result that is the reverse of the normal situation.

Morgan examined the chromosomes of the exceptional females and found that their two X chromosomes, instead of being separate as they are normally, were permanently joined at or near their centromeres. This attachment, she further observed, prevented the two chromosomes from separating from one another at metaphase. Finally, she noted that females with such **attached X** chromosomes (X̂X) also carried a Y chromosome.

By knowing the karyotype of the exceptional females we can interpret their unusual pattern of gene transmission. First, it is clear that they should produce two types of egg in equal numbers: X̂X eggs and Y eggs, as shown in Figure 4-19. When the females are crossed with normal males, therefore, four types of zygote are possible.

1. Fertilization of an X̂X egg by a Y-bearing sperm will produce an X̂XY zygote which should develop into a normal female (recall that the Y plays no role in sex determination in flies).
2. Fertilization of an X̂X egg by an X-bearing sperm will produce an X̂XX zygote which is usually inviable or matures into a peculiar female fly with distinctive characteristics.
3. Fertilization of a Y egg by an X-bearing sperm will produce a normal XY male.
4. Fertilization of a Y egg by a Y-bearing sperm will produce a YY zygote, inviable because the missing X chromosome carries genes that are essential for survival.

These possibilities are summarized in Figure 4-19.

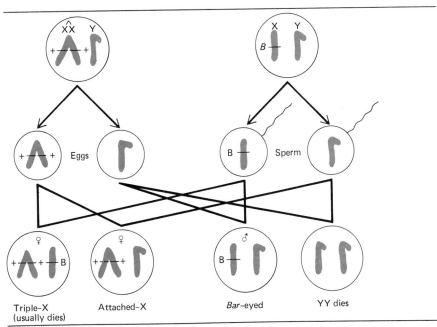

Triple-X (usually dies)　Attached-X　*Bar*-eyed　YY dies

Figure 4-19 A cross in *Drosophila* between an exceptional X̂XY female and a *Bar*-eyed male.

Thus it is clear that among the four possible types of offspring only two, namely X̂XY females and XY males, are viable. In a cross between wild X̂XY females and *Bar*-eyed males, therefore, only wild females and *Bar*-eyed males are expected, and this is what Morgan observed. In other words, these experiments show a direct relationship between a visible chromosomal abnormality and the transmission of specific hereditary traits, and they therefore offer an unambiguous demonstration that chromosomes are indeed the vehicles of gene transmission; their behavior at meiosis determines the patterns of inheritance.

References

Segregation And Independent Assortment

Bateson, W. *Mendel's Principles of Heredity.* Cambridge: Cambridge University Press, 1909.

*Correns, C. (translated by L. K. Piternick). "G. Mendel's Law concerning the behavior of progeny of varietal hybrids," *Genetics* **35** (Pt. 2):33–41 (1950). (Original paper in German, 1900). [Reprinted in L. Levine, *Papers on Genetics,* St. Louis: Mosby, 1971.]

Demerec, M., Ed. *The Biology of Drosophila.* New York: Wiley, 1950.

Elliot, A. M., Ed. *Biology of Tetrahymena.* Stroudsburg: Dowden, Hutchinson and Ross, 1973.

Fincham, J. R. S., and P. R. Day. *Fungal Genetics.* Oxford:Blackwell, 1963.

Gillies, C. B. "Synaptonemal complex and chromosome structure." *Ann. Rev. Genetics* **9:**91–109 (1975).

Iltis, H. (translated by E. and C. Paul). *Life of Mendel.* New York: Norton, 1932.

John, B. "Myths and mechanisms of meiosis." *Chromosoma* **54:**295–325 (1976).

McKusick, V. A. *Human Genetics,* 2nd ed. Englewood Cliffs, N. J.: Prentice-Hall, 1969.

McKusick, V. A. *Mendelian Inheritance in Man,* 3rd ed. Baltimore: Johns Hopkins Press, 1971.

*Mendel, G. "Experiments in plant hybridization" (English translation of "Versuche uber Pflanzen Hybriden," (1966). [Reprinted in J. A. Peters, Ed. *Classical Papers in Genetics.* Englewood Cliffs, N. J.: Prentice-Hall, 1959.]

Rasmussen, S. W. "The meiotic prophase in *Bombyx mori* females analyzed by three-dimensional reconstructions of synaptonemal complexes." *Chromosoma* **54:**245–293 (1976).

Rimoin, D. L., T. J. Merimee, and V. A. McKusick. "Growth-hormone deficiency in man: an isolated, recessively inherited defect," *Science* **152:**1635–1637 (1966). [Reprinted in L. Levine, *Papers on Genetics.* St. Louis: Mosby, 1971.]

Roth, R. "Temperature-sensitive yeast mutants defective in meiotic recombination and replication," *Genetics* **83:**675–686 (1976).

Sinnot, E. W., L. C. Dunn, and T. Dobzhansky. *Principles of Genetics,* 5th ed. New York: McGraw-Hill, 1958.

Sonneborn, T. M. "Paramecium aurelia." In: *Handbook of Genetics* **2,** R. C. King, Ed. New York: Plenum, pp. 469–594 (1974).

Stern, C. *Principles of Human Genetics.* San Francisco: Freeman, 1960.

Stern, C., and E. R. Sherwood, Eds. *The Origin of Genetics, a Mendel Source Book.* San Francisco: Freeman, 1966.

*Denotes articles described specifically in the chapter.

Sturtevant, A. H. *A History of Genetics.* New York: Harper and Row, 1965.

Sturtevant, A. H., and G. W. Beadle. *An Introduction to Genetics.* Philadelphia: Saunders, 1940.

*Sutton, W. S. "The chromosomes in heredity," *Biol. Bull.* **4:**213–251 (1903). [Reprinted in J. A. Peters, Ed. *Classical Papers in Genetics.* Englewood Cliffs, N.J.: Prentice-Hall, 1959.]

Zimmering, S., L. Sandler, and B. Nicoletti. "Mechanisms of meiotic drive," *Ann. Rev. Genetics* **4:**409–436 (1970).

Sex Chromosomes and Sex Linkage

*Ford, C. E., K. W. Jones, P. E. Polani, J. C. de Almeida, and J. H. Briggs. "A sex-chromosome anomaly in a case of gonadal dysgenesis," *The Lancet* **1:**711–713 (1959). [Reprinted in L. Levine, *Papers on Genetics,* St. Louis: Mosby, 1971.]

Hess, O., and G. F. Meyer. "Genetic activities of the Y chromosome in *Drosophila* during spermatogenesis," *Adv. Genetics* **14:**171–223 (1968).

*Jacobs, P. A., and J. A. Strong. "A case of human intersexuality having a possible XXY sex-determining mechanism," *Nature* **183:**302–303 (1959). [Reprinted in L. Levine, *Papers on Genetics,* St. Louis: Mosby, 1971.]

*Morgan, L. V. "Non-criss-cross inheritance in *Drosophila melanogaster,*" *Biol. Bull.* **42:**267–274 (1922).

*Morgan, T. H. "Sex limited inheritance in *Drosophila,*" *Science* **32:**120–122 (1910). [Reprinted in L. Levine, *Papers on Genetics,* St. Louis: Mosby, 1971, and in J. A. Peters, Ed. *Classical Papers in Genetics,* Englewood Cliffs, N.J.: Prentice-Hall, 1959.]

Ohno, S. *Sex Chromosomes and Sex-Linked Genes.* Berlin: Springer-Verlag, 1967.

Rich, C. M., and G. C. Hanna, "Determination of sex in *Asparagus officinalis,*" *Am. J. Bot.* **30:**711–714 (1943). [Reprinted in L. Levine, *Papers on Genetics.* St. Louis: Mosby, 1971.]

Westergaard, M. "The mechanism of sex determination in dioecious flowering plants," *Adv. Genetics* **9:**217–281 (1958).

Wilson, E. B. "The chromosomes in relation to the determination of sex in insects," *Science* **22:**500–502 (1905). [Reprinted in L. Levine, *Papers on Genetics.* St. Louis: Mosby, 1971.]

(a)   (b)

Thomas Hunt Morgan (a) and Lilian Vaughan Morgan (b) were pioneers in establishing the genetics of *Drosophila melangaster.* (Photographs were generously provided by the Morgan family.)

Questions and Problems

1. (a) A mutant strain of *Chlamydomonas* having paralyzed flagella (*pf*) was crossed to the motile, wild-type strain and 100 tetrads were analyzed. Each tetrad gave a one-to-one ratio of paralysis to motility. Explain this result in terms of the behavior of a pair of homologous chromosomes during meiosis. (b) Instead of analysis of 100 tetrads, as in (a), assume that one meiotic product was recovered at random from each of the 100 tetrads. What would be the ratio of paralyzed to motile?

2. A cross was made between two *Drosophila* that had the wild phenotype. Their progeny were found to have the same phenotype. A sample was taken of 200 of the progeny, each of which was crossed with a fly having a purple eye color. Half of the crosses gave only wild-type flies and the other half gave 50 percent wild-type and 50 percent purple-eyed progeny. What were the genotypes of the original pair of wild-type flies?

3. A cross was made between wild-type (+) *Neurospora* and a mutant strain that cannot synthesize the vitamin thiamin (*thi*). Tetrad analysis showed a one-to-one segregation for +: *thi*. It was also found that 20 percent of the tetrads showed second-division segregation. How were these tetrads recognized and what are the events in meiosis that lead to second-division segregation?

4. Human albinism is the absence of pigment from the hair, skin, and eyes. Determine from the pedigree given below whether it is a dominant or a recessive gene and whether it is sex-linked.

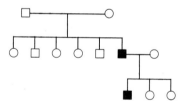

5. A marriage between two brachydactylous heterozygotes, as postulated in the text, has in fact never been reported, and thus it is not known whether *B/B* homozygotes would be brachydactylous or would instead suffer from more severe disabilities. If *B/B* homozygotes are in fact inviable and die at an early embryonic stage, then what proportion of pregnancies would you expect to end in abortion in the proposed mating? Of the couple's viable children, what would be their expected genotypes and the proportions of each?

6. List the nine different genotypes emerging from the cross summarized in Table 4-4.

7. In *Chlamydomonas* a mutant gene *arg* assorts independently of a mutant

gene *pf*. What types of tetrads are expected from the cross *arg* + $\times$ + *pf*. How many different genotypes will there be and in what ratio?

8. The tetrads formed in *Neurospora* are ordered as a result of meiosis in the ascus. Assume that one mating type of this organism carries the mutant gene *a* and the wild-type allele of mutant gene *b*, while the other mating type carries the wild-type allele of the mutant gene *a* and the mutant gene *b*. A cross, *a* + $\times$ + *b*, gave the following kinds and numbers of ordered tetrads:

A	B	C	D	E	F
a +	*a* *b*	*a* +	*a* *b*	*a* +	+ +
a +	*a* *b*	*a* *b*	*a* +	+ +	*a* +
+ *b*	+ +	+ +	+ *b*	*a* *b*	+ *b*
+ *b*	+ +	+ *b*	+ +	+ *b*	*a* *b*
134	132	105	108	11	10

(a) Show how the data reveal that *a* and *b* are assorting independently.

(b) Classify each tetrad by name and show with diagrams the *simplest* origin for each. Indicate for each tetrad whether segregation occurs at either the first or second division for the *a* locus and *b* locus.

(c) Are all the expected tetrad classes present? If not, which are missing and in what proportions would you expect them to appear?

(d) Instead of analyzing 500 tetrads, assume that one ascospore was recovered at random from each of 1000 asci. How many genotypes would you find and how many of each kind?

9. A cross between two wild-type flies gave progeny all of which were wild type. When they were test crossed to $\dfrac{vg\ e}{vg\ e}$ flies the following results were obtained:

(a) 1/4 of the test crosses gave wild type, *vestigial ebony, vestigial,* and *ebony* in a 1:1:1:1 ratio.

(b) 1/4 of the test crosses gave all wild type.

(c) 1/4 of the test crosses gave *vestigial* and wild type in a 1:1 ratio.

(d) 1/4 of the test crosses gave *ebony* and wild type in a 1:1 ratio.

What are the genotypes of the original pair of wild-type flies?

10. Assume that the gene *r* and its + allele show **incomplete dominance** with respect to flower color such that +/+ is red, +/*r* is pink, and *r*/*r* is white and that the gene *s* and its allele show incomplete dominance with regard to seed color such that +/+ has red-black seeds, +/*s* pink seeds, and *s*/*s* white seeds. Give the phenotypes and their ratios expected among the progeny in the following crosses:

(a) $\dfrac{+\quad s}{+\quad s} \times \dfrac{r\quad +}{r\quad +}$

(b) $\dfrac{+\quad +}{r\quad s} \times \dfrac{+\quad +}{r\quad s}$

(c) $\dfrac{+\quad+}{r\quad s} \times \dfrac{r\quad s}{r\quad s}$

(d) $\dfrac{r\quad+}{r\quad s} \times \dfrac{+\quad s}{r\quad s}$

11. For the *Neurospora* cross:

list all the tetrads that could be obtained from a *single* crossover event.

12. Thalassemia is a type of human anemia rather common in Mediterranean populations, but relatively rare in other peoples. The disease occurs in two forms, minor and major; the latter is much more severe. Persons with thalassemia major are homozygous for an aberrant recessive gene; mildly affected persons (with thalassemia minor) are heterozygous; persons normal in this regard are homozygous for the normal allele. The following four questions relate to this situation:

(a) A man with thalassemia minor marries a normal woman. With respect to thalassemia, what types of children, and in what proportions, may they expect (let t = the allele for thalassemia minor and T = its normal allele)?

(b) Both father and mother in a particular family have thalassemia minor. What is the chance that their baby will be severely affected? Mildly affected? Normal? Diagram the possible germ-cell unions in this family.

(c) An infant has thalassemia major. According to the information so far given, what possibilities might you expect to find if you checked the infant's parents for anemia?

(d) Thalassemia major is almost always fatal in childhood. How does this fact modify your answer to (c)?

13. In humans the most frequent type of albinism (itself quite rare) is inherited as a simple recessive characteristic. Standard symbols are C = normal pigmentation; c = albino. Assume that the genes for thalassemia and albinism assort independently.

(a) A husband and wife, both normally pigmented and neither with severe anemia, have an albino child who dies in infancy of thalassemia major. What are the probable genotypes of the parents?

(b) If these people have another child, what are its chances of being phenotypically normal with respect to pigmentation? Of having entirely normal (that is, nonthalassemic) blood? Of being phenotypically normal in both regards? Of being homozygous for the normal alleles of both genes?

14. Suppose the analysis of 100 ordered tetrads from a *Neurospora* cross gave the data below (top and bottom of tetrads are not distinguished, i.e., the orders *AaAa* and *aAaA* are considered to be identical.)

No. tetrads	85	2	3	2	3	3	1	1
Tetrad	*Ab*	*Ab*	*Ab*	*AB*	*AB*	*Ab*	*Ab*	*aB*
type	*Ab*	*AB*	*AB*	*Ab*	*Ab*	*aB*	*aB*	*Ab*
	aB	*ab*	*aB*	*ab*	*aB*	*Ab*	*aB*	*Ab*
	aB	*aB*	*ab*	*aB*	*ab*	*aB*	*Ab*	*aB*

(a) Are the *A* and *B* loci linked or unlinked? Explain.
(b) What are the genotypes of the parents?
(c) At what stage of meiosis did the crossovers occur which produced each tetrad?

15. The following is a pedigree of a fairly common sex-linked trait

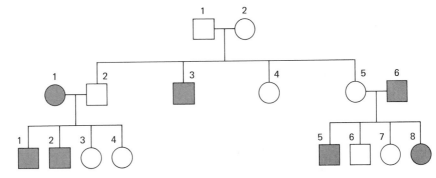

(a) Is the gene which causes the trait dominant or recessive? Explain.
(b) Does this data prove that the gene is sex-linked?
(c) Designate the genotype for I 1, 2; II 2, 5, 6; III 1, 8.

16. A cross was made between two wild-type *Drosophila*. Their progeny were 187 *raspberry* eye color males, 194 wild-type males, and 400 wild-type females. Is *ras* a sex-linked gene? Explain. What are the parental genotypes? What are the genotypes of the F_1 wild-type females and what is their ratio?

17. A cross was made between a female heterozygous for the recessive genes *ct* (*cut* wings) and *se* (*sepia* eye color) and a *sepia* male. Among their female progeny the phenotypes were 1/2 wild type and 1/2 *sepia*. Among their male progeny the phenotypes were 1/4 wild type, 1/4 *cut*, 1/4 *sepia*, and 1/4 *cut* and *sepia*. Are either of the genes sex-linked? What are the genotypes of the parents and their offspring?

18. If a woman having normal color vision has a color-blind father, what is the probability that her sons will be color blind if she marries a man with normal color vision? What genotypes are possible among her male and female offspring? What is the probability of her having a color-blind child if she marries a color-blind man, and is this probability different depending on whether the child is a male or a female?

19. Neither Tsar Nicholas II nor his wife, Empress Alexandra, nor their daughter, Princess Anastasia, had the disease hemophilia. However, their son, the Tsarevich Alexius, did have the disease. Can one automatically assume that Anastasia was a carrier? Why or why not?

20. How could you distinguish from pedigree studies between a sex-linked recessive factor and a dominant autosomal factor expressed only in males?

21. In *Drosophila*, the gene for *red* eye is dominant to its *white* allele and the gene for *long* wing is dominant to its *vestigial* allele. The $+/w$ locus is on the X chromosome; the $+/vg$ locus is not on a sex chromosome. Two *red*-eyed *long*-winged flies bred together produced the following offspring:

Females: 3/4 *red long*, 1/4 *red vestigial*
Males: 3/8 *red long*, 3/8 *white long*, 1/8 *red vestigial*,
1/8 *white vestigial*

What are the genotypes of the parents?

CHAPTER
5

DNA Replication and the Transmission of Prokaryotic and Viral Chromosomes

THE PROKARYOTIC BACTERIAL CELL

5.1 BACTERIAL CELL AND CHROMOSOME STRUCTURE

The bacteria *E. coli, Bacillus subtilis, Salmonella typhimurium,* and pneumococcus represent the prokaryotic species most widely used in genetic research. While each bacterial species has its own distinctive properties, all can be said to be small cells (perhaps 1–2 μm long and 0.5 μm wide) surrounded by one or more membranes and walls. The bacteria contain a ribosome-filled cytoplasm and DNA-containing regions known as **nucleoids** (Figure 5-1) with an apparent volume of 0.1 μm³. The DNA is so densely packed into a nucleoid that most cytoplasmic particles are excluded, but the nucleoid is not surrounded by a membrane, nor does the cytoplasm contain the other organelles that typify eukaryotic cells (compare Figure 5-1 with Figure 2-1).

The DNA in *E. coli* and *B. subtilis* is largely in the form of a single **main chromosome** containing about 1100 μm of DNA and having a molecular weight of about 2.6×10^9 daltons. This DNA differs from eukaryotic chromosomal DNA in that it does not associate with histones, nor does it form regular nucleosome structures (Section 2.11). It is also distinctive in that it takes the form of a giant circle: the duplex chromosome comes back on itself and seals covalently so that if one were to scan along the DNA molecule it would be impossible to locate its "beginning" or "end." In addition to this major chromosome, bacterial cells often possess one or more minor chromosomes, each called a **plasmid,** which may contain from

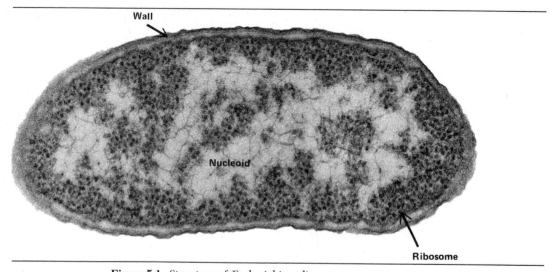

Figure 5-1 Structure of *Escherichia coli.*

0.5 to 2 percent of the DNA of the cell. A plasmid may contain 4 μm up to 35 μm of duplex DNA and this, too, is free of histone and circles about itself to form a small ring (plasmids are considered in detail in Chapter 14).

An interesting property of circular chromosomes is best described by illustration. If two pieces of stiff laboratory tubing are held together and twisted at both ends, a reasonable facsimile of the secondary structure of a DNA duplex is produced. If one now attempts to join the two ends of this twisted structure into a ring, new tertiary kinks develop, and instead of forming a single circle a twisted multilooped configuration is observed. A circular duplex DNA molecule exhibits similar tertiary kinks that are called **superhelical twists.** Figure 5-2 illustrates such twists in the small circular chromosome of a bacteriophage. The twisting phenomenon is relevant when we consider how an *E. coli* cell folds into its nucleoid a chromosome that is approximately 1000 times longer than itself: the chromosome adopts a floral structure, with the superhelical twists looping in and out of the nucleoid center.

5.2 THE BACTERIAL CELL CYCLE

The cell cycle of *E. coli* at slow growth rates ($>$ 60 min per generation) consists of three periods: preparation for initiation of DNA replication, DNA replication, and a period between termination of DNA replication and cell division. At rapid growth rates, a second round of DNA replication may be initiated before the first has been completed; therefore, the 3 periods become blurred and the synthesis of DNA can be said to be essentially continuous. No events resembling mitosis are observed during the bacterial cell cycle, and the mechanism of chromosome segregation during bacterial division is unknown. Since cell division is accomplished by an ingrowing shelf of bacterial membrane called the **mesosome,** the simplest model of bacterial chromosome segregation states that a bacterial chromosome attaches to the membrane at the start of replication via an **attachment point,** that a "daughter" attachment point is synthesized during the course of replication, and that the ingrowing mesosome separates these two attachment points spatially, thereby segregating the daughter chromosomes into daughter cells. Consistent with this model are reports showing that *E. coli* DNA indeed binds to specific membrane proteins at the time of its replication.

An *E. coli* cell will divide as rapidly as once every 20 minutes (yeast, one of the fastest growing eukaryotes, divides every $1\frac{1}{4}$ hours under comparable growth conditions). This rapid **generation time** is, of course, an attractive feature of *E. coli* for genetic research, since it means that large numbers of genetically identical organisms can be generated from a single cell in a very short period of time.

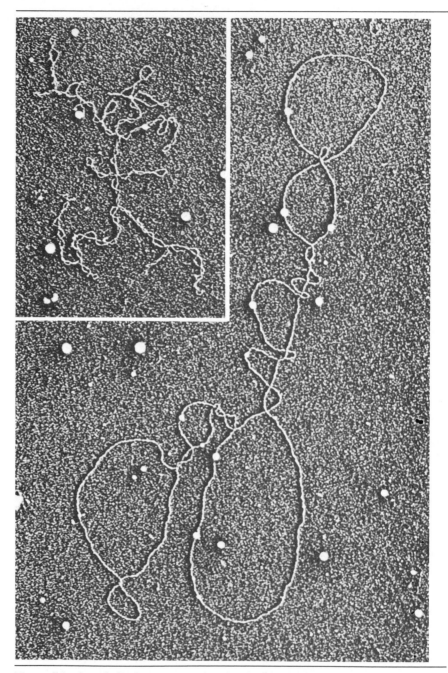

Figure 5-2 Superhelical twists in a circular duplex chromosome. Illustrated are two chromosomes from bacteriophage λ; the variation in the number of twists is caused by differences in temperature and ionic environment. (With permission from V. C. Bode and L. A. MacHattie, *J. Mol. Biol.* **32**:673, 1968. Copyright by Academic Press Inc. (London) Ltd.)

5.3 · DNA REPLICATION IN *E. COLI*

Molecular mechanisms for chromosomal DNA replication have been elegantly worked out for *E. coli* and its phages. These details, in and of themselves, are of only passing interest to the geneticist, for while it is clearly essential to normal gene transmission that chromosomes be copied correctly, genetic analysis can be said to begin with events that change the normal genotype—be it by mutation or recombination—rather than with those events that conserve it. It turns out, however, that many of the enzymes and mechanical processes involved in chromosome replication operate during mutation and recombination as well. DNA replication is therefore considered here with emphasis on information that is of relevance to molecular genetics.

DNA replication is scrutinized in *E. coli* because many insights into this complex biochemical phenomenon have come through the analysis of *E. coli* mutant strains. A mutation affecting an enzyme essential for chromosome replication will ordinarily be lethal. This problem can be circumvented, however, by isolating **conditional mutations** which affect the organism when it is maintained under one set of growth conditions but not under another. Most common in DNA-replication studies are **temperature-sensitive (ts)** conditional strains. In these strains, a mutant gene product can function (and therefore permit the strain to survive) at the **permissive temperature** (for example, 37°C) but is unable to function at the **nonpermissive temperature** (for example, 47°C). You will recall (Box 2.1) that enzyme activity is critically dependent on the spatial configuration adopted by enzyme molecules. A given *ts* mutant will produce, for example, mutant DNA polymerases that retain their ability to carry out DNA replication at 37°C. When the temperature is shifted to 47°C, however, these enzymes become unstable, lose their essential structural features, and can no longer catalyze DNA polymerization. Specific studies of *ts* strains are described after an overview of DNA replication is presented.

5.4 THE SEMICONSERVATIVE MODE OF DNA REPLICATION

You will recall from Chapter 1 that the Watson-Crick model of DNA structure suggested that once chromosome replication is initiated, the two original polynucleotide strands of the helix will unwind, at least locally, so that each can serve as a template for a new strand (Figure 1-10). An immediate prediction follows from this proposal: both duplexes that result from replication should be **hybrid** in nature, each containing an *old* strand derived from the original molecule and a *new* strand which has been formed during the replication process. This prediction is diagrammed in Figure 5-3, in which original DNA strands are shown in gray and new DNA

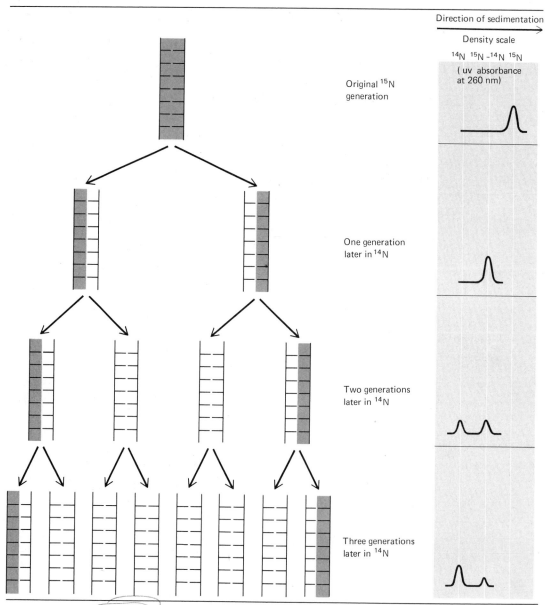

Figure 5-3 The semiconservative replication of DNA and its demonstration in the Meselson-Stahl experiment. Gray color indicates "old" (^{15}N-labeled) DNA; black color indicates "new" (^{14}N-labeled) DNA.

in black. Figure 5-3 also outlines what would be predicted if these two hybrid duplexes went on to replicate themselves. Four duplexes would result, two of which would still contain a single strand derived from the original chromosome and two of which would contain a single strand of totally new DNA. Similarly, a third round of replication would result in eight duplex chromosomes, with two still containing an original strand.

Experimental support for this mode of DNA replication came in 1958 from M. Meselson and F. Stahl. They utilized the fact that if cells are grown in a medium containing ^{15}N, the heavy isotope of nitrogen, the DNA of the cells will become heavier than ordinary DNA and that this density difference can be distinguished by centrifugation in a CsCl gradient (see Box 5.1). By growing *E. coli* in the ^{15}N isotope for several generations, Meselson and Stahl obtained a population of cells that contained fully ^{15}N-labeled molecules. The DNA from these cells was shown to exhibit the buoyant density expected of ^{15}N-labeled DNA (Figure 5-3). The cells were then transferred to a medium containing only ^{14}N as a nitrogen source. It was found that after one round of replication the DNA in the daughter cells had neither the original ^{15}N nor the pure ^{14}N density. Instead, a single species with the intermediate density of a ^{15}N-^{14}N hybrid was found (Figure 5-3), exactly as one would expect from the mode of replication proposed by Watson and Crick. If the daughter cells were then allowed to divide again, two species of DNA appeared in the gradient: roughly half of the DNA showed ^{15}N-^{14}N hybrid density while the rest showed pure ^{14}N density (Figure 5-3). As the cells continued to divide in the ^{14}N medium a greater and greater proportion of the DNA in the population showed ^{14}N density, but for many generations a faint band was detectable at the ^{15}N-^{14}N DNA position in the gradient (Figure 5-3). Again, these results are compatible with the mode of replication proposed by Watson and Crick.

Box 5.1
CsCl GRADIENT
CENTRIFUGATION

If a concentrated solution of CsCl is centrifuged at high speeds for many hours, centrifugal forces will induce the CsCl to sediment toward the bottom of the tube, a process that will be counteracted by back-diffusion of the CsCl toward the top of the tube. Eventually, at equilibrium, the opposing processes will balance and a stable CsCl concentration gradient will be established in which there is progressively more and more CsCl as one scans from the top to the bottom of the tube. If pieces of DNA are mixed with the CsCl and both are centrifuged to equilibrium, each piece will tend to move and become concentrated into that region of the gradient in which the density of the solution matches the **buoyant density** (symbolized by ρ and expressed as g/cm^3) of the DNA (see accompanying illustration). Thus if all the pieces in a sample have the same net buoyant density, a single band of DNA will form in the gradient, whereas if the pieces are markedly different in density, two or more bands may form.

To visualize the position and width of the various bands, advantage is taken of the fact that nucleic acids strongly absorb ultraviolet light at 260 nm and can be detected by optical devices. The ρ-values of each class of fragment can then be calculated.

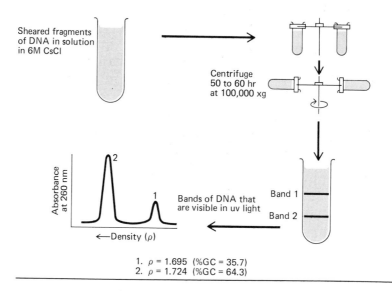

1. $\rho = 1.695$ (%GC = 35.7)
2. $\rho = 1.724$ (%GC = 64.3)

The Meselson-Stahl experiment is particularly valuable in that it not only establishes the mode of replication of DNA but it also eliminates the possibility of other modes. For example, it might be postulated that replication is **conservative:** the original duplex serves as a template for a new duplex but it also remains intact such that at the end of one round of replication an old and a new double helix are observed. Hybrid ^{15}N-^{14}N DNA should not form in this case. The experimental result, however, disproves this. A **dispersive** mode of DNA replication is also ruled out by the experiment: this mode predicts that no pattern of transmission exists and that parental DNA strands break at random during the replication process in such a way that the DNA duplexes in daughter cells contain varying amounts of both old and new DNA. In this case a wide spectrum of DNA densities should be detected after one round of duplication rather than a single ^{15}N-^{14}N species.

The mode by which DNA replicates has been termed **semiconservative** to indicate that the parental strands of DNA *are* conserved (as opposed to dispersed) but that as replication proceeds they wind up, as it were, in two different helices.

5.5 VISUALIZING REPLICATION

Replicating *E. coli* chromosomes have been visualized by autoradiography, a technique described in Box 2.2. The first such study was made by J. Cairns, who grew cells in the presence of ^{3}H-thymidine for about two generations, carefully isolated their DNA, and made autoradiographs from the isolated preparations. The most favorable autoradiograph, shown in Figure 5-4, revealed an intact circular structure with two diverging segments. Cairns counted the number of exposed silver grains along the various lengths of the molecule and found that certain segments were roughly twice as radioactive as others. From these observations he argued convincingly that the chromosome is replicated as an intact circle and that its DNA replication is semiconservative.

The distribution of radioactivity in the Cairns autoradiograph indicates that the replicating chromosome possesses but a single **growing point,** meaning that DNA replication seemed to initiate at a fixed position and then to proceed in one direction, moving within a **replication fork** in which the original strands progressively separate and new strands are synthesized as this separation takes place. More recent studies by M. Masters and P. Broda have shown that the replication of the *E. coli* chromosome is more commonly **bidirectional:** after initiation two growing points travel in opposite directions around the circular chromosome, each copying 50 percent of the genome and both meeting at the opposite side of the chromosome. Since the entire *E. coli* chromosome can be copied *in vivo* in 40 minutes, and since the chromosome is 1100μ long, $15\ \mu$ of chromosome must be copied per minute along each replication fork. This corresponds to 45,000 base pairs added to each growing point per minute, and far exceeds the $0.5\ \mu$ per minute rate estimated for DNA replication in mammals (Section 2.3).

5.6 INITIATION OF DNA REPLICATION

The bidirectional replication of the *E. coli* chromosome always begins at the same unique site, the **replication origin** which, following the convention used to designate position on the *E. coli* circular chromosome, is located at 86 minutes (Figure 5-5). At the molecular level, the sequence of nucleotides at the initiation site is presumably unique and signals the events critical for initiation.

As noted earlier, one of these initiating events *in vivo* is likely to be the attachment of the replication origin to the *E. coli* cell membrane. Subsequent events must include the first "opening up" of the parent helix and the laying down of the first daughter nucleotides opposite the parent strands. Setting this initiation sequence up requires the products of two *E. coli* genes, *dna A* and *dnaC-D,* both of which have been identified by tempera-

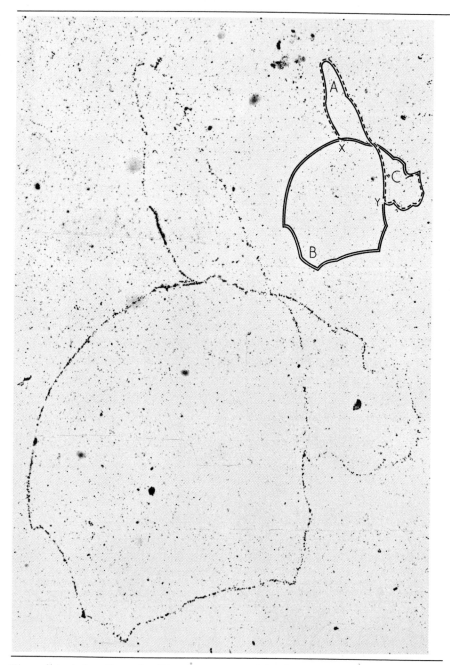

Figure 5-4 Autoradiograph of the chromosome of *E. coli* K12. Inset, the same structure shown diagrammatically and divided into three sections (A, B, and C) that arise at the two forks (X and Y). (From J. Cairns, *Cold Spring Harbor Symp. Quant. Biol.* **28**:44, 1963.)

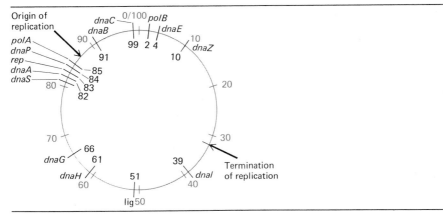

Figure 5-5 A map of the circular *E. coli* chromosome showing only the location of the genes identified as playing a role in DNA replication. Positions on the map are given in *minutes*, each minute representing the length of DNA transferred during one minute of sexual conjugation. The total map covers 100 minutes; each minute therefore corresponds to about 110 μm of DNA. Details of how such maps are constructed are given in Chapter 11.

ture-sensitive mutations. When cells carrying the mutant *dna A* gene, for example, are shifted from the permissive to the nonpermissive temperature, ongoing DNA replication continues but no new rounds of replication are initiated. The locations of the *dna-A* and *dnaC-D* genes in the *E. coli* chromosome are shown in Figure 5-5, where it is seen that the *dna A* gene lies intruigingly close to the replication origin.

5.7 ENZYMES PARTICIPATING IN CHAIN ELONGATION

The enzymes required for the elongation of daughter strands of DNA undoubtedly function during initiation, but their modes of action are more easily studied during the elongation process. The properties and genetic origin of these enzymes are described below.

DNA Polymerase III DNA polymerase III, specified by the *dnaE* gene (Figure 5-5), is the true **replicase** in *E. coli,* a replicase being defined as a DNA polymerase that acts within the replication fork to copy parental DNA strands into daughter strands. The polymerization reaction is diagrammed in Figure 5-6 and described in the legend to Figure 5-6. The essence of the reaction is that the replicase recognizes a base on the template DNA strand, chooses a complementary nucleoside-5'-triphosphate monomer from the cytoplasm according to Watson-Crick base-pairing rules (Chapter 1), and attaches the monomer to the growing daughter strand by a 3', 5' phosphodiester bond, releasing two of the three phosphates in the process.

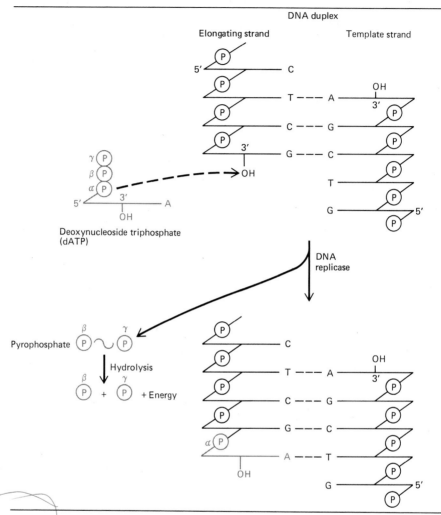

Figure 5-6 DNA polymerization. A dATP nucleotide is shown being added to an elongating strand opposite a thymine (T) base in the template strand, a Watson-Crick base recognition event catalyzed by the DNA replicase. The α phosphate of the dATP forms a 3′, 5′ phosphodiester bond with the free 3′-OH of the elongating strand, and a molecule of pyrophosphate (PNP) is simultaneously released. Pyrophosphate contains a "high-energy" or "~" bond, meaning that when the released pyrophosphate is hydrolyzed into two phosphate molecules, energy is liberated which drives the polymerization process forward. The resultant polymerization will always proceed in a net 5′ ⟶ 3′ direction, meaning that the nucleotide at the 3′ end is always the most recently added to the chain.

Two features of the *E. coli* (and all other) DNA replicase enzyme have important consequences for the mode of DNA replication. First, all DNA replicases thus far characterized are incapable of initiating DNA synthesis: they can only add nucleotides onto the 3'-OH end of a DNA strand that is hydrogen-bonded to a template strand, as drawn in Figure 5-6. This raises an obvious question: how do the first nucleotides in a daughter strand get laid down? In other words, how does the initiation of DNA polymerization occur? This question, which has plagued biochemists for many years, appears to have been answered by studies of *in vitro* DNA replication conducted by A. Kornberg and associates and by S. Wickner. They find that an essential enzyme for replication, the product of the *dnaG* gene (Figure 5-5), acts to make a short **primer** strand of either DNA or RNA. This primer is then elongated by the DNA polymerase III in the fashion depicted in Figure 5-6. Once initiation is accomplished, the primer sequences are enzymatically removed from the replication-origin region.

The second important feature of all DNA replicase enzymes is that they can add deoxyribonucleotides only to the 3'-OH end of a hydrogen-bonded primer, never to its 5'-P end. This creates a conceptual dilemma. As noted in Chapter 1, the two strands of a DNA duplex are oriented in opposite directions (Figure 1-5); we now learn that DNA polymerases can only elongate chains from the 3'-OH end and therefore in a 5' $\longrightarrow$ 3' direction. Thus a single growing point in the *E. coli* chromosome contains two parent strands of opposite polarity being replicated at the same time and in the same direction by an enzyme capable only of unidirectional synthesis. How is this accomplished?

Figure 5-7 diagrams a current view of how elongation occurs. The parent helix locally unwinds as DNA replication is proceeding, in the allowed 5' $\longrightarrow$ 3' direction, along the upper strand (Figure 5-7a). A primer fragment (shown as a jagged line in Figure 5-7b) is next synthesized opposite the lower strand, and this is elongated, again in the allowed 5' $\longrightarrow$ 3' direction (Figure 5-7c). The parental helix, meanwhile, unwinds further so that its upper strand is copied continuously, while the lower strand experiences new primer synthesis (Figure 5-7d) and the generation of another short piece of DNA (Figure 5-7e). These pieces of DNA, it should be noted, are known as **Okasaki fragments** after R. Okasaki, who first reported that discontinuous pieces of DNA, perhaps 1000 nucleotides long, are generated during DNA replication. Finally, primer sequences are degraded (Figure 5-7f) and the discontinuous pieces associated with the lower strand are joined together enzymatically (Figure 5-7g).

DNA Polymerase I DNA polymerase I, specified by the *pol* A gene (Figure 5-5) is a single polypeptide but has three enzymatic activities: it polymerizes DNA chains from 3'-OH primers, although at a slower rate than DNA polymerase III, and it has two **exonuclease** activities, one of which allows it to digest DNA strands in a 5' $\longrightarrow$ 3' direction and the other of which

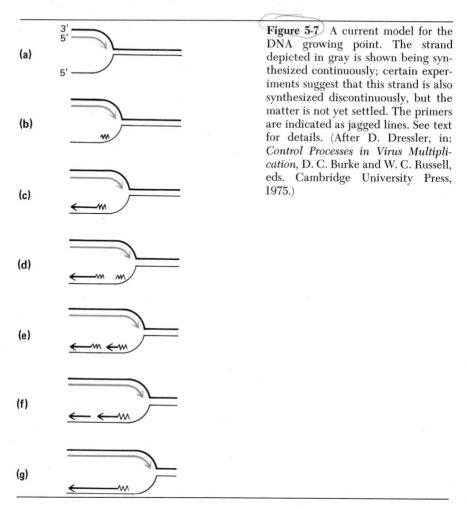

(a)

(b)

(c)

(d)

(e)

(f)

(g)

Figure 5-7 A current model for the DNA growing point. The strand depicted in gray is shown being synthesized continuously; certain experiments suggest that this strand is also synthesized discontinuously, but the matter is not yet settled. The primers are indicated as jagged lines. See text for details. (After D. Dressler, in: *Control Processes in Virus Multiplication*, D. C. Burke and W. C. Russell, eds. Cambridge University Press, 1975.)

catalyzes a 3′ ⟶ 5′ digestion (see Figure 5-8 and its legend for a description of exonucleases).

DNA polymerase I enzymes appear to be responsible for "cleaning up" the primer ends of Okazaki pieces. In the wake of the advancing replication fork, a DNA polymerase I molecule is thought to recognize a primer-daughter strand juncture, chew up the primer sequences using its 5′ ⟶ 3′ exonuclease activity, and fill in the gap it creates with deoxyribonucleotides using its polymerase activity.

DNA Polymerase I also serves as a **proofreader** or **editor** in the wake of the replication fork: it "checks out" the base pairs formed by the polymerase III enzymes, and clips out any incorrect nucleotides that may be present

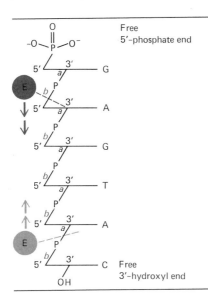

Free
5'-phosphate end

Free
3'-hydroxyl end

Figure 5-8 Exonuclease action. An exonuclease must begin its attack from a **free end** of a polynucleotide. Therefore, depending on the specificity of the enzyme, an exonuclease will either begin at a free 3'—OH end of a polynucleotide and progressively cleave the bonds on the 3'—OH side of the phosphodiester backbone (labeled *a*) or it will begin at a free 5'—P end and cleave the polynucleotide from the 5' side (labeled *b*). A 3'-exonuclease is depicted in light gray, a 5'-exonuclease in dark gray.

using its 3' ⟶ 5' exonuclease activity, replacing the clipped-out nucleotides with the correct nucleotides. Such replication errors will ordinarily be rare: all known polymerases, including DNA polymerase III, are extraordinarily accurate in following the Watson-Crick rules of base-pairing during the DNA polymerization reaction (Figure 5-6). Since more than 3 million nucleotides are polymerized for each round of *E. coli* chromosome replication, however, even a low error rate could produce one wrong base per chromosome and that one wrong base could prove lethal to a daughter cell. Therefore, the editing function of polymerase I is vital for correct chromosome transmission.

About 400 copies of polymerase I are present in an *E. coli* cell, in contrast to only about 10 copies of polymerase III. In part because of this difference in abundance, polymerase I was considered to be the true replicase for many years and its *in vitro* properties were intensively studied, notably by A. Kornberg and his associates. In 1969, however, P. DeLucia and J. Cairns isolated a mutant strain of *E. coli*, *pol A*, that contained no active form of the "Kornberg enzyme" and appeared capable of normal growth and DNA replication; a search for replicase activity in *polA* strains subsequently revealed the existence of polymerase III. It has since been found that the 5' ⟶ 3' exonuclease activity of DNA polymerase I is not abolished by the *polA* mutation and that if this exonuclease activity is made temperature-sensitive by a second mutation, cells cannot survive at the nonpermissive temperature. It therefore seems clear that while DNA polymerase III alone is capable of replicating DNA—perhaps because it is able to associate with essential initiation factors—both DNA polymerase enzymes perform vital functions for an *E. coli* cell.

DNA Ligase The existence of Okazaki fragments can be demonstrated by exposing cells to ³H-thymidine for short periods of time, isolating their DNA, and showing that most of the radioactivity is found in low-molecular-weight DNA strands. If after the short "pulse" of ³H-thymidine the cells are washed, given unlabeled thymidine, and allowed to continue replication for an additional short period (a so-called **pulse-chase experiment**), the label is now found in high-molecular-weight strands of DNA. This joining-together of Okazaki pieces to form continuous daughter strands is effected by the enzyme **DNA ligase,** which is specified in *E. coli* by the *lig* gene (see Figure 5-5). Mutants defective in ligase activity join Okazaki pieces very slowly. The mechanism of DNA ligase action is diagrammed in Figure 5-9 and detailed in the Figure legend. As noted earlier, this reaction is thought to be preceded during DNA replication by the concerted excision-synthesis activities of DNA polymerase I, which excise and replace primer sequences.

DNA Unwinding Enzyme and Unwinding Protein DNA replication clearly requires that a duplex parental molecule be "unwound" so that its internal

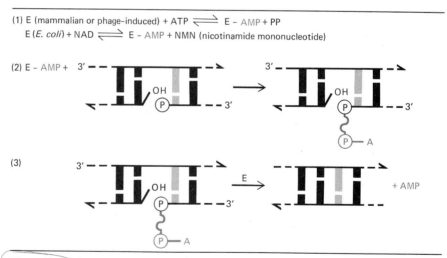

(1) E (mammalian or phage–induced) + ATP $\rightleftharpoons$ E – AMP + PP

    E (*E. coli*) + NAD $\rightleftharpoons$ E – AMP + NMN (nicotinamide mononucleotide)

(2) E – AMP +

(3)

Figure 5-9 DNA ligase action. (1) The enzyme (E) forms an enzyme-adenylate (E-AMP) complex. (2) The E-AMP reacts with a nicked DNA chain, generating a pyrophosphate bond linking the 5'—P terminus of the DNA to the P group of the AMP. (3) The enzyme mediates the formation of a phosphodiester bond with the release of AMP. (Reproduced with permission, from "Enzymes in DNA Metabolism," *Annual Review of Biochemistry* **38:**795–840. Copyright © 1969 by Annual Reviews, Inc. All rights reserved.)

bases are available to the replication enzymes. Unwinding activity in *E. coli* appears to be mediated by the **rep protein,** identified first in the *rep* mutant strain (Figure 5-5), which hydrolyzes ATP while it actively forces the DNA helix apart.

Once opened up, the exposed single strands are stabilized by a second protein known as the **DNA unwinding protein.** This protein has not yet been defined in terms of a temperature-sensitive mutation, but similar proteins have been shown to be essential for DNA replication in phages (for example, the P32 protein of phage T4). The unwinding protein exhibits no interaction with duplex DNA, nor does it actively force open a double helix. Once a helix has opened to reveal its single strands, however, a molecule of unwinding protein readily associates with the single-stranded region; additional proteins proceed to bind much more readily than the first (a phenomenon known as cooperativity) so that the single-stranded state is rapidly stabilized by a "coating" of unwinding proteins. DNA unwinding protein is thought to be responsible for holding open a double helix in advance of the replication fork. A single unwinding protein is long enough to cover 8 nucleotides of DNA, meaning that the 400 unwinding proteins present in each cell can stabilize substantial lengths of both replication forks.

Other Proteins Participating in Replication   A complete catalogue of proteins required for DNA replication in *E. coli* include the following: 1) a **relaxing enzyme** (the **ω protein**) which is thought to remove the superhelical twists and kinks generated in a circular duplex chromosome as it unwinds for replication; 2) at least 2 **elongation factors** required for DNA replication in vitro; and 3) the product of the *dnaZ* gene (Figure 5-5), whose role in elongation is not yet clear.

5.8   OVERVIEW OF DNA REPLICATION

Several stages of *E. coli* DNA replication have been defined, and these can now be summarized. **Initiation,** which is least well understood, occurs at a fixed site in the chromosome, requires at least 3 gene products and an initial unwinding of the helix, probably occurs in association with the membrane, and includes the synthesis of the first primer strands. **Elongation** requires polymerase III acting in conjunction with, and probably in physical association with, several other replication enzymes and factors. Elongation must be preceded by a continued unwinding of the helix and a stabilization of the opened state by unwinding proteins, and relaxing enzyme must ultimately act to relieve the resultant twists and avoid a chromosomal tangle. Finally, a **synthesis-excision-joining** reaction must be effected by DNA polymerase I, DNA ligase, and perhaps other enzymes to proofread, patch up, and join together the Okazaki pieces created by the replication apparatus.

Studies of DNA replication in viruses and in eukaryotes reveal many of the major molecular features found for *E. coli.* Some viruses replicate rod-shaped chromosomes and others replicate circular chromosomes; some "borrow" heavily from the stock of replication enzymes present in their host cells, while others rely heavily on their own replication enzymes. Yet the fundamental patterns of initiation, elongation, and synthesis-excision-joining are followed, and the virus-specified enzymes usually act in much the same way as the host enzymes they supplant. In eukaryotes, DNA synthesis within a replicon (Section 2.3) appears to be bidirectional, beginning at a focal origin and advancing outward in opposite directions to two termini (see Figure 2-3) in much the same fashion as *E. coli* replication. Semiconservative replication of eukaryotic DNA has also been demonstrated, as has the existence of Okazaki pieces and at least 3 DNA polymerase enzymes. References to articles on viral and eukaryotic chromosome replication are given at the end of the chapter.

5.9   MODIFICATION AND RESTRICTION IN *E. COLI*

As an *E. coli* cell replicates its chromosome, certain classes of **methylating enzymes** often attach methyl groups to the nucleotides found in particular sequences along the daughter strands. Thus, for example, one class of methylating enzymes recognizes the sequence $\frac{5'pGpApApTp\,Tp\,C}{CpTp\,Tp\,ApApGp5'}$ and transforms it into $\frac{5'pGpApAp\overset{*}{A}pTp\,Tp\,C}{CpTp\,Tp\,ApApGp5'}$ where the asterisks denote methyl groups. The resultant DNA is said to have undergone **modification,** and genes specifying such modification enzymes are found both in the main chromosome and in certain plasmids.

The "purpose" of modification becomes apparent when a second set of bacterial enzymes, the **restriction endonucleases,** are described. An **endonuclease** resembles an exonuclease (Figure 5-9) in that it attacks one or the other side of the phosphodiester linkage in a nucleic acid, but an endonuclease reacts only with those bonds that occur along the internal length of a polynucleotide chain. If the polynucleotide chain is single-stranded, such an attack will obviously cut the chain into two pieces. If the polynucleotide strand is a member of a DNA double helix, then two results are possible: a cut may be made across both strands to sever the duplex in two, or alternatively, only one strand may be cut, in which case the helix is then said to contain a **nick** (Figure 5-10).

Restriction endonucleases constitute a special class of endonuclease enzyme in that they are restricted to attacking DNA that contains particular sequences of nucleotides. Different restriction enzymes attack different sequences, moreover, as described in some detail in Section 7.1. Here we can cite as an example the enzyme known as *Eco*RI, which makes the single-strand cuts indicated by the two arrows whenever it encounters the se-

quence $\overset{\downarrow}{\underset{CpTp\ Tp\ ApA\underset{\uparrow}{p}Gp5'}{5'pG\overset{\downarrow}{p}ApApTp\ TpC}}$. This is precisely the sequence we just iden-

tified as the substrate for a particular methylating enzyme. Obviously a restriction enzyme will not encounter this sequence if the cell has successfully methylated all such sequences during replication. Therefore, the "purpose" of a particular class of methylating enzyme is apparently to provide self-protection from its restriction-enzyme counterpart.

What, then, is the "purpose" of the restriction enzymes? They appear to be a line of defense against foreign DNA, poised to degrade the nucleic acids of infecting bacteriophages before these have a chance to replicate within the bacterial cell. Thus if a phage has developed in a cell in which the above-described methylating enzyme was not present, this phage is likely to have a "naked" (unprotected) $\frac{GAATTC}{CTTAAG}$ sequence within its chromosome. If it now infects a cell with the modification-restriction system we are considering, the phage will likely be "restricted" the moment it enters the cell. It should be clear that the phage chromosome would also be a substrate for the methylating enzyme of the host, but it seems that unmodified phage DNA is usually restricted before it has a chance to be modified, while replicating bacterial DNA is usually modified before it has a chance to be restricted.

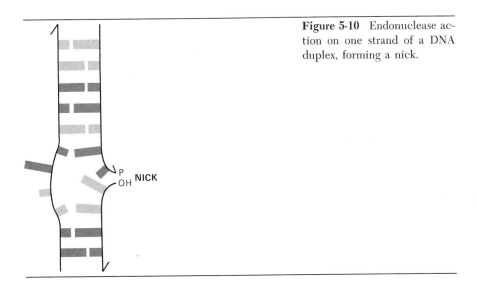

Figure 5-10 Endonuclease action on one strand of a DNA duplex, forming a nick.

Restriction enzymes have become critically important in the analysis of chromosome structure and are considered again in Section 7.1.

BACTERIOPHAGES, VIRUSES, AND THEIR LIFE CYCLES

It is meaningless to speak of a virus as a prokaryote or a eukaryote since a virus has no cellular structure; instead, viruses are identified according to the type of cell they infect, with bacteriophages infecting bacteria and animal and plant viruses infecting eukaryotic animals and plants. All forms of virus are exceedingly simple as compared with most living organisms: each **virion** (individual virus), contains a single type of nucleic acid—either DNA or RNA—and a protein coat which serves in the infection process and in protecting the virion from the external environment. The nucleic acid contains genetic information necessary for the construction of new viruses but little or none of the apparatus necessary for such construction (for example, ribosomes or enzymes). Viruses must therefore infect living cells and divert these cells' biosynthetic apparatus into the business of synthesizing virus progeny.

Bacteriophages have evolved a number of strategies for infecting and diverting the metabolism of their bacterial hosts. The events that occur from the time of phage adsorption to the time progeny phages are released from the host cell are collectively known as a phage's **infectious cycle.** The following sections describe the infectious cycles of various bacteriophages and viruses, focusing on those events that will be of relevance to the viral genetics presented in later chapters. Pertinent physical properties of these viruses are summarized in Table 5-1, and several are illustrated in Figure 5-11.

5.10 VIRULENT PHAGES

Virulent phages cause the death of the host, and a general model for a virulent infectious cycle is given by that of T7 (Figure 5-12) and its close relative T3. Infection begins when a T7 phage **adsorbs** to the E. coli cell wall via its tail and injects its rod-shaped duplex DNA chromosome into the bacterial cytoplasm. Three sets of genes are then sequentially expressed: the **early gene** products essentially inhibit all E. coli RNA synthesis; the **DNA metabolism** gene products include nucleases that digest the E. coli chromosome to provide free nucleotides for T7 DNA synthesis and the necessary replication enzymes for producing many new individual copies of the T7 chromosome; and the **late genes** dictate viral structure and assembly proteins, including coat and tail components. These three sets of genes are considered in some detail in Chapter 17.

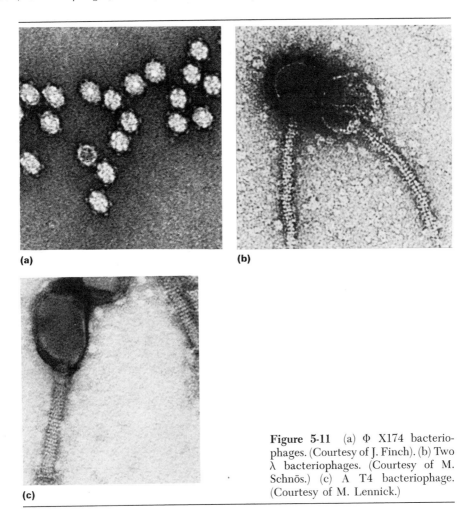

(a)

(b)

(c)

Figure 5-11 (a) Φ X174 bacteriophages. (Courtesy of J. Finch). (b) Two
λ bacteriophages. (Courtesy of M.
Schnös.) (c) A T4 bacteriophage.
(Courtesy of M. Lennick.)

Following their synthesis, the segregation of daughter T7 chromosomes
from one another is accomplished by the coat proteins: these have the
inherent ability to **self-assemble** in a specific fashion around individual
chromosomes. Since each T7 chromosome measures 11.5 μm in length and
each head measures only 0.065 μm, considerable folding and condensation
of the T7 chromosome must occur during the packaging process. Finally the
bacterial cell is lysed, partially because of the production of a phage-specified enzyme known as **lysozyme** which digests the bacterial cell walls.
Perhaps 250 new phages are released in what is called the **burst,** and these
are then free to find new bacterial hosts.

An important variation on the above theme occurs when an *E. coli* cell is
infected by two or more phages that differ at one or more of their genetic

Table 5-1 Properties of Representative DNA and RNA Viruses

Category	Virus Name	Host	Weight of Virion (m.w. × 10⁻⁶)	Weight of Nucleic Acid (m.w. × 10⁻⁶)	Length of Chromosome (μm)	%GC in Chromosome	Chromosome Description
Duplex DNA Phages	T2, T4, T6	*E. coli*	220	110	56	35	Circularly permuted, rod
	T7	*E. coli*	38	24	12	48	Unique sequence, rod
	λ	*E. coli*	66	30	17	49	Rod with sticky ends
	Ø80	*E. coli*	—	29	14	53	Rod with sticky ends
	P2	*E. coli*	—	20	10	—	Rod with sticky ends
	P22	*Salmonella*	—	26	14	48	
	SP29	*B. subtilis*	—	11	5.8	—	
Single-Stranded DNA Phages	ØX174	*E. coli*	6.2	1.6	1.8	A:25, T:33 G:24, C:18	Ring
	fd	*E. coli*	11.0	1.7	—	A:24, T:34 G:20, C:22	Ring
Duplex DNA Animal Viruses	Herpes	Human	—	100	53	68	
	Pox	Human	3200	160	89	37	
	Polyoma	Mouse	25	2.9–3.4	—	48	Supercoiled ring
	SV40	Human	17	2.3–2.5	—	41	Supercoiled ring

Single-Stranded RNA Phages	Qβ	E. coli	4.2	1.5	1.4	A:22, U:29 G:24, C:25	
	R17	E. coli	3.6	1.1	1.1	A:25, U:23 G:27, C:25	
Small Single-Stranded RNA Animal Viruses	Poliomyelitis	Human	5.5	2	—	A:28, U:25 G:25, C:22	
	Foot and mouth disease	Cattle	6	3.1	—	A:26, U:22 G:24, C:28	
	Coxsackie A	Human	7	2	—	A:27, U:25 G:28, C:20	
Large Single-Stranded Animal and Plant RNA Viruses	Tobacco mosaic	Tobacco	50	2.1	—	A:30, U:26 G:25, C:19	
	Influenza	Human	300	2.9	—	A:23, U:33 G:20, C:24	Several pieces
	Newcastle disease	Fowl	800	7.5	—	A:26, U:22 G:25, G:27	Several pieces
	Murine leukemia	Mouse	220	13	—	A:25, U:23 G:25, C:27	Several pieces
Double-Stranded RNA Viruses	Reovirus	Mammals	>70	17	8.3	A:38, U:28 G:17, C:17	Several pieces
	Wound tumor	Plant	—	>10	—	A:31, U:32 G:19, C:19	Several pieces

From A. K. Kleinschmidt, in *Handbook of Cytology* (A. Lima-de-Faria, Ed.). Amsterdam: North-Holland Pub. Co., 1969.

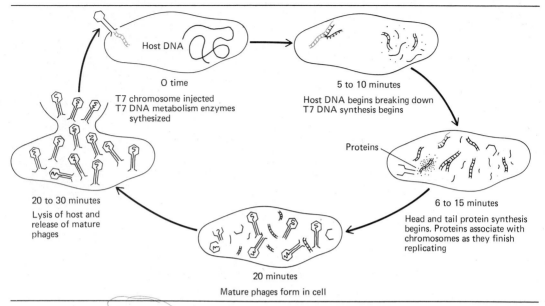

Figure 5-12   The life cycle of the *E. coli* bacteriophage T7 at 30°C.

loci (a **mixed infection**). In this case, the two types of homologous chromosomes will occasionally participate in recombination so that the burst will contain a mixture of *parental-type phages* (for example, + + + and *a b c*) as well as *recombinant phages* (for example, + *b c* and *a* + +). A detailed consideration of recombination during mixed infection is given in Chapter 10.

The basic virulent infection cycle scheme exemplified by T7 is followed by many other phages, the major distinctions between them being in their mode of chromosome replication and segregation. While the details of such cycles lie beyond the scope of this text, we can summarize a few pertinent examples.

T4 and T2   The closely related phages T2 and T4 possess a long (56 μm), rod-shaped chromosome (Figure 5-13). The initial products of T4 chromosome replication are enormously long DNA molecules—many times longer than a T4 chromosome—called **concatamers.** The concatamers appear to be converted into chromosome-length rods by the so called **headful mechanism:** DNA is pulled into a forming phage head until there is no more room, at which point it is cleaved.

φX174, M13, fd, and fl   These phages possess small, single-stranded circular DNA (Table 5-1) as their genetic material, and carry out the unique se-

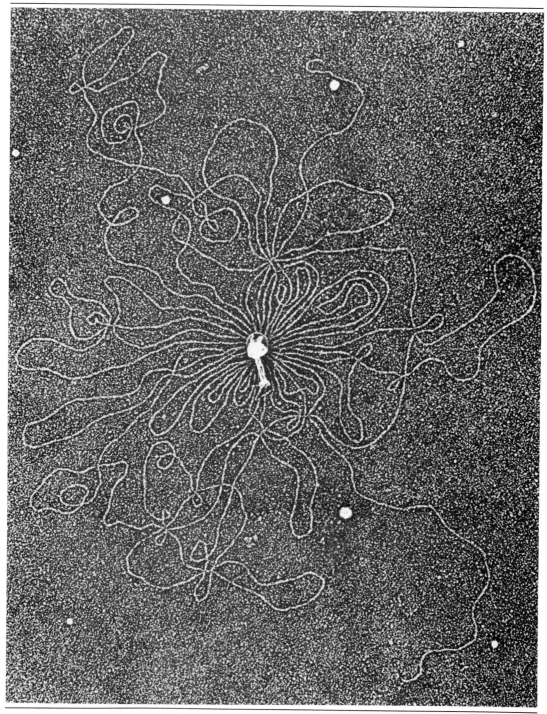

Figure 5-13 T2 bacteriophage showing its duplex DNA chromosome (released by osmotic shock) with two free ends. (From A. K. Kleinschmidt et al., *Biochim. Biophys. Acta* **61:**857, 1962.)

quence of replication events illustrated in Figure 5-14. The infecting parental chromosome (the **positive strand**) is copied by the *E. coli* replication-enzyme apparatus to produce a complementary **negative strand,** and the two strands are then covalently linked to produce a double-stranded circle, the **parental replicative form** (RF). This undergoes several rounds of semi-conservative replication to produce perhaps 60 **daughter RF** molecules. Finally, the negative strands of the RF molecules serve as templates for positive strand synthesis, and the resulting positive strands are packaged into protein coats specified by phage late genes.

R17, MS2, f2, and Qβ These phages, which contain single-stranded RNA chromosomes, can only infect host cells that bear long, thin surface ap-

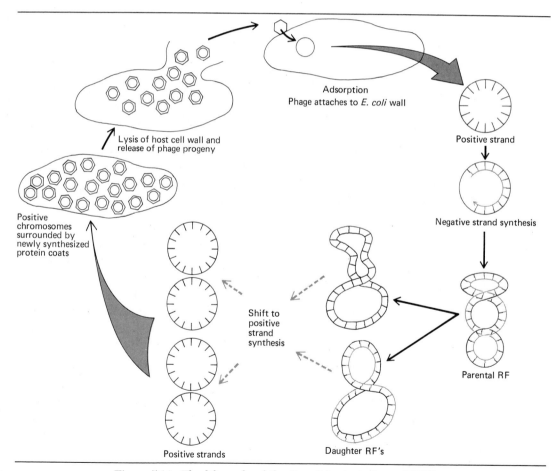

Adsorption
Phage attaches to *E. coli* wall

Positive strand

Negative strand synthesis

Parental RF

Daughter RF's

Shift to positive strand synthesis

Positive strands

Lysis of host cell wall and release of phage progeny

Positive chromosomes surrounded by newly synthesized protein coats

Figure 5-14 The life cycle of the *E. coli* bacteriophage ΦX174.

pendages called **sex pili** (singular, **pilus**) or **sex fimbria** (see Figure 11-2). In
E. coli, such appendages are carried only by cells of the relatively rare
"male" sex (Chapter 11 includes a detailed description of *E. coli* sexuality);
therefore, the RNA phages are valuable for identifying male bacterial strains.
Following adsorption to a pilus, an RNA phage injects its single strand of
RNA into the pilus and the RNA is transferred, by an unknown mechanism,
into the cell. A single early gene dictates the production of a replicase
protein. This protein combines with other *E. coli* proteins to form an
enzyme which proceeds to generate many copies of the infecting chromosomes. Two late genes then dictate a **maturation protein** (the A protein) and
a **coat protein** required for assembly of mature virions. No lysozyme is
specified, and the mechanism of cell lysis by RNA phages is not understood
except that it appears to be mediated by the coat protein of the virus.

5.11 TEMPERATE PHAGES

The phages λ, φ80, and 21 have two options—a **lytic** pathway and a **lysogenic** pathway—to their infectious cycles. These are summarized in Figure
5-15 and described below.

When λ infects an *E. coli* cell it will often direct a lytic life cycle very
much like the virulent cycles of T7 or T4. Under other conditions of infection, however, λ directs what is known as a **lysogenic** cycle: the infecting λ
chromosome is not replicated as an independent entity but instead becomes physically inserted into a special region of the host chromosome, in
which state it is replicated along with the host chromosome for many
generations and is called a **prophage.**

An *E. coli* carrying such a λ prophage is said to be lysogenic for λ and
usually grows at the same rate as a normal bacterium. Very infrequently
(perhaps once in 10,000 divisions of a lysogenic bacterium), the prophage is
spontaneously released from the host chromosome, whereupon it proceeds
to direct a lytic cycle and lyses its host. Such release can be stimulated in the
laboratory by irradiating lysogenic bacteria with ultraviolet light or by
exposing them to certain chemicals, such manipulations being known as
induction. There are thus two occasions when λ may initiate a lytic cycle,
one immediately after it infects an *E. coli* cell and the other after it leaves
the lysogenic state.

Viruses such as λ, with both lytic and lysogenic options to their infectious
cycles, are termed **temperate,** in contrast to virulent phages such as T4 and
T7. The strategic advantage to temperance should be evident: once λ DNA
sequences are introduced into the chromosome of an *E. coli* cell, these
sequences are transmitted to subsequent generations of *E. coli* with each
cell division, so that billions of cells harboring λ prophages are generated in

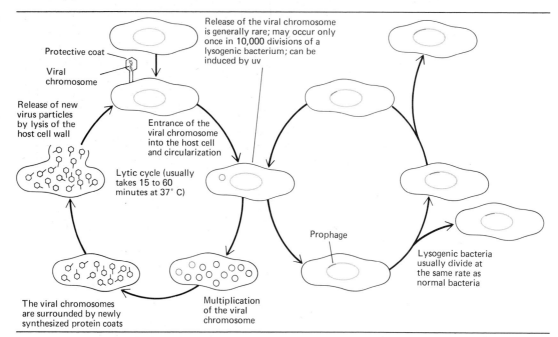

Figure 5-15 The life cycle of the *E. coli* bacteriophage λ, showing its lytic (left) and lysogenic (right) cycles. Phage chromosomes are depicted in dark gray and black and host chromosome in light gray. (From J. D. Watson, *Molecular Biology of the Gene*, 3rd ed. Copyright © 1976 by J. D. Watson; W. A. Benjamin, Inc., Menlo Park, Calif.)

a short period of time. The transmission of phage information in space and in time is therefore not solely dependent on the infectious process, a process which requires the correct meeting of virus with host. As the host invades new habitats, the phage effectively goes along for the ride.

Temperate phages have been shown to insert into and exit from the *E. coli* chromosome by a specialized type of recombination described in detail in Section 13.16. An important feature of such exit and entry is that an occasional phage will make a mistake and will include a sector of the bacterial chromosome in the length of DNA that leaves the chromosome. Specifically, the λ prophage lies in the *E. coli* genome between the genes *gal* (*gal*actose) and *bio* (biotin), so that the lysogen has the sequence *gal* λ *bio*. A mistake during exit may produce the **defective transducing phage** λ *dgal*. This phage is defective in that it may have left behind some λ genes, and it is transducing in that its chromosome contains the bacterial gene *gal* which the phage can carry over (transduce) into any new *E. coli* cell that it infects. Such transducing phages prove to be invaluable for analyzing bacterial genes, as will become apparent in later chapters.

5.12 TUMOR VIRUSES

The final variety of infectious cycle considered here is exhibited by the tumor virus SV40 and its close relative polyoma, both of which infect animal cells and can cause tumors. Like λ, SV40 can conduct two types of infections, one leading to host death and the other to a physical insertion of viral DNA into the host genome. Thus if SV40 virions are presented to cultured monkey cells, they will carry out a **productive infection** which is formally analogous to a lytic cycle in bacteriophages. Several hours after infection three SV40-specific molecules are detected, of which the **T-antigen** is most significant: it is required for viral DNA replication, for stimulation of host-cellular enzymatic activities, and for initiating the transcription of late viral genes. Some 12-20 hours after infection, SV40 chromosome replication commences; viral structural proteins then begin to appear in the nucleus of the host cell, where they assemble to form mature progeny viruses. By 2 days after infection approximately 10,000 new virus particles have been released. The host cell is killed in the process although a lytic "burst" does not necessarily occur.

If rather than being presented to monkey cells, SV40 is presented instead to cultured mouse cells, an alternate mode of infection is possible. Following infection, T antigen appears in the host cell nucleus but no viral DNA or capsid protein synthesis occurs. After 18-24 hours, the host cell itself divides a few times and acquires novel properties: it can grow without being attached to a solid substratum (it acquires **anchorage independence**); it will proliferate in culture media depleted of its usually high levels of calf serum (it acquires a so-called **serum independence**); and it can be caused to clump to other cells when presented with carbohydrate-binding lectin proteins such as concanavalin A (it acquires **lectin agglutinability**). Most mouse cells so altered will return to normal after a few days and are said to have been abortively transformed, but some remain anchorage- and serum-independent and lectin-agglutinable and are said to be **stably transformed.** If such transformed cells are then injected into the peritoneal cavity of a young mouse, they will divide uncontrollably and eventually kill the animal. In other words, stably transformed cells have the potential to cause tumors.

The SV40 chromosome in a stable transformant remains in the nucleus of its host cell and appears to become covalently linked to the host's chromosomal DNA. It continues to dictate the synthesis of T antigen but late gene expression almost never occurs, apparently because mouse ribosomes cannot be diverted into synthesizing late proteins. In certain respects, therefore, the transformed state is reminiscent of the lysogenic stage in phage λ. If a transformed cell is now fused with a "permissive" monkey cell so that their two cytoplasms blend together (cell-fusion techniques are discussed in Chapter 12), then mature infectious virions can be produced, showing that a complete copy of the virus chromosome must have been present in the mouse transformant. It might go without saying that intensive research is presently focused on SV40 transformation mechanisms.

References

General Reviews

Dressler, D. "The recent excitement in the DNA growing point problem," *Ann. Rev. Microbiol.* **29:**525–559 (1975).

Goulian, M., P. Hanawalt, and M. Fox. *DNA Synthesis and its Regulation.* Menlo Park, California: Benjamin, 1976.

Jovin, T. M. "Recognition mechanisms of DNA-specific enzymes," *Ann. Rev. Biochem.* **45:**889–920 (1976).

Kornberg, A. *DNA Synthesis.* San Francisco: Freeman, 1975.

Leibowitz, P. J., and M. Schaechter. "The attachment of the bacterial chromosome to the cell membrane," *Int. Rev. Cytol.* **41:**1–28 (1975).

Bacterial Chromosomes and DNA Replication

*Cairns, J. "The chromosome of *Escherichia coli,*" *Cold Spring Harbor Symp. Quant. Biol.* **28:**43–46 (1963). [Reprinted in E. A. Adelberg, *Papers on Bacterial Genetics,* 2nd ed. Boston: Little, Brown, 1966.]

*De Lucia, P., and J. Cairns. "Isolation of an *E. coli* strain with a mutation affecting DNA polymerase," *Nature* **224:**1164–1166 (1969).

*Gefter, M. L., Y. Hirota, T. Kornberg, J. A. Wechsler, and C. Barnoux. "Analysis of DNA polymerase II and III in mutants of *Escherichia coli* thermosensitive for DNA synthesis," *Proc. Natl. Acad. Sci. U.S.* **68:**3150–3153 (1971).

Gottesman, M. M., M. L. Hicks, and M. Gellert. "Genetics and function of DNA ligase in *Escherichia coli,*" *J. Mol. Biol.* **77:**531–547 (1973).

*Gudas, L. J., R. James, and A. B. Pardee. "Evidence for the involvement of an outer membrane protein in DNA initiation," *J. Biol. Chem.* **251:**3470–3479 (1976).

*Horiuchi, T., and T. Nagata. "Mutations affecting growth of the *Escherichia coli* cell under a condition of DNA polymerase I-deficiency," *Molec. Gen. Genet.* **123:**89–110 (1973).

*Konrad, E. B., P. Modrich, and I. R. Lehman. "Genetic and enzymatic characterization of a conditional lethal mutant of *Escherichia coli* K12 with a temperature-sensitive DNA ligase," *J. Mol. Biol.* **77:**519–529 (1973).

Konrad, E. B., and I. R. Lehman. "Novel mutants of *Escherichia coli* that accumulate very small DNA replicative intermediates," *Proc. Nat. Acad. Sci. U.S.* **72:**2150–2154 (1975).

Livingston, D. M., and C. C. Richardson. "Deoxyribonucleic acid polymerase III of *Escherichia coli.* Characterization of associated exonuclease activities," *J. Biol. Chem.* **250:**470–478 (1975).

*Masters, M., and P. Broda. "Evidence for the bidirectional replication of the *Escherichia coli* chromosome," *Nature New Biol.* **232:**137–140 (1971).

McPherson, A., I. Molineux, and A. Rich. "Crystallization of a DNA-unwinding protein: Preliminary X-ray analysis of fd bacteriophage gene 5 product," *J. Mol. Biol.* **106:**1077–1081 (1976).

*Meselson, M., and F. W. Stahl. "The replication of DNA in *Escherichia coli,*" *Proc. Natl. Acad. Sci. U.S.* **44:**671–682 (1958). [Reprinted in G. L. Zubay, *Papers in Biochemical Genetics.* New York: Holt, Rinehart and Winston, 1968.]

*Denotes articles described specifically in the chapter.

*Okazaki, R. T., K. Okazaki, K. Sakabe, K. Sugimoto, and A. Sugino. "Mechanism of DNA chain growth. I. Possible discontinuity and unusual secondary structure of newly synthesized chains," *Proc. Natl. Acad. Sci. U.S.* **59**:598-605 (1968).

*Portalier, R., and A. Worcel. "Association of the folded chromosome in the cell envelope of *E. coli:* Characterization of the proteins at the DNA-membrane attachment site," *Cell* **8**:245-255 (1976).

Wake, R. G. "Circularity of the *Bacillus subtilis* chromosome and further studies on its bidirectional replication," *J. Mol. Biol.* **77**:569-575 (1973).

Wake, R. G. "Termination of *Bacillus subtilis* chromosome replication as visualized by autoradiography," *J. Mol Biol.* **86**:223-231 (1974).

*Weiner, J. H., L. L. Bertsch, and A. Kornberg. "The deoxyribonucleic acid unwinding protein of *Escherichia coli.* Properties and functions in replication," *J. Biol. Chem.* **250**:1972-1980 (1975).

*Wickner, S. "Mechanism of DNA elongation catalyzed by *Escherichia coli* DNA polymerase III, DNA protein, and DNA elongation factors I and III," *Proc Nat Acad Sci U.S.* **73**:3511-3515 (1976).

*Worcel, A., and E. Burgi. "On the structure of the folded chromosome of *Escherichia coli*," *J. Mol. Biol.* **71**:127-147 (1972).

Modification and Restriction

Arber, W. "DNA modification and restriction," *Prog. Nucl. Acid. Res. Mol. Biol.* **14**:1-37 (1974).

*Dugaiczyk, A., J. Hedgpeth, H. W. Boyer, and H. M. Goodman. "Physical identity of the SV40 deoxyribonucleic acid sequence recognized by the *Eco*RI restriction endonuclease and modification methylase," *Biochemistry* **13**:503-512 (1974).

Meselson, M., R. Yuan, and J. Heywood. "Restriction and modification of DNA," *Ann Rev Biochem* **41**:447-466 (1972).

Modrich, P., and D. Zabel. "*Eco*RI endonuclease. Physical and catalytic properties of the homogeneous enzyme," *J. Biol. Chem.* **251**:5866-5874 (1976).

Nathans, D., and H. O. Smith. "Restriction endonucleases in the analysis and restructuring of DNA molecules," *Ann. Rev. Biochem.* **44**:273-293 (1975).

Viruses and Viral Chromosome Replication

Dalton, A. J., and F. Haguenau, Eds. *Ultrastructure of Animal Viruses and Bacteriophages: An Atlas.* New York: Academic Press, 1973.

Eisenberg, S., J. F. Scott, and A. Kornberg. "Enzymatic replication of viral and complementary strands of duplex DNA of phage ØX174 proceeds by separate mechanisms," *Proc. Natl. Acad. Sci. U.S.* **73**:3151-3155 (1976).

Ellis, E. L., and M. Delbrück. "The growth of bacteriophage," *J. Gen. Physiol.* **22**:365-384 (1939). [Reprinted in G. S. Stent, *Papers on Bacterial Viruses,* 2nd ed. Boston: Little, Brown, 1966.]

Fenner, F., B. R. McAuslan, C. A. Mims, J. Sambrook, and D. O White. *The Biology of Animal Viruses.* New York: Academic Press, 1974.

Germond, J. E., B. Hirt, P. Oudet, M. Gross-Bellard, and P. Chambon. "Folding of the DNA double helix in chromatin-like structures from simian virus 40," *Proc. Natl. Acad. Sci. U.S.* **72**:1843-1847 (1975).

Ikeda, J., A. Yudelevich, and J. Hurwitz. "Isolation and characterization of the protein coded by gene A of bacteriophage ØX174 DNA," *Proc Natl Acad Sci U.S.* **73**:2669-2673 (1976).

Levine, A. J. "SV40 and adenovirus early functions involved in DNA replication and transformation," *Biochim. Biophys. Acta* **458**:213-241 (1976).

McGeoch, D., P. Fellner, and C. Newton. "Influenza virus genome consists of eight distinct RNA species," *Proc. Natl. Acad. Sci. U.S.* **73**:3045-3049 (1976).

Modrich, P., and C. C. Richardson. "Bacteriophage T7 deoxyribonucleic acid replication *in vitro*," *J. Biol. Chem.* **250**:5508-5514; 5515-5522 (1975).

Moise, H., and J. Hosoda. "T4 gene 32 protein model for control of activity at replication fork," *Nature* **259**:455-458 (1976).

*Streisinger, G., J. Emrich, and M. M. Stahl. "Chromosome structure in phage T4 III. Terminal redundancy and length determination," *Proc. Natl. Acad. Sci. U.S.* **57**:292-295 (1967).

Takahashi, S. "Role of genes *O* and *P* in the replication of bacteriophage λ DNA," *J. Mol. Biol.* **94**:385-396 (1975).

Thomas, C. A., Jr., and L. A. MacHattie. "The anatomy of viral DNA molecules," *Ann. Rev. Biochem.* **36**:485-518 (1967).

Zinder, N.D., Ed. *RNA Phages.* Cold Spring Harbor Monographs Series, 1975.

Eukaryote Chromosome Replication

Bollum, F. J. "Mammalian DNA polymerases," *Prog. Nucl. Acid Res. Mol. Biol.* **15**:109-144 (1975).

Edenberg, H. J., and J. A. Huberman. "Eukaryotic chromosome replication," *Ann Rev. Genet.* **9**:245-284 (1975).

Genta, V. M., D. G. Kaufman, W. K. Kaufmann, and B. I. Gerwin. "Eukaryotic DNA replication complex," *Nature* **259**:502-503 (1976).

Kriegstein, H. J., and D. S. Hogness. "Mechanism of DNA replication in *Drosophila* chromosomes: Structure of replication forks and evidence for bidirectionality," *Proc. Natl. Acad. Sci. U.S.* **71**:135-139 (1974).

Loeb, L. A. "Eucaryotic DNA polymerases." In: *The Enzymes, vol. X.* (P. D. Boyer, Ed.) New York: Academic Press, 1974, pp. 174-210.

Sheinin, R. "Preliminary characterization of the temperature-sensitive defect in DNA replication in a mutant mouse L cell," *Cell* **7**:49-57 (1976).

Weissbach, A. "Vertebrate DNA polymerases," *Cell* **5**:101-108 (1975).

(a) (b) (c)

B. Singer (a) and (b) H. Fraenkel-Conrat (University of California, Berkeley) have studied the genetic properties of viral RNA chromosomes. (c) Sue Wickner (National Institutes of Health) uses mutant strains of *E. coli* to dissect DNA replication *in vitro.*

Questions and Problems

1. Diagram the uv absorbance patterns expected in a Meselson-Stahl-type experiment if replication were conservative and the experiment followed 3 generations (as in Figure 5-3).

2. Diagram the structure of an *E. coli* chromosome in a rapidly growing cell in which a second round of DNA replication is initiated before the first has been completed.

3. (a) Diagram how segregation of daughter DNA chromosomes is thought to be accomplished during the *E. coli* cell cycle. (b) Describe the properties you would expect of a mutant strain with a temperature-sensitive attachment point.

4. Describe what you would expect to happen to DNA replication when an *E. coli* strain carrying a temperature-sensitive *dnaG* mutation is shifted to nonpermissive temperature.

5. Would you expect Okasaki fragments to continue being synthesized for some minutes after a shift to nonpermissive temperatures in rapidly growing cultures of *dnaA* strains? *dna C-D* strains? *dnaG* strains? *dnaE* strains? a strain carrying a temperature-sensitive *lig* mutation? Explain for each case.

6. Distinguish between the activities of the unwinding protein, the unwinding enzyme, and the ω protein in opening up a parental duplex for replication.

7. The DNA polymerase II enzyme of *E. coli*, encoded by the *polB* gene, has both exonuclease and polymerization activities. How might you use mutant strains to learn what role this enzyme plays in the cell?

8. You wish to look for a strain of *E. coli* carrying a mutation in the gene coding for the methylating enzyme described in the chapter. You begin with a strain having the gene for the *Eco*RI enzyme. Would it be necessary to isolate a conditional mutant strain? Explain.

9. Examine the sequence recognized by the *Eco*RI enzyme. What pattern is evident in the sequence? Can you speculate why the enzyme might recognize such a pattern? Would you expect the enzyme to make restriction cuts in an MS2 chromosome? An M13 chromosome? Explain.

10. SV40 DNA associates with histones. What structures might you expect to see if SV40 chromosomes are isolated under gentle conditions?

11. You wish to learn whether the T antigen of SV40 is responsible for any of the novel properties of transformed cells. Outline how you would ask such a question with mutant strains of SV40.

12. List the differences between phage, bacterial, and eukaryotic chromosomes in terms of size, appearance, and behavior during the cell or infectious cycle.

13. What kind(s) of enzymes or proteins (e.g., replicase, polymerase, etc.) would you expect to find associated with the following: (a) closing the parental RF of ØX174; (b) cleaving a T4 concatamer to form a headful; (c) reducing the number of superhelical twists in a λ chromosome (Figure 5-2); (d) insertion of a λ chromosome into an *E. coli* chromosome. Explain your reasoning for each case.

14. A pure preparation of nonmethylated λ chromosomes is presented with purified *Eco*RI enzyme *in vitro*. Would you expect the DNA to be digested into individual nucleotides? If not, what would you expect the test tube to contain following incubation? How might you utilize CsCl centrifugation to test these predictions? Would you expect different results using λ and λ*dgal*? Explain.

15. You isolate a mutant strain of λ which can be induced to leave the lysogenic state and direct a lytic cycle by shifting the temperature to 45°C. How might you explain this phenotype?

16. Decay of ^{32}P produces β-particles. When added prior to replication, ^{32}P phosphate can be incorporated into the phosphodiester bonds of the genome of an organism. This has been done extensively with different bacteriophages, where it has been shown that labelled phages lose viability during storage. This phenomenon, called suicide, can be expressed by a constant, k, which represents the efficiency of "killing" per radioactive disintegration. For the T- series phages, the value is about 0.1, whereas with ØX174 the k value is about 1.0. Explain the difference in k between the two phages.

CHAPTER
6

Mutation and Repair

6.1   DEFINITIONS AND TERMS

A **mutation** occurs when the sequence or the number of nucleotides in a nucleic acid is altered and the new sequence or number is passed from parent to offspring. Mutations are broadly classified as being either **point mutations** affecting short regions of a chromosome (for example, the **substitution** of one base for another or the **deletion** or **addition** of one or several bases), or as being **chromosomal mutations** or **aberrations** affecting larger pieces of the chromosome or the total chromosome number of the species. Clearly many mutations will be intermediate between these extremes and the classifications are simply for convenience.

In defining a mutant organism, reference is usually made to some "normal" standard with which the mutant organism is compared and contrasted. As noted in Section 3.15, geneticists have adopted the term **wild type** to refer to such a standard. The term wild type originally indicated the sort of organism found in nature—in the wild—as opposed to the type produced in the laboratory or kept under domestication. In fact, however, the term is often applied to the stock maintained in the laboratory from which mutant stocks are derived. With years of laboratory existence the original stock may change; hence its wild-type appellation is often quite arbitrary. Moreover, different laboratories working with the same species may use different wild-type strains. The wild-type term is thus often loosely defined, but is nonetheless useful.

With wild type as a reference, point mutations are commonly classified as being **forward mutations** if they proceed in the wild-type $\longrightarrow$ mutant direction and as **reverse mutations** or **reversions** if they proceed in the mutant $\longrightarrow$ wild-type direction. Reversion was encountered in Chapter 1 as the process that caused *rough* pneumococcus to resume the capsule characteristics of the *smooth* strain from which it derived. Strictly speaking, a mutant virus or organism is said to have undergone a reversion when a second mutation restores exactly the base sequence of its parental strain. Thus if the parent is CATCAT and the mutant is CAACAT, the revertant will be CATCAT. Reversion is often, however, mimicked by **suppression.** Suppression results when a second mutation occurs at a site in the chromosome different from the first mutation site but is in some way able to mask, or suppress, the phenotypic expression of the first mutation. Thus suppression will produce an organism that appears to be reverted but is in fact doubly mutant. The two types of mutations can be distinguished in genetic crosses, as described in later chapters.

Chromosomal mutations are less delicate than point mutations and therefore are far less likely to be revertable or suppressible in the strict sense. The major kinds of chromosomal changes are **deficiencies** or **duplications** of genetic material, **translocations** of a piece of one chromosome onto the end of another, and **inversions** of a chromosomal segment so that

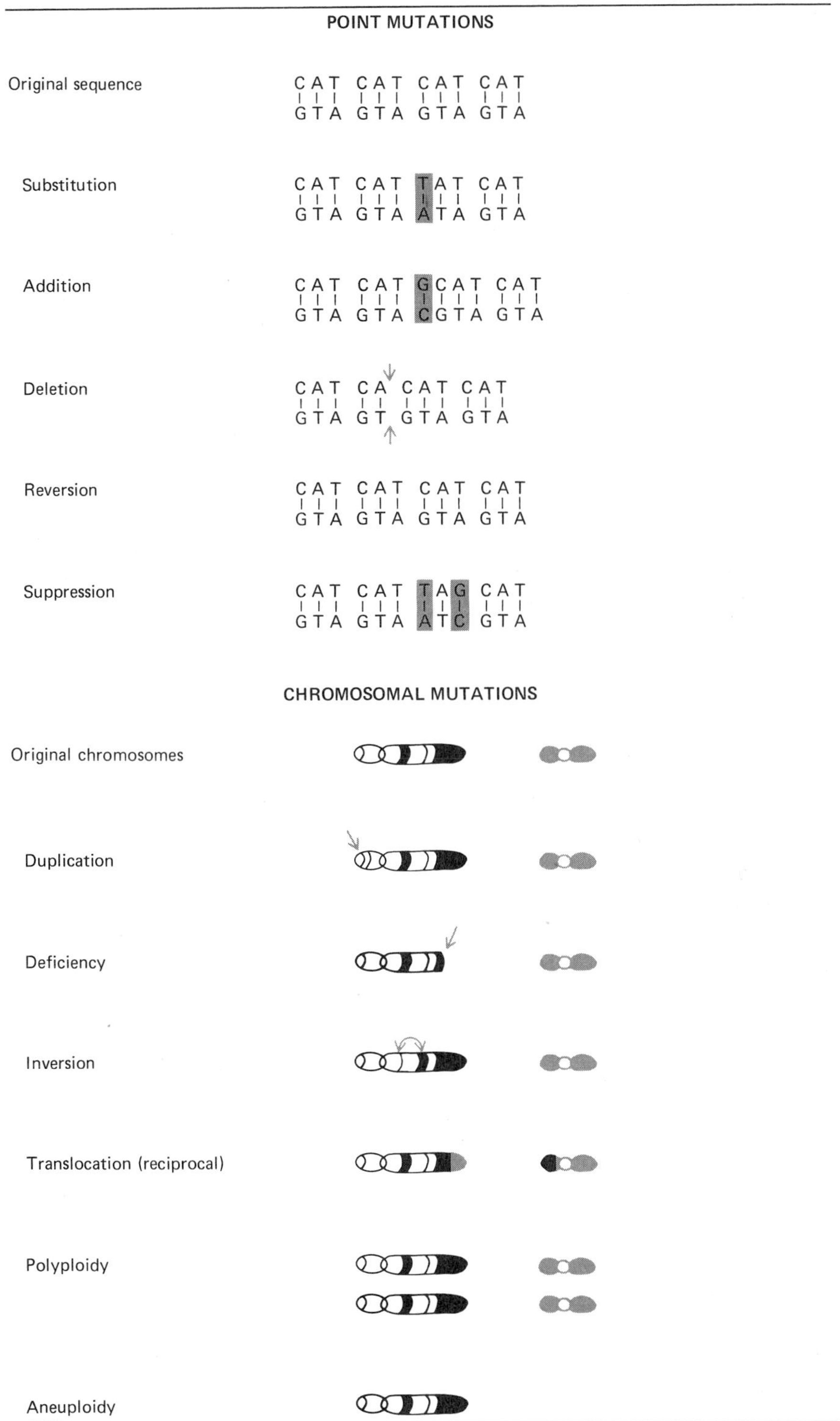

POINT MUTATIONS

Original sequence
```
C A T   C A T   C A T   C A T
| | |   | | |   | | |   | | |
G T A   G T A   G T A   G T A
```

Substitution
```
C A T   C A T   T A T   C A T
| | |   | | |   | | |   | | |
G T A   G T A   A T A   G T A
```

Addition
```
C A T   C A T   G C A T   C A T
| | |   | | |   | | | |   | | |
G T A   G T A   C G T A   G T A
```

Deletion
```
              ↓
C A T   C A   C A T   C A T
| | |   | |   | | |   | | |
G T A   G T   G T A   G T A
          ↑
```

Reversion
```
C A T   C A T   C A T   C A T
| | |   | | |   | | |   | | |
G T A   G T A   G T A   G T A
```

Suppression
```
C A T   C A T   T A G   C A T
| | |   | | |   | | |   | | |
G T A   G T A   A T C   G T A
```

CHROMOSOMAL MUTATIONS

Original chromosomes

Duplication

Deficiency

Inversion

Translocation (reciprocal)

Polyploidy

Aneuploidy

Figure 6-1 Point and chromosomal mutations.

its genes are in the reverse order. Changes in chromosome number arise either by **polyploidy** (a 2*n* organism becomes 3*n* or 4*n*) or by **aneuploidy** wherein a single chromosome is gained (2*n* + 1) or lost (2*n* − 1).

Figure 6-1 summarizes the various kinds of point and chromosomal mutations that have been described. All are known to occur **spontaneously,** and all can be **induced** by a variety of agents and chemicals known collectively as **mutagens.** This chapter begins by describing how organisms carrying mutations are identified and classified; we then discuss how mutagens act to produce various kinds of point mutations. Since many point mutations prove to be consequences of the faulty repair of DNA lesions, repair mechanisms are included in some detail. Finally, we consider chromosomal mutations and the mutagenic/carcinogenic effects of agents present in our environment.

SCREENING PROCEDURES

Screening procedures are essential for two related types of study. In the first, the geneticist wishes to isolate a mutant strain with a particular phenotype. Since the mutation process is random, the geneticist is unable to control which gene will become mutant. Techniques must therefore be devised to screen a large number of potentially mutant organisms for the mutant phenotype of interest. In the second, the geneticist wishes to determine whether a given chemical or agent is mutagenic; the geneticist therefore exposes organisms of one phenotype to the putative mutagen and screens their descendents for mutants. In this case the important features of the mutant phenotype are that it be readily detectable and the result of a single mutational event, so that it is possible to calculate how many mutational **hits** are produced by a particular dose of mutagen.

The screening procedures described in the following sections are used for one or both of these purposes. Their description involves much of the genetic terminology and most of the organisms introduced in Chapters 2–5.

6.2   SCREENING FOR VISIBLE MUTATIONS

Visible mutations affect a morphological trait, and screening is done by inspection. The rare white-eyed fly among a group of red-eyed flies seen under a low-power microscope can be readily recognized and set aside, and thus visibles constitute ideal kinds of mutations for many sorts of genetic studies. In diploid organisms, a visible mutation is often recessive, so that it is expressed only in homozygotes. Some mutations, however, affect the phenotype of heterozygotes, an example being the *B* duplication in *Drosophila* which leads to the formation of *Bar*-shaped eyes (Figure 4-14). Recessive visibles will also, of course, be detected in male diploid organisms if the mutant gene resides in the X chromosome (Section 4.12).

Haploid microorganisms can be screened for changes in their colony morphology, and these mutations can also be classified as visibles. Wild-type *Neurospora*, for example, will produce the long fluffy hyphae depicted in Figure 6-2a while the strain *abn-1* produces minute hyphae under the same growth conditions (Figure 6-2b). Similarly, yeast colonies may be classified as normal or as *petite*, while *C. reinhardi* may lose, by mutation, the ability to synthesize chlorophyll in the dark, and will form yellow colonies instead.

Visible mutations are encountered even in the case of viruses. A petri plate containing agar and growth medium **(nutrient agar)** can be prepared and inoculated with *E. coli* to provide an even growth of bacteria that will cover the agar surface—a bacterial "lawn." If a dilute suspension of wild-type phage T2 is mixed with the bacteria at the time of inoculation, each virus will encounter, infect, and eventually lyse a single bacterium, releasing about 100 progeny. The progeny can, in turn, infect 100 adjacent bacteria, and so on, so that a clear, bacteria-free **plaque** soon forms within the lawn (Figure 6-3). Certain mutant strains of T2 lyse the bacterial cells so rapidly that during the time a wild-type (r^+) phage produces a small plaque with a characteristic fuzzy edge, these rapid-lysis or *r* mutants produce a large clear plaque (Figure 6-3). By scanning hundreds of petri dishes a single *r*-type

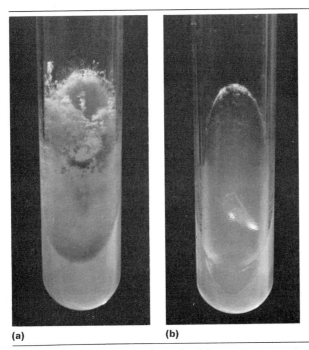

(a)   **(b)**

Figure 6-2 (a) Normal hyphae of wild-type *Neurospora crassa*. (b) Small hyphae produced by the mutant strain *abn-1*. (Courtesy of Dr. D. J. L. Luck).

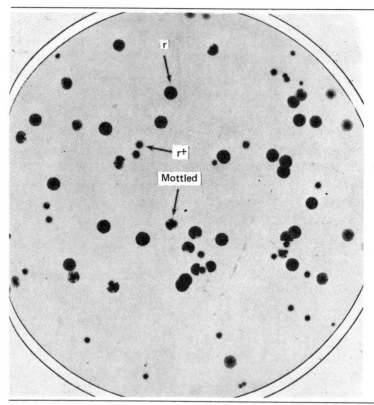

Figure 6-3 Plaques produced by wild-type (r^+) and mutant (r) T2 phages. The mottled plaques are produced when both r and r^+ phages grow simultaneously in the same plaque; phages that produce such plaques are described in Section 13.9. (From *Molecular Biology of Bacterial Viruses* by Gunther S. Stent. W. H. Freeman and Company. Copyright © 1963.)

phage can be distinguished from many thousands of wild-type phages by the abnormally large plaque it (and its progeny) produces.

Strains of bacteriophage λ that cannot carry out lysogeny (a process described in Section 5.11) can also be recognized visibly. The plaques produced when normal λ particles are placed on an *E. coli* lawn are turbid, since many of the bacteria infected are lysogenized rather than lysed (Figure 6-4). Mutant particles, on the other hand, will form a clear plaque on the lawn and can be spotted easily (Figure 6-4).

Specific Locus Test in Mouse   L. B. Russell and W. L. Russell constructed a mouse tester strain homozygous for the 7 recessive visible point mutations shown in Figure 6-5. Each mutant allele has a clear-cut effect on the phenotype and none has any adverse effect on viability. To estimate induced mutation rates, homozygous wild-type mice are exposed to a muta-

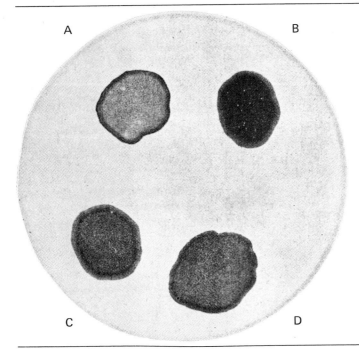

Figure 6-4 Turbid plaque produced by wild-type λ phage is shown at A. The other three clear plaques are produced by various mutant strains that cannot carry out lysogeny. (From A. D. Kaiser, Virology **3:**42, 1957.)

gen and crossed to the tester stock; the F_1 is then examined for mice that are phenotypically mutant (and therefore homozygous recessive) at any of the seven specific loci (this is effectively a test cross, as you should work out for yourself.) Typical results for control and X-irradiated mice indicate that the summed **spontaneous mutation rate** at these seven loci is about 8 mutations per locus per million tested male gametes while the **induced mutation rate** by X irradiation is at least 90 mutations per locus per million male gametes tested.

6.3 SCREENING FOR NUTRITIONAL MUTATIONS

A **nutritional** or **biochemical** mutation affects an organism's ability to produce a molecule (for example, an amino acid) essential for growth. Nutritional mutations are most commonly studied in haploid organisms such as *E. coli, Chlamydomonas, Neurospora,* and yeast, which can grow in the laboratory on well-defined media. Mutant strains that require a nutritional supplement for growth are termed **auxotrophic** in contrast to **proto-**

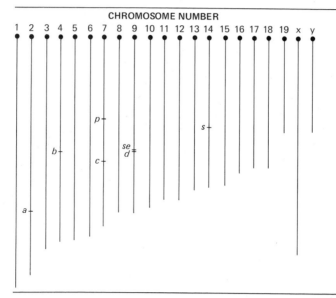

Figure 6-5 Mouse karyotype, showing approximate positions of specific loci used in mutagenesis experiments. Abbreviations: *a*, non-agouti; *b*, brown; *c*, chinchilla; *p*, pink-eye dilution; *d*, dilute; *se*, short ear; and *s*, piebald.

trophic wild-type or revertant strains. The replica-plating technique, described in Figure 4-2, is commonly used for detecting auxotrophs in these organisms.

Screening for mutant strains of microorganisms with metabolic abnormalities can sometimes be accomplished by the use of special media. Thus *E. coli* cells can be grown on a medium containing the sugar lactose and the brown indicator dye **2,3,5-triphenyl tetrazolium chloride.** Colonies that are able to ferment lactose (*lac*+) cause the pH of the medium to drop and thereby bleach the dye so that the colonies appear white, whereas mutant *lac*− colonies that cannot ferment lactose from their surroundings appear brown.

Enrichment for Auxotrophs Screening for rare mutant organisms in a population is facilitated considerably when an **enrichment procedure** is devised to increase the proportion of mutant to wild-type organisms present. Two examples can be given. First, auxotrophs of *E. coli* and other Gram-negative bacteria can be selectively enriched for by placing mutagenized cells in an unsupplemented medium that contains the antibiotic penicillin. Penicillin prevents the formation of cell walls; therefore, wild-type cells that attempt to divide in the presence of penicillin are promptly killed. Auxotrophs, on the other hand, cannot divide in the unsupplemented medium and thus remain alive (although they are slowly starving).

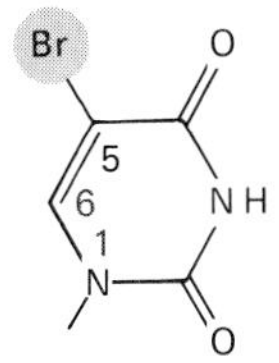

Figure 6-6 Structure of 5-bromodeoxyuracil (5BU).

The culture is then plated on to supplemented, penicillin-free medium and colonies are allowed to form; these are tested by replica-plating (Figure 4-2) to determine the specific nutritional requirements they carry.

A second enrichment procedure, utilized both for bacteria and for cultured eukaryotic cells, employs **5-bromodeoxyuracil** (5BU), an analogue of thymine (recall that thymine is 5-methyluracil) with a bromine atom in place of the methyl group on carbon 5 (Figure 6-6). A mutagenized culture is incubated in minimal medium and presented with 5BU. Wild-type cells incorporate the analogue into their daughter chromosomes whereas mutant, non-growing cells do not. The culture is then exposed to near-uv light (313 nm). 5BU molecules absurb uv much more strongly than do the usual bases; therefore, cells that have incorporated 5BU into their chromosomes are much more likely to be photolyzed (literally, "light-broken") by uv exposure than are 5BU-free cells. Thus the mutants survive while the wild-type cells die, an essential feature of most nutritional enrichment procedures.

Nutritional Revertant Assay A nutritional revertant assay is often used to calculate mutation rates. A microorganism auxotrophic for an amino acid (histidine, for example) is exposed to a mutagen; the cells are then plated onto histidine-free medium so that only "revertants" (either true revertants or suppressed mutants) will grow. If 10^7 cells are exposed to the mutagen and plated, and 10 colonies appear on the plate, a mutation rate of $10/10^7 = 10^{-6}$ can immediately be deduced; it is not necessary, as in most forward-mutation assays, to examine the phenotypes of thousands of organisms.

6.4 SCREENING FOR LETHAL MUTATIONS

A gene that has undergone a **lethal mutation** is typically incapable of producing an active form of an indispensible protein. In haploid organisms this means that the mutant itself or its immediate mitotic progeny will not survive. Mutations in "indispensible" genes must therefore be isolated in haploids as **conditional lethals:** as already noted in Section 5.3, an organism carrying a conditional mutation survives under permissive growth conditions where the gene product is functional, but dies under restrictive growth conditions where the gene product is nonexistent or nonfunctional.

Lethal mutations in diploids possess the ability to kill the organism directly or to prevent it from reproducing **(genetic death);** the latter organisms are more often known as **steriles.** A dominant lethal or sterile mutation in a diploid organism cannot be maintained past one generation; however, this is not true of recessive lethals. A new germ-line recessive lethal mutation in a diploid organism will not affect the viability of the original parent, nor will it affect the viability of the next generation since the gamete carrying the mutation (*l*) is almost certain to fuse with a gamete carrying its dominant allele (+) to produce a viable +/*l* heterozygote. Indeed, the recessive lethal can be maintained indefinitely in the heterozygous condition by appropriate crosses. It is only when +/*l* heterozygotes are crossed that the presence of the lethal is detected as an inviability for one-quarter of the progeny (those with an *l*/*l* genotype). If the lethal gene is carried by the X chromosome, moreover, a female with a +/*l* genotype will produce only half the expected number of sons. You should work these ratios out yourself using the procedures outlined in Sections 4.2 and 4.12.

Selection for Lethals Using Special Chromosomes Screening for sex-linked lethal mutations in *Drosophila* can be facilitated by the use of stocks carrying special chromosomal mutations. One such chromosome is the **Basc X** chromosome of *D. melanogaster,* so-named because it carries the dominent *Bar* allele *B,* the recessive *apricot* eye-color allele *w^a*, and an inversion involving the *scute* (*sc*) region of the chromosome. The *scute* inversion effectively prevents meiotic crossing over in the left arm of the X chromosome, presumably because homologues cannot readily undergo synapsis (Section 3.4) when one is inverted with respect to the other. Therefore, females heterozygous for a *Basc* X chromosome do not generate eggs carrying recombinant X chromosomes, and any progeny flies with *Bar* eyes can be automatically assumed to carry *Basc* chromosomes.

The *Basc* chromosome can be used to screen for recessive lethal mutations in the left arm of the X chromosome in the fashion diagrammed in Figure 6-7. Wild-type males are mutagenized and crossed with females homozygous for *Basc*. The F_1 females are collected as virgins (they are not allowed to mate with their brothers) and are individually crossed with wild-type males. Those females whose sons are all *Bar, apricot* (and never wild type) are identified as possessing one X chromosome that carries a recessive lethal.

6.5 MUTATIONS TO RESISTANCE AND DEPENDENCY

Mutant organisms that are resistant to a drug, to infection by a particular virus, or to other agents that are normally toxic are widely known. Similarly,

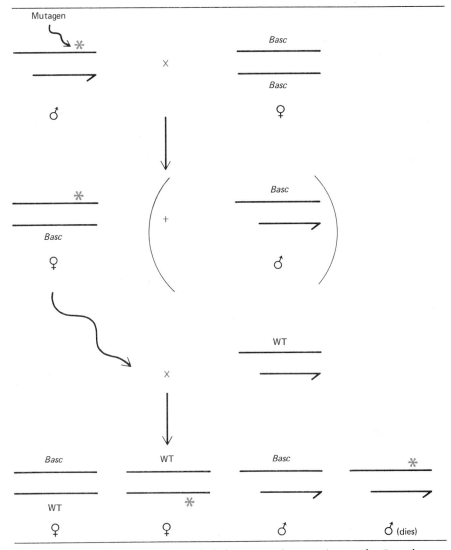

Figure 6-7 Screening for a recessive lethal mutation (gray star) using the *Basc* chromosome of *D. melanogaster*. Y chromosome symbolized by ⟶.

mutant organisms can exhibit a dependency on such agents for growth. Screening for these mutant strains is a simple matter: a population of cells or organisms is exposed to the agent and only those that are resistant, or those that are both resistant and dependent, will survive. The dependent strains can subsequently be recognized by their inability to grow in the absence of the agent.

Analysis of many drug-resistant strains has revealed the presence of two general classes: strains that have become impermeable to the drug and strains whose intracellular targets have become resistant to the effects of the drug. The latter class is of particular interest. Many drugs act by inhibiting essential enzymes such as polymerases, or by blocking the function of essential structures such as ribosomes. Thus a comparison of resistant and sensitive strains often yields important information about how these enzymes or structures function.

The Fluctuation Test When toxic—and potentially mutagenic—agents are used to screen for mutant strains, a fundamental question often arises: did the mutations preexist in the population of cells before exposure to the selective agent, or did the agent itself cause the mutations? If, for example, streptomycin-sensitive *E. coli* cells are mutagenized and plated on a medium containing streptomycin, are the streptomycin-resistant colonies that arise induced by the mutagen, or are they induced by the streptomycin in the growth medium?

Such a question is best answered by a **fluctuation test,** as first devised by S. Luria and M. Delbrück. In the experiments we describe, Luria and Delbrück were interested in learning whether the presence of the bacteriophage T1 elicited the formation of T1-resistant *E. coli.* They first inoculated a few cells into a flask containing liquid culture medium, allowed the culture to grow until it contained millions of cells, and then plated a number of 0.05 ml samples of the culture to agar media containing T1 phages. As seen in Table 6-1, the number of phage-resistant colonies that arose from each sample was about the same—an average or **mean** of about 51 with a variance of 27, the **variance** describing the way values are dispersed about a mean and here reflecting both the sampling variance and also miscellaneous factors such as inaccurate pipetting (recall chi square, Section 4.9).

With these data serving as controls, Luria and Delbrück next performed the following experiment: they inoculated each of nine different flasks with a few cells. Each culture was then allowed the same period of growth, and one sample from each flask was plated to the T1-containing medium. In this case the results were very different (Table 6-1). The average number of resistant colonies per sample happened to be about the same as in the control, but the variance was now 3498. The individual sample values, in other words, fluctuated greatly.

From such results one can argue as follows. If mutation to resistance is indeed *caused* by exposure to the phages on the plate, the bacteria in each plated sample should have the same chance of experiencing a mutational event, and the variance in the second experiment of Table 6-1 should be similar to the variance in the control experiment. Since this is not the observed result, the hypothesis of a virally induced mutation to resistance is not supported. If, on the other hand, mutation to resistance is a *random event* that can occur at any time during the growth period in liquid culture, the marked fluctuation observed in the second experiment is precisely what one would predict: in one flask a mutation might happen to "hit" a bacte-

Table 6-1 Comparison of the Number of Phage-Resistant Bacteria in Different Samples of the Same Culture and in Samples from Independent Cultures

Sample No.	Samples from Same Culture	Samples from Independent Cultures
	Resistant Bacteria	
1	46	30
2	56	10
3	52	40
4	48	45
5	65	183
6	44	12
7	49	173
8	51	23
9	56	57
10	47	51
Average per sample	51.4	62
Variance*	27	3959

*The variance $V = \dfrac{\Sigma D^2}{n-1}$ where $D =$ the amount by which each value differs from the mean, and $n =$ the number of values tabulated (Section 4.9).

After S. E. Luria and M. Delbrück, *Genetics* **28**:491 (1943), and W. Hayes, *The Genetics of Bacteria and Their Viruses,* 2nd ed. Oxford: Blackwell, 1968.

rium early in the growth period, so that a large clone of resistant descendants will appear in the final sample, whereas in a second flask a mutation might not happen to hit any bacteria until late in the growth period and only a few resistant bacteria will be present in the final sample.

DIRECT MUTAGENESIS

Direct mutagenesis alters the sequence or order of bases in a chromosome. The change may be induced in a parent chromosome, and become copied into an altered daughter chromosome, or it may occur at the time of chromosome replication itself. Either possibility is distinguished from **indirect mutagenesis** in which a **premutational lesion** is introduced into the chromosome by the mutagen and only later is this lesion converted into a mutation, usually as a consequence of faulty repair. Indirect mutagenesis is considered in the next major section of the chapter.

6.6   PERMANENT ALTERATIONS IN NUCLEOTIDE STRUCTURE

Transitions and Transversions   Most direct-acting mutagens transform the chemical structure of a nucleotide, giving it the base-pairing properties of another nucleotide. The resultant **mispairing** will give rise to two sorts of mutation, transitional and transversional. In the case of a **transition,** a purine is replaced by another purine or a pyrimidine is replaced by another pyrimidine (the structures of purines and pyrimidines are found in Figure 1-1). Consider, for example, a particular adenine on one strand of a DNA duplex: a mutagen may transform this into the modified molecule A* which has the base-pairing properties of guanine (Figure 6-8a). At replication, then, this A* will mispair with cytosine (Figure 6-8b), and the complementary daughter strand will carry a cytosine at this position. One round of replication later, the cytosine will pair with a guanine; thus the granddaughter strand, which should be identical to the original, will carry the purine guanine where it should carry the purine adenine, and an A ⟶ G transition is said to have occurred (Figure 6-8c). Similar sequences of events produce G ⟶ A, T ⟶ C, and C ⟶ T transitions.

The second kind of mutation that can result from mispairing is a **transversion,** in which a purine is replaced by a pyrimidine or vice-versa (for example, A ⟶ C, G ⟶ C, and so on). Of the many transversion-inducing mispairings one might try to visualize, however, most fit very

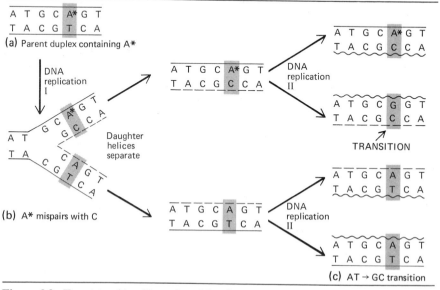

Figure 6-8   Transition (A→G) produced by the mispairing of a modified adenine (A*) with cytosine.

poorly within the confines of a double helix; in the case of purine-purine mispairs, for example, the sugar-phosphate backbone would need to undergo considerable distortion to accomodate both large bases. A few mutagens are, nonetheless, believed to induce transversions directly, although their mode of action remains a matter of speculation.

Permanent alterations in nucleotide structure can be effected by a variety of agents. Of these, four of the most widely cited are nitrous acid (HNO_2), hydroxylamine (NH_2OH), a group of compounds known as the alkylating agents, and heat. Figure 6-9 summarizes what is known of these agents' mutagenic effects on nucleotide structure.

Nitrous Acid   Nitrous acid deaminates nucleotides, removing amino ($—NH_2$) groups and substituting instead a keto group ($=O$). As shown in Figure 6-9a, deamination of cytosine yields uracil and deamination of adenine yields an unusual base, **hypoxanthine (HX).** Uracil is seen in Figure 6-9a to pair with adenine (recall that uracil replaces thymine in RNA), and thus the deamination of cytosine can produce a C $\longrightarrow$ T transition. Hypoxanthine resembles guanine at positions 1 and 6 on its ring and can therefore pair with cytosine, as shown in Figure 6-9a. Again a transition results, this time in the A $\longrightarrow$ G direction.

Hydroxylamine   Hydroxylamine reacts only with the pyrimidines of DNA, and probably only its reaction with cytosine is mutagenic. The way in which hydroxylamine is thought to alter cytosine to produce mutations is shown in Figure 6-9b. Note that hydroxylamine attacks the amino group in position 4 of cytosine, converting it to a hydroxylimine ($=N—OH$); the resulting base (N^4-hydroxycytosine) preferentially pairs with adenine to produce a C $\longrightarrow$ T transition.

Alkylating Agents   The alkylating agents are a diverse group of highly reactive chemicals that introduce alkyl groups ($CH_3—$, $CH_3CH_2—$, and so on) into nucleotides at numerous positions. Of the various alkylated products, O^6-alkylguanine and O^4-alkylthymine are the most likely to undergo mispairing. As drawn in Figure 6-9c, O^6-alkylguanine can pair with thymine to generate G $\longrightarrow$ A transitions, while O^4-alkylthymine can pair with guanine to produce T $\longrightarrow$ C transitions.

Alkylating agents are the largest group of mutagens and include mustard gas, epoxides, dimethyl- and diethylsulfonate, methyl- and ethylmethane sulfonate (**MMS** and **EMS**) and N-methyl-N'-nitroso-N-nitroguanidine (**MNNG**). Most alkylating agents prove to exert their mutagenic effects by triggering misrepair, as described in Section 6.10.

Heat   Heat has only recently been accorded attention as an important mutagen. It clearly brings about the deamination of cytosine to form uracil, much as does nitrous acid, thereby bringing about C $\longrightarrow$ T transitions;

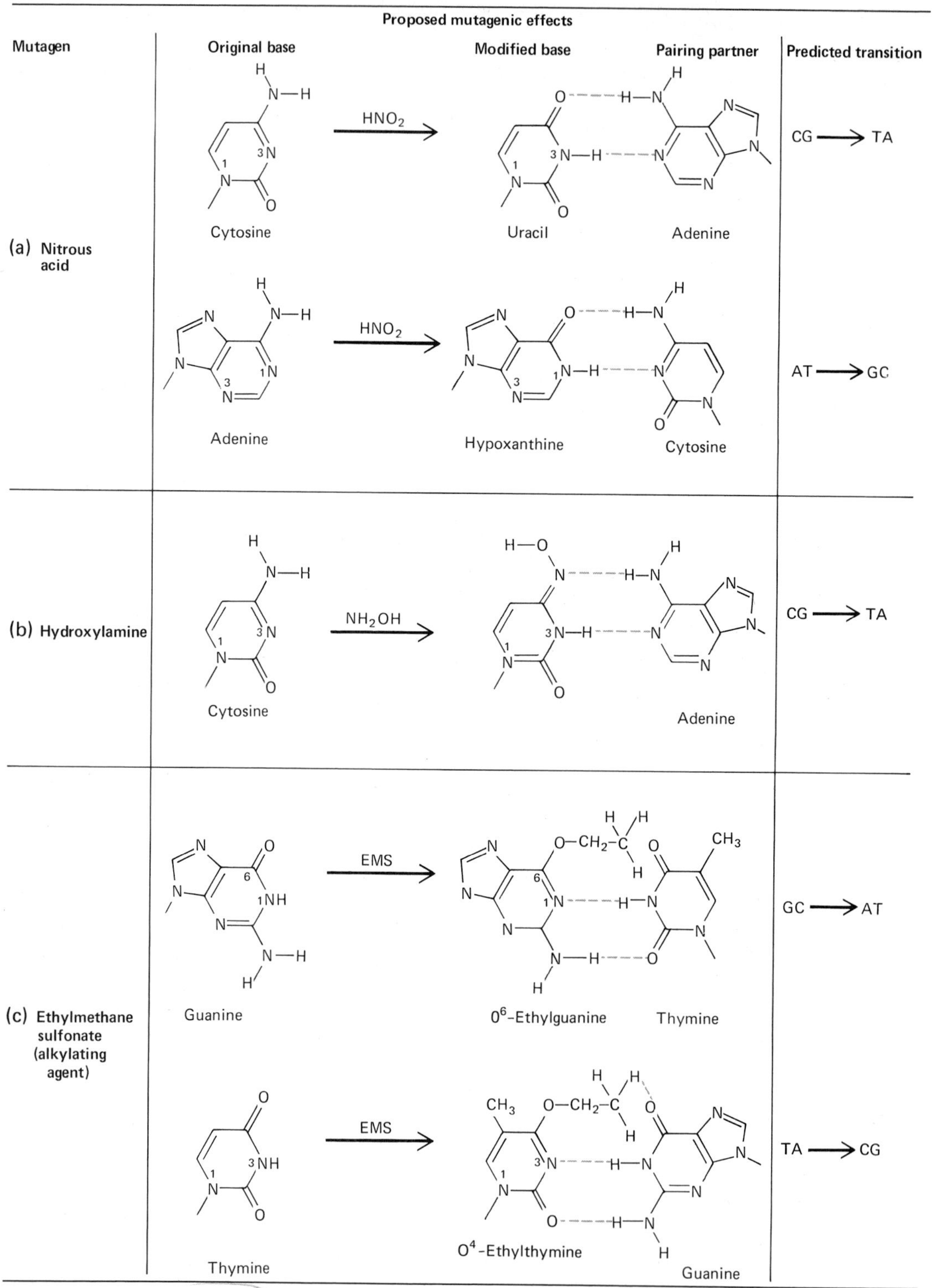

Figure 6-9 Three types of mutagens that modify the chemical structure of bases.

heat also causes G $\longrightarrow$ C transversions by an as yet unknown mechanism. It has been estimated that over 100 heat-induced mutations occur in a typical human cell each day, but the vast majority of these are no doubt later repaired.

6.7 ADDITIONAL LESIONS CAUSED BY MUTAGENS

Chemical mutagens and heat are capable of producing mispairing modifications in specific nucleotides which give rise to mutations directly, as shown above. These agents, and radiations as well, also produce a variety of additional alterations in nucleotides and nucleic acids, which can be classified as nonhereditary, inactivating, and premutational alterations.

Nonhereditary alterations in nucleotide structure are without effect either on DNA replication or on the transfer of genetic information. Thus the major reaction product of many alkylating agents is 7-alkylguanine, a nucleotide that appears to behave exactly like guanine. This means that measurements of 7-alkylguanine formation are poorly correlated with the mutagenic potential of an alkylating agent.

Inactivating alterations prevent the transmission of the altered genome from parent to offspring. For example, the alkylation of A and C is often inactivating. Hydroxylamine decomposes, producing lethal peroxides which damage nucleic acids indiscriminately, ultimately killing the cell directly. Nitrous acid and certain alkylating agents may **cross-link** the strands of a DNA duplex so that they cannot separate for replication; and so on. These inactivating events should not be confused with lethal mutations. A lethal mutation is potentially heritable, provided some way is found to keep the mutant daughter organism alive, whereas in an inactivating event a daughter cell or virus is never even formed.

Inactivation becomes an important consideration during a mutagenesis experiment, for if the dose of mutagen is increased in the hope of obtaining large numbers of mutant organisms, there will be an increase in the number of inactivating "hits" as well as in mutagenic hits, until the point is reached at which most of the organisms being tested will be killed by the treatment. Experimental organisms (or viruses) are therefore first treated with a range of mutagen concentrations and the number of surviving progeny is plotted against concentration. From such a plot a dose that will give some reasonable number of survivors is selected, and this dose is used to induce mutations.

The final category of additional events induced by mutagens is a **premutational lesion.** An individual nucleotide or a region of DNA may be sufficiently altered or damaged that under most circumstances it produces inactivation. Such a lesion may, however, be repaired by the cell either prior to, during, or following replication. In some cases the repair is faultless: the damaged region is replaced by nucleotides identical to those in the original sequence and neither mutation nor inactivation occurs. In other cases the repair is faulty: nucleotides that are not the same as, or complementary to, the parent DNA sequence are inserted into the damaged strand or into a daughter strand. This, of course, produces a mutation, and the damage eliciting such **misrepair** is correctly termed a premutational lesion.

REPAIR AND MISREPAIR

During the course of its lifetime, chromosomal DNA is subjected to a variety of damages: endogenous nucleases may nick it; breaks may occur in it during packaging into a phage head or during mitotic segregation; endogenous cellular chemicals and heat, or uv light and even white light from the sun, may alter it; and noxious environmental chemicals may interact with it. Therefore, even without the added stress of deliberate mutagenesis, the integrity and survival of a genome is critically dependent on the existence of DNA repair mechanisms.

This section describes two broad classes of repair. We first consider two repair systems—**photoreactivation** and **excision repair**—that function with remarkable efficiency and are virtually **error-free.** We then consider post-replication repair systems that are **error-prone** and produce mistakes that are frequently immortalized into DNA as mutations. Finally, we review enzymatic **"proofreading"** mechanisms that operate during DNA replication to overcome both spontaneous and induced mutations, and consider what happens when these mechanisms function imperfectly.

6.8   PHOTOREACTIVATION

Ultraviolet radiation produces several effects on DNA, one being the formation of chemical bonds between two adjacent pyrimidine molecules in a polynucleotide, particularly between adjacent thymine residues, as shown in Figure 6-10. As the two residues associate, or **dimerize,** their position in the DNA helix becomes so displaced that they can no longer form hydrogen bonds with the opposing purines and thus the regularity of the helix becomes distorted. Such dimers are inactivating (and lethal) unless repaired.

A **photoreactivating enzyme** can convert such a thymine dimer ($\widehat{TT}$) into two thymine monomers and thereby eliminate the lesion from the parental strand. The enzyme is so named because, although it can associate with a dimer in the dark, it must absorb a photon of visible light before it can bring about monomerization (Figure 6-10a). Photoreactivation enzymes are present in both prokaryotes and eukaryotes, including humans; indeed, certain humans suffering from the hereditary disease **xeroderma pigmen-**

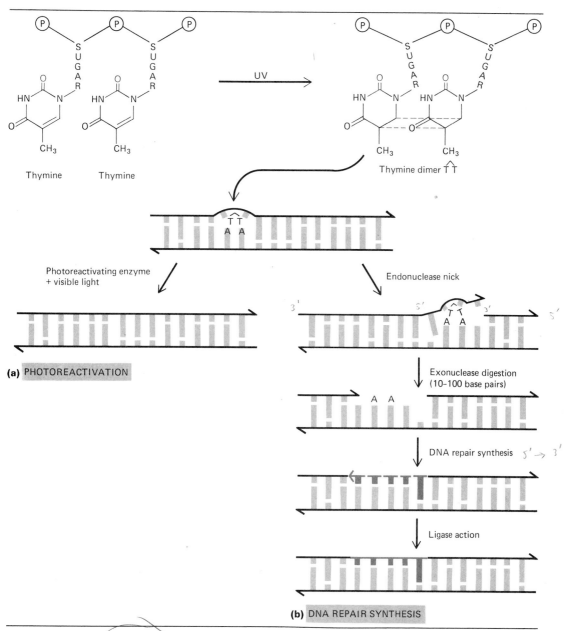

Figure 6-10 Thymine dimer created by uv irradiation and its removal by photoreactivation (a) or by excision repair (b).

tosum,—which produces severe skin sensitivity to the ultraviolet rays in sunlight—have very low levels of photoreactivation enzyme. Thus photo-reactivation is an important line of defense against a major type of DNA damage caused by the sun. It is limited, however, to the repair of pyrimidine dimers.

6.9   EXCISION REPAIR

In describing DNA replication in Section 5.7 we noted that DNA polymer-ase I possesses a 3' $\longrightarrow$ 5' exonuclease activity which serves to "edit out" of newly synthesized daughter strands any nucleotides that are mispaired with the parent template; correct nucleotides are then put into the gaps, the polymerases taking instructions from the template. A similar sequence of events can occur in the dark following the production of thymine dimers: the dimer-containing region of the chromosome is physically removed (excised) from the DNA duplex and a new, dimer-free section of polynu-cleotide is put in its place. Figure 6-10b diagrams the sequence of events that is thought to occur during such **excision repair.** An endonuclease, perhaps recognizing the presence of the dimer by its distortion of the helix, first introduces a nick in the dimer-containing strand, somewhere in the vicinity of the dimer. An exonuclease then enters the gap created by the nick and digests a portion of the strand, including the dimer, in a 5' $\longrightarrow$ 3' direction. A DNA polymerase next proceeds to synthesize a replacement piece of DNA, starting from the exposed 3'—OH end, proceeding in a 5' $\longrightarrow$ 3' direction, and copying the intact strand in a complementary manner. Finally, when the newly synthesized piece of DNA reaches the 5'—P end of the broken-and-digested strand, the two ends are joined by a DNA ligase.

Mutant strains of E. coli (Figure 6-11) have aided in sorting out the

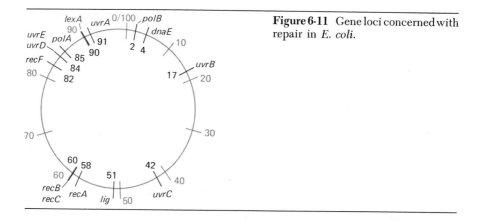

Figure 6-11 Gene loci concerned with repair in *E. coli.*

enzymes involved in excision repair in much the same fashion as they have clarified the process of DNA replication (Section 5.7 and Figure 5-5). Thus synthesis of the T͡T-specific endonuclease has been found to require the gene products of the *uvrA* and *uvrB* loci; mutations in the *pol A* gene leave *E. coli* cells defective in their ability to repair the gaps that result from dimer excision, implicating DNA polymerase I as a major contributor to the DNA-synthesis stage of the repair process; and *lig* mutants show greatly reduced levels of excision repair. In addition, the genes *uvrC* and *uvrD* control as yet unidentified steps in the repair process. In yeast, at least 7 *rad* genes appear to be involved in excision repair, and in humans some patients with xeroderma pigmentosum are reportedly defective in their ability to excise thymine dimers. You will recall that defects in photoreactivation are also reported for this disease; the clinical syndrome called xeroderma pigmentosum may well be caused by a variety of inherited recessive mutations, each of which blocks repair processes in different ways.

Excision repair is readily detected in eukaryotic cells as the occurence of **unscheduled DNA synthesis:** cells in the G_0 or G_1 stage of the cell cycle (Section 2.2) are placed in the dark, exposed to irradiation or chemical treatments, and then given ^{3}H-thymidine and subjected to autoradiography (Box 2.2). The repair is found to be extremely efficient and effective: thousands of dimers can be excised from an *E. coli* chromosome without error. Furthermore, excision repair systems possess broad specificities: numerous types of DNA damage (in addition to pyrimidine dimers) can be corrected by this mechanism.

6.10   POST-REPLICATION MISREPAIR

Despite the efficient lines of defense provided by photoreactivation and excision repair, occasional lesions undoubtedly remain in chromosomes at the time of DNA replication (Figure 6-12a). When the replication apparatus encounters such a lesion it will presumably "stall" since the lesion does not resemble any known base. It is at this juncture that **post-replication misrepair** appears to occur. As diagrammed in Figure 6-12b, the replication apparatus is thought to skip past the lesion and recommence synthesis on the other side, some 10^3 residues along. Filling of the resultant daughter-strand gap is then initiated. Several models have been proposed to describe how this repair process occurs. Depicted in Figure 6-12c is the addition of several bases opposite the lesion by a DNA-synthesizing enzyme that does not require template DNA for its activity. An example of such an "untemplated" enzyme is **terminal deoxynucleotidyltransferase,** which simply adds nucleotides onto the ends of polynucleotide chains. The nature of the enzymes actually involved, however, is not known. The net effect, in any case, is that the daughter strand carries one or several nucleotides that have not been dictated by the parental strand (Figure 6-12d, jagged lines);

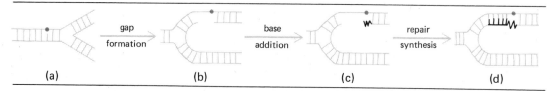

Figure 6-12 Model of a post-replication misrepair mechanism, where the gray circle depicts an unrepaired lesion and the jagged line an untemplated DNA synthesis. An alternate model for this repair process, not shown here, calls for an error-prone recombination event between one of the parental strands and the gap-containing strand.

relatively error-free repair synthesis then takes over to complete the rest of the daughter strand (Figure 6-12d). Among the mutations that result from this repair mechanism, transitions and transversions would both be expected, as would the addition or deletions of bases. All such mutations are in fact found to result from post-replication misrepair.

Ultraviolet light, ionizing radiations, and numerous alkylating agents all produce in DNA premutational lesions that are converted into mutations by error-prone post-replication repair. Such a link between misrepair and mutagenesis was, in fact, first postulated by E. Witkin. She discovered that the *E. coli* mutant *lexA* (Figure 6-11) was *less* prone than wild-type *E. coli* to undergo ultraviolet mutagenesis, but was *more* prone than wild-type *E. coli* to be killed by ultraviolet irradiations. From this she proposed that ultraviolet radiation triggers (induces) an error-prone repair pathway, that the *lexA* mutation prevents the induction of this pathway, and that *lexA* cells are subsequently more inactivated by, and less mutagenized by, ultraviolet radiation. The *rev3* and *rad6* mutations in yeast have effects similar to the *lexA* mutation in *E. coli*. Other mutations (for example, *uvrE* in *E. coli*) have since been found to enhance misrepair mutagenesis, and the *recA* gene product, described in Section 13.4, plays a major role in the process. A complex genetic control of post-replication misrepair will undoubtedly be revealed within the next few years.

6.11 MUTATOR AND ANTIMUTATOR MUTATIONS

The normal replication apparatus is remarkably accurate (Section 5.7): mistakes are only very rarely made by the replicase in copying the template strand, and any errors are normally detected and corrected by the "proofreading" activities that function in the wake of the replication fork. Should this mechanism falter, of course, then frequent mutations will result. Such is the consequence of lesions known as **mutator mutations.** In phage T4, for example, the phage-specified DNA replicase is encoded by gene 43. A number of mutations in gene 43 have the effect of increasing spontaneous mutation rates at many T4 loci, producing deletions, additions, transitions, and transversions. When the replicases specified by such gene-43 mutants are isolated, their ratio of exonuclease to polymerase activity is typically found to be much lower than the ratio in the wild-type enzyme. In other words, the ability of the mutant enzymes to excise base-mispairings in the T4 DNA is defective, a defect that can at once be correlated with their mutator activity *in vivo*. Similar mutator mutations occur in *dnaE*, the gene for *E. coli* DNA polymerase III, and numerous mutator genes have been described in yeast and in other eukaryotes.

Antimutator mutations that decrease spontaneous mutation rates have also been reported in gene 43 of T4, and *in vitro* testing of the resulting enzymes often reveals that the "antimutator polymerases" have a high exonuclease/polymerase ratio compared to their wild type counterparts, meaning that *in vivo*, these enzymes are unusually efficient in mispair recognition and excision.

The above examples should not be taken to mean that the exonuclease activities of DNA polymerases are the sole guardians of accurate DNA replication. Other activities also undoubtedly function to assure that the genome is transmitted intact, and these may also be altered by mutator or antimutator mutations.

6.12 SPONTANEOUS MUTATIONS AND REPAIR

In the preceding section the term **spontaneous mutation** was used, and this should be defined more carefully. If we imagine for the moment that the environment is mutagen-free (which it is not), then a spontaneous mutation can be defined as a mutation that arises without any deliberate application of mutagens (recall the data in Table 6-1). From the preceding sections we can surmise that spontaneous mutations will arise from two major sources: either the "vigilante enzymes" in the replication fork fail to function correctly and allow a mispaired base to remain unexcised, or else a "spontaneous" lesion is not excised or photoreactivated and induces post-replication misrepair.

The rate at which spontaneous mutations occur is found to be remarkably constant throughout the biological kingdom, being about 0.005 mutations **per genome** per DNA replication. This means, of course, that spontaneous mutation rates **per base pair** decrease markedly as one moves from the small-genome bacteriophages to the large-genome eukaryotes. In other words, the eukaryotes appear to have evolved more sophisticated or efficient means for avoiding spontaneous mutational events.

Faulty Repair and Aging An interesting theory of aging put forward by L. E. Orgel proposes that senescent cells are increasingly prone to mistakes in macromolecular synthesis and that the ensuing faulty cell metabolism leads to an **error catastrophe.** R. Holliday and associates are applying this theory to the enzymes of DNA metabolism in the cultured human fibroblast cell line called MRC-5. L. Hayflick had observed some years ago that after such cells have been transferred ("passaged") from growth medium to growth medium about 65 times, the cells invariably die. The Holliday group therefore isolated DNA polymerase enzymes from early- and late-passage cells and tested the ability of these enzymes to copy artificially synthesized DNA templates *in vitro*. They found that the enzymes from late-passage cells are significantly more error-prone than are early-passage enzymes. The Holliday group also observed that rates of replicon elongation (Section 2.3) are far slower in the senescent phase of cell life, again suggesting faulty polymerase activity. Finally, they found that spontaneous mutation rates to glucose-6-phosphate dehydrogenase deficiency are far higher in passage-60 cells than in passage-16 cells.

Together, these observations give support to the concept that "Orgel's error catastrophe" may have as one of its origins an increasingly faulty system of DNA synthesizing and repair enzymes (perhaps the enzymes have incurred mutator mutations) which produce increasing levels of "spontaneous" **somatic mutations** (somatic mutations being transmitted to the immediate mitotic progeny of an affected cell but not to the germ cells of an organism and hence not to the F_1 offspring). Somatic mutations are presently also believed to be a major agent in carcinogenesis, as we describe at the end of this chapter. Furthermore, since cancer is primarily (although not exclusively) a disease of the elderly, somatic mutation, aging, and carcinogenesis are thought by some (although not by all) geneticists to be intimately related phenomena.

6.13 DELETIONS AND ADDITIONS

A **deletion** is represented by the sequence ABDEFGH and an **addition** by the sequence ABXCDEFGH where X represents any base or group of bases (Figure 6-1). Deletions and additions occur spontaneously and are among the mutations elicited by γ-irradiation and MMS, agents that act *via* post-replication misrepair. One would indeed expect error-prone enzyme systems to add too few or too many bases to a daughter strand as they fill in untemplated gaps.

An interesting model for deletion/addition formation has been proposed by G. Streisinger and colleagues and is shown and explained in Figure 6-13. Essentially, the model proposes that following strand breakage, either the broken strand or the intact strand may "buckle" and form base-pairs with inappropriate sectors of its complement; DNA repair then creates strands that are either too long (Figure 6-13a) or too short (Figure 6-13b).

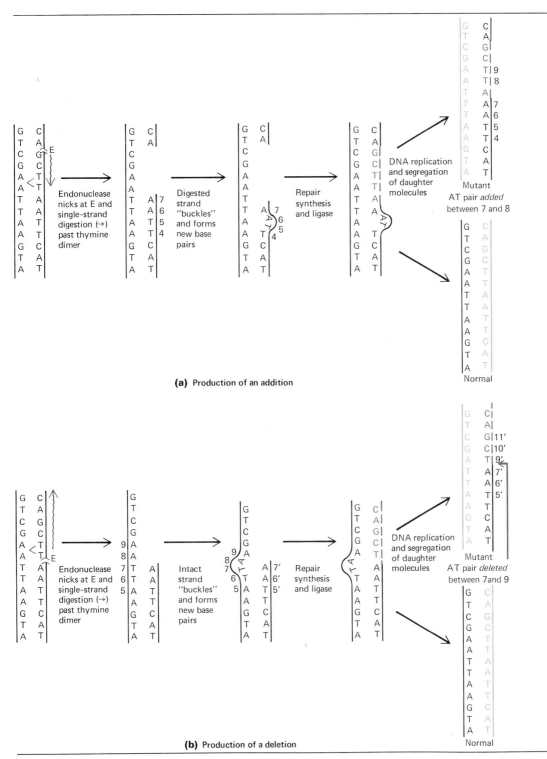

Figure 6-13 Proposed sequence of events leading to the formation of an addition (a) and a deletion (b). Thymine dimers are shown as T̂T̂. (After G. Streisinger et al., *Cold Spring Harbor Symp. Quant. Biol.* **31**:77, 1966.)

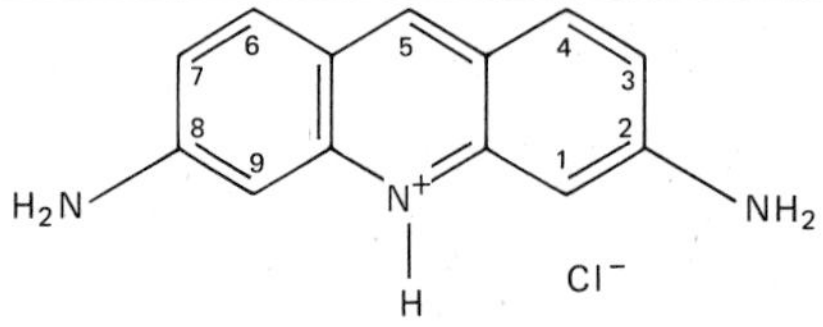

Figure 6-14 The structure of proflavin, an acridine mutagen.

Proflavin (chloride): (2, 8–diamino acridine hydrochloride)

Additions and deletions are also efficiently induced by a class of mutagens known as **acridines,** the most widely used of which is **proflavin** (Figure 6-14). Acridines are flat aromatic molecules that interact with DNA in such a way that they become wedged, or **intercalated,** between the stacked bases of a double helix. Streisinger has proposed that acridines may stabilize the postulated buckled-out regions of DNA that predicate additions and deletions (Figure 6-13), but their precise mutagenic mode is unknown.

Additions and deletions often produce an effect on the "reading" of the genetic message known as a **frameshift.** Frameshift mutations are described in detail in Section 9.9 where we consider experiments by Crick and his associates that provided important insights into the nature of the genetic code.

CHROMOSOMAL MUTATIONS

6.14 DEFICIENCIES AND DUPLICATIONS

A large deletion in a eukaryotic chromosome, in contrast to a small deletion produced by an acridine, is usually referred to as a **deficiency** (Figure 6-1). A large addition is known as a **duplication.** The occurrence of either aberration can sometimes be verified by examining chromosomes with the light microscope: the chromosome appears either longer or shorter than normal. Additions and deletions can also be detected in polytene chromosomes of the Diptera (Section 2.9), in which one or several bands may be missing or duplicated. Finally, it is possible to detect even very small deficiencies and duplications in G-banded or Q-banded chromosomes (Section 2.7). Such analyses have, for example, shown an association between the presence of a deficiency in the short arm of human chromosome 5 and the **cri du chat** syndrome, which includes severe mental retardation and a characteristic catlike cry in very young infants.

Malignant cells are sometimes found to possess specific deficiencies or duplications. The white blood cells of patients with chronic myelogenous leukemia frequently exhibit a karyotype in which a portion of the long arm of chromosome 22 is deleted (producing what is known as the **Philadelphia**

chromosome) and the missing piece is translocated to the end of the long arm of chromosome 9. Patients with a variety of lymphomas often display a duplication of bands in the long arm of chromosome 14. Whether such changes are a cause or a result of the malignancy is unknown.

The presence of a deficiency or a duplication is particularly obvious in a Dipteran organism heterozygous for the aberration. As noted in Section 2.9, somatic pairing occurs between homologous polytene chromosomes in the Diptera. Since pairing cannot take place along a deleted region in the heterozygote, the undeleted homologue simply buckles, as diagrammed in Figure 6-15 and as illustrated in Figure 6-16. A Dipteran heterozygous for a duplication will also exhibit a buckled region; in this case, however, the buckled region will contain bands that are duplicates of adjacent regions of the chromosome, the adjacent regions being properly paired with the homologue.

Duplications and deficiencies produce a change in the amount of genetic material present in a chromosome. Higher organisms homozygous for a deficiency are usually inviable (exceptions include certain strains of corn), whereas organisms heterozygous for a deficiency are frequently viable. These results indicate that a single dose of certain genes may be sufficient for development and self-maintenance, whereas the total loss of those genes cannot be supported. Duplications, on the other hand, are not generally as harmful as deficiencies and may, under evolutionary pressures, even become advantageous. An organism carrying a duplication has, in effect, an extra portion of genetic material that is not essential to its development and reproduction. Mutations in this extra genetic material are thus much less likely to be deleterious to the organism than are mutations in its essential genes. Moreover, the duplicated region can undergo repeated mutation without being subjected to the usual negative selective pressures, and in this sense it enjoys a kind of independent existence within the

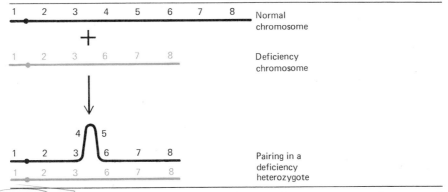

Figure 6-15 A chromosome carrying a deficiency, showing pairing with its normal homologue.

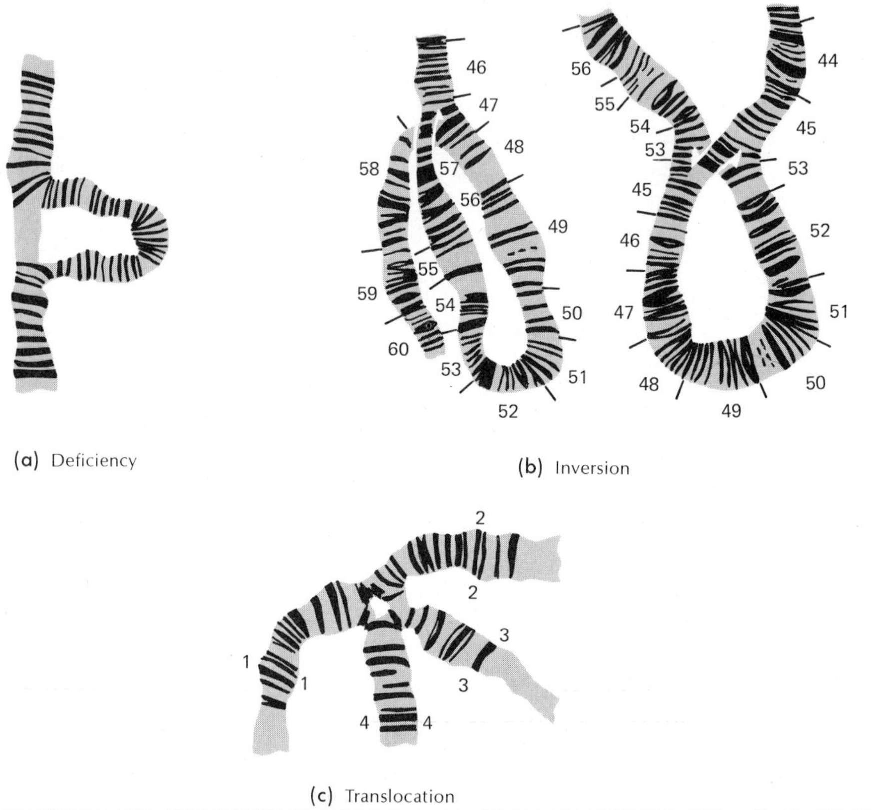

(a) Deficiency

(b) Inversion

(c) Translocation

Figure 6-16 Appearance of chromosomal mutations in polytene chromosomes of *Drosophila*. Numbers designate particular bands on particular chromosomes. (After T. S. Painter, *J. Hered.* **25**:464–476, 1934.)

genome. Should accumulated mutations produce a gene whose protein product is in some way beneficial to an organism, the organism will be provided with a selective advantage and the new gene—for such it can be called—may become increasingly prevalent in a population. Specific examples of gene duplications are given in Sections 15.8–15.17.

6.15   INVERSIONS

An **inversion** changes the *arrangement* of the genetic material in a chromosome rather than the amount. It is produced when a portion of the chromosome is broken, the broken piece assumes a reversed position, and the break is repaired (Figure 6-1). Should the inversion involve an internal segment of a chromosome, as it usually and perhaps always does, two breaks must occur, one at either end of the inverted segment. An inversion will, of course, cause a reversal in the band sequence of a polytene chromosome or a banded eukaryotic chromosome.

In an inversion heterozygote, synapsis between the inverted portion of a chromosome and its homologue can occur only if one chromosome loops back to invert itself in the affected region (see Figure 6-17). Thus in pachytene preparations of inversion heterozygotes, the length and position of an inversion can frequently be estimated by examining the size of the loops that are present. Similar loops also form during somatic pairing of dipteran polytene chromosomes, in which the inverted region can be identified quite precisely by its banding pattern (Figure 6-16b). It is not uncommon for a second inversion to occur within a single, or simple, inversion, in which case a secondary loop will form at synapsis. In certain inversion-carrying strains, therefore, extremely complicated synaptic patterns can be found.

An inversion may be **pericentric** (containing a centromere) or **paracentric** (not containing a centromere). If pericentric, it may produce a chromosome quite different from the original in appearance. A metacentric chromosome, for example, can be transformed into an acrocentric chromosome by a pericentric inversion that includes unequal lengths of the chromosome to the right and left of the centromere. This is illustrated in Figure 6-18. Much evolution of the karyotype has presumably occurred by such chromosomal alterations.

We noted when we considered the *Basc* X chromosome (Section 6.4) that inversions tend to suppress meiotic recombination. In addition to the mechanical difficulties in the synapsis process itself, crossovers within inversion synapses usually fail to generate recombinant chromosomes, an effect best understood by studying Figure 6-19. Thus the genes contained within an inversion tend to be inherited together as a package, rather than

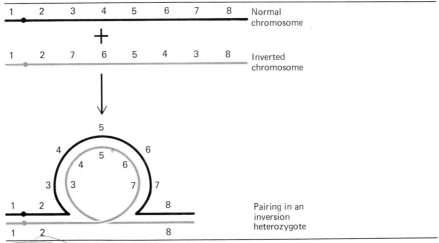

Figure 6-17 A chromosomal inversion, showing pairing with its normal homologue.

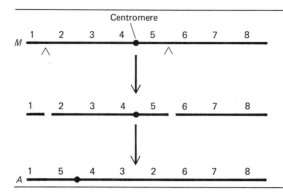

Figure 6-18 The transformation of a metacentric chromosome (M) into an acrocentric chromosome (A) by a pericentric inversion.

being separated from one another by crossing over. This protective effect of inversions appears to be utilized by *Drosophila*, at least, in adapting to different kinds of environmental situations (Chapter 20).

6.16 TRANSLOCATIONS

A second process by which the arrangement of genetic material can be altered is **translocation.** When a translocation occurs, a portion of one chromosome becomes physically associated with (usually) a nonhomologous chromosome (Figure 6-1). Translocations are commonly (but not invariably) **reciprocal,** meaning that each chromosome both donates and receives a piece of chromosomal material (Figures 6-1 and 6-20).

Following a meiotic translocation, the next generation includes individuals heterozygous for the translocation. Meiotic cells from such an individual will exhibit cross-shaped chromosomal configurations at the time of synapsis as the translocated regions attempt to pair. Figure 6-16c illustrates the conformation assumed by reciprocal translocation chromosomes in polytene cells, and a detailed view of how such configurations are established is given in Figure 6-20.

6.17 ANEUPLOIDY

As noted at the start of this chapter, the chromosome complement of an aneuploid cell is increased or decreased by one or more chromosomes (Fig 6-1). Thus if the normal diploid chromosome number of an organism is $2n$, an aneuploid organism might be $2n - 1$ **(monosomic),** $2n + 1$ **(trisomic),** $2n + 2$ **(tetrasomic,** or **double trisomic),** and so on.

Aneuploidy is a common mutation. In humans, for example, probably almost a third of all spontaneous abortions involve a fetus with an abnormal number of chromosomes, and among these more than 60 percent are aneuploids, most being either trisomics or monosomics. Some relevant data are summarized in Table 6-2.

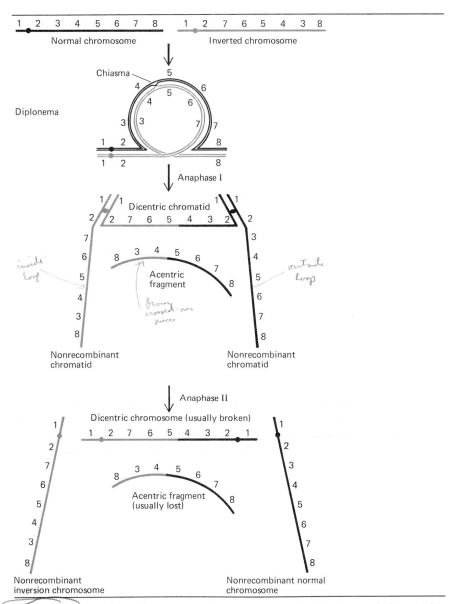

Figure 6-19 The effects of crossing over between a normal and an inverted chromatid in an inverted region that does not contain a centromere (a paracentric inversion).

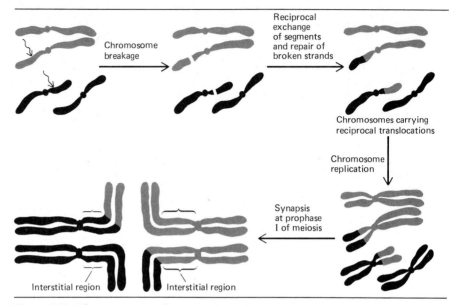

Figure 6-20 The occurrence of reciprocal translocations between nonhomologues and its effect on synapsis during meiosis.

Aneuploidy is usually produced by **primary nondisjunction** at meiosis: two holomogues fail to disjoin at the first meiotic division and two of the resulting gametes carry a double dose of the chromosome; the other two gametes lack the chromosome entirely. When these abnormal gametes are fertilized by normal gametes, trisomic and monosomic zygotes result. In

Table 6-2   Frequency of Selected Chromosomal Aberrations in Humans

	Spontaneous Abortions	Live-Born
Monosomy X (XO, "Turner's syndrome")	1/18	1/3,000 females
Trisomy chromosome 16	1/33	Almost zero
Trisomy chromosome 8	1/33	1/14,500
Trisomy chromosome 21	1/40	1/600
Trisomy chromosome 18	1/200	1/4,500
Trisomy, sex chromosomes XXY, "Klinefelter syndrome"	Not found	1/600 males
XYY	Not found	1/1000 males
Triploidy	1/22	Almost zero
Trisomy, X chromosome (XXX)	Not found	1/1,600 females

From V. A. McKusick, *Human Genetics.* Englewood Cliffs, N.J.: Prentice-Hall, © 1969, Table 2.3, and P. A. Gerald, *New England J. Med.* **294:**706 (1976).

addition, one or more chromosomes may be eliminated from any type of cell if some mechanical difficulty prevents one or more chromosomes from migrating to a pole at anaphase. Clearly, this may occur during mitosis as well as meiosis, and it is not an uncommon occurrence in cells maintained in tissue culture or in diseased cells.

Monosomic organisms are usually inviable for the same reason that large deficiencies are usually lethal, even in heterozygotes: a great many genes in diploid organisms must apparently be present in two doses if proper development is to occur. Trisomy, like duplication, may be less severely deleterious. In the plant *Datura stramonium* 12 trisomic types have been identified, one for each of the 12 chromosome types, and each plant type has a different morphology (Figure 6-21). In most animals, however, trisomy can be tolerated only for certain chromosomes, the other combinations being

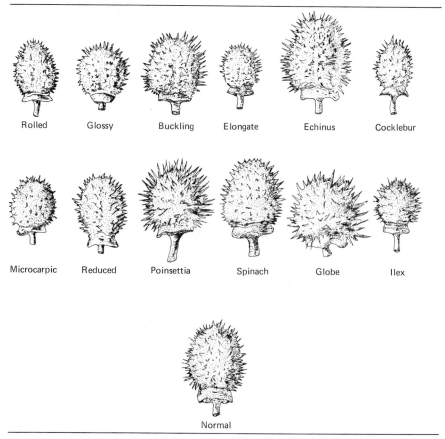

Rolled   Glossy   Buckling   Elongate   Echinus   Cocklebur

Microcarpic   Reduced   Poinsettia   Spinach   Globe   Ilex

Normal

Figure 6-21 The trisomic strains of the plant *Datura* (Jimson weed). (From E. Sinnott, L. Dunn, and T. Dobzhansky, *Principles of Genetics*. New York: McGraw-Hill, 1958, after A. F. Blakeslee. Used with permission of McGraw-Hill Book Company.)

lethal during the course of embryogenesis. Why this should be so is not known, but the differentiation of a diploid animal zygote is such a delicately balanced process that the presence of an extra chromosome's worth of genetic information might create developmental problems at critical embryonic stages. Supporting this line of reasoning is the observation that live-born human trisomic individuals carry extra copies of chromosomes 8, 13, 18, 21, Y, or X (Table 6-2). The first five of these chromosomes contain large blocks of constitutive heterochromatin, which is presumed to be genetically inert (Section 2.13); thus their presence in an extra dose might do less to upset the postulated balance than an extra dose of fully euchromatic chromosomes. Extra X chromosomes in aneuploid cells also become heterochromatic, as described in Section 18.7.

Trisomy for any of the above-listed chromosomes except the Y usually produces an abnormal individual, the nature and severity of the syndrome being dependent upon the chromosome involved. The best-known example in humans is trisomy for chromosome 21, which produces **Down's syndrome** (also known by the misnomer Mongolian idiocy). This condition occurs in about one in every 600 newborns, and it is clear from Figure 6-22 that its frequency increases significantly when mothers (but not fathers) are more than 35 years of age. This suggests that nondisjunction occurs more often in older egg cells, and when it is recalled from Section 3.6 that the human egg cell remains in diplonema from the fetal life of a woman onward, it is perhaps not unexpected that mechanical difficulties become more prevalent by the time the cell is, say, 45 years old. In this context it is relevant to note that male mammals are increasingly more apt to produce mutation-bearing sperm with increasing paternal age. Since these are more often point than chromosomal mutations, one theory holds that misreplication in dividing stem spermatogonia (Section 3.12) becomes more prevalent with age.

Trisomy for the Y chromosome produces XYY males, and these individuals are characteristically taller than average but have normal fertility and

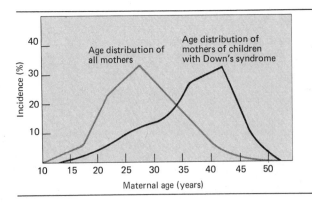

Figure 6-22 Effect of increased maternal age on the incidence of Down's syndrome. (From V. A. McKusick, *Human Genetics*, 2nd ed., Englewood Cliffs, N.J.: Prentice-Hall, © 1969.)

produce either X or Y sperm. Initial speculation concerning the existence of an "XYY syndrome" of antisocial tendencies in such men has recently been countered by a large-scale study of Danish men in whom no such syndrome is apparent. Aneuploidy for X chromosomes has been described in Section 4.11.

Partial trisomy refers to an individual who is a **translocation heterozygote:** one chromosome in the karyotype contains a translocated portion of a heterologous chromosome, so that the individual carries 3 copies of a certain sector of the genome. Partially trisomic individuals can usually be identified only by examining their banded chromosomes, but such individuals prove to be quite common: partial trisomies have been described for nearly all the human chromosomes, and about 10% of Down's syndrome individuals carry a portion of an extra chromosome 21 translocated to some other chromosome.

6.18   EUPLOIDY

Most species of eukaryotes have either predominantly haploid or predominantly diploid phases to their life cycles. Many haploid organisms go through a brief diploid stage as zygotes, whereas diploid organisms usually produce short-lived haploid gametes (Sections 3.12–3.14): these normal haploid-diploid fluctuations are not considered euploidy. If, however, an egg is induced to undergo embryogenesis without first being fertilized by a sperm, a **haploid** individual who is considered to be euploid will result.

Haploid embryos rarely develop normally in animals, just as monosomic organisms rarely survive. In plants, however, haploids frequently reach maturity, although they are usually weak and delicate. They are also usually sterile, a fact that can readily be appreciated in trying to visualize a haploid meiosis in which the pairing of homologues cannot occur and segregation of chromosomes to daughter cells becomes a random process.

Polyploid individuals ($2n$ if the original chromosome number was n and $3n$, $4n$, and so on, if the original number was $2n$) are often viable and may be larger than their haploid or diploid counterparts. Polyploidy is common among natural populations of plants, particularly the grasses, and plant breeders frequently create polyploid lines because of their greater vigor. Even-numbered polyploids, and tetraploid plants in particular, do not usually experience difficulties during meiosis, since each chromosome and its homologue are represented equally. Plants with an uneven chromosome number (triploid, pentaploid, and so on), on the other hand, are usually sterile in that one chromosome set is without homologous partners and (as noted for haploid meioses) aneuploid gametes are produced. Sterility is sometimes desirable for the fruit breeder, since inedible seeds do not form and the plant can still be propagated vegetatively by cuttings. Therefore such commercial plants as the banana are triploids.

Polyploid higher animals are rarely fertile, even when tetraploid, an exception being polyploid silkworms. The difference between plants and animals is believed to rest, at least in part, with the sex chromosomes. Most genera of higher plants do not appear to possess differentiated sex chromosomes and many, although not all, are bisexual (or, in botanical terms, **monoecious**). In higher animals, on the other hand, sex and fertility depend on a critical balance of sex-determining chromosomes, as discussed in Chapter 18. Polyploid individuals that have lost this balance can therefore be expected to be sterile.

While triploids are well known in plants, in *Drosophila,* and even in amphibians, human triploids cannot survive: nearly 1 percent of all human conceptions are triploid, but most die within the first 3 months after fertilization and the few live births die within the neonatal period. As with most aneuploidy, it appears that the very stringent developmental schedule of a human pregnancy is incompatible with such an increase in gene dosage.

Polyploids can arise in a number of ways: a meiotic aberration may result in a failure of reduction division during gametogenesis; an egg may be fertilized by more than one sperm; or chromosomal separation at anaphase may be blocked by **colchicine** or **colcemid,** agents that disrupt microtubules and hence meiotic/mitotic spindle formation. Polyploids also frequently arise if gametes of two different species are induced to fertilize one another. The resultant hybrid will ordinarily be sterile, for the usual difficulties will be encountered when homologous pairing is unsuccessfully attempted at meiosis. If, however, tetraploidy can be achieved by any of the germ-line or gametophytic cells, each chromosome will have a homologue and viable 2*n* gametes can form. In monoecious higher plants, in which self-pollination is possible, these 2*n* gametes can unite to form 4*n* zygotes. Thus in one generation an artificially produced hybrid can give rise to a fertile, self-perpetuating line that may be considered a new species. The formation of new species by this process, called **allopolyploidy** because the two chromosome sets are of different origin, has occurred in natural populations, and the karyotype of the two parental species can occasionally be recognized in the karyotype of a polyploid hybrid species.

6.19 INDUCTION OF CHROMOSOMAL MUTATIONS

Chromosomal aberrations occur spontaneously. They can also be induced by any agent that affects the physical integrity of chromosomes, the mechanics of chromosome movements at meiosis, or both. Thus any agent that is toxic to cells is, at least potentially, mutagenic, and it is perhaps not surprising to find that many environmental pollutants, pesticides, food additives, hallucinogens, narcotics, and the like have been implicated as inducing chromosomal aberrations.

The experimental induction of chromosomal mutations is frequently accomplished with X rays. It was once thought that X rays broke chromosomes by a direct fission of the phosphodiester bonds in nucleic acids, but it now appears that this kind of direct damage is rare. Instead, X rays appear to induce the intracellular formation of free radicals (molecules containing an atom with an unpaired electron), especially when oxygen is present. These highly reactive molecules can apparently wreak havoc with chromosome structure by causing single-stranded or double-stranded breaks. The broken ends may reanneal in a grossly incorrect manner during repair to produce deficiencies or duplications in much the same way that proflavin-mediated deletions and additions are thought to be produced (Figure 6-13). Alternatively, the broken pieces of chromosomes may unite to produce translocations.

The graph in Figure 6-23 demonstrates an important property of X rays: their mutagenic effects are directly proportional to their dosage (measured in **roentgen units, r**), at least in the low-dosage range. X-ray dosage is cumulative, meaning that exposure to several low doses over a long period is fully as mutagenic as a single exposure at a high dose. It is for this reason that all unnecessary exposure to X rays is to be avoided.

The free radicals produced by X-irradiation are also capable of bringing about chemical changes in individual bases, especially the pyrimidines, and these altered bases trigger misrepair (Section 6.10). In other words, X rays

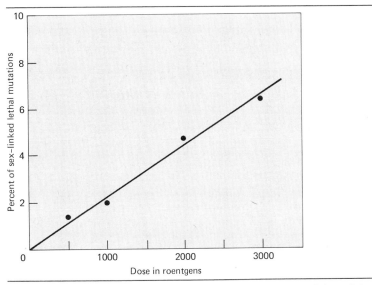

Figure 6-23 Increase in sex-linked lethal mutations in *Drosophila* with increasing X-ray dosage. Roentgen units are based on the number of ionizations produced in one cubic centimeter of air under standard conditions.

can produce revertible point mutations as well as chromosomal mutations; indeed, the first X-ray induced mutations studied by H. J. Muller (1927) were point mutations. In this context it should be mentioned that the chemical mutagens considered earlier in this chapter are all capable of producing chromosome breaks and other aberrations, particularly if they are administered in high doses, and, in general, no mutagen is known to produce one type of alteration of the genotype exclusively. Once mutant organisms have been experimentally induced, the nature of the mutations they carry can be determined only by careful genetic tests.

Sister-Chromatid Exchanges   Deficiencies, duplications, inversions, and translocations must all ultimately entail chromosome breakage. A sensitive test for the ability of various mutagens to induce chromosome breakage has recently been developed, the test involving the cytological detection of exchanges between mitotic sister chromatids.

In the sister-chromatid-exchange (**SCE**) assay, cultured cells are allowed to undergo two rounds of mitosis in the presence of 5BU (Figure 6-6). During the second cell cycle, a mutagen is presented to experimental but not to control cells. At the time of the second mitosis, all cells are treated with colcemid so that all reach, but cannot pass, the metaphase stage. The cells are then treated with the Giemsa stain or with a fluorochrome known as **Hoechst 33258** and examined with the light or fluorescence microscope.

In the control cells it is observed that the two sister chromatids are stained differently (Figure 6-24a): as a consequence of semiconservative replication (Section 5.4), one sister chromatid possesses one 5BU-containing strand while the other chromatid possesses two such strands, and the staining reagents interact more strongly with the doubly-halogenated than the singly-halogenated chromatid. Many of the control cells contain, in addition, sister-chromatid pairs that have a checkerboard pattern (Figure 6-24, arrows). While the molecular basis for such exchanges is poorly understood, they unquestionably require reciprocal breaks in the two chromatids.

When mutagenized cells are now examined, the SCE frequency is frequently enhanced (Figure 6-24b). Figure 6-25 summarizes data on the induction of SCEs by various mutagens; it is evident that **mitomycin C** and MNNG are particularly effective. While it is presently unclear whether a mutagen effective in generating SCEs will necessarily be effective in causing chromosomal abberations or the reverse, the ability to generate SCEs usually correlates well with the overall mutagenicity of a given reagent.

6.20   CARCINOGENS

Carcinogens are agents that cause normal cells to become malignant such that they divide uncontrollably and possess the ability to spread and establish cell lines in inappropriate parts of the body (**metastasis**). Well-

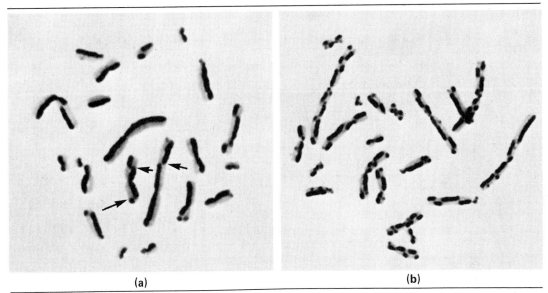

(a) **(b)**

Figure 6-24 (a) Metaphase cell from the Chinese hamster ovary (CHO) cell line after 2 rounds of replication in the presence of 5BU followed by staining with Hoechst 33258 plus Giemsa. Only one chromatid of each sister-chromatid pair is stained. Arrows point to several sister-chromatid exchanges (SCEs); a total of 12 SCEs are present. (b) As in Figure 6-24a, but exposed to nitrogen mustard (HN$_2$) at 3×10^{-6} M for the two cell cycles before sampling. The frequency of SCEs has increased approximately tenfold relative to Figure 6-24a. (From P. Perry and H. J. Evans, *Nature* **258:**121, 1975.)

established carcinogens for humans include vinyl chloride and asbestos; some component(s) of cigarette smoke are also clearly carcinogenic. Present estimates hold that 80–90 percent of all human cancers are generated by carcinogens present in the human environment; therefore, identifying the carcinogenic agents and understanding their mode of action is a key goal of cancer research.

The relevance of carcinogenesis to genetics lies in the fact that while potent mutagens are not necessarily strong carcinogens (MMS being one example), carcinogens are usually mutagens (see below). This fact allows at least two interpretations, and since "cancer" is actually many hundreds of different types of related diseases, each interpretation may prove to be correct for different malignancies. One theory holds that carcinogens are such highly reactive molecules that even though their cancer-generating activity is unrelated to mutagenesis, mutations are likely to be generated

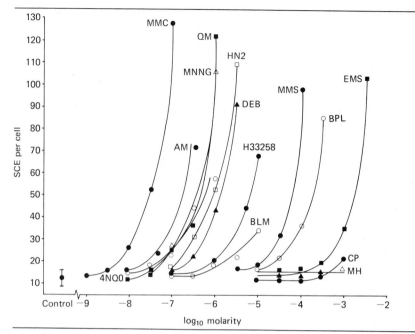

Figure 6-25 Dose-response curves for SCE incidence in cultured Chinese Hamster Ovary (CHO) cells against log of initial concentration. MMC, mitomycin C; AM, adriamycin; MNNG, N-methyl-N-nitro-N-nitrosoguanidine; QM, quinacrine mustard; 4NQO, 4-nitroquinoline 1-oxide; HN2, nitrogen mustard; DEB, diepoxy-butane; H33258, Hoechst 33258; BLM, bleomycin; MMS, methylmethane sulfonate; BPL, β-propio-lactone; EMS, ethyl methane sulfonate; CP, cyclophosphamide; MH, maleic hydrazide. (From P. Perry and H. J. Evans, *Nature* **258**:121–125, 1976.)

once they enter a cell. The alternate **somatic-mutation theory of cancer** holds that mutations are themselves able to precipitate malignancy.

One line of support for the somatic theory of cancer comes from the phenotype of persons suffering from the heterogeneous group of diseases collectively known as xeroderma pigmentosum (XP). As noted in Sections 6.9 and 6.10, XP persons are unable to repair ultraviolet-induced damage to their DNA and are presumably therefore highly susceptible to spontaneous mutation by misrepair. It turns out that XP persons are also highly suscepti-ble to numerous forms of skin cancer (squamous cell carcinomas, malignant melanomas, and many others), suggesting that the accumulation of such mutations eventually transforms a skin cell into a malignant cell.

Regardless of theory, the fact that carcinogens are usually mutagens (and vice-versa) can be exploited to screen environmental chemicals for their mutagenicity, with those found to be mutagenic given high priority for direct carcinogenicity tests. Of the many screening tests that have been devised, that of B. Ames and his associates is particularly well developed. Strains of *Salmonella typhimurium* are constructed to carry a cell wall mutation that permits most chemicals to enter the cell, plus a *uvr* mutation that abolishes most types of excision repair. The test organisms also carry the plasmid pKM101 which, for as yet unknown reasons, exerts mutator activity in *Salmonella* so that DNA damage is converted into mutations with high frequency. Finally, the bacterial cells are auxotrophic for histidine, carrying one of several possible *his* mutations that can be reverted either by base-pair substitutions or by addition-deletion (frameshift) mutations. Since many environmental chemicals are not carcinogenic to mammals until they have been taken up by the body and metabolized by a class of liver enzymes known as **mixed function oxidases,** a "microsomal" cell fraction from rat liver is also included. The number of revertant *his*$^+$ colonies on the test plates is then ascertained (see Section 6.3).

The **Ames test,** as it has come to be called, has been applied to more than 300 chemicals and the results are summarized in Table 6-3. Of 179 compounds known to be carcinogenic in animals, 157 (87.7 percent) are able to revert the *Salmonella* strains, and 101 of 117 (86 percent) of the compounds believed to be noncarcinogenic are unable to revert the bacterial strains. Additional tests performed since Table 6-3 was compiled have raised the "detection frequency" to about 95%. Such high predictive values strongly commend the test as a "prescreen" for the many thousands of chemicals and drugs presently in the human environment and for the many compounds proposed to be introduced each year. An example of the use of the test can be given. In 1975, undergraduates at the University of California at Berkeley subjected several hundred commercial products to the Ames test as

Table 6-3 Correlation of animal carcinogenicity and bacterial mutagenicity

Group of Compounds	Carcinogens Detected as Bacterial Mutagens	Non-Carcinogens Not Mutagenic to Bacteria	Compounds of Uncertain Carcinogenicity Detected as Mutagens
A Aromatic amines etc.	23/25	10/12	5/7
B Alkyl halides, etc.	17/20	1/3	1/1
C Polycyclic aromatics	26/27	7/9	1/1
D Esters, epoxides, carbamates, etc.	13/18	5/9	0/1
E Nitro aromatics and heterocycles	28/28	1/4	0/2
F Miscellaneous organics	1/6	13/13	0/1
G Nitrosamines	20/21	2/2	1/1
H Fungal toxins and antibiotics	8/9	5/5	—
I Mixtures (cigarette smoke condensate)	1/1	—	—
J Miscellaneous heterocycles	1/4	7/7	—
K Miscellaneous nitrogen compounds	7/9	2/4	—
L Azo dyes and diazo compounds	11/11	2/3	3/3
M Common laboratory biochemicals	—	46/46	—
Total	157/178	101/117	11/17

From McCann, J., E. Choi, E. Yamasaki, and B. N. Ames, *Proc. Nat. Acad. Sci.* **72:**5135, 1975.

a laboratory exercise and discovered that most hair dyes were highly mutagenic. The dyes are now being tested for their ability to produce tumors in mice. Since each carcinogenicity test takes many months, sacrifices many hundreds of mice, and requires expensive animal facilities and personnel, it would have been prohibitive to subject the many commercial products to such animal tests, but in light of the prescreen results, the hair-dye tests become imperative. An increasing number of industries are utilizing Ames tester strains to assay the mutagenicity of the compounds they produce.

References

Screening Procedures

Childs, B. "Genetic Screening," *Ann Rev Genet.* **9**:67–89 (1975).

*Edgar, R. S., and I. Lielausis. "Temperature-sensitive mutants of bacteriophage T4D: their isolation and genetic characterization," *Genetics* **49**:649–662 (1964).

*Lederberg, J., and E. M. Lederberg. "Replica plating and indirect selection of bacterial mutants," *J. Bacteriol.* **63**:399–406 (1952). [Reprinted in E. A. Adelberg, *Papers on Bacterial Genetics,* 2nd ed. Boston: Little, Brown, 1966.]

*Lederberg, J., and N. Zinder. "Concentration of biochemical mutants of bacteria with penicillin," *J. Amer. Chem. Soc.* **70**:4267 (1948).

*Luria, S. E., and M. Delbrück. "Mutations of bacteria from virus sensitivity to virus resistance," *Genetics* **28**:491–511 (1943). [Reprinted in E. A. Adelberg, *Papers on Bacterial Genetics,* 2nd ed. Boston: Little, Brown, 1966.]

Suzuki, D. T. "Temperature-sensitive mutations in *Drosophila melanogaster,*" *Science* **170**:695–706 (1970).

Point Mutations

*Bingham, P. M., R. H. Baltz, L. S. Ripley, and J. W. Drake. "Heat mutagenesis in bacteriophage T4: the transversion pathway," *Proc. Nat. Acad. Sci.* **73**:4159–4163 (1976).

Budowsky, E. I. "The mechanism of the mutagenic action of hydroxylamines," *Proj. Nucl. Acid Res.* **16**:125–188 (1976).

Conkling, M. A., J. A. Grunau, and J. W. Drake. "Gamma-ray mutagenesis in bacteriophage T4," *Genetics* **82**:565–575 (1976).

Drake, J. W., and R. H. Baltz. "The biochemistry of mutagenesis," *Ann. Rev. Biochem.* **45**:11–38 (1976).

Kohn, H. I. "X-ray induced mutations, DNA and target theory," *Nature* **263**:766–767 (1976).

*Muller, H. J. "Artificial transmutation of the gene," *Science* **66**:84–87 (1927). [Reprinted in J. A. Peters, Ed. *Classical Papers in Genetics.* Englewood Cliffs, N.J.: Prentice-Hall, 1959.]

Russell, L. B., W. L. Russell, R. A. Popp, C. Vaughan, and K. B. Jacobson. "Radiation-induced mutations at mouse hemoglobin loci," *Proc. Nat. Acad. Sci.* **73**:2843–2846 (1976).

Searle, A. G. "Mutation induction in mice," *Adv. Rad. Biol.* **4**:131–207 (1974).

Singer, B. "The chemical effects of nucleic acid alkylation and their relation to mutagenesis and carcinogenesis," *Prog. Nucl. Acid Res.* **15**:219–284 (1975).

*Stadler, L. J. "Mutations in barley induced by X-rays and radium," *Science* **68**:186–187 (1928).

*Denotes articles described specifically in the chapter.

*Streisinger, G., Y. Okada, J. Emrich, J. Newton, A. Tsugita, E. Terzaghi, and M. Inouye. "Frameshift mutations and the genetic code," *Cold Spring Harbor Symp. Quant. Biol.* **31**:77–84 (1966).

Waring, M. "Variation of the supercoils in closed circular DNA by binding of antibiotics and drugs: evidence for molecular models involving intercalation," *J. Mol. Biol.* **54**:247–279 (1970).

Repair and Misrepair

*Bessman, M. J., N. Muzyczka, M. F. Goodman, and R. L. Schnaar. "Studies on the biochemical basis of spontaneous mutation. II. The incorporation of a base and its analogue into DNA by wild-type, mutator and antimutator DNA polymerases," *J. Mol. Biol.* **88**:409–421 (1974).

Boram, W. R., and H. Roman. "Recombination in *Saccharomyces cerevisiae*: A DNA repair mutation associated with elevated mitotic gene conversion," *Proc. Nat. Acad. Sci. U.S.* **73**:2828–2832 (1976).

*Braun, A., and L. Grossman. "An endonuclease from *Escherichia coli* that acts preferentially on uv-irradiated DNA and is absent from the *uvrA* and *uvrB* mutants," *Proc. Nat. Acad. Sci. U.S.* **71**:1838–1842 (1974).

*Bridges, B. A., R. P. Mottershead, and S. G. Sedgwick. "Mutagenic DNA repair in *Escherichia coli*. III. Requirement for a function of DNA polymerase III in ultraviolet-light mutagenesis," *Molec. Gen. Genet.* **144**:53–58 (1976).

Cleaver, J. E., and D. Bootsma. "Xeroderma pigmentosum: Biochemical and genetic characteristics," *Ann. Rev. Genetics* **9**:19–35 (1975).

d'Ambrosio, S. M., and R. B. Setlow. "Enhancement of postreplication repair in Chinese hamster cells," *Proc. Nat. Acad. Sci. U.S.* **73**:2396–2400 (1976).

*Fornace, A. J., K. W. Kohn, and H. E. Kann. "DNA single-strand breaks during repair of uv damage in human fibroblasts and abnormalities of repair in xeroderma pigmentosum," *Proc. Nat. Acad. Sci. U.S.* **73**:39–43 (1976).

Grossman, L., A. Braun, R. Feldberg, and I. Mahler. "Enzymatic repair of DNA," *Ann. Rev. Biochem.* **44**:19–43 (1975).

Hanawalt, P. C., and R. B. Setlow, Eds. *Molecular Mechanisms for the Repair of DNA.* New York: Plenum, 1975.

*Hastings, P. J., S-K. Quah, and R. C. von Borstel. "Spontaneous mutation by mutagenic repair of spontaneous lesions in DNA," *Nature* **264**:719–722 (1976).

*Heddle, J., and K. Athanasiou. "Mutation rate, genome size and their relation to the *rec* concept," *Nature* **258**:359–361 (1975).

Ho, K. S. Y., and R. K. Mortimer. "Two mutations which confer temperature-sensitive radiation sensitivity in the yeast *Saccharomyces cerevisiae*," *Mut. Res.* **33**:157–164 (1975).

Lawrence, C. W., and R. Christensen. "UV mutagenesis in radiation-sensitive strains of yeast," *Genetics* **82**:207–232 (1976).

*Linn, S., M. Kairis, and R. Holliday. "Decreased fidelity of DNA polymerase activity isolated from aging human fibroblasts," *Proc. Nat. Acad. Sci. U.S.* **73**:2818–2822 (1976).

*Maher, V. M., L. M. Ouellette, R. D. Curren, and J. J. McCormick. "Frequency of ultraviolet light-induced mutations is higher in xeroderma pigmentosum variant cells than in normal human cells," *Nature* **261**:593–595 (1976).

*Mortelmans, K., E. C. Friedberg, H. Slor, G. Thomas, and J. E. Cleaver. "Defective thymine dimer excision by cell-free extracts of xeroderma pigmentosum cells," *Proc. Nat. Acad. Sci. U.S.* **73**:2757–2761 (1976).

*Sutherland, B. M., M. Rice, and E. K. Wagner. "Xeroderma pigmentosum cells contain low levels of photoreactivating enzyme," *Proc. Nat. Acad. Sci. U.S.* **72**:103–107 (1975).

Vogel, F., and R. Rathenberg. "Spontaneous mutation in man," *Adv. Hum. Genetics* **5**:223–318 (1975).

Witkin, E. M. "Ultraviolet-induced mutation and DNA repair," *Ann. Rev. Genetics* **3**:525–552 (1969).

Chromosomal Mutations Involving Rearrangements

Cleland, R. E. "The cytogenetics of *Oenothera*," *Adv. Genetics* **11**:147–237 (1962).

Creighton, H. B., and B. McClintock. "A correlation of cytological and genetical crossing-over in *Zea mays*," *Proc. Natl. Acad. Sci. U.S.* **17**:492–497 (1931). [Reprinted in L. Levine, *Papers on Genetics*. St. Louis: Mosby, 1971.]

Ford, C. E., and H. M. Clegg. "Reciprocal translocations," *Brit. Med. Bull.* **25**:110–114 (1969). [Reprinted in L. Levine, *Papers on Genetics*. St. Louis: Mosby, 1971.]

*Goto, K., T. Akematsu, H. Shimazu, and T. Sugiyama. "Simple differential Giemsa staining of sister chromatids after treatment with photosensitive dyes and exposure to light and the mechanism of staining," *Chromosoma* **53**:223–230 (1975).

*Hollaender, A., W. K. Baker, and E. H. Anderson. "Effect of oxygen tension and certain chemicals on the X-ray sensitivity of mutation production and survival," *Cold Spring Harbor Symp. Quant. Biol.* **16**:315–325 (1951).

*Kihlman, B. A. "Sister chromatid exchanges in *Vicia faba*. II. Effects of thiotepa, caffeine, and 8-ethoxycaffeine on the frequency of SCE's," *Chromosoma* **51**:11–18 (1975).

Latt, S. A., and J. C. Wohlleb. "Optical studies of the interaction of 33258 Hoechst with DNA, chromatin, and metaphase chromosomes," *Chromosoma* **52**:297–316 (1975).

McClintock, B. "The stability of broken ends of chromosomes in *Zea mays*," *Genetics* **25**:234–282 (1941).

Muller, H. J. "The nature of the genetic effects produced by radiation." In *Radiation Biology* (A. Hollaender, Ed.). New York: McGraw-Hill, 1954, p. 351–473.

Novitski, E. "The genetic consequences of anaphase bridge formation in *Drosophila*," *Genetics* **37**:270–287 (1952).

Painter, T. S. "A new method for the study of chromosome rearrangements and plotting of chromosome maps," *Science* **78**:585–586 (1933). [Reprinted in J. A. Peters, Ed. *Classical Papers in Genetics*. Englewood Cliffs, N.J.: Prentice-Hall, 1959.]

Roberts, P. A. "The genetics of chromosome aberration." In: *The Genetics and Biology of Drosophila*. (M. Ashburner and E. Novitski, Eds.) New York: Academic Press, 1976. pp. 67–184.

Chromosomal Mutations Affecting Ploidy

*Åkesson, H. O., and H. Rorssman. "A study of maternal age in Down's syndrome," *Ann. Hum. Genetics* **29**:271–276 (1966).

*Blakeslee, A. F. "New Jimson weeds from old chromosomes," *J. Hered.* **25**:80–108 (1934).

Clausen, R. E., and T. H. Goodspeed. "Interspecific hybridization in *Nicotiana*. II. A tetraploid glutinosa-tabacum hybrid, and experimental verification of Winge's hypothesis," *Genetics* **10**:278–284 (1925). [Reprinted in L. Levine, *Papers on Genetics*. St. Louis: Mosby, 1971.]

*Fukuhara, S., S. Shirakawa, and H. Uchino. "Specific marker chromosome 14 in malignant lymphomas," *Nature* **259**:210–211 (1976).

Klush, G. S. *Cytogenetics of Aneuploids*. New York: Academic Press, 1973.

*Lejeune, J., M. Gautier, and R. Turpin. "Study of the somatic chromosomes of nine mongoloid idiot children," *Compt. Rend. Acad. Sci.* **248**:1721–1722 (1959). [Reprinted in S. H. Boyer, *Papers on Human Genetics*. Englewood Cliffs, N.J.: Prentice-Hall, 1963.]

Little, T. M. "Gene segregation in autotetraploids. II," *Bot. Rev.* **24**:318–339 (1958).

Neervath, P., K. DeRemer, B. Bell, L. Jarvik, and T. Kato. "Chromosome loss compared with chromosome size, age, and sex of subjects," *Nature* **225**:280–281 (1970).

*Philip, J., C. Lundsteen, D. Owen, and K. Hirschhorn. "The frequency of chromosome aberrations in tall men with special reference to 47,XYY and 47,XXY," *Am. J. Hum. Genet.* **28**:404–411 (1976).

Sparrow, A. H., and A. F. Nauman. "Evolution of genome size by DNA doublings," *Science* **192**:524–529 (1976).

Carcinogens

*Ames, B. N., H. O. Kammen, and E. Yamasaki. "Hair dyes are mutagenic: Identification of a variety of mutagenic ingredients," *Proc. Nat. Acad. Sci. U.S.* **75**:2423–2427 (1975).

Bridges, B. A. "Short term screening tests for carcinogens," *Nature* **261**:195–200 (1976).

Fishbein, L. "Atmospheric mutagens. I. Sulfur oxides and nitrogen oxides," *Mut. Res.* **32**:309–330 (1976).

Heddle, J. A., Ed. International Symposium on Genetic Hazards to Man from Environmental Agents. *Mut. Res.* **33**:1–106 (1975).

Huberman, E., L. Sachs, S. K. Yang, and H. V. Gelboin. "Identification of mutagenic metabolites of benzo[a] pyrene in mammalian cells," *Proc. Nat. Acad. Sci. U.S.* **73**:607–611 (1976).

Kirkland, D. J., and S. Venitt. "Cytotoxicity of hair colourant constituents: chromosome damage induced by two nitrophenylenediamines in cultured Chinese hamster cells," *Mut. Res.* **40**:47–56 (1976).

Montesano, R., and H. Bartsch. "Mutagenic and carcinogenic N-nitroso compounds: possible environmental hazards," *Mut. Res.* **32**:179–228 (1976).

*Perry, P., and H. J. Evans. "Cytological detection of mutagen-carcinogen exposure by sister chromatid exchange," *Nature* **258**:121–125 (1975).

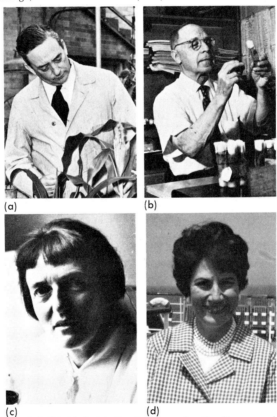

(a) (b)

(c) (d)

(a) L. J. Stadler and (b) H. J. Muller independently discovered the mutagenic effects of X-rays in plants and animals, respectively. (c) Charlotte Auerbach performed early studies on the mutagenic effects of certain chemicals. She is now at the University of Edinburgh. (d) Evelyn Witkin, of Rutgers University, discovered misrepair mutagenesis in *E. coli.*

Questions and Problems

1. What is the ratio of phenotypes and genotypes expected in the progeny of a cross between two organisms heterozygous for the same lethal gene?

2. Cite six human phenotypic traits that you would classify as wild type (that is, highly invariant) and six traits that commonly appear in a number of variant forms in the human population.

3. Devise an experiment utilizing a colony-forming haploid microorganism such as *C. reinhardi* that would allow you to screen for a mutant strain carrying two mutations, one leading to an acetate (carbohydrate) requirement and the other to a thiamin (vitamin) requirement.

4. After mutagenesis with uv, organisms must be left in the dark for at least 12 hours. Why is this so?

5. When *C. reinhardi* cells are exposed to a chemical mutagen at a particular concentration, 70 percent of the cells cannot form colonies. Describe the kind(s) of effects these cells have suffered.

6. What molecular similarities exist between adenine and cytosine that render them susceptible to the same mutagens? What is similar about thymine and guanine in the same regard?

7. Explain why mutagenic transitions are easier to explain at a molecular level than are transversions.

8. Diagram a sequence of molecular events leading to the reversion of the addition shown in Figure 6-13a and a reversion of the deletion shown in Figure 6-13b.

9. A portion of a gene contains the sequence $\xrightarrow[\text{GTAAC}]{\text{CATTG}}$. For each of the mutagens listed below, show a probable series of events by which this sequence would become mutated after one or two rounds of replication: (a) nitrous acid; (b) hydroxylamine; (c) EMS; (d) uv irradiation.

10. Diagram somatic pairing in the salivary gland cells of a Dipteran heterozygous for a duplication.

11. Nondisjunction can occur during meiotic anaphase I when homologues fail to separate; it can also occur at meiotic anaphase II when centromeres fail to divide. It is unlikely that both events will occur during the same meiosis. (a) Diagram the consequences of each type of nondisjunction on the sex chromosome constitution of sperm and egg cells formed during spermatogenesis and oogenesis (recall that males are XY and females XX). (b) State whether the following gamete types can be produced by nondisjunction at anaphase I, at anaphase II, or either way: a YY sperm; an XY sperm; an XX egg; a sperm carrying no sex chromosome; an egg carrying no sex chromosome. (c) Describe the kinds of chromosomal mutations carried by human zygotes formed when the above gametes are fertilized by an X-bearing gamete.

12. Hemophilia is a sex-linked recessive trait. A son born to phenotypically

normal parents has Klinefelter's syndrome and hemophilia. In which meiotic division of which parent did nondisjunction of the X chromosome most likely occur? Explain and state your assumptions.

13. Describe how you would set up a carcinogen test in which human skin fibroblasts were monitored for the induction of unscheduled DNA synthesis. What would be the rationale of such a test?

14. Describe how you would set up a screening procedure for temperature-sensitive *dna* strains of *E. coli* using the 5BU enrichment procedure.

15. Why are the F1 female flies in Figure 6-7 not allowed to inbreed? How would this change the experiment?

16. You wish to isolate a large deficiency in the *a* (*apricot* eye) region of the *D. melanogaster* X chromosome. Show how you would do this utilizing the *Basc* chromosome. Does it matter whether or not the deficiency is lethal?

17. Post-replication misrepair appears to be **induced** in *E. coli,* meaning that the enzymes involved do not exhibit activity unless unrepaired lesions persist into the replication phase of the cell cycle. Why would it be advantageous to *E. coli* to be able to mobilize these enzymes "on call" rather than having them active at all times?

18. It is reported that the alkylating agent MNNG selectively mutates replicating regions of bacterial chromosomes, producing double or multiple mutations in close proximity to one another. MNNG is highly reactive in alkylating thiol (sulfur-containing) groups in proteins. Devise a hypothesis on the mechanism of MNNG mutagenesis based on these observations.

19. The purines and pyrimidines of DNA have the stable chemical structures shown in Figure 1-1, and the accuracy of Watson-Crick base-pair formation depends on the stability of these configurations. Watson and Crick proposed a theory of mutagenesis wherein, on rare and short-lived occasions, certain hydrogen atoms in a purine or pyrimidine ring would migrate to new unstable positions (a **tautomeric shift**). The resultant tautomers are drawn below.

Rare enol form of thymine (T*)

Rare imino form of cytosine (C*)

Rare imino form of adenine (A*)

Rare enol form of guanine (G*)

(a) Show with diagrams how T* forms base-pairs with guanine, C* with adenine, A* with cytosine, and G* with thymine (use Figure 1-6 as a model).

(b) Show with diagrams similar to Figure 6-8 the mutagenic effect if a cytosine undergoes a tautomeric shift during DNA replication I but resumes its stable configuration at DNA replication II. (c) Would you expect tautomeric shifts to produce transitions? Transversions? Deletions/additions? Explain.

20. Mutagens known as **base analogues** have chemical structures analogous to naturally occurring bases but carry critical modifications. They are mutagenic only if they are presented to cells at the time of chromosome replication. 5BU is an analogue of thymine and usually pairs with adenine. The bromine atom in 5BU (Figure 6-6) so alters the charge distribution of the molecule, however, that it tautomerizes to the 5BU* form quite frequently, in which case it possesses the hydrogen-bonding properties of cytosine. (a) Show with diagrams similar to Figure 6-8 how a chromosome given 5BU nucleoside triphosphates at DNA replication I could undergo a G $\longrightarrow$ A transition after 2 more rounds of replication. (b) Show in a similar fashion how 5BU could induce an A $\longrightarrow$ G transition.

21. Because base analogues can induce both G $\longrightarrow$ A and A $\longrightarrow$ G transitions, they are able to cause reversions of the mutations they induce. (a) Would you expect hydroxylamine to be able to cause reversions in hydroxylamine-induced mutations? Explain. (b) Would you expect 5BU to be able to cause reversions in hydroxylamine-induced mutations? Explain.

22. Discuss the following statement: The real purpose of the Ames test is not the prevention of cancer but the protection of future generations.

23. The drug adenine arabinoside is able to combat such herpes virus infections as chicken pox, shingles, and herpes encephalitis without killing infected humans. How do you think the drug might act? How could you test this prediction with adenine-arabinoside-resistant herpes mutants?

24. (a) When Chinese hamster ovary cells are cultured in undiluted Burpee's Cola and then examined, the average cell exhibits 20 SCEs. On the basis of this test, would you recommend that the cola be removed from the market? (b) Male strains of mice were fed a steady diet of Burpee's Cola and subjected to the specific locus test. Of 65,548 offspring, 40 showed mutations at one of the 7 loci. What is the mutation rate per locus per 10^6 gametes? On the basis of this test, would you recommend that the cola be removed from the market?

25. A normal woman and man have a child with severe birth defects and mental retardation. Banded karyotyping reveals that the mother is heterozygous for a reciprocal translocation between portions of chromosomes 8 and 19. The father's karyotype is normal. (a) Draw the possible karyotype(s) of the afflicted child. (b) Why is the woman normal and the child abnormal? (c) What proportion of normal: abnormal children would this couple expect to have?

26. An enrichment procedure known as **tritium suicide** involves exposing mutagenized cells to ^{3}H-adenine, harvesting the cells, and storing them in the cold for weeks to allow the radioactive decay within the tritium-containing cells to "burn them out." What kind(s) of mutations would such a procedure enrich for?

27. (a) Indicate the proper method(s) to isolate *his*⁻, streptomycin (*str*)-resistant

and *str* independent *E. coli* mutants, and explain the defects of the improper method(s).

1. *E. coli* $\longrightarrow$ grown on minimal medium with penicillin added $\longrightarrow$ plated to minimal medium with histidine and streptomycin added $\longrightarrow$ transferred to his-supplemented, streptomycin-free medium.

2. *E. coli* $\longrightarrow$ grown on minimal medium + streptomycin $\longrightarrow$ plated to minimal medium $\longrightarrow$ transferred to plate with penicillin and all nutritional substances added, with the exception of histidine.

3. *E. coli* $\longrightarrow$ grown on minimal medium with ^{3}H-adenine added $\longrightarrow$ harvest all the cells and store them in the cold for weeks $\longrightarrow$ plate the cells to minimal medium with histidine added $\longrightarrow$ transfer to his-supplemented penicillin-free medium.

4. *E. coli* $\longrightarrow$ plated on complete medium lacking histidine but with penicillin added $\longrightarrow$ his-supplemented, streptomycin-added minimal medium $\longrightarrow$ replicate plate to minimal medium; use this result to choose which colonies to transfer to his-supplemented minimal medium.

(b) Devise a method for isolating a bacterial auxotroph which requires both histidine and arginine but cannot ferment lactose. (Show your isolation procedures stepwise and concisely, e.g., *E. coli* treatment(s) $\longrightarrow$ genotypes isolated.)

28. In the process of studying a stock of Florida *Drosophila* with a high frequency of sex-linked lethal mutations, Demerec inbred the stock for two generations. In doing this, a mutability factor (a recessive mutator gene on the 2nd chromosome of Drosophila associated with a high frequency of sex-linked recessive lethal mutations) located on the second chromosome became homozygous in some of the F_2 flies, and consequently they expected a higher frequency of visible mutations (in addition to lethal mutations) in the F_3 generation than had been found in the original stock. The visible mutations actually observed in the F_3 generation among 15,000 individuals were:

yellow—24 times	*black*—2 times
forked—3 times	*blistered*—1 time
lozenge—2 times	*dwarfish*—1 time
vermilion—2 times	*curled wing*—1 time

Knowing that these genes are scattered throughout the entire length of the X chromosome, indicate the conclusions which are consistent with the above data, and explain your choices.

(a) The mutability factor is affecting the entire X chromosome equally, as the mutations are randomly distributed. (b) The mutability factor is affecting the entire chromosome equally, but some genes contained more sites for lethal mutations than others. (c) The region of the chromosome controlling the *yellow* phenotype is much larger than the other regions scored. (d) The mutability factor cannot be affecting the whole chromosome equally, because among 15,000 progeny, one would certainly have seen more loci represented. (e) List one other conclusion that is consistent with the data presented.

29. A mouse strain is heterozygous for a paracentric inversion. (a) Diagram the meiotic synapsis configuration adopted by this strain, following Figure 6-17 but showing all 4 chromatids. (b) Diagram how a single crossover between positions 4 and 5 on two nonhomologous chromatids will generate a **dicentric** (two centromeres) chromatid and an **acentric** (no centromeres) fragment. (c) What do you think will happen to the dicentric chromatid at anaphase I? What will be the fate of the acrocentric fragment? (d) Based on the above, explain why inversion heterozygotes typically produce low numbers of viable offspring, and why the viable offspring that do arise are usually non-recombinant. In other words, explain why inversions are effective crossover suppressors.

30. Follow the instructions and questions for the preceding problem, but assume that the inversion chromosome is pericentric and that a crossover occurs between positions 3 and 4 following meiotic synapsis in an inversion heterozygote.

$$\underline{1 \quad 2 \quad 3 \quad 4 \quad \bullet \quad 5 \quad 6 \quad 7 \quad 8} \quad \text{normal}$$

$$\underline{1 \quad 2 \quad 6 \quad 5 \quad \bullet \quad 4 \quad 3 \quad 7 \quad 8} \quad \text{inverted}$$

31. A *Drosophila* strain is homozygous for a reciprocal translocation involving noncentromeric portions of chromosomes 2 and 3. (a) Diagram schematically chromosomes 2 and 3 of a normal strain and the translocation strain. Assign arbitrary genes (2A, 2B, etc.) to each chromosome so the effect of the translocation is clear. (b) Diagram the gametes produced by each strain. (c) Diagram meiotic prophase in an F_1 fly produced by crossing these two strains (recall Figure 6-20).

32. Diploid karyotypes are often denoted in a shorthand that describes the total number of autosomes and sex chromosomes present. Thus a normal human female would be 44A, XX and a normal male 44A, XY. (a) How would you denote a person with Turner's syndrome? Klinefelter's syndrome? Down's syndrome? (b) A diploid species with the karyotype female 10A, XY; male 10A, XX, developed a stable tetraploid subpopulation. Give the karyotype of this population. (c) Demonstrate how this population will yield equal numbers of male and female offspring without producing large numbers of inviable zygotes. (d) Would it give fertile offspring in matings with the members of the original population? Why?

33. Phage lysates of T2 and T4 were uv irradiated separately and the surviving fraction was determined at 10 second intervals, on assay plates incubated either in the dark or in the light (permitting photoreactivation (PhR).) The results are given on the next page.

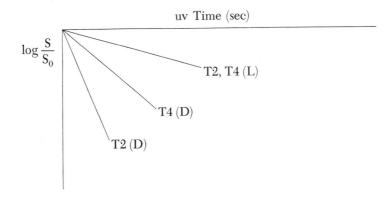

D = incubated in the dark

L = incubated in the light

$\dfrac{S}{S_0}$ = surviving fraction, where S = survivors, S_0 = original number irradiated

Which is the best explanation? Explain your choices.

A. PhR is more efficient in T2 than in T4.

B. T4 has a dark repair mechanism which eliminates damages not susceptible to PhR. T2 lacks this mechanism.

C. T4 has a dark repair mechanism which eliminates the same class of damages as PhR. T2 lacks this mechanism.

D. The T2 genome is larger than the T4 genome hence it is more susceptible to uv.

CHAPTER
7

Molecular Biology of Chromosomes

INTRODUCTION

The broad scientific discipline known as molecular biology seeks to understand biological processes in molecular terms. The molecular biologist may use the methods and analytical tools of other disciplines (biochemistry, biophysics, x-ray crystallography) or may use methods and tools unique to molecular biology; in either case the focus is on the molecules themselves. It is in this sense that molecular biology can be distinguished from molecular genetics, for the molecular geneticist begins with a genetic phenomenon (mutation, repair, recombination, protein variants) and seeks to understand its molecular basis.

This chapter considers observations made by molecular biologists who have focused their attention on chromosomal DNA (and RNA). They have asked such questions as: How much DNA is present? What base sequences are present and how are they organized in the chromosome? How does chromosomal DNA differ from species to species? In asking these questions they have in some cases started to provide molecular answers to such long-standing genetic questions as: What does a gene look like? What is special about constitutive heterochromatin? Does all the DNA in a nucleus code for protein? In other cases they have described molecular properties of chromosomes which cannot as yet be related to known genetic phenomena, this being particularly the case for eukaryotic DNA. In all cases it is clear that their findings have profoundly influenced the way in which all geneticists—molecular, "classical," and population geneticists alike—think about chromosomes and genes. The findings of molecular biologists are presented in this textbook on genetics because much present-day research in genetics is predicated on this material. Readers with a background in molecular biology will undoubtedly need to give this chapter less attention than those without such background.

MOLECULAR STRUCTURE OF PHAGE AND VIRAL CHROMOSOMES

7.1 SV40 RESTRICTION MAPS

Restriction endonucleases, described in Section 5.9, act as bacterial lines of defense against foreign DNA. Many species of bacteria produce such enzymes; Table 7-1 summarizes the properties of some of these enzymes that have been isolated and characterized. Like the *Eco*RI described in Chapter 5, all the enzymes listed make site-specific cuts in DNA. The specific sequences they recognize are given in the Table 7-1.

Purified restriction enzymes have become major research tools of molecular biology because the selective substrate specificity of each enzyme can be used to transform a duplex chromosome into a collection of homo-

Table 7-1 Class II site-specific restriction endonucleases

Strain	Enzyme[a]	Sequence (5'→3')	Number of Cleavage Sites		
			λ	Ad2	SV40
Escherichia coli (end I⁻, R⁺, RI)	EcoRI	G↓AATTC	5	5	1
Escherichia coli (end I⁻, R⁺, RII)	EcoRII	↓CCTGG	>35	>35	16
Haemophilus aegyptius (ATCC 11116)	HaeIII	GG↓CC	>50	>50	18
Haemophilus influenzae serotype d	HindII	GTPy↓PuAC	34	>20	7
	HindIII	A↓AGCTT	6	11	6
Haemophilus parainfluenzae	HpaI	GTT↓AAC	11	7	4
	HpaII	C↓CGG	>50	>50	1
Anabaena variabilis	AvaI	CGPu↓PyCG	—	—	—

[a] Additional, less well-characterized enzymes and the strains from which they are derived are as follows: AvaII, Anabaena variabilis; AluI, Arthobacter luteus, ATCC 21606; HaeII, Haemophilus aegyptius, ATCC 11116; HgaI, Haemophilus gallinarium, ATCC 14385; HhaI, Haemophilus hemolyticus, ATCC 10014; HinH-1, Haemophilus influenzae H-1; HphI, Haemophilus parahaemolyticus; MboI, MboII, Moraxella bovis, ATCC 10900; SmaI, Serratia marcescens; SacI, SacII, Streptomyces achromogenes ATCC 12767; SalI, SalII, Streptomyces albus ATCC 3004; XamI, Xanthomonas amaranthicola ATCC 11645, HapI, Haemophilus aphrophilus, ATCC 19415.
From D. Nathans and H. O. Smith, Ann. Rev. Biochem. **44**:273, 1975.

geneous fragments that are far easier to study than the intact chromosome. The SV40 chromosome, for example, is a circular duplex molecule with a molecular weight of about 3×10^6. When exposed to EcoRI, a single cut is made through the chromosome (Table 7-1) meaning that there exists but a single 5'GAATTCC / 3'CTTAAGG sequence in the SV40 chromosome. The single cut converts the circle into a chromosome-length rod of duplex DNA, as can be ascertained by electron microscopy. When SV40 chromosomes are instead exposed to EcoRII restriction enzymes, sixteen 5'CCTGG / 3'GGACC sequences are encountered (Table 7-1) and the chromosome is cleaved into 16 discrete **restriction fragments.** These can be separated from each other by **gel electrophoresis** (see Box 7.1).

K. J. Danna, G. H. Sack, and D. Nathans were among the first to use restriction enzymes to produce a **restriction map** of SV40. They began with the enzyme Hin from Haemophilus influenzae, and found that it converted the SV40 chromosome into the eleven fragments shown in Figure 7-1, with fragment Hin-A the largest (slowest-migrating) and Hin-K the smallest

Box 7.1
GEL
ELECTROPHORESIS

The technique of electrophoresis involves placing molecules in a high-voltage field such that they move toward the positive or the negative pole, depending on their net charge. Movement takes place within a solid substrate permeated with a buffered aqueous solution. The pH of the buffer, the strength of the electric field, the length of time that the field is applied, and the nature of the solid substrate can all be varied, and when the appropriate combination of conditions is chosen molecules differing only slightly in their net charge can be separated from one another.

Polyacrylamide gel electrophoresis is suitable for the separation of **oligonucleotides** (short DNA or RNA fragments) as well as larger polynucleotides and the polypeptide chains of proteins (described in Chapter 9). A narrow column of polyacrylamide gel is prepared, and the sample is layered on top. The gel is then placed in an electric field. When low concentrations of polyacrylamide are present in the gel (for example, 2.2 percent), relatively large molecules can move into the gel; at higher concentrations (for example, 7.5 percent) large molecules may be excluded. Thus the size range of the molecules under examination can be selected. Following electrophoresis, the gel can be stained with a dye such as toluidine blue so that its component bands are visible, or it can be scanned with ultraviolet light and its absorbance plotted. The curves to the right show the effect of various gel concentrations on the separation of tumor cell RNA molecules. Molecular weights are obtained by performing standard gel runs with marker RNAs of known

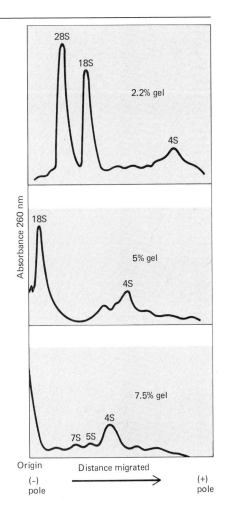

molecular weight. (Gel scans from J. N. Davidson, *The Biochemistry of Nucleic Acids*. Norfolk, England: Cox and Wyman, 1972, Figure 5.3.)

Many extensions of this basic technique have been developed. For example, oligonucleotides or polypeptides can be labeled with such radioisotopes as ^{32}P or ^{14}C and subjected to electrophoresis in **slab gels** in which the polyacrylamide gel is molded to be flat and paper-thin. The gel is then dried and sandwiched next to an X-ray film for **autoradiography** (see Box 2.2). The black lines on the resultant autoradiogram reveal the positions of the labeled macromolecules. A second extension, very valuable for separating a complex array of macromolecules, is called **two-dimensional polyacrylamide electrophoresis:** a radioactively-labeled sample is subjected to electrophoresis under one set of conditions (e.g. an 8% gel at pH 3.5 in 6 M urea); the gel is then laid on top of a second slab gel (e.g. 16% polyacrylamide at neutral pH) and its contents, which were partially separated from one another in the "first dimension," move into the second gel and become even better separated under the second set of conditions.

(fastest-migrating) fragment in the gel. A number of approaches were then used to reconstruct the original order of the 11 *Hin* fragments within the SV40 chromosome. One strategy was to use two restriction endonucleases simultaneously or in succession. For example, if an SV40 *Hin* digest was exposed to *Eco*RI and then subjected to electrophoresis, fragment *F* was no longer found in its characteristic position on the gel; instead, two new smaller fragments were present, the summed sizes of which correspond to the size of fragment *F*. Thus fragment *F* was identified as carrying the *Eco*RI cleavage site. Alternatively, the SV40 chromosomes were first digested with the enzyme *Hpa*I, which generates three large fragments, *Hpa-A*, *Hpa-B*, and *Hpa-C*. Each of these was then eluted separately from a gel, exposed to *Hin,* and rerun by electrophoresis. When, for example, the *Hpa*-C fragments

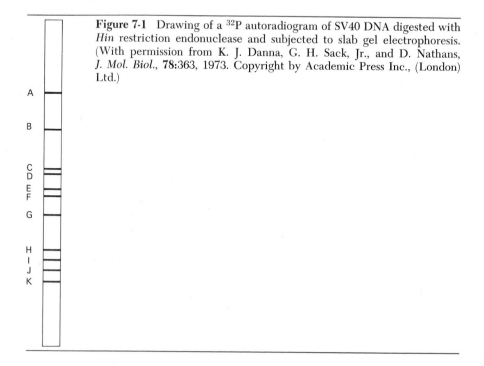

Figure 7-1 Drawing of a ^{32}P autoradiogram of SV40 DNA digested with *Hin* restriction endonuclease and subjected to slab gel electrophoresis. (With permission from K. J. Danna, G. H. Sack, Jr., and D. Nathans, *J. Mol. Biol.,* **78**:363, 1973. Copyright by Academic Press Inc., (London) Ltd.)

were exposed to Hin, bands identical in mobility to the *Hin-B* and *Hin-I* were found on the resultant gels, indicating that the *B* and *I* fragments lie next to one another in the chromosome. By such approaches, Nathans and colleagues were able to produce the **physical map** of SV40 shown in Figure 7-2, where the single *Eco*RI cleavage site is arbitrarily assigned a "O" position and other sites map between O and 1 on the circle.

More elaborate physical maps of SV40 have since been generated (Figure 7-3). These have utilized restriction enzymes that recognize short nucleotide sequences (for example, tetranucleotides). Since it is more probable that a particular tetranucleotide will be present in a chromosome than, say, a hexanucleotide, such restriction enzymes cleave the chromosome into numerous small fragments. The small fragments are difficult to separate from one another on the basis of size; therefore, they are most reliably identified using one of the elegant techniques presently available for determining the **DNA sequence** of small restriction fragments (see Box 7.2). Once the nucleotide sequences of particular fragments are known, **overlapping sequences** can be sought: for example, if one strand of the fragment X generated by *Hae* III has the sequence 5'CCGACTA<u>AGCTTG</u> . . . 3' and fragment Y generated by *Hind* III has the sequence 5'<u>AGCTTG</u>CTATA . . . 3', then the underlined region of overlap indicates that fragments X and Y derive from a sector of the chromosome with the sequence 5'CCGACTA<u>AGCTTG</u>CTATA3', and a map order 5' X — Y 3' is deduced.

It should be obvious that a careful comparison of numerous overlapping and sequenced restriction fragments will eventually yield the nucleotide sequence of an entire chromosome. This has in fact been accomplished

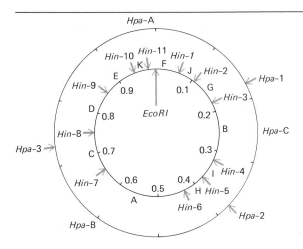

Figure 7-2 A cleavage map of the SV40 genome. Map units are given as

$$\frac{\text{distance from } Eco\text{RI site}}{\text{length of SV40 DNA}}$$

in the direction F-J-G-B . . . Arrows illustrate the sites of restriction endonuclease cuts (*Hin-1*, *Hin-2*, *Hpa-1*, . . .) and letters indicate the fragments generated, with the inner circle showing the position of the eleven *Hind* II fragments (A-K) and the outer circle the 3 *Hpa* fragments (A-C). (With permission from K. J. Danna, G. H. Sack, Jr., and D. Nathans, *J. Mol. Biol.*, **78**:363, 1973. Copyright by Academic Press Inc., (London) Ltd.)

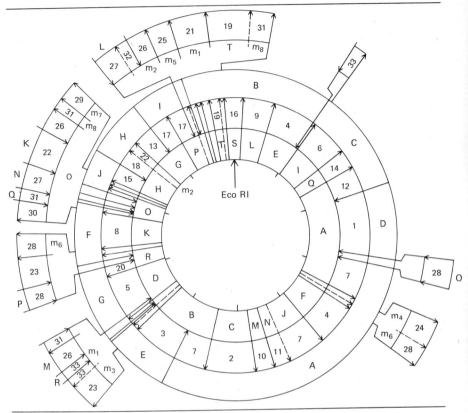

Figure 7-3 Alu-Hae cleavage map of SV40 DNA. Inner circle: the *Eco* RI site is shown. Inner ring: the location of the 32 *Alu* (*Arthrobacter luteus*) sites (and fragments) are indicated. *Alu* sites which are also recognized by *Hind* III are shown by dashed arrows. Middle ring: the 50 fragments produced by double digestion of SV40 DNA with *Alu* and *Hae* III (*Haemophilus aegyptius*) enzyme are indicated. These *Alu-Hae* fragments are identified by an arabic numeral; fragments moving with the same mobility have the same number. Outer ring: the 18 *Hae* sites (and fragments) are shown. For clarity, several regions are shown on an expanded scale on the outside. Outward arrows are *Alu* sites (dashed arrows are *Alu* or *Hind* III sites), and inwards arrows *Hae* III sites. *Alu* fragments are indicated on the inside and *Hae* fragments on the outside (From R. Yang, A. van de Voorde, and W. Fiers. *Eur. J. Biochem.* **61**:119–138, 1976).

| Box 7.2 DNA SEQUENCING | The DNA sequencing procedure devised by W. Gilbert and A. Maxam is described below (see also art on facing page). An alternate technique, developed by F. Sanger and A. R. | Coulson, involves the primed synthesis of highly labeled complementary strands.

1. A pure preparation of DNA (for example, a 64-base-pair *Alu* I restric- |

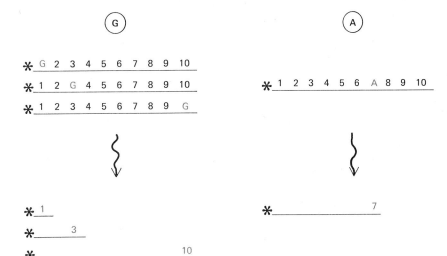

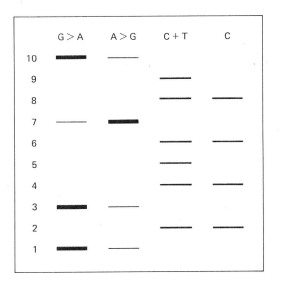

tion fragment from λ chromosomal DNA) is labeled enzymatically with ^{32}P at the 5' ends of both strands.

2. The strands of the duplexes are separated by electrophoresis into two single-strand fractions known as "heavy" and "light" (see Section 8.4). Considered below is the heavy-strand sample; identical procedures can be performed on the light-strand sample to provide a complementary sequence.

3. The heavy-strand sample is divided into 4 aliquots, one of which will indicate the position of every G in the strand, one of which will indicate every A, one every T, and one every C. These are considered in turn in the next four steps.

Ⓖ The aliquot is treated with dimethyl sulfate under conditions that result, on the average, in the methylation of one purine per single strand. G is methylated about 5 times faster than A. Methylation causes the strands to break in the presence of warm alkali, generating ^{32}P-labeled fragments (fragments 1, 3, and 10 in the diagram on the preceding page, where the asterisk denotes radioactivity). A smaller quantity of labeled fragments broken at methylated A positions is also generated. Electrophoresis and autoradiography of this sample gives the **G > A** pattern, with strong bands at guanine breaks and weak bands at adenine breaks.

Ⓐ The purines in this aliquot are methylated as in the previous step, but the sample is then exposed to cold dilute acid, which preferentially causes breaks at A positions. Electrophoresis and autoradiography yields the **A > G** pattern with dark bands at adenines and light bands at guanines.

Ⓣ This aliquot is treated with hydrazine under conditions that result, on the average, in the hydrazinolysis of one pyrimidine per strand, with T and C equally sensitive. The reacted pyrimidines are rendered susceptible to strand scission by the reagent piperidine. Electrophoresis and autoradiography result in the **C + T** pattern, with bands of equal intensity corresponding to cleavages at cytosines and thymines.

Ⓒ This aliquot is exposed to hydrazine and piperidine but in the presence of high salt, which preferentially suppresses the hydrazinolysis of thymines. The resultant **C** pattern displays bands that derive solely from breakages at cytosines.

4. Comparison of the 4 autoradiographic patterns allows a direct determination of the DNA sequence. In the sample gel on the preceding page, for example, the sequence can be directly read as

5' G C G C T C A C T G 3'
  1 2 3 4 5 6 7 8 9 10

---

by F. Sanger, G Brownlee, and their associates for the 5375 nucleotides comprising the RF chromosome of φX174; examples of this work are described in Chapters 8 and 9. A complete sequence of the 5000 nucleotides in the SV40 chromosome will probably also be available by the time this text is published: a major portion of the HinK fragment of SV40 has, for example, already been sequenced in W. Fiers' laboratory.

7.2 THE MS2 CHROMOSOME

The MS2 chromosome, with 3569 ribonucleotides and 3 genes (for replicase, A-protein, and coat protein, as described in Section 5.10) is among the smallest genomes known. **RNA sequencing** methods, developed chiefly by F. Sanger and his co-workers (see Box 7.3), have been utilized by W. Fiers and colleagues to generate the entire nucleotide sequence of the MS2 chromosome. Thus it is possible to write down $\begin{smallmatrix}5'p1\ 2\ 3\ 4\ 5\ 6\ 7\ 8\ 9\\ GGGUGGGAC\dots\end{smallmatrix}$ and continue through to $\begin{smallmatrix}3565\ 3566\ 3567\ 3568\ 3569\\ \dots A\quad C\quad C\quad C\quad A\end{smallmatrix}$ 3'OH.

Box 7.3 **RNA** **SEQUENCING**	Sequencing a large RNA molecule such as the MS2 chromosome usually begins by exposing purified RNA to a **ribonuclease (RNase)** called T_1. Digestion conditions are controlled, moreover, such that the only sequences the enzyme "sees" are those exposed at the tips of hairpin loops (see text). As a result, each chromosome is cut into a number of discrete RNA fragments varying in chain length from 30 to 260 nucleotides. The next step is to separate these from one another in such a way that, for example, all the 35-nucleotide-length fragments containing the bases 1–35 are separated from another collection of 35-nucleotide fragments containing the bases 2000–2035. This is best accomplished by various modes of gel electrophoresis (Box 7.1).

Each set of discrete fragments is now sequenced by a technique known as **fingerprinting.** If we consider the fragment set containing bases 1-35, then one portion of this sample is digested completely with T_1, which cleaves the ester bond between the phosphate on the 3′ side of G and 5′-OH of its neighbors, while a second portion is digested completely |

with RNase A, which cleaves only after a U or a C. Two sets of very small fragments result—the T_1 oligonucleotides and the RNase oligonucleotides—and these are again separated into discrete subclasses by gel electrophoresis. Finally, each oligonucleotide subclass is degraded by a combination of endonucleases and exonucleases (Figures 5-8 and 5-10) to yield, eventually, single ribonucleotides; these are readily identified by various forms of electrophoresis. Since the oligonucleotides in the T_1 and RNase A digests usually contain overlapping sequences, a laborious comparison of the fragments will yield the nucleotide sequence of the entire 1–35 base fragment.

The final step is to orient this fragment into the rest of the chromosome. This is accomplished by taking MS2 chromosomes and digesting them very gently with T_1 to produce very large pieces which might, for example, include a set of fragments with the nucleotides 1–47. Fingerprints of these larger fragments are compared with the smaller fragments and the entire sequence is eventually pieced together.

When such a primary sequence is examined, certain segments are found to exhibit **internal complementarity,** that is, they would be expected to fold back on themselves like a **hairpin** and form interstrand base pairs. For example, a strand 5′ . . . AAUGUACGXXXCGUACAUU . . . 3′ (where X is any nucleotide) would be most stable if its secondary structure were to assume

the hairpin form

$$5' . . .AAUGUACG$$
$$\quad\quad\quad | | | | | | | |\quad X$$
$$\quad\quad\quad\quad\quad\quad\quad\quad\quad\quad X.$$
$$3' . .UUACAUGC$$

In the case of the MS2 chromosome, many such internally complementary sequences have been identified, and although there is no reason to assume that all of these form hairpin structures *in vivo,* it is likely that many of them do. Each gene is therefore thought to loop and bend like a many-petaled flower, the entire chromosome resembling a bouquet. Such a "flower model" for the secondary structure of the coat protein gene is shown in Figure 7-4; the proposed secondary structures for the larger A-protein and replicase genes entail even more elaborate configurations.

The folded secondary structure of the MS2 chromosome has several consequences. First, the "packaging problem" (Section 5.10) of fitting a 1 μm length of RNA into a 25 nm capsid is simplified if the chromosome exists naturally as a compact molecule. Second, the sensitivity of the chromosome to ribonucleases present in an *E. coli* host cell is greatly

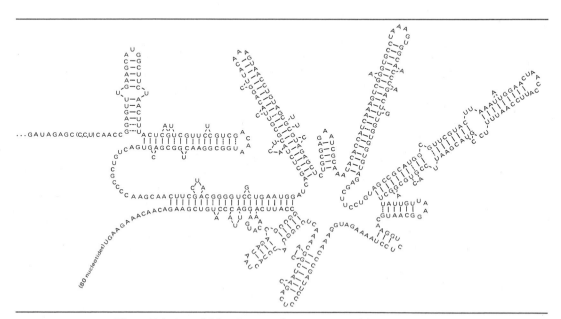

Figure 7-4 A model for the secondary structure of the coat protein gene. (From W. Fiers *et al., Nature* **237**:83, 1972.)

reduced if much of the RNA is "tucked inside" the structure. Finally, some of the nucleotide sequences prominently displayed on the exterior of the "bouquet" may play important roles in terms of associating with ribosomes, initiating transcription, seeding capsid proteins, and so on. In other words, phage chromosomes may conduct their infectious cycles not only by virtue of the genes encoded in their chromosomes but also by virtue of the physical topology of the chromosomes themselves.

7.3 THE "STICKY ENDS" OF λ

The λ chromosome contains 46,500 base pairs or about 9 times the DNA of ΦX174; the sequencing of this and the other large phage chromosomes will therefore be ambitious undertakings. Meanwhile, molecular information is available on particular features of these chromosomes.

The λ chromosome is rod-shaped within the virion but soon after infection closes on itself to form a covalently sealed ring (Figure 5-15). This circularization process depends on a unique feature of the chromosomes of λ and its temperate-phage relatives, namely, that at the 5' terminus of each chromosomal strand, 12 additional nucleotides protrude out in single-stranded form beyond the long double helix. These 12 nucleotides have been sequenced and prove to be exactly complementary to one another, as diagrammed in Figure 7-5a. These nucleotides do not pair with one another inside the phage head, but when the chromosome enters an *E. coli* cell the ends undergo base pairing and form a ring (Figure 7-5a) which is then covalently sealed by ligase enzymes.

How are the **sticky ends** of a λ chromosome generated? It turns out that the initial products of λ chromosome replication are giant concatamers, very much like those of T4 (Section 5.10); these are then cut into "unit length" chromosomes and packaged into phage heads. Length determination is not made for λ by the "headful" mechanism described in Section 5.10, however. Instead, a λ-specified endonuclease called **ter** (for *terminus*-generating activity) recognizes in the concatamer a sequence of bases known as *cos* (Figure 7-5b). Following recognition, *ter* makes two cuts, one in each of the two opposed strands and each twelve nucleotides apart, to generate the single-stranded cohesive ends (Figure 7-5b). Thus *ter* clearly acts in the same fashion as a restriction enzyme; its biological role, however, is quite different from the restriction enzymes of a bacterium.

7.4 REPETITIONS AND PERMUTATIONS IN T4

The phage T4 chromosome is a rod-shaped duplex of about 165,000 nucleotide pairs which does not form rings either within the virus particle or

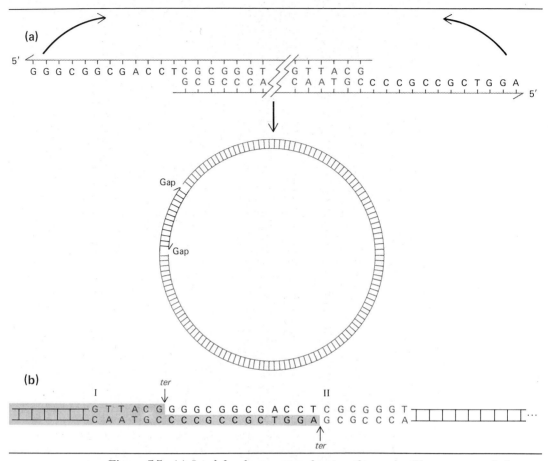

(a)

5' G G G C G G C G A C C T C G C G G G T ∥ G T T A C G
      G C G C C C A ∥ C A A T G C C C C G C C G C T G G A 5'

Gap

Gap

(b)

         I   *ter*   II

G T T A C G G G G C G G C G A C C T C G C G G G T
C A A T G C C C C G C C G C T G G A G C G C C C A ...

                               ter

Figure 7-5 (a) Lambda chromosome showing the nucleotide sequence of the two "sticky" ends in gray. The two ends become paired as the chromosome circularizes, leaving two gaps in the molecule that are sealed by ligases to form a covalently closed duplex ring. (b) A *cos*-sequence region of a giant concatamer formed during λ chromosome replication. Arrows show the location of *ter* nicks which generate two sticky-ended λ chromosomes, I and II, from the concatamer.

within the host cell. Since rod-shaped chromosomes have two ends, we might reasonably expect the order of genes along the T4 chromosome to be ABCDEFG, with the two "end" genes, A and G, defined as those farthest apart. When T4 is mapped by genetic crosses (Chapter 10), however, chromosomal ends cannot be found. Instead, the hypothetical gene G behaves as though it were as close to A as it is to F. The simplest explanation for this is to propose that the chromosome exists as a circle: D E F G A · C B

Such a circle, of course, will have no ends.

The real explanation for the rod-shaped chromosome and circular map of T4 is in many ways bizarre. It has been found that the sequence of nucleotides in all T4 chromosomes is **terminally repetitious,** meaning that the genetic "text" begins over again at the end of the chromosome. Using our previous analogy, a T4 chromosome from one phage particle might read ABCDEFGAB. Unlike λ, however, in which terminal repetition represents approximately 12 nucleotides that are exposed as single strands, terminal repetition in T4 may represent 2000 to 6000 nucleotides, all of which are base-paired in a DNA duplex. The DNA from a second T4 particle is not necessarily the same as that from the first; instead, it may be terminally repetitious for different parts of the chromosome; for example, CDEFGABCD. A third chromosome might read FGABCDEFG, and so on. All of these combinations occur with equal frequency in a T4-phage population. Stated more formally, each phage chromosome is a different **circular permutation** of a common nucleotide sequence, ABCDEFG, and each is terminally repetitious for the particular region of the sequence in which the chromosome happens to begin. For this reason, when a large population of T4 is examined, as in a mapping experiment, it appears as if any one gene in the sequence is adjacent to the next gene in the sequence, and no ends are detected.

7.5 DISSOCIATION-REASSOCIATION ANALYSIS

The circularly permutated state of T4 chromosomes is most convincingly demonstrated in an experiment involving the dissociation and reassociation of T4 DNA, a technique first developed by J. Marmur and associates. Since this technique is essential to many of the experiments described in this chapter, it is described in some detail.

If a solution containing duplex DNA molecules is either heated or treated with alkali, the hydrogen bonds holding each duplex together become increasingly unstable and the helix starts to come apart (**denaturation** or **melting**). If at this point the solution is cooled or neutralized, the hydrogen bonds will reform and the helix is said to **renature,** as illustrated in Figure 7-6.

If the solution is instead maintained at high temperature or high alkalinity, the two strands of each helix will separate completely and diffuse away from one another, a phenomenon known as DNA **dissociation.** When a solution of dissociated DNA is incubated under appropriate conditions, the formation of new base pairs eventually causes **reassociation** or **reannealing** of the DNA: two complementary sequences encounter each other in register and form a duplex molecule. Since the two strands in the new duplex are very unlikely to have been associated with one another in the original collection of DNA molecules—they almost certainly derive from different duplexes—the term **re**association is somewhat misleading but is commonly used nonetheless.

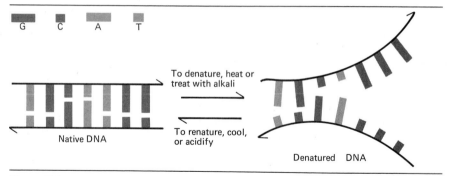

Figure 7-6   Denaturation and renaturation of duplex DNA.

Dissociation-Reassociation of T4 Chromosomes   We can now return to the molecular structure of the T4 chromosome and describe an experiment performed by C. Thomas. Thomas subjected rod-shaped T4 chromosomes to dissociation and reassociation, examined the resulting molecules with the electron microscope, and found molecules predominantly in the form of circles. The explanation for this result is diagrammed in Figure 7-7. Three hypothetical T4 DNA molecules are drawn, each a circular permutation of the next, and each a rod-shaped duplex. When these molecules are dissociated and the separated strands are allowed to reassociate, a given strand may find complementary sequences on any portion of another strand. Once such base pairing begins, it will continue around both strands and a circle will form.

MOLECULAR COMPARISONS BETWEEN PROKARYOTIC AND EUKARYOTIC CHROMOSOMAL DNA

The prokaryotic bacterial chromosome is similar to a haploid set of eukaryotic chromosomes, and distinct from a viral chromosome, in that it contains all the genetic information necessary to perpetuate the species. Despite this similarity, we have already noted (Section 5.1) that an *E. coli* chromosome differs from a eukaryotic chromosome in its circular shape, its absence of histones, and its apparent absence of coiling and uncoiling during the cell cycle. In this section we examine molecular differences between prokaryotic and eukaryotic chromosomal DNA. Molecular biologists have been able to define major differences in the amount, the chemical composition, and the sequence organization of the DNA in prokaryotes *versus* eukaryotes which are particularly intriguing in that their significance is as yet fundamentally unknown. It seems likely, however, that these differences must influence the basic prokaryote-eukaryote "split" in the biological kingdom, and that they relate in some way to the vastly greater genetic complexity attained by the eukaryotes.

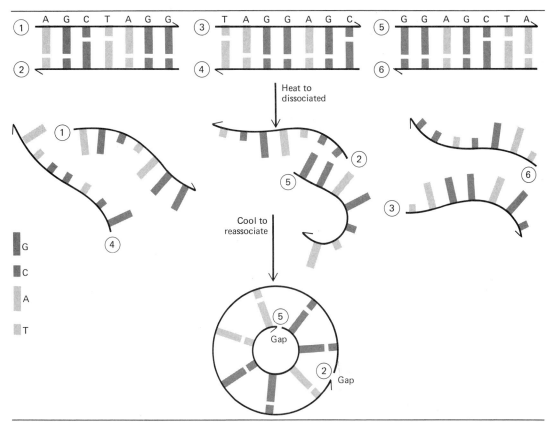

Figure 7-7 Diagram of a dissociation-reassociation experiment with a circularly permuted set of chromosomes from bacteriophage T4. (After C. A. Thomas, *J. Cell Physiol.* **70, Suppl. 1:**13–34, 1967.)

7.6 TOTAL AMOUNT OF DNA COMPARED

The first obvious difference between prokaryotic and eukaryotic chromosomes lies in their DNA quantities. As already discussed in Chapter 5, a typical eukaryotic nucleus contains perhaps a thousand times more DNA than a bacterium (and a hundred thousand times more than a virus). The actual numbers are shown in Table 5-1 and are summarized in Figure 7-8. Table 5-1 and Figure 7-8 also indicate that eukaryotes vary markedly among themselves with regard to the amount of DNA they possess. Thus a fungus such as yeast contains not much more DNA than *E. coli*, whereas the nuclei of salamanders and other tailed amphibia contain much more DNA than human nuclei. Although it is perhaps not difficult to imagine why humans might possess more DNA than a fungus, it is not at all clear why salamanders should have more DNA than humans.

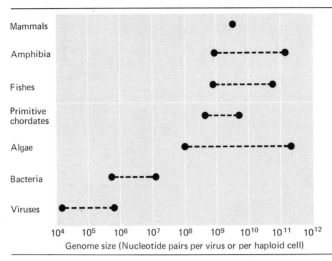

Figure 7-8 Amount of DNA per cell in organisms of increasing evolutionary complexity. For each group of organisms, the diagram indicates the maximum and minimum amounts of DNA that have been recorded in various species. For higher forms germ cells are used to calculate the haploid amount. Mammals are all seen to have about the same amount. (After "Repeated Segments of DNA" by R. J. Britten and D. E. Kohne. Copyright © 1970 by Scientific American, Inc. All rights reserved.)

7.7 CHEMICAL COMPOSITION (GC CONTENT) OF DNA COMPARED

Eukaryotic and prokaryotic DNAs also differ in their chemical compositions. As we saw in Chapter 1, Chargaff and others have demonstrated that the DNA of a species has a characteristic A + T/G + C ratio which can also be expressed as percent G + C or, less precisely but more commonly, as **percent GC** or the **GC content.** Thus to say that an organism has a GC value of 30 percent is to say that G = 15 percent, C = 15 percent, A = 35 percent, and T = 35 percent of the total DNA. When different species of bacteria are compared, the percent GC varies from as little as 24 to as much as 75 percent (Table 1-3). A similar variability is found in the base composition of various viral DNAs. As one moves up the evolutionary scale, however, the overall GC content becomes less variable, so that, for mammals, it rests at an average of about 40 percent GC. It is not yet understood why mammals have a relatively fixed GC content compared with prokaryotes, but the difference is a basic one.

Determining Percent GC In Chargaff's experiments (Table 1-3), the total GC content was determined by digesting DNA into its component nucleotides and analyzing their quantities biochemically. An alternative method utilizes density gradient centrifugation in CsCl, a technique described in

Box 5.1. It turns out that a fragment rich in GC pairs will equilibrate in a denser region of a CsCl gradient than will a GC-poor (that is, an AT-rich) fragment, and if the two fragment classes are sufficiently different, each will form a discrete band in the gradient. The buoyant density (ρ) value of each band is then compared to the ρ values obtained when DNA fragments of known GC percent (determined by the Chargaff method) are centrifuged to equilibrium; in this way, the percent GC of the DNA in the various bands can be deduced.

7.8 CsCl BANDING PATTERNS COMPARED

When a culture of *E. coli* cells is harvested and many millions of *E. coli* chromosomes are isolated by rather gentle procedures, each of these chromosomes breaks at random into fragments averaging about 10^5 nucleotide-pairs in length. These fragments will all move at equilibrium to a single position in a CsCl gradient, forming a band whose ρ-value corresponds to 51 percent GC. The formation of only a single band in such an experiment means that, overall, the *E. coli* genome has a **uniform base composition:** its random fragmentation does not yield certain chromosome fragments that have significantly different base compositions than other fragments.

When the DNA of most mammalian cells is subjected to the same analysis, on the other hand, a quite different picture is obtained. The bulk of the DNA fragments equilibrate at a single position in the CsCl gradient, one that corresponds to the percent GC (roughly 40 percent) that is obtained by chemical analysis. In addition to this **main band** DNA, however, **satellite bands** are often (although not always) found, having buoyant densities indicating either a higher or lower GC content than that of the main band. Some eukaryotes may exhibit only one satellite band whereas others may exhibit several. Additional, "**cryptic satellites**" can, moreover, be resolved by alternate gradient systems; for example, if **actinomycin D** is added to the gradient, it preferentially binds to GC pairs, thereby magnifying density differences between DNAs that are only slightly different in their GC content.

When satellite DNA was first detected, some confusion existed regarding its localization in the eukaryotic cell. It is now clear that certain classes of satellite DNA correspond to DNA that resides in cytoplasmic organelles. This can be clearly seen in *Chlamydomonas reinhardi,* whose "total DNA" separates into a main (α) and a satellite (β) band in CsCl gradients (Figure 7-9a). If, however, a preparation of *C. reinhardi* chloroplasts is made and the DNA is extracted from this preparation, a great enrichment for the β-band DNA is observed (Figure 7-9b), with only a relatively minor peak of contaminating α-DNA. Organelle chromosomes, such as those in *C. reinhardi* chloroplasts, are considered separately in Chapter 14; for the present it is

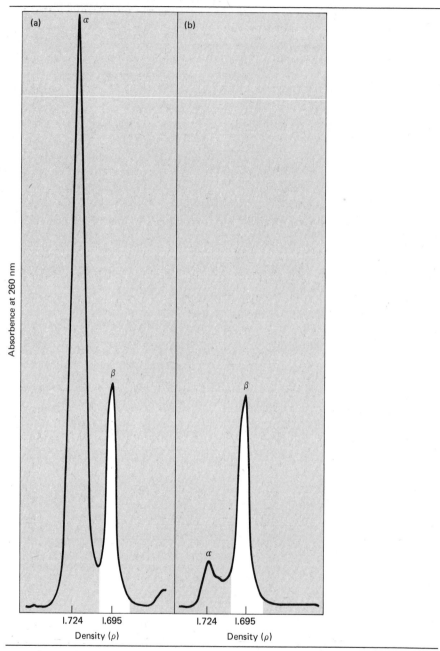

Figure 7-9 The α and β DNA fractions from *C. reinhardi* as prepared in CsCl density gradients and visualized by uv absorbance at 260 nm. In whole cells (a) most of the DNA belongs to the α class ($\rho = 1.724$), whereas in chloroplast preparation (b) most of the DNA is β ($\rho = 1.695$). (Courtesy of J. D. Rochaix.)

sufficient to note that such chromosomes characteristically (although not inevitably) have a different GC value than the nuclear chromosomes of the same species of organism.

Satellite bands corresponding to organelle DNAs are now commonly designated as mitochondrial or chloroplast bands, and the term **satellite DNA** has instead come to refer to a second class of divergent DNAs that are localized within the nucleus itself. Table 7-2 lists the properties of some of these nuclear satellite DNAs. It is clear that some are GC-rich and others AT-rich with respect to the main band. Moreover, great variability exists in the percentage of an organism's total DNA that has satellite properties: in the mouse it is perhaps 6 percent of the total, while in the crab, up to 30 percent. Variability in the percent GC of satellite DNAs appears not to have a significant effect on the total base composition of the mammalian nucleus since, as we noted earlier, most mammals possess DNA with a total GC content of about 40 percent.

Satellite DNAs are considered in more detail in Sections 7.11 and 7.13. Here we stress that the existence of satellite DNA does, in general, distinguish eukaryotes from prokaryotes, exceptions being two species of *Halobacterium* which have satellite DNAs representing from 10 to 20 percent of their total DNA (Table 7-2).

Table 7-2 Base Composition of Some Satellite DNAs.

Organism	Main Component (%GC)	Satellite (%GC)	Percent of Total DNA Represented by Satellite
Bacteria			
Halobacterium salinarium	67	58	20
H. cutirubrum	68	59	10
Invertebrates			
Cancer antennarius (crab)	40	3	30
Cancer magister (crab)	41	3	12
Cancer productus (crab)	41	3	31
Balanus nubilis (barnacle)	47	55	–
Vertebrates			
Mouse	41	34	11
Guinea pig	41		
Type I		39	5
Type II		45	4
Human	41		
Type I		33	0.5
Type II		35	2
Type III		39	1.5

From B. J. McCarthy, in *Handbook of Cytology* (A. Lima-de-Faria, Ed.). Amsterdam: North-Holland, 1969, Table 4. Copyright 1971 by the American Association for the Advancement of Science.

7.9 REASSOCIATION KINETICS (C_0T PLOTS) COMPARED

At this point we can cite three characteristic properties that distinguish eukaryotic DNA from prokaryotic DNA: there is much more of it per cell; it has a more stable average GC content; and it is typically nonhomogeneous, for it exhibits satellite DNA fragments in CsCl gradients. The fourth and final molecular difference between eukaryotic and prokaryotic DNA considered here is perhaps the most intriguing and most complex. This relates to the different **reassociation kinetics** exhibited by prokaryotic DNA on the one hand and by eukaryotic DNA on the other. Again we must describe the biochemical technique used to measure this phenomenon before we can discuss its significance. This technique was first developed and exploited by R. Britten and D. Kohne.

Analyzing Reassociation Kinetics We have already described the process of dissociation and reassociation of DNA with regard to viral chromosomes (Figure 7-7). In a kinetic study of reassociation one simply measures the time course of the reassociation process. The extent of reassociation can be monitored by electron microscopy, as described for the phage experiments, but more often it is estimated by passing samples of the reaction mixture over **hydroxyapatite** columns, since hydroxyapatite crystals, under appropriate ionic conditions, selectively bind only duplex DNA, and the amount of reassociated DNA formed in a given time period can be readily determined. To compare reassociation kinetics from one experiment to another and from one organism to another, certain standard conditions are introduced. The DNA is fragmented into small pieces of relatively uniform size (usually in the range of 250 to 450 nucleotide pairs), and the temperature and ionic conditions of the reaction are adjusted so that they are comparable from one experiment to another.

An idealized reassociation experiment follows **second-order reaction kinetics,** meaning that the rate-limiting step in the reaction is the "correct" collision of strand pairs such that a mutual region of complementarity is recognized and base-pair formation begins. Second order reactions are known to follow the equation:

$$\frac{C}{C_0} = \frac{1}{1 + kC_0 t}$$

where C_0 is the total DNA concentration, C the concentration of DNA remaining in single-stranded form at time t, and k the reassociation rate constant. In DNA reassociation experiments, it has proven most convenient to express this relationship by plotting C/C_0 (the fraction of DNA strands left single-stranded) as a function of the logarithm of the term $C_0 t$ (C_0 multiplied by the time, t, that the reassociation has been allowed to proceed). $C_0 t$ is a convenient experimental parameter, since it takes account of the reciprocity of concentration and time in second-order reactions. A $C_0 t$ **plot,** therefore, compensates for the effect on reaction kinetics of differ-

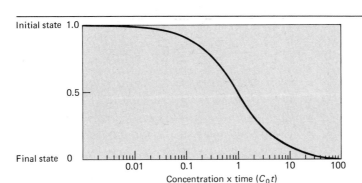

Figure 7-10 Ideal time course for DNA reassociation as seen in a C_0t plot. In the initial state all DNA is single-stranded; in the final state all DNA has reassociated into duplex form. (After R. J. Britten and D. E. Kohne, Carnegie Institution of Washington Year Book 66, Washington D.C., 1967. Courtesy of the Carnegie Institution of Washington.)

ences in initial DNA concentrations and allows a comparison of reactions that differ by large factors in their overall rates.

A C_0t plot for an idealized DNA reassociation experiment, shown in Figure 7-10, follows the simple sigmoidal-curve pattern of a second-order reaction. This curve is based on an important assumption: that all the single-stranded DNA fragments in the initial reaction mixture have an equal probability of forming a reassociated duplex. Obviously, if one class of such fragments is capable of reassociating much faster than another, then the

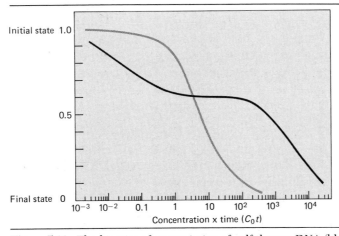

Figure 7-11 The kinetics of reassociation of calf thymus DNA (black curve) and *E. coli* DNA (gray curve), as seen in a C_0t plot in which the DNA is all single-stranded in the initial state and double helical in the final state. (After R. J. Britten and D. E. Kohne, Carnegie Institute of Washington Year Book 66, Washington D.C., 1967. Courtesy of the Carnegie Institution of Washington.)

smooth S-shaped curve of Figure 7-10 will be replaced by a skewed curve that reflects the occurrence of several different kinds of reactions going on at the same time.

Figure 7-11 shows the actual renaturation kinetics obtained for *E. coli* (gray curve) and calf (black curve) DNA fragments. The *E. coli* C_0t plot has the same symmetrical form as the idealized curve plotted in Figure 7-10. The calf C_0t curve, on the other hand, is clearly skewed, indicating that the population of DNA fragments is heterogeneous with respect to its kinetics of reassociation. In particular, it is evident that some reassociation of calf DNA can occur at very low C_0t values, meaning that a fraction of the calf DNA "finds partners" very rapidly. It is also evident that the fraction of the calf DNA which reassociates with the same kinetics (slope) as the *E. coli* DNA does so only at relatively high C_0t values.

Heterogeneous renaturation kinetics, then, represents yet another hallmark of eukaryotic chromosomal DNA. In the following and final sections of this chapter we consider what is known about the various "kinetic classes" of DNA in eukaryotic chromosomes.

THE "KINETIC CLASSES" OF EUKARYOTIC DNA

7.10   SINGLE-COPY SEQUENCES

In Chapters 3 and 4, Mendelian genetics was presented with the assumption that a gene specifying a simple trait at a particular locus is represented but once per haploid genome. Thus a homozygous *round* pea plant is assumed to contribute one copy of its *R* gene to a zygote (and not two or more); a homozygous *wrinkled pea* plant is assumed to generate pollen having only one *r* gene per gamete; and so on. If instead, the haploid pea genome were to contain two copies of *R* rather than one, then *R* ⟶ *r* mutations would be expected to be extremely rare since both dominant *R* genes would have to be mutated in the same cell to yield a recessive (*rr*) gamete. It is therefore widely assumed that most genes will be found in **single-copy** (also called **unique**) DNA.

Single-copy DNA should have the following property: when fragmented, dissociated, and incubated under reassociation conditions, a given strand should be able to reassociate with very few other strands in the DNA mixture; more specifically, only its original partner or a comparable fragment from some other genome will be perfectly complementary. Therefore, a relatively long time will be required for the reassociation reaction to go to completion, since most collisions will produce incorrect pairings. Single-copy DNA, in other words, is predicted to have the kinetic properties of the slowest-reassociating kinetic class of DNA in eukaryotic C_0t plots (Figure 7-11).

In most eukaryotes, single-copy DNA represents about 70 percent of the chromosomal DNA. While most genes are believed to reside in this DNA fraction, this does not necessarily mean that 70 percent of the eukaryotic genome contains genes. If all single-copy DNA were to represent genes, then a typical mammalian haploid genome, containing 3 pg of DNA, would contain about half a million genes. Most estimates put the number of genes in a mammalian haploid genome at a maximum of 50,000 (Chapter 18). Thus the single-copy fraction of eukaryotic DNA appears to contain at least 10-fold more DNA than is needed for genetic coding. An analogous excess may be present in the *E. coli* genome as well. The function of this extra single-copy DNA is not known.

7.11 HIGHLY REPETITIVE SEQUENCES

At the opposite extreme from the slowest-reassociating DNA in eukaryotic C_0t plots is the rapid reassociating DNA which forms duplexes at low C_0t values (Figure 7-11). When such rapid-reassociating fragments are isolated and analyzed, they prove to represent the **highly repetetive** DNA of the eukaryotic genome.

Highly repetetive DNA typically represents from 5-10 percent of the eukaryotic genome (but may be absent as, for example, in the insect *Chironomus*). In most cases this DNA is composed of a short sequence of bases repeated again and again, with from 10^5 to 10^7 repeats per genome. Since this highly reiterated fraction is usually (but not inevitably) somewhat different in its average base composition from the bulk of the DNA in the eukaryotic genome, highly repetitive DNA often bands in a satellite position in a CsCl gradient (Table 7-2); therefore, satellite DNA and highly repetetive DNA are often synonymous.

A number of eukaryotic satellite DNAs have been isolated and sequenced (Table 7-3). Guinea pig satellite DNA, for example, has as its basic repeating unit the sequence

CCCTAA
GGGATT,

which is repeated **in tandem** along long stretches of duplex DNA. Over the course of time, mutations have accumulated within this satellite DNA so that an occasional "module" of the DNA may carry a

CGCTAA
GCGATT

sequence, while another may contain a

CCATAA
GGTATT

sequence. Nonetheless, a basic six-base (hexamer) repeat pattern is con-

Table 7-3   Sequenced Satellite DNA's

Organism	Base Pairs per Repeat	Location	Sequence of One Strand (Satellite Designation)
Drosophila melanogaster (fruit fly)	5	Arms of Y chromosome and centric heterochromatin of chromosome 2 only; also distal end of 2L	AGAAG (polypurine) ATAAT (1.672)
	7		ATATAAT (1.672)
	10	Centric heterochromatin of all chromosomes and tip of 2L	AATAACATAG (1.686) AGAGAAGAAG (1.705)
Drosophila virilis	7	Centric heterochromatin	ACAAACT (I) ATAAACT (II) ACAAATT (III)
Cancer borealis (marine crab)	2	?	AT
Pagurus pollicaris (hermit crab)	4	?	ATCC
	3	?	CTG
Cavia poriella (guinea pig)	6	Centric heterochromatin	CCCTAA (α)
Dipodomys ordii (kangaroo rat)	10	Centric heterochromatin	ACACAGCGGG (HS-β)

From Tartoff, K., *Ann. Rev. Genet.* **9:**355–385.

served. Other satellite DNAs exhibit tetrameric units (as in the hermit crab), heptameric units (in *Drosophila*) and so on (Table 7-3), while bovine satellite I DNA appears to repeat only every 1400 bases.

Highly repetetive DNA exhibits rapid reassociation kinetics because the chances that a fruitful collision will occur are obviously very good if from 5-10 percent of the DNA in a mixture is of the same kind.

The chromosomal location of highly repetitive DNA is considered in Section 7.14 of this chapter.

7.12   MIDDLE-REPETITIVE SEQUENCES

Middle-repetitive DNA represents perhaps 1 to 30 percent of the eukaryotic genome and contains sequences that are repeated from a few times to hundreds of thousands of times. It is a far more heterogeneous class of DNA than the highly repetitive fraction. When sheared into fragments and dissociated, a middle-repetitive strand encounters a **related family** of sequences with which to reassociate: these are not perfectly complementary but are at least sufficiently similar that they can combine to form imperfect duplexes, as illustrated in Figure 7-12. Such duplexes are more sensitive to heat than are perfectly base-paired duplexes, but they are sufficiently stable

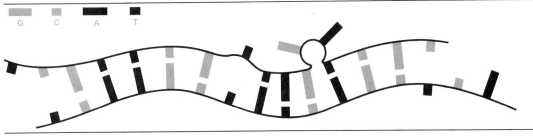

Figure 7-12 Imperfect DNA duplex. The two strands are generally complementary but occasional bases do not match. This reduces the stability of the molecule, as measured by its dissociation temperature. (After "Repeated Segments of DNA" by R. J. Britten and D. E. Kohne. Copyright © 1970 by Scientific American, Inc. All rights reserved.)

to bind to hydroxyapatite and score as reassociated DNA. Therefore, the middle repetitive class of DNA reassociates with kinetics intermediate between the highly repetetive and the single-copy DNA classes.

7.13   FOLD-BACK DNA

The final class of eukaryotic DNA, known variously as **fold-back, snap-back, palindromic,** or **inverted-repeat** DNA, in fact never features in a C_0t plot at all. When a chromosome sector carrying such DNA is reassociated, the single strands of the DNA find their inverted complements on their own strands and simply fold back and pair with these, exactly as we have seen for regions of internal complementarity in the MS2 chromosome earlier in this chapter (Figure 7-4). Since such folding back requires no fragment-to-fragment collisions and therefore occurs extremely rapidly, fold-back DNA can be isolated in a dissociation-reassociation experiment as that fraction of DNA that becomes duplex (that is, binds to hydroxyapatite columns) at zero time (that is, seconds after the beginning of reassociation). When individual *Drosophila* fold-back DNA molecules are examined with the electron microscope, 80 percent have the structure diagrammed in Figure 7-13: two long complementary stretches on the strand are separated by a stretch of noncomplementary DNA which loops out. The length of the loop ranges widely as does the length of the inverted-complement region, and these units are apparently interspersed with the other kinetic classes of DNA.

In both the human and *Xenopus* genome there are an estimated 120,000 fold-back sequences. This is a proper order of magnitude for postulating that these sequences associate with genes and function, for example, in transcription control or in recombination. However, such postulates are as yet without direct evidence. Several laboratories have also reported that fold-back sequences have the intriguing property of moving about in the

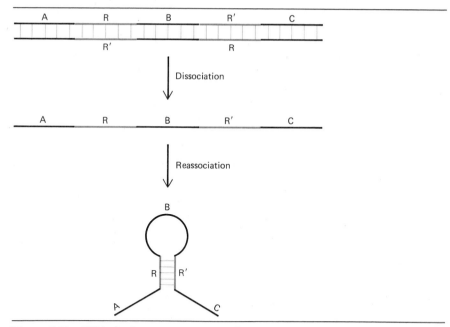

Figure 7-13 DNA duplex carrying inverted repeat (fold-back) sequences R and R′ separated by a nonrepeated sequence of length B. Dissociation and reassociation allows single strands to "snap back" on themselves and form base pairs in the R-R′ sector. (After C. Schmid, J. Manning, and N. Davidson, *Cell* **6**:159, 1975. Copyright © MIT.)

eukaryotic genome, earning them the name **jumping genes.** The significance of these observations is, again, unknown.

LOCALIZING CLASSES OF DNA IN CHROMOSOMES

7.14 *IN SITU* HYBRIDIZATION

Granted the existence of four "kinetic classes" of eukaryotic DNA, one next wishes to learn how these classes are distributed within the chromosomes. An effective technique for addressing this queston makes use of **nucleic acid hybridization,** which must first be described.

In a hybridization experiment DNA duplexes are dissociated and the resultant single strands are challenged with RNA or single-stranded DNA fragments that derive from another source, as, for example, another animal or another species. The DNA mixture is then allowed to incubate to permit duplex formation to occur. If the fragments and the dissociated DNA carry homologous base sequences, "hybrid" DNA-DNA or DNA-RNA duplexes will form. Where no homology exists, no hybridization will occur.

Nucleic acid hybridization can be performed in the test tube, and many important experiments in molecular genetics have utilized this technique, as seen in later chapters. With our focus now on chromosomes, we describe a technique known as *in situ* **hybridization,** in which dissociated DNA is left in place inside the cell nucleus, and then subjected to hybridization. Cells are first preserved by chemical fixatives and mounted on a glass slide suitable for light microscopy; they are treated with ribonuclease to remove RNA, and are then briefly exposed to warm alkali (NaOH) to denature the DNA; finally, they are incubated in a solution containing single-stranded radioactively labeled fragments. These may be DNA fragments collected from a particular band in a CsCl gradient. More commonly they are purified RNA species or RNA copies (**cRNA**) of particular DNA fragments, the copies being made *in vitro* by using DNA-dependent RNA polymerase (Section 8.3). The temperature of the preparation is adjusted to permit hybridization of the labeled strands with the denatured DNA exposed in the cell nuclei. After a suitable period the slide is washed so that only the hybridized, radioactively labeled nucleic acid remains associated with the cellular DNA. When cRNA is used, the slide is further exposed to ribonucleases which digest all the RNA molecules that are not part of DNA-RNA complexes. As a last step, the slide is subjected to autoradiography and stained for light microscopy.

When this procedure makes use of mouse cells and moderately repetitive or single-copy mouse DNA, the radioactive label is distributed uniformly throughout all the chromosomes, indicating that these DNA fractions are made up of base sequences that derive from all parts of the mouse genome. In contrast, when mouse repetitive satellite DNA is used as the labeled fragment, the label appears exclusively over constitutive heterochromatin (Section 2.13) which, in the mouse, is located adjacent to each centromere (Figure 7-14). Thus mouse satellite DNA, which constitutes 6 percent of the total DNA of the mouse cell and is comprised of repeated sequences of a $\frac{\text{GAAAAATGA}}{\text{CTTTTTACT}}$ module, is neither dispersed throughout the mouse genome nor confined to a single chromosome. Instead, this DNA occupies specific sites in *all* of the mouse chromosomes, with the possible exception of the Y chromosome.

In situ hybridization of satellite DNAs from a variety of plants and animals has been performed, and the general pattern seen is similar to that of the mouse: hybridization occurs over blocks of DNA in the pericentromeric regions, in portions of the Y chromosome, and often at the **telomeres** (chromosome ends)—all common locales of constitutive heterochromatin (Table 7-3). Since there is no evidence that satellite DNA codes for protein and since heterochromatin appears to be genetically inert (Chapter 18), it is generally assumed that if highly repetitive DNA has any function, it lies in the rather ill-defined category of "chromosome mechanics," possibly including chromosome folding at mitosis, meiotic pairing, meiotic chromosome separation, and the like.

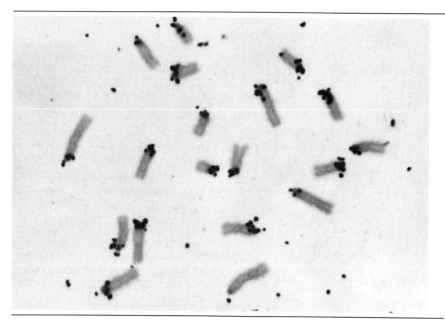

Figure 7-14 Autoradiograph of a mouse metaphase cell hybridized with radioactive cRNA copied *in vitro* from mouse satellite DNA. (From M. L. Pardue and J. G. Gall, *Chromosomes Today* **3**:47, 1971.)

Such speculations are tempered by the fact that chromosomes apparently lacking both constitutive heterochromatin and highly repetitive DNA appear able to fold, pair, and separate as well as their satellite-bearing counterparts. This situation is well illustrated by the human karyotype. Humans possess four species of satellite DNA. J. Gosden and collaborators have demonstrated that when these are individually hybridized to C-banded (Section 2.13) human chromosomes in which constitutive heterochromatin can be visualized microscopically, only about half the heterochromatic pericentromeric regions present in the nuclear DNA show any hybridization. Moreover, different chromosomes show very different degrees of affinity for the cRNAs, with an intensely reactive site being found at the C-band on chromosome 9, a less intensely reacting site in the long arm of the Y, a strongly responsive site at the centromeric regions of the acrocentric chromosomes 15, 21, and 22, less responsive sites on chromosomes 16, 17, and 20, and so on. Finally, individual satellites localize differently, with chromosome 13 hybridizing to satellites I, III, and IV, but not to II, chromosome 16 hybridizing only to satellite II; et cetera. Since some human chromosomes appear to be altogether devoid of satellite sequences, it is difficult to maintain that these DNAs play a major role in human chromosome maintenance. Why, then, are they so ubiquitous in the eukaryotic kingdom? This question does not, at present, have an answer.

7.15 SEQUENCE ORGANIZATION OF MIDDLE REPETITIVE AND UNIQUE DNA

Both middle-repetitive and unique DNA, as noted above, hybridize uniformly in *in situ* hybridization experiments, requiring that their chromosomal organization be deduced by other means. Recent work, notably by R. Britten, E. Davidson, and their associates, has revealed an ordered arrangement of these two classes of DNA: a short, middle-repetitive DNA sequence, with an average length of 300 nucleotide pairs, is followed by a single-copy sequence of about 1200 nucleotide pairs, which is followed by another 300-base-pair repeat, and so forth along the major portion of the chromosomal DNA length. This pattern can be detected by several types of experimental protocols, one being to shear human DNA into long (perhaps 8,000 base-pair) fragments, dissociate these, allow them to reassociate at intermediate $C_o t$ values (for example, 10^{-1}) where reannealing of repetitive-DNA is strongly favored, and examine the reassociated DNA by electron microscopy. When this is done, "double fork" structures are found; these consist of a central duplex region, about 300 nucleotides in length, from which two single-strands emerge at either end.

Approximately 50 percent of the middle-repetitive DNA of most eukaryotes is thus interspersed between 1200 base-pair sectors. It should be noted, however, that perhaps 20 percent of the mid-repetitive DNA lies between longer, 4000 base-pair segments; the remaining mid-repetitive DNA occurs in blocks longer than 300 nucleotide pairs, with its overall distribution unknown. It should also be noted that while a very wide range of organisms, including the silkmoth and the housefly, exhibit such a sequence organization (known as the "*Xenopus*" pattern), the insects *Drosophila* and *Chironomus* exhibit the "*Drosophila* pattern," in which middle-repetitive DNA exists not in short 300 base-pair units but in a heterogenous collection of segments ranging from a few hundred to more than 10,000 nucleotide pairs. This DNA is interspersed with long lengths of single-copy DNA, typically on the order of 13,000 base pairs.

Thus at present it can be said that the kinetic classes of eukaryotic DNA represent not merely experimental oddities but rather classes of DNA that have discrete localizations within the eukaryotic genome. They therefore, presumably have discrete but as yet undefined functions in the controlled expression and transmission of genetic information.

References

Methods of Characterizing DNA and RNA

*Eigner, J., and P. Doty. "The native, denatured and renatured states of deoxyribonucleic acid," *J. Mol. Biol.* **12**:549–580 (1965).

*Denotes articles described specifically in the chapter.

Gilham, P. T. "RNA sequence analysis," *Ann. Rev. Biochem.* **39**:227–250 (1970).

Kennell, D. E. "Principles and practices of nucleic acid hybridization," *Progr. Nucl. Acid Res.* **11**:259–302 (1971).

*Marmur, J., R. Rownd, and C. L. Schildkraut. "Denaturation and renaturation of deoxyribonucleic acid," *Progr. Nucl. Acid Res.* **1**:231–300 (1963).

*Maxam, A. M., and W. Gilbert. "A new method for sequencing DNA," *Proc. Nat. Acad. Sci. U.S.* **74**:560–564 (1977).

*Meselson, M. S., F. W. Stahl, and J. Vinograd. "Equilibrium sedimentation of macromolecules in density gradients," *Proc. Natl. Acad. Sci. U.S.* **43**:581–588 (1957).

Parish, J. H. *Principles and Practice of Experiments with Nucleic Acids.* New York. Wiley, 1972.

Sanger, F., and A. R. Coulson. "A rapid method for determining sequences in DNA by primed synthesis with DNA polymerase," *J. Mol. Biol.* **94**:441–448 (1975).

*Wetmur, J. G., and N. Davidson. "Kinetics of renaturation of DNA," *J. Mol. Biol.* **31**:349–370 (1968).

*Yang, R. C., A. van de Voorde, and W. Fiers. "Specific cleavage and physical mapping of Simian-virus-40 DNA by the restriction endonuclease of *Arthrobacter luteus*," *Eur. J. Biochem.* **61**:119–138 (1976).

Molecular Structure of Phage and Viral Chromosomes

*Danna, K. J., G. H. Sack, and D. Nathans. "Studies of Simian virus 40 DNA. VII. A cleavage map of the SV40 genome," *J. Mol. Biol.* **78**:363–376 (1973).

*Fiers, W., R. Contreras, F. Duerinck, G. Haegeman, D. Iserentaut, J. Merregaert, W. Min Jou, F. Molemans, A. Raeymaekers, A. Van den Berghe, G. Volckaert, and M. Ysebaert. "Complete nucleotide sequence of bacteriophage MS2 RNA: primary and secondary structure of the replicase gene," *Nature* **260**:500–507 (1976).

Jeppesen, P. G. N., L. Sanders, and P. M. Slocombe. "A restriction cleavage map of ØX174 DNA by pulse-chase labelling using *E. coli* DNA polymerase," *Nucleic Acids Res.* **3**:1323–1339 (1976).

*MacHattie, L. A., D. A. Ritchie, C. A. Thomas, and C. C. Richardson. "Terminal repetition in permuted T2 bacteriophage DNA molecules," *J. Mol. Biol.* **23**:355–363 (1967).

Nathans, D., and H. O. Smith. "Restriction endonucleases in the analysis and restructuring of DNA molecules," *Ann. Rev. Biochem.* **44**:273–293 (1975).

*Sanger, F., G. M. Air, B. G. Barrell, N. L. Brown, A. R. Coulson, J. C. Fiddes, C. V. Hutchison III, P. M. Slocombe, and M. Smith. "Nucleotide sequence of bacteriophage ØX174 DNA," *Nature* **265**:687–695 (1977).

Wadsworth, S., R. T. Jacob, and B. Roizman. "Anatomy of herpes simplex virus DNA II. Size, composition, and arrangement of inverted terminal repetitions," *J. Virol.* **15**:1487–1497 (1975).

*Weigel, P. H., P. T. Englund, K. Murray, and R. W. Old. "The 3'-terminal nucleotide sequences of bacteriophage λ DNA," *Proc. Nat. Acad. Sci. U.S.* **70**:1151–1155 (1973).

*Wu, R., and E. Taylor. "Nucleotide sequence analysis of DNA II. Complete nucleotide sequence of the cohesive ends of bacteriophage λ DNA," *J. Mol. Biol.* **57**:491–511 (1971).

Satellite DNA

Botchan, M. R. "Bovine satellite I DNA consists of repetetive units 1400 base pairs in length," *Nature* **251**:288–292 (1974).

Cooke, H. "Repeated sequence specific to human males," *Nature* **262**:182–186 (1976).

Eckhardt, R. A. "Cytological localization of repeated DNAs." In: *Handbook of Genetics V.* (R. C. King, Ed.) New York: Plenum Press, 1976, pp. 31–54.

Gall, J. G., and D. D. Atherton. "Satellite DNA sequences in *Drosophila virilis*," *J. Mol. Biol.* **85**:633–664 (1974).

Southern, E. M. "Long range periodicities in mouse satellite DNA," *J. Mol. Biol.* **94**:51–69 (1974).

Straus, N. A. "Repeated DNA in eukaryotes." In: *Handbook of Genetics* v. 5. R. C. King, ed. N. Y. Plenum Press, 1976 pp. 3–30.

Tartof, K. D. "Redundant genes," *Ann. Rev. Genet.* **9**:355–385 (1975).

Kinetic Analysis and Sequence Organization of Chromosomal DNA

*Britten, R. J., and D. E. Kohne. "Repeated sequences in DNA," *Science* **161**:529–540 (1968).

Britten, R. J., and E. H. Davidson. "Studies on nucleic acid reassociation kinetics: Empirical equations describing DNA reassociation," *Proc. Natl. Acad. Sci. U.S.* **73**:415–419 (1976).

Cavalier-Smith, T. "Palindromic base sequences and replication of eukaryotic chromosome ends," *Nature* **250**:467–470 (1974).

*Davidson, E. H., G. A. Galan, R. C. Angerer, and R. J. Britten. "Comparative aspects of DNA organization in metazoa," *Chromosoma* **51**:253–259 (1975).

Lee, A. S., R. J. Britten, and E. H. Davidson. "Interspersion of short repetetive sequences studied in cloned sea urchin DNA fragments," *Science* **196**:189–192 (1977).

Manning, J. E., C. W. Schmid, and N. Davidson. "Interspersion of repetitive and nonrepetetive DNA sequences in the *Drosophila melanogaster* genome," *Cell* **4**:141–155 (1975).

*Schmid, C. W., J. E. Manning, and N. Davidson. "Inverted repeat sequences in the *Drosophila* genome," *Cell* **5**:159–172 (1975).

In situ Hybridization

*Gosden, J. R., A. R. Mitchell, R. A. Buckland, R. P. Clayton, and H. J. Evans. "The location of four human satellite DNAs on human chromosomes," *Exp. Cell Res.* **92**:148–158 (1975).

Henderson, A. S., D. Warburton, and K. C. Attwood. "Localization of ribosomal DNA in the human chromosome complement," *Proc. Natl. Acad. Sci. U.S.* **69**:3394–3398 (1972).

Hennig, W., I. Hennig, and H. Stein. "Repeated sequences in the DNA of *Drosophila* and their localization in giant chromosomes," *Chromosoma* **32**:31–63 (1970).

*Pardue, M. L., and J. G. Gall, "Chromosomal location of mouse satellite DNA," *Science* **168**:1356–1358 (1970).

Rae, P. M. M. "The distribution of repetetive DNA sequences in chromosomes," *Adv. Cell. Mol. Biol.* **2**:109–149 (1972).

Timmis, J. N., B. Deumling, and J. Ingle. "Localization of satellite DNA sequences in nuclei and chromosomes of two plants," *Nature* **257**:152–155 (1975).

(a)   **(b)**

(a) Mary Lou Pardue developed, with Professor J. Gall at Yale, the technique of *in situ* hybridization. She is now at the Massachusetts Institute of Technology. (b) Charles A. Thomas, Jr. utilized reassociation studies to elucidate the structures of several phage chromosomes. He teaches at the Harvard Medical School.

Questions and Problems

1. The DNAs from two different cell types have identical GC content. Does this mean that the two DNAs have the same nucleotide sequence? The same nucleotide composition? The same length? The same buoyant density? The same reassociation kinetics? Explain your reasoning.

2. Does a fast-reassociating class of DNA necessarily have a satellite buoyant density? Explain your answer.

3. If the fast-reassociating fraction of guinea pig DNA is sheared into fragments, dissociated, reassociated, and examined with the electron microscope, what would you expect to see? What would you expect of the slow-reassociating fraction in a similar experiment?

4. Outline *in situ* hybridization experiments you would perform to determine whether the Type I human satellite DNA (33% GC) is similar in sequence to the repetitive satellite DNA from mouse (34% GC).

5. The experiment of C. Thomas on phage T4 (Figure 7-7) was repeated using the rod-shaped chromosomes of phage T7, and almost all the resultant duplexes were rod-shaped. What does this experiment reveal about the T7 chromosome?

6. When DNA is isolated from *D. melanogaster* and the related species *D. virilis* and subjected to CsCl density gradient centrifugation, the patterns shown on the facing page (top) are obtained, where the black lines represent DNA from brain tissues and the gray lines represent DNA from salivary glands. (a) How would you determine whether any of the *virilis* satellites are similar to the *melanogaster* satellites? (b) In what respect would you predict the S phase of a *Drosophila* brain cell cycle to differ from a salivary-gland cell cycle, based on the above patterns? (c) Would you expect polytene chromosomes to have much constitutive heterochromatin? Explain your answer.

7. Why does the single-copy DNA of calf thymus reassociate at a higher C_0t value than the *E. coli* DNA in Figure 7-11 (Hint: recall the relative genome sizes of calf and *E. coli*).

8. The *ter* enzyme cuts the λ chromosome only once. Describe the properties of a hypothetical mutant λ chromosome that would be cut twice by wild-type *ter*. Explain why such a mutational lesion would presumably be lethal. Would it help to search for a temperature-sensitive mutant? Explain.

9. Circular duplex DNA from an animal virus is digested by the restriction enzymes *EcoR* 1 and *Hpa* to produce the sets of gels shown on the bottom of the facing page (numbers refer to molecular weight $\times 10^6$, as determined by standard gels). (a) What is the molecular weight of the active chromosome? (b) Draw a restriction map of the chromosome, showing the positions of sectors A–F. (c) What would you expect to find if each of the 4 *Eco* fragment classes were individually cut from gels and digested from *Hpa*?

10. The 5 histone proteins (Chapter 2) are very similar to one another in all

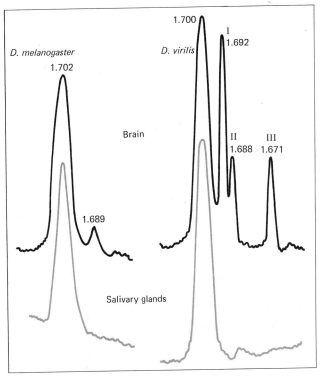

Buoyant density (ρ) (g/cm³)

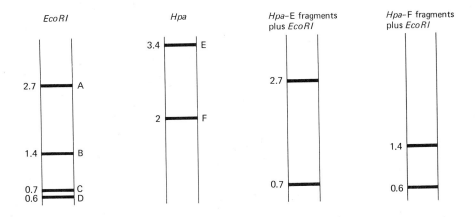

eukaryotes; the histone genes are repeated a number of times in the eukaryotic genome; such "histone DNA" from sea urchin bands in a distinct region of a CsCl gradient so that it can be isolated and purified. Using the above information, describe how you would locate the chromosomal location of histone genes in *Drosophila* and in humans.

11. A strain of bacteria was cultured in ^{15}N ammonium ion for many generations. When the DNA from this organism is sedimented, a single band appears at a density of 1.714. The organism is then removed from its ^{15}N source, and replicated in ^{14}N for exactly one cell division. Now its DNA bands at two places in equal amounts, 1.709 and 1.705. After a second round of ^{14}N replications, a fourth band appears at 1.700. This band becomes more intense after each further cell division in ^{14}N. Explain this data in terms of the strands' base composition. Hint: is the base composition of each strand of DNA necessarily equal?

12. Lambda DNA has been shown to have the structure

5'_____3'
   3'_____5'

How could this have been distinguished from:

5'_____3'
3'_____5' ?

13. Diagram the double-fork structures described in Section 7.15. Do you expect the four single strands to be of equal or unequal length? Explain.

CHAPTER
8

Genes and Gene Transcripts

INTRODUCTION

This chapter and the chapter that follows are broadly concerned with a key activity of the genetic material: its ability to direct protein synthesis. This process is conveniently thought of as occurring in two major stages, **transcription** and **translation,** which are summarized diagrammatically in Figure 8-1. Many of the details of this figure will become meaningful as the two chapters progress.

The present chapter describes the molecular biology of transcription and, in the process, the molecular structure of the genes that are transcribed. The chapter also describes those salient features of gene transcripts that are essential for their biological activity. Throughout the chapter, heavy emphasis is placed on the features of these subjects that have a direct relevance to genetics.

MOLECULAR BIOLOGY AND GENETICS OF TRANSCRIPTION

8.1   GENERAL FEATURES OF TRANSCRIPTION

The transcription of genes is superficially similar to DNA replication (Section 5.3): a polymerase enzyme matches complementary nucleotides along a DNA template according to the Watson-Crick base-pairing rules, and the bases are then polymerized to form a polynucleotide copy of the original strand. The polynucleotide transcripts are usually single-stranded RNA molecules; the enzyme that forms these copies is thus called **DNA-dependent RNA polymerase,** (abbreviated as **RNA polymerase** in this text) and ribonucleotides, rather than deoxyribonucleotides, are the substrates for the polymerization reaction. The base uracil replaces thymine in RNA, as noted in Chapter 1, but otherwise the DNA strand is faithfully copied by the polymerase enzyme. As with DNA replication, polymerization proceeds in a 5' $\longrightarrow$ 3' direction and the transcript has opposite polarity to the template, so that the sequence 3' TACAAC 5' in DNA is transcribed as 5' AUGUUG 3' in RNA.

The "purpose" of transcription is to allow gene expression. Genes must remain in chromosomes where their replication, repair, and transmission are assured, yet they must also be able to direct the activities of the cell, most notably protein synthesis. They therefore produce transcripts which diffuse away from the chromosomes and participate in protein synthesis. The transcripts do not, therefore, remain hydrogen-bonded to the template the way daughter chromosomes do; instead, they "peel off" the chromosome, as illustrated in Figure 8-1. They are not, moreover, endowed with the same kind of permanence as the original genetic material; instead, they are typically degraded by cellular ribonucleases once their functional usefulness has been spent.

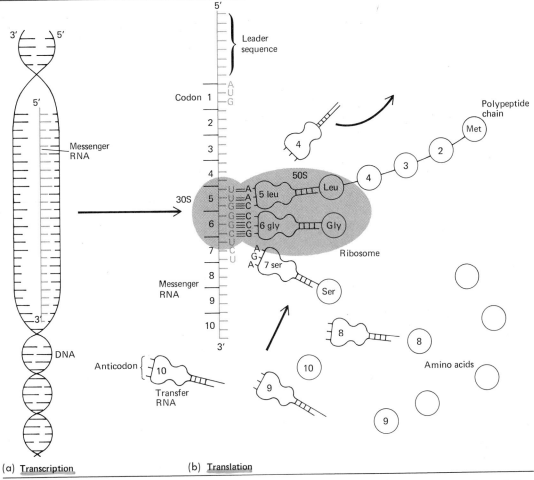

(a) Transcription (b) Translation

Figure 8-1 A schematic view of the principal features of transcription and translation. (a) During transcription a strand of the DNA helix serves as a template for the synthesis of a complementary RNA copy or transcript (in this case, messenger RNA). This then "peels off" the DNA template. (b) During translation, three classes of RNA interact with a variety of enzymes and proteins to generate the formation of a new polypeptide chain. Ribosomal RNA is a component of the ribosomes that serve as a kind of scaffolding for the process of polypeptide synthesis. The ribosomes contain a large (50S) and a small (30S) subunit. Transfer RNA interacts with amino acids and mediates their correct insertion into the growing polypeptide chain. Messenger RNA carries the information contained in a gene to the ribosome. The information is encoded as groups of three nucleotides, with each specifying a particular amino acid. Each codon is recognized by a complementary anticodon on a transfer RNA molecule which has previously associated with that particular amino acid. In the figure most of the amino acids are represented by numbered circles; the amino acid glycine has just been bound to its site on the ribosome by the corresponding transfer RNA. It will form a peptide bond with leucine, thereby extending the growing polypeptide chain. The ribosome then moves the length of the codon along the messenger RNA and so comes in position to bind the transfer RNA carrying serine. (From *Ribosomes* by M. Nomura, copyright © October 1969 by Scientific American, Inc. All rights reserved.)

see fig. 9-6
p. 313

8.2   DEFINING GENES AND GENE TRANSCRIPTS

A **structural gene** is defined as a sequence of nucleotides which specifies the amino acid sequence (that is, the structure) of a polypeptide. Associated with the beginning and end of each structural gene are nucleotide sequences known as **controlling elements** which are involved in the regulation of transcription. The interactions between controlling elements, RNA polymerase enzymes, and other regulatory proteins are considered in Chapters 17 and 18. Here it is sufficient to note the existence of two ubiquitous types of controlling elements: a **promoter** sequence to which the RNA polymerase must bind if it is to transcribe the gene into mRNA, and a **terminator** sequence which signals that the polymerase should dissociate from the template.

We must note at this point a special arrangement of structural genes known as an **operon.** In viruses and bacteria, groups of two or more related structural genes often share a single promoter and a single terminator and can thus be denoted as promoter . . . gene 1 . . . gene 2 . . . . gene *n* . . . terminator. Other controlling elements, such as **operators,** may be associated with this unit as well, as described in detail in Chapter 17. In the case of operons, then, several genes will be copied into a single long RNA transcript.

The initial transcript of most (and perhaps all) structural genes (and operons) must be modified before it acquires full biological activity. Such **post-transcriptional processing** may involve adding nucleotide sequences onto the original transcript, cleaving sequences from the original transcript, or modifying existing bases by, for example, methylation. The final **processed** transcript of a structural gene is known as **messenger RNA (mRNA)**; for operons, the transcript is sometimes specified as a multigenic mRNA. The mRNA molecules are then translated into polypeptide chains under the aegis of the **translation apparatus** (Figure 8-1 and Chapter 9) of the cell.

Structural genes represent most of the genes in a chromosome, and when "genes" are considered in this and subsequent chapters, usually the term will refer to structural genes. Bacteria and eukaryotes possess, in addition, several kinds of genes whose RNA transcripts serve not as messenger intermediaries but as the final products of gene expression. Included here are the genes that specify **ribosomal RNA (rRNA)** and **transfer RNA (tRNA)** molecules, both of which perform critical functions in the translation of mRNA (Figure 8-1). The rRNA and tRNA genes are associated with promoter and terminator controlling elements, and their initial RNA transcripts undergo an elaborate processing before they assume their biologically active form. The tRNA and rRNA transcripts are not, however, translated into polypeptide chains. The tRNA and rRNA genes are mentioned here because, as is evident in later sections of the chapter, their analysis has figured prominently in our present understanding of gene structure.

8.3 THE RNA POLYMERASES

An RNA polymerase enzyme must have the following properties: 1) it must be able to recognize promoter controlling elements in the double-stranded state of DNA; 2) it must be able to "burrow into" the DNA duplex at the proper promoter region and unwind the initial sequence of the gene for transcription; 3) it must copy the gene accurately; and 4) it must stop transcribing when it encounters and recognizes terminator controlling elements.

The *E. coli* RNA Polymerase The most widely studied RNA polymerase enzyme, that from *E. coli,* possesses 5 distinct polypeptides—**α, β, β′, ω** and **σ**—the α polypeptide being represented twice (Figure 8-2).

A combination of genetic and biochemical approaches has helped elucidate the roles these various *E. coli* polypeptide subunits play in gene transcription. The antituberculosis drug **rifampicin,** for example, was found to inhibit *E. coli* RNA polymerase activity in vitro and to kill *E. coli in vivo.* Rifampicin-resistant mutants were therefore isolated (Section 6.5) and were found to possess rifampicin-resistant polymerase activity. W. Zillig and coworkers separated the subunits from both the rifampicin-resistant and rifampicin-sensitive polymerase enzymes, mixed them in all combinations, and allowed "hybrid" enzymes to reassemble (a **mixed reconstitution** experiment). Of these hybrid enzymes, only the enzyme reconstituted with the β subunit from the rifampicin-resistant enzyme proved to be rifampicin-resistant *in vitro.* Since rifampicin interferes with the initiation of RNA transcripts, these experiments are consistent with the concept that the β subunit is involved, at least in part, in the initiation of transcription. Other studies have shown that if the σ subunit becomes dissociated from the enzyme, the remaining "core enzyme" (Figure 8-2) begins transcription at random along the *E. coli* DNA template and copies both strands but that when σ is added back, transcription begins at true promoter sites and copies only one strand. The σ subunit, then, is critical for promoter recognition.

The genes for the β and β′ subunits of the *E. coli* enzyme, denoted as *rpoB* and *rpoC* (for **R**NA **po**lymerase), appear to lie adjacent to one another in the *E. coli* chromosome at 88.5 minutes on the genetic map (Figure 8-3); the gene for α (*rpoA*) is not, however, found in this grouping but at 72 minutes (Figure 8-3), while the *rpoD* gene for σ maps at 66 minutes.

Viral RNA Polymerases Phages and viruses employ various strategies for copying their genes. Most appear to make use of an unmodified version of existing host enzymes early in their infection cycle. Some add phage-coded polypeptide subunits onto the host enzyme in place of sigma, presumably so that the resulting polymerase acquires a strong preference for phage promoters and therefore preferentially transcribes the phage genome.

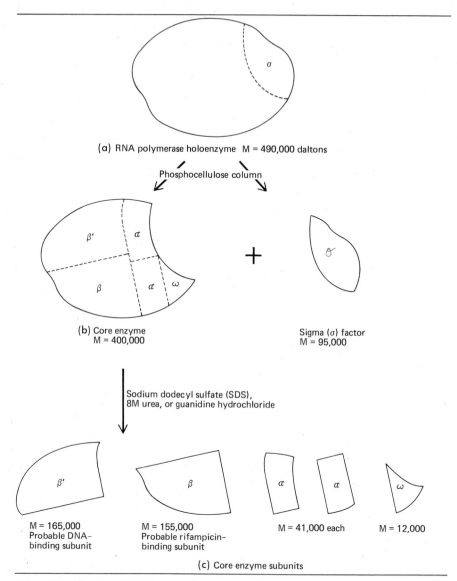

(a) RNA polymerase holoenzyme M = 490,000 daltons

Phosphocellulose column

(b) Core enzyme
M = 400,000

Sigma (σ) factor
M = 95,000

Sodium dodecyl sulfate (SDS),
8M urea, or guanidine hydrochloride

M = 165,000
Probable DNA–
binding subunit

M = 155,000
Probable rifampicin-
binding subunit

M = 41,000 each

M = 12,000

(c) Core enzyme subunits

Figure 8-2 The *E. coli* DNA-dependent RNA polymerase. All subunit shapes are highly schematic; the correct molecular structures are not yet known. (a) The intact molecule or holoenzyme. (b) The separated core enzyme and sigma units. (c) Subunits of the core enzyme.

Finally, some specify an RNA polymerase that functions independently of any host proteins. Mutations in all of these phage-coded proteins have been described, and details of their role in the phage infection cycle are found in Chapter 17.

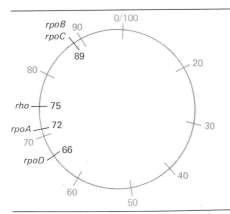

Figure 8-3 Genes involved in DNA transcription in *E. coli*.

Eukaryotic RNA Polymerases Three RNA polymerases, **I, II** and **III,** are present in eukaryotic nuclei. As summarized in Table 8-1, these enzymes transcribe different classes of genes: polymerase I, for example, resides in the nucleolus, is specialized for the synthesis of the large rRNA species known as 18S and 28S (described later in this chapter), and is resistant to the inhibitor **α-amanitin** (derived from the mushroom *Amanita phalloides*), whereas polymerase II is highly sensitive to α-amanitin. All three types of enzyme are composed of multiple subunit polypeptides, and the mutant approach to their analysis (for example, the characterization of RNA polymerases from α-amanitin-resistant cultured cells) is now beginning to be utilized.

8.4 TRANSCRIPTION OF THE SENSE STRAND OF A GENE

The statement that the sequence of nucleotides in a gene defines the amino acid sequence of a polypeptide refers to only one strand, the **sense strand,** of a duplex gene. Thus if the DNA base sequence 5′ . . . . . ATGTTTCAGACT . . . . 3′ codes for a sequence of several amino acids in a protein, then the

Table 8-1 Location and General Functions of Animal Cell RNA Polymerases

Enzyme Class	Subnuclear Localization	Cellular Gene Transcripts	Viral Gene Transcripts
I	nucleolus	18S, 28S rRNAs	none identified
II	nucleoplasm	hnRNA, mRNA	mRNA precursors
III	nucleoplasm	tRNAs, 5SRNA	low molecular weight RNAs

base sequence of the complementary DNA strand, 3′ . . . TACAAAGTCTGA . . . 5′ is dictated by the Watson-Crick base-pairing rules followed by DNA polymerase enzymes and not by any dictum that it, too, should contain a "meaningful" sequence. In fact, the sole function of this **antisense strand** is to generate a complementary sense strand for use by the next generation of a virus or cell. (Of course, the sense strand of a gene will, for its part, give rise to an antisense strand during DNA replication). The complete structure of a given gene can therefore be depicted as:

During transcription, then, only the sense strand of a gene is transcribed. In the case of small viruses such as the SP8 phage of *Bacillus subtilis*, all of the genes are located on one strand of the viral chromosome and this strand alone is transcribed; in the case of viruses such as T4, on the other hand, one strand of the duplex is the sense strand for certain genes while the other strand bears the sense sequences for other genes. We describe below experiments that demonstrate these features of gene transcription—experiments that depend on an important technique in gene transcription studies, known as DNA-RNA hybridization.

DNA-RNA Hybridization, first developed by B. Hall, D. Gillespie, and S. Spiegelman, is identical in principle to the *in situ* hybridization method described in Section 7.14 except that the dissociated DNA strands obtained from chromosomes are immobilized on a membrane filter rather than being left in the cell; the filter is then incubated with radioactively-labeled RNA transcripts, is washed, exposed to a ribonuclease that destroys all non-annealed RNA, and "counted" by a **scintillation counter** that assesses its level of radioactivity. Stable hybrids will form only between the RNA transcripts and those segments of the DNA strands that bear base sequences complementary to the transcripts.

J. Marmur and colleagues applied this technique to the SP8 phage of *Bacillus subtilis,* taking advantage of the fact that two strands of the phage chromosome can be separated from one another in a CsCl gradient (Box 5.1) because one strand (the heavy or **H strand**) contains relatively more purines than the light or **L strand,** giving each strand a discrete buoyant density ($\rho = 1.762$ for the *H* strand and 1.756 for the *L* strand as shown in Figure 8-4). Marmur and his colleagues immobilized the *H* strands of the SP8 phage on one set of membrane filters and the *L* strands on a second. They then allowed SP8 phages to infect *B. subtilis* cells in the presence of ^{32}P, isolated the mRNA that formed, and performed hybridization studies with each set of filters. Significant levels of radioactivity were found to be associated *only* with the filters bearing the *H* strands. Thus in SP8, only the *H* strand is transcribed into RNA and it is, therefore, defined as the sense strand.

Experiments by R. Jayaraman and E. Goldberg are among those demonstrating that, for T4, the sense strand corresponds to the *H* strand for some

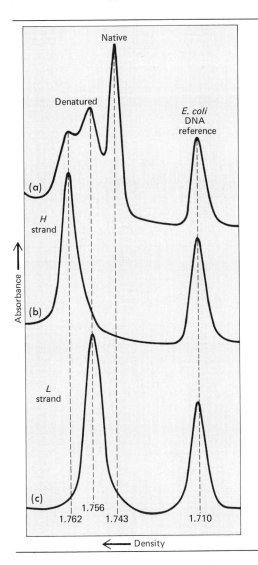

Figure 8-4 Buoyant densities of native, heat denatured, and fractionated strands of SP8 DNA, using *E. coli* DNA as a reference standard. (a) Native and denatured SP8 DNA. (b) *H* strand after purification through an MAK column. (c) Purified *L* strand. (From J. Marmur et al., *Cold Spring Harbor Symp. Quant. Biol.* **28**:191, 1963.)

genes and to the *L* strand for other genes. Previous workers had established that during the T4 infection cycle (Section 5.10) certain genes are transcribed early whereas other genes are transcribed late. Jayaraman and Goldberg therefore began by isolating "early" and "late" RNA transcripts from T4-infected *E. coli* cells. They then isolated wild type T4 chromosomes, dissociated these into *H* and *L* strands, and prepared *H*-strand filters and *L*-strand filters in much the same way as did Marmur in the experiments described above. Each RNA sample was next hybridized with each filter, and after hybridization each preparation was treated with an endonuclease that would digest all of the DNA in the filter preparations except those segments that had formed stable DNA-RNA hybrids. In other words, the lengths of

DNA remaining on the filter represented those DNA segments that had been transcribed: if the *H* strand contains a gene that is transcribed early, this gene will be protected from nuclease digestion when hybridization is carried out with early RNA but not when hybridization is carried out with late RNA.

The final step in Jayaraman and Goldberg's experiment was to determine which of the 4 sets of protected DNA fragments (which we can designate as *H*/early-RNA, *H*/late-RNA, *L*/early-RNA and *L*/late-RNA) contained early genes and which contained late genes. This was accomplished by gene recombination experiments. Bacterial cells (*Aerobacter aerogenes*) were treated in such a way that they could be induced to take up both T4 chromosomes and DNA fragments and to mediate recombination between the two kinds of DNA. Four cultures of these bacteria were first presented with each of the 4 sets of protected DNA fragments along with T4 chromosomes carrying mutations in an early gene known as *rIIb*. Recombinant

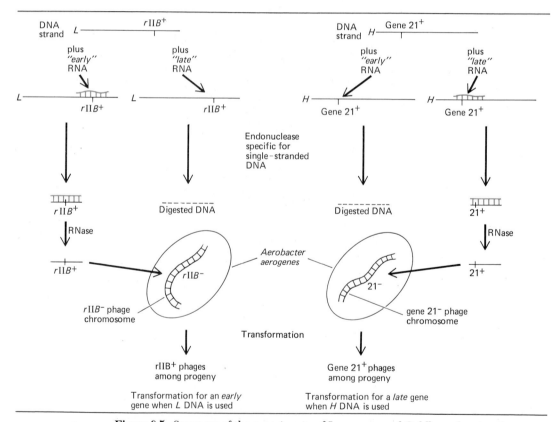

Figure 8-5 Summary of the experiments of Jayaraman and Goldberg showing that genes on the different strands (*H* and *L*) of T4 DNA are transcribed either "early" or "late" in the phage growth cycle.

rIIb⁺-containing chromosomes proved to emerge only from those bacteria harboring *L*/early fragments. The experiment was then repeated, this time using T4 chromosomes carrying mutations in a late gene known as gene 21. In this case, *gene-21⁻* chromosomes could be converted to *gene-21⁺* wild-type chromosomes only when *H*/late RNA fragments served as the source of DNA. These results are diagrammed in Figure 8-5.

The Jayaraman-Goldberg experiments, in summary, indicate that the information for at least one early gene is encoded in the *L* strand of the T4 chromosome, whereas at least one late gene is encoded in the *H* strand. Other experiments have expanded this conclusion: it appears that most of the early genes are encoded in the *L* strand and most of the late genes are in the *H* strand of the T4 genome. This means that in visualizing the chromosome of T4 we can think of a given single strand of DNA as having long sequences of bases containing meaningful information, followed by long sequences of bases that do not code directly for protein and presumably code for nothing at all.

8.5 STRUCTURE OF A SENSE STRAND AND ITS TRANSCRIPT

We can now focus more closely on a sector of a chromosome that carries a structural gene. Since chromosome is transcribed in a 5′ ⟶ 3′ direction and since the resultant mRNA is translated in a 5′ ⟶ 3′ direction (Figure 8-1), it follows that the beginning of the gene will lie to the 3′ end of the coding sequence and the conclusion of the gene will lie at the 5′ end:

As indicated in Figure 8-1 and detailed fully in Chapter 9, the first amino acid in a polypeptide chain is usually methionine, which is specified by the sequence 5′-AUG-3′ in mRNA. We can therefore be more specific:

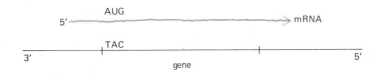

You will note that the AUG is not drawn at the 5′ end of the mRNA. The nucleotides that lie between the 5′ end and the AUG constitute what is called a **leader sequence**, and a complementary "anti-leader" sequence will, of course, be present in the DNA. Moreover, the initial nucleotide at the 5′-end of an RNA transcript is found to be a purine (5′-pppA or 5′-pppG), meaning that the DNA will carry an **initiating pyrimidine for transcription** (C or T). Incorporating this information, we have

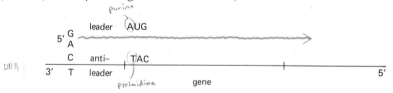

Finally, we must add the promoter, defined as the site at which an RNA polymerase attaches to DNA and initiates transcription, and the terminator, where transcription is concluded. A **trailer** sequence may be transcribed between the conclusion of the gene and the terminator sequence.

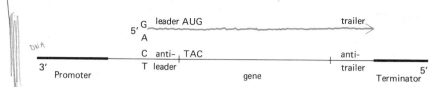

The above diagram of a DNA strand and its mRNA transcript allows us to imagine an RNA polymerase enzyme binding to the promoter at the left and "flowing" into the gene. Such a picture of transcription has led to the useful terms **upstream** and **downstream**: the promoter is said to be up-stream from the gene; the antitrailer sequence is said to be downstream from the gene; and so on.

We must at this point deal directly with a somewhat difficult matter of notation. Many biologists have become accustomed to reading nucleotide sequences with the 5′ nucleotide on the left-hand side of the page and the 3′ nucleotide on the right-hand side. This is, of course, the logical way to write the sequence of an mRNA molecule since it parallels the direction of mRNA translation. The difficulty comes in deciding how to denote the sequence of a gene. In the above paragraphs the **transcribed** strand of the gene has been written with the 3′ end on the left, a convention we shall generally continue to use throughout the text. Most publications, however, depict the **non-transcribed strand** of a gene with the 5′ end on the left, a sequence identical to the sequence of the mRNA transcript of that gene (except, of course, that thymines replace uracils).

8.6   LEADER SEQUENCES

A number of mRNA species have been purified and subjected to RNA sequencing (Box 7.3), starting at the 5′ end. The AUG translational start signal defines the beginning of the amino-acid-coding portion of the

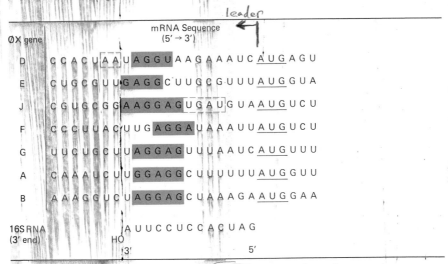

Figure 8-6 Leader sequences of ΦX174 mRNAs for the indicated genes, where sequences complementary to the 3' end of 16S rRNA are boxed. Broken lines indicate further complementarity if some nucleotides are looped out or not matched. Ribosome binding to mRNA has been demonstrated in these regions for genes J, F, G, and B. (From F. Sanger et al., *Nature* 265:687, 1977.)

molecule, and the nucleotides between the 5'-P and the adenine of the AUG constitute the leader.

The length of a leader varies considerably from one species of mRNA to the next. Of the mRNAs specified by the phage ΦX174, for example, one has a leader of 18 nucleotides and another has a leader of over 175 nucleotides. The significance of this variation is not yet known. Examination of all sequenced mRNA leaders reveals, however, a common feature: all carry a sequence of nucleotides, just upstream from the AUG, which is homologous to the 3' end of 16S rRNA (see Section 8.12). These homologies, some of which are diagrammed in Figure 8-6, are thought to have functional significance. The 16S rRNA sequence is believed to be exposed on the ribosome surface, and base-pairing between the 16S and the leader sequence is thought to be a key event in binding the mRNA to the ribosome for translation.

8.7 PROMOTER SEQUENCES

Several approaches have been taken to identify promoter sequences and learn how they might be recognized by polymerase enzymes. One approach, pioneered by H. Schaller, involves incubating bacterial or viral chromosome fragments with RNA polymerases under *in vitro* conditions in which transcription cannot proceed. The mixture is then exposed to pancreatic DNase. The DNase digests all the DNA except that which is pro-

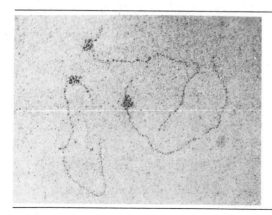

Figure 8-7 RNA polymerases bound to 3 Pr promoters of phage λ, located 100 base pairs from the ends of 3 *Hae* III restriction fragments. (From J. Hirsh and R. Schleif, in *RNA Polymerase*, R. Losick and M. Chamberlin, Eds. Cold Spring Harbor Laboratory, 1976.)

tected by its association with RNA polymerase (see Figure 8-7) and these "polymerase binding sites" are then isolated and subjected to DNA sequencing (Box 7.2).

Figure 8-8 shows 4 such polymerase-protected sequences plus several other promoter-containing regions identified by other means. There is no

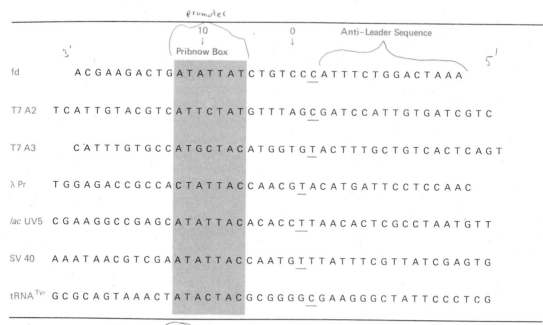

Figure 8-8 DNA sequences (reading 3′ ⟶ 5′ from left to right) that include putative 7-nucleotide promoter sequences, (often called **Pribnow boxes** after D. Pribnow, who first noted them). The initiating pyrimidines are underlined, and anti-leader sequences are given. Transcription would proceed from left to right. A2 and A3 are two T7 early promoters, Pr is the rightward early promoter of λ, and UV5 is a mutant high-level promoter of the *lac* operon; all are considered in Chapter 17. (After W. Gilbert, in *RNA Polymerase*, R. Losick and M. Chamberlin, Eds. Cold Spring Harbor Laboratory, 1976.)

reason to assume, of course, that the 41–44 nucleotides in each polymer-ase-protected fragment will all be part of the promoter; the promoter itself might be very short, but the bulky polymerase enzyme bound to it would shield several additional turns of the helix from pancreatic DNase digestion (Figure 8-7). Indeed, as shown in Figure 8-8, the initiating pyrimidine (underlined) and portions of anti-leader sequences are present in all of the protected fragments that have been studied.

A box is drawn in Figure 8-8 around a set of 7 bases, generally homolo-gous with the sequence 3'-ATAPyTAPy-5' (where Py is either T or C), which are centered about 10 nucleotides upstream from the initiating pyrimidine. That these 7 bases may constitute at least a portion of the promoter is most convincingly argued by studies of strains carrying promoter mutations. A **promoter mutation** either blocks or enhances the transcription of a partic-ular gene, and all promoter mutations that have been sequenced to date carry alterations in one of these 7 nucleotides. The current hypothesis, therefore, is that the polymerase binds tightly to the DNA at the boxed promoter region and proceeds to initiate RNA synthesis at a site 10 nucleo-tides (or one helical turn) downstream.

RNA CHAIN ELONGATION AND TERMINATION

Once RNA synthesis is initiated it appears to proceed at a uniform rate—from 40 to 50 nucleotides per second in *E. coli* at 37°C—presumably in concert with a "melting" of the template approximately one turn-of-the-helix in front of the polymerase and a "closing" of the helix in the wake of the enzyme. Growth of the chain is always in the 5' ⟶ 3' direction so that a pppA or a pppG initial purine marking the 5' end of the molecule exists as a ribonucleoside 5' triphosphate.

Termination of transcription occurs when the polymerase comes to the **terminator sequence** at the end of a gene or operon. Most prokaryotic transcriptional units that have been sequenced are found to end with a string of from 4–12 contiguous GC base pairs followed by a string of AT pairs. Similarly, most prokaryotic mRNAs terminate with the sequence 5'-UUUUUUA-3'. Since the GC-rich region of the helix would be expected to be "tight," having 3 hydrogen bonds per base pair (Figure 1-6), while the AT-rich region would be relatively "loose," a prevalent model for the termination event holds that an RNA polymerase is in some way signaled by this tight-loose progression to dissociate from the template.

Termination in *E. coli* is abetted by the protein **rho** whose gene (*rho*) maps at 75 minutes on the *E. coli* chromosome (Figure 8-3). When rho is added to many *in vitro* transcription systems, termination is far more accurate than when it is absent; however, the mode of action of rho in termination is not yet known.

Having such an overall picture of the transcription process, we can now consider specific genes and gene transcripts.

TRANSFER RNA AND tRNA GENES

8.8   ROLE OF TRANSFER RNA IN PROTEIN SYNTHESIS

Transfer RNA (tRNA) molecules play a key role in protein synthesis. Each of these molecules possesses the capacity to combine specifically with only one amino acid in a reaction mediated by a set of amino acid-specific enzymes called **aminoacyl-tRNA synthetases.** For example, a tRNA that binds to leucine (and is thus referred to as tRNA^{Leu}) does so in the presence of leucyl-tRNA synthetase, while a tRNA^{Val} binds to valine in the presence of valyl-tRNA synthetase, and so on. There often exists, moreover, more than one kind of tRNA for a given amino acid; thus *E. coli* possesses two species of tyrosine tRNA, tRNA$_1^{Tyr}$, and tRNA$_2^{Tyr}$, which are encoded by distinct genes and have distinct structures but which both bind to tyrosine. Such tRNAs are known as **isoaccepting tRNAs.**

Once the tRNAs are "charged" with their appropriate amino acids, they migrate to specific sites on the ribosome and interact with the ribosome-associated messenger RNA, as diagrammed in Figure 8-1b. Messenger RNA contains a series of **codons,** each codon being a sequence of three nucleotides which dictates that a certain amino acid must be placed in a certain position in the growing polypeptide chain. For this purpose, each mRNA codon is recognized by a complementary sequence of three nucleotides, called the **anticodon,** in a specific tRNA molecule. Thus the anticodon of a tRNA^{Gly} will match up with the mRNA codon specifying the amino acid glycine, whereas the anticodon on a tRNA^{Ser} must find an mRNA codon that specifies serine. Once the correct matching has occurred, the amino acid on the tRNA molecule becomes incorporated, via a peptide bond, into the polypeptide that is being assembled on the ribosome. The tRNA is then released from the ribosome and is free to combine with another amino acid.

It should be apparent from this description and from Figure 8-1 that a tRNA molecule performs highly complex functions: it interacts with a specific synthetase enzyme, possesses a binding site for a specific amino acid, possesses a second site for interacting with a ribosome, and contains an anticodon that must be exposed to the mRNA codon.

8.9   GENES FOR tRNA

At least 40 different kinds of tRNA are believed to be present in an *E. coli* cell and perhaps 60 different kinds are present in a eukaryotic cell; therefore, there must be a minimum of 40 or 60 tRNA genes present in every species' genome. It turns out that this minimum is exceeded: in both prokaryotes and eukaryotes, there is more than one gene present in the genome for most, if not all, tRNAs.

tRNA Genes in *E. Coli* In *E. coli*, information for certain tRNA species may be present only once per genome, but there is increasing evidence for what is called tRNA **gene redundancy.** In some cases the redundant genes may lie side by side (a **redundant tandem cluster**): for example, at 94 min on the genetic map of *E. coli* (Figure 8-9), 2–3 closely linked identical copies of the gene *glyV* are present, each coding for tRNA$_3^{Gly}$. In other cases the redundant genes may lie in different parts of the chromosome: thus at 42 min on the *E. coli* genetic map, gene *glyW* is present, and this also specifies tRNA$_3^{Gly}$. Finally, a given cluster of tRNA genes need not contain redundant sequences: at position 88 min on the genetic map of *E. coli* there exists the tandem series *tyrU glyT thrT* which code, respectively, for tRNA$_2^{Tyr}$, tRNA$_2^{Gly}$ and tRNA$_3^{Thr}$; such a grouping can be described as a **nonredundant tandem cluster.**

Figure 8-9 shows the map positions of the known tRNA genes in *E. coli*. These map positions have been determined in one of two ways. In some cases a tRNA gene will undergo mutation and give rise to **suppressor tRNAs** that have anomalous properties in protein synthesis, as described in Section 9.20; the map position of the gene giving rise to this suppressor phenotype can then be determined by genetic crosses. The chromosomal location of other tRNA genes has been deduced from their presence in specialized transducing phages (Section 5.11) of λ or φ80. Thus the phage λ*rif*^{d} 18 was selected because its "picked-up" *E. coli* DNA included the *rif* gene (now called *rpoB*), which codes for the β subunit of RNA polymerase and maps at 88.5 min (Figure 8-3). When λ*rif*^{d}18 chromosomes were cleaved with restriction endonucleases, one of the resultant fragment classes proved to contain a

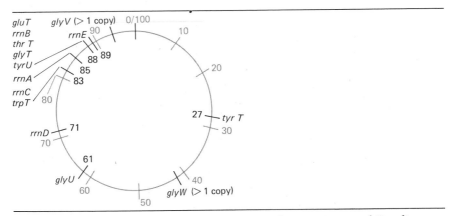

Figure 8-9 Known tRNA and rRNA (*rrn*) genes in the genetic map of *E. coli*.

totally unexpected sequence for $tRNA_2^{Glu}$. This tRNA gene, therefore, is at once localized close to *rpoB* on the genetic map.

tRNA Genes in Eukaryotes   Estimates of tRNA gene redundancy in eukaryotes have been made by hybridizing eukaryotic DNA and a total-tRNA preparation from the same species. While such experiments can at best yield only approximations, it is found, for example, that about 0.08 percent of the yeast genome will hybridize with tRNA; knowing the yeast genome size, this extrapolates to give about 400 tRNA genes in the yeast genome, or an average of from five to seven sequences for each of the 60 tRNA sequences present in the cell. By similar calculations, *Drosophila* is believed to carry 8 copies of each tRNA sequence and the toad *Xenopus laevis* is believed to carry 200 copies of each gene per haploid genome.

The tRNA genes of eukaryotes also exist in tandem clusters. Because the percent GC of these clusters may deviate from the average percent GC of the genome, it is possible in some cases to isolate tRNA gene clusters by virtue of their satellite buoyant density in CsCl gradients (see Chapter 7.8). One such cluster from *X. laevis,* for example, contains information for $tRNA_1^{Met}$ and for at least one other (unidentified) tRNA species. This cluster is therefore of the nonredundant variety; whether redundant clusters exist in eukaryotes, as they do in prokaryotes, is not yet established.

8.10   tRNA PROCESSING

tRNA genes are transcribed into oversized precursor molecules **(pre-tRNA)** which, as noted at the beginning of this chapter, must undergo processing to acquire the final, mature tRNA structure. For *E. coli,* some pre-tRNA sequences have the structure 5'-leader-tRNA-trailer-3' where leader and trailer sequences are destined to be clipped off. In other cases, a tandem tRNA cluster may have a single promoter and two or more tRNA sequences, in which case a stretch of nucleotides called **spacer DNA** occupies the interval between one tRNA sequence and the next. This spacer is included in the transcript (and is therefore denoted as **transcribed spacer**) so that the pre-tRNA transcripts from such clusters have the general structure 5'-leader-(tRNA-spacer)$_n$-tRNA-trailer-3' where the transcribed spacer is also fated for processing.

Processing of pre-tRNA begins with endonucleolytic cuts that eliminate leader, spacer, and trailer sequences. Temperature-sensitive mutants of *E. coli* that produce defective **RNase P** enzymes are unable to make such cuts, and large pre-tRNAs therefore accumulate in the cell. From analysis of the processing of such molecules *in vitro* it appears that RNase P and a second enzyme, **RNase P$_2$,** recognize base-paired loops (recall Figure 7-4) or other conformational properties of pre-tRNA molecules and make cuts in leader and spacer sequences, while a third enzyme, **RNase Q,** is involved in removing trailer sequences.

The next stage in tRNA processing, which occurs either on the precursor molecules or the cleaved products, is termed **nucleoside modification:** certain uridines, for example, are reduced to dihydrouridine or rearranged into a form known as pseudouridine; certain adenosines are deaminated to yield inosine; and methyl groups are added at certain positions to yield 5-methyl-cytosine or 1-methyl-guanosine. The structures of some of these **minor bases** are given in Figure 8-10; at least 50 have been described to date. The importance of nucleoside modification becomes apparent in the phenotype of the *hisT* mutant of *Salmonella.* The *hisT* mutation affects the production of a pseudouridylation enzyme; the consequent absence of pseudouridine in key positions on many tRNA molecules results in the defective biosynthesis of many proteins in *hisT* strains.

In the final tRNA processing step, a specific nucleotidyl transferase enzyme adds the sequence 5'CCA-OH3' onto the 3' terminus generated by RNase Q cleavage. This step is not required in cases in which the CCA sequence is an internal part of the transcript and becomes exposed after RNase Q action. The final CCA terminus is essential for the amino-acid acceptor and transfer functions of all tRNA molecules.

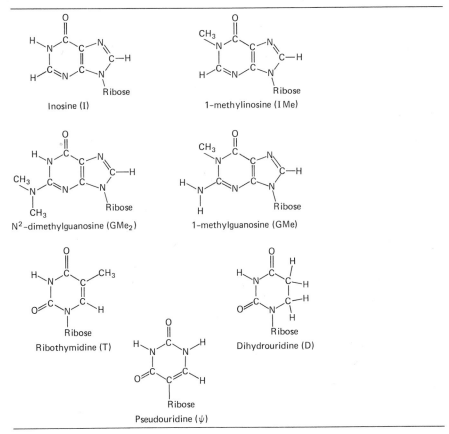

Figure 8-10 Some of the minor bases found in tRNA.

8.11   THE STRUCTURE OF MATURE tRNA

The mature tRNA molecule contains 70 to 80 nucleotides. A number of methods have been devised for obtaining individual species of tRNA in pure form, and some 75 tRNAs have now been sequenced. Regions of internal complementarity are evident in these sequences, much as with the MS2 chromosome (Figure 7-4), but many of the modified bases of tRNA carry substitutions or alterations in those positions that usually participate in hydrogen bonding. Consequently, the model builder is forced to construct several nonbase-paired loops in the molecule. Figure 8-11 shows the resulting **cloverleaf structure** proposed by R. Holley for yeast tRNA^{Ala}, the first

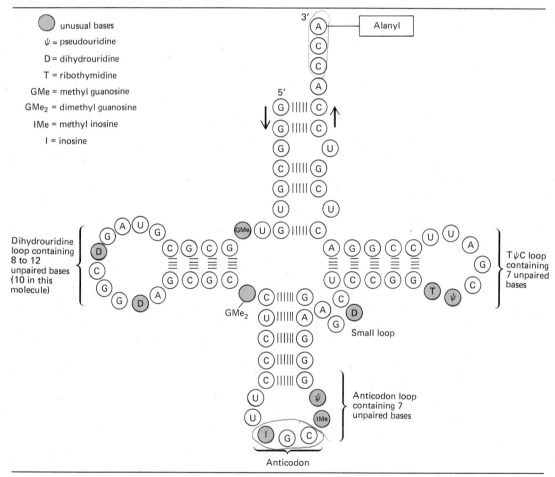

Figure 8-11   The cloverleaf configuration of a charged yeast alanine tRNA. (From James D. Watson, *Molecular Biology of the Gene*, 3rd ed., copyright © 1976 by J. D. Watson; W. A. Benjamin, Inc., Menlo Park, California.)

tRNA to be completely sequenced. The structure exhibits three large loops, one small loop, and an arm containing the terminal CCA segment. Unusual bases are prominent in the formation of each of the loops.

The cloverleaf model accomodates several of the known functions of tRNAs, one of which, it will be recalled, is to carry the anticodon in a prominent position, so that it can base-pair with its complementary codon in mRNA. When the cloverleaf configuration of tRNA^{Ala} was first constructed, the anticodon for alanine (3'-CGI-5') was found exposed on one of the cloverleaf loops, as shown in Figure 8-11. Each subsequent tRNA to be analyzed also contained its specific anticodon in a comparable position in a comparable loop, strongly suggesting that it is at this position in the molecule that the interaction with mRNA takes place.

The cloverleaf model of tRNA is in two dimensions, whereas the tRNA-synthetase interaction occurs between molecules that are, of course, three-dimensional. The tertiary structure of yeast tRNA^{Phe} has been determined by X-ray diffraction studies, and Figure 8-12 depicts a model that has been offered for this tertiary structure. It is presumed that aspects of this

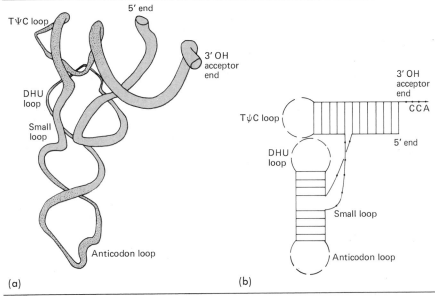

(a)   (b)

Figure 8-12 (a) Molecular model of the three-dimensional configuration of yeast phenylalanine tRNA. The polynucleotide chain is represented as a continuous coiled tube. The different shading represents the various loops and stem regions of the tRNA molecule. The TψC and DHU loops (see Figure 8-11) come in very close contact. (b) The way that the cloverleaf representation must be transformed in order to show the physical connections between various parts of the molecule. Two double-stranded helical regions are seen, each oriented at right angles to the other to produce an L-shaped structure. (From S. H. Kim, et al., *Science,* **179,** 285–288, January 19, 1973. Copyright 1973 by the American Association for the Advancement of Science.)

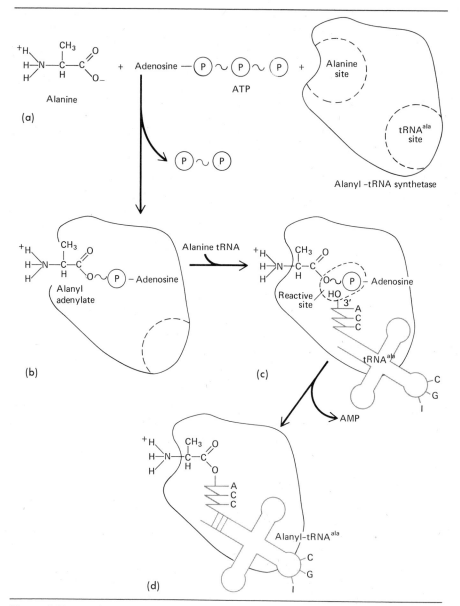

Figure 8-13 A schematic view of the joining of an amino acid (alanine) to its tRNA molecule as mediated by its synthetase enzyme.

structure are recognized by a phenylalanyl-tRNA synthetase enzyme; the enzyme then proceeds to "charge" the tRNA with phenylalanine, producing an **aminoacyl-tRNA** that is ready to participate in protein synthesis. The steps that accompany such a charging reaction are diagrammed in Figure

8-13, and the participation of aminoacyl-tRNAs in protein synthesis is described in Chapter 9.

RIBOSOMAL RNA AND rRNA GENES

8.12 ROLE OF RIBOSOMAL RNA IN PROTEIN SYNTHESIS

Ribosomal RNA (rRNA), like tRNA, is an essential component of the cellular machinery for translating genetic messages into protein and, like tRNA, it is needed by the cell in large quantities. Every known cell contains three kinds of rRNA which interact with some 50 species of ribosomal protein. This interaction produces the discrete particles known as **large and small ribosomal subunits.** At the time protein synthesis is initiated the large and small subunits combine to form an intact ribosome (see Figure 8-1b). The ribosomal surface is constructed to produce sites for messenger RNA and tRNA binding. During protein synthesis the ribosome mediates the codon-anticodon recognition event discussed earlier; it also permits the formation of peptide bonds between amino acids and mediates the movement of messenger RNA so that its codons are exposed to tRNA anticodons in a sequential fashion. Little is now known as to how the ribosome performs these various functions, but the problem is under active investigation.

The rRNA species contained in a ribosome are readily characterized by **sucrose gradient centrifugation** (Box 8.1). By analytic procedures that need not concern us here, it is possible to calculate from such sucrose gradient analysis a **sedimentation coefficient,** expressed in terms of **Svedberg units (S),** for individual RNAs; sedimentation coefficients can also be determined for ribosomal subunits and for ribosomes themselves. As a general rule, the larger the macromolecule, the larger the **S value.** Therefore, in the case of *E. coli,* the large ribosomal subunit (50S) can be shown to contain a 23S and a 5S rRNA, and the small subunit (30S) is found to contain a 16S rRNA. Eukaryotes tend to have larger high-molecular-weight rRNAs: thus the large ribosomal subunit contains, generally speaking, a 28S rRNA (the sedimentation constant varies between 25S and 30S, depending on the species) and a 5S rRNA, whereas the small subunit contains, generally speaking, an 18S rRNA. Partly for this reason, intact eukaryotic ribosomes are larger and sediment in sucrose faster than prokaryotic ribosomes; the former are generally denoted as **80S** and the latter as **70S ribosomes.**

Box 8.1 **SUCROSE** **GRADIENT** **CENTRIFUGATION**	(a) A centrifuge tube containing a continuous gradient of sucrose concentrations ranging from 5 percent at the top to 25 percent	at the bottom (other ranges may also be chosen) is prepared. (b) A small sample of RNAs is carefully layered on top of the gradi-

ent, and the tube is centrifuged at high speeds for several hours. Like-sized RNA molecules tend to move with similar velocities.

(c) Centrifugation is stopped. RNAs of similar molecular weights are located in discrete regions of the gradient.

(d) The bottom of the tube is punctured with a hypodermic needle and successive samples are collected with a fraction collector. The absorbance of each sample is then measured at 260 nm, an ultraviolet absorption maximum for nucleic acids. Samples with little RNA will show little absorbance, those with more RNA more absorbance. A plot of these absorbances generates a profile of RNA concentrations within the gradient. In such a profile heavier RNAs will be found near the bottom of the gradient and lighter RNAs near the top.

This technique differs from the CsCl density gradient procedure we have described for DNA (Box 5.1) in two respects. (1) The sucrose gradient is *preformed*, whereas the CsCl gradient is produced only after many hours of centrifugation; therefore the sucrose gradient approach is faster. (2) The CsCl gradient contains an *equilibrium* distribution of macromolecules, whereas the sucrose gradient contains molecules that are in the process of moving to the bottom of the tube; the centrifugation is stopped while they are still in transit.

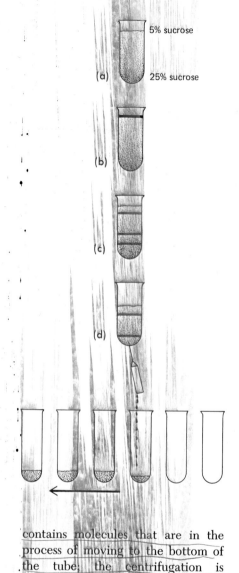

8.13   GENES FOR rRNA IN E. COLI

Genes for rRNA (*rrn*) in *E. coli* are present in 3 mapped positions in the *E. coli* genome (Figure 8-9). From 5–10 copies of *rrn* are estimated to be present per chromosome, a value deduced from rRNA-DNA hybridization studies; it is not known whether tandem redundancy exists or whether more map locations will be found. Each *rrn* gene has a single promoter from which is copied a 30S **pre-rRNA** transcript containing 16S, 4S (tRNA), 23S, and 5S sequences (as noted earlier, a tRNA sequence often—and perhaps always—lies between 16S and 23S). Processing of this transcript involves initial cuts in the transcribed spacer, followed by secondary trimming of leader and trailer sequences, much as we saw for pre-tRNA. Details are given in Figure 8-14.

About 0.4 percent of an *E. coli* chromosome is devoted to carrying rRNA sequences, whereas about 80 percent of the RNA in an *E. coli* cell is rRNA. Two explanations can be given for this discrepancy in percentages. First, rRNA is greatly stabilized and protected by ribosomal proteins, and thus any rRNA molecule that is synthesized is expected to have a far longer lifetime than other RNA species. Second, in contrast to most genes that may be transcribed only a few times or perhaps not at all during the life of a cell, the rRNA genes are in a state of perpetual transcription. Indeed, after a polymerase has attached to an *rrn* promoter site and has moved along the gene a certain distance a second polymerase attaches and begins to read the gene, then a third, and so on. As a result at least 30 polymerase molecules are working on a given gene at any one time, and strands of rRNA of progressively longer lengths are observed coming off the template. This process has

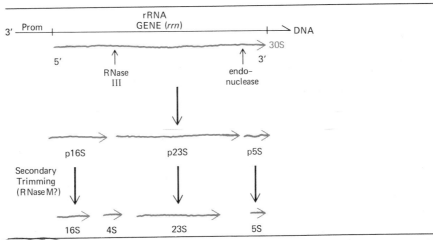

Figure 8-14 Processing of the *rrn* 30S transcript into mature rRNA molecules in *E. coli*. A cleavage site for RNase III exists in the transcribed spacer between the 16S and 28S sequences. Normally this cut is made during the course of transcription so that a full-length 30S transcript is rarely formed; the full-length transcript is, however, recovered in an RNase III-deficient mutant strain. Secondary trimming may involve a second enzyme, RNase M or maturase, and generates a 4S tRNA from either the p16S or the p23S precursor. Nucleoside modifications, notably methylations, occur at an early stage of processing for the 23S molecule and later for the 16S species.

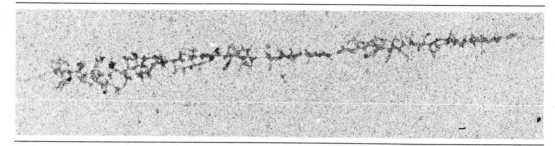

Figure 8-15 Electron micrograph of rRNA transcription in *E. coli*. (From O. L. Miller, Jr. and B. A. Hamkalo, *Int. Rev. Cytol.* **33**:1, 1972.)

been observed directly with the electron microscope, as shown in Figure 8-15. By such an arrangement, 60 molecules of 16S rRNA can be generated per minute from each gene, a rate sufficient to account for the number of 16S molecules known to be present in a rapidly growing *E. coli* cell (perhaps 20,000 molecules).

8.14   GENES FOR 18S AND 28S rRNA IN EUKARYOTES

Each eukaryotic nucleus possesses 1 or more nucleoli (Figure 2-1); these represent sites where newly synthesized rRNA accumulates and becomes associated with the ribosomal proteins that are synthesized in the cytoplasm and then migrate back into the nucleolus for assembly. Experiments by D. Brown and J. Gurdon were among the first to establish that the nucleolus is not only a storage site for rRNA but also the site of rRNA synthesis.

The Brown and Gurdon experiments focused on the mutation *O-nu* (for *zero-nucleolus*) in *Xenopus laevis*. Normal *Xenopus* cells (+/+) contain 2 nucleoli whereas heterozygous toads (+/*O-nu*) exhibit only one nucleolus in each of their nuclei. Such heterozygotes are viable, but when they are crossed with one another about one-quarter of their progeny die at the swimming larva stage. You will recall from Chapter 4 that whenever two heterozygotes are crossed one-quarter of their progeny are expected to be homozygous for a mutant allele. Therefore we can predict that the inviable progeny in the *X. laevis* crosses represent *O-nu/O-nu* homozygotes. This expectation is confirmed when cytological specimens are prepared from embryo tail tips: about one-quarter of the embryos exhibit no nucleoli in their nuclei.

The swimming larva stage of *X. laevis* embryogenesis is preceded by a period when the embryos have used up the store of ribosomes inherited from their mothers and are called on to synthesize their own rRNA. Brown and Gurdon therefore compared the ability of homozygous and heterozygous larvae to synthesize rRNA during this period. They exposed embryos to $^{14}CO_2$ and examined the amount of radioactivity incorporated into rRNA. The results are shown in Figure 8-16, in which the absorbance at 260 nm is shown in gray and the radioactivity is shown in black. It is clear that the control larvae have much more 28S and 18S RNA than the anucleolate mutants, as measured by absorbance. Even more striking is the fact that the anucleolate embryos incorporate virtually no radioactivity into either species of rRNA, indicating that the 28S and 18S rRNA that they do possess is of maternal origin.

The mutation carried by *O-nu* toads proves to be a deletion of the **nucleolar organizer (NO),** a sector of chromosome #12 in *Xenopus laevis.* This sector can be shown by *in situ* hybridization studies with 18S and 28S rRNAs to be the exclusive site of 18S and 28S gene sequences. RNA-DNA hybridization studies reveal that 500 copies of the 28S and 18S sequences are present in each *X. laevis* nucleolar organizer. A similar level of NO gene redundancy is found for all other eukaryotes examined, but there is some variation with regard to the distribution of these genes: *Drosophila*

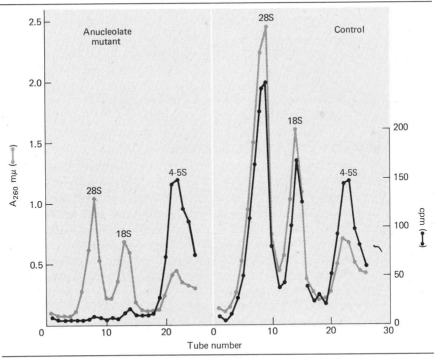

Figure 8-16 Sucrose gradient centrifugation of total RNA isolated from anucleolate (*O-nu/O-nu*) embryos and control (+/+ and +/*O-nu*) embryos. The bulk RNA is given by the absorbance at 260 nm (gray tracing). The RNA synthesized between the neurula and muscular response stage is represented by the radioactivity measurements (black tracing). (From D. Brown and J. Gurdon, *Proc. Natl. Acad. Sci. U.S.* **51:**139, 1964.)

resembles *Xenopus* in having but one NO per haploid genome (in the X and Y chromosomes); humans have 5 NOs (in the short arms of the 5 acrocentric D and G chromosomes); while the field vole (a rodent) has *rrn* genes in almost all of its chromosomes.

Transcription of nucleolar organizer DNA can be readily visualized, as shown in Figure 8-17, where the DNA is represented by the long axial strands. Segments of the NO DNA are seen in Figure 8-17 to be in an active state of transcription in which multiple initiation sites produce multiple rRNA strands of progressively longer lengths (compare with Figure 8-15). Each such "feather region" is flanked on either side by DNA that is clearly not being transcribed and is appropriately termed **non-transcribed spacer.**

NO DNA from *Xenopus* can be isolated as a satellite band in CsCl gradients and has been subjected to extensive molecular analysis, notably in the laboratory of D. Brown; Figure 8-18 diagrams the structure of this DNA. The 18S and 28S sequences are copied as a single large 40S pre-rRNA transcript; processing of this transcript is detailed in Figure 8-18 and its legend. All 500 copies of the rRNA gene in the DNA appear to be identical to one another; the intervening nontranscribed spacer, on the other hand, is of variable length, even along the same piece of NO DNA (Figure 8-18). Sequence analysis of several of these nontranscribed spacers reveals an internal pattern of repeats, with certain "modules" differing from others by nucleotide substitutions. In other words, the nontranscribed spacer in NO DNA bears a strong structural resemblance to the nontranscribed "satellite DNAs" found in constitutive heterochromatin (Section 7.11). Since the function of both classes of DNA is unknown, it cannot be said whether or not this resemblance is fortuitous.

We should pause here to raise an intriguing question in molecular genetics: How do the hundreds of copies of rRNA genes in the NO resist mutational alterations? It is a simple matter to explain resistance to mutational change when a gene is present in only one copy: in such a case, it is argued that all changes have proved to be lethal. When, instead, a chromosome carries several copies of a gene and one sequence experiences a lethal change, the remaining sequences will presumably still be functional. Over the millions of years of a species' existence, we would expect mutations in redundant genes to accumulate, so that *X. laevis,* for example, would produce several different types of rRNA of the 18S variety instead of what appears to be a single species.

It might be argued that this does not occur because any change in the sequence of rRNA might lead to an abnormal population of ribosomes and that the cell's demand for ribosomes is so great that any reduction in its effective ribosome potential cannot be tolerated. This may well be the case for bacteria, with their rapid growth rate in a highly competitive environment, but it does not appear to be true for eukaryotes. In the $+/O\text{-}nu$ heterozygote of *X. laevis,* for example, the cells are found to compensate for the nucleolar deletion by synthesizing twice as much rRNA from their single

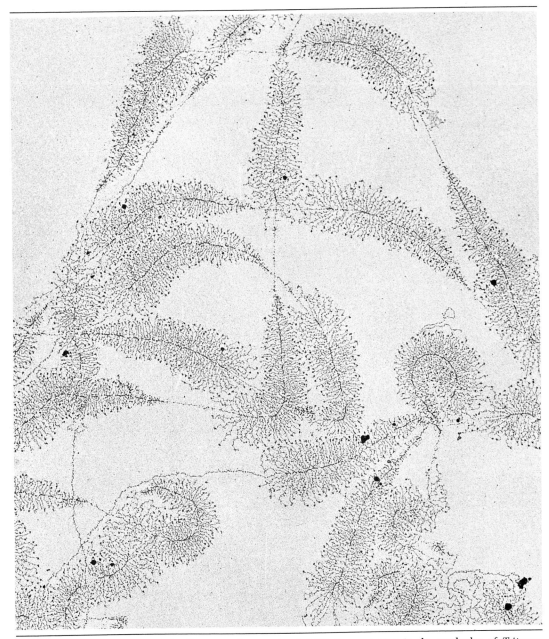

Figure 8-17 Electron micrograph of rRNA transcription in the nucleolus of *Triturus*. (From O. L. Miller and B. R. Beatty, *Science* **164**:955, 1969. Copyright 1969 by the American Association for the Advancement of Science.)

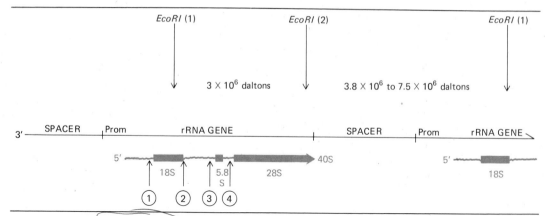

Figure 8-18 The structure and transcription of rRNA genes in *Xenopus*. Each gene is separated from the next by a non-transcribed spacer of variable length (as determined by the size of the indicated *Eco*RI restriction fragments). The 40S pre-rRNA transcript (45S for mammals) contains a 5′ leader sequence which is cleaved from the molecule at step ①. Cuts ② and ③ in the transcribed spacer occur next, followed by cut ④ adjacent to the 28S block. Secondary trimming produces the final 18S, 28S, and 5.8S rRNAs; all three then associate with ribosomal proteins. (After P. Wellauer, *et al.*, *J. Mol. Biol.* **105**:461, 1976, and R. Perry, *Ann. Rev. Biochem.* **45**:605, 1976.)

NO (a regulatory phenomenon called **dosage compensation** which we discuss again in Section 18.17). Thus *X. laevis* possesses the capacity, at least, to dispense with half its rRNA genes.

Since the non-transcribed spacer DNA between the rRNA genes of *X. laevis* has diverged both in length and in sequence, it seems probable that the rRNA genes undergo mutation as well but that some mechanism exists for correcting such mutations or dispensing with any mutant genes so that sequence identity is maintained in the cluster. This subject is considered again in Section 15.7.

8.15   GENES FOR 5S rRNA IN EUKARYOTES

It is apparent from Figure 8-16 that while 18S and 28S genes are not synthesized in *O-nu* homozygotes of *Xenopus laevis*, 4S RNA synthesis (which represents both tRNA and 5S ribosomal RNA) continues apace, suggesting that the genes for these RNAs are not removed by the *O-nu* deletion. This inference is confirmed by *in situ* hybridization, where 5S rRNA does not localize to the NO region but rather to the telomeres of many, if not all, of the 18 chromosomes of *X. laevis*. With the possible exception of yeast and *Dictyostelium*, eukaryotes generally do not have 5S genes near their 18S and 28S genes: in *Drosophila*, for example, 5S genes localize to bands 56F 1–9 of chromosome 2 rather than to the sex chromosomes, while in humans 5S genes are found in chromosome 1 and in a number of other chromosomes as well.

The 5S genes in *X. laevis* are very similar to the NO genes in their organization: highly conserved transcribed sequences, 120 nucleotides long, alternate with variable nontranscribed spacer sequences about 600 nucleotides long. An estimated 20,000 copies of 5S genes are present in the *X. laevis* genome; 2000 copies are believed to be present in the human haploid complement. Again, mechanisms for conserving rRNA sequences in the face of such redundancy are unknown.

STRUCTURAL GENES AND MESSENGER RNA

Ribosomal and transfer RNA genes make up less than 1 percent of the genome of both prokaryotes and eukaryotes, although they generate perhaps 95 percent of an *E. coli* cell's total RNA. The overwhelming proportion of the information in a genome takes the form of structural genes. We have already considered structural genes in Sections 8.2 and 8.5. Here we consider mRNA processing and metabolism, followed by the organization of the histone genes.

8.16   mRNA PROCESSING

Most structural-gene transcripts **(pre-mRNA)** have the general structure: leader-coding sequence-trailer. In eukaryotes, at least, the initial transcripts are usually processed so that the final mRNA that reaches the ribosome is smaller than the original (see Section 8.19). In addition, many (but not all) eukaryotic mRNAs have a 200-nucleotide tract of adenosine residues **(poly A)** attached post-transcriptionally to their 3' ends, and a 7-methylated guanidine residue may be attached to the 5' purine (an mRNA **cap**); the functional significance of either modification is not clear at the present time. We can thus write the structure of the final mRNA transcript as 5'-untranslated sequence-coding sequence-untranslated sequence-3', with the understanding that either or both untranslated sequences may be processed versions of the original leader and trailer sequences.

8.17   SYNTHESIS AND TURNOVER OF mRNA

Figure 8-19 shows structural-gene transcription in *E. coli*, where ribosomes are seen to associate with each mRNA transcript even as it peels off the template. Thus, in *E. coli*, transcription and translation are **tightly coupled** to one another. The ribosomes must work fast to translate the message,

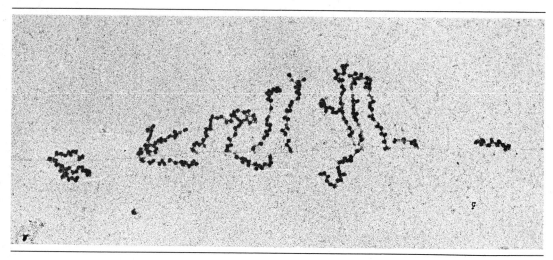

Figure 8-19 Electron micrograph of mRNA transcription in *E. coli;* ribosomes are associated with the mRNA. (From O. L. Miller, B. A. Hamkalo, and C. A. Thomas, Jr., *Science* **169:**392, 1970. Copyright 1970 by the American Association for the Advancement of Science.)

moreover; since cellular ribonucleases are highly active against mRNA, and the lifetime of an average *E. coli* mRNA molecule is on the order of only 3 minutes. The short lifetime of mRNA in *E. coli* is characteristic of rapidly growing organisms whose genomes are governed by a labile regulatory system: since *E. coli* cells destroy most of their mRNA molecules every few minutes, a new spectrum of mRNAs can replace an old spectrum with great rapidity.

In eukaryotes, mRNA lifetime is dependent on the state of differentiation of the cell. For dividing human HeLa cells in tissue culture, for example, about one-third of the mRNA has a half-life of 7 hours and the remaining two-thirds has a half-life of 24 hours. Eukaryotic mRNA, then, is far more stable than its *E. coli* counterpart, even when differences in the length of the cell cycle are taken into account. Much of this stabilization is probably due to the fact that eukaryotic mRNA associates with protein in the nucleus to form a **ribonucleoprotein particle** which subsequently migrates to the cytoplasm for translation.

Particularly long-lived mRNA is characteristically produced when a cell becomes committed to synthesizing one or a few kinds of protein. For example, more than 90 percent of the protein synthesis in mammalian reticulocyte cells is devoted to hemoglobin synthesis, and even though reticulocytes lose their nuclei spontaneously in the course of maturation, so becoming unable to synthesize more mRNA, hemoglobin synthesis continues for another two days. In extreme cases, such as in the state of dormancy adopted by many animal eggs and plant seeds, mRNA is maintained in a stable form for months or even for years.

8.18 THE HISTONE GENES

Most of what is known about structural genes comes from genetic analysis; the remaining chapters of this text, therefore, are almost entirely concerned with the nature of structural genes. M. Birnstiel and colleagues have, however, recently undertaken a molecular analysis of the genes coding for the five histone proteins (Section 2.11) that associate in a one-to-one fashion with chromosomal DNA. These are unusual structural genes for eukaryotes in that they are found in clusters, very much like tRNA. Their repetitive nature gives histone-gene-containing fragments of DNA a satellite buoyant density, thus allowing their isolation from the rest of the genome. Histone mRNA also proves to be readily isolated, for it has a characteristic small size (9S), appears only during the S phase of the cell cycle (Section 2.2), and is synthesized in bulk (constituting some 70 percent of the total mRNA synthesis) in rapidly-dividing sea urchin embryos. Each of the five histone mRNAs has been separated from the others by electrophoresis (Box 7.1), and each has been used to drive cell-free protein synthesis; analysis of the resultant polypeptides reveals that the mRNA with the greatest electrophoretic mobility codes for H1, the slowest for H4, and so on.

By subjecting the isolated histone gene-containing DNA fragments to a series of restriction endonucleases (Section 7.1) and hybridizing the various cleaved fragment classes with known histone mRNAs, a map order for each histone gene cluster in sea urchins has been determined. This has turned out to be H1- H2A- H3- H2B- H4, the entire unit being about 6000 base-pairs long. The histone unit is repeated from 10-20 times in *Xenopus laevis* and up to 300-1000 times in sea urchins. A "rationale" for such repetitiveness can be found in the unusually high demand for histone biosynthesis during the cleavage stage of embryogenesis: in contrast, most, if not all, structural genes are believed to be present in only a single copy per haploid genome in both prokaryotes and eukaryotes (Section 7.10).

Isolated histone-gene clusters can also be subjected to **denaturation mapping** in which the DNA is heated just enough to induce AT pairs to come apart; the resulting preparation is then examined with an electron microscope. As seen in Figure 8-20, lengths of nondenatured (GC-rich) DNA are flanked by lengths of denatured (AT-rich) DNA. Since histone mRNA is known to be very GC-rich, the nondenatured regions can be equated with the histone genes (they have the expected length as well) and the AT-rich regions can be identified as nontranscribed spacers. The spacer DNA, as by now one comes to expect, is found to have diverged considerably in both its length and sequence.

Histone mRNAs from sea urchin hybridize to the 3DE bands on the left arm of chromosome 2 of *Drosophila*, thus localizing the *Drosophila* histone genes to one sector of the genome. That hybridization occurs at all in this experiment attests to the extreme conservation of the histone genes

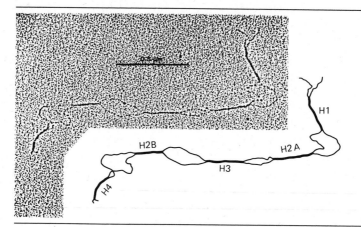

Figure 8-20 Partially denatured DNA from the histone gene cluster of the sea urchin. The genes (H1, H2A, H2B, H3, and H4) are relatively GC-rich and therefore remain in duplex form at 61°C, whereas the AT-rich spacers are denatured. (From R. Portmann, W. Schaffner, and M. Birnstiel, *Nature* **264**:31, 1976.)

throughout the eukaryotic kingdom: amino-acid sequence studies show very few differences between histones of a given class derived from widely separated species, the histones H3 and H4 being particularly highly conserved.

8.19 HETEROGENEOUS NUCLEAR RNA

The fourth and final class of RNA, considered in this chapter, known as **heterogeneous nuclear RNA (hnRNA),** is found only in eukaryotes. It shares several properties with mRNA: it commonly has a GC content similar to that of the DNA in the genome of a species; it is heterogeneous in size; it hybridizes with a considerable fraction of the genome; and it may carry a poly-A "tail" on its 3' end. Unlike mRNA, however, a large fraction of hnRNA appears never to leave the nucleus, and the participation of this fraction in cytoplasmic protein synthesis is considered unlikely. Moreover, the half-life of hnRNA, on the order of 30 minutes or less in many cases, is much shorter than that of most eukaryotic mRNA. The hnRNA also varies in size from 10S to 200S, whereas most mRNA has a sedimentation coefficient between 6S and 30S. Finally, certain studies suggest that much hnRNA contains internally redundant nucleotide sequences that are not noted in mRNA. In short, the eukaryotic nucleus synthesizes numerous, enormous, and heterogeneous RNA molecules that are degraded shortly after they are synthesized and do not engage in protein synthesis. What, then, is their function?

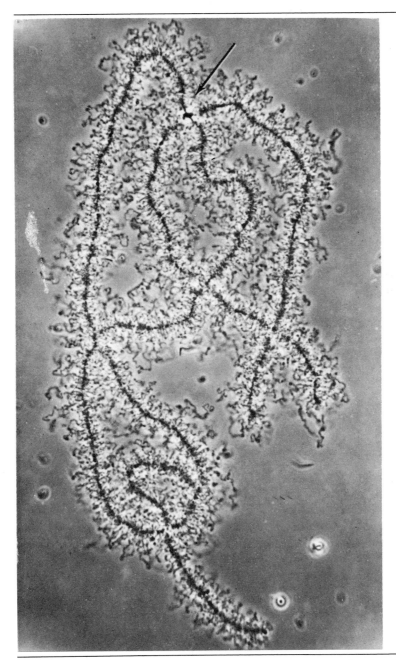

Figure 8-21 Lampbrush chromosomes from a *Triturus* oocyte. Two homologous are present, each being made up of two chromatids that cannot be individually resolved here. Extending from each chromatid pair is an array of lateral loops which reminded early cytologists of a lampbrush. Each loop is believed to be a single strand of nucleosomes (Figure 2-15) surrounded by newly transcribed RNA (Figure 8-22). (From J. Gall.)

A possible answer comes when one considers two occasions in which enormous primary transcripts are generated by eukaryotic chromosomes. The first occurs during the diplotene stage of meiotic prophase in vertebrate and invertebrate oocytes and gametes: in this stage, the so-called **lampbrush chromosomes** of these cells (Figure 8-21) generate long loops of DNA from their chromomeres, and giant RNA transcripts peel off this DNA. The second occasion in which giant primary transcripts are created occurs in larval cells of many Diptera: during certain periods of larval life, or in response to such stimuli as hormones, certain polytene chromosome bands in these larval cells lose their compact form and appear to separate into their component strands, forming what is aptly termed a **puff** (Figure 8-22). RNA isolated from these individual puffs proves to be very large (75S in one case). Thus it is <u>attractive to imagine that these large transcripts are "pre-mRNA," that messenger-sized (6S to 30S) pieces of RNA are cleaved from the pre-mRNAs and stabilized as ribonucleoprotein, and that the remaining RNA rapidly breaks down with the characteristic kinetics of hnRNA.</u>

Strong support for this notion comes from studies of mouse β-globin, one of the polypeptide constituents of hemoglobin (Sections 9.3 and 15.9). The mRNA for β-globin is a 10S species which is readily purified. This mRNA can then be copied into a radioactively labeled **complementary DNA (cDNA)** species using the enzyme **reverse transcriptase** (which makes DNA copies

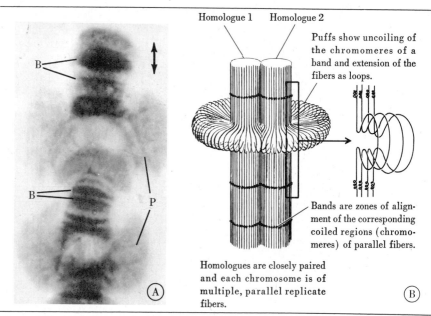

Homologue 1   Homologue 2

Puffs show uncoiling of the chromomeres of a band and extension of the fibers as loops.

Bands are zones of alignment of the corresponding coiled regions (chromomeres) of parallel fibers.

Homologues are closely paired and each chromosome is of multiple, parallel replicate fibers.

(A)   (B)

Figure 8-22 (A) Micrograph showing two chromosome puffs (P) and unpuffed polytene bands (B) from the salivary gland of *Chironomus*. (From U. Clever.) (B) Schematic representation of chromosomal puffing. (From A. B. Novikoff and E. Holtzman, *Cells and Organelles*. New York: Holt, Rinehart and Winston, 1970, Figure IV-25.)

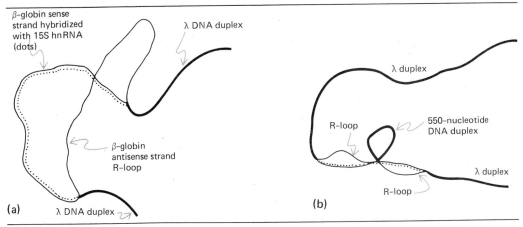

β-globin sense strand hybridized with 15S hnRNA (dots)

λ DNA duplex

β-globin antisense strand R-loop

(a) λ DNA duplex

λ duplex

R-loop

550-nucleotide DNA duplex

λ duplex

R-loop

(b)

Figure 8-23 R-loop mapping of β-globin DNA (see text for details). (Courtesy S. Tilghman, P. Curtis, D. Tiemeier, P. Leder, and C. Weissman.)

of RNA chains), and the cDNA probe can be used to hunt for globin-like sequences in hnRNA derived from globin-producing cells. When this is done, a 15S hnRNA species is found to hybridize well with the cDNA. Therefore, a globin gene DNA ⟶ 15S hnRNA ⟶ 10S mRNA ⟶ β-globin progression is indicated, with the 15S ⟶ 10S conversion ~~begin~~ being accomplished by RNA processing enzymes.

Intervening Sequences and hnRNA Processing In mid-1977, several laboratories made a most unexpected discovery about the structure of several eukaryotic genes; we consider here mouse β-globin gene studies from P. Leder's laboratory. Using recombinant-DNA technology (described in Section 14.6), a λ phage chromosome was constructed which contained the mouse β-globin gene. Replication of such recombinant phages generates abundant copies of pure genomic β-globin DNA sequences for experimental use. When such DNA is partially denatured, presented with the 15S globin hnRNA described above, subjected to renaturation conditions, and examined with the electron microscope, structures such as that shown in Figure 8-23a are observed. The λ DNA sequences are found to have renatured normally, but in the globin-gene region, the 15S hnRNA has hybridized with the sense strand of the gene, preventing the reannealing of the anti-sense DNA strand. As a result, the anti-sense DNA strand bulges out, forming what is known as an **R-loop.** Thus the 15S hnRNA is clearly implicated as the primary transcript of the β-globin gene.

The unexpected result comes when this experiment is repeated using the 10S mRNA species rather than the 15S species. As drawn in Figure 8-23b, two R-loops are present, between which is a sector of about 550 DNA base-pairs that is not homologous to the 10S species and therefore does not participate in R-loop formation. In other words, it appears that during the 15S ⟶ 10S processing, a sector of RNA in the **middle** of the 15S transcript is excised and the two broken pieces resealed to form the 10S species. Such excision and rejoining events must clearly occur with great precision so that a continuous and correct β-globin message is generated for translation.

Turning to the β-globin gene itself, these experiments reveal that the sense strand of the gene is interrupted by about 550 nucleotides that do not code for globin at all, after which the globin-encoding sequence resumes. What, if any, "purpose" such **intervening sequences** might serve in the expression or transmission of the eukaryotic genome remains a matter of speculation as this text goes to press.

References

RNA Polymerases

*Burgess, R. R. "Separation and characterization of the subunits of ribonucleic acid polymerase," *J. Biol. Chem.* **244:**6168–6176 (1969).

*Gilbert, W. "Starting and stopping sequences for the RNA polymerase." In: *RNA Polymerase.* (R. Losick and M. Chamberlin, Eds.) Cold Spring Harbor Laboratory, 1976, pp. 193–206.

Hayward, R. S., and J. G. Scaife. "Systematic nomenclature for the RNA polymerase genes of prokaryotes," *Nature* **260:**646–647 (1976).

Lobban, P., and L. Siminovitch. "The RNA polymerase II of an α-amanitin-resistant Chinese hamster cell line," *Cell* **8:**65–70 (1976).

*Losick, R., and M. Chamberlin, Eds. *RNA Polymerase.* Cold Spring Harbor, N.Y.: Cold Spring Harbor Laboratory, 1976. (Contains numerous relevant articles and reviews.)

Transfer RNA

Altman, S. "Biosynthesis of transfer RNA in *Escherichia coli*," *Cell* **4:**21–29 (1975).

Chambers, R. W. "On the recognition of tRNA by its aminoacyl-tRNA ligase," *Prog. Nucl. Acid Res.* **11:**489–525 (1971).

Clarkson, S. G., and V. Kurer. "Isolation and some properties of DNA coding for tRNA^{Met} from *Xenopus laevis*," *Cell* **8:**183–195 (1976).

Grigliatti, T. A., B. N. White, G. M. Tener, T. C. Kaufman, and D. T. Suzuki. "The localization of transfer RNA$_5^{Lys}$ genes in *Drosophila melanogaster*," *Proc. Nat. Acad. Sci. U.S.* **71:**3527–3531 (1974).

*Holley, R. W., J. Apgar, G. A. Everett, J. T. Madison, M. Marquisee, S. H. Merrill, J. R. Penswick, and A. Zamir. "Structure of a ribonucleic acid," *Science* **147:**1462–1465 (1965).

Perry, R. P. "Processing of RNA," *Ann. Rev. Biochem.* **45:**605–629 (1976).

Quigley, G. J., and A. Rich. "Structural domains of transfer RNA molecules," *Science* **194:**796–806 (1976).

Smith, J. D. "Transcription and processing of transfer RNA precursors," *Prog. Nucl. Acids Mol. Biol.* **16:**25–73 (1976).

Squires, C., B. Konrad, J. Kirschbaum, and J. Carlson. "Three adjacent transfer RNA genes in *Escherichia coli*," *Proc. Nat. Acad. Sci. U.S.* **70:**438–441 (1973).

*Denotes articles described specifically in the chapter.

Ribosomal RNA

Benhamou, J., and B. R. Jordan. "Nucelotide sequence of *Drosophila melanogaster* 5S RNA: Evidence for a general 5S RNA model," *FEBS, Lett.* **62**:146–149 (1976).

*Brown, D. D., and J. B. Gurdon. "Absence of ribosomal RNA synthesis in the anucleolate mutant of *Xenopus laevis*," *Proc. Natl. Acad. Sci. U.S.* **51**:139–146 (1964).

Gargano, S., and F. Graziani. "Increase of rDNA redundancy in *bb* females of *Drosophila melanogaster*," *Molec. Gen. Genet.* **145**:255–258 (1976).

Hsu, T. C., S. E. Spirito, and M. L. Pardue. "Distribution of 18 + 28S ribosomal genes in mammalian genomes," *Chromosoma* **53**:25–36 (1975).

Procunier, J. P., and K. D. Tartoff. "Restriction map of 5S RNA genes of *Drosophila melanogaster*," *Nature* **263**:255–257 (1976).

*Ritossa, F. M., K. C. Atwood, and S. Spiegelman. "A molecular explanation of the *bobbed* mutants of *Drosophila* as partial deficiencies of ribosomal DNA," *Genetics* **54**:819–834 (1966).

*Ritossa, F. M., and S. Spiegelman. "Localization of DNA complementary to ribosomal RNA in the nucleolus organizer region of *Drosophila melanogaster*," *Proc. Natl. Acad. Sci. U.S.* **53**:737–745 (1965).

Stambrook, P. J. "Organization of the genes coding for 5S RNA in the Chinese hamster," *Nature* **259**:639–641 (1976).

Tartof, K. D., and J. B. David. "Similarities and differences in the structure of X and Y chromosome rRNA genes of *Drosophila*," *Nature* **263**:27–30 (1976).

Wellauer, P. K., I. B. Dawid, D. D. Brown, and R. H. Reeder. "The molecular basis for length heterogeneity in ribosomal DNA from *Xenopus laevis*," *J. Mol. Biol.* **105**:461–486 (1976).

Messenger RNA

Brawerman, G. "Characteristics and significance of the polyadenylate sequence in mammalian messenger RNA," *Prog. Nucl. Acid Res.* **17**:117–148 (1976).

*Brenner, S., F. Jacob, and M. Meselson. "An unstable intermediate carrying information from genes to ribosomes for protein synthesis," *Nature* **190**:576–581 (1961). [Reprinted in G. S. Stent, *Papers on Bacterial Viruses*, 2nd ed. Boston: Little, Brown, 1965.]

Daneholt, B. "Transcription in polytene chromosomes," *Cell* **4**:1–9 (1975).

*Gillespie, D., and S. Spiegelman. "A quantitative assay for DNA-RNA hybrids with DNA immobilized on a membrane," *J. Mol. Biol.* **12**:829–842 (1965).

Greenberg, J. R. "Messenger RNA metabolism of animal cells," *J. Cell Biol.* **64**:269–288 (1975).

*Gross, K., E. Probst, W. Schaffner, and M. Birnstiel. "Molecular analysis of the histone gene cluster of *Psammechinus miliaris*. I. Fractionation and identification of five individual histone mRNAs," *Cell* **8**:455–469 (1976).

*Hall, B. D., and S. Spiegelman. "Sequence complementarity of T2-DNA and T2-specific RNA," *Proc. Natl. Acad. Sci. U.S.* **47**:137–146 (1961). [Reprinted in L. Levine, *Papers on Genetics*. St. Louis: Mosby, 1971.]

Jacob, E., G. Malacinski, and M. L. Birnstiel. "Reiteration frequency of the histone genes in the genome of the amphibian, *Xenopus laevis*," *Eur. J. Biochem.* **69**:45–54 (1976).

*Jayaraman, R., and E. B. Goldberg. "A genetic assay for mRNA's of Phage T4," *Proc. Natl. Acad. Sci. U.S.* **64**:198–204 (1969).

Kedes, L. H. "Histone messengers and histone genes," *Cell* **8**:321–331 (1976).

Lewin, B. "Units of transcription and translation: The relationship between heterogeneous nuclear RNA and messenger RNA," *Cell* **4**:11–20 (1975).

*Marmur, J., C. M. Greenspan, E. Palecek, F. M. Kahan, J. Levine, and M. Mandel. "Specificity of the complementary RNA formed by *Bacillus subtilis* infected with bacteriophage SP8," *Cold Spring Harbor Symp. Quant. Biol.* **28**:191–199 (1963).

Marotta, C. A., B. G. Forget, S. M. Weissman, I. M. Verma. R. P. McCaffrey, and D. Baltimore. "Nucleotide sequences of human globin messenger RNA," *Proc. Natl. Acad. Sci. U.S.* **71**:2300–2304 (1974).

*Miller, O. L., Jr., B. R. Beatty, B. A. Hamkalo, and C. A. Thomas, Jr. "Electron microscopic visualization of transcription," *Cold Spring Harbor Symp. Quant. Biol.* **35**:505–512 (1970).

*Portman, R., W. Schaffner, and M. Birnstiel. "Partial denaturation mapping of cloned histone DNA from the sea urchin *Psammechinus miliaris*," *Nature* **264**:31–34 (1976).

Proudfoot, N. J. "Sequence analysis of the 3' non-coding regions of rabbit α- and β-globin messenger RNAs," *J. Mol. Biol.* **107**:491–525 (1976).

Ross, J. "A precursor of globin messenger RNA," *J. Mol. Biol.* **106**:403–420 (1976).

Rozenblatt, S., R. C. Mulligan, M. Gorecki, B. E. Roberts, and A. Rich. "Direct biochemical mapping of eukaryotic viral DNA by means of a linked transcription-translation cell-free system," *Proc. Natl. Acad. Sci. U.S.* **73**:2747–2751 (1976).

(a)

(b)

(a) Joan Steitz (Yale University) has sequenced the leader ends of mRNA molecules and RNA bacteriophage chromosomes. (b) Barbara Hamkalo and Oscar Miller, Jr., photographed during their collaborative studies on eukaryotic gene transcription. Dr. Hamkalo is presently at the University of California, Irvine, and Dr. Miller at the University of Virginia.

Questions and Problems

1. In what respect do the histone genes resemble the mammalian X chromosome (Section 4.11)? How might this similarity be explained?

2. You wish to learn which mRNA sequences are "shielded" from nuclease digestion when mRNA first associates with ribosomes. Outline how you would perform this experiment. Which nucleotides would you predict would be shielded?

3. Review how actinomycin D interacts with DNA (Section 7.8), and then explain why actinomycin D is a widely used inhibitor of transcription. Would you expect actinomycin D to interact preferentially with histone genes or with histone spacer? Do you expect that actinomycin D-resistant strains of *E. coli* could be isolated and, if so, what would most likely be altered in these strains (review Section 5.5)?

4. A series of strains of *D. melanogaster* carry recessive mutations known as *bobbed* which, in homozygous form, cause slow development, reduced growth and fertility, and poor viability. Some *bobbed* mutations have a stronger effect than others on development and viability. Describe DNA/RNA hybridization experiments that would ascertain whether the various *bobbed* mutations represent more or less extensive NO deletions.

5. It has been proposed that the DNA carrying rRNA sequences in *E. coli* is in a "melted" state at all times. What considerations have led to this suggestion?

6. You are told that certain RNA samples have the following properties: (a) a larger precursor size than final size; (b) an additional sequence added onto the original transcript; (c) a homogenous molecular weight; (d) a short lifetime. In each case, state whether the RNA could reasonably be tRNA, mRNA, rRNA, and/or hnRNA (one or more than one may apply), and give your reasons.

7. Compare Figures 5-5 and 8-3. How would you characterize the relative complexity of transcription and replication based on a comparison of the two maps? Speculate as to why transcription lacks an elaborate error-correcting mechanism (Hint: Compare the genetic consequences of a single transcription error *versus* a single replication error).

8. The promoter for the wild-type *lac* operon of *E. coli* is termed a low-level promoter in that RNA polymerases bind to it only very infrequently unless a regulatory protein known as CAP is present. Several mutations have been isolated that convert this into a high-level, CAP-independent promoter, one being the *uv5* promoter shown in Figure 8-8. The Pribnow-Box sequence of the wild-type *lac* promoter region reads 3′ATACAAC5′. (a) What mutation(s) have occurred to produce the *uv5* promoter? (b) Write out the duplex sequences of the two promoters. Which would be more readily "melted" (denatured) by a polymerase? Explain. How might this difference explain the different properties of the two promoters?

9. Is a cDNA copy of an mRNA molecule equivalent to the transcribed strand of a gene? If not, how would you distinguish them?

10. Most species of eukaryotic mRNA can be purified from a cell extract by passing total cellular RNA through a column containing long polymers of thymine linked to a solid support. What is the molecular basis for this purification procedure?

11. The cellular RNase activity responsible for the short lifetime of bacterial mRNA acts against free, single-stranded RNA. Why are bacterial rRNA and tRNA relatively resistant to this nuclease?

12. Is the promoter to the left or the right in Figure 8-15? Explain with diagrams.

13. The X and Y chromosomes of *D. melanogaster* each carry an NO. Mutant X chromosomes are known that carry either an NO duplication or an NO deficiency. (a) Diagram the karyotypes of flies you would construct with one, two, three, and four "doses" of the NO. What is the sex in each case? (b)

Would you expect a difference in the ability of DNA extracted from these 4 strains to hybridize with 18S rRNA? 28S rRNA? 5S rRNA? If so, what would the differences be?

14. A common property of tandem-repeat genes is that they undergo meiotic synapsis "out of register" and then undergo crossing over to produce a deficiency and a duplication chromosome. Diagram how such an event could generate the X chromosomes described in Problem 13.

15. What might you expect to find if you compared the restriction fragments generated from β-globin gene DNA, from 15S cDNA, and from 10S cDNA by a given restriction enzyme (assume the cDNA to be duplex)?

CHAPTER
9

Structural Gene Expression and the Genetic Code

This chapter is primarily concerned with the translation of mRNA into protein and the nature of the genetic code, but we begin with a brief review of protein structure and an overall account of the mechanics of protein synthesis. Throughout these first sections emphasis is placed on those aspects of protein synthesis that are important to an understanding of genetic phenomena, and details that are more appropriately considered in a biochemistry text are omitted.

PROTEIN STRUCTURE

9.1   AMINO ACIDS

The basic subunits of proteins are **amino acids.** Figure 9-1 shows the 20 amino acids found in naturally occurring proteins. All of these amino acids except proline have as a common denominator the structure drawn in the shaded region in Figure 9-1. This consists of a carbon atom, known as the **α-carbon,** to which is bonded an amino group (the α—NH_2), a carboxyl group (the α—COOH), and a proton. At the pH found in most cells the α—NH_2 and the α—COOH groups of free amino acids exist in solution as —NH_3^+ and —COO^-. The remaining portion of each amino acid is uniquely distinctive and is referred to as an **R group** (R standing for radical). Thus the R group for alanine is CH_3—, that for valine, $(CH_3)_2CH$—, and so forth (Figure 9-1). The R groups of amino acids ultimately determine most of the chemical and configurational properties of a protein.

Amino acids are commonly classified according to the chemical properties of their R groups. Amino acids are thus characterized as being either relatively nonpolar (Figure 9-1a) or relatively polar (Figures 9-1b, c, and d). The polar amino acids can be further characterized by the net charge they tend to carry under conditions of physiological pH: most polar amino acids are uncharged (Figure 9-1b), but five are charged (Figures 9-1c and d). Lysine, for example, carries an —NH_2 residue in its R group, and this —NH_2 group tends to accept protons, becoming —NH_3^+. Consequently, lysine is called a **basic amino acid,** as noted in Chapter 2.11; the two other basic amino acids are arginine and histidine (Figure 9-1c). Similarly, aspartic and glutamic acids tend to lose protons from the —COOH residues located in their R groups and are thus called **acidic amino acids** (Figure 9-1d). Other categories of amino acids, not designated as such in Figure 9-1, include the aromatic amino acids, which possess unsaturated carbon rings in their R groups, and the two sulfur amino acids, which carry a sulfur atom.

9.2   PEPTIDE AND POLYPEPTIDE STRUCTURE

Amino acids can become linked in the following way. The α—NH_2 group of one amino acid interacts with the α—COOH of a second, and a total of one molecule of water is removed, as shown in Figure 9-2a. The resulting —C—N— linkage is known as a **peptide bond** (Figure 9-2b). When two or several amino acids become so linked, the resulting polymeric structure is known as a **peptide** (as for example, a dipeptide, having two amino acids, or a tripeptide, having three). A peptide possesses a zigzagging backbone of nitrogen and carbon atoms (Figure 9-2c) that can be analogized with the sugar-phosphate backbone of a polynucleotide, with R groups projecting outward in alternating fashion. A peptide containing many amino acids is called a **polypeptide.**

The amino acid sequence of a purified polypeptide (or peptide) chain can be characterized by several biochemical procedures; here we outline a procedure frequently used in genetic studies. The amino acid at one end of a polypeptide (the **N-terminal amino acid**) possesses a free α—NH_2 group, while the amino acid at the other end possesses a free α—COOH group (the **C-terminal amino acid**), as shown in Figure 9-2b. These amino acids can be selectively cleaved from the ends of the chain and can be identified by chromatography, a technique described in Box 9.1. Samples of the polypeptide are then exposed to two different enzymes, **trypsin** and **chymotrypsin.** Each enzyme cleaves peptide bonds only between particular amino acids. Trypsin, for example, will cleave the bond formed between the α—COOH of lysine or arginine and the α—NH_2 of any other amino acid, whereas chymotrypsin generally cleaves the same bond, provided that the α—COOH is donated by phenylalanine, tryptophan, or tyrosine rather than lysine or arginine. Two collections of small peptides result; one set is composed of **tryptic peptides,** the other of **chymotryptic peptides.** The tryptic peptides are then separated by chromatography or by electrophoresis (Box 7.1), and each separate peptide class is subjected to an **Edman degradation,** a procedure outlined in Figure 9-3. This yields the complete amino acid sequence of the peptide. The process of separation and sequencing is performed similarly with the chymotryptic peptides.

At the end of all of these analyses, the amino acid sequences of numerous trypic peptides and numerous chymotryptic peptides is known. Because the two enzymes cleave the polypeptide at different positions, many of these amino acid sequences will overlap one another: a tryptic peptide NH_2—Ala—Gly—Gly—Lys—COOH may correspond to the chymotryptic fragment NH_2—Ala—Gly—Gly—Lys—Ser—Phe—COOH, and so on. By a laborious comparison of these sequences the overall amino acid sequence for an original polypeptide can eventually be generated, much as described for the sequencing of nucleic acids in Chapter 7. An automated apparatus is now available that subjects entire polypeptide chains to a sequential Edman degradation, thereby determining amino-acid sequences without enzymatic hydrolysis.

9.3 POLYPEPTIDE AND PROTEIN CONFORMATIONS

The **primary structure** of a polypeptide chain is simply the sequence of its amino acids. This single chain forms characteristic twists and turns upon itself (the **secondary structure**) and eventually assumes a characteristic

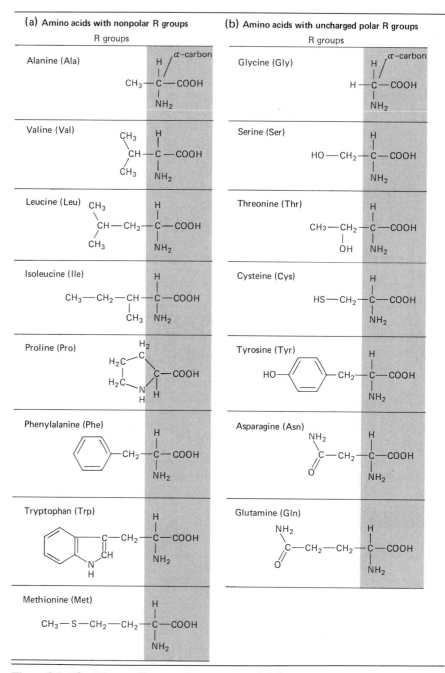

Figure 9-1 The 20 naturally occurring amino acids. The invariant regions are shaded in gray. (From A. L. Lehninger, *Biochemistry*. New York: Worth, 1970.)

Amino acids with charged polar groups at pH 6.0 to 7.0.

(c) Basic amino acids (positively charged at pH 6.0)

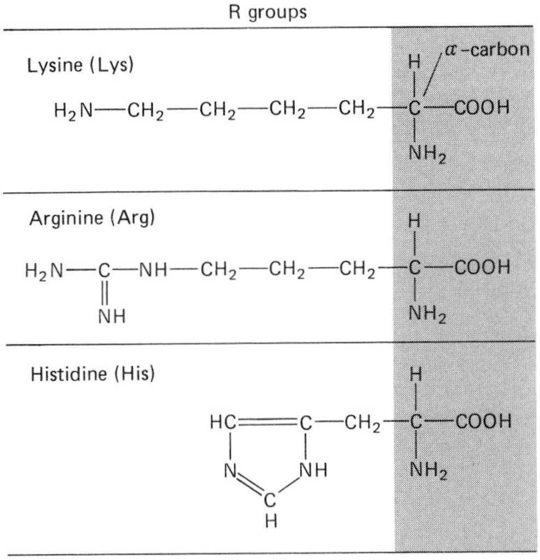

R groups

Lysine (Lys)

$H_2N-CH_2-CH_2-CH_2-CH_2-\overset{H}{\underset{NH_2}{C}}-COOH$ α-carbon

Arginine (Arg)

$H_2N-\overset{NH}{\underset{\parallel}{C}}-NH-CH_2-CH_2-CH_2-\overset{H}{\underset{NH_2}{C}}-COOH$

Histidine (His)

$HC=C-CH_2-\overset{H}{\underset{NH_2}{C}}-COOH$
$\underset{\underset{H}{C}}{N}\diagdown NH$

(d) Acidic amino acids (negatively charged at pH 6.0)

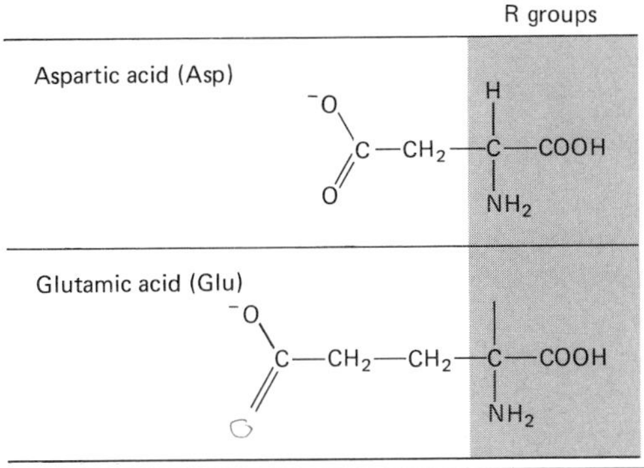

R groups

Aspartic acid (Asp)

$\overset{^-O}{\underset{O}{C}}-CH_2-\overset{H}{\underset{NH_2}{C}}-COOH$

Glutamic acid (Glu)

$\overset{^-O}{\underset{O}{C}}-CH_2-CH_2-\overset{H}{\underset{NH_2}{C}}-COOH$

Box 9.1
CHROMATOGRAPHY

Chromatography is the general term used to describe a variety of techniques that separate molecules according to their size and charge. The techniques are similar to electrophoresis (described in Box 7.1) in that molecules are allowed to move through a solid supporting material that is permeated with a solvent; the nature of the material and the solvent can be selected so that separations are effected between molecules having the size range of interest. Unlike electrophoresis, however, no electric field is applied. The term chromatography derives from the fact that in certain early procedures the separated molecules were visualized by appropriate stains. It is now common practice to localize certain molecules (for example, nucleotides) by their uv absorption.

Chromatographic procedures that are widely used by molecular geneticists are described below.

Paper Chromatography Amino acids and small peptides are frequently identified by this procedure. A sample is applied to one end of a piece of filter paper and known amino acids or peptides are applied beside it. The paper is then dipped into a solution containing water and organic solvents (for example, isopropanol or butanol). As the solvent slowly moves up the paper by capillary action, the amino acids (or peptides) in the sample will migrate at characteristic speeds, depending on how strongly they are adsorbed to the water-saturated fibers in the paper relative to their affinity for the more mobile organic solvent. When the solvent front reaches the top of the paper, the paper is dried, sprayed with **ninhydrin** (an amino acid stain), and heated. The amino acids or peptides in the sample appear as purple dots and can be identified by comparison with the positions of the known samples.

Ion Exchange Chromatography This technique utilizes columns filled with synthetic resins that carry fixed charge groups. For the analysis of amino acids the column contains particles with sulfonic acid groups attached, and these acid groups are equilibrated with NaOH so that all are fully charged with Na^+ ions. An amino acid mixture is now applied to the top of the column. The amino acids are in solution at pH 3, at which pH they all carry a net positive charge. As they move down the column, particular amino acids have a greater or lesser tendency to displace the Na^+ and interact with the sulfonic acid groups. Specifically, the basic amino acids will be bound most tightly (and so move most slowly); the acid amino acids are bound the least tightly (and therefore travel fastest). The column is now washed (eluted) with buffers of increasing pH and Na^+ concentration, and because the amino acids move at different rates they appear in the **effluent** (at the bottom of the column) at different times and can be collected in separate fractions. A device known as an **amino acid analyzer** performs all these operations automatically so that the amino acid composition of a protein can be determined overnight.

Molecular-Exclusion (-Sieve) Chromatography This technique is appropriate for separating a mixture of proteins. A column is packed with small particles, one widely used preparation being known as **Sephadex.** Pores exist *between* the particles through which proteins can pass. In addition, pores exist *within* the particles, since each particle consists of a meshwork of long carbohydrate polymers. Therefore, when a protein solution is placed on top of the column and allowed to flow through by gravity, smaller proteins will tend to move into and be retarded by the small intraparticle pores, whereas larger proteins will be able to flow only through the large interparticle pores. As a result, large proteins will appear in the effluent sooner than the small, and proteins can be effectively separated by size.

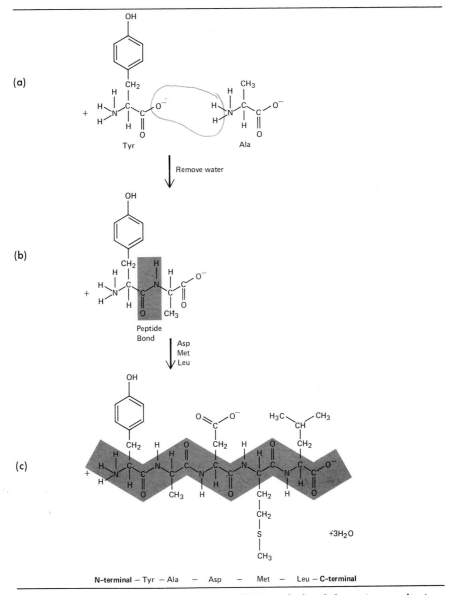

Figure 9-2 (a) Two amino acids interact. (b) Peptide bond formation results in a dipeptide. (c) A pentapeptide, where the basic backbone structure of the molecule is shaded. (From James D. Watson, *Molecular Biology of the Gene,* 3rd ed., © 1976 by J. D. Watson; W. A. Benjamin, Inc., Menlo Park, California.)

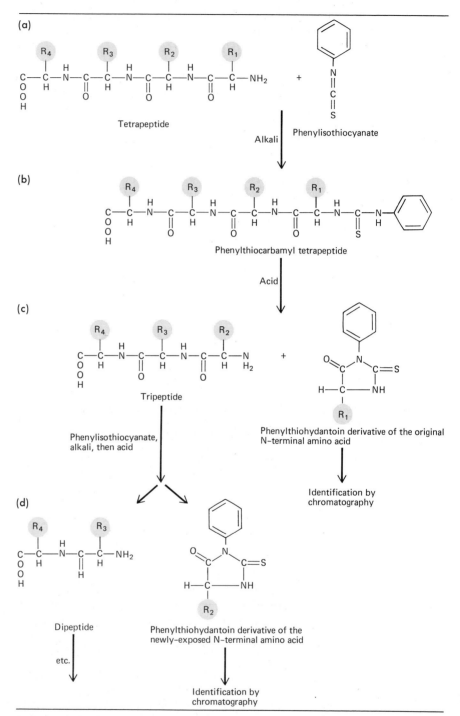

Figure 9-3   The Edman degradation to determine the amino acid sequence of a tetrapeptide. (After A. L. Lehninger, *Biochemistry*. New York: Worth, 1970.)

overall shape in space (the **tertiary structure**) (Figure 9-4). Two, three or many polypeptides may also proceed to associate with one another, the resulting protein being said to possess a **quaternary structure.** Perhaps the best known example of such an **oligomeric protein** is hemoglobin, which contains four separate polypeptide chains called globins—two α chains and two β chains—elaborately fitted together (Figure 9-5). Another example of an oligomeric protein that we have already encountered is the RNA polymerase of *E. coli,* which is made up of six polypeptide subunits (Figure 8-2).

All of these interactions—secondary, tertiary, and quaternary—are ultimately determined by the primary structure of the polypeptide chain(s). For example, during the folding process in a single chain, nonpolar R groups of the amino acids tend to orient toward the interior of the polypeptide chain, whereas polar groups, and particularly charged groups, tend to become localized on the polypeptide's outer surface. Sulfur-containing amino acids, when they come into contact, often form **disulfide (S-S) linkages** that tend to stabilize the protein. Since the conformation of a protein ultimately determines its activity, amino-acid changes in the primary sequence of a

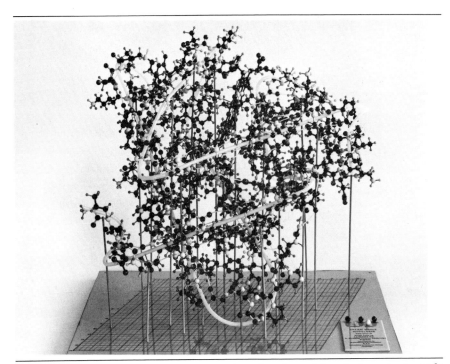

Figure 9-4 Structure of myoglobin deduced from data derived from high-resolution (2 Å) X-ray diffraction. The secondary structure of the polypeptide is formed by the twists and turns taken by the single chain, many of which are α-helical. The tertiary structure of the polypeptide—its overall topology—is indicated by its overall shape in space. (Courtesy Dr. J. C. Kendrew, Cambridge University)

Figure 9-5 Structure of hemoglobin as deduced from data derived from X-ray diffraction. The α chains are light gray; β chains are black; heme groups are shown as gray discs. Each heme group binds oxygen, and there are four per molecule. (From "The Hemoglobin Molecule," by M. F. Perutz. Copyright © November 1964 by Scientific American, Inc. All rights reserved.)

protein may cause it to assume an altered secondary or tertiary conformation, and hence to lose its biological activity. Such amino-acid changes, of course, are ultimately elicited by gene mutations; therefore, as we examine "wild" and "mutant" phenotypes throughout the rest of this text, we shall in fact usually be referring to the effect exerted by a change in the primary sequence of a polypeptide on the overall topology of a protein.

PROTEIN SYNTHESIS

We have already considered certain features of protein synthesis in Chapter 8, features that are summarized in Figure 8-1b. In the sections that follow, we first review the overall process of synthesis and then give a detailed analysis of individual events occurring during synthesis. In the course of these descriptions, for the sake of completeness, certain statements are made about the genetic code. Evidence for these statements appears later in the chapter.

9.4 AN OVERALL VIEW OF PROTEIN SYNTHESIS

A polypeptide is synthesized in a stepwise fashion, starting with the N-terminal amino acid and ending with the C-terminal amino acid. In a like manner the mRNA dictating the sequence of the polypeptide is read by the protein-synthesizing apparatus in a stepwise fashion, one codon at a time, starting at or near the 5' end and working toward the 3' end. Each codon in an mRNA molecule consists of three nucleotides (a **triplet**). A given codon is complementary, in the Watson-Crick sense, to an anticodon carried by a tRNA molecule. Each tRNA is also charged with its specific amino acid to form an aminoacyl-tRNA (see Figures 8-11 and 8-13). As succeeding codons of the mRNA are exposed to a specific ribosomal site, successive aminoacyl-tRNA molecules participate in an anticodon-codon recognition event which is presumed to involve the familiar hydrogen bond pairing of adenine with uracil and guanine with cytosine. Thus a codon UUG (written in a 5' $\longrightarrow$ 3' direction) on mRNA will be recognized by the AAC (written in a 3' $\longrightarrow$ 5' direction) anticodon of a tRNA molecule, as illustrated in Figure 8-1. The UUG codon specifies the amino acid leucine, and the tRNA bearing the AAC anticodon is charged with a leucine molecule. The next codon in the mRNA illustrated in Figure 8-1 is a 5'—GGC—3' codon specifying glycine. A glycyl-tRNA will pair with this codon via its 3'—CCG—5' anticodon; and so forth. As successive amino acids are thus brought into position on the ribosomes, peptide bonds form between them and a polypeptide is synthesized (Figure 8-1).

9.5 INITIATING POLYPEPTIDE SYNTHESIS

The first amino acid to be incorporated into a polypeptide is apparently always methionine. As the polypeptide matures, this N-terminal methionine may or may not be cleaved off, so that the final protein may or may not have methionine as its N-terminal amino acid. Methionine is also found as an internal amino acid in polypeptides.

The first methionine is brought to the ribosome by the **initiator tRNA,** known as tRNA$_f^{Met}$. The **f** in this designation stands for N-formyl methionine because in prokaryotes and in organelles, the α—NH$_2$ group of the initial

$$\overset{\text{O}}{\underset{\|}{}}$$

methionine carries a formyl (—C—H) group; this is not the case in the eukaryotic cytoplasm, but the term persists for all initiator tRNAs.

The anticodons of methionyl-(or N-formyl methionyl-) tRNAs ordinarily recognize one of two **initiator codons** in mRNA, namely, 5'-AUG-3' or 5'-GUG-3'. Recognition of these triplets triggers an intricate series of events. A small ribosomal subunit, a methionyl-tRNA$_f^{Met}$, and an initiator codon (AUG, for example) first interact (Figure 9-6a) to form a subunit · methionyl-tRNA$_f^{Met}$—AUG complex called the **initiation complex** (Figure 9-6b). This interaction requires at least three protein initiation factors designated IF-1, IF-2, and IF-3, and involves the hydrolysis of a molecule of GTP. A recognition and base-pairing interaction between the leader sequence of the mRNA and the 3' end of the 16S rRNA in the small ribosomal subunit (Figure 8-6) may also be critical in establishing the initiation complex.

Once formed, the initiation complex associates with a large ribosomal subunit to form a functional 70S (or 80S in eukaryotes) ribosome. The large subunit possesses two sites, a **peptidyl site** and an **aminoacyl site,** whose actual spatial configurations are unknown. As the two ribosomal subunits associate, the methionyl-tRNA$_f^{Met}$ becomes bound to the peptidyl site in a reaction that also requires GTP; at the same time, the mRNA becomes sequestered in a groove formed by the apposition of the two ribosomal subunits. These relationships are diagrammed in Figure 9-6c.

9.6   ELONGATION OF THE POLYPEPTIDE CHAIN

The alignment of the initiator codon of mRNA with the anticodon of the methionyl-tRNA$_f^{Met}$ in the peptidyl site of the ribosome fixes the alignment of the next codon of the mRNA, and consequently, the alignment of the aminoacyl site. In Figure 9-6d, in which we have diagrammed this stage of polypeptide chain elongation, the next codon specifies valine, and a valyl-tRNA^{Val} is seen to bind valine at its aminoacyl site and also to pair with the mRNA codon. This binding again requires GTP and is mediated in *E. coli* by a complex of two protein-elongation factors called EF-T$_s$ and EF-T$_u$.

At this point the peptidyl site still contains a methionyl group covalently linked to its tRNA$_f^{Met}$ by the α—COOH group of methionine. A displacement now occurs so that the α—COOH group of methionine leaves the tRNA linkage and forms a peptide bond with the α—NH$_2$ of the valyl-tRNA sitting in the adjacent aminoacyl site (Figure 9-6e). This exchange of one bond for another, called the **peptidyl transferase reaction,** is mediated by peptidyl transferase, an enzyme bound to the large ribosomal subunit. At the completion of this step the tRNA^{Val} carries a dipeptide residue (NH$_2$-met-val-), while the tRNA$_f^{Met}$ has given up its amino acid.

The next step in polypeptide synthesis, called **translocation,** involves two simultaneous events that are mediated by GTP hydrolysis, by the elongation factor EF-G, and in all likelihood by a pronounced conformational change in the structure of the ribosome. In the first event the dipeptidyl-tRNA (in Figure 9-6e, the dipeptidyl-tRNA^{Val}) moves from the aminoacyl to the

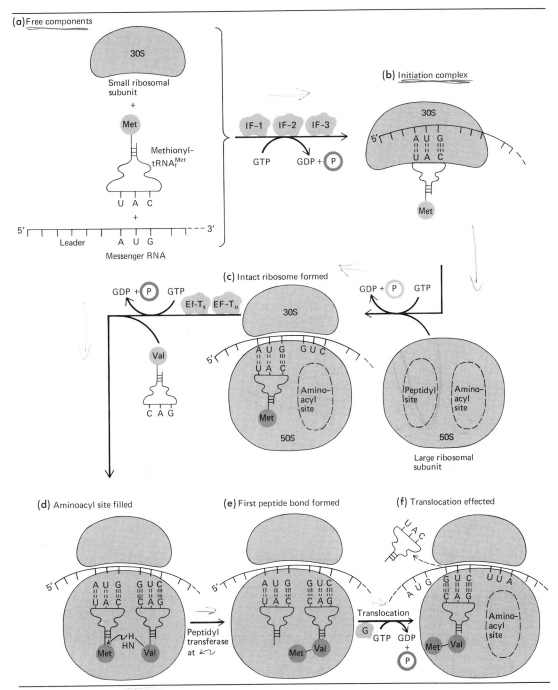

Figure 9-6 Steps in the synthesis of a polypeptide chain, with molecular interactions often indicated in highly schematic form. (See text for details.)

peptidyl site on the ribosome, dislodging the empty tRNA$_f^{Met}$ in the process. The second and simultaneous event is that the whole ribosome moves along the mRNA a distance equivalent to three nucleotides (one codon). As a result, the valine codon remains associated with the anticodon of the dipeptidyl-tRNA^{Val}, and a new codon moves into register with the amino-acyl site. The net result of these translocation events is shown in Figure 9-6f. In Figure 9-6f the new codon specifies leucine, and the cycle described by steps (c), (d), and (e) can now repeat itself: leucyl-tRNA^{Leu} will move into and bind to the aminoacyl site; a peptidyl transferase reaction will occur; a translocation of the tripeptidyl-tRNA^{Leu} to the peptidyl site will take place; and a fourth codon will be exposed to the aminoacyl site. As the ribosome moves in a 5' $\longrightarrow$ 3' direction along the mRNA strand, 8–15 amino acids will be incorporated into the growing polypeptide every second. Since the correct aminoacyl-tRNAs apparently find the correct codons by a random trial-and-error diffusion process, this rate of chain elongation seems truly remarkable.

After a ribosome has translated perhaps 25 codons of an mRNA, the 5' end of the mRNA becomes free to form a second initiation complex and a second ribosome begins moving along the mRNA, mediating the synthesis of a second polypeptide chain. A third ribosome follows, and so on. The resulting structure, called a **polyribosome** or **polysome,** consists of an mRNA molecule which is being simultaneously translated by several ribosomes into several polypeptide chains. Polysomes from E. coli are visualized in Figure 8-19.

9.7 TERMINATION OF POLYPEPTIDE SYNTHESIS

The synthesis of a protein is concluded when a given ribosome in a polysome encounters a genetic signal encoded in the mRNA which specifies that the C-terminal amino acid of a polypeptide has been added to the chain. At least one of the termination signals is the presence of one or more **terminator codons** in the mRNA. A terminator codon is not recognized by the anticodons of any of the normally occurring aminoacyl-tRNAs, and its presence in the aminoacyl site of the ribosome therefore precludes the addition of any further amino acids to the chain.

In E. coli and its phages, and probably in all eukaryotes as well, the RNA triplets UAA, UGA, and UAG all function as terminator codons, a fact we explore fully in the final sections of this chapter. When a terminator codon moves into an aminoacyl site (Figure 9-7a), the following events are thought to occur. The terminator codon first interacts with one of two release factors: RF-1, which appears to recognize UAA or UAG, or RF-2, which recognizes UAA or UGA. The resultant RF · terminator codon · ribosome complex effectively blocks further chain elongation (Figure 9-7b). With the aminoacyl site so clogged, the completed polypeptide remains esterified to

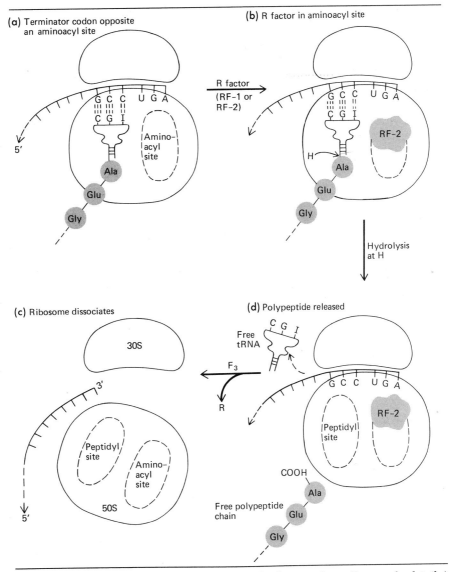

(a) Terminator codon opposite an aminoacyl site

(b) R factor in aminoacyl site

R factor (RF–1 or RF–2)

(c) Ribosome dissociates

(d) Polypeptide released

Figure 9-7 Stages in the termination of polypeptide chain synthesis. (See text for details.)

the final tRNA occupying the peptidyl site. This linkage is then broken by hydrolysis in a reaction mediated by still another protein factor, and both a free tRNA molecule and a complete polypeptide are released from the ribosome (Figure 9-7c). The ribosome then dissociates into its large and small subunits (Figure 9-7d). The dissociated subunits are now free to form new initiation complexes and participate in another round of polypeptide synthesis.

GENETIC SPECIFICATION OF THE TRANSLATION APPARATUS

The translation apparatus in *E. coli* consists of more than 140 distinct macromolecular species, including at least 60 types of tRNA, 20 types of aminoacyl-tRNA synthetase, 30 proteins in the large ribosomal subunit (proteins L1–L30), 19 proteins in the small ribosomal subunit (S1–S19), 3 rRNAs (23S, 16S, and 5S), and 9 factors for initiation (IF), elongation (EF), and release (RF). These multiple components must fit together and interact in a highly specific manner because mutations affecting the structures of many of these components often prove to be highly deleterious or lethal: cells are unable to translate accurately their mRNA into protein. There is therefore considerable interest in the genes that specify these components. We have already considered in detail the genes encoding tRNA and rRNA information (Chapter 8); here we describe the genes for the protein elements of the translation apparatus and consider how these genes might be regulated.

9.8   TRANSLATION-APPARATUS GENE CLUSTERS

Some of the genes for ribosomal proteins in *E. coli* can be identified by screening for mutants resistant to antibiotics that ordinarily block bacterial protein synthesis (Section 6.5). Just as mutations that produce rifampicin-resistance affect the structure of an RNA polymerase subunit (Section 8.3), so do mutations that produce resistance to streptomycin, spectinomycin and erythromycin affect the structural genes that code for particular ribosomal-subunit proteins: in one streptomycin-resistant mutant, for example, the protein S12 is found to be altered in its electrophoretic mobility and in its ability to bind streptomycin.

M. Nomura and others have selected defective transducing phages (Section 5.11) that have "picked up" certain genes involved in antibiotic resistance. The phages λ*dspc1* and λ*dspc2* (*spc* for spectinomycin resistance), for example, were isolated and identified as carrying *spc* genes. Nomura and colleagues next investigated what other genes had been picked up along with the *spc* gene,—that is, what genes were very closely linked to *spc*. To do this, they irradiated *E. coli* cells with doses of ultraviolet light sufficient to totally abolish the effectiveness of the *E. coli* chromosome as a template for transcription. They then infected such cells with λ*dspc1* in the presence of ^{35}S-methionine, and next determined, by autoradiography of polyacrylamide gels (Box 7.1), which proteins were synthesized following phage infection. They found that the phages dictated the synthesis of 14 of the L ribosomal proteins of *E. coli* and 13 of its S ribosomal proteins, indicating that the genes for these proteins lie in a very small region of the *E. coli* chromosome, one that is readily picked up and carried by a λ phage.

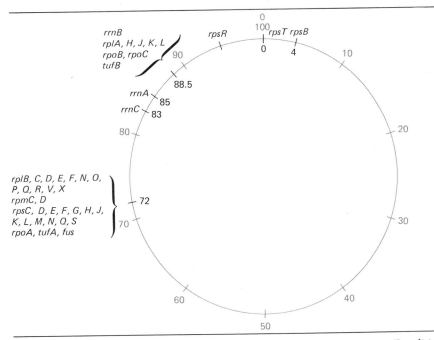

Figure 9-8 Genes specifying components of the translational apparatus in *E. coli* (see also Figures 8-3 and 8-9). Abbreviations: *rrn*, ribosomal RNA gene cluster; *rpl*, ribosomal protein of the large subunit; *rpm*, the *rpl* genes continued past *rplZ*; *rps*, ribosomal protein of the small subunit; *rpo*, RNA polymerase; *tuf*, EF-T$_u$ gene; *fus*, EF-G gene.

Mapping experiments place the *spc* locus at 72 min on the *E. coli* map (Figure 9-8). It thus appears that a cluster of ribosomal-protein genes occupies this site. An unexpected windfall was the finding that certain λdspc strains also carry the gene *tufA* for the elongation factor EF-T$_u$, the gene *fus* for EF-G, and the gene *rpoA* for the α subunit of RNA polymerase (Figures 8-2 and 8-3). Similar analysis of the λrif^{d}18 strain, which carries the 88.5 min sector of the *E. coli* chromosome, has revealed that this phage not only carries the RNA polymerase genes *rpoB* and *rpoC* (Figure 8-3), but also carries genes for 5 of the L ribosomal proteins, a cluster of rRNA genes, a tRNA gene, and a second gene for EF-T$_u$ called *tufB* (Figure 9-8). The remaining genes for translation-apparatus proteins appear to be dispersed elsewhere in the chromosome (Figure 9-8).

9.9 CONTROL OF TRANSLATION-APPARATUS GENES

Whenever a bacterial geneticist learns that genes functionally related to one another are clustered in a small sector of a chromosome, the possibility arises that these genes constitute an operon (Section 8.2) with a single promoter governing their transcription. Although there is as yet no genetic evidence that the genes at 72 min or the genes at 88.5 min in *E. coli* are transcribed as a unit, there is considerable evidence that many and perhaps most of the *E. coli* genes that specify translation-apparatus components are sensitive to similar kinds of controls. Thus during steady-state growth of *E. coli*, the synthesis of all ribosomal proteins and rRNAs is coordinately regulated—none appears to "get ahead of" the others—even though their genes are located in diverse parts of the chromosome (Figure 9-8).

Stringent Control   A particularly elaborate form of translation control known as the **stringent response** occurs when auxotrophic *E. coli* cells are starved for an essential amino acid. The stringent response is apparently triggered when an mRNA codon (for example, GUC for valine) moves into the aminoacyl site on a ribosome but the corresponding valyl-tRNA^{Val} species is not available because the cell (a valine auxotroph) is starved for valine. Under these circumstances, an "uncharged" tRNA^{Val} instead attaches to the ribosome. A protein called the **stringent factor** responds to such uncharged tRNA binding by associating with the ribosome and catalyzing the formation of two unusual nucleotides, **ppGpp (MSI)** and **pppGpp (MSII)**. These nucleotides inhibit (whether directly or indirectly is not yet known) the transcription of all of the tRNA and rRNA genes in the chromosome and probably other structural genes as well.

Although the stringent response is most readily studied in auxotrophic *E. coli,* wild-type cells presumably utilize a similar mechanism to modulate their translational output. The adaptive significance of this response should be apparent: cells do not squander their energy on producing a useless translation apparatus when amino-acid supplies become limiting. A mutation in the *rel*$^+$ gene of *E. coli,* known as *relA,* prevents the production of the stringent factor: *relA* **(relaxed)** strains are therefore unresponsive to amino-acid depletion and continue their normal rates of RNA synthesis. Interestingly, and with the exception of yeast, all eukaryotes that have been studied have the relaxed phenotype: rates of RNA synthesis are not responsive to amino-acid levels. The importance of this kind of control over translation does not, therefore, appear to extend to "higher" organisms.

"CRACKING" THE GENETIC CODE

By the late 1950s it was clear to a number of investigators that the ultimate products of genes were polypeptides, that the synthesis of a polypeptide involved the ordering of amino acids into a defined sequence, and that this information might be encoded by the sequence of nucleotides in a gene. In this and the following sections we present a largely historical account of how the genetic code was deciphered in the early 1960s. The experimental approaches taken involved a brilliant mix of genetics and biochemistry, and the results unquestionably revolutionized the biological sciences.

9.10   THE TRIPLET NATURE OF THE CODE

Those trying to deduce the nature of the genetic code quickly realized that if a single nucleotide in an mRNA specified a single amino acid in a protein, proteins could contain only four amino acids, whereas in fact they contain 20. Similarly, a doublet code made up of all possible pairs of the four nucleotides could generate only 16 combinations, still not enough to specify the 20 amino acids. Therefore, the simplest code that could be envisioned as biologically useful was a triplet code. When all triple combinations of the four nucleotides are made, 4^3, or 64, different sequences or codons are possible. Since, in theory, only 20 are necessary the possibility was entertained that the code is **degenerate, meaning that an amino acid can be specified by more than one triplet**. Of course, another theoretical possibility is that as many as 44 of the putative triplets are never found in natural mRNAs, or that they occur but do not specify amino acids. In the experiments described below, F. Crick, L. Barnett, S. Brenner, and R. Watts-Tobin were able to establish both the triplet nature and the degeneracy of the code. Since these experiments illustrate several important genetic principles, they are described in detail.

In their experiments, Crick and co-workers made use of r-mutant strains of the T4 phage that exhibit two distinct phenotypes, as you may recall from Section 6.2: they cannot grow on strain K of E. coli when this strain is lysogenic for phage λ, and they bring about a rapid atypical lysis of E. coli strain B so that their plaque morphology is distinctive (Figure 6-3). The wild-type allele of the r gene dictates the synthesis of a T4 protein that inserts into the E. coli cell membrane during the course of infection, and the insertion of this protein is apparently necessary for a normal lytic response.

It is possible to classify r strains of T4 as being **leaky** or **nonleaky.** A leaky strain is capable of partial wild-type function; in this case it is capable of producing plaques on E. coli strain B that somewhat resemble wild-type plaques. Mutagenesis of T4 with base analogs such as 5BU (Figure 6-6), which characteristically generate GC $\rightleftharpoons$ AT transitions (see problem 20 of Chapter 6), typically produces leaky r strains. In contrast, a nonleaky r strain shows no growth on E. coli strain K and produces unmistakably large, clear r plaques on strain B. The r strains induced by acridine dyes (Figure 6-14), which generate deletions or insertions of one or a few nucleotides in the T4 chromosome (Chapter 6.13), are characteristically nonleaky.

Crick and co-workers pursued these observations with a series of carefully designed experiments that established molecular distinctions between acridine-induced and base-analog-induced mutations. They began with a nonleaky acridine-induced r strain which they designated FC0. During successive cycles of infection of E. coli B with this T4 strain they occasionally came upon a plaque that was indistinguishable, or nearly so, from a wild-type plaque. Moreover, the phages isolated from these plaques could successfully infect strain K. At first sight, therefore, these phages seemed to represent strains in which the *FC0* mutation had reverted to wild type, but Crick and his associates were able to demonstrate that each new strain was in fact a **pseudo-wild** strain carrying *two* mutations, the original *FC0* mutation plus a second mutation. They found that if they crossed each new phage strain with a wild-type strain by mixed infection (Chapter 5.10), they could isolate rare recombinant strains carrying either *FC0* alone or a new mutation alone. This outcome, diagrammed in Figure 9-9, would not have occurred had *FC0* undergone reversion. Therefore each new strain was in fact doubly mutant, and each could generate a pseudo-wild lytic response only because it carried, in addition to its original *FC0* mutation, a second mutation that acted to **suppress** the effect of the original.

When Crick and his co-workers studied the properties of the recombinant phages that carried only the new suppressor mutations (Figure 9-9), they found in almost all cases that each produced a nonleaky *r* phenotype. In other words, each suppressor mutation generated a phage with the phenotype of an acridine-induced *r* strain.

These observations led Crick and his co-workers to make the following proposals. They first postulated that acridines produce *either* deletions *or* insertions in DNA. Specifically, they suggested that the original *FC0* strain carried an insertion, designated as +, and that each of the *FC0* suppressor mutations consisted of a spontaneous deletion in the *r* gene, designated as −. They next argued that if the reading of a genetic message begins from

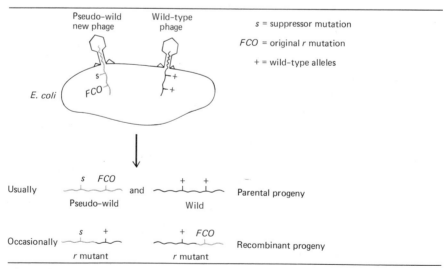

Figure 9-9 Demonstration of the doubly mutant nature of a pseudo-wild strain as isolated by Crick et al. The new strain is allowed to infect *E. coli* along with wild-type phages (a mixed infection). Occasional recombinant progeny emerge carrying either FC0 alone or the new suppressor mutation(s) alone.

a fixed point and proceeds sequentially from that point, one codon at a time, a + mutation would have the effect of shifting the reading frame of a transcript (a **frameshift mutation**) in a forward direction; for example, if the correct reading frame is GAG · GAG · GAG · GAG · . . . , the insertion of a C would produce the reading GAG · GAC · GGA · GGA · . . . . Similarly, a −

GAG · GAC · GGA · GGA
         ↑

mutation involving the deletion of an A would shift the frame in a reverse direction and produce a GAG · G↓GG · AGG · AGG · . . . message. In either case it is clear that a single deletion or insertion, particularly if it occurs near the beginning of a gene, would lead to a disruption in the reading of most of the genetic message and a nonfunctional (nonleaky) protein would almost certainly be synthesized. In contrast, they argued, a base analog-induced transitional mutation in a gene would disrupt only a single base in an mRNA, resulting, at most, in the insertion of a single incorrect amino acid into a polypeptide. This mutant polypeptide might well still be capable of partial functioning; that is, it might well be able to sustain a leaky phenotype.

Finally, Crick and his co-workers proposed that the − mutations were able to exert their suppressor effects because they were of opposite sign from the FC0 mutation. They argued that when both + and − mutations are included within nearby portions of the same gene they have the effect of canceling each other out: the reading frame of the transcript from such a doubly mutant gene would first be shifted out of and then back into the correct phase. Looking at our GAG polynucleotide, the insertion of a C (↑) followed by the deletion of an A (↓) would have this effect:

GAG · GAC · GGA · GGA · GG↓G · GAG · GAG . . . .
    o.k.   ↑   o.k.   o.k.

In this sequence, only four triplets are wrong instead of the entire message, and a pseudo-wild phenotype might well be achieved by the resultant protein.

Based on these arguments, Crick and his co-workers designed and performed a series of experiments that established the reading frame used in translating messenger RNA. They first infected *E. coli* B cells with the − suppressor strains of T4 that they had collected. These strains were, as we have said, of the *r* phenotype, but pseudo-wild plaques were occasionally encountered, much as in the experiments with FC0. These plaques again proved to be formed by suppressed strains carrying double mutations, but in this case the new second mutations were suppressors of − mutations and were thus of the + class. In this way Crick and his co-workers eventually accumulated a store of about 80 different *r* strains, all designated (+) or (−).

Crosses were then made between the various + and − strains. Most combinations of + and − produced recombinants with a pseudo-wild phenotype, just as the original combination of FC0 and its − suppressor gave pseudo-wild progeny (Figure 9-9). In other words, most − mutations were found to be suppressors of + mutations and vice versa. In contrast, phages carrying combinations of any two + mutations or any two − mutations still had the nonleaky r phenotype. Thus, two mutations of the same sign could not suppress each other.

The final step was to create strains that carried **three** mutations of the same sign, + + + or − − −. All exhibited a pseudo-wild phenotype. Thus three mutations of the same sign were found to suppress one another where two could not. The simplest interpretation of this result is to propose that the **coding ratio** is three: if each codon consists of a triplet of nucleotides, three insertions or three deletions would have the effect of restoring the correct reading frame. Again, by example with poly · GAG, three + + + insertions of C would produce

GAG · GAG · GAC · GGA · GCG · ACG · GAG · GAG.

Four incorrect codons are included in this message, but the bulk is back in phase.

9.11   THE DEGENERACY OF THE CODE

Besides indicating that a genetic message is read in successive frames of three nucleotides, the Crick experiments suggested that the code is degenerate. You will recall that many of the + − combinations in T4 were found to yield a pseudo-wild phenotype. This indicates that a reasonably functional protein was often synthesized by such doubly mutant strains and suggests that the wrong triplets found between a + and a − mutation (or between + + + or − − − mutations) *do* code for amino acids, even though they are probably in most cases the wrong amino acids. Since the wrong triplets are created by random mutational events, they presumably include most of the 64 possible triplet combinations of the four nucleotides, suggesting that many of the 20 amino acids are specified by more than one triplet.

A more direct demonstration that the code is degenerate was made shortly after the 1961 publication of the Crick paper, when several laboratories succeeded in deciphering the genetic code by biochemical means.

9.12   DECIPHERING THE CODE: *IN VITRO* CODON ASSIGNMENTS

Three approaches were taken to establish, *in vitro,* which nucleotide triplets specify which amino acids. This work was done most notably in the laboratories of M. Nirenberg and G. Khorana. In the first approach synthetic polyribonucleotides that contained one, two, or three different types of base were constructed. They were then presented to a complete **cell-free protein-synthesis system** containing ribosomes, the full complement of aminoacyl-tRNAs, and the protein factors necessary for polypeptide synthesis. The composition of the polypeptides synthesized by the *in vitro* system was then analyzed. In cases in which the artificial messenger contained only one repeating nucleotide, the results were straightforward. Poly U, for example, was found to direct the synthesis of polyphenylalanine. Assuming a triplet code, this experiment indicated that UUU is a codon for phenylalanine. Similarly, poly C directs the synthesis of polyproline and poly A, that of lysine, indicating that CCC codes for proline and AAA for lysine.

When mixed copolymers are made, one can calculate all the possible triplets they contain assuming that the bases are incorporated randomly into the molecule; for example, when poly AC is synthesized from a mixture containing equal proportions of A and C, eight triplets should occur: CAC, AAC, AAA, ACA, ACC, CCC, CCA, and CAA. Using poly AC it was found that six amino acids were incorporated into polypeptides: asparagine, glutamine, histidine, lysine, threonine, and proline. The next step was to vary the ratio of A and C, and it was found that when the polymer contained more A than C the ratio of asparagine to histidine in the polypeptide increased. By such experiments the *composition* of the bases in a certain triplet, although not their absolute order, could often be deduced.

The second approach taken was similar to the first except that methods were devised to synthesize polyribonucleotides of a known sequence; for example, a regular copolymer of C and U (CUCUCU . . .) was constructed and was found to direct the synthesis of regularly alternating copolypeptides of leucine and serine. Polymers of repeating tri- and tetranucleotides were also elaborated and the polypeptides they specified were analyzed. In this way amino acid assignments could readily be made for a number of triplets.

We should note at this point that magnesium concentrations must be kept artificially high in these *in vitro* experiments if the translation of artificial messengers is to occur. High magnesium concentrations allow initiation without an initiation codon, and an AUG or GUG codon will, of course, be absent from most of the artificial polyribonucleotides that might be synthesized *in vitro*. The artificial initiation process occurs at random along a synthetic messenger, reading in all possible frames: thus a regularly repeating poly AAG stimulates the synthesis of three kinds of homopolypeptide: polyarginine (AGA), polylysine (AAG), and polyglutamic acid (GAA).

A third and quite different approach to solving the genetic code was made by synthesizing trinucleotides of known base sequences, associating them with ribosomes, and testing the ability of the resulting complexes to stimulate the binding of specific aminoacyl tRNAs. For example, the trinucleotide GCC was found to be active in promoting the binding of alanyl-tRNA^{Ala} but *not* of any other aminoacyl-tRNA, and it was therefore concluded that GCC is a codon for alanine.

None of these *in vitro* methods is in itself without ambiguity, but taken together they established the codon assignments for the amino acids with considerable certainty. The full genetic code so derived is summarized in Table 9-1. Each codon is written as it would appear in mRNA, reading in a $5' \longrightarrow 3'$ direction; the corresponding codons in DNA will, of course, be both complementary to these mRNA codons and written in the reverse order on a $5' \longrightarrow 3'$ strand (Section 8.5). Similarly, the bases in the corresponding tRNA anticodons will be both complementary and antipolar to the mRNA codons.

An examination of Table 9-1 reveals several features of the genetic code. First, the code is indeed seen to be degenerate, there being, for example, six codons that specify serine. Only two amino acids are represented by a single codon: tryptophan and methionine. Methionine, as we have learned, is signaled by the special codon AUG which also specifies the initiation of translation. The three codons UAA, UAG, and UGA are often called **nonsense codons** because they fail to stimulate aminoacyl-tRNA binding. They serve, as already noted, a punctuation function as chain-terminator signals, and are considered more extensively in Section 9.18.

9.13 PATTERNS TO THE GENETIC CODE

Two notable patterns emerge when the genetic code is tabulated as in Table 9-1. First, amino acids with similar structural properties tend to have related codons. Thus the aspartic acid codons (GAU and GAC) are similar to the glutamic acid codons (GAA and GAG); similarly, the codons for the aromatic amino acids phenylalanine (UUU, UUC), tyrosine (UAU, UAC), and tryptophan (UGG) all begin with uracil. This feature of the code is thought to have evolved to minimize the consequences of mistakes made during translation or of mutagenic base substitutions. Thus, if an amino acid in a protein is erroneously replaced by one with similar properties, the protein may still be functional.

The second pattern to the genetic code is that for many of the synonym codons specifying the same amino acid the first two bases of the triplet are constant, whereas the third can vary; for example, all codons starting with CC specify proline (CCU, CCC, CCA, and CCG) and all codons starting with AC specify threonine. This flexibility in the third nucleotide of a codon may well help to minimize the consequences of errors, and F. Crick has offered a molecular explanation for its occurrence. He suggests that as an aminoacyl-tRNA molecule is lining up to form base pairs with an mRNA codon on a ribosome the *initial* codon-anticodon interaction is between the nucleotide at the 5' end of the codon and the nucleotide at the 3' end of the

Table 9-1 The Genetic Code

First Position (5' End of mRNA) (Read Down)	Second Position (Read Across)				Third Position (3' End) (Read Down)
	U	C	A	G	
U	Phe	Ser	Tyr	Cys	U
	Phe	Ser	Tyr	Cys	C
	Leu	Ser	Stop	Stop	A
	Leu	Ser	Stop	Trp	G
C	Leu	Pro	His	Arg	U
	Leu	Pro	His	Arg	C
	Leu	Pro	Gln	Arg	A
	Leu	Pro	Gln	Arg	G
A	Ile	Thr	Asn	Ser	U
	Ile	Thr	Asn	Ser	C
	Ile	Thr	Lys	Arg	A
	Met (start)	Thr	Lys	Arg	G
G	Val	Ala	Asp	Gly	U
	Val	Ala	Asp	Gly	C
	Val	Ala	Glu	Gly	A
	Val (start)	Ala	Glu	Gly	G

anticodon, while the second interaction takes place between the two middle nucleotides. He proposes that once correct base pairs have formed at these two positions some **wobble** will be permissible at the third position. For example, the purine inosine (I) is similar to guanine and will normally pair with cytosine (C), as shown in Figure 9-10a; when it is in the third position of an anticodon, however, it may be free to shift its position so that it can form base pairs with a U (Figure 9-10b) or an A (Figure 9-10c) in the third position of the codon, such pairs being called **wobble base pairs.** Therefore, as illustrated in Figure 9-10, a seryl-tRNA[Ser] molecule with the anticodon 3'—AGI—5' will be able to interact with the serine codons UCC, UCU, and UCA. Similarly, a U at the wobble position will be able to pair with an A or a G.

Because of the proposed wobble base-pairing, one tRNA species is thought to be able to recognize more than one codon for the same amino acid. As noted in Section 8.10, moreover, several distinct tRNA species often exist for a given amino acid. In some cases these distinct species respond to the same codon or codon family, but in other cases they will respond only to different codons. In short, there is no simple one-to-one relationship between the number of amino acid codons that exist in a cell and the number of tRNA species possessed by the cell.

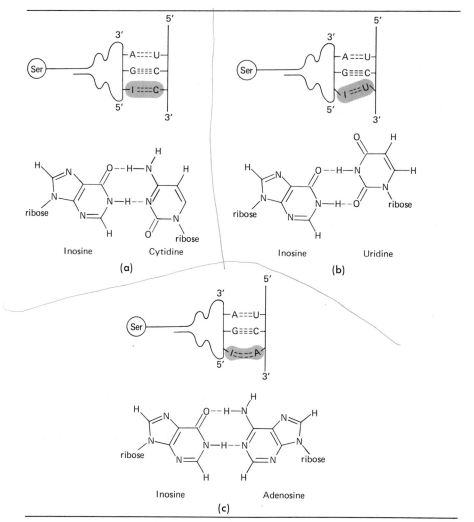

Figure 9-10 (a) Normal base pairing between inosine (I) and cytidine in the third position of anticodon-codon recognition (shaded). (b) Wobble base pairing between I and uridine. (c) Wobble base pairing between I and adenosine.

9.14 DECIPHERING THE CODE: *IN VIVO* CODON ASSIGNMENTS

The code summarized in Table 9-1 was established with synthetic molecules and cell-free protein-synthesizing systems. Experiments designed to determine whether this same code is also used *in vivo* soon followed. One approach compared the amino acid sequences of proteins from wild-type and mutant strains of various organisms.

Early experiments by C. Yanofsky and co-workers, for example, focused on the enzyme tryptophan synthetase of *E. coli*. In one chymotryptic peptide derived from the wild-type enzyme, position 8 was occupied by glycine. Several base-analog-induced mutant strains had substitutions for glycine at position 8, and the new amino acids were always either arginine or glutamate, as shown in Figure 9-11a. When the arginine-containing strains were again treated with base analogs, revertants were recovered which had glycine at position 8. New mutant strains were also found, and these carried either serine, isoleucine, or threonine at position 8 (Figure 9-11a). In contrast, when the glutamate-containing strains were again mutagenized with base analogs, new mutant strains carrying either valine or alanine at position 8 were obtained in addition to the glycine revertants (Figure 9-11a). By assuming that the original glycine at position 8 was coded for by GGA and that each mutational event involves a single base change, it was found that the observed amino acid substitutions could be explained by a unique set of codons (Figure 9-11b), all selected from the *in vitro* codon assignments. This outcome strongly suggests (but does not prove) that the *in vitro* code is followed *in vivo*.

Similar experiments were performed by E. Terzaghi, G. Streisinger, A. Tsugita, and their associates. This group used a different protein and a different mutagen than Yanofsky and colleagues. The experiments concerned the enzyme lysozyme, which is specified by T4 and digests bacterial cell walls. In this case the complete amino acid sequence of the enzyme was known, and comparisons were thus possible between the wild-type sequence and the sequence specified by pseudo-wild strains that carried two acridine-induced, + − frameshift mutations. One pseudo-wild enzyme was found to carry a cluster of five amino acids that differed from those in the wild-type enzyme, as shown in Figure 9-12, thus confirming the prediction of Crick and his colleagues that pseudo-wild strains should

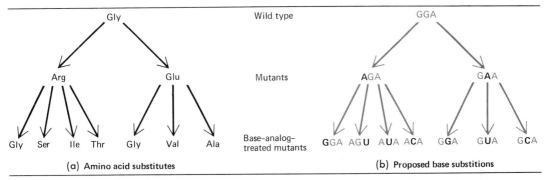

(a) Amino acid substitutes

(b) Proposed base substitions

Figure 9-11 (a) Amino acid substitutions observed at position 8 of a chymotryptic peptide from tryptophan synthetase A of *E. coli*, as described in the text. (b) Proposed base substitutions (boldface) leading to the amino acid substitutions seen in (a).

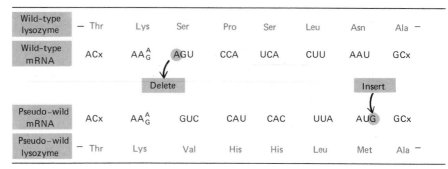

Figure 9-12 Frameshift mutations in the lysozyme gene of T4 phage and their effect on the primary structure of the enzyme, as shown by E. Terzaghi and colleagues. The lysine codon is shown as AA_G^A to indicate the possible variability at the third position. X = U, C, A, or G. (After C. Yanofsky, *Ann. Rev. Gen.* **1**:117, 1967.)

dictate the synthesis of proteins that carry several wrong amino acids in tandem. When *in vitro* codon assignments were written out for the five wild-type amino acids, it was found that if a single nucleotide were first deleted and then inserted, the *in vitro* codon assignments for the five mutant amino acids could be generated, as illustrated in Figure 9-12. The amino acid sequences of several other pseudo-wild strains can be similarly made to fit the *in vitro* code by assuming simple frameshift events.

The most unambiguous validation of the *in vitro* code has been made possible by comparing the amino acid sequences of various proteins with the nucleotide sequences of the mRNAs that specify them. Initiator codons in mRNA can be identified by the fact that they are protected from ribonuclease digestion when mRNA is associated with ribosomes under *in vitro* conditions that disallow translocation (Figure 9-6). Starting from these codons, one can mark off successive triplets on mRNA and determine which amino acids they should specify according to the *in vitro* code; this theoretical sequence is then compared with the known sequence of amino acids in the protein. A number of "proofs" of the *in vitro* code have been made in this fashion.

9.15   INITIATOR CODONS

Sequencing studies have revealed that GUG as well as AUG can serve as an initiator codon *in vivo*. The *in vitro* code (Table 9-1) states that GUG should code for valine, yet the N-terminal amino acid inserted in response to an initiator GUG is (N-formyl)-methionine. This aberrant finding is explained by the observation that methionyl-tRNA$_f^{Met}$ appears to obey unique codon-anticodon recognition rules: you will recall (Figure 9-10) that most aminoacyl-tRNAs appear to make initial contact with the first nucleotide in the triplet codon, the third nucleotide being the "wobble base;" methionyl-tRNA$_f^{Met}$, on the other hand, makes initial contact with the third nucleotide (G) of the codon, the first nucleotide becoming the "wobble base". Therefore, methionine is inserted in response to AUG and GUG codons (and probably UUG and CUG codons as well).

Why doesn't N-formyl methionyl-tRNA$_f^{Met}$ recognize internal AUG and GUG codons and insert N-formyl methionine in these positions? It appears that the elongation-factor complex EF-T$_s$ · EF-T$_u$ (Figure 9-6) is unable to interact with N-formyl methionyl-tRNA$_f^{Met}$ and therefore never brings this species into the aminoacyl site of the bacterial ribosome. Hence, internal AUGs and CUGs are accessible only to methionyl-tRNA^{Met} and valyl-tRNA^{Val} species. As is evident in Figure 9-13, tRNA$_f^{Met}$ and tRNA^{Met} have the same 5'-CAU-3' anticodon but are otherwise very different in their base sequence; presumably, therefore, they assume quite different tertiary structures (see Figure 8-12) and are recognized by quite different sets of proteins in the translation apparatus.

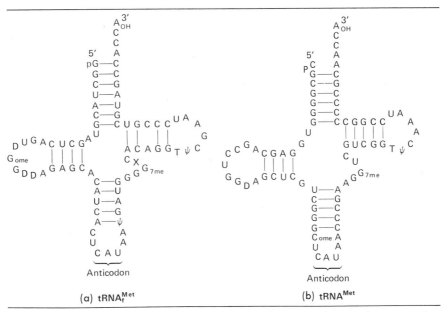

Figure 9-13 (a) The tRNA that places N-formyl methionine as the initial amino acid in a polypeptide. (b) The tRNA that places methionine in the interior (middle) of a polypeptide. Abbreviations are the same as in Figure 8-11 plus G$_{7me}$, 7-methylguanylic acid; G$_{ome}$, 2'-0-methyl-guanylic acid; C$_{ome}$, 2'-0-methylcytidylic acid. X represents an unidentified nucleotide.

9.16 THE UNIVERSALITY OF THE CODE

The experiments described above involve *E. coli* and its phages, but similar experiments have demonstrated that eukaryotes use the same type of genetic code. F. Sherman and collaborators, for example, have made considerable use of yeast strains that have abnormal levels of a functional iso-l-cytochrome c (as estimated by its absorption spectrum). Amino-acid sequencing in these strains has established the existence of an alanine in position 12 of the wild-type cytochrome; X ray-induced mutations have caused this alanine to be replaced by proline, and further mutagenesis changes this proline to either serine or threonine. The inferred sequence of events— GC__(Ala) $\longrightarrow$ CC__(Pro) $\longrightarrow$ AC__ or UC__ (Thr or Ser)— is consonant with the code established *in vitro*. (The third base of each triplet in this series is in all cases degenerate and is thus indicated by a blank.)

Amino acid substitutions have also been determined for a number of spontaneously arising mutant forms of human hemoglobin, as shown in Table 9-2. Again, each substitution can be accounted for by assuming the structure of the *in vitro* codons and postulating the occurrence of a transition or transversion of a single base pair.

That the genetic code is indeed universal can be demonstrated quite directly by presenting an *E. coli in vitro* protein-synthesizing system with, for example, purified mRNA from poliovirus (which is normally translated by human cells) and observing the synthesis of virus protein. A second approach utilizes an *in vitro* system described earlier in the context of "cracking" the code. In this approach, synthetic trinucleotides of known sequence are complexed with *E. coli* ribosomes, and their binding affinity for *E. coli* aminoacyl-tRNAs is compared with their affinity for aminoacyl-tRNAs from other organisms. In general, such studies demonstrate that different types of organisms are similar, if not identical, in the codons that their various aminoacyl-tRNAs recognize, although some may have little or no aminoacyl-tRNA that corresponds to certain codons.

It is speculated, therefore, that the code became fixed in its present form when the first kinds of cellular life evolved, from one to three billion years ago. Whether the code originated essentially all at once or whether it began with only a few meaningful codons and gradually expanded to include all 64 codons is not, of course, known, but apparently once it was established, it specified such essential biological information that any further variations did not survive evolutionary pressures.

NONSENSE MUTATIONS AND CHAIN TERMINATION

9.17 PROPERTIES OF NONSENSE MUTATIONS

As stated earlier in the chapter, three RNA codons are not recognized by any of the aminoacyl-tRNAs normally present in cells. These are UAG (also called **amber** because an investigator who studied the properties of this

Table 9-2 Correlation between Amino Acid Replacements and Codon Assignments in some Variants of Human Hemoglobin.

Hemoglobin Variant	Affected Chain	Position	Amino Acid Substitution	Codon Alteration	Mutational Mechanism
Hikari	β	61	Lys $\longrightarrow$ Asn	AAA $\longrightarrow$ AAU AAG $\longrightarrow$ AAC	Transversion
I	α	16	Lys $\longrightarrow$ Glu	AAA $\longrightarrow$ GAA AAG $\longrightarrow$ GAG	Transition
D Ibadan	β	87	Thr $\longrightarrow$ Lys	ACA $\longrightarrow$ AAA ACG $\longrightarrow$ AAG	Transversion
G Philadelphia	α	68	Asn $\longrightarrow$ Lys	AAU $\longrightarrow$ AAA AAC $\longrightarrow$ AAG	Transversion
O Indonesia	α	116	Glu $\longrightarrow$ Lys	GAA $\longrightarrow$ AAA GAG $\longrightarrow$ AAG	Transition
G Chinese	α	30	Glu $\longrightarrow$ Gln	GAA $\longrightarrow$ CAA GAG $\longrightarrow$ CAG	Transversion
San Jose	β	7	Glu $\longrightarrow$ Gly	GAA $\longrightarrow$ GGA GAG $\longrightarrow$ GGG	Transition
G Galveston	β	43	Glu $\longrightarrow$ Ala	GAA $\longrightarrow$ GCA GAG $\longrightarrow$ GCG	Transversion
S	β	6	Glu $\longrightarrow$ Val	GAA $\longrightarrow$ GUA GAG $\longrightarrow$ GUG	Transversion
Mexico	α	54	Gln $\longrightarrow$ Glu	CAA $\longrightarrow$ GAA CAG $\longrightarrow$ GAG	Transversion
Shimonoseki	α	54	Gln $\longrightarrow$ Arg	CAA $\longrightarrow$ CGA CAG $\longrightarrow$ CGG	Transition
K Ibadan	β	46	Gly $\longrightarrow$ Glu	GGA $\longrightarrow$ GAA GGG $\longrightarrow$ GAG	Transition
Norfolk	α	57	Gly $\longrightarrow$ Asp	GGU $\longrightarrow$ GAU GGC $\longrightarrow$ GAC	Transition
Seattle	β	76	Ala $\longrightarrow$ Glu	GCA $\longrightarrow$ GAA GCG $\longrightarrow$ GAG	Transversion
M Milwaukee	β	67	Val $\longrightarrow$ Glu	GUA $\longrightarrow$ GAA GUG $\longrightarrow$ GAG	Transversion
L Ferrara	α	47	Asp $\longrightarrow$ Gly	GAU $\longrightarrow$ GGU GAC $\longrightarrow$ GGC	Transition
G Accra	β	79	Asp $\longrightarrow$ Asn	GAU $\longrightarrow$ AAU GAC $\longrightarrow$ AAC	Transition
M Boston	α	58	His $\longrightarrow$ Tyr	CAU $\longrightarrow$ UAU CAC $\longrightarrow$ UAC	Transition
Kenwood	β	95	Lys $\longrightarrow$ Glu	AAA $\longrightarrow$ GAA AAG $\longrightarrow$ GAG	Transition
Zurich	β	63	His $\longrightarrow$ Arg	CAU $\longrightarrow$ CGU CAC $\longrightarrow$ CGC	Transition
Horse hemoglobin	α	24	Phe $\longrightarrow$ Tyr	UUU $\longrightarrow$ UAU UUC $\longrightarrow$ UAC	Transversion

From A. Sadgopal, *Adv. Genetics* **14**:325–404 (1968), Table 3.

codon belonged to the Bernstein family, and Bernstein means "amber" in German); UAA (also called *ochre*), and UGA (*opal*). One or more of these three codons is used profitably by the cell to signal the natural end of translation of a particular polypeptide. If, however, a mutation occurs such that one of these codons appears in the interior of a genetic message, incomplete polypeptides are released from the ribosome. Such a mutation is called a **nonsense** mutation, as contrasted to the **missense** mutations we have been considering in which a mutant codon specifies an alternate amino acid.

Strains carrying nonsense mutations can be recognized by four criteria. First, they often produce peptide fragments that can be detected experimentally. The length of the fragment will depend, of course, on the position in the gene in which the nonsense codon appears. A given codon may be **proximal** to (near) or **distal** to (far from) the promoter region of its gene, and a proximal nonsense codon will exert its effect near the N-terminus of a protein to produce a small polypeptide fragment, whereas a distal mutation will have a less severe effect on the protein length.

The second, and correlative, property of a nonsense mutation is that it usually produces a nonleaky phenotype, as might be expected for an incomplete protein. Only when the mutation occurs in the distal part of the gene and the untranslated portion of the mRNA is not essential does the polypeptide product show any activity at all. Frequently, so little of the protein is synthesized that extracts of the mutant strains will not even cross react, in the precipitin test, with antibody made against the wild-type protein (the precipitin test is described in Box 9.2.)

The third characteristic of nonsense mutations, called **polarity,** is observed when a nonsense mutation lies in a gene proximal to the promoter of an operon. In such a case, the nonsense mutation prevents expression not

Box 9.2 PRECIPITIN AND IMMUNO- DIFFUSION TESTS	A highly specific attribute of a protein is its **antigenic** properties. When a protein is injected into a foreign vertebrate animal—for example, an *E. coli* protein into a rabbit—the rabbit will respond by producing an **antibody** to that protein. The blood serum from the rabbit (the **antiserum**) is then collected. If this is mixed with a small amount of the *E. coli* antigen, a precipitate is formed known as the **precipitin.** Precipitin is formed because each antibody has two binding sites for the antigen (as described in Section 15.12) so that an antibody-antigen network is created. A version of this test known as **double immunodiffusion** requires only small amounts of material. Antiserum and antigen are placed in separate wells in a gel matrix and allowed to diffuse toward each other. A solid band of precipitin forms where the antibody and antigen meet. No such band appears when the antiserum instead meets a protein against which the rabbit has not been immunized.

only of the promoter-proximal gene but also of all genes in the operon that are distal to the mutation. Such polar effects have been found to be "relieved" when *E. coli* cells also carry the mutant gene *suA* (*supressor*). Analysis of *suA* strains has revealed that these strains carry a mutation in their structural gene for rho factor, cited in Section 8.8 as a participant in the termination of transcription. It appears, therefore, that when an *E. coli* ribosome falls off a message at a nonsense codon, rho blocks further transcription of that message and no genes distal to the nonsense codon are copied. Mutant versions of rho perform this activity less efficiently, and polarity is thereby less severe in those *suA* strains that carry nonsense mutations in an operon.

The fourth feature of a nonsense mutation is that its phenotypic expression is usually conditional. Certain genes in bacteria and other cells suppress the expression of nonsense codons in a manner quite distinct from the suppression of polarity. The molecular basis of nonsense suppression is discussed in Section 9.19; meanwhile its basis may occur to you.

9.18 ESTABLISHING THE CHAIN-TERMINATION CODONS

The behavior of UAG, UAA, and UGA triplets *in vitro* made them likely candidates as the sources of nonsense mutations. It was observed, for example, that neither UAG, UAA, nor UGA was active as a trinucleotide in promoting the binding of aminoacyl-tRNAs to ribosomes. Moreover, when synthetic polynucleotides were synthesized such that any one of these triplet combinations appeared in the message, these polynucleotides were characteristically translated into short peptide fragments rather than long polypeptides. Such observations indicated—but did not prove—that these three codons were involved both in nonsense chain termination and in natural chain termination.

The approaches taken to analyze nonsense mutations *in vivo* resemble the experiments described earlier for the tryptophan synthetase gene. In one approach followed by A. Garen and colleagues, *E. coli* was mutagenized with base analogs and strains were isolated that exhibited no alkaline phosphatase activity whatsoever, and were thus likely to be carrying nonsense mutations in the alkaline phosphatase gene. When these strains were again mutagenized with base analogs, "cured" strains having at least partial alkaline phosphatase enzyme activities were recovered. The amino acid sequences of the wild-type and the cured alkaline phosphatases were then compared. It was found that the wild-type protein had a tryptophan (coded for by UGG) in a certain position and that the cured proteins from several strains of *E. coli* carried a spectrum of seven amino acid substitutions in its place, including the original tryptophan. By writing out the codons for these seven amino acids it could be deduced that the original nonsense mutation must have converted the UGG to a UAG, for the substituted codons could

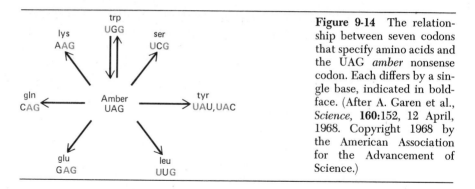

Figure 9-14 The relationship between seven codons that specify amino acids and the UAG *amber* nonsense codon. Each differs by a single base, indicated in boldface. (After A. Garen et al., *Science*, **160**:152, 12 April, 1968. Copyright 1968 by the American Association for the Advancement of Science.)

all be related to UAG by single nucleotide changes. This result is diagrammed in Figure 9-14. The study of other cured strains led, by similar reasoning, to the implication of UAA and UGA as nonsense codons.

SUPPRESSOR MUTATIONS

9.19 GENERAL NATURE OF SUPPRESSORS

Suppressor mutations were defined in Section 6.1 as second mutations that restore the original (wild) phenotype of a mutant strain. Such a second mutation may occur within the gene carrying the first mutation **(intragenic)** or may lie in a different gene **(intergenic).** We have already encountered several examples of intragenic suppression in the course of this chapter: base substitutions in *E. coli* have been shown to convert missense or nonsense codons into other missense codons that restore protein activity (Figures 9-11 and 9-14), while a suppressor frameshift mutation occurring close to the position of a first mutation can produce a functional lysozyme or *r* gene product in T4 (Figures 9-9 and 9-12).

Intergenic suppression can originate in two different ways; one indirect and one direct. During **indirect intergenic suppression** a second gene mutation has the effect of circumventing the expression of the original mutant phenotype without in any way altering the original mutant gene product. For example, certain mutant strains of *Neurospora* are known to have an abnormally high sensitivity to intracellular arginine. When these strains also carry the *s* suppressor gene, they synthesize very little arginine and thus the arginine sensitivity, although still present, is not exhibited. Additional examples of genes that affect the expression of other genes are given in Chapters 17 and 18. Of more immediate interest here is the phenomenon of direct intergenic suppression.

Direct intergenic suppression is most easily defined after an example is given to illustrate it. We first imagine an *E. coli* cell carrying an *amber* mutation in its tryptophan synthetase gene such that the mRNA for trypto-phan synthetase carries a UAG codon near its proximal end. Such a cell, as we have learned, would synthesize no tryptophan synthetase and would normally be inviable. Now let us imagine that a second mutation occurs, this time in a gene that specifies tRNA^{Leu}. This mutation has no effect on the affinity of the tRNA^{Leu} molecule for leucine, and thus leucyl-tRNA^{Leu} complexes form normally. The mutation does, however, change the anti-codon of the tRNA^{Leu} so that instead of being 3'—AAC—5' and pairing with a 5'—UUG—3' leucine codon, it is 3'—AUC—5'. This mutant tRNA has the capacity to recognize a 5'—UAG—3' nonsense codon and insert leucine in response to it. Therefore, when the mutation in the tRNA gene, called a suppressor mutation, is included in the cell along with the mutation in the tryptophan synthetase gene, a tryptophan synthetase of normal size will be synthesized. This will contain one wrong amino acid (leucine), but it will usually have at least some activity and the cell can survive.

In general, then, direct intergenic suppressor mutations can be said to affect genes whose products function in the translation of mRNA into protein. The mutant suppressor molecules permit a mutant codon to be translated incorrectly but beneficially. Most often the suppressor molecules are tRNA molecules, as in the foregoing example. In *E. coli* nine different tRNA genes (Figure 8-9) have been observed to mutate so that their tRNA transcripts perform a mutant activity: each inserts its normal amino acid in response to a nonsense triplet. As in the preceding example, the amino acid inserted is usually wrong and the protein synthesized in the presence of the suppressor is therefore usually still mutant; however, it is almost always able to function more effectively than the original nonsense fragment.

9.20   NONSENSE SUPPRESSOR GENES IN E. COLI

One of the first isolated and most intensively studied **suppressor strains** of *E. coli* carries the mutant gene known as *su3*. A cell carrying *su3* plus an *amber* mutation in its alkaline phosphatase gene will demonstrate partial or pseudo-wild enzymatic activity. Similarly, if this cell is infected with a T4 strain carrying an amber mutation in a head protein gene, some infective progeny phages will be produced.

H. Goodman and colleagues analyzed the tRNA present in wild-type and *su3* cells. Both proved to have identical amounts of tRNA$_2^{Tyr}$. The wild-type cells also possessed tRNA$_1^{Tyr}$, whereas *su3* cells produced a reduced amount of tRNA$_2^{Tyr}$ and, in addition, a new tyrosine-inserting tRNA species not present in the wild type cells. When the sequence of this new tRNA was determined, it was found to be identical to tRNA$_2^{Tyr}$ except at the anticodon: thus, whereas the wild type anticodon reads 5'-G*UA-3' (where G* is a modified guanine), the mutant anticodon reads 5'-CUA-3'. The mutant tRNA is thus able to read 5'-UAG-3' codons, insert tyrosine, and eliminate the chain-terminating effects of nonsense triplets, exactly as in our earlier hypothetical example.

The *su3* mutation maps to 27 min on the *E. coli* chromosome (Figure 8-9), thus localizing the gene for tRNA$_2^{Tyr}$. Further analysis has revealed that two closely linked copies of this gene (*tyrT*) occupy the locus; therefore, *su3* *E. coli* cells carry one *tyrT* and one *tyrTsu3* gene, explaining how they are able to synthesize both normal and suppressor versions of tRNA$_2^{Tyr}$.

Other *amber* suppressors in *E. coli* (Figure 8-9) insert glutamine, leucine, lysine or serine in response to UAG. The codons for these amino acids all differ from UAG by a single base (Table 9-1), and in a number of these strains a suppressor tRNA with a mutant anticodon has been identified. Ochre and opal suppressors are also known in *E. coli,* as are suppressor tRNAs with base changes in non-anticodon regions of the molecule. For example, one suppressor tRNA carries a mutation in the dihydrouridine loop (Figures 8-11 and 8-12) which changes its translation properties.

Strains of *E. coli* carrying efficient *amber*-suppressor genes are capable of the same rates of growth as wild-type cells, a puzzling and unexplained fact since one would expect that in such strains many natural chain-terminator codons would be suppressed and overly-long proteins would be synthesized by **read-through** into ordinarily nontranslated sectors of mRNA molecules such as trailer or spacer sequences.

Temperature-sensitive Amber Suppressor   Since a nonsense mutation typically abolishes all protein activity, *amber* mutations cannot ordinarily be isolated in essential genes: if a non-suppressor strain is used to select for such a mutation then the mutant cell will die, whereas if a suppressor strain is used the cell will not express a mutant phenotype during the screening procedure. This dilemma is circumvented by the *SupD^{ts}* mutant of *E. coli,* which produces an efficient serine-inserting amber suppressor at 14°C but has virtually no suppressor activity at 43°C.

9.21   NONSENSE SUPPRESSOR GENES IN YEAST

Of the eukaryotes, yeast is presently the most well-characterized with respect to intergenic nonsense suppressors. Suppressor genes can be detected by their ability to suppress nutritional nonsense mutants and by their ability to restore near-normal iso-l-cytochrome c levels to strains carrying proximal *amber* or *ochre* mutations in the *cyc1* structural gene. Of the large number of *amber* or *ochre* suppressor (*SUP*) mutations screened for by such procedures by F. Sherman and colleagues, all prove to map to 8 loci in the yeast genome (Figure 12-16), with alleles at each locus for both *amber* and *ochre* suppressors: thus the gene SUP-2a and the gene SUP-2o are alleles at the SUP-2 locus which specify, respectively, *amber* and *ochre* suppressors.

In at least one case—the SUP-5a strain—the *amber* suppressor has been identified as a tRNA^{Tyr} species with a mutant anticodon (5'-CTA-3') that inserts tyrosine in response to UAG. Interestingly, all the *amber* and *ochre* suppressors mapping to the 8 loci also insert tyrosine. Since wild-type yeast possesses only one species of tRNA^{Tyr}, these 8 loci may prove to represent 8 redundant copies (Section 8.10) of this tRNA gene.

References

Protein Synthesis

"The mechanism of protein synthesis." *Cold Spring Harbor Symp. Quant. Biol.,* Vol. 34. New York: Cold Spring Harbor Laboratory, 1970.

Anfinsen, C. B. *Aspects of Protein Biosynthesis.* New York: Academic, 1970.

Champney, W. S., and S. R. Kushner. "A proposal for a uniform nomenclature for the genetics of bacterial protein synthesis," *Molec. Gen. Genetics* **147:**145–152 (1976).

Lucas-Lenard, J., and F. Lipmann. "Protein biosynthesis," *Ann. Rev. Biochem.* **40:**409–448 (1971).

Nomura, M., Ed. *Ribosomes.* Cold Spring Harbor, N. Y. Cold Spring Harbor Laboratory, 1974.

Weissbach, H., and S. Ochoa. "Soluble factors required for eukaryotic protein synthesis," *Ann. Rev. Biochem.* **45:**191–216 (1976).

Genes Specifying the Translation Apparatus

Furano, A. V., and F. P. Wittel. "Effect of the *RelA* gene in the synthesis of individual proteins in vivo," *Cell* **8:**115–122 (1976).

Haseltine, W. A., R. Block, W. Gilbert, and K. Weber. "MSI and MSII made on ribosome in idling step of protein synthesis," *Nature* **238:**381–384 (1972).

Jaskunas, S. R., L. Lindahl, M. Nomura, and R. R. Burgess. "Identification of two copies of the gene for the elongation factor EF-Tu in *E. coli,*" *Nature* **257:**458–462 (1975).

Ratner, D. R. "Evidence that mutations in the *suA* polarity suppressing gene directly affect termination factor rho," *Nature* **259:**151–153 (1976).

Yamamoto, M., W. A. Strycharz, and M. Nomura. "Identification of genes for elongation factor Ts and ribosomal protein S2 in *E. coli,*" *Cell* **8:**129–138 (1976).

The Genetic Code

"The genetic code." *Cold Spring Harbor Symp. Quant. Biol.,* Vol. 31. New York: Cold Spring Harbor Laboratory, 1966.

Barrell, B. G., G. M. Air, and C. A. Hutchison III. "Overlapping genes in bacteriophage ØX174," *Nature* **264:**34–38 (1976).

Clegg, J. B., D. J. Weatherall, and P. F. Milner. "Haemoglobin Constant Spring—A chain termination mutant?" *Nature* **234:**337–340 (1971).

*Crick, F. H. C. "Codon-anticodon pairing; the wobble hypothesis", *J. Mol. Biol.* **19:**548–555 (1966). [Reprinted in G. L. Zubay, *Papers in Biochemical Genetics.* New York: Holt, Rinehart and Winston, 1968.]

*Crick, F. H. C., L. Barnett, S. Brenner, and R. J. Watts-Tobin. "General nature of the genetic code for proteins," *Nature* **192:**1227–1232 (1961). [Reprinted in G. S. Stent, *Papers on Bacterial*

*Denotes articles described specifically in the chapter.

Viruses, 2nd ed. Boston: Little, Brown, 1966, and in G. L. Zubay, *Papers in Biochemical Genetics.* New York: Holt, Rinehart and Winston, 1968.]

Hoffman, G. W. "On the origin of the genetic code and the stability of the translation apparatus," *J. Mol. Biol.* **86:**349–362 (1974).

Khorana, H. G. "Polynucleotide synthesis and the genetic code," *Harvey Lectures* **62:**79–105 (1966–67).

*Nirenberg, M., and P. Leder. "RNA code words and protein synthesis," *Science* **145:**1399–1407 (1964). [Reprinted in G. L. Zubay, *Papers in Biochemical Genetics.* New York: Holt, Rinehart and Winston, 1968.]

*Nirenberg, M., and J. H. Matthaei. "The dependence of cell-free protein synthesis in *E. coli* upon naturally occurring or synthetic polyribonucleotides," *Proc. Natl. Acad. Sci. U.S.* **47:**1588–1602 (1961). [Reprinted in G. L. Zubay, *Papers in Biochemical Genetics.* New York: Holt, Rinehart and Winston, 1968.]

*Ocada, Y., S. Amagase, and A. Tsugita. "Frameshift mutation in the lysozyme gene of bacteriophage T4: demonstration of the insertion of five bases, and a summary of *in vivo* codons and lysozyme activity," *J. Mol. Biol.* **54:**219–246 (1970).

Sadgopal, A. "The genetic code after the excitement," *Adv. Genetics* **14:**325–404 (1968).

Stewart, J. W., F. Sherman, N. A. Shipman, and M. Jackson. "Identification and mutational relocation of the AUG codon initiating translation of iso-l-cytochrome c in yeast," *J. Biol. Chem.* **246:**7429–7445 (1971).

*Streisinger, G., Y. Okada, J. Emrich, J. Newton, A. Tsugita, E. Terzaghi, and M. Inouye. "Frameshift mutations and the genetic code," *Cold Spring Harbor Symp. Quant. Biol.* **31:**77–84 (1966).

Wong, J. T. "The evolution of a universal genetic code," *Proc. Nat. Acad. Sci. U.S.* **73:**2336–2340 (1976).

*Yankofsky, C. "Amino acid replacements associated with mutation and recombination in the A gene and their relationship to *in vitro* coding data," *Cold Spring Harbor Symp. Quant. Biol.* **28:**581–588 (1963). [Reprinted in E. A. Adelberg, *Papers on Bacterial Genetics,* 2nd ed. Boston: Little, Brown, 1966.]

Nonsense and Suppression

*Brenner, S., A. O. W. Stretton, and S. Kaplan. "Genetic code: the 'nonsense' triplets for chain termination and their suppression," *Nature* **206:**994–998 (1965). [Reprinted in G. L. Zubay, *Papers in Biochemical Genetics.* New York: Holt, Rinehart and Winston, 1968.]

Brody, S., and C. Yanofsky. "Suppressor gene alteration of protein primary structure," *Proc. Natl. Acad. Sci. U.S.* **50:**9–16 (1963). [Reprinted in G. L. Zubay, *Papers in Biochemical Genetics.* New York: Holt, Rinehart and Winston, 1968.]

Garen, A. "Sense and nonsense in the genetic code," *Science* **160:**149–159 (1968).

*Gilmore, R. A., J. W. Stewart, and F. Sherman. "Amino acid replacements resulting from super-suppression of nonsense mutants of iso-l-cytochrome c from yeast," *J. Mol. Biol.* **61:**157–173 (1971).

*Goodman, H. M., J. Abelson, A. Dandry, S. Brenner, and J. D. Smith. "Amber suppression: A nucleotide change in the anticodon of a tyrosine transfer RNA," *Nature* **217:**1019–1024 (1968).

Gorini, L. "Informational suppression," *Ann. Rev. Genetics* **4:**107–134 (1970).

*Liebman, S. W., F. Sherman, and J. W. Stewart. "Isolation and characterization of amber suppressors in yeast," *Genetics* **82:**251–272 (1976).

Nagata, T., and T. Horiuchi. "Isolation and characterization of a temperature-sensitive amber suppressor mutant of *Escherichia coli* K12," *Molec. Gen. Genet.* **123:**77–88 (1973).

Piper, P. W., M. Wasserstein, F. Engbaek, K. Kaltoft, J. E. Celis, J. Zeuthen, S. Liebman, and F. Sherman. "Nonsense suppressors of *Saccharomyces cerevisiae* can be generated by mutation of the tyrosine tRNA anticodon," *Nature* **262:**757–761 (1976).

Rasse-Messenguy, F., and G. Fink. "Temperature-sensitive nonsense suppressors in yeast," *Genetics* **75:**459–464 (1973).

(a)   (b)   (c)

(a) Masayasu Nomura (University of Wisconsin) studies the genetic specification and assembly of ribosomes. (b) Charles Yanofsky (Stanford University) has demonstrated relationships between amino acid substitutions and the genetic code in *E. coli.* (c) Luigi Gorini contributed much to our understanding of the molecular nature of suppressor mutations.

Questions and Problems

1. In what ways is a polyribosome similar to an *rrn* gene (Sections 8.14 and 8.15) in the process of being transcribed?

2. Crick et al. found that they could mixedly infect *E. coli* cells with a + and a − *r* strain and could recover + − recombinants. In most cases these had a pseudo-wild phenotype, but several continued to exhibit an *r* phenotype, particularly in cases where both the + and − mutations occurred near the beginning of the *r* gene. Give an explanation for this result, using diagrams.

3. The following repeating polyribonucleotides were used in a cell-free system to direct amino acid incorporation into polypeptides: (a) poly AC; (b) poly UAC; (c) poly AAG; (d) poly GAU; (e) poly UAUC. Which amino acids would you expect to be incorporated in each case?

4. When U occupies the third (5′) position of an anticodon, wobble base-pairing can occur with either A or G in the codon, whereas a C in this position can pair only with a G. Predict whether the following mutant anticodons can suppress *amber* (5′-UAG-3′) or both *amber* and *ochre* (5′-UAA-3′) codons: (a) 5′-UUA-3′; (b) 5′-CUA-3′.

5. The RNA chromosome of the R17 bacteriophage, which contains 3 genes, also serves as phage messenger RNA. The R17 coat protein, specified by the middle gene in the chromosome, has the C-terminal amino-acid sequence

Asn-Ser-Gly-Ile-Tyr. The R17 replicase, specified by the gene at the 3′ end of the chromosome, has the N-terminal sequence Met-Ser-Lys-Thr-Thr-Lys.

The following R17 RNA sequence is found to be protected by ribosomes from ribonuclease digestion:

1	2	3	4	5	6	7	8	9	10	11	12	13	14

5′ . . . AAC UCC GGU AUC UAC UAA UAG AUG CCG GCC AUU CAA ACA UGA ~

15	16	17	18	19	20

GGA UUA CCC AUG UCG AAG . . . 3′

(a) Which codon specifies the C-terminus of the coat protein? (b) What is significant about the two codons that follow the codon identified in (a)? Would you expect an *ochre* suppressor mutation to confuse the termination of coat-protein synthesis? Explain. (c) Explain how the AUG #18 and not the AUG #8 is identified as the initiator codon for replicase synthesis. (d) If AUG #8 were to initiate protein synthesis, write out the sequence of amino acids that would be polymerized.

6. A bacterial operon has the gene sequence promoter-*a b c d*-terminator, where genes *a-d* specify polypeptides A-D, respectively. (a) Write out a general structure for the mRNA transcribed from such an operon, assuming that it has not undergone processing. Indicate the polarity of the molecule and include the following features: leader, trailer, and transcribed spacer sequences; initiator and terminator codons; initial purine. (b) The C-terminus of polypeptide C is found to be . . . Thr-Arg-Arg; the N-terminus of polypeptide D is found to be Met-Leu-Ser. The sequence of a ribosome-protected fragment of the operon mRNA is found to be . . . ACUCGCCGCUGAUGCUAUCA . . . Define and comment on the reading frames used to translate the *c* and *d* sequences in this mRNA. How long is the transcribed spacer?

7. When "multigenic" mRNAs transcribed from bacterial operons are translated, equimolar quantities of each polypeptide chain encoded by the message are synthesized in the case of the *trp* (*tryptophan*-synthesis) and *gal* (*galactose*-utilization) operon mRNAs of *E. coli*. When *lac* operon mRNA is translated, however, the enzyme β-galactosidase (specified by the first coding sequence at the 5′ end of the message) is synthesized 3-5 times more frequently than the transacetylase enzyme (specified by the third and final coding sequence at the 3′ end of the message). In what region(s) of the *lac* operon mRNA would you look for sequences that might explain this discrepancy?

8. A mutation called *glyUsuA36* in *E. coli* converts a tRNA^{Gly}, which recognizes the GGA glycine codon, into a tRNA^{Gly} which recognizes the AGA arginine codon. (a) Write out the probable anticodons for these two tRNAs and specify the nature of the mutational event. (b) The missense mutation *trpA36* of *E. coli*, which alters amino acid #211 in the active site of the

tryptophan synthetase enzyme, is suppressed in *glyUsuA36* strains. What amino acid would you expect to find at position #211 of tryptophan synthetase in wild-type cells? In *trpA36* cells? In cells carrying both *trpA36* and *glyUsuA36*?

9. Wild-type yeast are sensitive to the inhibitor canavanine (*can^s*). The *ochre* mutation *canl-100* causes yeast to be canavanine-resistant (*can^r*). When *canl-100* cells are mutagenized, two types of *can^s* strains can be recovered which we can call A strains and B strains. The mutations carried by the A strains map to the *canl-100* locus; the mutations carried by the B strains are not linked to *canl-100*. (a) Describe the probable nature of the type-A and the type-B mutations. (b) Describe how you would use the new *can^s* strains to isolate temperature-sensitive *ochre* suppressors.

10. Describe how you would screen for an *amber* mutation in an essential gene of *E. coli* using the *SupD^ts* strain. Assume you have an antiserum available that is directed against the protein product of this gene.

11. A suppressor tRNA is discovered that will suppress certain frameshift mutations if they are of the +1 addition variety but not if they are +2 or −1 or −2. Describe a likely conformation assumed by its anticodon loop.

12. A mutation changes a 3′TCT5′ DNA sequence to an ACT, which creates a mutant phenotype in strain Y of *E. coli*. Strain Y is now mutagenized, and three strains (A, B, and C) are recovered that have an apparently wild phenotype. Sequence analysis of this position of the gene reveals that strain A carries a GCT sequence, strain B a TCT sequence, and strain C a TTT sequence. Explain why each of these strains has an apparently wild phenotype.

13. A nonsense suppressor mutation is found to map to the *rpsR* gene of *E. coli* (Figure 9-8). Speculate on how the product of this gene might function in translation.

14. A temperature-sensitive suppressor mutation is found to relieve UAA and UAG nonsense mutations but is without effect on UGA mutants. tRNA extracts from the suppressor strain demonstrate no suppressor activity in an *in vitro* translation assay, nor do washed ribosomes, but the soluble protein fraction of the cells is highly active. (a) Design an *in vitro* translation assay that could be used for the assays described above. (b) What component(s) of the translation system are likely to be affected by this mutation? Explain your reasoning.

15. Is iso-1-cytochrome *c* indispensible for growth in yeast? Explain.

16. Transversions can arise by 4 general mutational pathways (AT ⟶ CG, AT ⟶ TA, GC ⟶ CG, GC ⟶ TA) and transitions can arise by 2 mutational pathways (GC ⟶ AT, AT ⟶ GC) (Chapter 6). You wish to learn which pathway(s) are followed by a new mutagen. Outline how you would accomplish this using the following tester strains: a serine-missense T4 mutant; a tyrosine-missense T4 mutant; a leucine-missense T4 mutant; and *amber, ochre,* and *opal* suppressor strains of *E. coli*. The diagram on the next page will assist your answer.

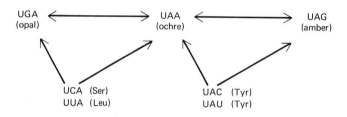

17. It is observed that heat induces reversion of *amber* but not *ochre* mutations. Explain how this observation can be used to argue that heat causes guanine transversions.

18. The phage ØX174 genome is remarkable in that, in two cases, the same stretch of DNA codes for two polypeptides which are translated in different reading frames. Specifically, the complete E sequence is contained within the D gene region, while the complete B sequence is contained within the A gene region:

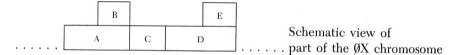

   Schematic view of part of the ØX chromosome

The mRNA specifying the D and E polypeptides (which function, respectively, in phage DNA replication and host lysis) has the following sequence:

D start
↓
5′G . . . 15 bases . . . AAUAGGUAAGAAAUCAUGAGUCAAGU . . . 152

E start
↓
bases . . . UUGAGGCUUGCGUUUAUGGUACGCUGGA . . .

E end   D end
↓   ↓
250 bases . . . GCGGAAGGAGUGAUGUAA . . . 3′

(a) Is the reading frame for D and E shifted by one or two nucleotides? Explain. (b) What sector(s) of this mRNA would you expect to be protected by *in vitro* ribosome binding? (Recall Figure 8-6) (c) What is the expected C-terminal amino acid of the D protein? The E protein? (d) Would you expect an *amber* mutation in gene E to affect the translation of gene D? Why or why not? (e) Speculate as to how this arrangement of genes might have evolved.

19. Sequencing of the ØX174 chromosome (Section 7.2) has allowed a determination of the frequency with which the 64 possible codons are actually used in the phage genome. The results are tabulated on the next page.

Phe	TTT	39	Ser	TCT	35	Tyr	TAT	36	Cys	TGT	12
	TTC	26		TCC	9		TAC	15		TGC	10
Leu	TTA	19		TCA	16	Ter	TAA	3	Ter	TGA	5
	TTG	26		TCG	14		TAG	0	Trp	TGG	16
Leu	CTT	36	Pro	CCT	34	His	CAT	16	Arg	CGT	40
	CTC	15		CCC	6		CAC	7		CGC	29
	CTA	3		CCA	6	Gln	CAA	27		CGA	4
	CTG	24		CCG	21		CAG	34		CGG	8
Ile	ATT	45	Thr	ACT	40	Asn	AAT	37	Ser	AGT	9
	ATC	12		ACC	18		AAC	25		AGC	5
	ATA	2		ACA	13	Lys	AAA	47	Arg	AGA	6
Met	ATG	42		ACG	19		AAG	31		AGG	1
Val	GTT	53	Ala	GCT	64	Asp	GAT	44	Gly	GGT	38
	GTC	14		GCC	17		GAC	35		GGC	28
	GTA	10		GCA	12	Glu	GAA	27		GGA	13
	GTG	11		GCG	12		GAG	34		GGG	3

(a) Are the codons used randomly or nonrandomly? Explain with examples. (b) The "most popular" codon for each amino acid has a distinctive feature. What is it? How might this relate to the wobble hypothesis? (c) Referring to Table 1-3, would you expect that a preference for particular codons might prove to be a distinctive feature of a species? How might an analysis of the tRNA species present in a cell help to document your answer?

20. A hemoglobin variant called **Constant Spring** is known in humans. Whereas the normal length of the hemoglobin α chain (Figure 9-5) is 141 amino acids, the length of the Constant Spring α polypeptide is 172 amino acids. Describe how this phenotype could arise either via a chain-terminator mutation or a frameshift mutation. Would you expect α-hemoglobin mRNA to contain "trailer sequences" based on this information? Explain.

21. Explain why a frameshift mutation does not lead to the incorrect expression of all the genes on a chromosome.

CHAPTER
10

Mapping Viral Chromosomes

INTRODUCTION

The position of genes along a chromosome is expressed by a **genetic map** of the chromosome. Such a map also indicates the order of genes along a chromosome. It further indicates the relative distances that separate the various genes. Thus if a group of genes is thought of as being physically linked along a chromosome, neighboring genes will be closely linked and widely separated genes will be loosely linked.

The principles that govern the construction of genetic maps were established, in large part, with eukaryotic organisms, notably *D. melanogaster*. As early as 1911 T. H. Morgan conceived the idea that variations in the degree of gene linkage, as measured by recombination frequencies, correlated with differences in the distances separating genes. A. H. Sturtevant proceeded to utilize this concept to determine the order and the relative spacing of five genes of *D. melanogaster* located in the X chromosome. Since that time genetic maps of the chromosomes of numerous organisms have been published, with maps of viral and bacterial chromosomes being relatively recent.

The principles of genetic mapping are presented here using bacteriophages—despite the fact that this means reversing historical precedents—because the principles of mapping are readily visualized with viruses. Each virus possesses only one chromosome, so that all of its genes appear within a single linkage group. Furthermore, the molecular nature of the viral chromosome is well known (Chapter 5), in contrast to the mystery presently surrounding the organization of the eukaryotic chromosome (Chapter 2), and genetic data can in certain cases be correlated with molecular data. Finally, each viral chromosome possesses a rather small number of genes so that the ordering of all or most of the genes on the chromosome can actually be visualized.

AN OVERVIEW OF GENETIC MAPPING

10.1 THE IMPORTANCE OF MAPPING

There are a number of reasons why genetic maps are constructed. Fundamentally, maps represent a summary of a species' genetic information. Thus for a virus whose genetic map is well populated with genes we can tell at a glance the kinds of information the virus possesses; we can recognize the blank spaces on the map in which genetic functions have still to be identified; and we can learn whether the genes are ordered in a circular array or whether two ends of the chromosome exist.

Any gene that has been mapped can be identified by its location in a particular chromosome, and this is presently the best single way to distinguish it from all other genes. Other questions that can be answered by an analysis of gene order can, in addition, be cited. First, genetic mapping can often give an indication of the kind of mutation a chromosome has undergone. A large deletion that removes several genes from a chromosome shows different results in mapping experiments than a point mutation affecting a single nucleotide. If a mutant strain is used to argue that a certain gene determines the production of a certain protein, it is clearly essential to demonstrate that the strain does not carry a deletion affecting several genes, any one of which might conceivably be responsible for dictating the synthesis of the protein in question.

It is also possible by genetic mapping to determine whether independent mutations affect the same or different genes. Suppose that, following mutagenesis and an appropriate screening procedure, 11 different strains of a virus or organism appear to have the same mutant phenotype. Do these 11 strains carry mutations affecting the same gene or is more than one gene involved in producing the same phenotype? One way to answer this question is to locate each mutation on the genetic map: those that can be shown to occupy the same locus or position in the chromosome can be readily distinguished from those that map to loci that are far from one another on the chromosome. Thus, if the 11 strains are found to carry mutations at three distinct loci, we can assume that at least three genes will be of interest in studying the production of the phenotype in question.

Another important question that can be answered by genetic mapping is closely related to the last. Returning to our example, let us suppose that four mutations affecting a phenotype all map within a very small region on the chromosome. Do these mutations actually occur within a single gene or is the chromosome so organized that several genes concerned with the phenotype are closely clustered in a small portion of the genome? Genetic mapping can identify the existence of these gene clusters within a chromosome.

The foregoing catalog is by no means complete, but a final application for genetic mapping is that it should allow, in theory, the localization, and hence the identification, of every gene that makes up an organism's chromosome(s). This goal has already been realized for certain small bacteriophages that have only a few genes, and it is probably close at hand for such well-studied middle-sized phages as λ.

10.2 SOME TERMINOLOGY DEFINED

Before we describe how genetic mapping is done, we should make clear what is meant by the terms **gene, marker,** and **locus.** These terms are frequently used during genetic analysis, and they are often confused.

We have already offered a definition of a gene at the start of Chapter 8, and by now you should feel comfortable with the concept of a gene as existing in a number of different forms, one usually denoted as wild type and the others as mutant. Ordinarily a gene cannot be mapped, and, indeed, its existence cannot ordinarily be known, until a strain carrying a mutation in that gene is somehow procured. Thus we might rightly infer that the chromosome of a bacteriophage must contain genes specifying bacterio-phage head protein, but until it is possible to devise a means of obtaining mutant phages that are unable to direct the synthesis of this head protein, it will be difficult to learn much about the chromosomal location or the properties of the normal gene.

Genetic markers are therefore sought, these being mutations that mark the existence of a given gene. The marker is not, of course, synonymous with the gene itself, any more than a substituted base could be considered synonymous with a gene. One reason that the two are often confused is that the *name* given to a marker often becomes loosely used as the *name* of the gene; for example, when a mutant strain of phage that is defective in head protein synthesis is isolated, the mutation it carries might arbitrarily be denoted *hp* for *head protein*. In this case, then, *hp* would serve as a genetic marker. Subsequent genetic analysis might localize the marker at a certain position on the phage's map, leading to the conclusion that a gene speci-fying the structure of head protein resides at that position. Unfortunately this gene is often called *hp* when it should strictly be called "the gene marked by the *hp* mutation." The latter terminology is obviously more cumbersome and the former is therefore often used, even though it can result in misunderstanding.

Locus is a convenient term, introduced in Section 3.15, that denotes a physical position within a chromosome. Thus, if two markers are said to "map to the same locus," this indicates that they occupy similar positions. The term can be used quite generally to denote a certain chromosomal region or it can be used more specifically to refer to small sectors of a chromosome. The term **gene locus** is also used to denote that portion of a chromosome's nucleotide sequence that constitutes a certain gene. Two **allelic genes** (defined in Section 3.15) will, of course, occupy the same chro-mosomal (and gene) locus.

We advise that you become familiar with the above terms before pro-ceeding in this chapter. If you understand what is being manipulated in a genetic cross, then an analysis of the outcome of the cross becomes straightforward.

10.3 THE GENERAL CASE FOR MAPPING BY RECOMBINATION ANALYSIS

As noted in Chapters 5 and 9 and illustrated in Figure 10-1, a bacterium can be subjected to **mixed infection** by two or more phages carrying different genetic markers—for example, by one phage with the chromosome consti-tution $a+$ and another with $+b$. During the ensuing lytic cycle exchanges

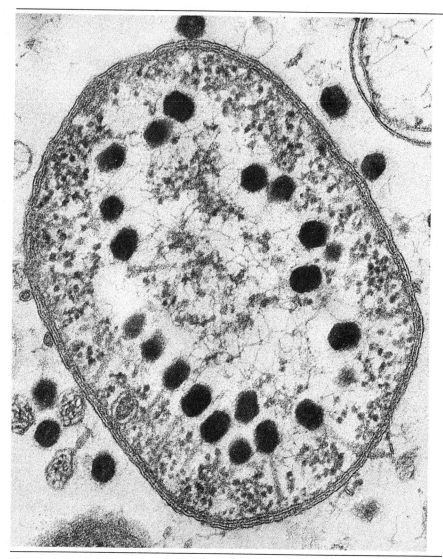

Figure 10-1 Electron micrograph of an *E. coli* B cell infected with T2 bacteriophages (black hexagons). The small black particles are ribosomes; the thin fibrous material is DNA. (From L. D. Simon, *Virology* **38**:287, 1969.)

will occasionally occur so that recombinant + + and *ab* progeny phages appear in the burst that follows lysis of the host cell. In an analogous process known as **transfection,** specially treated bacterial cells can be caused to take up "naked" copies of phage chromosomes, thus bypassing the need for the phage injection apparatus; such chromosomes also proceed to replicate and recombine within the host.

The principles underlying genetic mapping in such situations are diagrammed in Figure 10-2. It is first assumed that genetic exchanges occur with comparable frequencies at all points along a chromosome's length. Granted this assumption, then the farther apart two genes are on a chromosome, the more likely it is that they will be separated by a genetic exchange. To give an extreme example, we can assume that the *a* and *b* loci lie at opposite ends of a chromosome, as shown in Figure 10-2a. In this case almost any exchange occurring in an internal position on the chromosome will have the effect of bringing the markers into new combinations. The observed **recombination frequency** between the markers will therefore be large; in other words, a relatively large recombination frequency between two markers suggests that the markers are far apart on the chromosome. In contrast, two other loci, *c* and *d*, might lie right next to each other in the chromosome, in which case there will be a small probability that any particular exchange will happen to occur in the interval separating the two markers. The observed recombination frequency for these two markers will, therefore, be small (Figure 10-2b). By comparing the recombination frequencies obtained with a number of different marker combinations (for example, *ab* X ++ crosses and *cd* X ++ crosses), the four markers can be arranged in a linear order (Figure 10-2c) such that their relative distances

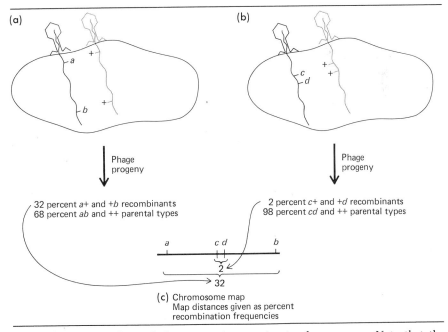

(a)

(b)

Phage progeny

Phage progeny

32 percent *a*+ and +*b* recombinants
68 percent *ab* and ++ parental types

2 percent *c*+ and +*d* recombinants
98 percent *cd* and ++ parental types

(c) Chromosome map
Map distances given as percent recombination frequencies

Figure 10-2 Principles underlying genetic mapping in phage crosses. Note that the placement of *c* and *d* between *a* and *b* cannot be assumed on the basis of the two crosses shown alone; other crosses must also be performed.

from one another correspond to the recombinational distances between them. This procedure forms the basis of **genetic mapping** in viruses and in all other organisms as well.

THE ANALYSIS OF PHAGE LAMBDA CROSSES

We now turn to a specific problem, the construction of a genetic map for bacteriophage λ, to illustrate how genetic crosses are performed and the data analyzed.

10.4 KAISER'S MUTANT STRAINS: MARKER SUITABILITY

Some of the earliest genetic mapping experiments with phage λ were published by A. D. Kaiser in 1955. Starting with a wild-type strain, which we can designate $+++++$, he obtained five mutant strains by ultraviolet irradiation. Each mutant strain produced a variant type of plaque morphology (Section 6.2): the s (small) strain created a small plaque; mi (minute), a minute plaque; c (clear), a completely clear plaque; co_1 (cocarde), a clear plaque except for a ring of colonies in the center; and co_2, a plaque in which the central ring is denser than for co_1. The last three gene mutations can be immediately recognized as interfering with a proper lysogenic response: whereas wild-type λ will frequently lysogenize cells on the bacterial lawn and will thus create a turbid plaque containing viable lysogens (Figure 6-4), the clear and cocarde strains lyse virtually every cell they meet.

Before mutant strains are used for genetic mapping purposes they are scrutinized to make sure that the markers they carry are suitable. Suitability is usually defined by five criteria; these five criteria govern the choice of genetic markers regardless of the organism under study.

First, the mutant phenotypes should be clearly distinguishable from the wild type and from one another, so that the progeny of each can be easily classified. In the case of the λ strains, for example, Kaiser established that the s and mi plaques could be readily differentiated and that each plaque type could easily be singled out from a wild-type plaque. Mutations that are leaky (a term defined in Section 9.9) make poor genetic markers.

Second, each mutant strain should have a reliable phenotype to prevent ambiguity in classification. For bacteriophages, for example, a mutation would be chosen whose expression is not highly dependent on the age of the bacteria in the lawn or the humidity in the Petri dish.

Third, each mutant strain should ideally reproduce itself as effectively as the wild type. In other words, for phages, the **burst size**—meaning the number of progeny emerging from a bacterium infected by a single phage particle—following a lytic cycle of growth should be comparable for both the wild and mutant strains. This requirement ensures that during a mixed infection equivalent numbers of both types of chromosome will be present within the cell, an important factor indeed if exchanges are to occur between *unlike* as often as between *like* chromosomes.

The fourth requirement of a marker used for mapping experiments is a low frequency of reversion. If, for example, it can be shown that when cells are infected by phages carrying the *mi* mutation no more than one phage in 10^4 or 10^5 has a wild phenotype arising from a reversion, the marker is suitable for mapping. If, however, reversion to wild type occurs more frequently, then during a mixed infection a wild-type plaque that arose by reversion will be confused with a wild-type plaque that arose by recombination; indeed, the two cannot be distinguished, and recombination frequencies will appear erroneously high.

The fifth requirement of a mutant strain is that, ideally, it should differ only by a single mutation—at least with respect to the phenotype being considered—from both the wild-type and other strains being examined. In plaque-morphology-mutant strains of λ this contingency can be investigated with **one-factor crosses,** so-called because only one phenotypic trait is followed during any one cross.

10.5 ONE-FACTOR CROSSES

To perform a one-factor cross two types of phage, one wild and one mutant, are allowed to infect a suspension of *E. coli* cells. As an example, we can cross the λ wild type with the *small* strain and simplify the notation by focusing only on the markers of interest. Thus the cross becomes $+ \times s$. The infection is adjusted so that a dense suspension of *E. coli* is presented with an even denser suspension of the two types of phage, and, on the average, about 10 phages (five of each type) infect each cell at the same time. Stated more formally, the **multiplicity of infection** (m.o.i.) is adjusted to have a value of approximately 10; 10^9 phages and 10^8 bacteria are present. Phage multiplication is then allowed to proceed until lysis occurs; the progeny phages in the lysate are diluted to appropriate concentrations for plating onto bacterial lawns; and the morphologies of the plaques that form on the lawns are individually scored.

Kaiser found that among 4390 plaques examined after a cross of $+ \times s$, 2050 had the *s* phenotype and 2340 had the $+$ phenotype. No other plaque morphologies were detected on any of the plates. Similarly, for $+ \times co_1$, $+ \times co_2$, $+ \times mi$, and $+ \times c$, only the two parental types were found among the progeny and they were present in roughly equal numbers, as shown in Table 10-1. Thus the one-factor crosses indicate that each of the five mutant strains differs by only a single factor from the wild type: no additional kinds of plaque phenotype emerge when a cross is performed and the phages are given ample opportunity, by a high multiplicity of infection, to undergo genetic exchange.

Table 10-1   One-Factor Crosses Involving Phage λ

The numbers of progeny plaques examined are given for each cross. Only parental types were recovered.

Parents	Progeny	
s $\times$ +	2050 s	2340 +
co_1 $\times$ +	761 co	707 +
mi $\times$ +	923 mi	736 +
c $\times$ +	5900 c	6600 +
co_2 $\times$ +	2100 co_2	1500 +

From A. D. Kaiser, *Virology* **1:**424 (1955).

Let us explore what the result would be for such a cross if a mutant strain did in fact carry a complex mutation. Let us suppose that the *small* phenotype is actually the composite effect of *two* gene mutations, a and b, affecting plaque formation. When such a *small* strain is crossed to wild type at high multiplicity, we would predict that genetic exchange would occasionally separate a and b so that they would come to reside in different progeny chromosomes. In other words, returning to our original notation, the original *small* strain might be a b, in which case recombination during a cross with wild type would be expected to yield a + and + b progeny. It is usually the case for viruses and haploid organisms that recombinant progeny carrying only one mutation will have phenotypes different from either their doubly mutant parent or their wild-type parent. Thus we would expect that plaques other than wild and *small* would appear among the progeny; for example, intermediate-sized plaques or plaques with a distinctively new morphology might turn up. When such additional plaque types appear, it is clear that the cross is not a one-factor cross: the two strains clearly differ by mutations at more than one locus.

10.6   TWO-FACTOR CROSSES

The hypothetical cross we have just outlined was, in fact, a **two-factor cross,** in which a doubly mutant parent, a b, was crossed with a wild-type parent, + +. When, as in this case, both mutant genes are carried by one parent and their wild-type alleles are carried by the other parent, the cross is said to be **in coupling**—the two markers are coupled in the same chromosome. In genetic shorthand the cross is written + + $\times$ a b, and recombination between the a and b loci will produce singly mutant progeny designated + b and a +. The alternative in a two-factor cross is that the markers be **in repulsion,** in which case each parent carries only one of the two markers being followed in the cross (+ b $\times$ a +).

Returning now to Kaiser's experiments, two-factor crosses in repulsion were carried out with his various mutant strains of λ. The genetic notation for such crosses is as follows. One parental strain, for example, *mi*, is denoted + *mi*, indicating that its chromosome is wild type (at least with respect to plaque morphology) in all but the *mi* locus. The other parent, for example, *c*, is denoted *c* + to indicate that it is mutant only at the *c* position and to stress that it is wild type at the *mi* position. Thus the cross is written + *mi* X *c* +, and if any recombination occurs between *c* and *mi* recombinant progeny will be either wild type (+ +) or doubly mutant (*c mi*). The doubly mutant phages will in this case form plaques that are both *clear* and *minute;* they can be distinguished from the two parental types and from the recombinant wild-type plaques.

Kaiser performed two-factor crosses in repulsion with many combinations of his mutant strains. He then isolated some of the doubly mutant recombinant progeny and crossed them with wild-type phages to yield data for crosses in coupling. The results of these crosses are shown in Table 10-2.

Table 10-2 Two-Factor Crosses Involving Phage λ

Numbers of progeny plaques, classified according to genotype, are given opposite each cross. Percent recombination is

$$\frac{\text{sum of two recombinants}}{\text{sum of all types}} \times 100$$

In the upper section several examples of the same cross are given to show the amount of variation between experiments. In the lower sections, only total numbers, obtained by adding together the results of individual experiments, are given.

Parents	Number of Experiments	Progeny				Recombination, %
co_1 + X + *mi*	1	5162 *co* +	6510 + *mi*	311 + +	341 *co mi*	5.3
	1	459	398	17	25	4.7
	1	720	672	44	46	6.1
co_1 *mi* X + +	1	36	30	795	620	4.4
	1	74	56	1005	956	6.2
s + X + co_1	2	7101 *s* +	5851 + *co*	145 + +	169 *s co*	2.4
s co_1 X + +	2	46	53	1615	1774	2.8
s + X + *mi*	1	647 *s* +	502 + *mi*	65 + +	56 *s mi*	9.5
s mi X + +	3	1024	1155	13083	13253	7.6
s + X + *c*	1	808 *s* +	566 + *c*	19 + +	20 *s c*	2.8
c + X + *mi*	1	1213 *c* +	1205 + *mi*	84 + +	75 *c mi*	6.2
c + X + co_1	3	6000 *c* +	6000 + *co*	14 + +	— *c co*[a]	0.1
co_2 + X + *mi*	1	1477 co_2 +	1949 + *mi*	109 + +	131 co_2 *mi*	6.6

[a] Type *c co* not distinguishable from *c* +.

From A. D. Kaiser, *Virology* **1**:424 (1955).

Looking first at the uppermost section of the table, data are presented for three separate $co_1 + \times + mi$ crosses and for two separate $co_1\ mi \times + +$ crosses. For each experiment the **percent recombination R** between co_1 and mi is calculated as

$$R = \frac{\text{sum of two recombinant types}}{\text{sum of all types}} \times 100$$

Thus for the first experiment

$$R_{co_1mi} = \frac{311 + 341}{5162 + 6510 + 311 + 341} \times 100$$
$$= \frac{652}{12,324} \times 100$$
$$= 0.053 \times 100$$
$$= 5.3$$

The results in Table 10-2 clearly indicate that co_1 and mi are separated by recombination during a phage cross about 5 percent of the time, whether the markers are initially in coupling or in repulsion.

The second section of Table 10-2 reveals that recombination between co_1 and s is half as frequent (2.6 percent) as between co_1 and mi, again whether the cross is made with markers in coupling or repulsion. This result suggests that s is much closer to co_1 than is mi. Given this information alone and using the average percent recombination values as map distances, we might conclude that the three markers bear one of two relationships to one another on the chromosome:

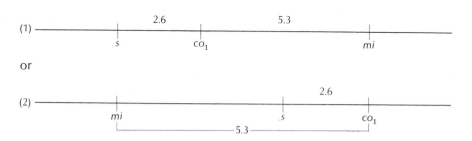

or

Clearly, more information is needed to establish the marker order.

The third section of Table 10-2 provides further information. Here we see that the percent recombination between s and mi averages 8.5. Thus it can be concluded that s and mi are farther apart than are either s and co_1 or mi and co_1, since exchanges occur between these two latter marker pairs with greater frequency. Given this fact, it is most likely that co_1 lies somewhere between s and mi on the genetic map, giving a marker order of $s\ co_1\ mi$ (or $mi\ co_1\ s$, since the choice of left and right ends of the chromosome is at this point an arbitrary one). Thus possibility (2) can be eliminated.

Immediate support for this conclusion comes from the observation that the sum of the R values for $s - co_1$ and $co_1 - mi$ is $5.3 + 2.6 = 7.9$, which is approximately equivalent to the value of 8.5 for R between s and mi. Thus we can rewrite the map with some assurance as illustrated below, where the bracket indicates the percent recombination between the two **outside markers** on the map.

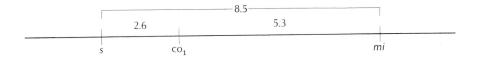

The bottom section of Table 10-2 gives data that allow us to position marker c with respect to s and mi. It is seen that the percent recombination between s and c is 2.8, while that between c and mi, is 6.2. The sum of these two R values is 9, again approximately equivalent to the 8.5 value for R between s and c is 2.8, while that between c and mi is 6.2. The sum of these s and mi, and that it is closer to s than to mi. Indeed, these data indicate that c must lie relatively close to co_1 on the chromosome, since both appear to be comparably positioned with respect to s and mi. This deduction is affirmed by the results of the cross $c + \times + co_1$ given in Table 10-2: the percent recombination (0.1) in this cross is small, which indicates that a genetic exchange seldom occurs in the interval separating c and co_1.

We are still unable to position c and co_1 with respect to the outside markers; the map can only be written $s - (c,\ co_1) - mi$, where the parentheses indicate the uncertainty of the $c - co_1$ order. Nor have we any information on the position of co_2. The last entry in Table 10-2, the cross $co_2 + \times + mi$, indicates that co_2 is separated from mi by 6.6 recombination units on the map. With this information alone we cannot determine whether co_2 lies within the segment of the chromosome we have been considering, namely the segment bracketed by s and mi, or whether it lies 6.6 units outside the segment. For example, either of the two map sections shown below might be correct:

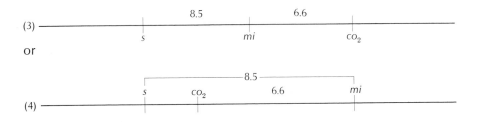

To explore these possibilities every mutant phage strain must be crossed with every other. Fortunately, a simpler and more precise determination of map order is possible by using the **three-factor crosses** described below.

10.7   THREE-FACTOR CROSSES

In a three-factor cross, as you might by now anticipate, the parental phages are distinguished by three traits. Laboratory phage stocks carrying three mutations can be obtained by crossing double-mutant strains with single mutant strains and isolating the recombinants. Three-factor crosses are written, for example, as $s\, co_1\, mi \times +\,+\,+$ or $s + mi \times + co_1 +$. The principles underlying genetic mapping that we have been following so far are still observed for these triply mutant strains, but an additional consideration is introduced. This is the occurrence of **double exchanges,** in which two chromosomes participate in not one but two exchange events.

In the one-factor and two-factor crosses we have been discussing so far many of the progeny have in fact undoubtedly participated in double (and higher order) exchanges, but it was not possible to detect their occurrence; for example, in a two-factor cross $e f \times +\,+$, if two exchanges occur within the e-f interval, the resulting phages will remain ef and $+\,+$ and be genetically nonrecombinant (parental), even though they are physically hybrid. This outcome is diagrammed in Figure 10-3a. If, on the other hand, a single exchange occurs within the e-f interval and a second outside the interval, the second exchange will have no effect on the marker relationships being followed and the resultant phages will be e + and + f single recombinants, as shown in Figure 10-3b. At least three markers must therefore be followed in order to recognize double recombinants. Here, for $e f g \times +\,+\,+$ the double recombinants will be e + g and + f +, which are readily distinguishable from the singly recombinant and parental-type progeny (Figure 10-3c).

So far in this chapter we have been assuming that a single exchange will occur with equal probability at any point along a chromosome's length. A second assumption we now make, one that holds for many purposes in the analysis of prokaryotic recombination, is that once an exchange has occurred at one position on a chromosome it does not preclude or influence the probability that a second, third, or fourth exchange will occur at some other position on the same chromosome. In other words, we assume for the present that all exchange events are independent of one another.

As already noted for the χ^2-test (Section 4.10), the probability that two independent events will occur simultaneously is the product of their individual probabilities. A familiar example of this principle occurs with coin-flipping: given that the probability of flipping heads or tails is 50–50 (0.5) for one coin, the probability that two coins flipped simultaneously will come up heads is $(0.5) \times (0.5) = 0.25$, or one-quarter of the double flips.

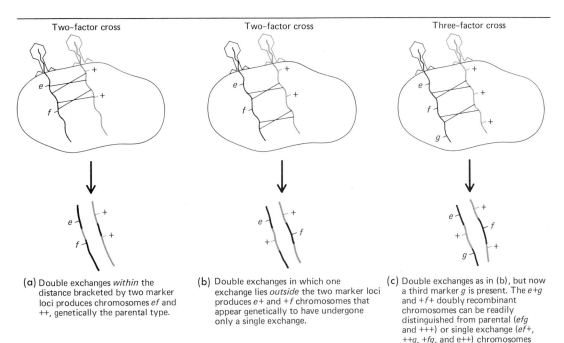

Two-factor cross	Two-factor cross	Three-factor cross

(a) Double exchanges *within* the distance bracketed by two marker loci produces chromosomes *ef* and ++, genetically the parental type.

(b) Double exchanges in which one exchange lies *outside* the two marker loci produces *e*+ and +*f* chromosomes that appear genetically to have undergone only a single exchange.

(c) Double exchanges as in (b), but now a third marker *g* is present. The *e*+*g* and +*f*+ doubly recombinant chromosomes can be readily distinguished from parental (*efg* and +++) or single exchange (*ef*+, ++*g*, +*fg*, and *e*++) chromosomes

Figure 10-3 Effect of double exchanges (each exchange shown as ✕) in two-factor crosses (a) and (b) and in a three-factor cross (c) where one exchange occurs between each marker pair.

Thus, if the recombination frequency between markers *a* and *b* is 0.05 (so that $R = 0.05 \times 100 = 5$) and the recombination frequency between *b* and *c* is 0.03 (so that $R = 3$), the expected probability that *two* exchanges will occur between *a* and *c* in a given cross is the product of the two individual frequencies; that is, $(0.05)(0.03) = 0.0015$ (so that $R = 0.15$). Clearly, this product will always be a much smaller number than either individual frequency alone, an outcome that makes intuitive sense: it seems logical that any one chromosome would be far more likely to participate in one exchange than in two; more likely to take part in two than in three; and so forth, just as it would be a better bet to wager that a given coin flip will be heads than that when two coins are flipped both will be heads.

The infrequent nature of double exchanges can be profitably applied to three-factor crosses when gene order is being investigated, for of the six possible recombinant classes among the progeny the two rarest classes can be taken as the doubly recombinant classes. Once these two classes are recognized the map order of the three markers in the cross can be deduced directly.

The data in Table 10-3 illustrate this principle for three-factor crosses of the mutant strains of λ. Looking at the first line of the table (in which the

Table 10-3 Three-Factor Crosses Involving Phage λ

Numbers of progeny plaques are classified by genotype. The total number of progeny plaques examined is given in the last column.

Parents	Progeny								Total
	+ + +	s co mi	s + +	+ co mi	s co +	+ + mi	s + mi	+ co +	
$s\ co_1\ mi$ × + + +	975	924	30	32	61	51	5	13	2091
$s + mi$ × $+ co_1 +$	38	23	273	318	112	121	6389	5050	12324

From A. D. Kaiser, *Virology* **1**:424 (1955).

order of markers in the parental stock is given initially as an arbitrary sequence), we find that the cross $s\ co_1\ mi \times +\ +\ +$ yields, as expected, the two parental types in greatest numbers. Among the recombinant types the two reciprocal classes $s + mi$ and $+ co_1 +$ are clearly present in the fewest numbers, and thus it is assumed that they are derived from a double exchange between $s\ co_1\ mi$ phages and $+\ +\ +$ phages. The marker order must therefore be as we have arbitrarily written it: $s\ co_1\ mi$. Were it instead, for example, $s\ mi\ co_1$, then the double recombinants in the cross would be $s + co_1$ and $+ mi +$, which is clearly not the experimental result.

In the second cross given in Table 10-3, the same markers are followed, only the parental genotypes are $s + mi \times + co_1 +$. Again the two rarest classes indicate that the marker order must be $s\ co_1\ mi$.

How is the marker order deduced when the correct sequence is not provided by the textbook? Suppose, for example, you know only that one parent carries co_1, mi, and $+$, whereas the other parent carries c, $+$, and $+$; when these are crossed, you learn that one of the rarest recombinant types carries the co_1, c, and $+$ markers while the other carries mi, $+$, and $+$. The most foolproof way to proceed is to write out all the possibilities and then select the correct one, in the following manner:

Possible Marker Order	Expected Double Recombinants
(1) co_1 mi $+$	co_1 $+$ $+$
$+$ $+$ c	$+$ mi c
(2) co_1 $+$ mi	co_1 c mi
$+$ c $+$	$+$ $+$ $+$
(3) $+$ co_1 mi	$+$ $+$ mi
c $+$ $+$	c co_1 $+$

Of the three possible marker orders, only (3) yields expected double recombinants whose genotypes match those of the two rarest types in the actual cross; therefore, a $c\ co_1\ mi$ marker order can be determined. Try this yourself: from a cross $c\ co_2 + \times +\ + mi$, the rarest progeny are $c\ mi +$ and $co_2 + +$. What is the relative order of c, co_2, and mi?

The map of the λ chromosome generated by Kaiser on the basis of such crosses is drawn in Figure 10-4, where recombination frequencies are translated as **map distances.** Additional markers are evident on the map, including m_6, m_5, and g_1 (plaque-morphology markers) and h (extended *host* range). The distance between m_6 and mi, expressed as R units, is about 15, and these two markers are now known to lie near either end of the λ chromosome map.

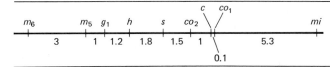

Figure 10-4 An early map of the phage λ chromosome. Marker symbols are given in the text. Map distances (R units) represent percent recombination × 100 in two-factor crosses. (From A. Kaiser, *Virology* **1:**424, 1955.)

10.8 MAP DISTANCES FROM A THREE-FACTOR CROSS

We should note that data from three-factor crosses can be used not only to order genetic markers but also to calculate map distances between markers. This is done as follows:

1. Classify the progeny of the cross according to whether they have experienced no exchange, a single exchange, or a double exchange in the three-marker interval. Thus in a cross $a\,b\,c \times +\,+\,+$, where $a\,b\,c$ represents the determined gene order, there emerge eight kinds of progeny that can be grouped into four classes: $a\,b\,c$ and $+\,+\,+$ progeny represent the **parental** or **nonrecombinant** class; $a\,+\,+$ and $+\,b\,c$ progeny arise from a single exchange between the a and b loci and represent what we can call the **single-recombinant-I** class; $a\,b\,+$ and $+\,+\,c$ progeny result from an exchange between b and c and constitute the **single-recombinant-II** class; and finally, the rare $a\,+\,c$ and $+\,b\,+$ progeny make up the **double-recombinant** class.

2. Count up the number of progeny in each class and the grand total scored. Thus with the data below, the calculations are as follows:

Genotype	Number	Class	Class Total
$+\,+\,+$	975 ⎫	Nonrecombinant	1899
$a\quad b\quad c$	924 ⎭		
$a\,+\,+$	30 ⎫	Single I	62
$+\quad b\quad c$	32 ⎭		
$a\quad b\,+$	61 ⎫	Single II	112
$+\,+\quad c$	51 ⎭		
$+\quad b\,+$	5 ⎫	Double	18
$a\,+\quad c$	13 ⎭		
		GRAND TOTAL	2091

3. Calculate the **frequency** of each class of recombinant progeny as follows:

Recombinant Class	Frequency	Percent of Total Progeny
Single I	62/2091 = 0.029	2.9
Single II	112/2091 = 0.053	5.3
Double	18/2091 = 0.0086	0.86

4. For simplicity, designate the three percentages so obtained as α, β, and γ. Thus

α = percent of total progeny of the Single I class = 2.9
β = percent of total progeny of the Single II class = 5.3
γ = percent of total progeny of the Double class = 0.86

5. The values for R_{ab}, R_{bc}, and R_{ac} can now be calculated directly. The markers a and b are clearly separated in the Single I progeny (a + + and + b c). They are also separated in the Double progeny (+ b + and a + c). Therefore,

$$R_{ab} = \alpha + \gamma$$

Similarly,

$$R_{bc} = \beta + \gamma$$

while

$$R_{ac} = \alpha + \beta$$

Work out for yourself how these relationships yield the map

6. Examination of this map reveals an apparent anomaly: the distance between the two outside markers (8.2) is less than the sum of the a-b distance plus the b-c distance (3.8 + 6.2 = 10). That the two numbers are not the same is due to the occurrences of double crossovers. If two crossovers occur in the a-c interval then, of course, the one cancels the other out and no recombination appears to have occurred when the R_{ac} frequency is being calculated; each of these crossovers is, however, likely to be scored when R_{ab} and R_{bc} frequencies are being determined. In general, therefore, as genes lie increasingly farther apart on a chromosome, their apparent distance in two-factor map intervals becomes increasingly foreshortened by the occurrence of double crossovers.

10.9 THE ANALYSIS OF SUPPRESSOR-SENSITIVE STRAINS OF λ

Returning to the specific goal of mapping the λ chromosome, numerous three-factor crosses generated the Kaiser map shown in Figure 10-4. The markers utilized by Kaiser had a common feature: none of the mutations prevented λ from carrying out a lytic cycle. In order to obtain markers in genes essential to the production of infective phage progeny, it is, of course, necessary to obtain conditional mutants (Section 6.3) which exert their effects only under certain conditions.

A profitable approach, taken by A. Campbell, was to seek *sus* (*suppressor-sensitive*) strains of λ that carry *amber* nonsense mutations (Section 9.16) and can therefore give rise to mature progeny when they grow on *amber*-suppressor strains of *E. coli* (see Section 9.18 for a review of suppression). A collection of *sus* strains (sus_1, sus_2, sus_3, and so on) was obtained by screening for phages that will grow in suppressor but not in nonsuppressor hosts (the terms **permissive** and **nonpermissive** hosts are also sometimes used). If these strains are crossed with one another in appropriate combinations, recombinant strains carrying three genetic markers can be isolated; for example, a $sus_1 + sus_3$ strain and a $+ sus_2 +$ strain can be constructed. By crossing these with one another the relative marker order can be deduced in the manner we have outlined above for three-factor crosses.

The kind of map that results is shown in Figure 10-5, in which 130 different λ *sus* mutants have been ordered with respect to one another. This particular map was in fact constructed by Campbell not from three-factor crosses but by a technique called deletion mapping described later in this chapter. The same marker order has, in other experiments, been deduced with three-factor crosses; Campbell's data are used here because they illustrate particularly well several important considerations in map construction.

The first question that can be asked of two *sus* mutant strains is whether they carry mutations at identical sites within a gene. You will notice in Figure 10-5 that certain loci on the map are marked by only one *sus* mutation (sus_{11}, sus_{19}, or sus_{32}, for example), whereas other loci are marked by several mutations (for example, the segment containing sus_{36}, sus_{109}, and sus_{121}). When two mutations map at the same locus, there are two possible explanations for the result: they may actually be duplicates, so that the *same nucleotide position* is affected in both cases, or they may represent two distinct mutations that lie so close to each other that the probability is extremely small that a recombinational event will occur between them. To distinguish between the two possibilities, experiments must be designed in which many thousands of progeny from a cross are examined so that if very rare recombinants arise they can be detected. We consider such experiments in Chapter 13.

A second important kind of question to be asked of the *sus* mutant strains mapped in Figure 10-5 is whether the various strains carry mutations in the same or different genes. We can imagine two extreme situations. One is that the entire portion of the λ chromosome from sus_{11} to *h* represents a single gene controlling a single function. We know that this function must be vital for the production of infectious phages, so let us suppose that the

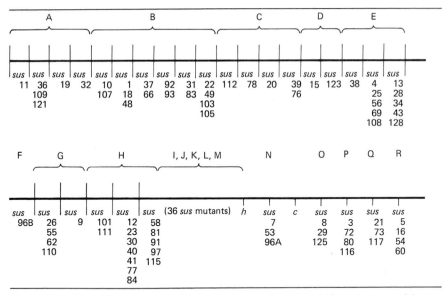

Figure 10-5 Location of *sus* mutations on the genetic map of phage λ. Capital letters designate complementation classes. (From A. Campbell, *Virology* **14**:22, 1961.)

entire sus$_{11}$-h segment represents the structural gene for a single head protein and that all the *sus* markers represent *amber* mutations that have occurred at various sites within this gene. Stated in the terminology used in Chapters 3–5, one possibility is that all the *sus* markers are alleles of a single gene locus. At the other extreme we can imagine that the 61 separate loci between *sus*$_1$ and *h* represent 61 distinct genes, each of which has undergone a single amber mutation. Each gene, by definition, would control a different genetic function. The most likely possibility lies, of course, between these two extremes; namely, that the chromosomal segment contains some number between one and 61 genes, that some of the closely linked *sus* mutations lie within a single gene and are therefore allelic, but that the more widely separated markers lie in different genes and are therefore nonallelic. How are the gene boundaries established?

COMPLEMENTATION AND GENE BOUNDARIES

10.10 THE COMPLEMENTATION TEST

Gene boundaries are most often established by a **complementation test.** The principle of complementation is as follows:

If two homologous chromosomes carry mutations within different genes—let us say one chromosome is mutant in gene P and the other in gene Q—and these two chromosomes come to reside in the same cell, one chromosome will direct the synthesis of a mutant P gene product but a normal Q product and the other will direct the synthesis of a normal P but a mutant Q product. The important point is that the cell will come to contain both normal P and normal Q products and thus normal function can occur. In such a case the two mutant chromosomes are said to **complement** each other, and it is inferred that the two mutations lie within the boundaries of two different genes. If, on the other hand, the two chromosomes carry mutations in the same gene—for example, both carry mutations in gene P (although not necessarily at identical positions within P)—then neither chromosome will be able to direct the synthesis of a normal P product, and even when both chromosomes come to reside in the same cell normal P function will not be observed. The two mutant chromosomes in this case do *not* complement each other, and it can be inferred that both mutations are allelic, that is, that they lie within the boundaries of the same gene.

The actual design of a complementation test depends on the organism and on the kinds of mutations being studied. For *amber* mutant strains of a bacteriophage, one approach is to mixedly infect a suspension of non-suppressor bacteria with equal multiplicities of two mutant phage strains, allow sufficient time for a single lytic cycle to occur, and count the total number of phages produced by plating the lysate onto a lawn of suppressor bacteria. When such experiments are performed with *sus* mutant strains of λ, it is found that if the two strains can complement each other, 100 to 250 phages (a normal burst) will emerge per infected cell, whereas if the two strains are noncomplementary then fewer than 0.1 phage will emerge per infected cell. In other words, for every cell that produces one phage nine produce none, or, more realistically, for every cell that releases a normal burst of at least 100 phage progeny, 99 cells produce no phage progeny at all.

The few successful bursts that do occur in a mixed infection between noncomplementing strains could, of course, take place because one of the two mutations in question has undergone reversion. They could also take place because recombination has occurred. Two noncomplementing (allelic) strains that carry mutations at two close, but distinct, sites within the same gene will undergo recombination with a very low frequency to produce phages with a normal nucleotide sequence. These rare recombinants can direct a normal lytic cycle and produce normal infective progeny in the nonpermissive host. As long as the experiment is limited to one lytic cycle, the occurrence of these rare recombinational events does not obscure the results. If, however, the complementation test is not properly designed and a number of lytic cycles occur before the lysate is plated and counted, the original rare recombinants might give rise to a sufficient number of descendants to make it *appear* that complementation has occurred. Should there be any doubt, there is, of course, a simple check: phages picked from

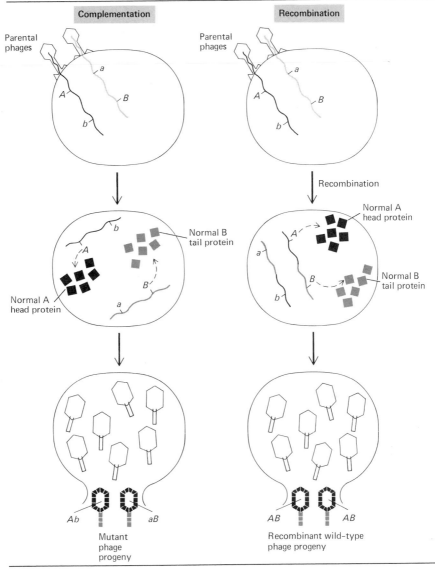

Figure 10-6 Salient differences between complementation and recombination as illustrated for two mutant phage strains. One strain, *A b*, carries a normal gene for the A head protein but a defective tail protein gene *b*. The other strain, *a B*, carries a normal B tail protein gene but an abnormal head protein gene *a*. During complementation, normal A and B proteins are synthesized under the direction of the complementing chromosomes, and normally constructed phages emerge that are mutant (*A b* and *a B*) in genotype. During recombination, normal proteins are also synthesized and normally constructed phages emerge, this time carrying normal (*A B*) genotypes. It should be obvious that the phenomenon of complementation will operate in a cell where recombinant chromosomes have also arisen.

single plaques can be tested for their ability to direct a lytic cycle in a nonpermissive host; if they can, they are recombinants (or revertants, for we cannot distinguish between the two in such an experiment), but if they cannot they originated by complementation.

It is very important to keep complementation and recombination distinct. They are two quite different processes, as summarized below and illustrated in Figure 10-6.

1. Complementation involves the interaction of gene **products** (usually proteins) to produce a normal phenotype, whereas wild-type recombinants produce a normal phenotype because a normal sequence of nucleotides has actually been created along a chromosome via genetic exchange (Figure 10-6).
2. The ability to generate a normal phenotype is genetically transmitted from parent to offspring when chromosomes arise by recombination, but progeny produced as a result of complementation remain individually defective in genotype (Figure 10-6).
3. As a general rule, recombination can occur within a gene, whereas complementation occurs only between one gene and another. The exceptional case, in which a special kind of complementation can occur within a gene, is detailed in the next section.

10.11   INTRAGENIC COMPLEMENTATION

There is one major exception to the rule that **complementation occurs only between one gene and another.** The exception, known as **intragenic complementation,** is illustrated schematically in Figure 10-7. In the example two identical polypeptides, both called α and both specified by the same gene, assume opposite orientations with respect to each other and so form an enzyme $\alpha\alpha$ (Figure 10-7a). In wild-type cells the region of the enzyme important to its catalytic activity (the active site) contains an arginine and a proline contributed by one α subunit and an arginine and a proline contributed by the other α subunit (Figure 10-7a). The interaction of all four amino acids is essential to catalytic activity. The arginine, we can imagine, resides at position No. 86 on the α polypeptide and the proline at position No. 37. We now suppose that a GCG $\longrightarrow$ GAG mutation occurs and that leucine rather than arginine appears at position No. 86 in one mutant strain. This mutant polypeptide, which we can call α_{86}, will not form an active enzyme when it interacts with a second α_{86} polypeptide, perhaps because the four amino acids cannot interact in such a way as to form an active catalytic site (Figure 10-7b). We can similarly suppose that a second strain has undergone a GGG $\longrightarrow$ GTG mutation and produces an α_{37} polypeptide in which histidine instead of proline appears at position No. 37. Again,

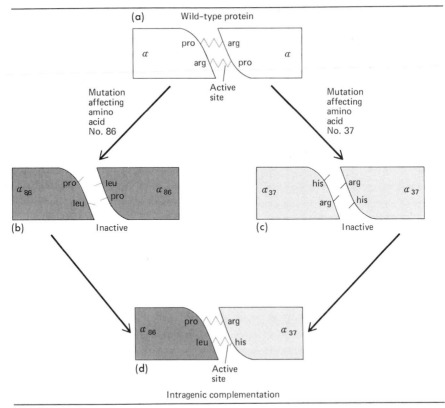

(a) Wild-type protein

(b) Inactive

(c) Inactive

(d) Active site

Intragenic complementation

Figure 10-7 Diagram of intragenic complementation between two different mutant subunits of a dimeric protein. (See text for details.)

$\alpha_{37}\alpha_{37}$ dimers are found to be inactive (Figure 10-7c). When, however, the two mutant polypeptides are allowed to interact in a single cell during a complementation test, then $\alpha_{86}\alpha_{37}$ **heterodimers** can form, and the presence of the proline, leucine, arginine, and histidine residues in the active site results in a configuration, or perhaps a net charge, that is compatible with catalytic activity (Figure 10-7d). Thus *two mutations occurring within the same gene can occasionally give rise to complementary gene products when the products interact to form a multisubunit protein.*

10.12 COMPLEMENTATION GROUPS IN THE λ CHROMOSOME

Returning now to the various *sus* strains that we have been considering, Campbell found that the strains *sus*$_{11}$ through *h* (Figure 10-5) could be placed in 13 different **complementation groups,** designated A to M in Figure 10-5. Thus none of the *sus* strains with mutations in group A was able to complement one another, but all were able to complement strains in groups B, C, D, and so on. Five other complementation groups, N to R, were also identified to the right of *h*; these are considered later in the chapter.

In the experiments leading to the generation of the map in Figure 10-5 the relative order of groups I to M was not determined. More recent experiments, notably those of J. Parkinson, have indicated that their order is M, L, K, I, J, with the *h* mutation lying within the J region. Parkinson also demonstrated the existence of five additional complementation groups, T to W and Z, that lie within the A-J segment; their existence was not detected in the earlier experiments because *sus* mutations in these groups did not happen to be included in the original collection of 130 *sus* strains. An updated version of this **left arm** of the λ chromosome is found in Figure 10-8, which also shows the order of **middle** and **right-arm** genes on the λ map.

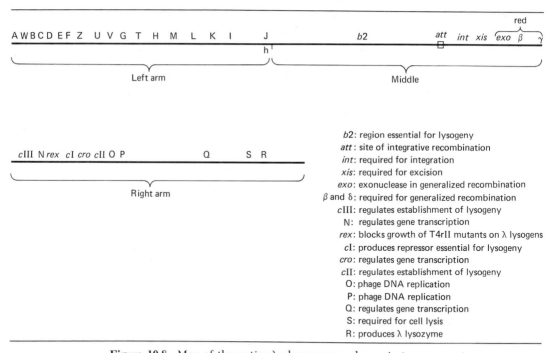

Figure 10-8 Map of the entire λ chromosome, drawn in two segments.

10.13 *IN VITRO* COMPLEMENTATION

How does one determine the function of a collection of newly defined genes? For the left-arm genes of λ, J. Weigle was able to demonstrate an involvement with the synthesis of proteins that form the head capsule and the tail structure of the phage. Weigle exploited the fact that *sus* mutations in these genes do not interfere with the establishment of lysogeny, nor do they prevent an integrated λ chromosome from leaving the host chromosome upon induction with ultraviolet light (Section 5.11). Therefore, when nonsuppressor *E. coli* cells that are lysogenic for, say, strain sus_{20} of λ—a strain defective in gene C function—are irradiated, the induced prophages will proceed to direct the synthesis and assembly of most of the components required for the growth and maturation of phage progeny. Because gene C is defective, however, the cycle cannot be completed and the cells instead fill up with putative λ components. Weigle found that if such cells are lysed artificially with chloroform and the lysate is plated onto a permissive bacterial lawn, no plaques will form because no infective phage particles are present. Numerous plaques will form, however, if the lysate is first mixed with an equal quantity of a second lysate derived in similar fashion from cells lysogenic for sus_9 (mutant in gene G) and the two lysates are incubated together *in vitro* several hours before plating. In other words, lysates from sus_{20} can provide something that will complement, *in vitro,* lysates from strain sus_9.

Weigle repeated his ***in vitro* complementation** experiments with other pairwise mixtures of artificial lysates and found that if lysates from *sus* strains defective in genes A to F and W are mixed together, they cannot restore one another's infectivity, nor can mixtures between lysates in which G to M, U to V, or Z are defective. Lysates of the first class will, however, readily restore infectivity to lysates of the second class. These results are shown in Table 10-4.

The explanation of these results becomes clear when the various lysates are examined by electron microscopy. Lysates from the A to F and W defective classes are found to contain normal λ tails but incomplete or otherwise abnormal heads, whereas the G to M, U to V, and Z lysates contain normal heads but abnormal or absent tails. When a lysate mixture includes both normal heads and normal tails, phage halves can unite spontaneously by self-assembly to produce whole infective phages. It can therefore be concluded from these experiments that the left arm of the λ chromosome possesses a group of seven genes which carry information required for the synthesis and assembly of λ heads and a group of 11 genes involved in the construction of tails.

The **head genes** and **tail genes** of λ are identified on the expanded left-arm map in Figure 10-9. Also indicated in Figure 10-9 is an interesting feature of the arrangement of these genes: proteins that appear to interact most directly during the assembly of the phage particle are encoded by genes most closely linked to one another in the phage chromosome.

Table 10-4 Lysate Complementation Between *sus* Mutant Strains of Phage λ Defective in Genes *A* Through *M*

	A	B	C	D	E	G	H	I	J	K	L	M
A	0.03	0.03	0.06	0.06	0.12	950	40	670	550	1000	1800	550
B	—	0.002	0.03	0.04	0.10	84	16	130	100	280	800	120
C	—	—	0.03	0.07	0.12	180	8	160	110	170	360	50
D	—	—	—	0.04	0.13	260	24	180	210	210	440	50
E	—	—	—	—	0.09	330	9	180	170	120	120	25
G	—	—	—	—	—	0.37	0.39	0.45	0.46	0.68	0.10	0.56
H	—	—	—	—	—	—	0.03	0.96	0.14	0.55	0.70	0.23
I	—	—	—	—	—	—	—	0.07	0.17	0.56	0.75	0.27
J	—	—	—	—	—	—	—	—	0.10	0.52	0.78	0.30
K	—	—	—	—	—	—	—	—	—	0.58	0.60	0.55
L	—	—	—	—	—	—	—	—	—	—	0.68	0.88
M	—	—	—	—	—	—	—	—	—	—	—	0.20

Defective lysates of the various mutants were diluted in the following way: to 0.8 ml of medium 0.1 ml of each of two lysates was added. The control contained 0.1 ml of a single lysate. The mixtures were incubated for 1 hr at room temperature and the titer of active phage determined. The numbers given in the table are titers divided by 10^3. Lysates prepared on different days varied in their activity and thus the numbers given are indicative of orders of magnitude only.

From J. Weigle, *Proc. Natl. Acad. Sci. U.S.* **55**:1462 (1966).

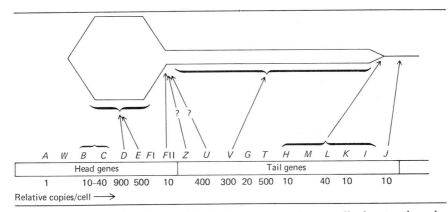

Figure 10-9 Map of the left arm of the λ chromosome, schematically showing the order of the morphogenetic genes and the likely points of action of those gene products where known. The approximate relative number of copies of the various gene products made during a productive infection are shown below the map. (With permission from S. Casjens and R. Hendrix, *J. Mol. Biol.* **90**:20, 1974. Copyright by Academic Press Inc. (London) Ltd.)

DETECTING DELETIONS BY GENETIC MAPPING

10.14 GENERAL PROPERTIES OF DELETIONS

As detailed in Section 6.13, a deletion chromosome lacks at least one and more commonly a stretch of nucleotides. By definition, the sector of a chromosome lost by a deletion has specific properties in genetic crosses:

1. The deleted sector is not available to participate in recombination.
2. The deleted sector cannot ordinarily be reverted.
3. The deleted sector cannot complement defects in other genes.
4. The deletion cannot be corrected by a suppressor mutation.

A deletion is usually recognized genetically because it fails to recombine with several point mutations. Examples of this phenomenon are given in Section 10.18 of this chapter. Here we consider the properties of a deletion in the middle of the λ chromosome which cannot be defined in this fashion because the deleted sector is not marked by known point mutations.

10.15 THE λb2 DELETION

The left half of the central portion of the λ chromosome is designated b_2 (Figure 10-8), so named because in the mutant strain λb_2, this region of the chromosome is defective. Phenotypically, λb_2 is able to direct a lytic cycle but can establish only abortive lysogeny, meaning that the prophage does not establish a stable state of integration in the host chromosome and thus does not become replicated and passed on to future bacterial generations. Three-factor crosses involving λb_2 have indicated that b_2 lies between h and c on the λ map. The λb_2 chromosome has also been shown by CsCl density gradient centrifugation (Box 5.1) to contain 17 percent less DNA than the wild-type chromosome, which suggests that the b_2 marker represents a large deletion lying between h and c.

If b_2 is in fact a large deletion, then its presence should physically shorten the distance between h and c. Therefore, recombination frequencies between these two markers in a λb_2 strain should be *lower* than in a strain that does not carry b_2. More specifically, the interval between h and c represents roughly 34 percent of the λ chromosome; therefore, the removal of 17 percent of the chromosome in this interval should shorten the interval by half, and recombination between h and c should occur only half as often in λb_2 as in wild-type phage. This is exactly what is observed. In the cross $h\,b_2 + \times + b_2\,c$ the percent of recombination between h and c is 2.2, whereas it is 5 in the cross $h + + \times + + c$.

The most direct demonstration that b_2 represents a large deletion in the middle of the λ chromosome comes from an approach called **heteroduplex mapping.** B. Westmoreland, W. Szybalski, and H. Ris dissociated wild-type and b_2-carrying chromosomes to separate their individual polynucleotide strands, as described in Section 7.5. Reassociation permitted the formation of **heteroduplexes** between + strands and b_2-carrying strands; these heteroduplexes were then examined by electron microscopy. One of the resulting molecules, drawn in Figure 10-10, is a double-stranded duplex interrupted at a median position by a single-stranded loop corresponding in length to about 17 percent of the total chromosome length. Thus the presence of b_2 in one strand prevents the corresponding segment of the wild-type strand from participating in base pairing and forces it to loop out of the main duplex, in certain ways analogous to the looping that occurs when a eukaryotic chromosome attempts to establish synaptic pairing with its deficiency homologue (Figure 6-15).

GENETIC ANALYSIS OF PHAGE T4

10.16   CIRCULARITY OF THE T4 MAP

Although most of the principles of genetic mapping can be illustrated with phage λ, one important phenomenon cannot, namely, the linkage patterns exhibited by markers located in a genetically circular chromosome. Such patterns are therefore illustrated with phage T4.

Early genetic analyses of T4 were performed by G. Streisinger and his colleagues, who studied recombination during mixed infection and established that a number of markers were linked. These markers were ordered

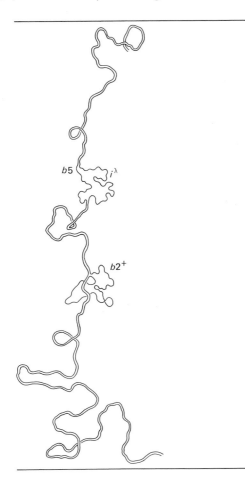

b5

i^λ

b2^+

Figure 10-10 Drawing of heteroduplex λ DNA, where one strand carries the *b2* deletion. The second discontinuity in the molecule is formed between two nonhomologous regions called *i*^λ and *b5*. (From B. Westmoreland, W. Szybalski, and H. Ris, *Science* **153**: 1345, March 21, 1969. Copyright 1969 by the American Association for the Advancement of Science.)

on a linear map as shown in Figure 10-11a. The map showed no signs of being unusual as long as the markers followed in the various crosses were fairly closely linked. Aberrant results were obtained, however, when three-factor crosses involving markers placed at the two "ends" of this linear map were made; for example, when the cross *r67 h42* + × + + *ac41* was performed, the rarest recombinant classes were found to be *r67* + + and + *h42 ac41*, indicating a map order of *r67-h42-ac41*, whereas the map shown in Figure 10-11a predicted that the *r* marker would be more closely linked to *ac* than to *h*. To resolve this dilemma Streisinger and his co-workers proposed that the ends of the linear array in fact closed on themselves to form a circle. Additional crosses, shown by the various arcs in Figure 10-11b, confirmed this explanation: all the markers could be assigned a unique order on a circular map. A more recent map of the T4 chromosome is found in Figure 10-12.

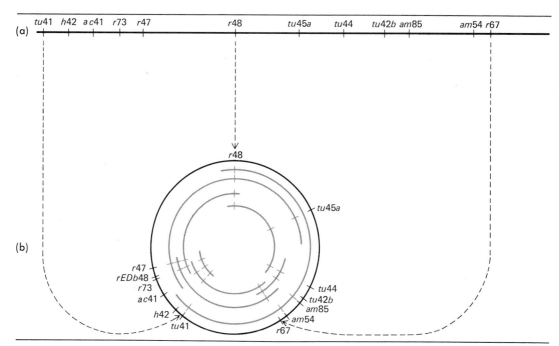

Figure 10-11 (a) A linear sequence of genetic markers in T4 established by recombinational analysis. (b) A circular genetic map of T4. The markers used for each three-or-four-factor cross are connected by an arc. (After G. Streisinger, R. Edgar, and G. Denhardt, *Proc. Natl. Acad. Sci. U.S.* **51**:775, 1964.)

You will recall from Section 7.5 that the T4 chromosome is, in fact, physically a rod; the circular map results from the circularly permuted set of terminally redundant chromosomes in a phage population. Conversely, the λ chromosome is physically a circle during much of its infective cycle, yet its map is linear. It is obvious, therefore, that a genetic map does not necessarily reveal the physical nature of a chromosome.

10.17 RECOMBINATION FREQUENCIES IN T4 AND λ

The T4 and λ chromosomes differ not only in their relative numbers of genes (compare Figures 10-8 and 10-12) and in their circular genetic behavior, but also in their recombination frequencies. During the course of a lytic cycle, a λ chromosome will, on the average, participate in fewer than one genetic exchange, meaning that many chromosomes will not undergo any exchanges, others will undergo one; and still fewer will undergo more than one. In contrast, the average T4 chromosome undergoes 20 exchanges during a lytic cycle. Moreover, **group matings** are thought to occur in which a T4 chromosome undergoes genetic exchange at two (or more) different

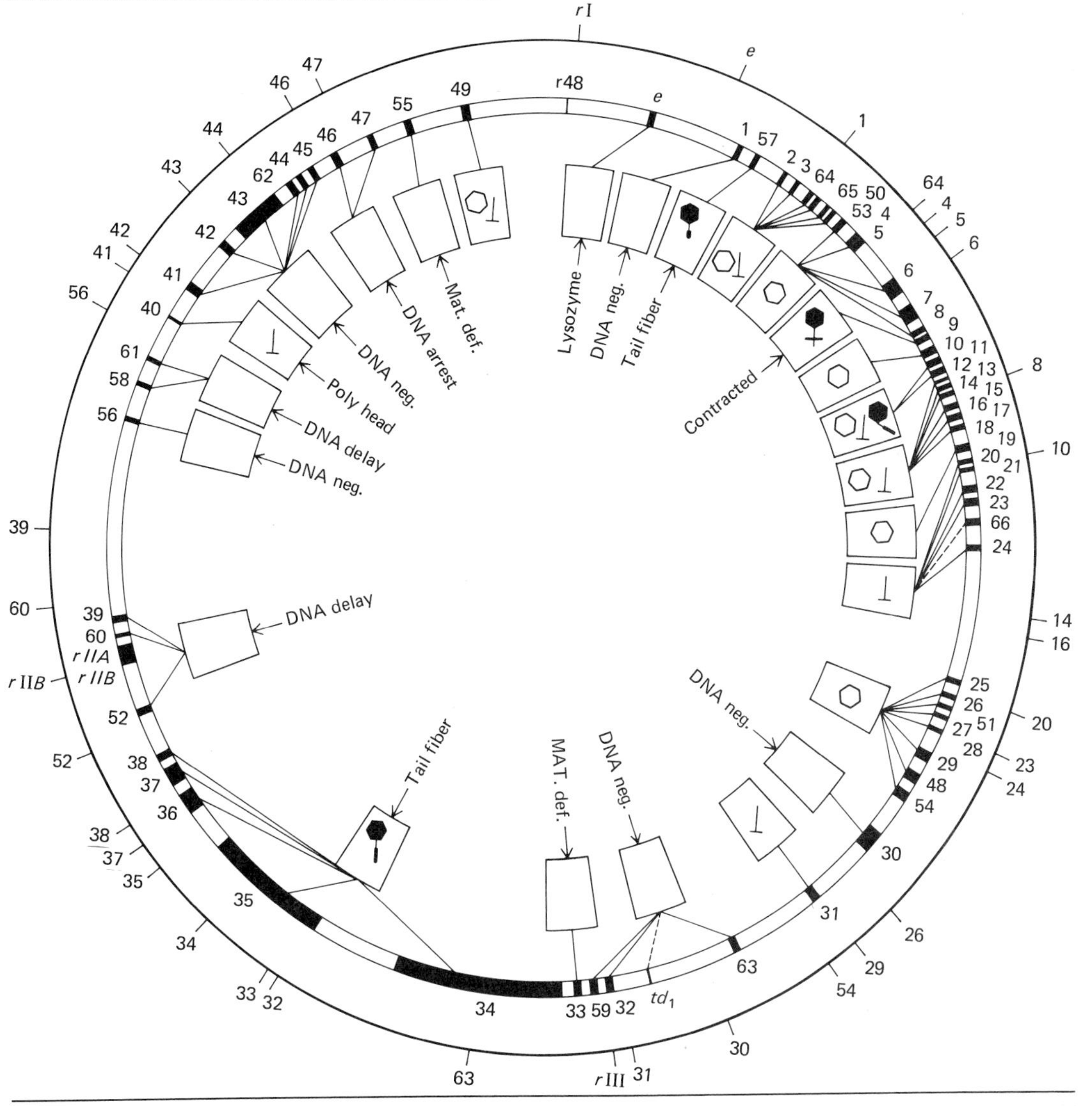

Figure 10-12 Map of the phage T4 chromosome. The inner circle shows the function of genes and their recombination distances. The outer circle represents proposed physical distances between markers, determined by isolating fragments of T4 DNA that have been prematurely packaged into phage particles and analyzing the markers that they contain. (From G. Mosig, *Adv. Genetics* **15**:1, 1970.)

loci, the exchange being with one chromosome at one locus and with another chromosome at the other. Because recombination is a much more frequent and complex event in T4, map distances based on recombination frequencies cannot correspond to the same units of physical distance as map distances derived for λ. In other words, if $R = 2.6$ between two markers on a T4 chromosome, this probably indicates that they are physically much closer together than two markers separated by a 2.6 R value on a λ chromosome would be. More generally, it is safe to assume that map distances based on recombination frequencies should not be freely translated into physical terms when comparing one organism with another.

10.18 DELETION MAPPING OF THE rII GENES OF T4

We conclude a consideration of T4 by describing some of the most important experiments conducted in genetics, namely, the mapping of the *rII* locus of T4 performed by S. Benzer in the late 1950's.

The *rII* locus is seen in Figure 10-12 to lie in the lower-left quadrant of the T4 map. Mutations in the *rII* locus have two consequences, as noted in Section 6.2: the mutant phages cannot conduct a lytic cycle in *E. coli* cells of strain *K* lysogenic for λ, and the growth of the mutant phages on strain *B* of *E. coli* is uncontrolled, leading to the rapid lysis of the host and an abnormal plaque on a bacterial lawn.

Benzer assembled a collection of thousands of different *rII* mutant strains of T4. When a number of these were subjected to complementation tests (Section 10.10), it became clear that they fell into two classes. Strains of class *A* could complement the function(s) missing in strains of class *B*, and vice versa, so that together the two could bring about a lytic cycle on *E. coli* strain *K* in the absence of any gene recombination. The *rII* region of T4 was therefore shown to contain two genes, *rIIA* and *rIIB*.

We should pause here to clarify a term. Benzer referred to the *A* and *B* genes as the *A* and *B* **cistrons**, and the term cistron has become widely used as a synonym for a gene as we have defined it in this text. A cistron derives its name from the **cis-trans test** for complementarity. If two mutations are located in the same chromosome, this corresponds to a *cis* configuration. A wild-type chromosome should be able to complement the missing functions of such a doubly mutant chromosome, regardless of whether the two mutations lie in the same or different genes; thus, complementation should be observed in mixed infections involving *cis* chromosomes and wild-type chromosomes. When the two mutations are instead in the *trans* configuration, each is located on a separate chromosome. Complementation in mixed infections involving two such *trans* chromosomes will occur *only* if the two mutations are located in different genes or, if one prefers, in different cistrons. In practice complementation studies usually do not include the *cis* part of the test, since it requires the construction of doubly mutant strains. The *trans* part of the test is identical to the complementation test described in Section 10.10.

Returning now to Benzer's experiments, attention was next focused on several *rII* strains that were never observed to revert to wild type and thus were excellent candidates as deletion mutants. Some of these strains behaved as if they carried a large deletion in the *A* gene in that they could not form wild-type recombinants when mixedly infected with a number of strains carrying point mutations in gene *A*. Others gave similar evidence that they carried deletions in the *B* gene or deletions that covered both genes. Benzer reasoned that if two strains carrying deletions that overlapped one another were crossed, then wild-type recombinants would never be found among the progeny, whereas if the two crossed strains carried deletions that did not overlap then wild-type recombinants would occasionally form. Figure 10-13 illustrates the principle of **mapping by overlapping deletions** and Figure 10-14 shows the deletions Benzer actually studied, each deletion being identified by a particular number (for example, 1272 and 1241). The ends of the various deletions in the collection delimit 47 different segments of the *A* and *B* genes, shown at the bottom of Figure 10-14 as *A1a*, *A1b1*, and so on. The "big seven" deletions shown at the top of Figure 10-14 cover particularly large regions of the *rII* sector.

With such a collection of deletion-mutant strains it becomes a relatively simple matter to map the *rII* genes. A strain carrying a point mutation at an unknown site is crossed first with the seven strains carrying the big seven deletions. If, for example, the point mutation lies in segment *A3e*, it will give no wild-type recombinants in crosses with strains *r1272*, *r1241*, or *J3*, but it will give wild-type recombinants with the remaining strains. The position of the point mutation within the region bracketed by the *J3* and *PT1* deletions can then be determined by crosses with the appropriate small deletion strains, in this case *r1231*, *r184*, *r250*, and *C33* (Figure 10-14). The *A3e* mutant strain should not yield wild-type recombinants with *r1231* but should with the others.

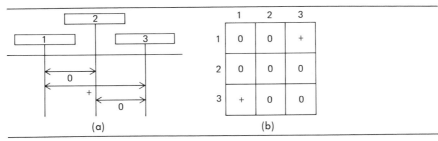

(a)   (b)

Figure 10-13 The method of mapping by overlapping deletions. (a) Three deletion mutations are shown, each affecting a different segment of a gene. Mutant strains carrying deletions number 1 and number 3 can recombine to produce wild-type recombinants (a result symbolized by the + below the arrow connecting number 1 and number 3), whereas neither strain can produce wild-type recombinants when crossed with strains carrying deletion number 2 (a result symbolized by the arrows labeled 0). (b) The same data as in (a) displayed in matrix form. The parental strains (numbers 1, 2, and 3) are drawn on the horizontal and vertical axes, and the result of a cross between any pair (for example, number 1 × number 3) is found in their intersecting box (in this case, a +). (From S. Benzer, in *The Chemical Basis of Heredity*, W. D. McElroy and B. Glass, Eds. Baltimore: Johns Hopkins Univ. Press, 1957.)

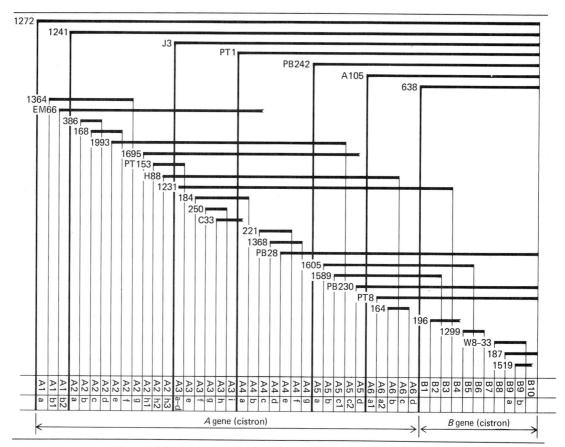

Figure 10-14 Deletions used to divide up the *rII* region of T4 into 47 small segments (shown as small boxes at the bottom of the figure). Some deletion ends have not been used to define a segment and are drawn fluted. The *A* and *B* genes, defined by independent complementation tests, are indicated. (From S. Benzer, *Proc. Natl. Acad. Sci. U.S.* **47**:410, 1961.)

Having localized the mutation to segment *A3e* by means of only 11 crosses, Benzer could now subject the strain to two-factor crosses with other strains that also carried point mutations within the *A3e* segment. In this way, Benzer was able to construct the first gene fine-structure map, shown in Figure 10-15. In addition to generating this map, Benzer went on to note with what frequency the various sites underwent mutation, scoring both $r^+ \longrightarrow r$ forward mutations and $r \longrightarrow r^+$ reversions. These results are included in Figure 10-15, in which each square represents one spontaneous mutational occurrence at each indicated site. It is obvious that the genes exhibit a number of "hot spots" in which spontaneous mutations occur preferentially. More than 500 mutational events are scored at the most

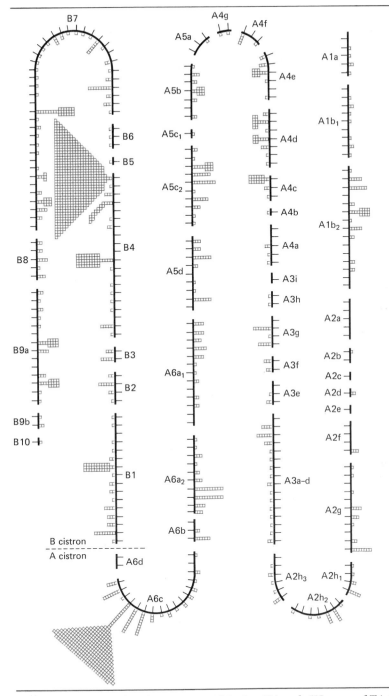

Figure 10-15 The order of mutations within the *rIIA* and *rIIB* genes of T4. Each square identifies the location in the chromosome of an independently occurring mutation. Where more than one mutation occurs, the number of squares indicates the number of mutations. (After S. Benzer, *Proc. Natl. Acad. Sci. U.S.* **47**:410, 1961.)

prominent hot spot, whereas only one mutation is noted at many other sites. Benzer constructed similar profiles to indicate the **topography** of mutations elicited by various mutagens, and these proved to be different both from one another and from the topography found for spontaneous mutations. Such evidence gave early support to the notion that different mutagens preferentially alter different spectra of nucleotides. Why there should be particularly "hot" mutational sites within a gene has not yet been explained, although one popular theory holds that the reactive nucleotides are those thrust into prominence by the particular tertiary structure assumed by a chromosome.

10.19   MAPPING PHAGE φX174

In this final section we present the various approaches taken to analyze the circular chromosome of phage φX174 (see Figure 5-14 for a review of the φX174 lytic cycle).

The "classical" genetic approaches described in preceding sections of this chapter have yielded considerable information about the φX174 genome. (1) Two-factor crosses of various nonsense and *ts* mutants isolated in R. Sinsheimer's laboratory have generated the map shown in Figure 10-16. (2) The genes affected by these mutations have been defined by complementation tests (Figure 10-16, genes A–J). (3) Comparison of the polypeptides synthesized during wild-type and mutant infections has permitted the identification of a specific polypeptide product for each gene (Table 10-5). (4) The relative size of each gene has been estimated from the calculated length of each polypeptide, giving the gene boundaries drawn in Figure 10-16.

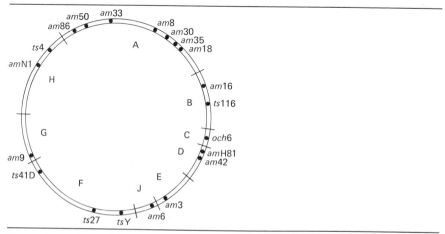

Figure 10-16 Recombination map of φX174, based largely on data of R. M. Benbow et al., *J. Virol.* **7**:549–558, 1971.

Table 10-5   Summary of the ϕX174 Genome

Gene	Function	Protein Molecular wt	Number of Nucleotides*
A	Replication	56,000	1536
B	Replication or packaging	13,800	(360)
C	Replication or packaging	7000	
D	Replication or packaging	16,800	456
E	Lysis	9900	(273)
J	Capsid	5000	114
F	Capsid	46,000	1275
G	Capsid	19,000	525
H	Capsid	36,000	984
Noncoding and C			485
Total			5375

*Values in parentheses are overlapping sequences and therefore not included in the addition to obtain the total length of DNA.
From F. Sanger et al., *Nature* **265**:687 (1977).

P. J. Weisbeek and his colleagues have more recently produced a map of the ϕX174 chromosome that is based on physical distances between mutant loci rather than on recombination frequencies. Their technique is based on the ability of restriction enzymes to cleave the chromosome into discrete fragments, exactly as we saw for SV40 (Section 7.1); thus the enzyme *Hind* II cleaves ϕX174 chromosomes into 13 fragments, *Alu* I produces 23 fragments, and so forth. The fragments are then analyzed by what is called **marker rescue.**

To illustrate the technique we can follow an experiment designed to localize the *am* 9 mutation. A *Hind* II digest of wild-type duplex DNA is prepared (Figure 10-17a). Each fragment class is then incubated separately with *am* 9 single-stranded chromosomes under dissociation-reassociation conditions (Section 7.5); as drawn in Figure 10-17b, partial duplexes form between the mutant chromosomes and complementary sequences present in the restriction fragments. Each DNA preparation is now used to transfect *E. coli* **spheroplasts** that have been so manipulated *in vitro* that they can take up and replicate naked DNA. Only those mutant chromosomes that carry with them a wild-type sequence covering the *am* 9 mutation are able to generate high titers of wild-type phages in this assay and, as drawn in Figure 10-17c, this capability is uniquely associated with the R9 fragment of the *Hind* II digest, localizing *am* 9 to the R9 sector of the genome. The results of many such experiments have produced the ϕX174 map shown in Figure 10-18, where the relative order of the genes around the circle remains

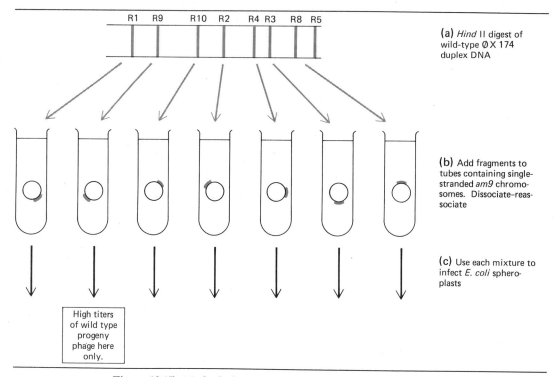

(a) *Hind* II digest of wild-type ∅X 174 duplex DNA

(b) Add fragments to tubes containing single-stranded *am9* chromosomes. Dissociate-reassociate

(c) Use each mixture to infect *E. coli* spheroplasts

High titers of wild type progeny phage here only.

Figure 10-17   Method of mapping φX174 using restriction-fragment "rescue" of mutant chromosomes.

as in Figure 10-16 but where gene boundaries and lengths are far more precisely defined than is possible by recombination.

The ultimate form that a chromosome map can take is, of course, its sequence of nucleotides (Box 7.2) and, as noted in previous chapters, a complete sequence for the φX174 chromosome has been published by F. Sanger and his associates. Such a DNA sequence allows gene lengths to be expressed with great precision (Table 10-5). In the case of φX174, moreover, the sequence revealed that gene B is contained within the boundaries of gene A and that gene E is contained within the D gene region, each gene being translated in a different reading frame. Details of this spectacular discovery are found in Problem 18 of Chapter 9 and in the publications from Sanger's laboratory cited in the Reference section of this chapter.

APPROACHES TO SOLVING MAPPING AND COMPLEMENTATION PROBLEMS

Here and at the conclusion of the next few chapters we consider approaches to solving genetics problems. It is **not** recommended that these

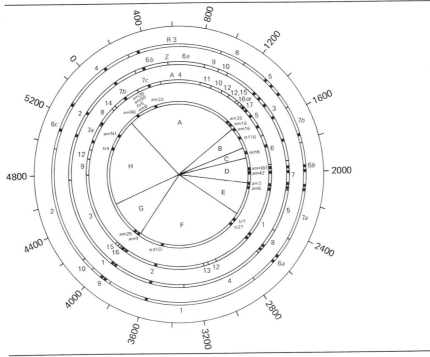

Figure 10-18 Restriction fragment map of φX174 (three middle circles), genetic map based on marker rescue (inner circle), and map distance in nucleotides (outer circle) starting at the gene A/H border. Fragment maps are for *Hind* II (R) (outermost), Hae III (Z), and *Alu* I (A) (innermost) fragments. (From P. J. Weisbeek et al., *Virology* 72:61–71, 1976.)

sample problems be memorized or applied step by step to other problems. The only reliable approach is to understand thoroughly the principles involved and then consider each question as a new puzzle to solve. The sample problems are instead presented as "warm ups," illustrating the ways that data are presented and the ways the data can be used.

10.20 TWO-FACTOR PHAGE CROSS PROBLEM

Problem When five mutant strains of a bacteriophage are crossed, they yield the following recombination frequencies (numbers indicate percent recombination):

Strain	A	B	C	D	E	Crossed with strain
	0	12	3	8	2	A
		0	9	4	6	B
			0	5	0	C
				0	2	D
					0	E

Draw a map showing the location of each mutation. What is unusual about results with strain *E*, and how could they be explained?

Solution   (a) Write out all recombination frequencies from the data:

$$
\begin{array}{lll}
R_{AB} = 12 & R_{BE} = 6 & R_{CE} = 0 \\
R_{AC} = 3 & R_{BD} = 4 & R_{DE} = 2 \\
R_{AD} = 8 & R_{BC} = 9 & \\
R_{AE} = 2 & R_{CD} = 5 &
\end{array}
$$

(b) Identify A and B as lying farthest apart, and put them at either ends of the map:

(c) Fill in the remaining distances, working always from large to small recombination frequencies, and omitting any anomalous data:

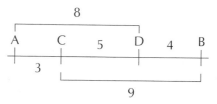

(d) Note that E is anomalous. If R_{BE} is 6, then R_{AE} should also be about 6; instead it is 2. C does not recombine with E at all. Conclusion: E is most likely a deletion extending to either side of C.

10.21   THREE-FACTOR PHAGE CROSS PROBLEM

Problem   Order the mutations *a*, *b*, and *c* (where *a b c* is not necessarily the marker order) on the basis of the results of the following three phage crosses and give the distances between the markers.

Cross	Percent wild type among progeny from the cross
(1) $a\,b\,+\ \times\ +\,+\,c$	8.75
(2) $a\,+\,c\ \times\ +\,b\,+$	3.75
(3) $+\,b\,c\ \times\ a\,+\,+$	1.25

Solution   (a) Note that the data are presented in an unusual fashion: only "percent wild type" is given. The reciprocal "triple-mutant" classes will, of course, figure equally among the progeny and must be included in the calculation of recombination frequencies. Therefore, the values given should first be multiplied by 2.

(b) Cross #3 produces the smallest numbers of + + + phages; this is therefore identified as the double-crossover class. Writing the marker order as + a + $\times$ b + c allows the generation of + + + (and b a c) progeny by double crossovers. The marker order is therefore assigned as b a c, and the frequency of doubles, $R_D = 2(1.25) = 2.5\%$ so that $\gamma = 2.5$.

(c) The Single I (α) and Single II (β) classes can now be identified as crosses #2 and #1 respectively. Therefore:

$$\alpha = 2(3.75) = 7.5$$
$$\beta = 2(8.75) = 17.5$$

(d) Recombination frequencies are therefore:

$$R_{ba} = \alpha + \gamma = 7.5 + 2.5 = 10.0$$
$$R_{ac} = \beta + \gamma = 17.5 + 2.5 = 20.0$$
$$R_{bc} = \alpha + \beta = 17.5 + 7.5 = 25.0$$

(e) Draw map as:

10.22 PHAGE COMPLEMENTATION PROBLEM

Problem Seven mutant strains of T4 can grow individually only in suppressor strains of *E. coli*. All possible pairwise mixed infections of non-suppressor *E. coli* cells are performed with the 7 strains to test for the ability to carry out a complete infection. In the results given, $0 =$ no phage growth; $+ =$ phage growth and lysis.

1	0						
2	+	0					
3	0	+	0				
4	0	0	0	0			
5	+	+	+	0	0		
6	0	+	0	0	+	0	
7	+	0	+	0	+	+	0
	1	2	3	4	5	6	7

What is the minimum number of genes that are represented in this collection of seven mutant strains?

Solution   (a) Examine the data for obvious anomalies. Strain 4 clearly cannot complement any other; it must therefore represent a long deletion or, less likely, a multi-point mutation.

(b) Proceed to assess the remaining strains. Strain 1 (first row of matrix) appears to be mutant in the same gene (no complementation) as strains 3 and 6. Call this gene A and write

(the relative order is, of course, arbitrary here). Strains 2, 5, and 7 are not mutant in gene A since they complement A-deficient strains. These are therefore assessed next.

(c) Strain 2 (second row of matrix) cannot complement strain 7. Call them B-deficient mutants:

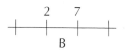

(d) Strain 5 complements all strains (except, of course, itself and the deletion strain). It is therefore mutant in a third gene C:

(e) Three genes, A–C, are represented in the collection.

References

Properties of Bacteriophages

Cairns, J., G. S. Stent, and J. D. Watson, Eds. *Phage and the Origins of Molecular Biology.* New York: Cold Spring Harbor Laboratory of Quantitative Biology, 1966.

Delbrück, M. "The growth of bacteriophage and lysis of the host," *J. Gen. Physiol.* **23**:643–660 (1940). [Reprinted in G. S. Stent, *Papers on Bacterial Viruses,* 2nd ed. Boston: Little, Brown, 1966.]

Delbrück, M., and W. T. Bailey. "Induced mutations in bacterial viruses," *Cold Spring Harbor Symp. Quant. Biol.* **11**:33–37 (1946). [Reprinted in L. Levine, *Papers on Genetics.* St. Louis: Mosby, 1971.]

d'Hérelle, F., F. W. Twort, J. Borclet, and A. Gratia. "Discussion on the bacteriophage (bacterio-lysin)," *Brit. Med. J.* **2**:289–297 (1927) (four articles). [Reprinted in G. S. Stent, *Papers on Bacterial Viruses,* 2nd ed. Boston: Little, Brown, 1966.]

Doermann, A. H. "The intracellular growth of bacteriophages. I. Liberation of intracellular bacteriophage T4 by premature lysis with another phage or with cyanide," *J. Gen. Physiol.* **35**:645–656 (1952). [Reprinted in G. S. Stent, *Papers on Bacterial Viruses,* 2nd ed. Boston: Little, Brown, 1966.]

Ellis, E. L., and M. Delbrück. "The growth of bacteriophage," *J. Gen. Physiol.* **22**:365–384 (1939). [Reprinted in G. S. Stent, *Papers on Bacterial Viruses,* 2nd ed. Boston: Little, Brown, 1966.]

Hayes, W. *The Genetics of Bacteria and Their Viruses,* 2nd ed. New York: Wiley, 1968.

Luria, S. E. "The frequency distribution of spontaneous bacteriophage mutants as evidence for the exponential rate of phage reproduction," *Cold Spring Harbor Symp. Quant. Biol.* **16**:463–470 (1951). [Reprinted in G. S. Stent, *Papers on Bacterial Viruses,* 2nd ed. Boston: Little, Brown, 1966.]

Stent, G. S. *Molecular Biology of Bacterial Viruses.* San Francisco: Freeman, 1963.

Physical Mapping

*Inman, R. B., and M. Schnös. Partial denaturation of thymine- and 5-bromouracil-containing λ DNA in alkali," *J. Mol. Biol.* **49**:93–98 (1970).

*Jeppesen, P. G. N., J. Argetsinger-Steitz, R. F. Gestland, and P. F. Spahr. "Gene order in the bacteriophage R17 RNA:5′-A protein-coat protein-synthetase-3′," *Nature* **226**:230–237 (1970).

Jeppesen, P. G. N., L. Sanders, and P. M. Slocombe. "A restriction cleavage map of φX174 by pulse-chase labelling using *E. coli* DNA polymerase," *Nucl. Acids Res.* **3**:1323–1339 (1976).

Parkinson, J. S., and R. W. Davis. "A physical map of the left arm of the lambda chromosome," *J. Mol. Biol.* **56**:425–428 (1971).

*Sanger, F., G. M. Air, B. G. Barrell, N. L. Brown, A. R. Coulson, J. C. Fiddes, C. A. Hutchison III, P. M. Slocombe, and M. Smith. "Nucleotide sequence of bacteriophage φX174," *Nature* **265**:687–695 (1977).

Simon, M. N., and F. W. Studier. "Physical mapping of the early region of bacteriophage T7 DNA," *J. Mol. Biol.* **79**:249–265 (1973).

*Smith, M., N. L. Brown, G. M. Air, B. G. Barrell, A. R. Coulson, C. A. Hutchison III, and F. Sanger. "DNA sequence at the C termini of the overlapping genes *A* and *B* in bacteriophage φX174," *Nature* **265**:702–705 (1977).

*Weisbeek, P. J., J. M. Vereijken, P. D. Baas, H. S. Jansz, and G. A. van Arkel. "Genetic map of bacteriophage φX174 constructed with restriction enzyme fragments," *Virology* **72**:61–71 (1976).

*Westmoreland, B., W. Szybalski, and H. Ris. "Mapping of deletions and substitutions in heteroduplex DNA molecules of bacteriophage lambda by electron microscopy," *Science* **163**:1343–1348 (1969).

Mapping Phage Lambda

*Campbell, A. "Sensitive mutants of bacteriophage λ," *Virology* **14**:22–32 (1961).

*Casjens, S., and R. Hendrix. "Comments on the arrangement of the morphogenetic genes of bacteriophage lambda," *J. Mol. Biol.* **90**:20–23 (1974).

Dove, W. F. "The genetics of the lamboid phages," *Ann. Rev. Genetics* **2**:305–340 (1968).

*Franklin, N. C. "Deletions and functions of the center of the φ80-λ phage genome: Evidence for a phage function promoting genetic recombination," *Genetics* **57**:301–318 (1967).

Gingery, R., and H. Echols. "Integration, excision, and transducing particle genesis by bacteriophage λ," *Cold Spring Harbor Symp. Quant. Biol.* **33**:721–727 (1968).

Hershey, A. D., Ed. *The Bacteriophage Lambda.* New York: Cold Spring Harbor Laboratory of Quantitative Biology, 1971.

*Jordan, E. "The location of the b_2 deletion of bacteriophage λ," *J. Mol. Biol.* **10**:341–344 (1964).

Jordan, E., and M. Meselson. "A discrepancy between genetic and physical lengths on the chromosome of bacteriophage lambda," *Genetics* **51**:77–86 (1965).

*Kaiser, A. D. "A genetic study of the temperate coliphage λ," *Virology* **1**:424–443 (1955).

*Parkinson, J. S. "Genetics of the left arm of the chromosome of bacteriophage lambda," *Genetics* **59**:311–325 (1968).

Parkinson, J. S., and R. J. Huskey. "Deletion mutants of bacteriophage lambda. I. Isolation and initial characterization," *J. Mol. Biol.* **56**:369–384 (1971).

Radding, C. M. "The genetic control of phage-induced enzymes," *Ann. Rev. Genetics* **3**:363–394 (1969).

*Denotes articles described specifically in the chapter.

Radding, C. M., and A. D. Kaiser. "Gene transfer by broken molecules of λDNA: Activity of the left half-molecule," *J. Mol. Biol.* **7:**225–233 (1963).

Signer, E. R. "Lysogeny: the integration problem," *Ann. Rev. Microbiol.* **22:**451–488 (1968).

Signer, E., H. Echols, J. Weil, C. Radding, M. Shulman, L. Moore, and K. Manley. "The general recombination of bacteriophage λ," *Cold Spring Harbor Symp. Quant. Biol.* **33:**711–714 (1968).

Signer, E. R., and J. Weil. "Site-specific recombination in bacteriophage λ," *Cold Spring Harbor Symp. Quant. Biol.* **33:**715–719 (1968).

*Weigle, J. "Assembly of phage lambda *in vitro*," *Proc. Natl. Acad. Sci. U.S.* **55:**1462–1466 (1966).

Mapping Other Phages

*Benbow, R. M., C. A. Hutchison, J. D. Fabricant, and R. L. Sinsheimer. "Genetic map of bacteriophage φX174," *J. Virol.* **7:**549–558 (1971).

*Benzer, S. "On the topology of the genetic fine structure," *Proc. Natl. Acad. Sci. U.S.* **45:**1607–1620 (1959).

*Benzer, S. "On the topography of the genetic fine structure," *Proc. Natl. Acad. Sci. U.S.* **47:**403–415 (1961).

Hershey, A. D., and M. Chase. "Genetic recombination and heterozygosis in bacteriophage," *Cold Spring Harbor Symp. Quant. Biol.* **16:**471–479 (1951).

Hershey, A. D., and R. Rotman. "Genetic recombination between host-range and plaque-type mutants of bacteriophage in single bacterial cells," *Genetics* **34:**44–71 (1949). [Reprinted in G. S. Stent, *Papers on Bacterial Viruses*, 2nd ed. Boston: Little, Brown, 1966.]

Levinthal, C. "Recombination in phage T2: its relationship to heterozygosis and growth," *Genetics* **39:**169–184 (1954).

Mosig, G. "Recombination in bacteriophage T4," *Adv. Genetics* **15:**1–53 (1970).

*Streisinger, G., and V. Bruce. "Linkage of genetic markers in phages T2 and T4," *Genetics* **45:**1289–1296 (1960).

*Streisinger, G., R. S. Edgar, and G. H. Denhardt. "Chromosome structure in phage T4. I. Circularity of the linkage map," *Proc. Natl. Acad. Sci. U.S.* **51:**775–779 (1964). [Reprinted in G. S. Stent, *Papers on Bacterial Viruses*, 2nd ed. Boston: Little, Brown, 1966.]

*Streisinger, G., J. Emrich, and M. M. Stahl. "Chromosome structure in phage T4. III. Terminal redundancy and length determination," *Proc. Natl. Acad. Sci. U.S.* **57:**292–295 (1967).

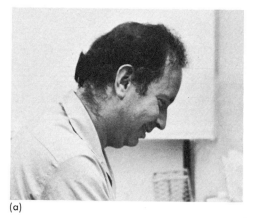

(a)   (b)

(a) George Streisinger (University of Oregon) has made a number of contributions to our present understanding of phage T4 genetics. (b) Seymour Benzer (California Institute of Technology) produced the first fine-structure maps of a phage gene.

Questions and Problems

1. Given the information in the problem presented in Section 10.21, what percent wild type would you expect from the cross $a + \times + c$?

2. A three-factor cross between 2 strains of a virus (abc and $+++$) was performed. The results are given as follows:

Genotype of progeny	Number of plaques
$+ + +$	1200
$a \ b \ c$	1100
$a + +$	280
$+ \ b \ c$	300
$a \ b +$	180
$+ + \ c$	160
$a + \ c$	85
$+ \ b +$	75

Determine the linkage order of these 3 genes and linkage distance between a–b, b–c, and a–c. Comment on your results.

3. Wild-type phage λ is able to multiply both in strain A and strain B of *E. coli*; *sus* mutant strains are able to multiply only in A, not in B. Multiplication in B does result after mixed infection of strain B by certain pairs of mutants. All 40 possible pairwise mixtures of eight different *sus* strains are tested for the ability to multiply cooperatively in strain B with the following results:

	1	2	3	4	5	6	7	8
1	0							
2	+	0						
3	0	+	0					
4	+	0	+	0				
5	0	+	0	+	0			
6	+	0	+	0	+	0		
7	+	+	+	+	+	+	0	
8	0	0	0	0	0	0	0	0

Mutant strain (rows); **Mutant strain** (columns)

(a) On the basis of these results, classify mutations 1 through 7 into genes.

(b) Give two possible explanations for the behavior of strain 8. How could you distinguish between these alternatives experimentally?

4. Traits A, B, C, D, E, F, G, H in T4 were mapped in a series of 2-factor crosses. The values turned out to work very consistently, with one exception.

(a) Draw a map (need not be on scale) of this part of the T4 genome

(b) Which recombination value is different from the one predicted from the map?

	A	B	C	D	E	F	G	H
A	0	10	17	1.5	3	6.5	10	8
B	—	0	0	6	9	0	2	4
C	—	—	0	11	14	4	7	9
D	—	—	—	0	0	0.5	4	2
E	—	—	—	—	0	3.5	7	5
F	—	—	—	—	—	0	0	0
G	—	—	—	—	—	—	0	2
H	—	—	—	—	—	—	—	0

5. A geneticist performed the following experiments with two suppressor sensitive mutations p and q. Plaques caused by $+\,+, p+$ or $+q$ and mixed yields could be distinguished.
   1. Simultaneously infect non-permissive host and plate on non-permissive host
   2. Simultaneously infect non-permissive host and plate on permissive host
   3. Simultaneously infect permissive host and plate on non-permissive host
   4. Simultaneously infect permissive host and plate on permissive host
   Predict the results of the experiments if:
   (a) p and q fail to complement and are 8 units apart.
   (b) p and q complement and are 10 units apart
   NOTE: Recombination rarely occurs before phage growth is underway.

6. One mutant of T4 is unable to make a head protein (H^-) while another mutant is unable to make tail fibers (T^-). Both of these strains can infect *E. coli* strain K but neither are able to produce progeny phage. However, if strain K is mixedly infected with both (H^-) and (T^-) strains of T4, phage progeny are produced in approximately the same amounts as cells infected with wild-type phage. What would the genotypes of the progeny phage be? Explain your choice.
   (a) Approximately equal amounts of wild type $(+\,+)$ and the double mutant (H^-T^-).
   (b) Mostly (H^-) and (T^-) in approximately equal amounts with a small proportion of wild type $(+\,+)$ and double mutant (H^-T^-).
   (c) Equal amounts of (H^-), (T^-), wild type $(+\,+)$, and the double mutant (H^-T^-).
   (d) Approximately equal amounts of wild type $(+\,+)$ and the double mutant (H^-T^-), with small amounts of (H^-) and (T^-).

7. Suppose the genome of T4 phage is represented as *ABCDEF*, and because of the circularly permuted DNA of T4, a typical sample from several particles yields sequences such as *ABCDEFAB, CDEFABCD, EFABCDEF*, etc.
   Now suppose deletion mutants are formed by deleting region *CD*, that these mutants are allowed to infect *E. coli* cells, and that the DNA of the progeny phage is then examined. Which of the sequences below ought to be found if the "headful" hypothesis (Section 5.10) is correct? Explain your choice.

   (a) *ABEFAB, EFABEF, FABEFA*
   (b) *ABEFABEF, EFABEFAB, FABEFABE*
   (c) *ABEFABEFA, EFABEFABE, FABEFABEF*

8. Two-factor crosses involving strains *a, b, c, d,* and *e* of a bacteriophage give the following results:

Cross	Recombination frequency (%)
$a \times b$	$<10^{-3}$
$a \times c$	2.0
$a \times d$	3.0
$a \times e$	1.0
$b \times c$	$<10^{-3}$
$b \times d$	1.0
$b \times e$	0.8
$c \times d$	1.1
$c \times e$	2.8
$d \times e$	3.8

Order the markers along an unbranched genetic map.

9. The following recombination frequencies are measured for two-factor bacteriophage crosses:

$$a\,b^+ \times a^+\,b \quad 3.0\%$$
$$a\,c^+ \times a^+\,c \quad 2.0\%$$
$$b\,c^+ \times b^+\,c \quad 1.5\%$$

   (a) What map order is suggested for the mutations *a, b,* and *c*? Why are the distances not additive?
   (b) Consider the three-factor cross $a\,b^+\,c \times a^+\,bc^+$. Which two recombinant types do you expect to be rarest?
   (c) Compute the recombination frequencies expected for each of the three reciprocal pairs of recombinant types emerging from the three-factor cross in (b).

10. Four mutant T4 strains are tested for complementation in *E. coli.* We get these results (+ = complementation, 0 = no complementation):

	1	2	3	4
1	0	0	+	+
2	0	0	0	+
3	+	0	0	+
4	+	+	+	0

   (a) Explain the above results in terms of the functional relationships between mutations 1 to 4.
   (b) Draw a map showing the relative locations of these mutations. What ambiguity remains?

11. From the data given for the first cross in Table 10-3, construct a map showing the distances between markers *s*, *co*, and *mi* in phage λ.

12. A three-factor cross yields the following kinds and numbers of progeny:

+ + +	235	*p q r*	270
p q +	62	*p* + +	7
+ *q* +	40	*p* + *r*	48
+ *q r*	4	+ + *r*	60

Total: 726

(a) What are the genotypes of the parental phages in this cross?

(b) What is the gene order?

(c) What are the map distances between the genes?

13. Four T4 phage deletion strains are tested for recombination by pairwise crossing in *E. coli*. We get these results (+ = recombination, 0 = no recombination):

	1	2	3	4
1	0	+	+	+
2	+	0	0	+
3	+	0	0	0
4	+	+	0	0

(a) Explain the above results in terms of a genetic map.

(b) A mutant phage, *m*, known to contain *no* deletions, recombines with mutants 1 to 4 as follows:

	1	2	3	4
m	0	+	0	+

Explain the nature of mutant phage *m* and locate it on the map.

14.

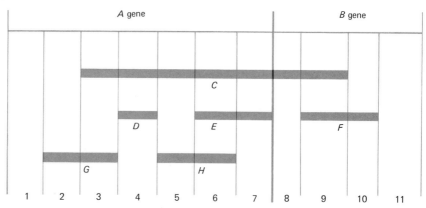

The ends of six deletion strains (*C, D, E, F, G,* and *H*) define 11 separate regions in the *rII* locus of phage T4, as shown above. When these strains are mixedly infected with four strains carrying mutations *K, L, M,* and *N*, the following results are obtained:

0 = no lysis
C = complementation
R = recombination

	C	D	E	F	G	H	K
K	0	R	R	C	R	0	0
L	0	C	C	R	C	C	C
M	0	0	0	C	R	0	0
N	0	R	0	R	R	0	0

In which regions of the map can you place mutations *K, L, M,* and *N*? What is the nature of mutations *M* and *N*? Give your reasoning.

15. The b_2 sector of the λ chromosome (Figure 10-8) is not marked by *sus* mutations. Why?

16. The λ chromosome exhibits a clustering of functionally related genes (Figure 10-8), with the left arm *morphological,* the right arm largely *regulative,* and the middle region *integrative* and *recombinational.* Is the T4 chromosome (Figure 10-12) similarly "logical"?

17. Phage chromosomes are often effectively analyzed by a denaturation map (Figure 8-20), one of which is shown on the next page.
   (a) To what position on the map does the arrow in the micrograph correspond? Explain.
   (b) What is the relative GC content of positions 8–10 and 12–14 microns on the chromosome relative to the rest of the chromosome? Explain.
   (c) Explain how such a map allows a sensitive estimate of the lengths of particular deletions.

18. A culture of *E. coli* B was mixedly infected with two *rIIB* mutants. An aliquot of the *B* lysate was diluted by 10^2 and a 0.1 ml sample was plated on *E. coli* strain K(λ). A second aliquot of the *B* lysate was diluted by 10^5 and a 0.1 ml sample was plated on *E. coli* strain *B* bacteria. A total of 20 plaques were counted on the *K* plates and 20 plaques were also counted on the *B* plates. What is the map distance between these two mutations?

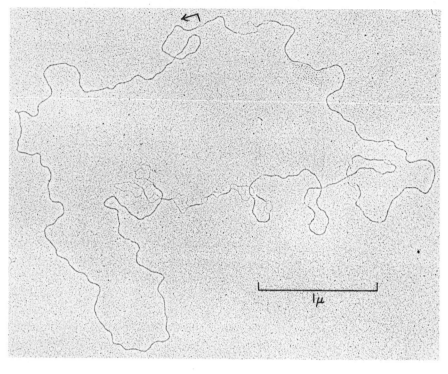

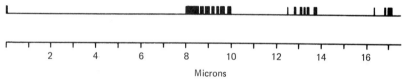

19. Benzer isolated a mutant phage strain Z with a single point mutation somewhere in the *rIIA* locus. He crossed mutant Z with three other deletion mutant strains A, B, and C which were deleted for various regions of the *rIIA* gene as indicated below.

	a	b	c	d	e
Normal *rllA* locus					
Strain A		b	c	d	e
Strain B			c	d	e
Strain C				d	e

The results of this series of crosses were as follows:

	Wild-type recombinants/ 10,000 progeny
Z × A	140
Z × B	35
Z × C	0

Benzer then crossed mutant strain Z with three strains I, II, and III possessing *point* mutations at sites I, II, and III, respectively, with the following results:

	Wild-type recombinants/ 10,000 progeny
Z × I	2
Z × II	8
Z × III	4

From this information, in what region (*a–b*, *b–c*, *c–d*, or *d–e*) of the *rIIA* locus is the mutation in strain Z located? Which of the three strains carrying point mutations have their mutant sites closest to that in strain Z, and what is the map distance separating them from Z?

20. The deletion mutants C, D, E, F, and G map in the following regions of the T4 *rII* genes.

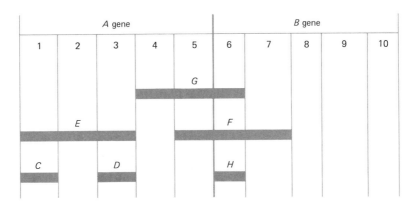

The following data for the unmapped mutants R, S, T, and U was obtained. (0 = no lysis; R = recombination; C = complementation.)

	C	D	E	F	G	H
R	R	R	0	R	R	C
S	R	R	R	R	0	C
T	C	C	C	0	R	R
U	C	C	C	R	R	R

Assign map positions to the 4 mutations.

CHAPTER
11

Mapping Bacterial Chromosomes

AN OVERVIEW OF BACTERIAL MAPPING

As we move from the bacteriophage (Chapter 10) to the bacterium, we make a major leap in the complexity of both genetic organization and gene transmission. A bacterium possesses perhaps a hundred-fold more DNA than most of its phages (Figure 11-1). The DNA of a bacterium, moreover, is found both in the single main chromosome and in the relatively small plasmids, (Section 5.1), so that frequently, the transmission of more than one genetic element from parent to offspring must be followed. Finally, a number of different avenues have evolved whereby the DNA from one bacterial cell can undergo genetic exchange with the DNA from another bacterial cell. Thus the DNA may be transmitted through a specialized sex pilus (**conjugation**); it may enter the cell as a duplex fragment (**transformation**); or it may be carried into and out of the cell by a bacteriophage (**transduction**). This chapter describes each of these modes of bacterial gene transmission, and demonstrates how each is used to construct bacterial— and, in some cases, phage—chromosome maps.

A MOLECULAR OVERVIEW OF BACTERIAL CONJUGATION

11.1 THE *F* ELEMENT AND $F^+ \longrightarrow F^-$ TRANSFER

A bacterial cell may contain one or more plasmids—small DNA molecules that usually maintain a distinct existence from the main chromosome and replicate independently of it. Plasmids are considered in some detail in Chapter 14, but one kind of plasmid, the **sex element** known as *F,* plays a key role in the sexuality of bacteria and is thus appropriately considered here as well.

Escherichia coli cells that carry an *F* element (the terms **F factor** and **F agent** are also used) are known as **F⁺**, and such cells are found, albeit quite rarely, among natural populations of the bacterium. In F^+ cells perhaps 2 percent of the cell's DNA is found in the *F* element, which takes the form of a covalently closed circular DNA molecule with a molecular weight of approximately 35×10^6. *E. coli* cells can be induced to carry two mutant *F* elements and subjected to complementation analysis (Section 10.10). These studies, described fully in Chapter 14, indicate that the *F* element is made up of at least 15 complementation groups or genes, nine of which are known to control the elaboration of **F pili**—long appendages that extend from the surface of F^+ cells (Figure 11-2). Cells that possess *F* pili are sensitive to infection by single-stranded RNA phages; **F⁻** cells, which lack an *F* element and are devoid of pili, are insensitive to such phage infection.

An F^+ cell will usually ignore another F^+ cell, whereas it will readily establish contact, probably via an *F* pilus, with an F^- cell. Once contact is

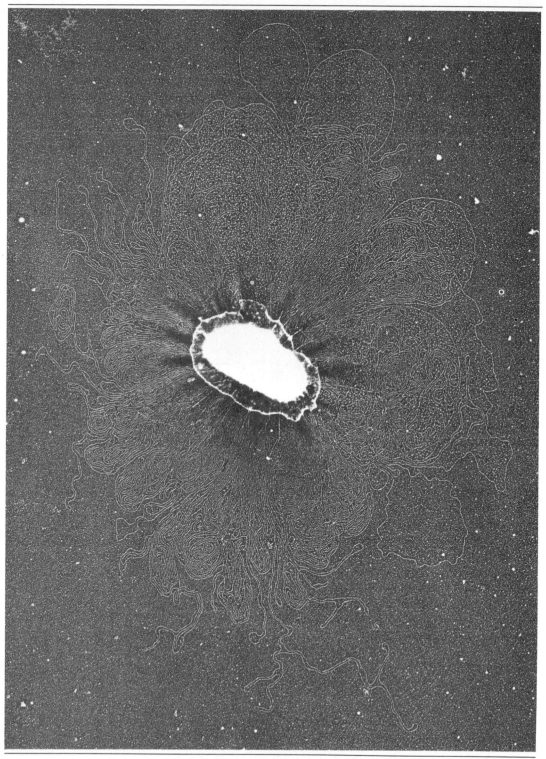

Figure 11-1 The bacterium *Hemophilus influenzae* after osmotic shock, revealing its long chromosome. (With permission from L. A. MacHattie et al., *J. Mol. Biol.* **11**:648, 1965. Copyright by Academic Press Inc. (London) Ltd.)

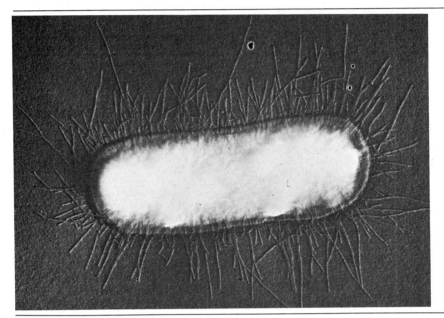

Figure 11-2 A bacterium with pili (also called fimbriae) extending from its surface. (From J. P. Duguid et al., *J. Path. Bact.* **29**:197, 1966.)

made the pilus is believed to become modified and to serve as a protoplasmic channel between the two cells; in this context it is often referred to as a **conjugation tube.** Under ordinary circumstances the only genetic element transferred through the tube is the F element itself. In a mixture of F^+ and F^- cells each F^+ donor will pass a copy of the F element on to an F^- recipient, while retaining at least one copy for itself; eventually, then, virtually every cell in such a mixed population becomes an F^+ cell.

The transfer of F is thought to proceed by a mode of DNA replication known as the **rolling circle** mechanism. In this mechanism, as shown in Figure 11-3, the 5' end of one strand of the F element is thought to be drawn into the recipient cell, where it is copied in the 5' ⟶ 3' direction. Meanwhile the second strand remains in the donor and serves as a rolling template for its own replication.

The infectious transmission of F elements from cell to cell is of interest in itself, but it yields no information about the genes in the main bacterial chromosome. Fortunately for bacterial genetics, however, the F element infrequently (about once in every 10,000 F^+ cells) becomes associated with the main bacterial chromosome in such a way that a copy of the main chromosome, instead of simply the F element, is transferred through the conjugation tube from donor to recipient.

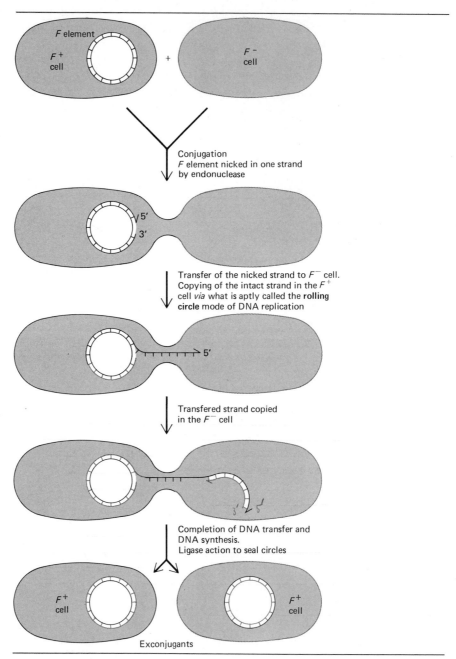

Figure 11-3 Diagram illustrating transfer of the F element from an F^+ to an F^- cell during conjugation in *E. coli*.

11.2 THE FORMATION OF *Hfr* CELLS AND *Hfr* ⟶ *F⁻* TRANSFER

The events that allow transfer of the main bacterial chromosome to an *F⁻* cell are diagrammed in Figure 11-4. In this process, an *F* element first inserts itself into the bacterial chromosome. The insertion process, which is in some ways analogous to λ chromosome insertion (Section 5.15), causes the circular *F* element to break at a particular point and to become a linear segment of the bacterial chromosome, as diagrammed in Figure 11-4a. An *F⁺*

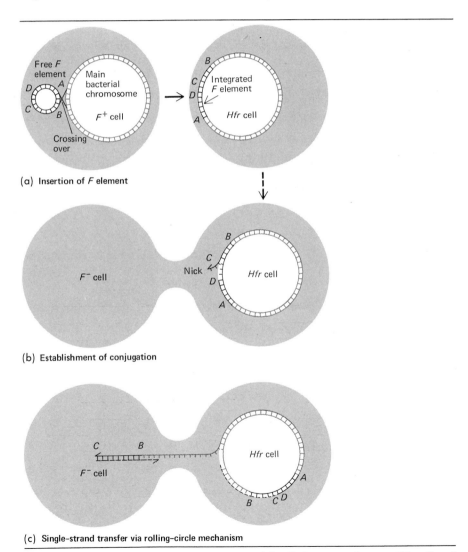

(a) Insertion of *F* element

(b) Establishment of conjugation

(c) Single-strand transfer via rolling-circle mechanism

Figure 11-4 Diagram illustrating integration of the *F* element to form an *Hfr* cell and transfer of *Hfr* chromosome to an *F⁻* cell during conjugation in *E. coli.*

cell that carries such an integrated *F* element is known as an **Hfr cell** for reasons that will become apparent shortly.

The integrated *F* element is ordinarily replicated passively along with the bacterial chromosome, very much like a λ prophage, and in this way is transmitted from one *Hfr* generation to the next. In other words, the integrated *F* element ordinarily shuts off its ability to replicate independently and behaves as an ordinary segment of the bacterial chromosome. When conjugation is initiated, however, the replication apparatus of the *F* element is somehow activated. The *F* DNA is nicked in one strand (Figure 11-4b), and the 5' end of the nicked strand is drawn into the conjugation tube, apparently by the impetus of a rolling-circle type of DNA replication (Figure 11-4c). This mode of transfer is similar to the round of replication that effects the transfer of an *F* element in an *F*$^+$ ⟶ *F*$^-$ conjugation (Figure 11-3). The difference, of course, is that the transferred strand is now covalently linked to the enormous main bacterial chromosome; one strand of this chromosome is therefore drawn into the *F*$^-$ cell as well (Figure 11-4c) and is copied in the 5' ⟶ 3' direction as it enters the *F*$^-$ recipient.

In Figure 11-4a the *F* element carries some arbitrary markers called *A, B, C,* and *D,* and the nick that initiates rolling-circle replication occurs between markers *C* and *D* (Figure 11-4b). This means that the *F* element is split during transfer, and only some of its genes enter the *F*$^-$ cell at the start of conjugation. The remaining *F* element genes (the *D-A* segment in Figure 11-4) will be transferred to the recipient cell only after approximately 1200 μ of chromosomal DNA has already passed through the conjugation tube. Usually the entering chromosome breaks at some intermediate position during transfer. Therefore the *F*$^-$ recipient cell usually inherits an incomplete copy of the *F* element during *Hfr* X *F*$^-$ conjugation, and remains *F*$^-$, (in contrast to an *F*$^+$ X *F*$^-$ interaction, in which the *F*$^-$ cell is converted to *F*$^+$). If transfer of the entire chromosome is in fact completed, then the recipient *F*$^-$ cell will inherit a complete *F* element and the *Hfr* property, and its descendants can then act as *Hfr* cells.

An *F*$^-$ cell that has received only a part of the donor chromosome is called a **partial zygote** or **merozygote.** A partial zygote is initially diploid for those genes that have been transferred, but the cell does not remain diploid. Instead, it takes part in genetic exchanges so that some of the donor DNA is included in the recipient chromosome (such exchanges are described from a molecular viewpoint in Chapter 13). All nonintegrated fragments of DNA are then lost from the cell line in subsequent divisions, and the clone emerges as haploid.

When donor and recipient chromosomes carry different genetic markers, the emergent cells are frequently recombinant. It is for this reason that cells capable of donating chromosomal material to recipient cells have come to be called *Hfr* for **high frequency of recombination.**

MAPPING BY BACTERIAL CONJUGATION

11.3   EARLY CONJUGATION STUDIES

Many features of bacterial sexuality were worked out in the 1950s by W. Hayes, F. Jacob, J. and E. Lederberg, and E. Wollman well before the molecular details of the process were known. Their studies revealed that recombination would occur when *Hfr* cells of one genotype were crossed with F^- cells of another genotype. We can, in particular, follow a series of crosses performed by Jacob and Wollman.

The *Hfr* strain used in the Jacob-Wollman cross had the following phenotype and genotype: it was capable of synthesizing the amino acids threonine (*thr⁺*) and leucine (*leu⁺*); it was sensitive to the metabolic inhibitor sodium azide (*azi-s*), and it was sensitive to the phage T_1 (T_1-s). It could ferment lactose (*lac⁺*) and galactose (*gal⁺*) as well as glucose and it was sensitive to the antibiotic streptomycin (*str-s*). The F^- strain used by Jacob and Wollman had the complementary genotype, that is, *thr⁻ leu⁻ azi-r* T_1-r *lac⁻ gal⁻ str-r* (F^-), where *resistance* is denoted by *-r*. The two cell types were mixed and allowed to interact for 60 minutes. The cells were then plated to various media and allowed to grow.

Jacob and Wollman began by plating the cells in the mating mixture to a minimal medium containing streptomycin. On such a medium all unmated *Hfr* cells, being streptomycin-sensitive, were killed, and any cells that required threonine or leucine, including unmated F^- cells, were unable to grow. This step therefore allowed only F^- cells with the recombinant genotype *thr⁺ leu⁺ str-r* to form colonies. Such recombinants emerged with very high frequency—some 10 percent of the total cells plated—meaning that genetic exchange had occurred frequently between the *thr⁺ leu⁺* markers contributed by the *Hfr* donor and the *str-r* marker contributed by the F^- recipient. This result, in other words, indicated that the *thr* and *leu* loci are quite loosely linked to the *str* locus.

The next step involved determining the rest of the genotype of the *thr⁺ leu⁺ str-r* recombinants. Jacob and Wollman replica-plated (Figure 4-2) the recombinant colonies to a minimal medium that contained either sodium azide or bacteriophage T_1 and determined the distribution of sensitivity and resistance to these agents among the colonies. They also replica-plated the *thr⁺ leu⁺ str-r* recombinants to special indicator media (such as that described in Section 6.2) to determine whether the various strains were able to ferment lactose or galactose. Typically, they found that among the *thr⁺ leu⁺ str-r* recombinants, 90 percent carried the *azi-s* marker of the *Hfr* donor, 80 percent carried the T_1-s marker, 40 percent carried the *lac⁺* marker, and 25 percent carried the *gal⁺* marker.

11.4   MAPPING BY INTERRUPTED MATING

At the time the Jacob-Wollman crosses were performed nothing was known about what was involved in bacterial conjugation, and thus the results were interpreted in many different ways. Indeed, the data were used to support models in which *E. coli* possessed a branched chromosome or multiple chromosomes. That conjugation in fact involves the transfer of a linear *Hfr* chromosome to the *F⁻* cell was established genetically by Jacob and Wollman in their subsequent **interrupted mating** experiments.

In these experiments *Hfr* and *F⁻* cells, having the same genotypes as before, were allowed to conjugate. At specific time intervals samples were taken and agitated in a Waring Blendor to separate mating cells. Samples that had been allowed to mate for 5 minutes, 10 minutes, and so on, were thus obtained. Each sample was then plated to a streptomycin-containing minimal medium to select, as before, recombinant *F⁻* cells that were *thr⁺ leu⁺ str-r*. The complete genotypes of these recombinant cells were then determined by appropriate platings.

The results are shown in Figure 11-5. Focus attention first on the order of appearance of the unselected markers in the recombinant cells. It is seen

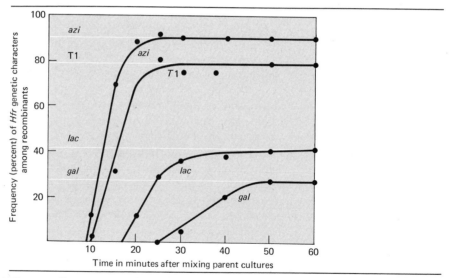

Figure 11-5   Kinetics of transfer of various unselected donor (*Hfr*) markers (*azi, T₁, lac,* and *gal*) among *F⁻* recombinants selected for *thr⁺ leu⁺ strʳ* markers following an interrupted mating experiment. For each marker, a plateau is observed after the frequency of transfer reaches a certain level. This corresponds to the rate of spontaneous cessation observed for that marker in noninterrupted mating experiments. (After F. Jacob and E. L. Wollman, *Sexuality and the Genetics of Bacteria.* New York: Academic, 1961.)

that *azi-s* is not transferred at all during the first nine minutes of mating, after which it is transferred readily. The T_1-*s* marker comes next, first appearing in recombinants at 10 minutes. The *lac*⁺ marker does not appear until about 18 minutes, whereas the *gal* locus makes its first appearance only after 25 minutes. The most direct way to interpret these results is to draw the donor chromosome as $\underrightarrow{(thr^+, leu^+)}$ with the arrowhead indicating the origin of entry, and to assume that this chromosome moves into the recipient in a linear, orderly fashion. Since *azi-s* begins to enter the recipient cells shortly after the selected *thr*⁺ *leu*⁺ markers, whereas *gal*⁺ does not enter for an additional 16 minutes, it can at once be deduced that *azi-s* must lie much closer to the *thr*⁺ *leu*⁺ loci than does *gal*⁺. Applying this logic to the remaining unselected markers, the order of genes in the donor chromosome can be deduced as $\underrightarrow{gal^+ \ lac^+ \ T_1\text{-}s \ azi\text{-}s \ (thr^+, leu^+)}$.

You will notice in Figure 11-5 that the percent recombination for each marker reaches a "plateau" value. These values prove to be the same as those obtained in the non-interrupted matings described in Section 11.3, with 90 percent for *azi*, 80 percent for T_1, and so on. Work out for yourself how these numbers arise, recalling that "natural" breaks in the donor chromosomes are increasingly likely to occur with increasing conjugation time, and assuming that once inside a recipient, all donated DNA is equally likely to participate in recombination with the *F*⁻ chromosome.

11.5   DIFFERENCES BETWEEN *Hfr* STRAINS

The *Hfr* strain used in the foregoing experiments was called *Hfr-H*. As different *Hfr* lines were isolated from *F*⁺ populations in various laboratories it became clear that each had distinctive properties. Thus a second strain, *Hfr-AB311*, when subjected to interrupted mating, transferred its markers to *F*⁻ cells in the order $\underrightarrow{thr \ leu \ azi \ T_1 \ lac \ gal}$; in this case the order of the markers is the reverse of the *Hfr-H* order. In a third strain, *Hfr-C*, the transmission order proved to $\underrightarrow{gal \ thr \ leu \ azi \ T_1 \ lac}$. Here, not only is the orientation the reverse of *Hfr-H* but also the linkages appear to differ from both *Hfr-H* and *Hfr-AB311*: *gal* and *lac* now appear to be at opposite ends of the chromosome, whereas earlier they had appeared next to each other. (Such a situation should sound familiar. You will recall from Section 10.15 that linkage relationships established in one phage T4 cross do not always hold for another.)

Once marker-transfer data were collected from a number of different *Hfr* strains it was possible to draw several conclusions about *Hfr* integration and the main bacterial chromosome. First, it became clear that the main chromosome can pass from *Hfr* to F⁻ cells either in one direction (let us say $A \longrightarrow Z$) or in the opposite direction ($Z \longrightarrow A$). Since *F* integration is now believed to be the stimulus for transfer, it is presumed that *F* itself can assume one or the other of opposite orientations within the main chromosome at the time it integrates, thus establishing the polarity of transfer.

The genetic data also reveal that although any one *Hfr* strain exhibits a unique, inheritable site of *F* integration, *F* does not always integrate at the same site in the chromosome. Instead, numerous sites have been demonstrated for different insertion events, many with opposite orientation, as shown in Figure 11-6.

An additional fact to emerge from the interrupted mating experiments is that the *E. coli* chromosome, like the T4 chromosome, is genetically circular. Thus the marker orders obtained from strain to strain are all circular permutations of one another, reading in the same or in opposite directions. This result is, of course, one we have come to expect, since we have been assuming a circular *E. coli* chromosome throughout the text.

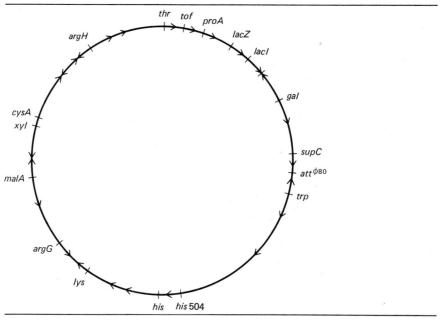

Figure 11-6 Sites of *F* element integration in the chromosome of *E. coli*. Each different point of insertion is symbolized by an arrowhead pointing in the direction of chromosome transfer, so that the gene behind the arrowhead is the first one transferred. (From A. Campbell, *Episomes*, New York: Harper and Row, 1969. Copyright © 1969 by Allan M. Campbell. Reprinted by permission of Harper & Row, Publishers, Inc.)

11.6   RAPID MAPPING BY CONJUGATION

B. Low has devised a particularly efficient method for conjugation mapping which depends on many of the principles we have described. This method utilizes 15 different *Hfr* strains of *E. coli;* each strain carries a different *F* integration site and therefore initiates the transfer of a different segment of the main chromosome; and each strain is sensitive to streptomycin. The 15 different *Hfr* colonies are replica plated onto a layer of *F⁻* streptomycin-resistant cells which carry an unmapped mutation. The plate is **recombinant-selective,** meaning, for example, that if the unmapped mutant strain is an acetate auxotroph, the medium is made acetate-free so that only re-combinants at the acetate locus will grow. Cells from each *Hfr* colony proceed to transfer their unique sectors of the main chromosome into the *F⁻* recipients on the plate. Since the probability of recombination decreases linearly with time because of spontaneous cessation of chromosome transfer, cells underlying a given *Hfr* strain are most likely to acquire genes transferred early in conjugation. After 30 minutes, the plate is sprayed with streptomycin to kill all *Hfr* cells, and is then incubated for 1-2 days. Heavy patches of growth are observed only in those areas where at least one *Hfr* cell has transmitted to a recipient mutant cell a wild-type gene which replaces its mutant allele. Thus simple inspection of the plate allows one to determine, for example, that *Hfr-AB311* cells are uniquely capable of con-verting the *F⁻* acetate mutants into prototrophs with high efficiency, and this localizes the acetate-utilization locus close to the *F* element integration site of the *Hfr-AB311* strain.

11.7   THE *E. COLI* CHROMOSOME MAP

Partial maps of the *E. coli* chromosome have been presented throughout the text; the complete genetic map of *E. coli,* as it is now known, is given in Figure 11-7. The map is divided into minutes, with 100 min representing the time taken for a complete chromosome to pass from *Hfr* donor to *F⁻* recipient during conjugation. This detailed map was not constructed with conjugation data alone, since transduction (Sections 11.11 and 11.15) must be used to establish linkage relationships between markers that are close to one another. In practice, therefore, a newly isolated mutant strain of *E. coli* carrying a mutation at some unknown position on the chromosome is first subjected to mapping by interrupted mating to determine the general location of the mutant gene; its precise location is then determined by transduction.

A striking feature of the *E. coli* chromosome is its nonrandom distribution of mapped genes: the 73 min and 83 min regions, for example, are densely populated with genes, whereas the 33 min region is a veritable genetic wasteland. One possibility is that the "empty" areas in fact contain genes that have not yet been marked by mutations, perhaps because we do not yet know how to select for such mutations. The alternative possibility is that the DNA in these regions does not contain genes and instead performs other functions for the cell. One intriguing suggestion is based on the fact that the 33 min "wasteland region" lies almost exactly 180° from the 86 min

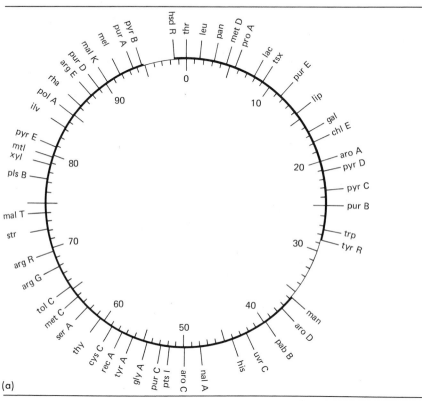

(a)

Figure 11-7 (a) Circular reference map of *E. coli* K-12. The large numbers refer to map position in minutes, relative to the *thr* locus. The 52 loci were chosen on the basis of greatest accuracy of map location, utility in further mapping studies, and/or familiarity as long-standing landmarks of the E. coli K-12 genetic map. The two thin portions of the circle represent the only two map intervals that are not spanned by a continuous series of P1 cotransduction linkages. (b) Linear scale drawings (pp. 409–10) representing the circular linkage map of E. coli K-12. The time scale of 100 min, beginning arbitrarily with zero at the *thr* locus, is based on the results of interrupted conjugation experiments. Parentheses around a gene symbol indicate that the location of that marker is not well known, sometimes having been determined only within very wide limits. An asterisk indicates that a marker has been mapped more precisely but that its position with respect to adjacent markers is not known. Arrows above genes and operons indicate the direction of messenger RNA transcription of these loci. Genetic symbols are defined in Bachmann, B. J., K. B. Low, and A. L. Taylor, *Bact. Rev.* **40**:116, 1976.

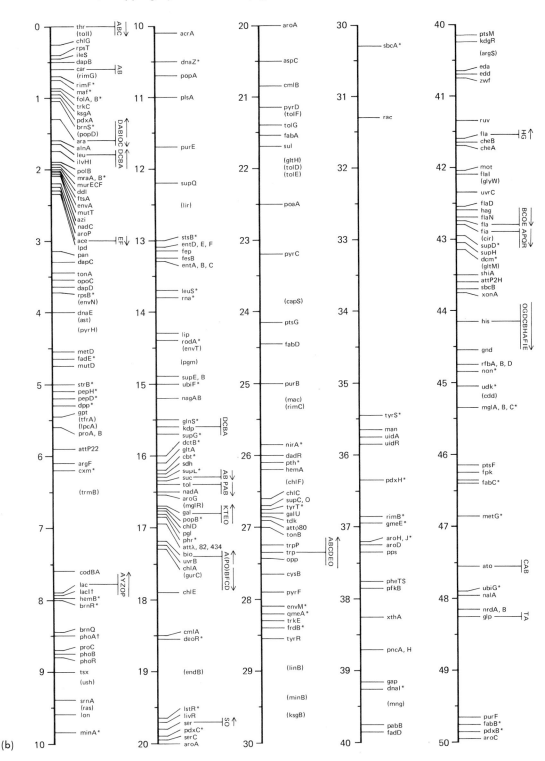

(b)

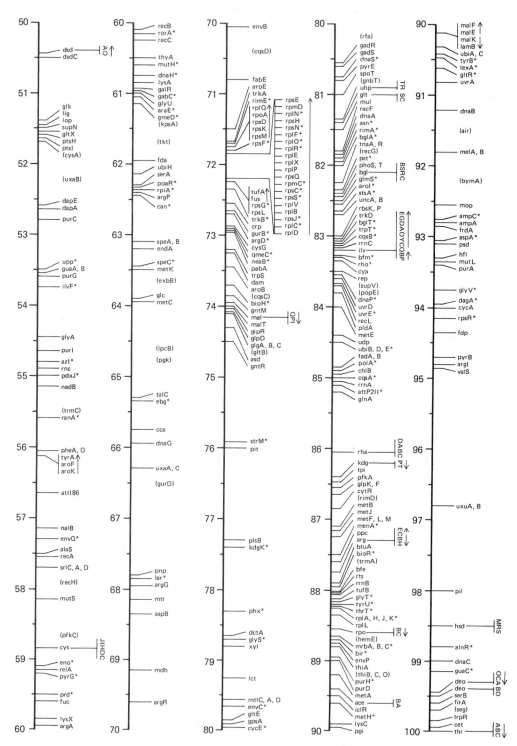

(b)

site where bidirectional chromosome replication in *E. coli* originates (Section 5.6); it has been proposed, therefore, that the 33 min DNA might attach to the membrane at the time of replication and that the late replication of this DNA somehow provides the signal for mesosome ingrowth and cell division.

BACTERIAL TRANSFORMATION

11.8   A MOLECULAR OVERVIEW OF BACTERIAL TRANSFORMATION

Transformation was described in Chapter 1 in the context of classical experiments with *D. pneumoniae* (pneumococcus), where DNA isolated from *smooth* (virulent) cells was taken up by *rough* (nonvirulent) cells and incorporated in such a way that information for *smooth* capsule formation was transmitted to progeny bacteria (Figure 1-12). Transformation is formally analogous to conjugation: in both situations, **donor** DNA is transferred to a **recipient** or **host** cell where it undergoes genetic exchange with the host chromosome to produce a recombinant bacterium. The processes are distinct, however, in that transforming DNA passes from donor to recipient in a free exposed state rather than through the shield of a conjugation tube. Moreover, in transformation, the recipient cell ordinarily receives random fragments of donor chromosomes, in contrast to the orderly transmission of a fixed gene sequence during conjugation.

A number of bacteria, including *D. pneumoniae, Hemophilus influenzae,* and *Bacillus subtilis,* are readily induced to undergo transformation in the laboratory, and at least low levels of transformation probably take place in their natural habitats. A bacterium such as *E. coli,* on the other hand, can undergo transformation only under special laboratory conditions: the cells must carry mutations abolishing exonuclease I and V activity; the cells must be treated with highly concentrated $CaCl_2$ solutions to render their membranes permeable to DNA; and they must be exposed to very high concentrations of donor DNA. In the paragraphs that follow, transformation is described as it is believed to occur in the more "natural" situations; the *in vitro* transformation of *E. coli* has, however, important applications for genetic research.

In any large population of bacteria such as *D. pneumoniae,* a few cells are probably always present that are capable of being transformed. A variety of laboratory procedures have, however, been developed to maximize the presence of transformable or **competent** cells so that transformation proceeds with high efficiency. These procedures include growing cells under particular culture conditions, harvesting them at a particular phase in their growth cycle, and so forth. Competent cells appear to elaborate a surface protein called **competence factor** which is involved in an energy-dependent binding of donor DNA fragments to the cell surface.

DNA fragments from the donor cell must also be in a particular state if high-efficiency transformation is to occur: they must be large, relatively intact, and double-stranded, having a molecular weight in the 0.3 to 8×10^6 dalton range. The requirement for duplex DNA presumably relates to the way in which donor DNA enters the cell: as the end of a DNA fragment crosses the membrane, an intracellular DNase hydrolyzes one of its strands. The energy for pulling in the intact strand subsequently derives from the hydrolysis of the degraded strand. Mutant strains of *D. pneumoniae* which lack this DNase activity (also called **DNA translocase**) cannot be transformed.

Once single-stranded DNA fragments have successfully been pulled into a cell, they are capable of inserting into homologous regions of the host chromosome. Obviously, any one piece of DNA will be homologous only to a certain segment of the recipient cell genome, and most of the pulled-in pieces will not be genetically marked. If, however, donor DNA derives from an erythromycin-resistant *ery*^{r} strain and the host strain is erythromycin-sensitive (*ery*^{s}), then an occasional host cell will pull in a donor DNA fragment containing a copy of the *ery*^{r} gene. If this donor DNA succeeds in inserting into the *ery*^{s} locus of the recipient, then a transformed, erythromycin-resistant cell will emerge. Models for how such insertions might take place are presented in Section 13.8; here we are concerned with how transformation is used to establish genetic linkage.

11.9 ESTABLISHING GENE LINKAGE BY TRANSFORMATION

To establish gene linkage by transformation the following considerations are made. If two genes, *E* and *F*, are so distantly linked that the probability is remote that both will ever be included within the same DNA fragment, then double transformants—cells transformed for both *E* and *F*—can arise only when a cell happens to receive and integrate two separate pieces of DNA, one carrying gene *E* and the other gene *F*. On the other hand, if *E* and *F* are closely linked, then double transformants can arise either because, as mentioned in the preceding sentence, a cell receives two separate DNA fragments, one with *E* and the other with *F*, or because a cell receives a single fragment carrying both *E* and *F*.

An experiment is therefore constructed in which a culture of competent cells carrying markers *e* and *f* is divided among several flasks and each subculture is presented with progressively more dilute concentrations of DNA fragments derived from donors carrying *E* and *F*. The recipient cells from each flask are then plated and the numbers of single (*E* or *F*) and double (*EF*) transformants are scored. Looking first at the single transformants, one finds, as expected, that as the concentration of DNA is decreased fewer and fewer single transformants appear among the progeny, with the probability of transformation being about the same for either marker. This

result is shown in Figure 11-8, Curve A. Looking next at double transformants, one predicts that if the genes *E* and *F* are linked, the concentration
dependence for producing double transformants (Figure 11-8, Curve B)
should have approximately the same slope as the concentration dependence for single transformants (Curve A). If, on the other hand, *E* and *F* are
unlinked, then, as DNA concentrations decrease, the occurrence of double
transformations will decrease with a much greater slope (Figure 11-8, Curve
C), since the occurrence of such double transformations will be totally
dependent on the rapidly decreasing probability that two appropriate
fragments will enter the same cell.

Stated somewhat more theoretically, the principle of establishing linkage
by transformation is as follows: The frequency of transformation for gene *E*
alone and for gene *F* alone is first established. Then, if genes *E* and *F* are
unlinked, the probability that double transformants will arise should equal
the product of the individual probabilities, and this will be a much smaller
number than either individual frequency alone. If, instead, the number of
observed double transformants greatly exceeds such expectations and, in
fact, approaches the frequency for an individual transformation event, it is
concluded that the two genes are closely linked and often take part in the
same transformation event.

Obviously this approach is dependent on the assumption that the DNA
molecules in the transformation mixture have a known, fairly uniform size.
When, for example, the mean molecular weight of transforming DNA is
5×10^6, we may find that genes *E* and *F* are linked whereas gene *G* is not
linked to either *E* or *F*. If the DNA is then isolated more gently and larger
fragments are acquired, linkage between *G* and either *E* or *F* may be
observed. A comparison of such overlapping data often allows relative
intergenic distances to be deduced.

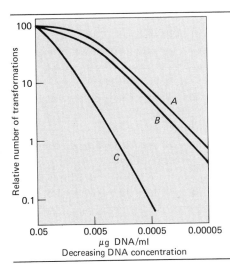

Figure 11-8 Idealized curves illustrating
the effect of decreasing concentration of
transforming DNA on the relative number of single and double transformants
for markers *E* and *F*. Curve A: Number of
single transformants of either the *E* or *F*
type. Curve B: Number of double transformants of type *EF* where *E* and *F* are
closely linked. Curve C: Number of double transformants of type *EF* when *E* and
F are unlinked and therefore carried by
different DNA fragments. (From W.
(Hayes, *The Genetics of Bacteria and Their
Viruses*, 2nd ed. New York: John Wiley,
1968.)

11.10   TRANSFORMATION MAPPING USING *BACILLUS SUBTILIS*

The transformation method just described can readily define whether two markers are linked. Moreover, by determining that marker *A* is linked to *B* and that *B* is linked to *C* a marker order of *A-B-C* can be deduced. The ordering of genes along a chromosome by transformation proceeds much more rapidly, however, in the kinds of experiments we now describe for *Bacillus subtilis.*

The initiation of the chromosome replication cycle in *B. subtilis* can be **synchronized**—meaning that every chromosome in a culture of cells begins being replicated at the same time—by procedures developed by N. Sueoka, H. Yoshikawa, and colleagues. Cells can, for example, be grown in a deuterated (D$_2$O) medium and then transferred to an aqueous (H$_2$O) medium; the transfer itself stimulates a synchronous onset of DNA replication. Such methods of initiating synchronous chromosome replication have been combined with transformation in the following way: the heavy (deuterated) DNA is induced to replicate in light (H$_2$O) medium so that the resulting daughter DNA duplexes are hybrid in density (Section 5.4). At intervals during the synchronous replication cycle, cells are removed from the culture and their DNA is extracted, fragmented, and subjected to CsCl density centrifugation (Box 5.1) to separate the hybrid DNA fragments from the heavy, unreplicated material. The fragments containing newly replicated DNA can then be used to transform competent recipient cells. By choosing recipient cells that carry numerous auxotrophic markers and donor cells carrying their prototrophic alleles, the genes that are included in the newly replicated DNA at each sampling time can be identified.

The resulting data indicate that synchronous chromosome replication in *B. subtilis* begins at a unique position such that the gene *purA,* involved with adenine synthesis, appears in early samples of hybrid DNA. At the other extreme the *metB* gene, involved with methionine synthesis, does not appear among the hybrid transforming fragments until the replication cycle is almost complete, thus localizing the *met* gene near the terminus of the replicating chromosome. The time of appearance of intermediate genes can also be scored and a map constructed in which the map units relate marker positions relative to the *met* terminal marker. Thus it is not only possible to establish by transformation that certain markers must be closely linked to one another, but also to map an entire chromosome.

Figure 11-9 shows a recent map of the *B. subtilis* chromosome where the map is seen to be in the form of a single, circular linkage group. The relative marker order was established by using the transformation methods we have just described. Such methods, however, do not reveal the relationships between very closely linked markers. The more sensitive technique of transduction has therefore been used to fill in the details of this map; we noted earlier that transduction is also needed to complete the

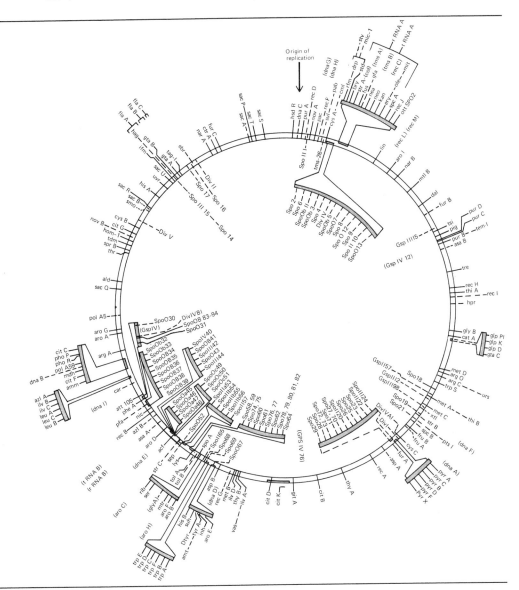

Figure 11-9 Genetic map of the chromosome of *B. subtilis*. Map distances are based on transduction frequencies with phage PBS1. (From F. E. Young and G. A. Wilson, in *Spores*, VI, P. Gerhardt, R. N. Costilov, and H. L. Sadoff, eds. Washington, D.C., American Society of Microbiologists, 1975.)

map of *E. coli* (Figure 11-7) after its general pattern is deduced by interrupted mating.

GENERALIZED TRANSDUCTION

11.11   A MOLECULAR OVERVIEW OF GENERALIZED TRANSDUCTION

Generalized transduction is one of three types of transduction known to occur in bacteria; the other two (specialized and *F*-mediated) are described in subsequent sections. During generalized transduction, a piece of a donor bacterial chromosome becomes incorporated into a phage head and is transduced (carried over) to a recipient bacterium, where it may insert into the recipient genome, much as a DNA fragment becomes incorporated during transformation.

Generalized transduction is performed by phage P1 of *E. coli*, P22 of *Salmonella,* and SP10 and PBS1 of *B. subtilis*. Towards the ends of their lytic cycles, while phage chromosomes are being packaged into protein coats and the host chromosome is being degraded, an occasional piece of host DNA, usually of the same length as a phage chromosome, becomes mistakenly packaged in a phage coat and is liberated along with the other phage progeny in the burst. This **transducing particle** is now able to infect a second bacterium, and the particle's DNA, which shares a homologous sequence with a portion of the infected cell's chromosome, can be inserted into the chromosome. When the inserted DNA carries donor genes that differ from recipient markers, recombinants will result.

We might pause to ask why it is that *all* virulent bacteriophages do not mediate generalized transduction, that is, why bacterial rather than phage DNA is not mistakenly packaged within, for example, the coats of T viruses. The answer appears to be a physiological one. The packaging process for some generalized transducing phages may be less selective about the DNA incorporated than is the same process for nontransducing phages. Alternatively, the DNA of the host may break into fragments of the appropriate size at just the right time in the transducing phage's lytic cycle. In the case of P22, this is apparently not a matter of chance: a phage-specified protein, coded by gene 3 and most likely the endonuclease responsible for cutting P22 concatamers into "headful" lengths during virion maturation (Section 5.10), apparently recognizes particular signals (base sequences?) in the *Salmonella* chromosome as well and cuts these into "headful" lengths. As a result, P22 will transduce certain *Salmonella* markers far more frequently than others, these transduced markers presumably lying between preferred cutting sites. High-transducing (HT) mutants of P22 continue to package phage-length pieces of *Salmonella* DNA into transducing particles but these now contain a random collection of markers: the marker-specificity of wild type is almost completely abolished. The mutations carried by HT strains, therefore, apparently alter the cutting specificity of the phage endonuclease.

11.12 MAPPING BY GENERALIZED TRANSDUCTION

Generalized transducing particles are formed very infrequently (one out of perhaps every 10^6 phages emerging during the lysis of a bacterial culture). In the analysis of transducing data, therefore, it is not necessary to be concerned (as one is with transformation) about the possibility that an apparent linkage of two markers has in fact arisen because a cell has become doubly recombinant for two separate pieces of DNA. Even if a given bacterium is infected with a multiplicity of infection as high as 10, the probability that one of the infecting particles will prove to be a transducing particle—and that this particle will generate a singly recombinant cell—is very small; therefore the probability that 2 of the 10 infecting particles are effective transducing particles becomes as low as 1 in 10^8, which is of the same order as the probability that one of the genetic markers being followed will undergo a reversion.

It follows that direct inferences can be drawn from generalized transduction data: if two markers are cotransduced with high frequency they are considered to be closely linked, whereas if they are never cotransduced they are probably separated by a length of DNA that is at least as long as one phage chromosome. This length is equivalent to perhaps 6×10^7 daltons of DNA for phage P1, and as much as 1.7×10^8 daltons for phage PBS1; in contrast, the average DNA fragment in a transformation preparation is in the 5×10^6 to 1×10^7 dalton range. Thus generalized transduction proves to be not only a more direct approach to establishing genetic linkage than transformation; it is also a means of examining linkage between both closely-linked and more widely separated markers.

When two markers are followed during generalized transduction as described above, a **two-factor transduction** is said to take place. Such crosses, as they are called, permit the establishment of **relative cotransduction frequencies,** and if marker A is cotransduced with B, B is cotransduced with C, and A is not cotransduced with C, it is concluded that the markers assume an A-B-C order along the chromosome. A far more informative cross is provided by a **three-factor transduction** which, like the three-factor crosses described in Sections 10.7 and 10.8 for bacteriophages, allows both marker order and map distance to be established directly in a single cross.

As an example of a three-factor transduction cross we can cite a typical experiment in which the donor E. coli cells have the genotype $a^+b^+c^+$ and the recipients have the genotype $a^-b^-c^-$. The donor is infected with P1, the P1 progeny are used to infect recipient cells, and the recipient cells are then plated and subjected to selection for the presence of *one* of the donor markers (for example, the presence of a⁺). Such recombinant recipient cells are said to have been transduced for at least one marker, and they are now tested by replica plating to see whether they are transduced (recombinant) for one or both of the other markers as well.

It is here that we apply reasoning similar to that for three-factor crosses of bacteriophages: *the rarest class of transductants should represent the most unlikely transduction event,* and the most unlikely event is for a recipient marker to be flanked by two donor markers; this requires the occurrence of four crossovers, as shown in Figure 11-10. Therefore, if *abc* is, in fact, the

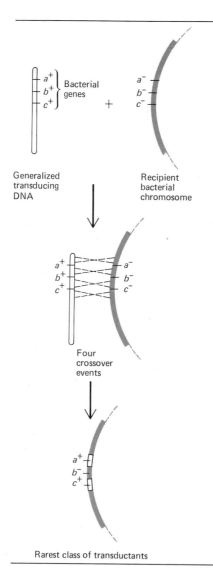

Figure 11-10   Formation of a rare transduced bacterium by four crossovers between the DNA of a transducing particle and a bacterial chromosome.

correct gene order, then $a^+b^-c^+$ cells should represent the rarest class of transductants; if *bac* is the correct gene order, then $b^+a^-c^+$ cells should be the rarest, and so on.

Three-factor transduction data can supply cotransduction frequencies as well as gene order; to derive these frequencies, the total number of times the selected marker appears with an unselected marker in the recipient cells is calculated. To give some actual data, an experiment was performed by E. Signer, J. Beckwith, and S. Brenner with $trpA^+$ $supC^+$ $pyrF^+$ donor cells and $trpA^-$ $supC^-$ $pyrF^-$ recipient cells of *E. coli*, where *trpA* is a gene involved in tryptophan biosynthesis, *supC* is an *ochre*-suppressor gene (Section 9.18), and *pyrF* is a gene involved in pyrimidine biosynthesis. P1-mediated transductants for $supC^+$ were initially selected.

These transductants could be classed as follows:

		Number
1.	$supC^+$ $trpA^+$ $pyrF^+$	36
2.	$supC^+$ $trpA^+$ $pyrF^-$	114
3.	$supC^+$ $trpA^-$ $pyrF^+$	0
4.	$supC^+$ $trpA^-$ $pyrF^-$	453
		603 total

The marker order is at once recognized as being *supC trpA pyrF,* based on the genotype of the rarest class of transductants (class 3). The $supC^+$ and $trpA^+$ markers are both transduced in classes (1) and (2) but not in (3) or (4); therefore, the *supC-trpA* cotransduction frequency is calculated as $36 + 114/603 = 150/603 = 0.25$. Similarly, the cotransduction frequency for *supC* and *pyrF* is seen to be equivalent to the class (1) frequency, namely, $36/603 = 0.06$.

Cotransduction frequency is inversely related to the distance between genes. Thus, if two markers are close together, they will almost always be transduced together and their cotransduction frequency will approach one. If, on the other hand, two markers are never or almost never included in the same piece of transducing DNA, their cotransduction frequency will approach or be zero. A mathematical expression of this relationship is given as:

$$d = L(1 - \sqrt[3]{x})$$

where d = the physical distance between two markers on a chromosome, L = the average length of transducing DNA, and x = the cotransduction frequency of two markers.

(For a derivation of this formula, refer to the paper by T. T. Wu cited in the References.) In other words, when the average length of the transducing particle's DNA is known, the molecular distances separating nearby genes can be estimated. When a value for L is not known, it can simply be considered as some constant.

As stated earlier, the kind of resolution that can be achieved with generalized transduction in *E. coli* is difficult or impossible to achieve with interrupted mating experiments; on the other hand, a generalized transduction experiment is most easily designed when one has some idea of the general location of a given mutation. In practice, therefore, when a new mutant strain is isolated in *E. coli,* for instance, its mutant gene is introduced into *Hfr* strains and its general chromosomal position is determined by conjugation. Its position is then pinpointed by generalized transduction, through the use of appropriately marked donor and recipient strains.

SPECIALIZED TRANSDUCTION

11.13 FORMATION OF SPECIALIZED TRANSDUCING PHAGES

Specialized transduction is mediated by temperate bacteriophages. You will recall from Section 5.11 that temperate viruses such as λ can exist in either an autonomous or an integrated state. They therefore qualify as **episomes**—genetic elements that can move into or out of the main chromosome of a cell. Most temperate phages normally occupy a fixed, invariant site within the bacterial chromosome, in which location they are referred to as **prophages.** Thus the λ prophage site (att^λ) in the *E. coli* chromosome is flanked on one side by the *galK, T,* and *E* genes of the *galactose* operon of the *E. coli* cell and on the other side by a block of genes involved with the synthesis of biotin, *bioA, B, F, C,* and *D.* This site occupies the 17 minute sector of the *E. coli* map (Figure 11-7). The ϕ80 prophage, on the other hand, specifically occupies a site ($att^{\phi 80}$) flanked by a cluster of *trp* genes on one side and two suppressor genes, *supC* and *supF,* on the other, as found in the 27 minute sector of Figure 11-7.

When prophages are induced to leave their chromosomal sites, an occasional error occurs so that the emergent piece of DNA contains some bacterial genes covalently linked to some phage genes. This hybrid DNA packaged within a phage coat is known as a **specialized transducing particle:** it is capable of transferring bacterial genes from one cell to another, and the transduction is specialized because only the genes adjacent to the prophage attachment site are transferred. Thus λ will specifically transduce *gal* or *bio* genes and ϕ80 will transduce *trp* or *sup* genes.

The total length of packaged DNA in a specialized transducing particle is roughly the same as that of a normal phage chromosome, presumably because it could otherwise not be packaged properly into a phage head. A specialized transducing particle carrying a sizable piece of the bacterial genome will therefore usually lack a corresponding length of phage genome. The missing phage genes are taken away from the end of the prophage chromosome opposite the one to which the bacterial genes are added. Therefore, as illustrated in the next paragraph, a λ particle transducing a certain number of *gal* genes will typically lack a corresponding block of genes in the J region of its chromosome, whereas a *bio*-transducing λ particle will lack genes in the N region. For this reason such specialized transducing particles are denoted as **defective** in phage genes so that, for example, λ*dgal* is defective and *gal*-transducing, while λ*dbio* is defective and *bio*-transducing. We have already encountered several examples of defective transducing phages, namely, those carrying *E. coli* RNA-polymerase genes (Section 8.3) and those carrying genes of the translation apparatus (Section 9.8).

A model accounting for the production of a λ*dgal* particle is presented in Figure 11-11. Figure 11-11a shows the prophage in its integrated state between *gal* and *bio*. As the prophage is looping out in preparation for excision (Figure 11-11b), a segment of the phage chromosome somehow interacts with a segment of the bacterial chromosome. The molecular basis for this interaction is unknown, but it is not likely to be based on homologous nucleotide sequences, since it can apparently occur at numerous positions along either chromosome. In any case, following the interaction, a genetic exchange produces a defective bacterial chromosome carrying phage genes (Figure 11-11c) and a complementary defective phage carrying

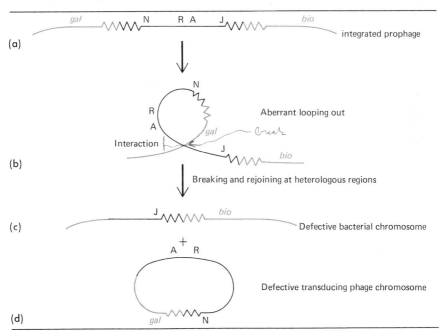

Figure 11-11 Model of the formation of a λ*dgal* specialized transducing phage. Bacterial DNA is represented in gray and phage DNA in black. Zigzag lines represent the *att* DNA normally involved in phage integration and excision; a detailed description of this DNA is given in Section 13.16. (From A. Campbell, *Episomes.* New York: Harper and Row, 1969. Copyright © 1969 by Allan M. Campbell. Reprinted by permission of Harper & Row, Publishers, Inc.)

bacterial genes (Figure 11-11d). This aberrant event happens quite infrequently, perhaps once for every 10^6 normal phage excisions, but when a culture of gal^+ *E. coli* cells lysogenic for λ is induced with ultraviolet light, (Section 5.11), the resulting phage lysate is likely to contain a few λ*dgal*$^+$ (and a few λ*dbio*$^+$) particles among many millions of normal phages.

11.14 TRANSDUCTION OF MARKERS TO HOST BACTERIA

If a λ phage lysate containing λ*dgal*$^+$ phages is used to infect gal^- bacteria, there arise transductants that are able to ferment galactose. These transductants are almost invariably lysogenic for λ and, when they are examined closely, it appears that they are formed in a unique fashion.

The entering λ*dgal*$^+$ DNA first circularizes (Figure 11-12a). Its gal^+ region then pairs with the homologous gal^- region in the recipient chromosome

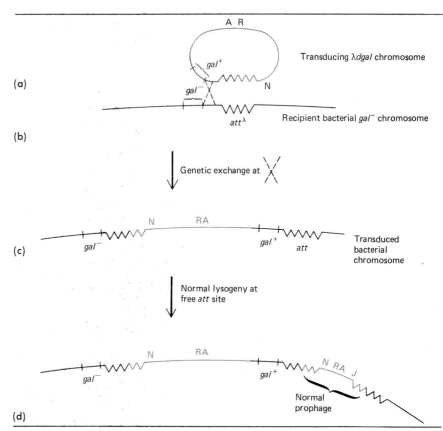

Figure 11-12 Course of events during specialized transduction of a gal^- bacterium by a λ*dgal*$^+$ transducing particle. Bacterial DNA is represented in black and phage DNA in gray. Zigzag lines represent the *att* DNA normally involved in phage integration and excision (Section 13.16).

(Figure 11-12b), but the donor genes do not displace the recipient genes (in contrast to the situation in transformation or in generalized transduction). Instead, recombination occurs such that the λ*dgal*⁺ DNA is inserted into the site where pairing took place (Figure 11-12c). Meanwhile, the usual λ attachment site in the *E. coli* chromosome remains unoccupied, and it ordinarily becomes filled by a second, normal λ phage that infects the cell at the same time as the λ*dgal*⁺ particle (Figure 11-12d). Thus the resulting transduced bacterium carries two copies of most (but not all) λ genes and two copies of a few bacterial genes (Figure 11-12d).

With respect to the bacterial genes, the cell is a **partial diploid,** *gal*⁺/*gal*⁻. It and its descendants are able to ferment galactose because the *gal*⁺ gene is dominant to the *gal*⁻ gene—meaning, in molecular terms, that the normal gene product carries out the activity that comes to define the phenotype: in this case, the ability to ferment galactose. The abnormal gene product specified by the *gal*⁻ gene is still synthesized in the cell, but its presence is without effect on the phenotype and can be detected only if it is isolated biochemically.

When partial diploids of the sort illustrated in Figure 11-12d are treated with ultraviolet light to induce λ excision, the resulting lysate will contain approximately half normal phages and half λ*dgal* transducing phages. Such a lysate, known as a **high-frequency transduction (HFT)** lysate, will clearly be greatly enriched for bacterial genes involved in galactose utilization.

MAPPING BY SPECIALIZED TRANSDUCTION

11.15 MAPPING BACTERIA BY SPECIALIZED TRANSDUCTION

Bacterial genes that flank the attachment sites of temperate phages—*gal* and *bio,* or *trp* and *supC*—are readily identified by specialized transduction as being closely linked to one another. As described in the final section of this chapter, moreover, it is now possible to introduce a wide variety of bacterial genes next to λ or φ80 attachment sites and thus to obtain specialized transducing phages carrying a number of *E. coli* chromosomal sectors. The value of such phages for establishing linkage has already been outlined in Sections 8.3 and 9.8, where we saw, for example, that several λ transducing phages carry a large number of linked ribosomal protein genes (Figure 9-8).

11.16 DELETION MAPPING OF BACTERIOPHAGES BY SPECIALIZED TRANSDUCTION

Specialized transducing particles are of added value in that they allow a rapid and sensitive genetic analysis of the temperate phages themselves. Specialized transducing particles usually carry deleted versions of phage chromosomes and, as stressed in Section 10.14, deletions represent stable and very useful genetic markers: they define a particular sector of a chromosome, and they are distinctive in that they fail to recombine with point mutations lying within the deletion-marked region.

For phage mapping, a collection of deletion strains is first assembled and characterized. The first step is to determine which genetic functions have been destroyed by each deletion, a determination made by pairwise complementation tests (Section 10.10). Thus, for example, if λdgal strains 1 and 2 cannot direct lytic cycles alone but can cooperate, when mated together, to generate a normal virulent infection, then the strains clearly complement one another and must carry defects in different genes. If, on the other hand, a similar mating between strains 1 and 3 fails to yield any normal plaques, then strains 1 and 3 must be deleted in the same gene or genes.

The second step is to determine how large a deletion is carried by each strain. This is accomplished by subjecting each deletion strain to mixed infections with strains carrying **point** mutations at **known** positions on the phage chromosome: if wild-type recombinants emerge, then the deleted region does not include the locus of the point mutation. Strain 1, for example, may produce wild-type phages when crossed with a strain carrying a known *amber* mutation in gene *P,* whereas strain 3 may fail to recombine with this strain. The strain 3 deletion is in this way identified as extending farther into the *P*-gene region of the λ chromosome than the strain 1 deletion.

A collection of such characterized λdgal strains, each carrying a slightly less extensively deleted version of the phage chromosome than the last, is shown in Figure 11-13. Suppose one now isolates a new mutant strain λ and wishes to characterize it genetically. Complementation tests with the collection of deletion strains will quickly establish which genetic function(s) are lost by the new mutation, and crosses with these strains will readily determine in which segment of the λ chromosome the unknown mutation resides. It was, in fact, this approach that Campbell employed when he ordered the various *sus* mutations in the left arm of the λ chromosome, as described in Section 10.9.

11.17   *F*-MEDIATED SEXDUCTION

The *F* element of *E. coli* (Sections 11.1 and 11.2) also qualifies as an episome: it is capable of an independent existence, and it can integrate into a chromosomal site that becomes specific for a given *Hfr* strain. Within any *Hfr* population certain cells are found to be *F*⁺, meaning that they transfer *F* elements (and not main chromosomes) to *F*⁻ cells. Such *F*⁺ cells are created when integrated *F* elements undergo excision and resume their status as autonomous plasmids. On rare occasions an error occurs during excision and genes from the main bacterial chromosome become included in the *F* element, which is now called an **F' element.** The *F'* element is unlike a λd

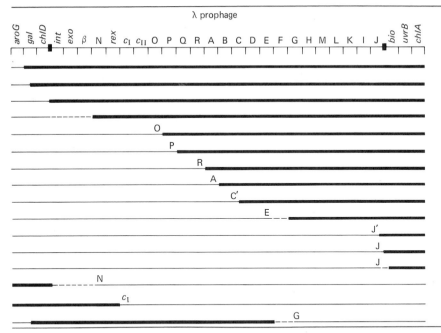

Figure 11-13 Terminal deletion mutations of prophage λ. The dark solid line represents the deleted segment, and the dashed lines indicate regions whose presence or absence cannot be determined from the genetic analyses that have been done. C′ and J′ indicate that the end points of the deletions lie between known *sus* markers of the C and J genes. (From S. Adhya, P. Cleary, and A. Campbell, *Proc. Natl. Acad. Sci. U.S.* **61**:956, 1968.)

particle in that rarely, if ever, is any of its own *F* genetic information deleted (perhaps because such deletions would invariably disallow sexual transfer). Instead, the *F* element is simply expanded in length by the number of main genes added to it. Since the *F′* element does not become encased in coat protein, this extra length does not present a packaging problem, and the element continues its autonomous existence within the bacterial cell.

The original *Hfr* cell giving rise to an *F′* element remains haploid and is usually viable; its genes are distributed somewhat differently, but all are still present in the cell in a single copy. When, however, its *F′* element is infectiously transmitted to an *F⁻* cell of differing genetic constitution, the recipient cell and its descendants become partial diploids for the bacterial genes introduced by the *F′* element. Continued cycles of replication and infection by *F′* elements can eventually convert an entire population into *F′* partial diploids. In addition, the bacterial genes in the *F′* element will not infrequently take part in exchanges with the recipient's chromosomal genes to produce true recombinants, a phenomenon commonly referred to as **sexduction** or **F-duction.**

Genetic studies of *E. coli* make frequent use of *F′* elements. The various *F′*

elements that arise independently within a single *Hfr* strain can be characterized, and the frequencies with which different genes are transmitted together can be calculated. These frequencies are similar to cotransduction frequencies and are used in constructing genetic maps in much the same way that cotransduction frequencies are used (Section 11.12).

F' elements also allow the study of gene interactions in bacteria. One can ask, for example, whether a mutant gene is dominant or recessive to its wild-type allele by bringing them together in partial diploids. One can also ask whether a mutant bacterial gene carried on the *F'* element can complement a mutation carried on the bacterial chromosome; if it can, then the two mutations are tentatively assigned to different genes. Finally, an *F'* element allows one to ask whether particular genes must be located in the same physical piece of DNA in order to be transcribed in a controlled manner, a problem considered in Chapter 17.

Since different *Hfr* strains have different integration and excision sites for the *F* element, each will give rise to *F'* elements that carry different arrays of bacterial genes. In Figure 11-14 the various arcs within the *E. coli* chromosome indicate some of the gene blocks that are included in known *F'* elements. The range in both the number and the kinds of genes that are picked up is apparent.

ISOLATION OF THE *lac* OPERON

This final section of this chapter describes a series of experiments, performed in the late 1960s by J. Beckwith and his colleagues, which led to the isolation of *lac* operon DNA which contains 3 genes (*lacZ, lacY,* and *lacA*) involved in lactose utilization by *E. coli*. These rather elaborate experiments are presented for several reasons. First, they illustrate the many uses to which carefully designed transduction experiments can be put. Second, the isolated *lacZ* gene, (with its promoter (*P*) and operator (*O*) sequences, proved useful in elaborating certain features of transcriptional control, as described in Chapter 17. Most important, perhaps, was the impact of these experiments: the gene-isolating feat brought to public attention the fact that the age of genetic engineering was upon us. As described in Chapter 14, this age has proved to be far nearer at hand than was imagined at the time the isolated *lac* operon was reported.

11.18   CONSTRUCTING *lac* TRANSDUCING PHAGES

The *lac* operon is not flanked by any attachment sites for temperate phages. To obtain *lac* genes in a transducing particle—an essential first step in the gene isolation procedure—it was therefore necessary to devise a means for

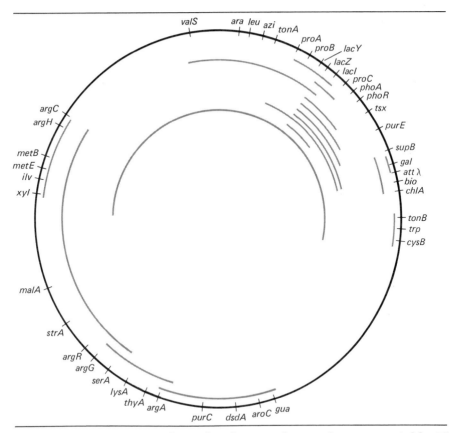

Figure 11-14 Genetic map of *E. coli* with arcs indicating the genes carried by *F'* elements in various partial diploid strains. (From A. Campbell, *Episomes*. New York: Harper and Row, 1969. Copyright © 1969 by Allan M. Campbell. Reprinted by permission of Harper & Row, Publishers, Inc.)

placing *lac* genes next to known *att* sites in the bacterial chromosome. This was accomplished by F. Cuzin, F. Jacob, J. Beckwith, and their collaborators.

The experiments began with an *Hfr* strain whose integrated *F* element was known to lie near the *lac* operon (Figure 11-15a). This strain gave rise to a clone harboring an autonomous *F'* element carrying *lac* genes (Figure 11-15b). The *F'* cells were then mutagenized and selection was made for still another clone whose *F'* element was temperature-sensitive in its DNA replication; specifically, the element could not be replicated at 45°C, although it could be replicated at 37°C. The mutant *F'* element was called $F_{ts}lac^+$ (Figure 11-15c).

As noted in Section 11.1, episomes such as *F* and *F'* are replicated independently of the main chromosome as long as they are in their autonomous state. When integrated, however, their replication proceeds concomitantly with the replication of the rest of the chromosome. Thus the

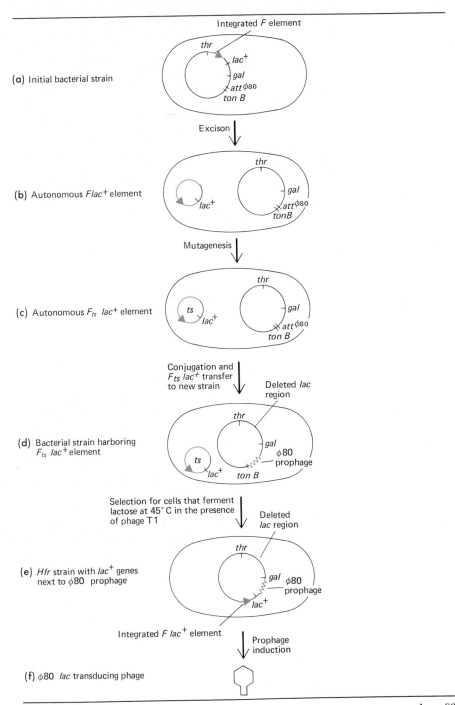

Figure 11-15 Procedure used to obtain *Hfr* cells with *lac* genes next to the *φ80* attachment site. (See text for details.) (After J. Beckwith, E. Signer, and W. Epstein, *Cold Spring Harbor Symp. Quant. Biol.* **31**:393, 1966.)

$F_{ts}lac^+$ element will not be transmitted to daughter cells at high temperatures so long as it exists in the autonomous state. If, however, it manages to undergo integration into the *Hfr* state at the time of the shift to high temperature, its transmission is ensured. In other words, the shift to high temperature becomes a strict selective agent for *Hfr* cells carrying an integrated $F_{ts}lac^+$ episome.

An *F* element may integrate at numerous sites within the bacterial chromosome, whereas an *F'* element preferentially pairs and interacts with homologous bacterial sequences in the main chromosome. To prevent this occurrence and thus maximize the probability that the $F_{ts}lac$ would integrate near a phage attachment site at the time the temperature shift occurred, the experiment was designed to include the $F_{ts}lac$ element in a lac^- strain whose *lac* operon was so extensively deleted that it shared little homology with the episome (Figure 11-15d).

A final selection procedure was used to ensure the isolation of *Hfr* strains whose $F_{ts}lac^+$ had integrated near the $\phi80$ attachment site. The $\phi80$ attachment site lies near or adjacent to the gene *tonB*, a gene that confers sensitivity to phage T1. Beckwith and colleagues reasoned that if the $F_{ts}lac^+$ episome integrated almost anywhere *within* this gene, the *tonB* sequence would be disrupted and the gene would be rendered nonfunctional. Such integrative events should produce bacteria that are both lac^+ at high temperature and *resistant* to T1 infection.

The experiments therefore involved the following steps. A culture of bacterial cells was grown at 37°C. Each cell had the constitution shown in Figure 11-15d: it was lysogenic for $\phi80$, it carried an extensively deleted *lac* operon in its main chromosome, and it possessed an autonomous $F_{ts}lac^+$ episome carrying a full complement of *lac* genes. The temperature of the experiment was then shifted to 45°C and, after several rounds of divisions, cells were plated to the special indicator medium for lactose utilization described in Section 6.2. The medium also contained T1 phages. Colonies that survived the T1 phages and also fermented lactose were then isolated (Figure 11-15e). Such colonies proved to arise very infrequently indeed, deriving from only one in every 10^9 cells of a bacterial population. They did, however, have the desired properties. When cells of the new strains were grown in large quantities and their $\phi80$ prophages were induced, *lac*-transducing particles emerged with low frequency (Figure 11-15f). Once these particles were used to transduce a second strain, HFT lysates (Section 11.11) containing a high proportion of *lac* genes could be derived.

Similar experiments permitted the isolation of *lac*-transducing particles of λ, and both $\phi80$ and λ *lac*-transducing strains were used in the experiments leading to the isolation of the *lac* genes. The strains used, it should be noted, are not defective for phage genes; each carries a complete set of phage genes plus a few bacterial genes, and each therefore has a slightly longer chromosome than a normal phage. Because such phages are plaque-forming, they are called $\phi80plac$ and $\lambda plac$.

11.19   UTILIZING *lac* TRANSDUCING PHAGES

Beckwith and colleagues focused their attention on two transducing phages, λ*plac*5 and *φ80plac*1, whose genetic constitutions are shown in Figure 11-16a. Consider first the λ*plac*5 strain. The bacterial genes are seen to be in the order *y z o p i* (where *i* is a closely linked regulatory gene) with respect to phage genes A and R; both *y* and *i* are present in deleted (incomplete) form, and gene *a* is not included at all. When the λ*plac*5 DNA is isolated and dissociated into its two strands, a heavy (*H*) strand can be separated from the complementary light (*L*) strand by centrifugation, the *H* strand having a greater proportion of G and C residues than the *L* strand (Section 8.4). When each strand is tested for its ability to form a hybrid with labeled mRNA carrying *lac* information (Section 8.4) only the *L* strand can do so. In other words, the *L* strand of the phage carries the sense nucleotide

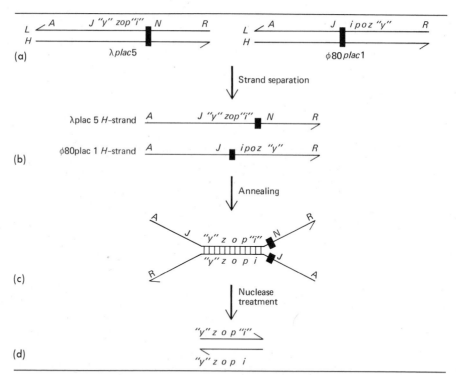

Figure 11-16 (a) Genetic constitution of transducing phages λ*plac*5 and *φ80plac*1. Genes in quotation marks have either been partially deleted or split into two parts by the integration of an F_{ts} *lac* episome. The genes A,J,N, and R are part of the phage chromosome, and the black rectangles indicate phage attachment sites. (b) Isolated heavy strands from the two chromosomes. (c) Structure formed when the two heavy strands anneal. (d) Isolated *lac* operon genes following nuclease treatment. (After J. Shapiro et al., *Nature* **224:**768, 1969.)

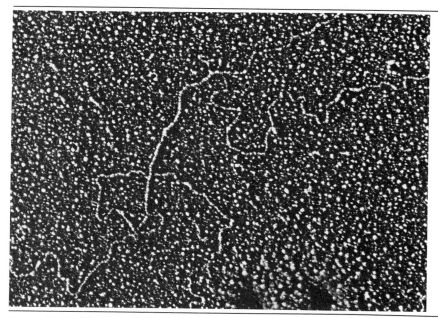

Figure 11-17 Heteroduplex formed by the noncomplementary heavy strands of phages λ*plac*5 and φ80*plac*1. The four single-stranded tails (Figure 11-16c) extend out from a thicker central duplex. (From J. Shapiro et al., *Nature* **224**:768, 1969; electron micrograph by L. MacHattie).

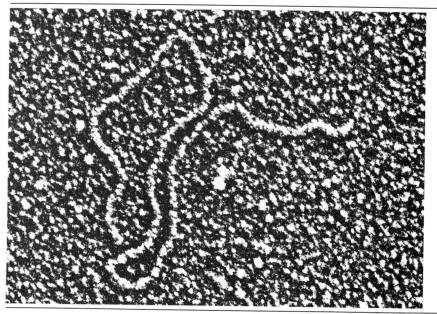

Figure 11-18 A single pure *lac* duplex. (From J. Shapiro et al., *Nature* **224**:768, 1969; electron micrograph by L. MacHattie.)

sequence of the *lac* genes and the *H* strand carries the complementary antisense sequence.

The second strain, ɸ80*plac*1, carries the reverse gene sequence, *i p o z y*, and only the *y* gene is incomplete (Figure 11-16a). Separation of this DNA into *H* and *L* strands can also be accomplished, and hybridization experiments indicate that in this phage it is the *H* strand that carries the sense *lac* sequences and the *L* strand that carries its noncoding complement. This would be expected if the phage arose in an *F* insertion event that was the inverse of that producing λ*plac*5 (Section 11.5).

It is known that ɸ80 and λ are closely related phages, having similar genetic maps and able to undergo gene recombination with one another in mixed infections. It is also known that both phages insert themselves into the bacterial chromosome during lysogeny with the same overall orientation. It is thus reasonable to assume that with respect to *phage* genes the *H* strand of the ɸ80*plac*1 chromosome should have a sequence of nucleotides very similar to the *H* strand of the λ*plac*5 chromosome, and that the two sequences should exhibit little or no complementarity. In contrast, with respect to *bacterial* genes the two *H* strands should be perfectly complementary and in the correct sense in terms of $5' \longrightarrow 3'$ polarity (Figure 11-16b). Therefore, when the two *H* strands are allowed to anneal in a hybridization experiment, a stable duplex should form *only* between the complementary *lac* sequences, with unpaired "tails" of similar phage sequences dangling from both ends of the helix, as diagrammed in Figure 11-16c. Figure 11-17 shows the structure actually observed with the electron microscope. The correspondence is strikingly clear.

The final step of Beckwith's experiment involved treating such structures with an endonuclease specific for single-stranded DNA, a procedure that should eliminate all dangling phage sequences (Figure 11-16d). One of the resulting duplexes, shown in Figure 11-18, is about 1.4 μ in length and contains a complete copy of the *lacz* plus the *p* and *o* regions of the *lac* operon. Sequence analyses of this sector of the *E. coli* genome have already been presented in Section 8.7, and control of its expression is described in Chapter 17.

APPROACHES TO SOLVING MAPPING AND COMPLEMENTATION PROBLEMS

11.20 DELETION MAPPING PROBLEM

Problem Five point mutants (*a* through *e*) were tested for wild recombinants with the seven deletion mutants drawn below. What is the order of point mutants on this deletion map (+ = recombination, 0 = no recombination)?

$$\underline{\hspace{2cm}} \quad \underline{\hspace{1cm}}$$

```
        1      2

      6        3
   _____  _____          deletion map
     5       4
   _____  _____

          7
   _____
```

deletions

	1	2	3	4	5	6	7
a	0	+	0	0	+	+	0
b	+	+	+	0	+	0	0
c	+	+	+	+	0	0	0
d	0	+	+	0	+	0	0
e	+	0	0	0	+	+	0

data — point mutants

Solution

a. Recognize anomalies. In this case #7 is probably an extensive deletion covering a whole region and is not useful.

b. Start with small deletion #5 and note that only c is unable to recombine with it: c . . . . . .

c. Similarly with #2, only e is unable to recombine with it: . . . . . e.

d. Deletion #6 covers b and d as well as c, whereas deletion #1 covers d. The map order in this region is therefore c b d . . . . .

e. Deletion #3 covers both a and e, while deletion #1 covers a and d. The map order here is therefore . . . d a e

f. The complete order is c b d a e.

11.21 TRANSDUCTION PROBLEM

Problem In a P1 transduction experiment, the donor bacteria were $synP^+$ $supM^+$ $trpZ^+$; the recipients were $synP^-$ $supM^-$ $trpZ^-$. Initial selection was for $supM^+$ transductants.

$$48 \text{ of these were } M^+ \, P^+ \, Z^+ \ (1)$$
$$120 \text{ of these were } M^+ \, P^+ \, Z^- \ (2)$$
$$500 \text{ of these were } M^+ \, P^- \, Z^- \ (3)$$
$$0 \text{ of these were } M^+ \, P^- \, Z^+ \ (4)$$

a. What is the marker order?

b. What is the cotransduction frequency between M and P? M and Z?

Solution

a. Recognize the rarest class. ($\#4$). The P^- recipient marker is in the middle in the order written; the marker order is therefore $M\ P\ Z$.

b. Grand total $= 48 + 120 + 500 = 668$ transductants.

c. Cotransduction of M and $P = \dfrac{48 + 120}{668} = 0.25$

d. Cotransduction of M and $Z = \dfrac{48}{668} = 0.072$

References
General

Brink, R. A., Ed. *Heritage from Mendel.* Madison, Wisc.: University of Wisconsin Press, 1967.

Campbell, A. *Episomes.* New York: Harper and Row, 1969.

Gunsalus, I. C., and R. Y. Stanier, Eds. *The Bacteria* (Vol. 5): *Heredity.* New York: Academic, 1964.

Hayes, W. *The Genetics of Bacteria and Their Viruses,* 2nd ed. New York: Wiley, 1968.

*Jacob, F., and E. L. Wollman. *Sexuality and the Genetics of Bacteria,* New York: Academic, 1961.

Susman, M. "General bacterial genetics," *Ann. Rev. Genetics* **4:**135–176 (1970).

Conjugation

Achtman, M. "Genetics of the F sex factor in *Enterobacteriaceae,*" *Curr. Top. Microbiol. Immunol.* **60:**79–123 (1973).

Adelberg, E. A., and S. N. Burns. "Genetic variation in the sex factor of *Escherichia coli,*" *J. Bacteriol.* **79:**321–330 (1960).

*Bachmann, B. J., K. B. Low, and A. L. Taylor. "Recalibrated linkage map of *Escherichia coli* K-12," *Bact. Rev.* **40:**116–167 (1976).

Curtiss, R. "Bacterial conjugation," *Ann. Rev. Microbiol.* **23:**69–136 (1969).

Curtiss, R., L. J. Charamella, D. R. Stallions, and J. A. Mays. "Parental functions during conjugation in *Escherichia coli* K-12," *Bact. Rev.* **32:**320–348 (1968).

Freifelder, D. "Studies on *Escherichia coli* sex factors. IV. Molecular weights of the DNA of several F' elements," *J. Mol. Biol.* **35:**95–102 (1968).

Gross, J. D., and L. G. Caro. "DNA transfer in bacterial conjugation," *J. Mol. Biol.* **16:**269–284 (1966).

Helmuth, R., and M. Achtman. "Operon structure of DNA transfer cistrons on the F sex factor," *Nature* **257:**652–656 (1975).

*Jacob, F., and E. Wollman. "Genetic and physical determinations of chromosomal segments in *Escherichia coli,*" *Symp. Soc. Exp. Biol.* **12:**75–92 (1958). [Reprinted in E. A. Adelberg, *Papers on Bacterial Genetics,* 2nd ed., Boston: Little, Brown, 1966.]

*Ohki, M., and J. Tomizawa. "Asymmetric transfer of DNA strands in bacterial conjugation," *Cold Spring Harbor Symp. Quant. Biol.* **33:**651–658 (1968).

*Low, B. "Rapid mapping of conditional and auxotrophic mutations in *Escherichia coli* K-12," *J. Bact.* **113:**798–812 (1973).

Skurray, R. A., H. Nagaishi, and A. J. Clark. "Molecular cloning of DNA from F sex factor of *Escherichia coli* K-12," *Proc. Nat. Acad. Sci. U.S.* **73:**64–68 (1976).

*Vapnek, D., and W. D. Rupp. "Asymmetric segregation of the complementary sex-factor DNA strands during conjugation in *Escherichia coli,*" *J. Mol Biol.* **53:**287–303 (1970).

*Denotes articles described specifically in the chapter.

Vapnek, D., and W. D. Rupp. "Identification of individual sex factor DNA strands and their replication during conjugation in thermosensitive DNA mutants of *Escherichia coli*," *J. Mol. Biol.* **60:**413–424 (1971).

Vielmetter, W., F. Bonhoeffer, and A. Schütte. "Genetic evidence for transfer of a single DNA strand during bacterial conjugation," *J. Mol. Biol.* **37:**81–86 (1968).

*Wollman, E. L., F. Jacob, and W. Hayes, "Conjugation and genetic recombination in *Escherichia coli* K-12," *Cold Spring Harbor Symp. Quant. Biol.* **21:**141–162 (1962). [Reprinted in E. A. Adelberg, *Papers on Bacterial Genetics,* 2nd ed. Boston: Little, Brown, 1966.]

Transformation

Archer, L. J. *Bacterial Transformation.* New York: Academic Press, 1973.

*Dubnau, D., D. Goldthwaite, I. Smith, and J. Marmer. "Genetic mapping in *Bacillus subtilis*," *J. Mol. Biol.* **27:**163–185 (1967).

*Ephrussi-Taylor, H. E. "Genetic aspects of transformations of pneumococci," *Cold Spring Harbor Symp. Quant. Biol.* **16:**445–455 (1951).

*Goodgal, S. H. "Studies on transformation of *Hemophilus influenzae*. IV. Linked and unlinked transformations," *J. Gen. Physiol.* **45:**205–228 (1961).

Gray, T. C., and H. Ephrussi-Taylor. "Genetic recombination in DNA-induced transformation of pneumococcus. V. The absence of interference, and evidence for the selective elimination of certain donor sites from the final recombinants," *Genetics* **57:**125–153 (1967).

Hotchkiss, R. D., and M. Gabor. "Bacterial transformation, with special reference to recombination processes," *Ann. Rev. Genetics* **4:**193–224 (1970).

*Hotchkiss, R. D., and J. Marmur. "Double marker transformations as evidence of linked factors in deoxyribonucleate transforming agents," *Proc. Natl. Acad. Sci. U.S.* **40:**55–60 (1954).

*Lacks, S., B. Greenberg, and M. Neuberger. "Role of a deoxyribonuclease in the genetic transformation of *Diplococcus pneumoniae*," *Proc. Nat. Acad. Sci. U.S.* **71:**2305–2309 (1974).

*Oishi, M., and S. D. Cosloy. "Specialized transformation in *Escherichia coli* K-12," *Nature* **248:**112–116 (1974).

Ravin, A. W. "The genetics of transformation," *Adv. Genetics* **10:**61–163 (1961).

Rosenthal, P. N., and M. S. Fox. "Effects of disintegration of incorporated ^{3}H and ^{32}P on the physical and biological properties of DNA," *J. Mol. Biol.* **54:**441–463 (1970).

Taylor, H. E. "Transformations récipriques des formes R et ER chez le pneumocoque," *Comptes Rend. Acad. Sci., Paris* **228:**1258–1259 (1949).

Tomaz, A. "Some aspects of the competent state in genetic transformation," *Ann. Rev. Genetics* **3:**217–232 (1969).

*Yoshikawa, H., and N. Sueoka. "Sequential replication of *Bacillus subtilis* chromosome, I. Comparison of marker frequencies in exponential and stationary growth phases," *Proc. Natl. Acad. Sci. U.S.* **49:**559–566 (1963).

Yoshikawa, H., and N. Sueoka. "Sequential replication of the *Bacillus subtilis* chromosome, II. Isotopic transfer experiments," *Proc. Natl. Acad. Sci. U.S.* **49:**806–813 (1963).

Transduction

*Campbell, A. M. "Episomes," *Adv. Genetics* **11:**101–145 (1962).

*Campbell, A. M. "Genetic recombination between λ prophage and irradiated λ*dg* phage," *Virology* **23:**234–251 (1964). [Reprinted in G. S. Stent, *Papers on Bacterial Viruses,* 2nd ed. Boston: Little, Brown, 1966.]

Garen, A., and S. Garen. "Complementation *in vivo* between structural mutants of alkaline phosphatase from *E. coli*," *J. Mol. Biol.* **7:**13–22 (1963). [Reprinted in G. L. Zubay, *Papers in Biochemical Genetics.* New York: Holt, Rinehart and Winston, 1968.]

Gottesman, M. M. "Isolation and characterization of a λ specialized transducing phage for the *Escherichia coli* DNA ligase gene," *Virology* **72:**33–44 (1976).

Ikeda, H., and J. Tomizawa. "Transducing fragments in generalized transduction by phage P1. III. Studies with small phage particles," *J. Mol. Biol.* **14:**120–129 (1965).

Jacob, F. "Transduction of lysogeny in *Escherichia coli*," *Virology* **1**:207–220 (1955). [Reprinted in G. S. Stent, *Papers on Bacterial Viruses,* 2nd ed. Boston: Little, Brown, 1966.]

*Jacob, F., and E. A. Adelberg. "Transfer of genetic characters by incorporation in the sex factor of *Escherichia coli*," *Comptes Rend. Acad. Sci., Paris* **249**:181–191 (1959). [Reprinted in E. A. Adelberg, *Papers on Bacterial Genetics,* 2nd ed. Boston: Little, Brown, 1966.]

Lennox, E. "Transduction of linked characters of the host by bacteriophage P1," *Virology* **1**:190–206 (1955).

*Morse, M. L., E. M. Lederberg, and J. Lederberg. "Transduction in *Escherichia coli* K-12" *Genetics* **41**:142–156 (1956). [Reprinted in E. A. Adelberg, *Papers on Bacterial Genetics,* 2nd ed. Boston: Little, Brown, 1966.]

Norkin, L. C. "Marker-specific effects in genetic recombination," *J. Mol. Biol.* **51**:633–655 (1970).

Ozeki, H., and H. Ikeda. "Transduction mechanisms," *Ann. Rev. Genetics* **2**:245–278 (1968).

*Signer, E. R., J. R. Beckwith, and S. Brenner. "Mapping of supressor loci in *Escherichia coli*," *J. Mol. Biol.* **14**:153–166 (1965).

Wollman, E. L., and F. Jacob. "Lysogeny and genetic recombination in *Escherichia coli* K-12," *Comptes Rend. Acad. Sci., Paris* **239**:455–456 (1954). [Reprinted in G. S. Stent, *Papers on Bacterial Viruses,* 2nd ed. Boston: Little, Brown, 1966.]

*Wu, T. T. "A model for three-point analysis of random general transduction," *Genetics* **54**:405–410 (1966).

Zinder, N. D. "Infective heredity in bacteria," *Cold Spring Harbor Symp. Quant. Biol.* **18**:261–269 (1953). [Reprinted in L. Levine, *Papers on Genetics.* St. Louis: Mosby, 1971.]

*Zinder, N. D., and J. Lederberg. "Genetic exchange in *Salmonella*," *J. Bacteriol.* **64**:679–699 (1952). [Reprinted in J. A. Peters, Ed. *Classical Papers in Genetics.* Englewood Cliffs, N.J.: Prentice-Hall, 1959.]

Gene Isolation

*Beckwith, J. R., E. R. Signer, and W. Epstein. "Transposition of the *lac* region of *E. coli*," *Cold Spring Harbor Symp. Quant. Biol.* **31**:393–401 (1966).

*Shapiro, J., L. MacHattie, L. Eron, G. Ihler, K. Ippen, and J. Beckwith. "Isolation of pure *lac* operon DNA," *Nature* **224**:768–774 (1969).

(a)   (b)

(a) Harriet Ephrussi-Taylor performed some of the early bacterial transformation experiments indicating linkage between genetic markers. (b) Jonathan Beckwith (Harvard Medical School) has utilized transduction and sexduction to elucidate many features of *E. coli* genetics.

Questions and Problems

1. Given the map of the 6 deletion mutants shown below, predict the results of crosses involving the 5 point mutants (a through e) with the 6 deletions (1 through 6). (Indicate + for recombination, 0, no recombination, in the table.)

(deletions)

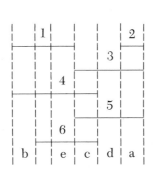

(point)	1	2	3	4	5	6
a						
b						
c						
d						
e						

2. A **complementation map** displays complementation relationships in the same fashion as a deletion map. Construct a complementation map from the following data derived from pairwise mixed infections of phage strains 1–5. Some or all of the strains may carry deletions. + for complementation, − for no complementation.

	1	2	3	4	5
1	0	0	0	+	+
2		0	0	0	0
3			0	+	0
4				0	0
5					0

3. The following groups of genes can be cotransduced by phage P1.
   1. A, H, and L
   2. H and S
   3. L and O
   4. M and O
   What is the order of the markers? Explain your logic.

4. Two labs calculated cotransduction frequencies for the genes *A* and *B*. Lab A found 0.63, the lab B found 0.47. Which lab reported the genes to be closer together?

5. In a generalized transduction experiment, donor *E. coli* cells have the genotype *trpC⁺ pyrF⁻ trpA⁻* and recipient cells have the genotype *trpC⁻ pyrF⁺ trpA⁺*. P1-mediated transductants for *trpC⁺* were selected and their total genotypes were determined with the following results:

Genotype			Number of Progeny
$trpC^+$	$pyrF^-$	$trpA^-$	274
$trpC^+$	$pyrF^+$	$trpA^-$	279
$trpC^+$	$pyrF^-$	$trpA^+$	2
$trpC^+$	$pyrF^+$	$trpA^+$	46

(a) Determine the order of the three markers. (b) Determine the cotransduction frequencies for $trpC$ and $pyrF$ and for $trpC$ and $trpA$. (*Note:* All cases in which the two markers are cotransduced must be scored.) (c) Calculate physical map distances between these markers assuming the P1 chromosome to be 10μ in length. (d) Calculate similarly the map distances between the $supC$, $trpA$, and $pyrF$ markers from the data in the text and construct a map for the region.

6. Four sus mutant strains of phage λ give the following results when crossed with the λ deletion strains (Figure 11-13). (0 = no recombination; $+$ = recombination).

		Deletion Strain							
		C_1	G	O	P	R	A	C'	E
sus	sus_1	+	0	+	+	+	+	+	+
strain	sus_2	+	0	0	0	0	0	0	+
	sus_3	+	+	0	0	0	0	0	0
	sus_4	+	0	0	0	0	+	+	+

(a) In which gene or genes (give all possibilities) does each sus mutation lie?
(b) How would you go about giving specific gene and map assignments to each sus mutant?

7. The gene order of closely linked *E. coli* genes can be determined from three-factor transduction experiments. Consider the cross $A^-B^+C^+$ (transducing chromosome) $\times$ $A^+B^-C^-$ (recipient chromosome) and its reciprocal cross $A^+B^-C^- \times A^-B^+C^+$. What is the possible gene order(s) if you find that the number of wild type recombinants $A^+B^+C^+$ are about the same in both crosses. Explain your answer.

8. Two bacterial strains were obtained with the following genotypes: Hfr, arg^-, leu^+, azi-s, str-r and F^- arg^+ leu^- azi-r str-s. You want to mate the two strains and after conjugation, detect and enrich for the F^- recombinant genotype arg^+ leu^+ azi-r. Which of the following media will accomplish this selection? Explain why the remaining media will not work.
(a) Minimal media with streptomycin.
(b) Minimal media with leucine and sodium azide.
(c) Minimal media with sodium azide.
(d) Enriched media without arginine or leucine, with streptomycin.
(e) Minimal media with streptomycin and sodium azide.

9. Suppose a strain of *E. coli* is found which only grows in minimal media when supplemented with methionine. When sexduced with an F′ element which is known to carry *ara+ leu+ azi-r* markers, the original strain can grow on unsupplemented media.
   Explain these results. Outline how you would test whether this explanation is correct.

10. Construct a complementation map from the following data derived from pairwise mixed infections of phage strains 1–5. Some or all of the strains may carry deletions. + for complementation, − for no complementation.

	1	2	3	4	5
1	0	+	+	0	0
2		0	0	0	+
3			0	+	0
4				0	+
5					0

11. How is the transfer of *F* similar to phage infection? How is it different?

12. A new mutation is introduced into an *Hfr* strain so that its general position can be determined by interrupted mating. Outline how you would carry out these steps.

13. The order of markers transferred by four *Hfr* strains to F^- cells is as follows:

Hfr Strain	Order of Markers Donated
1	Q W D M T $\longrightarrow$
2	A X P T M $\longrightarrow$
3	B N C A X $\longrightarrow$
4	B Q W D M $\longrightarrow$

Draw a map as in Figure 11-6 showing the sites and orientation of insertion of the *F* element in these four strains.

14. An *Hfr* strain of *E. coli* (A) transfers the *gal* operon early and with high efficiency and the *lac* operon late and with low efficiency to a *gal⁻ lac⁻ F⁻* strain (B) in an A × B cross. *Gal+* recombinants of strain B remain F^-. It is possible to isolate from strain A a variant called C that transfers the *lac* operon but not the *gal* operon early and with high frequency to strain B. In a C × B cross, *lac+* cells of strain B are generally F^+. What is the nature of strain C? How would you design experiments to isolate such a strain?

15. Compare the partial zygotes that result from *Hfr* $\longrightarrow$ F^- conjugation with the partial diploids that result from F' $\longrightarrow$ F^- conjugation.

16. An F' element including the maltose (*mal*) gene is discovered in *E. coli*. The episome marked with *mal+* is introduced into a *mal⁻ F⁻* strain. Most of the resulting cells transfer the original episome to other F^- cells. However,

occasional cells appear which transfer the entire *E. coli* chromosome, beginning with the maltose gene. These cells fall into two types: (a) those which transfer *mal*⁺ very early and *mal*⁻ very late and (b) those which transfer *mal*⁻ very early and *mal*⁺ very late.

Draw a diagram of the interaction between the episome and the chromosome which shows how cells of type (a) are most likely to have arisen. Do the same for the case of cells of type (b).

17. Genes *purB* and *purC* of *B. subtilis* are shown to be linked in transformation experiments. Outline three experimental approaches that could be used to demonstrate whether or not these genes are linked to the marker *lin* (*lincomycin-resistant*, lincomycin being an antibiotic).

18. Phage 105 is a lysogenic phage of *Bacillus subtilis*. It has recently been discovered that the gene order of the prophage map is identical to the gene order of the linear DNA in a phage 105 particle.

   (a) What does this finding imply about the location of the termini of the mature phage DNA and the prophage attachment site?

   (b) Why has it not been possible to isolate specialized transducing phage of phage 105?

CHAPTER 12

Mapping Eukaryotic Chromosomes

INTRODUCTION

This chapter parallels the two preceding chapters on mapping viruses and bacteria, but we now focus on nucleate organisms whose chromosomes undergo meiotic segregation and independent assortment (Chapters 3 and 4) as well as crossing over. We first consider "classical" early experiments demonstrating linkage in eukaryotes and correlating the occurrence of genetic recombination with the occurrence of physical exchanges between chromosomes. We next examine how genetic maps are generated in organisms with a sexual cycle, focusing attention on *Drosophila* and *Chlamydomonas*. Finally we examine mapping procedures utilizing "parasexual" organisms and cells, and conclude with a description of how human chromosomes are mapped *in vitro*.

CLASSICAL STUDIES ON LINKAGE AND RECOMBINATION

12.1   PARTIAL LINKAGE

The studies of Mendel (Box 3.1) on the garden pea, published in 1866, had little impact on the scientific world until 1900 when C. Correns and H. de Vries independently rediscovered his paper in the course of their own investigations on quantitative aspects of gene transmission. At about the same time investigators such as T. Boveri, C. Correns, and particularly W. Sutton were investigating the behavior of chromosomes at meiosis. It was Sutton in 1903 who clearly postulated a relationship between chromosomal behavior at meiosis and the segregation and independent assortment of genes, a relationship that forms the basis of the chromosome theory of inheritance (see Section 4.15).

Sutton and others realized that one gene could not possibly correspond to a whole chromosome: organisms clearly must possess more genes than chromosomes. Thus it was proposed that each chromosome carries many linked genes. Linkage was soon described experimentally, but it was found to be "partial" or "incomplete" linkage. For example, in experiments published in 1905 by W. Bateson, E. Saunders, and R. Punnett (and later expanded by Punnett in 1917) sweet peas with *purple flowers* and *long* pollen grains were crossed to *red* flowered plants with *round* pollen grains. Nothing was unusual about the F_1 progeny: all were *purple* and *long*, showing these to be the dominant traits. When the F_1 was inbred and each pair of alleles was examined separately, moreover, each proved to behave as expected of segregating Mendelian genes: *purple* and *red* flowers were present in a 3:1 ratio, as were the *long* and *round* traits.

It was when the traits were considered together that **partial linkage** was seen. Among 6952 F_2 plants 4831 were *purple, long,* 390 were *purple, round,*

393 were *red, long*, and 1338 were *red, round*. In other words, the parental types (*red, round* and *purple, long*) were present in excess of what would be predicted of independently assorting genes. Had the traits been located on independent chromosomes, perhaps 3915 *purple, long*, 1305 *purple, round*, 1305 *red, long*, and 435 *red, round* plants would have been expected in the F$_2$ in accordance with a 9:3:3:1 ratio (Section 4.8). The two traits did not, on the other hand, show complete linkage; had they done so, a 3:1 ratio, or 5214 *purple, long* and 1738 *red, round* progeny would have been expected (Section 4.2).

A number of theories were offered to explain partial linkage, but the correct interpretation was made in 1911 by T. H. Morgan based on analyses of crosses with *D. melanogaster.*

12.2 THE MORGAN CROSSES AND THE STURTEVANT MAP

Morgan began with the isolation of two spontaneously mutant male flies, one having white eyes and the other miniature wings. When these flies were crossed with normal females and the F$_1$ progeny were inbred, the mutant traits exhibited the sex-linked pattern described in Section 4.12. Morgan therefore postulated that the recessive genes *w* (*white*) and *m* (*miniature*) were both located in the X chromosome. By appropriate crosses he constructed a stock of female flies homozygous for both *w* and *m* and then set about to demonstrate that exchange could occur between these linked genes.

The results of his experiments are shown in Figure 12-1. The homozygous *white miniature* females were first crossed to wild-type males. In the F$_1$ the males were all *white miniature* and the females were wild in phenotype (and heterozygous in genotype), as is expected for sex-linked traits. When the F$_1$ were inbred, the original parental combinations (*white miniature* and wild type) were the most prominent phenotypes in the F$_2$. As we saw in the sweet pea experiments, **an excess of parental phenotypes in the F$_2$ strongly suggests the presence of linked genes.** The linkage was a relatively loose one in this case: 900/2441 or 36.9 percent of the progeny were recombinant (either *white* with wild-type wing or *miniature* with wild-type eye), as shown in Figure 12-1.

Morgan extended his studies to a third sex-linked recessive marker, *y*, which confers a *yellow* body instead of the normal gray. In crosses (similar to those of Figure 12-1) between *yellow*-bodied and *white*-eyed flies, 1.3 percent of the progeny were recombinant.

Morgan interpreted these experiments in the context of a study of salamander meiosis made by F. Janssens in 1909: Janssens had described chiasmata (Figure 3-4) and had suggested these might represent sites of physical exchange beween maternal and paternal homologues. Morgan therefore proposed that partial linkage was observed whenever two markers

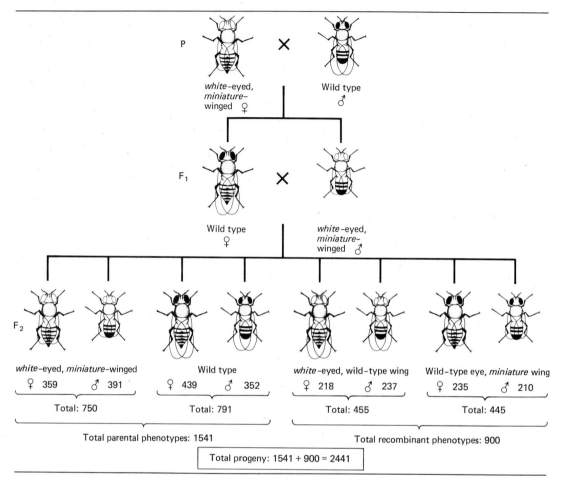

Figure 12-1 Morgan's crosses involving the markers *white* eye and *miniature* wing to demonstrate linkage and recombination in the X chromosome of *D. melanogaster.*

in the same chromosome were physically separated from one another by a chiasma (he coined the term **crossing over** for this process). To explain why 36.9 percent of the progeny were recombinant for *w* and *m* and only 1.3 percent were recombinant for *w* and *y*, Morgan proposed that *w* and *m* were relatively far apart on the X chromosome and therefore had a relatively good chance of being separated by crossing over, whereas *w* and *y* were close together and recombined quite infrequently.

It remained for A. Sturtevant, in 1911, to realize that the recombination frequencies from such two-factor crosses could be used to generate a chromosome map. Two more sex-linked markers, *v* (*vermillion eyes*) and *r* (*rudimentary wings*), were by this time known, and analysis of additional crosses had yielded slightly different recombination frequencies from those reported by Morgan. The five markers were therefore ordered and spaced as shown below, with *y* assigned an arbitrary value of 0 and the other map units expressing distances from *y*.

0	1.0		30.7	33.7		57.6
y	w		v	m		r

Such a use of recombination frequencies is, of course, identical to their more recent use in mapping viral and bacterial chromosomes, as described in detail in Chapters 10 and 11. Thus an idea conceived for *D. melanogaster* has since been found to be applicable to all chromosomes that undergo recombination.

12.3 CORRELATING CHIASMATA WITH GENE RECOMBINATION

Janssen's hypothesis that chiasmata lead to physical exchange and Morgan's hypothesis that physical exchange leads to gene recombination are both almost universally accepted. Experimental support for these hypotheses is in fact somewhat indirect, for it is based on a series of correlates rather than on direct evidence. Specifically, it rests on two observations: (1) the occurrence of chiasmata can be correlated with the physical exchange of chromosomal material; (2) the physical exchange of chromosomal material can be independently correlated with the occurrence of gene recombination. Therefore, it is argued that chiasmata must be correlated with gene recombination.

Correlation Between Physical Exchange and Chiasmata The correlation between physical exchange of chromosomal material and the occurrence of chiasmata has perhaps been demonstrated most clearly in the plant *Disporum sessile* by H. Kayano. The normal karyotype of *Disporum* is shown in Figure 12-2a, with homologues arranged above and below one another. Focusing on the A_1—A_2 pair of homologues, we see that A_1 carries an additional "tail" of chromatin, called a satellite, so that A_1 can be distinguished from A_2. The two are therefore called **heteromorphic homologues.** The homologues labeled *b* are, in this normal strain, identical in morphology. Figure 12-2b shows the chromosome complement of a mutant *Disporum* strain that has undergone a reciprocal translocation (Section 6.16) between A_2 and one of the *b* chromosomes: a small piece of material has broken off the *b* and is attached to the A_2, whereas a large piece has broken from A_2 and is attached to the *b*. The two translocation chromosomes so formed are called, respectively, *a* and B_2. Thus the karyotype of the mutant *Disporum* is particularly useful for study in that it carries chromosomes A_1, *a*, B_2, and *b*, all of which can be distinguished from one another.

Synapsis involving these four chromosomes produces the expected cross-shaped configuration (Figure 6-20) shown in Figure 12-2c. If no physical exchanges occur between the chromatids, then, when the homol-

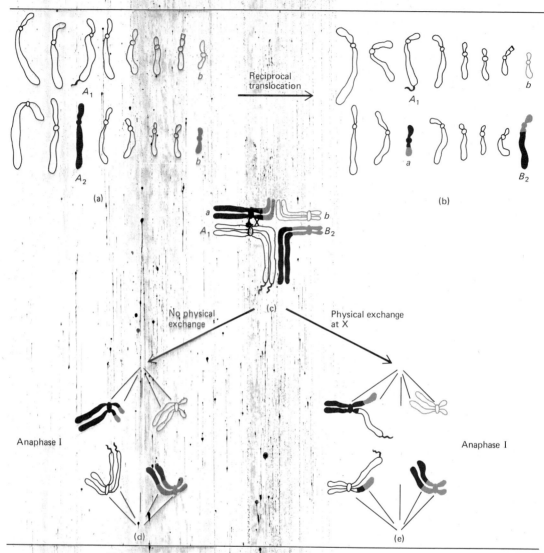

Figure 12-2 The effect of physical exchange in a translocation strain of *Disporum sessile* on the morphology of anaphase I chromosomes. See text for details. (After H. Kayano, *Cytologia* **25**:461–467, 1960.)

ogous centromeres separate at anaphase I, each centromere will be expected to carry a symmetrical pair of chromatids, as shown in Figure 12-2d. If, on the other hand, a physical exchange should occur within the lengths between the centromeres and the vertical axis of the cross (the so-called **interstitial regions**), then at anaphase I centromeres will carry chromatids of unequal length, one of the four arms being conspicuously longer than its counterpart. This is shown in Figure 12-2e.

The experiment therefore involves two sets of observations. Mutant

Disporum cells are first examined in meiotic prophase I, and the number of chiasmata within one or both of the interstitial regions is counted. The mutant cells in meiotic anaphase I (or II) are then examined, and the number of structures whose chromatid arms are of unequal length is counted. For *Disporum* both numbers prove to be approximately 30 percent; that is, 30 percent of the synapsing chromosomes have at least one interstitial chiasma and 30 percent of the separating translocation chromosomes have unequal arms. Similarly close values have been obtained in several other cases. Such results do not, of course, *prove* that chiasmata lead to physical exchanges, but they at least offer good support for the hypothesis.

Correlation Between Physical Exchange and Recombination Experiments that relate the occurrence of physical exchange between chromosomes to the occurrence of genetic recombination are best exemplified by C. Stern's experiments with *Drosophila*, published in 1931 and diagrammed in Figure 12-3. Stern first interbred various strains of flies that carried chromosomal abnormalities and mutant genes. He eventually obtained a stock of female flies that carried heteromorphic X chromosomes. One of these X's carried a translocated segment of a Y chromosome; the other X was broken into two segments, one segment remaining attached to its proper X centromere and the other being translocated to the tiny chromosome IV. The homologues, moreover, differed genetically at two different loci. The broken X carried mutations in two genes, called *car* and *B*, which affect the color (*carnation*) and shape (*Bar*) of the eye. The X with the attached Y segment carried normal alleles of each of these genes (both designated as + in Figure 12-3). The *B* gene is dominant to its + allele whereas *car* is recessive.

Such females were crossed with males whose X chromosome (recall that males have only one X, whereas females have two) carried *car* and the + allele of *B* (Figure 12-3). Female offspring of the cross, who would have acquired one X from their mother and one (always the recessive + car X chromosome) from their father, were then examined.

Four different types of female fly were found among the offspring (Figure 12-3): those with normal eyes, those with both *Bar*-shaped and *carnation*-colored eyes, those with *Bar*-shaped eyes, and those with *carnation*-colored eyes. The genotypes of these females can at once be recognized as $\frac{+\ +}{car\ +}$,

$\frac{car\ B}{car\ +}$, $\frac{+\ B}{car\ +}$, and $\frac{car\ +}{car\ +}$, as drawn in Figure 12-3. When the chromosome morphologies of these flies were examined, it was found that the normal flies had inherited an intact X chromosome with an attached Y segment from their mothers and that the doubly mutant flies had inherited the broken X. These two classes of progeny can therefore be said to have inherited unaltered versions of the maternal X complement. The two singly mutant flies, in contrast, each carried an X chromosome that was an altered

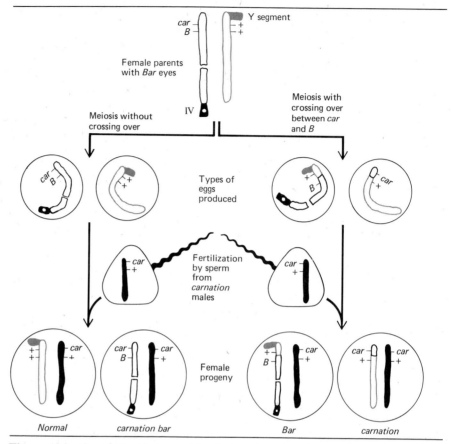

Figure 12-3 Correlation between physical exchange and recombinant progeny in *Drosophila*, as demonstrated in Stern's experiment with heteromorphic homologues.

version of the maternal X complement. Thus the flies with *Bar*-shaped but normal-colored eyes exhibited a broken X, one segment of which carried the Y translocation, whereas the *carnation*-eyed flies carried a normal-looking X (plus, in both cases, the normal X inherited from their fathers). As drawn in Figure 12-3, both new chromosome shapes can be generated by postulating that a reciprocal physical exchange of segments occurs between the two heteromorphic X's during meiosis. In other words, the altered chromosomes appear to represent recombinant chromosomes. Since the flies with normal eye color and shape and the flies with *carnation* and *Bar* eyes have parental genotypes, whereas those exhibiting either the *Bar* or the *carnation* trait alone have recombinant genotypes, this experiment clearly demonstrates a correlation between the inheritance of recombinant chromosomes and the inheritance of recombinant sets of genes.

MAPPING *DROSOPHILA* IN SEXUAL CROSSES

The sections that follow describe the steps taken in constructing and analyzing genetic crosses in *Drosophila*. Exactly the same principles hold for mapping all other sexually reproducing diploid eukaryotes; the problem section at the end of this chapter includes examples from other organisms.

12.4 TWO-FACTOR CROSSES

You will recall from Section 10.6 that the basic design of a two-factor cross in phages is to mixedly infect a bacterium with doubly marked phages ($+ +$ and $a\, b$, for example) and to look for recombinant phages ($+ b$ and $a +$) among the progeny. Exactly the same strategy holds in a eukaryotic two-factor cross, but the existence of diploidy and dominance in eukaryotes necessitate a few additional considerations.

The first step in a two-factor cross in *Drosophila* is to create flies with the general genotype $\dfrac{+\ +}{a\ b}$, heterozygous at the two loci of interest, where the continuous line connecting the markers denotes their linkage. This is usually accomplished by taking true-breeding $\dfrac{a\, b}{a\, b}$ flies and crossing them to true-breeding $\dfrac{+\ +}{+\ +}$ flies, thereby creating $\dfrac{+\ +}{a\ b}$ heterozygotes in the F_1 (or in the F_1 females in the case of sex-linked genes).

The next step is to allow these double heterozygotes to undergo gametogenic meiosis. Gametes produced without recombination will be either $+ +$ or $a\, b$, whereas gametes that are generated by crossing over will be either $+ b$ or $a +$.

The final step in the experiment is to devise some way of telling these gametes apart so that recombination frequencies can be deduced. This is usually accomplished by mating the double heterozygotes to homozygous recessive $\left(\dfrac{a\, b}{a\, b}\right)$ or, in the case of sex-linked genes, to hemizygous recessive $\left(\dfrac{a\, b}{=}\right)$ flies. Such flies produce either $a\, b$ or "silent" $\longrightarrow$ gametes, neither of which will obscure, via dominance, the genetic makeup of the gametes produced by the double heterozygote. In other words, going back to the terminology of Section 4.2, the double heterozygote is usually subjected to a test cross.

The test cross can be performed in one of two ways. In the 1911 crosses of Morgan described above, $\dfrac{w\, m}{w\, m}$ flies were crossed to $\dfrac{+\ +}{=}$ males to generate $\dfrac{w\, m}{+\ +}$ female heterozygotes in the F_1. In this case, the F_1 males were all $\dfrac{w\ m}{}$ hemizygotes and therefore suitable test-cross organisms, so that the F_1 was simply inbred and recombination frequencies were directly deduced from the F_2 phenotypes (Figure 12-1).

In other cases, the F_1 females must be taken from their vials while they are still virgin, to prevent their mating with their brothers, and allowed to mate with test-cross males. This can be illustrated by a cross designed to determine the recombination frequency between two autosomal markers, g (*glass* eye) and e (*ebony* body color). A *glass ebony* × *wild type* cross yields $\dfrac{+\ +}{gl\ e}$ F_1 heterozygotes in both females and males. Since the F_1 males have the potential to produce + + sperm, they are unsuitable for test crossing. The F_1 female virgins must therefore be collected and crossed to $\dfrac{gl\ e}{gl\ e}$ males, which yields the results shown in Table 12-1.

The crosses we have outlined utilized markers in coupling $\left(\dfrac{+\ +}{a\ b}\right)$, but the same results should occur if the markers are introduced in repulsion $\left(\dfrac{+\ b}{a\ +}\right)$. Thus the cross *glass* × *ebony* $\left(\dfrac{gl\ +}{gl\ +} \times \dfrac{+\ e}{+\ e}\right)$ will yield wild-type F_1 heterozygotes with a $\dfrac{gl\ +}{+\ e}$ genotype, and test-crosses of such females to $\dfrac{gl\ e}{gl\ e}$ males should again reveal that *gl* and *e* remain linked approximately 92 percent of the time and are recombined approximately 8 percent of the time.

It should be noted that female heterozygotes are used in all these analyses because, at least under laboratory conditions, crossing over does not occur in male *Drosophila*.

Table 12-1   Linkage and Recombination Between *gl* and *e* in *Drosophila*

$$\female \quad \frac{gl\ e}{+\ +} \times \frac{gl\ e}{gl\ e} \quad \male$$

	Number	Percent
$\dfrac{gl\ e}{gl\ e}$	1031	43.4
$\dfrac{+\ +}{gl\ e}$	1159	48.6
		92.0
$\dfrac{gl\ +}{gl\ e}$	92	3.8
$\dfrac{+\ e}{gl\ e}$	99	4.2
		8.0

12.5 THREE-FACTOR CROSSES

As in bacteriophages, two-factor crosses, if performed in enough combinations, can generate both the order and the spacing of markers on a chromosome. Three-factor crosses are preferable, however, since they can reveal marker orders directly by permitting the recognition of those rare recombinant classes that arise as a result of double crossovers (see Sections 10.7 and 10.8 for a review of the principles underlying three-factor crosses).

A typical cross for *D. melanogaster* that involves three sex-linked markers is outlined here. Females homozygous for *yellow* body (y), a rough eye known as *echinus* (ec), and *cut* wings (ct) are crossed, as usual, with wild-type males. The F_1 females will thus be phenotypically wild type and heterozygous for all three loci, $\frac{y\ ec\ ct}{+\ +\ +}$. These females are then inbred with their $\frac{y\ ec\ ct}{\longrightarrow}$ brothers; as noted earlier, these in effect are test crosses. We can predict that the females will produce eight different types of eggs: $y\ ec\ ct$, $+\ +\ +$, $y\ +\ +$, $+\ ec\ ct$, $y\ ec\ +$, $+\ +\ ct$, $y\ +\ ct$, and $+\ ec\ +$. We cannot, however, predict the number of each, since the gene loci are linked and therefore do not assort independently of one another.

The results of the cross are shown in Table 12-2. The eight progeny phenotypes fall into four classes, just as in the three-factor phage crosses outlined in Section 10.8. Using the same terminology introduced in Section 10.8, the parental class consists of *yellow echinus cut* and wild-type flies; the single-recombinant-I class consists of flies with the complementary

Table 12-2   Linkage and Recombination between *y*, *ec*, and *ct* in *Drosophila*

$$\frac{y\ ec\ ct}{+\ +\ +} \times \frac{y\ ec\ ct}{\longrightarrow}$$

Progeny Phenotypes	Number	Class	Class Total	Class Frequency
yellow, echinus, cut / wild type	1071 / 1080	Parental	2151	
yellow / echinus cut	78 / 66	Single I	144	$\alpha = 144/2880 = 0.050$
yellow echinus / cut	282 / 293	Single II	575	$\beta = 575/2880 = 0.199$
yellow cut / echinus	4 / 6	Double	10	$\gamma = 10/2880 = 0.0034$
		GRAND TOTAL	2880	

phenotypes *yellow* and *echinus cut;* the single-recombinant-II class consists of the complementary *yellow echinus* and *cut* flies; and the double-recombinant class consists of *yellow cut* and *echinus* flies. The parental class is, as expected, the largest; the double-recombinant class is the smallest. Assuming, as we did for prokaryotic genetics, that the rarest class arises as the result of double crossovers, the marker order can be assigned as *y ec ct.*

To determine map distances between the loci, the recombination frequencies for each class are calculated, as shown in Table 12-2, to yield values for α, β, and γ. The recombination frequencies between each pair of loci are then determined by familiar relationships:

$$R_{y,ec} = \alpha + \gamma = 5.3 \text{ percent}$$
$$R_{ec,ct} = \beta + \gamma = 20.2 \text{ percent}$$
$$R_{y,ct} = \alpha + \beta = 24.9 \text{ percent}$$

Converting recombination frequencies to map distances by multiplying by 100, we obtain the following map:

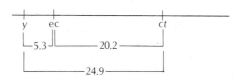

12.6   DOUBLE CROSSOVERS AND INTERFERENCE

The map drawn above reveals that the $y - ct$ interval of 24.9 map units is smaller than the $y - ec$ plus ec-ct interval (5.4 + 20.3 = 25.7 map units). As with phages (Section 10.8), this "foreshortening" of the outside markers is caused by the occurrence of double crossovers. Double crossovers in eukaryotes were considered in Section 3.6. As drawn in Figure 3-4, they may occur between the same two homologous chromatids in a bivalent **(two-strand doubles)** or they may involve three or four chromatids. In a **three-strand double** a chromatid engaged in crossing over with one homologous chromatid at one locus will establish a second chiasma with its second homologous chromatid at a different locus. In a **four-strand double** two homologous chromatids cross over at one level and the other two at a second level. Each type of double crossover yields a distinct ratio of parental: recombinant chromatids, as illustrated in Figure 12-4.

If we assume that the various double crossover configurations will occur with equal probability, then we would expect the ratio of two:three:four-strand double exchanges to be 1:2:1, the three-stranded events involving two possible acts of three chromatids and thus expected to occur twice as often. Three-strand events are in fact found to be deficient or two-strand events to be excessive, or both, but the observed ratio does not vary greatly from the ratio predicted. In genetic terminology it is therefore said that there

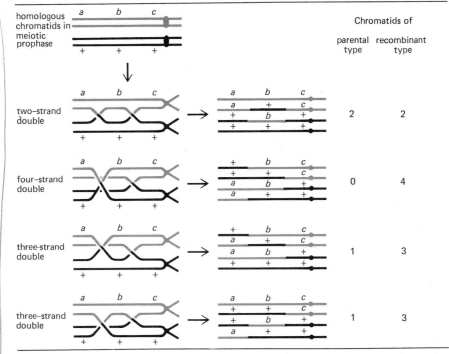

Figure 12-4 Various types of double crossover configurations and the types of recombinant chromosomes they produce. (After A. Srb, R. Owen, and R. Edgar, *General Genetics,* San Francisco: Freeman, 1965.)

is little if any **chromatid interference:** the fact that a chromatid has formed a chiasma with one homologous chromatid does not interfere with its ability to form a second chiasma with another chromatid.

In contrast, **there is often a marked interference exerted by one chiasma on the probability that a second chiasma will occur between the same two chromatids.** This can be seen by reexamining the data in Table 12-2. The double crossover class γ constitutes 0.34 percent of the progeny. As noted in Section 10.7, the probability that two independent events will occur at the same time is given by the product of their individual probabilities. Therefore, the frequency of double recombinants in the $y - ct$ interval would be expected to be the frequency of recombination between y and ec multiplied by the frequency of recombination between ec and ct. Thus one would expect (0.053) (0.202) = 0.011 or 1.1 percent of the progeny to be doubly recombinant. The discrepancy between the predicted (1.1 percent) and the observed (0.34 percent) figures indicates that the two crossover events are not independent of each other; in genetic terminology it is said that there is **interference** exerted by one chiasma on the occurrence of a second chiasma.

The nature and extent of interference can be conveyed by an expression known as the **coefficient of coincidence,** usually symbolized as s, where

$$s = \frac{\gamma}{(\alpha)\ (\beta)}$$

If the predicted frequency of double crossovers in an interval $[(\alpha)\ (\beta)]$ is the same as the observed frequency (γ), then $s = 1$; in this case, the occurrence of one exchange in the interval has no influence on the occurrence of a second, and it is said that there is no interference. If the number of double exchanges is less than predicted, then s will be some fractional number; in the y–ct example, $s = 0.34/1.1 = 0.31$, meaning that only 31 percent of the expected double crossovers in fact occur. Since in this case, one crossover appears to inhibit the second, the cross is said to exhibit **positive interference.** (When markers are very close together, particularly within the same gene, double exchanges occur more often than predicted and **negative interference** is said to take place. The molecular basis for negative interference is explored in Section 13.12).

Positive interference is not observed when markers are located on opposite sides of a centromere. For markers located in the same chromosomal arm, positive interference generally increases the closer the two markers are, an observation that suggests that chromatids experience some mechanical difficulty in establishing two chiasmata near each other. The actual molecular basis for positive interference is, however, unknown.

In summary, then, map distances derived from recombination frequencies must be corrected both for the shrinking effect of double crossovers and the compensatory effect of positive interference if they are to reflect the true physical distances between genes. Even with such corrections recombination frequencies cannot be relied on to give an accurate measure of distance, for it turns out that recombination does not occur with equal probability along the length of a eukaryotic chromosome. This will become apparent in a later section of this chapter.

CYTOLOGICAL MAPPING IN *DROSOPHILA*

An important and independent view of an organism's genome is gained from physical maps of chromosomes. Dipteran polytene chromosomes have proved to be particularly useful for this, since chromosomal aberrations will frequently alter both their banding patterns and their somatic pairing configurations (Figure 6-16), thus allowing the location and the extent of aberrations to be defined with some precision. When such physical disorders can be correlated with particular mutant phenotypes, it becomes possible to construct **cytological maps.**

Deficiencies (Section 6.14) are most commonly adopted for mapping purposes in *Drosophila.* Most deficiencies, even those involving only one or two bands, are lethal when carried in the homozygous or hemizygous condition and must be maintained in heterozygous flies. When a fly is heterozygous for a deficiency in one chromosome and for a recessive, nonlethal mutation at a comparable position on the homologous chromosome the recessive mutation will characteristically be expressed in the phenotype, a phenomenon aptly termed **pseudodominance.** It thus becomes possible to correlate the cytological location of a deficiency with the map locations of known genetic markers, as shown in the following examples.

12.7 CYTOGENETIC ANALYSIS OF THE *DROSOPHILA* X CHROMOSOME TERMINUS

Figure 12-5a shows the noncentromere end of the polytene X chromosome of *D. melanogaster.* You will recall that such a chromosome represents an amplified version of two homologous chromosomes which have come together by somatic pairing (Figure 2-17). Therefore, if a fly is made heterozygous for an X chromosome carrying a terminal deficiency, its polytene X chromosomes will resemble those shown in Figure 12-5b and 12-5c: that portion of the polytene structure deriving from the normal X will carry a full set of bands, whereas the portion deriving from the deficiency homologue will be foreshortened. Experiments by M. Demerec utilizing the chromosomes shown in Figure 12-5 offer a good example of cytological mapping by deficiencies. The deficiency chromosome in Figure 12-5b is seen to lack 11 bands present in the normal chromosome; in Figure 12-5a these bands are represented by the segment marked 260-1. Females carrying the recessive markers y (*yellow body*), ac (*achaete,* affecting the number and size of body bristles), and sc (*scute,* affecting the formation of scutellar bristles) can be made heterozygous for this deficiency, and all three recessive genes are expressed, thereby localizing these markers to the X terminus. When such triple mutants are instead made heterozygous for the deficiency shown in Figure 12-5c, a deficiency involving the loss of only the eight bands in the 260-2 region of Figure 12-5a, then only y and ac are expressed. This outcome localizes the sc locus to the few bands present in the second deficient chromosome (Figure 12-5c) but absent in the first (Figure 12-5b).

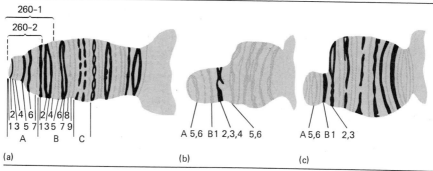

Figure 12-5 (a) Normal polytene X chromosome of *D. melanogaster.* (b) and (c) Polytene X chromosomes from deficiency heterozygotes. (From *General Genetics,* Second Edition, by A. M. Srb, R. D. Owen, and R. S. Edgar. W. H. Freeman and Company. Copyright © 1965, after Demerec.)

12.8 CYTOGENETIC ANALYSIS OF THE *ZESTE-WHITE* REGION OF THE *DROSOPHILA* X CHROMOSOME

B. Judd and collaborators have recently performed a particularly detailed genetic and cytogenetic analysis of the region of the X chromosome shown in Figure 12-6, extending from band 3A2 (the locus of z, the *zeste* eye marker) to band 3C2 (the locus of w, the *white* eye marker). Since these experiments are highly relevant to the question of the organization of genes in the eukaryotic chromosome, we consider them in some detail.

The first goal of the Judd experiments was to isolate *Drosophila* strains carrying mutations in every one of the genes in the *zeste-white* region. To screen for such mutations the Judd group employed the procedure outlined in Figure 12-7. Normal males were treated with mutagen and then crossed with females homozygous for the *Basc* chromosome (described in Section 6.4 and in Figure 6-7). The F₁ female progeny from this cross will therefore

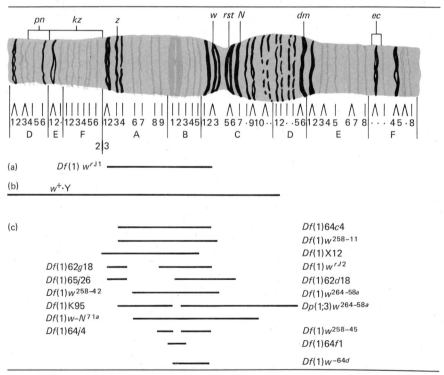

Figure 12-6 Portion of the polytene X chromosome of *D. melanogaster* analyzed in the experiments of Judd *et al.* The lines below designate (a) the extent of the deficiency in the *Df*(1)*w*ʳᴶ¹ chromosome used in screening; (b) the extent of the extra genes found in the *w⁺*Y chromosome used in screening; (c) the extents of various deficiencies used in complementation studies and in cytogenetic analysis. (From B. Judd, M. Shen, and T. Kaufman, *Genetics* **71**:139, 1972.)

inherit *Basc* chromosomes from their mothers and X chromosomes from their fathers.

Because the fathers were mutagenized, certain of the F_1 females would be expected to inherit a paternally derived X chromosome carrying a mutation in the *zeste-white* region (X*). Judd identified these carriers by crossing individual F_1 females to males carrying a deficiency in the 3A2-3C2 region of the X chromosome, a deficiency designated $Df(1)w^{rJ1}$ (Figure 12-6a). Such a deficiency is normally lethal in males, but the male strain was constructed to possess a Y chromosome carrying a translocated segment of X chromosomal material which includes all the genes absent in the deficient X chromosome. This chromosome is designated $w^+ \cdot Y$ (Figure 12-6b).

The possible F_2 progeny from such a cross are shown in Figure 12-7. Among the F_2 females half should carry a *Basc* X chromosome (and can be identified by their *Bar, apricot* eyes) and half should carry the original paternally derived X chromosome. If this chromosome carries a mutation (X*) in the *zeste-white* region and the mutation does not cause lethality, then the mutation will be expressed in the F_2 daughters since it will lie opposite the $Df(1)w^{rJ1}$ deficiency. Should the mutation be a recessive lethal, as proves most often to be the case, then all the F_2 females will be of

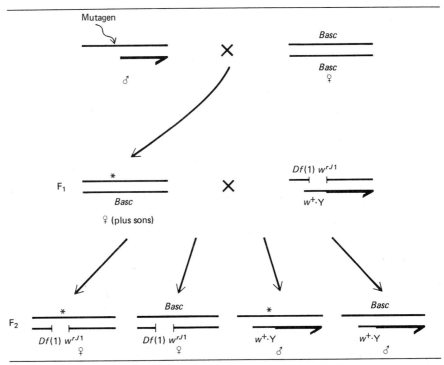

Figure 12-7 The mating scheme used to recover mutations in the X chromosome region 3A2-3C2. The X chromosome is represented by a thin line, the Y by a thicker line, and a mutation in the region of interest by °. (From B. Judd, et al., *Genetics* **71**:139, 1972.)

the *Bar, apricot* variety. Whenever Judd and his colleagues encountered an all *Bar-apricot*-female F_2, therefore, they isolated one of the $X^*/w^+ \cdot Y$ males in the vial and used it to start a stock of flies carrying the X^* chromosome in question. In this manner the Judd group was able to amass a collection of hundreds of stocks, each carrying an independently derived X^* chromosome. Such a collection is formally equivalent to the collection of *sus* strains of λ isolated by Campbell (Section 10.9) and, in like fashion, the next steps were to characterize the mutations genetically and to determine, by complementation, how many genes they marked.

Initial crosses between strains carrying X^* chromosomes revealed that 121 chromosomes behaved as if they carried point mutations (they could, for example, recombine to give wild-type recombinants). The approximate map locations of these 121 *zeste-white* mutations were first estimated by the deficiency mapping technique (Section 12.7): heterozygotes were constructed between a given mutant strain and one of the deficiency strains shown in Figure 12-6c, and the point mutation was expressed only when it occupied a position bracketed by a deficiency. The ability of a given pair of mutant chromosomes to complement each other was then assessed by crossing appropriate mutant strains and observing whether the resulting heterozygotes were normal or mutant (or absent, in the case of noncomplementing lethals). In this way the 121 point mutations were placed into 14 complementation groups, 12 lying between z and w and two lying to the left of z. Finally, pairwise crosses were performed to determine the recombinational distances between markers, and deficiencies covering known complementation groups were analyzed cytologically to determine which bands were missing.

The results of the Judd studies are summarized in Figure 12-8. The z locus is equated with band 3A3 and the w locus with band 3C2, a result obtained by other investigators as well. The 12 *zeste-white* (*zw*) loci between the z and w loci cover a region of the polytene chromosome estimated by Bridges to contain 12 bands. (Judd and colleagues were in fact able to identify positively only 11 bands in the *zeste-white* region, as shown in Figure 12-8). The Judd studies, in other words, demonstrate a correspondence between the presence of a chromomere (amplified as a polytene chromosome band) and the presence of a gene.

We are here confronted with a central question in cytogenetics: can chromomeres be equated with genes? Or is the chromomeric organization of chromatin independent of its polypeptide-coding functions? The question cannot yet be answered definitively (see also Section 2.12), but the following observations suggest the complexity of the situation. In some cases, several genes appear to localize to a region of a polytene chromosome where only one band can be identified. At the other extreme (the *zeste-white* region being one case), many chromomeres possess much more DNA than is necessary to specify the amino acid sequence of a polypeptide. It therefore appears necessary to postulate that some genes may not have a chromomeric organization whereas other genes associate with an excess of DNA (which possibly has a regulatory function).

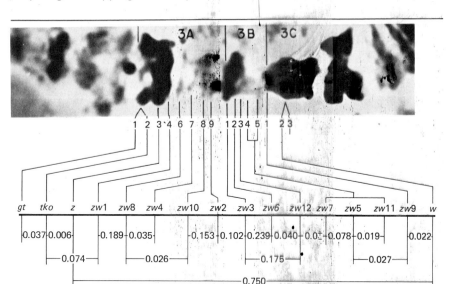

Figure 12-8 A micrograph of the 3A-3C region of the X chromosome, below which are the 16 complementation groups found in the region and their distances from one another on the genetic map, with distances expressed as map units. (From B. Judd et al., *Genetics* **71:**139, 1972.)

12.9 CORRELATING GENETIC AND CYTOLOGICAL MAPS

Over short distances, as in the Judd experiments, there appears to be good correspondence between physical distances in a polytene chromosome and distances on a genetic map (Figure 12-8). Over longer distances, however, the correspondence varies considerably, depending on the region of the chromosome being considered. It is seen in Figure 12-9, for example, that the X chromosome of *D. melanogaster* undergoes recombination more often in the center than at either end; markers at the ends therefore appear much closer together on the genetic map than they actually are on the physical chromosome. More specifically, there is as much recombination in the *w-fa* interval, which constitutes perhaps five polytene bands, as there is in the *y-w* interval, which has about 75 bands (Figure 12-9). Recombination is also rare in the immediate vicinity of centromeres, the sites of centromeric heterochromatin (Section 2.13), which again distorts the physical-genetic correspondence. Finally, as noted in Section 6.15, the presence of inversions and other chromosomal rearrangements will suppress recombination in certain regions.

Nonetheless, despite the fact that distances between markers are often disparate when cytological and genetic maps are compared, it is invariably the case that the *order* of markers deduced by both approaches is the same. This fact argues impressively for the linear arrangement of genes in chromosomes first proposed from recombinational data.

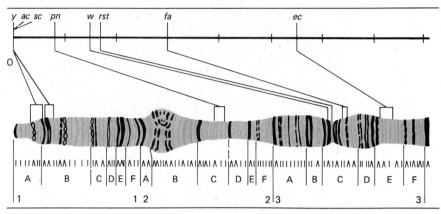

Figure 12-9   A portion of the left arm of the X chromosome of *D. melanogaster* and the corresponding genetic map of the region. (After C. B. Bridges.)

LINKAGE GROUPS AND CHROMOSOMES

12.10   DETECTING LINKAGE GROUPS

As more and more markers become available for study in a given eukaryotic species, a limited number of groups of linked genes, called **linkage groups,** begin to appear. **As a general rule, markers that assort independently are placed in different linkage groups and those that exhibit linkage are placed in the same group.**

The major qualification to this rule becomes apparent when the physical process of meiosis is recalled. As noted in Section 3.6, chiasmata form quite frequently during meiosis, typically at least once per bivalent. Taking as an example the *Drosophila* homologues marked by *gl* and *e* and by their + + alleles (Figure 12-10), if we were to examine 100 such bivalents we might expect to find that 84 have no chiasmata in the *gl-e* interval, while 16 have single chiasmata in this interval. It can thus be said that the overall **crossover frequency** in the *gl-e* interval is $16/100 = 0.16$ or 16 percent. By definition, however, only two of the four chromatids in a bivalent will participate in any single exchange event. Therefore, if we consider the $16 \times 4 = 64$ chromatids in the chiasma-bearing bivalents, only $\frac{1}{2} \times 64 = 32$ chromatids will be *gl* + or + *e* recombinants (Figure 12-10), and of the 400 chromatids present in the 100 bivalents, only $32/400 = 0.08$ or 8 percent will be recombinant. Thus the recombination frequency deduced from genetic analysis of random meiotic products will be half the crossover frequency.

It follows that when two markers on the same chromosome are so far from one another that they happen to be separated by a chiasma in virtually every meiotic bivalent, their crossover frequency will approach 100 percent and their recombination frequency 50 percent. Markers that appear to be linked in 50 percent of the offspring and unliked in the other 50 percent are

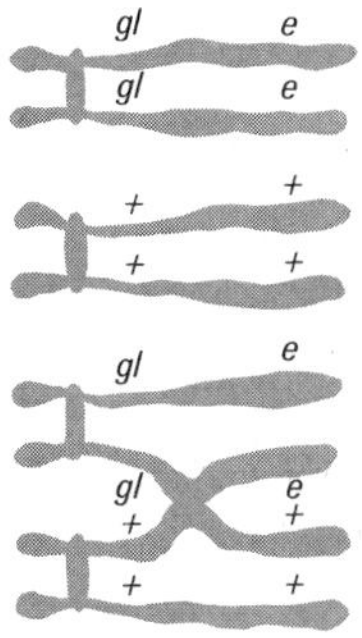

		Number of bivalents of each class	Number of chromatids in each class (bivalents × 4)	Number of recombinant chromatids
1) Non-Crossover Class		84	336	0
2) Crossover Class		16	64	32

Total bivalents = 100 Crossover frequency = 16/100 = 0.16	Total chromatids = 400	**Recombination frequency =** 32/400 = 0.08

Figure 12-10 The relationship between crossover frequency and recombination frequency, using the *glass ebony* data from Table 12-1 as an example.

indistinguishable from markers that assort independently. In other words, if two markers recombine about half the time, it is impossible to decide whether they lie in the same or different linkage groups.

To resolve this question the markers in question must be tested for their linkage to other markers in the relevant linkage group. Thus, for example, if the recombination frequency between *a* and *b* is about 50 percent, between *a* and *c* 20 percent, and between *b* and *c* 30 percent, then *a* and *b* are probably linked; if, however, *b* and *c* also recombine 50 percent of the time, then *a* and *b* are probably not linked.

The number of linkage groups possessed by a given organism should correspond to its number of chromosome pairs and thus to its haploid chromosome number. The number of linkage groups in *D. melanogaster* is four, and there are four pairs of chromosomes, the XX or XY pair, the large autosomes II and III, and the tiny autosome IV. Another *Drosophila* species, *D. pseudoobscura,* has five pairs of chromosomes and five linkage groups, corn has 10 linkage groups and a haploid chromosome number of 10, and so on. As we saw in Chapters 10 and 11, viruses and bacteria each have one major linkage group and one major chromosome. This correspondence is, of course, further evidence in support of the chromosome theory of inheritance.

12.11 ASSIGNING LINKAGE GROUPS TO CHROMOSOMES

In order to assign a particular linkage group to a particular chromosome, some sort of cytogenetic correlation must be made between the presence of a genetic marker carried by a linkage group and the presence of a distinctive chromosome. This is classically accomplished by chromosomal mutations: the presence of an identifiable deficiency in a chromosome (Section 12.7), for example, may allow the expression of a recessive trait in a known linkage group, thereby assigning that linkage group to that chromosome.

The assignment of the 4 linkage groups in *D. melanogaster* (Figure 12-11) to its 4 chromosomes was a straightforward affair: the chromosomes are easily distinguished from one another and have the landmark banding patterns in their polytenic state. Assigning the 10 linkage groups of maize (Figure 12-12) to its 10 haploid chromosomes was also greatly facilitated by "natural" landmarks: most of the chromosomes of maize bear characteristic arrays of chromomeres and "knobs", and the deletion or translocation of these features can be correlated with the acquisition of particular mutant phenotypes.

In the case of the mouse, 20 linkage groups were defined genetically (Figure 12-13) and 20 chromosomes were known to be present in the haploid karyotype, but it proved almost impossible to tell these small acrocentric chromosomes apart until "banding technology" was introduced (Box 2.3; see also Figure 2.10b for a banded mouse karyotype). O. J. Miller, D. A. Miller, and their coworkers proceeded to assign linkage groups to these chromosomes by studying **overlapping translocations.** For example, mouse stock T138Ca was known to carry a translocation that caused markers in linkage group (LG) II to be linked to markers in LG IX, while stock RB163H carried a translocation that caused LG II markers to associate with LG XII markers. The karyotype of stock T138Ca revealed a translocation chromosome with the banding patterns of chromosomes 9 and 17, while the translocation chromosome of stock RB163H proved to carry bands from both chromosomes 9 and 19. Thus it was deduced that LG II must be in chromosome 9, that LG IX corresponds to chromosome 17, and that LG XII is in chromosome 19.

MAPPING BY TETRAD ANALYSIS

As outlined in Section 4.1, the "lower" haploid eukaryotes can be subjected to tetrad analysis, wherein all four products of individual meioses are analyzed separately. Tetrad analysis also permits linkage and mapping studies.

12.12 MAPPING TWO LINKED GENES BY TETRAD ANALYSIS

When two markers assort independently (that is, when they are not linked), then one expects to find in tetrad analysis a near one-to-one ratio of PD to NPD tetrads (Section 4.1). **Linkage between two markers is signaled by a PD : NPD ratio that is considerably greater than unity.** For example, in an *arg* + X + *pab* (para-amino benzoic acid requirement) cross of *C. reinhardi,* the tetrads obtained are as follows (p. 466):

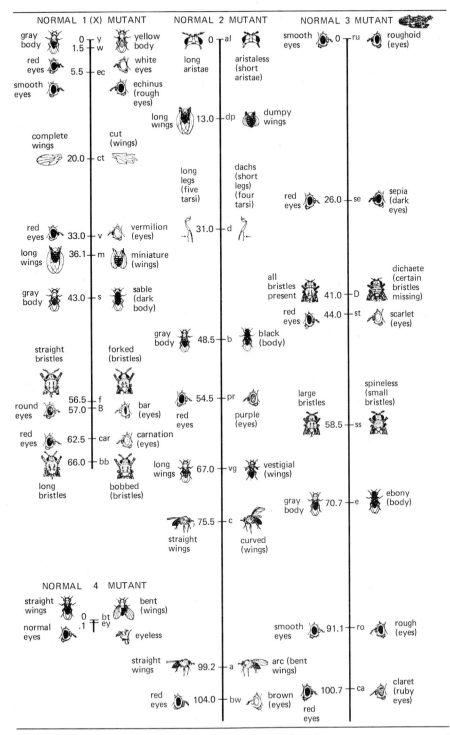

Figure 12-11 A portion of the genetic map of *Drosophila melanogaster*, showing some of the more widely used visible markers.

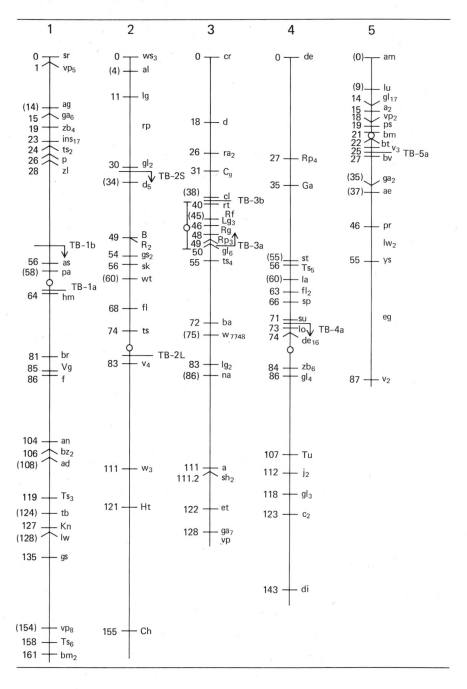

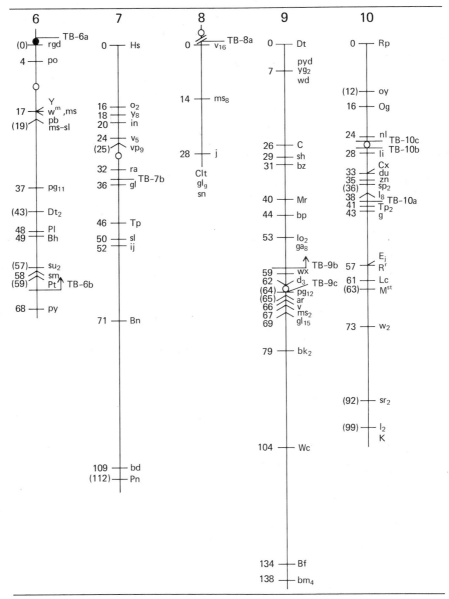

Figure 12-12 Linkage map of corn (*Zea mays*). Parentheses indicate probable position based on insufficient data, ○ indicates centromere position, and ● indicates organizer. Positions designated TB identify the genetic location of breakpoints of A-B translocations, which generate terminal deficiencies; ——→ indicates that the TB breakpoint is in that position or is some distance in the direction indicated. (From M. G. Neuffer and E. H. Coe, Jr. In: *Handbook of Genetics*, R. C. King, Ed. New York, Plenum Press, 1974.)

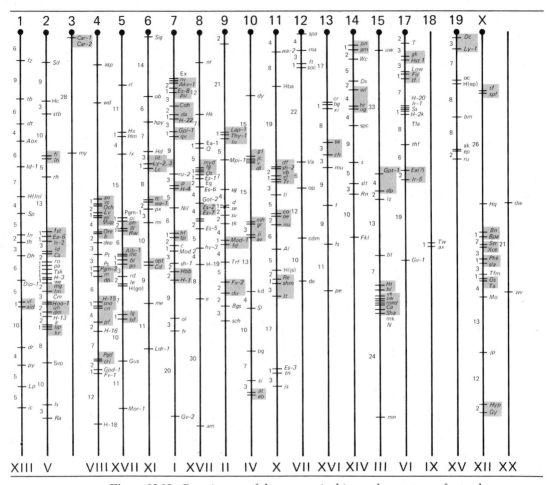

Figure 12-13 Genetic map of the mouse. Arabic numbers at top refer to chromosomes; Roman numerals at bottom refer to linkage groups. Loci whose order is uncertain are not italicized. Gray indicates that the order within the gray group has not been established. Knobs indicate the locations of the centromeres where they are known. (Reprinted with permission from M. Green, 46th Ann. Rep. Jackson Lab., Bar Harbor, Maine, 1975.)

PD	NPD	T
arg +	arg pab	arg +
arg +	arg pab	arg pab
+ pab	+ +	+ pab
+ pab	+ +	+ +
119	1	71

Total Tetrads: 191

The unequal ratio of PD to NPD is a major departure from independent assortment and indicates that the *arg* and *pab* loci tend to be inherited in their parental configurations rather than in any nonparental or recombinant configurations.

Assuming that *pab* and *arg* are linked, the origin of the three tetrad classes is diagrammed in Figure 12-14. The PD tetrads, of course, arise when the parental chromatids are transmitted intact to the meiotic products (Figure 12-14a). A PD tetrad might also arise after a two-strand double crossover, (Figure 12-4), but these would be expected to be relatively rare. The T tetrads form whenever a single crossover occurs between the *arg* and *pab* loci in two of the four chromatids in a bivalent (Figure 12-14b). Finally, the NPD tetrads, with four recombinant products, must result from the simultaneous occurrence of two crossovers in the *arg pab* interval involving all four chromatids; in other words, the NPD tetrads must result from a four-strand double crossover (Figure 12-14c). Once again the assumption is made that such double crossovers will be rare compared with single crossovers, and this assumption is borne out by the fact that only 0.5 percent of the tetrads in the present cross are of the NPD class, whereas 37.2 percent are of the T class and 62.3 percent are of the PD class.

Recombination frequencies can be calculated from such data in one of two ways. We can essentially follow the procedures set forth for *Drosophila* crosses by supposing that the progeny obtained in the *C. reinhardi* cross are in fact collected as random individuals rather than in their tetrad groupings. Since there are 191 tetrads, this means that there are 4×191 or 764 cells in the collection, of which $4 \times 1 = 4$ belong to the NPD class and $4 \times 71 = 284$ belong to T. All 4 NPD cells are recombinant *arg pab* or $+ +$ cells, whereas only half the T cells (that is, 142 cells) are *arg pab* or $+ +$ recombinants (recall Figure 12-10). Therefore, using the formula $R = \dfrac{\text{recombinant progeny}}{\text{total progeny}}$, the expression becomes

$$R_{arg,pab} = \frac{4 + 142}{764} = \frac{146}{764} = 0.191 \text{ or } 19.1 \text{ percent.}$$

A simpler calculation considers the tetrads in their class groupings and utilizes the expression

$$R = \frac{NPD + \frac{1}{2}T}{PD + NPD + T}$$

Thus the particular data are treated as

$$R_{arg,pab} = \frac{1 + \frac{1}{2}(71)}{191} = \frac{36.5}{191}$$

$$= 0.191 \text{ or } 19.1 \text{ percent}$$

As with *Drosophila,* this recombination frequency is translated as 19.1 map units.

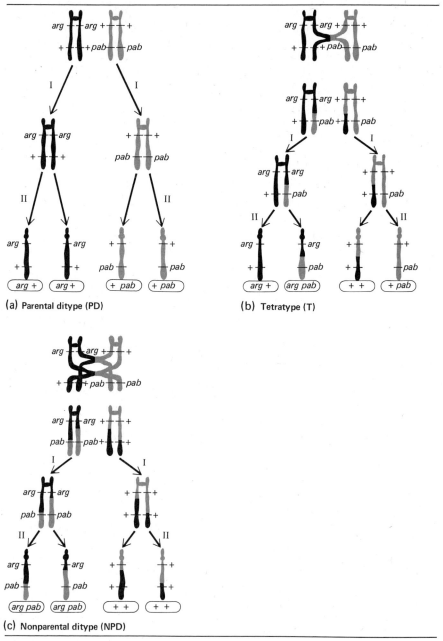

(a) Parental ditype (PD)

(b) Tetratype (T)

(c) Nonparental ditype (NPD)

Figure 12-14 Origin of tetrads in a cross of *arg* + X + *pab* in *Chlamydomonas*. I and II represent the two meiotic divisions.

12.13 MAPPING THREE LINKED GENES BY TETRAD ANALYSIS

bag

The use of a three-factor cross to determine both marker order and intra-marker map distances can be illustrated for *C. reinhardi* by a cross between an *arg* + + strain and a + *pab thi* strain, where *thi* indicates a requirement for thiamin. The kinds of tetrads obtained are shown in Table 12-3. The standard procedure with data of this kind is to classify each tetrad class as a PD, NPD, or T for each of the three pairwise marker combinations possible, namely, *arg pab, arg thi,* and *thi pab.* For example, tetrad (3) in Table 12-3 is a PD with respect to *arg* and *pab* but a T with respect to both *arg* and *thi* and *thi pab.* Tetrad (7) is a T with respect to *arg* and *pab,* NPD with respect to *arg* and *thi,* and T with respect to *thi* and *pab.* The total number of PD, NPD, and T tetrads is then determined for each marker combination, as shown in Table 12-4. The recombination frequencies calculated from these numbers are therefore

$$R_{arg,pab} \frac{1 + \frac{1}{2}(71)}{191}(100) = 19.1 \text{ percent}$$

$$R_{arg,thi} \frac{8 + \frac{1}{2}(167)}{191}(100) = 48 \text{ percent}$$

$$R_{thi,pab} \frac{2 + \frac{1}{2}(121)}{191}(100) = 32.8 \text{ percent}$$

Table 12-3   Kinds of Tetrads Obtained from the Cross *arg* + + × + *pab thi* in *Chlamydomonas*

(1)	(2)	(3)	(4)
arg + +	arg + +	arg + +	arg + +
arg + +	arg pab thi	arg + thi	arg pab +
+ pab thi	+ + +	+ pab +	+ + thi
+ pab thi	+ pab thi	+ pab thi	+ pab thi
14	54	103	2

(5)	(6)	(7)	(8)
arg + +	arg + thi	arg + thi	arg + thi
arg pab thi	arg pab +	arg pab thi	arg + thi
+ + thi	+ + +	+ pab +	+ pab +
+ pab +	+ pab thi	+ + +	+ pab +
6	3	6	2

(9)
arg pab +
arg pab thi
+ + +
+ + thi
1

Table 12-4 Number of PD, NPD, and T Tetrads from Data Given in Table 12-3

	arg-pab	arg-thi	pab-thi
PD	119	16	68
NPD	1	8	2
T	71	167	121
Total	191	191	191
Map distance	19.1	48.0	32.8

The marker order of *arg pab thi* can be deduced directly from these values: there is more recombination between *arg* and *thi* than between the other two sets of markers and, by our usual line of reasoning, *arg* and *thi* must therefore lie farthest apart. The marker order can also be deduced from the fact that the rarest tetrad classes (tetrads 4 through 9 in Table 12-3) can all be accounted for by assuming the occurrence of two-, three-, or four-strand double crossovers between chromatids bearing an *arg pab thi* sequence. If, for example, the marker order were instead *arg thi pab,* then tetrads of class 4 would arise by a single crossover and should be relatively frequent.

Extensive studies of *C. reinhardi* by tetrad analysis have disclosed the existence of 16 linkage groups (Figure 12-15). In *S. cerevisiae* (yeast) 18 linkage groups have been recognized (Figure 12-16), whereas *Neurospora crassa* has only seven (Figure 12-17). Most of the *N. crassa* linkage groups have been assigned to their respective chromosomes by correlating the inheritance of genetic translocations with the inheritance of chromosomes having altered morphologies.

12.14 GENE-CENTROMERE LINKAGES IN *NEUROSPORA*

The ordered tetrads of *N. crassa* (Figure 4-8) are particularly useful for determining the map distance separating a gene and its centromere. Except in certain abnormal cases, the centromeres of homologous chromosomes will always separate at the first meiotic division, meaning that they will always undergo first-division segregation. If a gene *b* is located very near a centromere, then *b* will also segregate from its + allele at the first division, and + + *b b* tetrads will emerge from all + × *b* crosses. If gene *b* instead is located some distance from the centromere so that crossing over takes place within the centromere-*b* interval, then every time an odd number of crossovers occurs a second-division-segregation tetrad will result (+ *b* + *b*). Thus the recombination frequency, and the map distance, between the gene and its centromere, is simply half the frequency of second-division segregation tetrads emerging from a cross. The one-half factor again derives from the fact that only two of the four chromatids in the relevant bivalent have participated in any one exchange.

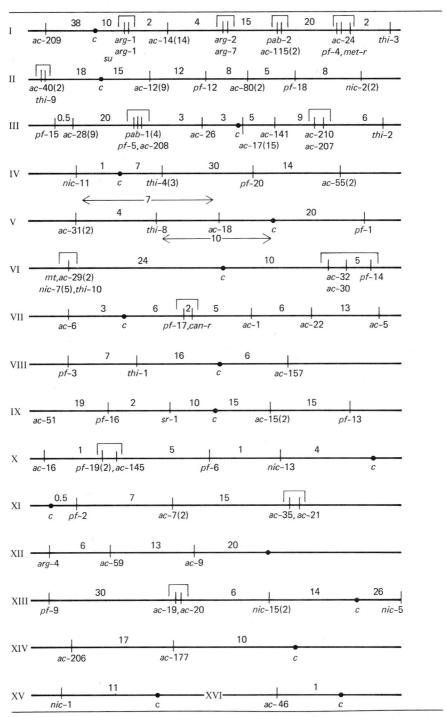

Figure 12-15 Linkage map of *Chlamydomonas reinhardi*. Figures in parentheses after the name of a locus indicate the number of alleles known at that locus. The bracket above a group of markers indicates that their relative positions are uncertain or unknown. *Abbreviations: ac*, acetate-requiring; *arg*, arginine-requiring; c, centromere; *can-R*, canavanine-resistant; *met-R*, methionine sulfoxamine-resistant; *mt*, mating type; *nic*, nicotinamide-requiring; *pab*, p-aminobenzoate-requiring; *pf*, paralyzed flagella; *SR*, streptomycin-resistant, *su^arg-1*, suppressor of *arg-1*; *thi*, thiamin-requiring.

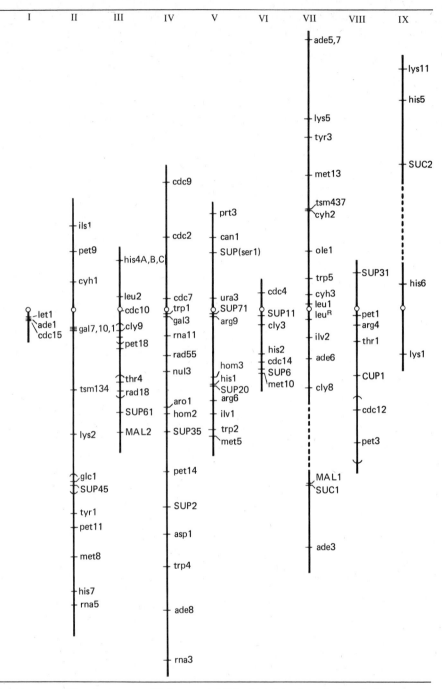

Figure 12-16 The genetic map of *Saccharomyces*. Linkages established by tetrad analysis, mitotic segregation, and trisomic analysis are denoted, respectively, by solid lines, dashed lines, and dotted lines. The order is unknown for the genes indicated within brackets. Gene symbols: nutritional requirements for adenine (*ade*), arginine (*arg*), aromatic amino acids phenylalanine, tryptophan, tyrosine, and *p*-aminobenzoate (*aro*), histidine (*his*), homoserine or threonine plus methionine (*hom*), isoleucine or isoleucine plus valine (*ilv*), leucine (*leu*), lysine (*lys*), methionine (*met*), oleate (*ole*), phenylalanine (*phe*), pyridoxine (*pdx*), serine (*ser*),

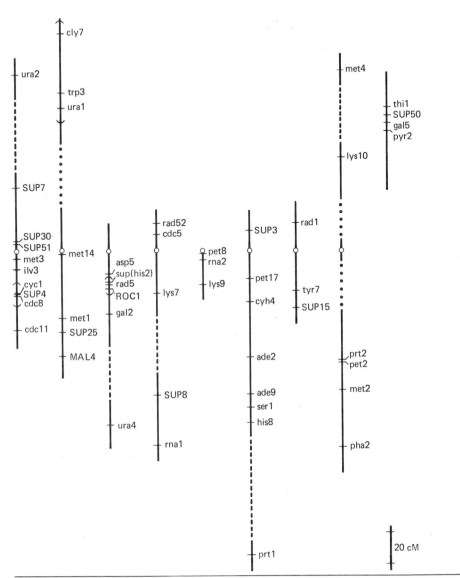

X   XI   XII   XIII   XIV   XV   XVI   XVII   F6

cly7

ura2

trp3
ura1

SUP7

SUP30
SUP51
met3
ilv3
cyc1
SUP4
cdc8
cdc11

met14

met1
SUP25

MAL4

asp5
sup(his2)
rad5
ROC1

gal2

ura4

rad52
cdc5

lys7

SUP8

rna1

pet8
rna2

lys9

SUP3

pet17

cyh4

ade2

ade9
ser1
his8

prt1

rad1

tyr7
SUP15

prt2
pet2

met2

pha2

met4

lys10

thi1
SUP50
gal5
pyr2

20 cM

thiamin (*thi*), threonine (*thr*), tryptophan (*trp*), tyrosine or tyrosine plus phenylalanine (*tyr*), and uracil (*ura*); defects of asparagine and aspartate metabolism (*asp*); fermentation of maltose (*MAL*) and sucrose (*SUC*); non-fermentation of galactose (*gal*) and its regulation (*i-gal*); conditional mutants affecting cell division (*cdc*), isoleucyl-tRNA synthetase (*ils*), methionyl-tRNA synthetase (*mes*), cell lysis (*cly*), protein synthesis (*prt*), RNA synthesis (*rna*), and unknown functions (*tsm*); lack of utilization of nonfermentable carbon sources (*pet*); deficiency of cytochrome *c* (*cyc*); electrophoretic mobility of esterases (*EST*); glycogen accumulation (*glc*); resistance to canavanine (*can*), copper salts (*CUP*), cycloheximide (*cyh*), and benzalkonium chloride, "Roccal" (*ROC*); sensitivity to radiation (*rad*); mating type (α) and nonmaters (*nul*); super-suppressors and specific suppressors (*SUP* or *sup*); recessive lethals (*let*). (From R. K. Mortimer and D. C. Hawthorne, *Genetics* **75**:459, 1973, and F. Sherman, unpublished observations.)

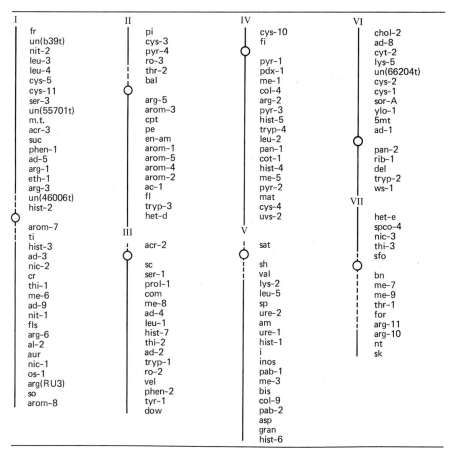

Figure 12-17 Linkage map of *Neurospora crassa*. The centromeres are shown in their most probable locations. Dashed lines indicate limits of uncertainty. Only unequivocally mapped markers are shown. A key to the symbols and a list of incompletely mapped markers is found in R. W. Barratt and A. Radford, *Handbook of Biochemistry*, 2nd ed., H. A. Sober, Ed. Cleveland, Ohio: Chemical Rubber Co., 1970.

12.15 MAPPING THE *CYC1* GENE OF YEAST

During the past 10 years, F. Sherman and his colleagues have been engaged in generating a fine-structure map of the *CYC1* gene of yeast. Located in linkage group X (Figure 12-16), the *CYC1* gene codes for the small heme protein, iso-1-cytochrome c, which localizes in the mitochondrion and is essential for respiration. The Sherman experiments can be said to be the equivalent of Benzer's experiments with the *rII* gene of T4 (Section 10.18) in the sense that they have provided the first detailed map of a eukaryotic gene.

Sherman and colleagues screened for *cyc1* mutations by first isolating yeast strains that could not respire, could not use lactate as a carbon source, and were therefore resistant to the toxic effects of chlorolactate. These mutant strains were then examined for the presence or absence of a protein with the spectral properties of iso-1-cytochrome c, and a total of 210 single-site mutants was isolated; these mutants either lacked iso-1-cytochrome c altogether or else produced a nonfunctional cytochrome (recall Section 9.20 on yeast nonsense mutations).

Sherman and colleagues proceeded to cross the various *cyc1* mutants against one another and calculated the frequency of *CYC1* wild-type recombinants. They also isolated a collection of strains carrying deletions that extended into the *CYC1* region, and subjected the point mutants to deletion mapping (Section 10.18). In this way they were able to assign the 210 point mutations to 47 mutational sites in the gene and to unambiguously order 60 of these 47 sites. We should pause here to note that in eukaryotic genetics, where the term allele has wide currency, the 210 mutant strains would usually be said to carry **allelic mutations,** or **multiple alleles,** or **heteroalleles** (the three terms are synonymous). Similarly, recombination between these strains would usually be spoken of as **heteroallelic recombination** or **interallelic recombination.**

Since the amino-acid sequence of iso-1-cytochrome c is known, Sherman next set out to construct a **gene product map** wherein the position of an amino-acid substition in a mutant polypeptide is correlated with the position of a mutational base change in a gene. Such a map could obviously not be constructed using the original *cyc1* mutants that produced no cytochrome. Sherman therefore proceeded to treat each of these non-producing strains with mutagens and isolated "revertant" clones that were able to grow on lactate and that synthesized a functional iso-1-cytochrome c. This cytochrome was then isolated and compared with the wild-type protein. Some of the "revertant" clones proved to be true revertants in that the amino-acid sequence of their cytochrome was the same as that of the wild type. Other clones, however, synthesized proteins with single amino-acid substitutions. Sherman reasoned that these latter clones arose by the same sequence of events as diagrammed in Figure 9-11 for bacteria: a codon in the original gene underwent a mutation such that a normal or stable cytochrome could not be produced; when the strain carrying this mutation was again mutagenized, a new codon resulted and this permitted an alternate, but acceptable, amino acid to appear in the protein. On the premise, then, that the positions of the substituted amino acids corresponded to the original lesions, he was able to produce the gene-product map shown in Figure 12-18. Italicized on this map are the amino-acid residues that have undergone substitutions in 15 mutant strains. Shown below each italicized amino acid is one of the mutant alleles (*cyc1*-13, *cyc1*-9, and so on) that incurred an amino-acid substitution at that position. The relative order of these alleles deduced by the gene-product map is identical to their relative order established by deletion mapping. In other

```
 −1   1   2   3   4   5           9  10     12  13        15
(Met) Thr-Glu-Phe-Lys-Ala-Gly-Ser-Ala-Lys-Lys-Gly-Ala-Thr-Leu-Phe-
      /   /   |   \       /   |   \
cyc1-13 cyc1-9 cyc1-31 cyc1-239  cyc1-179 cyc1-183 cyc1-6 cyc1-10
```

```
                    ┌─HEME─┐
     17             |      |            25              30
    Lys-Thr-Arg-Cys-Leu-Gln-Cys-His-Thr-Val-Glu-Lys-Gly-Gly-Pro-
        |                 |
      cyc1-15          cyc1-2
```

```
     31            35             40                45
    His-Lys-Val-Gly-Pro-Asn-Leu-His-Gly-Ile-Phe-Gly-Arg-His-Ser-
     |
cyc1-8
```

```
                  50              55              60
    Gly-Gln-Ala-Glu-Gly-Tyr-Der-Tyr-Thr-Asp-Ala-Asn-Ile-Lys-Lys-
```

```
              64 65  66            70  71            75
    Asn-Val-Leu-Trp-Asp-Glu-Asn-Asn-Met-Ser-Glu-Tyr-Leu-Thr-Asn-
                 |   |             |
          cyc1-166 cyc1-72       cyc1-76
```

```
       (CH3)3
        |          80             85              90
    Pro-Lys-Lys-Tyr-Ile-Pro-Gly-Thr-Lys-Met-Ala-Phe-Gly-Gly-Leu-
```

```
       93      95             100             105
    Lys-Lys-Glu-Lys-Asp-Arg-Asn-Asp-Leu-Ile-Thr-Tyr-Leu-Lys-Lys-
             |
          cyc1-140
```

```
       108
    Ala-Cys-Glu
```

Figure 12-18 The amino acid sequence of iso-1-cytochrome *c*. The residues related to the mutational lesions that were deduced from revertant iso-1-cytochromes *c* are shown in italics, directly above the corresponding *cyc1* mutant designation. Also shown are the sites of heme attachment at positions 19 and 22, the ϵ-N-trimethyllysine at position 77, and the amino-terminal residue of methionine that is excised from the normal protein. From F. Sherman et al., *Genetics* **81**:51 (1975).

words, there exists a strict **colinearity** between mutational sites in a gene as defined by these two independent mapping procedures. Similar colinearity demonstrations have been made by C. Yanofsky for the tryptophan synthetase genes of *E. coli;* together these studies provide strong evidence indeed for the concept of a structural gene as a linear sequence of nucleotides encoding a linear sequence of amino acids.

MAPPING *ASPERGILLUS* BY SOMATIC SEGREGATION AND RECOMBINATION

Sexual cycles that include gametic fusion and meiosis are the usual avenues by which segregation and recombination occur in eukaryotes. Occasionally, however, nuclei from somatic or vegetative cells can undergo fusion, and these nuclei may undergo processes akin to segregation and recombination. The study of such asexual processes has proved to be valuable for many kinds of genetic analyses.

12.16   HETEROKARYON FORMATION IN *ASPERGILLUS*

The organism whose somatic genetics is best known is the haploid filamentous fungus *Aspergillus nidulans*. This organism has a sexual cycle quite
similar to that of *N. crassa* (see Figure 3-9) and, like *N. crassa*, is able to
undergo an additional somatic process known as **heterokaryosis.** During
heterokaryosis cytoplasmic bridges form between two hyphae of different
genetic constitution, nuclei and cytoplasm pass freely across the bridges as
well as up and down the individual hyphae, and a **heterokaryon** that
contains a mix of nuclear types from the two parents results.

The nuclei of a heterokaryon of *N. crassa* always remain haploid and
usually remain haploid in an *A. nidulans* heterokaryon as well. Infrequently,
however, two haploid nuclei of differing genotype will fuse together in an
A. nidulans heterokaryon, and a diploid heterozygous nucleus results.
Repeated mitotic division of this nucleus generates a diploid sector in the
heterokaryon which can be recognized in several ways. One is by the
judicious use of genetic markers. In a heterokaryon formed between an
adenine (*ad*) auxotroph producing *yellow* (*y*) conidia and a *p-aminoben-
zoic acid* (*pab*) auxotroph producing *white* (*w*) conidia the nuclei in the
diploid sector will have a $\dfrac{+ \quad + \quad + \quad +}{ad \quad pab \quad y \quad w}$ genotype. These nuclei will dictate
the synthesis of enough normal (+) protein so that the sector will be
prototrophic and will produce wild-type (green) conidia at the hypha tips.
These green conidia, moreover, will be larger and will contain more DNA
than their haploid counterparts.

When diploid conidia are isolated and germinated, a growth of stable
diploid hyphae will form, but occasional sectors will appear that do not
have the same phenotype as the heterozygous parent; instead, these sectors
exhibit recessive mutant traits. Thus, in our example, a hypha bearing *white*
or *yellow* conidia may appear among the great majority bearing green
conidia. It would appear that in these sectors segregation has occurred such
that *w* or *y* are no longer accompanied by their + alleles and can thus be
expressed.

12.17   HAPLOIDIZATION AND MITOTIC CROSSING OVER

Genetic analysis, most notably by G. Pontecorvo and E. Käfer, has revealed
that segregant sectors can arise in stable diploid *Aspergillus* strains by two
mechanisms. The first is by a process known as **haploidization.** Presumably
because of mitotic nondisjunction (Section 6.17) or because one or more
chromosomes do not become effectively included in the spindle at metaphase, aneuploid cells are produced that are either hyperploid ($2n + 1$,
$2n + 2$, etc.) or hypoploid ($2n - 1$, $2n - 2$, etc.). The latter cells turn out to
be very unstable, exhibit poor growth, and continue to "throw off" chromosomes until they finally attain a stable haploid chromosome number. If,

during the process of haploidization, a chromosome that carries a + gene for which the cell is heterozygous is eliminated, the emerging cell line will display the phenotype of that gene's mutant allele.

The second way that segregant sectors can arise is by **mitotic crossing over** in the heterozygous diploids. During this process two homologous chromatids exchange segments at some time after the chromosomes have replicated. In the anaphase that follows, each centromere divides and one copy of each homologue passes to each daughter nucleus (recall that this is now **mitosis,** not meiosis).

Figure 12-19 diagrams two possible mitotic exchange events that might occur among the chromosomes we have been considering. In Figure 12-19a an exchange has occurred between the *pab* and the *y* loci. If we number the chromatids 1 to 4 and a nucleus receives chromatids 1 and 3, it will have the genotype $\frac{+\ +}{pab\ +}$ and will generate a normal phenotype. It will also generate a normal phenotype if it receives chromatids 2 and 3, or, of course, the nonrecombinant 1 and 4. Nuclei inheriting chromatids 2 and 4 will, however, have the genotype $\frac{+\ y}{pab\ y}$ and will thus permit the expression of the *y* mutant phenotype. Clones of such nuclei can produce segregant sectors.

Figure 12-19b diagrams an exchange between chromatids 2 and 3, but this time it occurs somewhere between the centromere and the *pab* locus. Nuclei inheriting chromatids 2 and 4 will in this case be homozygous for *pab* as well as for *y*.

These alternate outcomes can be stated more formally by introducing the

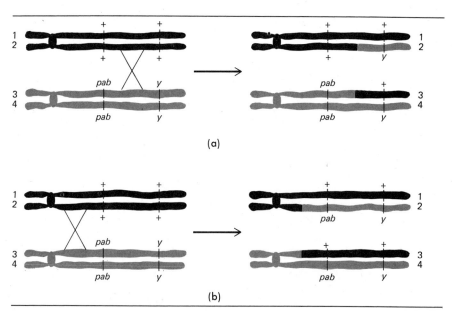

(a)

(b)

Figure 12-19 Mitotic crossing over in *Aspergillus nidulans.*

terms **proximal** and **distal** as referring to positions relatively close to and relatively distant from the centromere. Because the centromere divides at mitosis, any genes proximal to the site of a single genetic exchange in a given chromosome arm will always find themselves associated with the same alleles in daughter nuclei as they were in the parent nucleus. Thus in Figure 12-19a, where the parent nucleus is heterozygous for *pab* and its + allele, all possible daughter nuclei will also be heterozygous for *pab* and +, an outcome following directly from the usual mode of gene transmission accompanying mitosis. In contrast, all genes linked in coupling and distal to the site of a single exchange have the chance of being transmitted to a daughter nucleus in the homozygous condition. Thus in Figure 12-19a, *y* and its + allele are distal to the exchange and segregant nuclei can be homozygous for either allele; in Figure 12-19b both *y* and *pab* and their + alleles are distal to the exchange and segregant nuclei can be homozygous at both loci.

12.18   GENETIC ANALYSIS VIA THE PARASEXUAL CYCLE

The series of events we have been describing—the fusion of two haploid nuclei in a heterokaryon and the return of the resultant diploid nucleus to a recombinant haploid state via mitotic crossing over and/or haploidization—has been called the **parasexual cycle.** The parasexual cycle has been used for genetic analysis of *A. nidulans* in two ways.

First, it is possible to assign markers to linkage groups by following their patterns of inheritance during haploidization. A haploid segregant will inherit one copy of each chromosome type, and this collection of *n* chromosomes should represent a random sampling of the two parental sets; in other words, some of the chromosomes that are thrown off during haploidization will be of maternal origin and some will be of paternal origin. It follows that if two markers happen to be located on homologous chromosomes in a diploid nucleus, they will appear together in a haploid nucleus only if crossing over occurs in the appropriate region before the process of haploidization gets underway. Since this is an unlikely event, it can be stated that **maternal and paternal markers of the same linkage group will only very rarely be inherited together by a haploid nucleus during haploidization.** To give an example, we can imagine a diploid made between a certain *adenine*-requiring strain of *A. nidulans* known as ad_8 and a strain carrying *y, w,* and *lys* (*lysine-requiring*). The genotype of the diploid can be provisionally written as $\dfrac{ad_8 \; + \; + \; +}{+ \;\; y \;\; w \;\; lys}$. Haploid segregants of this diploid are often found to carry ad_8 and *w* or ad_8 and *lys* but almost never both ad_8 and *y.* It can thus be concluded that ad_8 and *y* are linked, so that the diploid's genotype is more correctly written as $\dfrac{ad_8 \; + \quad + \quad +}{+ \quad y \quad w \quad lys}$.

The second use of the parasexual cycle is to determine **mitotic recombination frequencies** between genes along a chromosome. We can take as an example a diploid A. *nidulans* strain heterozygous for *y, ad-14, pab-1, pro-1* (*proline-requiring*) and the + alleles of these genes. Haploidization data have indicated that the markers are all linked and the order of the markers has been determined by meiotic analysis to be *ad-14*, centromere, *pro-1*, *pab-1, y*. The four recessive alleles are introduced into the diploid in coupling (Figure 12-20). Segregants that bear *yellow* conidia are selected

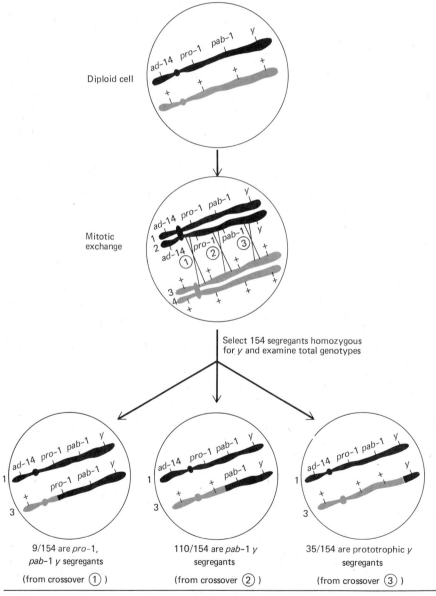

Diploid cell

Mitotic exchange

Select 154 segregants homozygous for *y* and examine total genotypes

9/154 are *pro-1*, *pab-1 y* segregants

(from crossover ①)

110/154 are *pab-1 y* segregants

(from crossover ②)

35/154 are prototrophic *y* segregants

(from crossover ③)

Figure 12-20 Genetic analysis utilizing mitotic recombination in *Aspergillus nidulans*.

first, these being the most easily detected and *y* being the most distal marker on the chromosome. The complete genotypes of the segregant nuclei are then determined by the phenotypes they produce. It is found (Figure 12-20) that among 154 segregants homozygous for *y*, 9 (5.5 percent) are homozygous for *pab-1* and *pro-1*, 110 (72 percent) are homozygous for *pab-1*, and 35 (22.5 percent) are prototrophic. Because almost none of the *y* segregants is found to be adenine-requiring, *ad-14* is most likely located on the other arm of the chromosome; from Figure 12-20 it should be apparent that mitotic crossing over in one arm of the chromosome is without effect on the segregation of alleles in the other arm.

Mitotic recombination frequencies cannot be converted into map distances in the same way that meiotic data are treated, for the total number of nuclei present in a hyphal mass (a number that corresponds to "total progeny" in conventional genetic analysis) is unknown. Moreover, it is not possible to recover all the products of a recombinational event, since only some will have a recognizable phenotype (in Figure 12-20, only the cells receiving chromatid 1 and a recombinant chromatid 3 are scored as *y* segregants). A mitotic map therefore simply utilizes the recombination percentages we have obtained above as map distances separating the markers. Obviously the scale established for one chromosome arm will not necessarily bear any relation to the scale established for any other arm.

In an organism such as *A. nidulans,* where both a sexual and a parasexual cycle are known, it becomes possible to compare mitotic and meiotic map distances by expressing the latter on a percentage basis similar to the former. Typical results, shown in Table 12-5 demonstrate little correspondence between the meiotic and mitotic distances. Thus there are undoubtedly fundamental differences between the processes of mitotic and meiotic crossing over, but their bases are unknown.

SOMATIC CELL GENETICS

12.19 PROPERTIES OF SOMATIC CELLS

When cells are taken from a sexually-reproducing "higher" eukaryote and are cultured under defined conditions *in vitro,* the resultant culture is said to represent a **line** of **somatic cells.** Cultured somatic cells are of two types. **Primary** somatic cells are taken directly from a plant or animal and maintained for a finite period; these are most likely to retain the chromosomal make-up of the parental organism. **Established cell lines,** on the other hand, have been selected for their ability to grow indefinitely in culture. Such cells frequently derive from malignant tissues, and their chromosomal makeup is typically different from that of the other tissues of the parental organism: chromosomes may be lost, added, deleted, translocated, and so on.

Somatic cells have been put to three genetic uses, the first two of which

Table 12-5   A Comparison of Meiotic and Mitotic Linkage Data in *A. nidulans*

Linkage group I, "left" arm (85 meiotic units total)

Marker	su		ribo1		an1		ad14	
Interval	I		II		III		IV	
Mitotic	23		7.4		6.2		63.4	
Meiotic	45.8		22.4		8.2		23.6	

Linkage group I, "right" arm (44 meiotic units)

Marker		pro1		paba1		y
Interval	V		VI		VII	
Mitotic	5.5		72.0		22.5	
Meiotic	46.1		18.0		35.8	

Linkage group II, "left" arm (46 meiotic units)

Marker	Acr1		w2	
Interval	VIII		IX	
Mitotic	14.7		85.3	
Meiotic	54.3		45.7	

From J. A. Roper, in *The Fungi*, Vol. 2, G. C. Ainsworth and A. S. Sussman, Eds. New York: Academic, 1968, p. 604.

are described here. First, they have been subjected to mutagenesis, selection, parasexual manipulations, and complementation analysis, the goal being to analyze the genome of higher eukaryotes under conditions that mimic the analysis of, say, *E. coli* or *Chlamydomonas*. Second, human cells have been fused with rodent cells; the resultant **hybrid cells** are found to eliminate human chromosomes preferentially, a phenomenon that has permitted the assignment of many human genes to particular chromosomes. Finally, cells in various states of differentiation have been fused in order to determine which differentiated state is "dominant" over the other.

12.20   MONOSPECIFIC SOMATIC CELL GENETICS

For a somatic cell line to be amenable to genetic analysis, it should be haploid or "functionally haploid" so that recessive mutations can be induced and expressed. This requirement is most easily met by higher plant cells, since cultures can be started from haploid pollen cells and, as noted in Section 6.17, the haploid state is generally tolerated by higher plants. Such clones can then be mutagenized and subjected to selection procedures much as those described in Chapter 6 for haploid microorganisms. Thus amino-acid and vitamin auxotrophs of cultured *Nicotiana tabacum* cells have been isolated by P. Carlson using the BuDR enrichment procedure

(Section 6.3), streptomycin-resistant lines of *Petunia* have been selected from streptomycin-containing media; and so on. Of direct agricultural interest are attempts to isolate mutant plant cell lines resistant to fungal or bacterial pathogens or to such adverse conditions as cold or wind; since it is possible to induce plant somatic cell lines to develop into complete plants that produce seed, such new cell lines are possible candidates for new and more vigorous crop plants.

True haploidy may be incompatible with animal cell viability in culture since attempts to produce haploid animal cell lines have so far been unsuccessful. The Chinese Hamster Ovary **(CHO)** cell line has, however, proven to be "functionally" haploid at many loci: although CHO cells possess a near-diploid amount of DNA and most of the original bands of the Chinese hamster karyotype remain, the chromosomes have been extensively rearranged and this appears to have resulted in the effective inactivation of many loci throughout the genome. Consequently, while CHO cells are demonstrably diploid for certain genes, they are "hemizygous" for many others, and these can be marked by recessive gene mutations. Thus F. Kao, T. Puck, and others have isolated CHO glycine, proline, and adenine auxotrophs, temperature-sensitive conditional lethals, cells with enzyme deficiencies, and so on.

After selection of an appropriate cell line, the next stage in the genetic analysis of somatic cells requires that the genomes of such mutant cells be brought together to allow some sort of interaction between them. The most common technique employed to date, called **somatic cell fusion,** was pioneered by G. Barski, B. Ephrussi, and H. Harris. Two marked cell types are cultured together in the presence of **Sendai virus** particles; the viral capsids are capable of causing cells to fuse together, and the resultant **heterokaryons** typically go on to fuse their nuclei and become **synkaryons.** The heterokaryons and synkaryons are usually selected from the background of unfused and fused homokaryotic cells on the basis of complementation between the two markers introduced. For example, two CHO glycine auxotrophs, *gly A* and *gly B,* may be grown and exposed to Sendai virus in a glycine-supplemented medium; the culture is then transferred to a glycine-free medium on which only the complementing heterokaryons can grow.

The final events in a fungal parasexual cycle—mitotic recombination and chromosome elimination—are as yet very difficult or impossible to manipulate in somatic cells. Detection of recombination awaits larger collections of linked markers, and chromosome elimination is very rare in CHO/CHO fused cells. Nonetheless, L. Chassin has been able to ascertain that the loss of the chromosome bearing the *gly A* marker is not accompanied by the loss of *gly B* and vice-versa, indicating that the two markers lie in different chromosomes; in the terminology of somatic cell genetics, *gly A* and *gly B* are not **syntenic** (that is, located in the same strand, the term "linked" being reserved for results of sexual crosses).

12.21 INTERSPECIFIC SOMATIC CELL GENETICS AND HUMAN CHROMOSOME MAPPING

The experiments described in the previous section concerned the fusion and interaction of cells of the same species. It is also possible to fuse cells of different species, forming **interspecific cell hybrids.** In the case of plant cells, development of this approach has great agronomic potential: the desirable features of two crop plants might, for example, be combined. Thus far, only two closely related tobacco species have generated a hybrid plant from a fused somatic cell, but an increasing number of laboratories are investigating such approaches.

For animal cells, interspecific somatic fusion has been found to be followed by a very efficient elimination of chromosomes. Usually, although not inevitably, the chromosomes of one species are lost in preference to the other. Specifically, in human-rodent hybrids, human chromosomes are preferentially eliminated, a phenomenon that has had far-reaching consequences for human genetics since it allows the assignment of human genes to human chromosomes. The procedures are outlined below.

Human-Rodent Cell Hybrids   Typical conditions for obtaining human-rodent cell hybrids are as follows. Primary human cells, usually leucocytes or fibrocytes, are obtained from donors with a known karyotype and genetic constitution. These are then mixed with CHO cells or with cells from established mouse lines that carry conditional mutations; for example, CHO glycine auxotrophs may be used. Following Sendai-mediated fusion, cells are transferred to glycine-free medium where unfused CHO cells fail to grow while hybrid cells, containing human wild-type genes for glycine prototrophy, continue to flourish. Unfused human leucocytes also fail to proliferate in culture and are gradually diluted out, leaving only human-CHO hybrids.

Chromosome Elimination   While it is not yet clear how chromosomes are lost from hybrid cells, it is found experimentally that after rodent-human hybrids have undergone about 30 rounds of mitosis, they contain a full or near-full complement of rodent chromosomes but an average of only 7 human chromosomes (the range extending from 1 to 20).

When the karyotypes of such cells are analyzed, it is generally found that the human chromosomes have been lost in a random fashion. In some cases, however, it is possible to assure that a particular human chromosome will be among those retained. This is done by including an additional selective feature to the growth medium. For example, if the parent mouse cell line lacks the enzyme **thymidine kinase** (TK) and the human parent is wild-type (Tk/Tk), then if the hybrid cells are plated to a medium known as **HAT** (for **h**ypoxanthine, **a**minopterin, and **t**hymidine), the aminopterin is lethal to the cells unless the Tk gene is present. All human-mouse hybrids that survive on HAT medium must therefore possess at least one copy of the human chromosome (or chromosome sector) bearing the Tk gene.

Detecting Human Traits in Hybrid Cells Once stable clones of hybrid cells have been obtained, the next step is to ask which clones express a particular human trait. If it is a **selected** human trait (that is, if the parent rodent cells lack the activity, as with the TK-deficient mouse line), then it should be expressed by all the surviving clones. If it is an **unselected** trait, on the other hand, then individual clones must be analyzed for the particular human gene product of interest. To give a specific example, the human form of lactate dehydrogenase B(LDH-B) has a different electrophoretic mobility pattern (Box 7.1) from the corresponding mouse or CHO enzymes and can be detected in gels by a specific staining reaction. Clones that carry the human form of the enzyme are therefore identified by subjecting samples of numerous hybrid clones to the electrophoresis assay.

Identifying Human Chromosomes in Hybrid Cells The final step in assigning the human gene to a chromosome is to determine which human chromosome is shared in common by all the clones exhibiting the particular human trait; this chromosome is then assumed to carry the relevant gene. Here, of course, banded chromosome preparations (Figure 2-10) are critical to assure correct identifications. By such analyses the human *Tk* gene has been assigned to chromosome 17 and the human *Ldh-a* gene to chromosome 11. Markers that segregate together, moreover, can be assumed to be syntenic: clones exhibiting human LDH-A activity, for example, prove to carry the human cell-surface antigen A_L as well, indicating that both gene loci lie on chromosome 11.

Translocations permit more finely-tuned assignments of markers to particular regions of chromosomes. One example, shown in Figure 12-21,

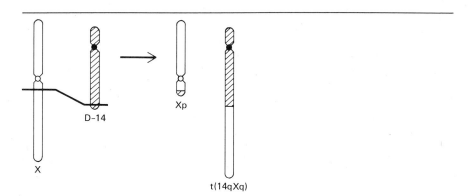

Figure 12-21 The *KOP X-14* translocation. The normal chromosomes X and 14 are shown on the left with the breakpoints indicated. The translocation products are shown on the right. Three X-linked genes, *Hprt*, *Pgk*, and *Gpd*, as well as the autosomally linked *Pnp*, were expressed in hybrids retaining the larger translocation product, indicating assignment of *Hprt*, *Gpd*, and *Pgk* to the long arm of the X chromosome, and the assignment of *Pnp* to chromosome 14. (From Ruddle, F. H., and Creagan, R. P., *Ann. Rev. Genetics* 9:407, 1975.)

involves a translocation break at band Xq12 (see Figure 2-12) and transposition of the long arm of the X chromosome to chromosome 14. This rearrangement causes three X-linked genes [*Hgprt* (hypoxanthine-guanine phosphoribosyl transferase [HGPRT]), *Pgk* (phosphoglycerate kinase), and *G6pd* (Glucose-6-phosphate dehydrogenase)] to become syntenic with the autosomal gene *pNp* (nucleoside phosphorylase). The translocation therefore localizes all three sex-linked genes to the Xq12 ⟶ Xq-terminus sector of the sex chromosome (and the *pNp* gene to chromosome 14). A second translocation, extending from Xq24 to the terminus, leaves the *Pgk* gene with the X but moves the *Hgprt* and *G6pd* genes. A still more terminal translocation leaves *Pgk* and *Hgprt* behind and moves *G6pd*. Thus the genes can be ordered as *Pgk Hgprt G6pd* in the long arm of the X chromosome (see Figure 12-22).

Human Chromosome Assignments and Linkages   Figure 12-22 and Table 12-6 summarize the present status of human chromosome mapping; much additional information will almost certainly be available by the time this text is in print. Some of the linkage assignments were made initially, or confirmed by, familial studies of inherited traits. As we saw in Section 4.11, X-chromosome assignments are readily made for all traits exhibiting a sex-linked pattern of inheritance; the *Hgprt* gene, for example, is marked in the human population by the *hgprt* mutation which causes an HGPRT deficiency in hemizygous males and leads to a purine-metabolism affliction known as the **Lesch-Nyhan syndrome.** Other familial studies permit a correlation between the inheritance of a certain trait and the inheritance of a particular chromosome aberration. For example, in one family, a structural variant of the heterochromatic region of chromosome 1 was found to be transmitted in parallel with a marker for the *Fy* (Duffy blood group system) locus, while in another family, loss of expression of acid phosphatase, red-cell type, is associated with a deletion in the short arm of chromosome 2, thus localizing the *Acp-1* gene to the 2p23 region.

12.22   GENE TRANSFER

A promising new approach to mapping animal chromosomes (and possibly higher plant chromosomes as well) is known as **gene transfer.** The various modes of gene transfer resemble bacterial transformation in that all involve isolating DNA from a genetically marked species and transferring that marker to a second species where it is both transmitted and expressed. The analogy with bacterial transformation may be incomplete in that there is presently no evidence that the donor DNA becomes integrated into the homologous sector of the host chromosome.

Claims for the occurrence of eukaryotic gene transfer were made for *D. melanogaster* by A. Fox and S. Yoon in 1966, when they found that

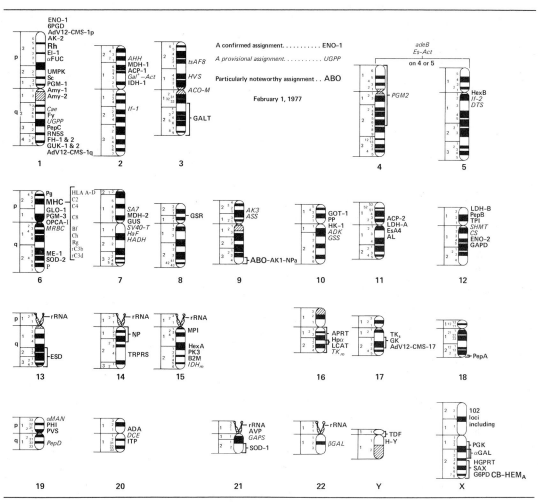

Figure 12-22 Genetic map of human chromosomes. Details of the banding patterns and numbering system are found in Figure 2-12. Key to symbols is found in Table 12-6. (From McKusick, V., and F. Ruddle, *Science* **196**:390, 1977.)

exposure of *vermilion brown* eggs to adult wild-type DNA would occasionally produce flies with wild-type traits. More recently, the yield of such "transformed" flies has been shown to be greatly enhanced if the DNA is physically injected into the egg. A comparable manipulaton is now possible with somatic cells: isolated metaphase chromosomes from one cell type can be taken up (probably phagocytically) by cultured somatic cells and the donor genes are both inherited and expressed by the recipients. Specifically, wild-type human metaphase chromosomes were presented to mouse A9 cells, which lack HGPRT, under conditions that select for HGPRT$^+$ cells. The

Table 12-6 Mapped Human Gene Loci

ABO	ABO blood group (chr. 9)	C4	Complement component-4 (chr. 6)	GAPS	Phosphoribosyl glycineamide synthetase (chr. 21)
ACO	Aconitase, mitochondrial (chr. 3)	C8	Complement component-8 (chr. 6)	Gc	Group-specific component (chr. 4)
ACO-S	Aconitase, soluble (chr. 9)	Cae	Cataract, zonular pulverulent (chr. 1)	GK	Galactokinase (chr. 17)
ACP-1	Acid phosphatase-1 (chr. 2)	CB	Color blindness (deutan and protan) (X chr.)	GLO-1	Glyoxylase I (chr. 6)
ACP-2	Acid phosphatase-2 (chr. 11)	Ch	Chido blood group (chr. 6)	GOT-1	Glutamate oxaloacetic transaminase-1 (chr. 10)
ADA	Adenosine deaminase (chr. 20)	CS	Citrate synthase, mitochondrial (chr. 12)	G6PD	Glucose-6-phosphate dehydrogenase (X chr.)
adeB	FGAR amidotransferase (chr. 4 or 5)	DCE	Desmosterol-to-cholesterol enzyme (chr. 20)	GSR	Glutathione reductase (chr. 8)
ADK	Adenosine kinase (chr. 10)	DTS	Diphtheria toxin sensitivity (chr. 5)	GSS	Glutamate-γ-semialdehyde synthetase (chr. 10)
AdV12-CMS-1p	Adenovirus-12 chromosome modification site-1p (chr. 1)	E1-1	Elliptocytosis-1 (chr. 1)	GUK-1 & 2	Guanylate kinase-1 & 2 (S & M) (chr. 1)
AdV12-CMS-1q	Adenovirus-12 chromosome modification site-1q (chr. 1)	E11S	Echo 11 sensitivity (chr. 19)	GUS	Beta-glucuronidase (chr. 7)
AdV12-CMS-17	Adenovirus-12 chromosome modification site-17 (chr. 17)	ENO-1	Enolase-1 (chr. 1)	HADH	Hydroxyacyl-CoA dehydrogenase (chr. 7)
AHH	Aryl hydrocarbon hydroxylase (chr. 2)	ENO-2	Enolase-2 (chr. 12)	HaF	Hageman factor (chr. 7)
AK-1	Adenylate kinase-1 (chr. 9)	Es-Act	Esterase activator (chr. 4 or 5)	HEM$_A$	Classic hemophilia (X chr.)
AK-2	Adenylate kinase-2 (chr. 1)	EsA4	Esterase-A4 (chr. 11)	Hex A	Hexosaminidase A (chr. 15)
AK-3	Adenylate kinase-3 (chr. 9)	ESD	Esterase D (chr. 13)	Hex B	Hexosaminidase B (chr. 5)
AL	Lethal antigen: 3 loci (a1, a2, a3) (chr. 11)	FH-1 & 2	Fumarate hydratase-1 and 2 (S and M) (chr. 1)	HGPRT	Hypoxanthine-guanine phosphoribosyltransferase (X chr.)
Amy-1	Amylase, salivary (chr. 1)	αFUC	Alpha-L-fucosidase (chr. 1)	HK-1	Hexokinase-1 (chr. 10)
Amy-2	Amylase, pancreatic (chr. 1)	Fy	Duffy blood group (chr. 1)	HLA	Major histocompatibility complex (chr. 6)
ASS	Argininosuccinate synthetase (chr. 9)	Gal+-Act	Galactose + activator (chr. 2)	Hpα	Haptoglobin, alpha (chr. 16)
APRT	Adenine phosphoribosyltransferase (chr. 16)	αGAL	α-Galactosidase (Fabry disease) (X chr.)	HVS	Herpes virus sensitivity (chr. 3)
AVP	Antiviral protein (chr. 21)	βGAL	β-Galactosidase (chr. 22)	H-Y	Y histocompatibility antigen (Y chr.)
Bf	Properdin factor B (chr. 6)	GALT	Galactose-1-phosphate uridyltransferase (chr. 3)	If-1	Interferon-1 (chr. 2)
β2M	β2-Microglobulin (chr. 15)	GAPD	Glyceraldehyde-3-phosphate dehydrogenase (chr. 12)	If-2	Interferon-2 (chr. 5)
C2	Complement component-2 (chr. 6)			IDH-1	Isocitrate dehydrogenase-1 (chr. 2)

IDH_m	Isocitrate dehydrogenase, mitochondrial (chr. 15)	P	P blood group (chr. 6)	SAX	X-linked species (or surface) antigen (X chr.)
ITP	Inosine triphosphatase (chr. 20)	PepA	Peptidase A (chr. 18)	Sc	Scianna blood group (chr. 1)
		PepB	Peptidase B (chr. 12)	SHMT	Serine hydroxymethyltransferase (chr. 12)
LCAT	Lecithin-cholesterol acyltransferase (chr. 16)	PepC	Peptidase C (chr. 1)		
		PepD	Peptidase D (chr. 19)	SOD-1	Superoxide dismutase-1 (chr. 21)
LDH-A	Lactate dehydrogenase A (chr. 11)	Pg	Pepsinogen (chr. 6)	SOD-2	Superoxide dismutase-2 (chr. 6)
LDH-B	Lactate dehydrogenase B (chr. 12)	PGK	Phosphoglycerate kinase (X chr.)	SV40-T	SV40-T antigen (chr. 7)
αMAN	Lysosomal α-D-mannosidase	PGM-1	Phosphoglucomutase-1 (chr. 1)	TDF	Testis determining factor (Y chr.)
MDH-1	Malate dehydrogenase-1 (chr. 2)	PGM-2	Phosphoglucomutase-2 (chr. 4)	TK_m	Thymidine kinase, mitochondrial (chr. 16)
MDH-2	Malate dehydrogenase, mitochondrial (chr. 7)	PGM-3	Phosphoglucomutase-3 (chr. 6)	TK_s	Thymidine kinase, soluble (chr. 17)
ME-1	Malic enzyme-1 (chr. 6)	6PGD	6-Phosphogluconate dehydrogenase (chr. 1)	TPI	Triosephosphate isomerase (chr. 12)
MHC	Major histocompatibility complex (chr. 6)	PHI	Phosphohexose isomerase (chr. 19)	TRPRS	Tryptophanyl-tRNA synthetase (chr. 14)
MPI	Mannosephosphate isomerase (chr. 15)	PK3	Pyruvate kinase-3 (chr. 15)	tsAF8	Temperature-sensitive (AF8) complementing (chr. 3)
MRBC	B-cell receptor for monkey red cells (chr. 6)	PP	Inorganic pyrophosphatase (chr. 10)		
		PVS	Polio sensitivity (chr. 19)	UGPP	Uridyl diphosphate glucose pyrophosphorylase (chr. 1)
NP	Nucleoside phosphorylase (chr. 14)	Rg	Rodgers blood group (chr. 6)	UMPK	Uridine monophosphate kinase (chr. 1)
NPa	Nail-patella syndrome (chr. 9)	Rh	Rhesus blood group (chr. 1)		
		rRNA	Ribosomal RNA (chr. 13, 14, 15, 21, 22)		
OPCA-1	Olivopontocerebellar atrophy I (chr. 6)	rC3b	Receptor for C3b (chr. 6)		
		rC3d	Receptor for C3d (chr. 6)		
		RN5S	5S RNA gene(s) (chr. 1)		
		SA7	Species antigen 7 (chr. 7)		

From: V. McKusick, and F. Ruddle, *Science* **196**:390, 1977.

resultant "transformants" proved to synthesize human HGPRT enzyme, identified both electrophoretically and immunochemically, but did not possess the closely linked genes *Pgk* and *Gpd* (Figure 12-22) or 18 tested autosomal markers. Thus it appears that only a small portion of the donor chromosomal DNA—perhaps 5×10^7 base pairs—is stably associated with the recipient nucleus. The linkage of two genes in such a short sector of the genome would presumably be easy to detect.

APPROACHES TO SOLVING MAPPING PROBLEMS IN EUKARYOTES

12.23 MULTIPLE MARKERS IN ONE CROSS

Problem The F_1 progency of a cross between an organism *AABBDD* and an organism *aabbdd* were backcrossed to the recessive parent and the following phenotypic frequencies were observed in the F_2:

ABD	20
abD	20
Abd	5
aBd	5
abd	20
ABd	20
aBD	5
AbD	5

What is the relationship of the three markers *ABD* to each other (assuming that they are on the same chromosome)?

Solution Problems of this sort are best solved by considering pairs of markers separately. Thus for the *AB* pair, we can classify the test-cross progeny (ignoring the recessive chromosome) as

$$
\begin{array}{lll}
AB & 20 + 20 = 40 \,\Big\} & \\
ab & 20 + 20 = 40 \,\Big\} & 80 \\
Ab & 5 + 5 = 10 \,\Big\} & \\
aB & 5 + 5 = 10 \,\Big\} & 20
\end{array}
$$

And can draw their linkage as A _____ 20 _____ B

For the *BD* pair we have

$$
\begin{array}{lll}
BD & 20 + 5 = 25 \,\Big\} & \\
bd & 5 + 20 = 25 \,\Big\} & 50 \\
Bd & 5 + 20 = 25 \,\Big\} & \\
bD & 20 + 5 = 25 \,\Big\} & 50
\end{array}
$$

The genes *B* and *D* are not linked genetically, even though they reside in the same chromosome

For the *AD* pair we have

AD	20 + 5 = 25	*A* and *D* are similarly
ad	5 + 20 = 25	not linked.
Ad	5 + 20 = 25	
aD	20 + 5 = 25	

12.24 DETERMINING PHENOTYPIC CLASSES

Problem In snails, pink shell color may be due to the homozygous condition of either or both of the recessive genes *a* and *b*. Brown shell color is dependent upon the presence of both dominant *A* and *B* genes. Pink male snails of genotype *AAbb* were crossed with pink females of genotype *aaBB*. The F_1 brown females were crossed with pink males of genotype *aabb*. There were 710 progeny of which 125 were brown and 585 were pink.
a. What ratio is expected for unlinked genes?
b. Based on the current data, what is the percent recombinants?
c. What is the map distance?
d. In what percent of the tetrads did a crossover between the two genes occur?

Solution

1. Write out all available information in genetic shorthand. Begin by assuming partial *AB* linkage

$$\text{Genotype of } F_1 \text{ brown females: } \frac{Ab}{aB}$$

$$\text{Genotype of testcross males: } \frac{ab}{ab}$$

Possible genotypes of testcross progeny:

$$\text{Parental: } \frac{Ab}{ab} \text{ and } \frac{aB}{ab}$$

$$\text{Recombinant: } \frac{AB}{ab} \text{ and } \frac{ab}{ab}$$

2. Recognize that the 125 brown progeny must correspond to the $\frac{AB}{ab}$ recombinant types since these are the only class with both *A* and *B* genes.

3. Assume reciprocal recombination, so that there should also be 125 $\frac{ab}{ab}$ pink snails among the progeny, giving a total of 250 recombinants.

4. The remaining $585 - 125 = 460$ pink testcross progeny should have the reciprocal parental genotypes, that is,

$$\left. \begin{aligned} \frac{Ab}{ab} &= 230 \\[1em] \frac{aB}{ab} &= 230 \end{aligned} \right\} \quad 460 \text{ parental}$$

5. Note that 460 parental + 250 recombinant yields the given total of 710.
6. Now proceed to answer the questions.
   (a) For unlinked genes, one would expect equal proportions of AB, ab, Ab, and aB progeny and a ratio of 75 percent pink, 25 percent brown. The excess of pinks (82 percent) signals linkage.
   (b) Percent recombination $= \dfrac{250}{710} = 35$ percent
   (c) Map distance = 35 map units
   (d) A crossover occurred in 70 percent of the tetrads.

References

Linkage in *Drosophila* and Early Linkage Studies

Bridges, C. B., and K. S. Brehme. *The Mutants of Drosophila melanogaster.* Washington, D.C.: Carnegie Institute of Washington Publication 552, 1944.

*Demerec, M., and M. E. Hoover. Three related X-chromosome deficiencies in *Drosophila, J. Heredity* **27:**206–212 (1936).

*Fox, A. S., S. D. Parzen, H. Salverson, and S. B. Yoon. "Gene transfer in *Drosophila melanogaster:* Genetic transformations induced by the DNA of transformed stocks," *Genet. Res.* **26:**137–147 (1975).

*Germeraad, S. "Genetic transformation in *Drosophila* by microinjection of DNA," *Nature* **262:**229–231 (1976).

Hilliker, A. J. "Genetic analysis of the centromeric heterochromatin of chromosome 2 of *Drosophila melanogaster.* Deficiency mapping of EMS-induced lethal complementation groups," *Genetics* **83:**765–782 (1976).

*Janssens, F. A. "Spermatogénèse dans les batraciens. V. La théorie de la chiasmatypie. Nouvelles interprétation des cinèses de maturation," *Cellule* **25:**387–411 (1909).

*Judd, B. H., M. W. Shen, and T. C. Kaufman. "The anatomy and function of a segment of the X chromosome of *Drosophila melanogaster,*" *Genetics* **71:**139–156 (1972).

Lindsley, D. L., and E. H. Grell. *Genetic Variations of Drosophila melanogaster,* Washington, D.C.: Carnegie Institution of Washington Publication 627, 1968.

*Morgan, T. H. "An attempt to analyze the constitution of the chromosomes on the basis of sex-limited inheritance in *Drosophila,*" *J. Exp. Zool.* **11:**365–414 (1911).

*Morgan, T. H. "Random segregation versus coupling in Mendelian inheritance," *Science* **34:**384 (1911). [Reprinted in L. Levine, *Papers on Genetics,* St. Louis: Mosby, 1971.]

Morgan, T. H., A. H. Sturtevant, H. J. Muller, and C. B. Bridges. *The Mechanism of Mendelian Heredity.* New York: Henry Holt, 1915.

*Denotes articles described specifically in the chapter.

*Painter, T. S. "A new method for the study of chromosome rearrangements and the plotting of chromosome maps," *Science* **78:**585–586 (1933). [Reprinted in J. A. Peters, Ed. *Classical Papers in Genetics,* Englewood Cliffs, N.J.: Prentice-Hall, 1959.]

*Punnett, R. C. "Reduplication series in sweet peas. II," *J. Genetics* **6:**185–193 (1917).

*Sturtevant, A. H. "The linear arrangement of six sex-linked factors in *Drosophila,* as shown by their mode of association," *J. Exp. Zool.* **14:**43–59 (1913).

*Sutton, W. S. "The chromosomes in heredity," *Biol. Bull.* **4:**231–251 (1903).

Linkage in Maize

Creighton, H. S., and B. McClintock. "A correlation of cytological and genetical crossing-over in *Zea mays,*" *Proc. Natl. Acad. Sci. U.S.* **17:**492–497 (1931). [Reprinted in J. A. Peters, Ed. *Classical Papers in Genetics,* Englewood Cliffs, N.J.: Prentice-Hall, 1959.]

*Darlington, C. D. "The origin and behavior of chiasmata. VII. *Zea mays,*" *Z. Vererbungslehre* **67:**96–114 (1934).

Kirk, D., and R. N. Jones. "Nuclear genetic activity in B-chromosomes of rye, in terms of the quantitative interrelationships between nuclear protein, nuclear RNA and histone," *Chromosoma* **31:**241–254 (1970).

*Rakha, F. A., and D. S. Robertson. "A new technique for the production of A-B translocations and their use in genetic analysis," *Genetics* **65:**223–240 (1970).

*Rhoades, M. M. "Meiosis in maize," *J. Hered.* **41:**58–67 (1950).

*Stadler, L. J. "The variability of crossing-over in maize," *Genetics* **11:**1–37 (1926).

Linkage in Other Higher Eukaryotes

Callan, H. G., and L. Lloyd. "Lampbrush chromosomes of crested newts *Triturus cristatus* (Laurenti)," *Phil. Trans. Royal Soc. Series B* **243:**135–219 (1960).

Lawler, S. D., and J. H. Renwick. "Blood groups and genetic linkage," *Brit. Med. Bull.* **15:**145–149 (1959). [Reprinted in L. Levine, *Papers on Genetics.* St. Louis: Mosby, 1971.]

*McKusick, V. A., and F. H. Ruddle. "The status of the gene map of the human chromosomes," *Science* **196:**390–405 (1977).

Miller, D. A., and O. J. Miller. "Chromosome mapping in the mouse," *Science* **178:**949–954 (1972).

*Miller, O. J., and D. A. Miller. "Cytogenetics of the mouse," *Ann. Rev. Genet.* **9:**285–303 (1975).

Linkage by Tetrad Analysis

Emerson, S. "Fungal genetics," *Ann. Rev. Genetics* **1:**201–220 (1967).

Fincham, J. R. S. "Fungal genetics," *Ann. Rev. Genetics* **4:**347–372 (1970).

Mortimer, R. K., and D. C. Hawthorne. "Yeast genetics," *Ann. Rev. Microbiol.* **20:**151–168 (1966).

Pontecorvo, G. *Trends in Genetic Analysis.* New York: Columbia University Press, 1958.

Fine-Structure Mapping

Moore, C. W., and F. Sherman. "Role of DNA sequences in genetic recombination in the iso-1-cytochrome c gene of yeast. I. Discrepancies between physical distances and genetic distances determined by five mapping procedures," *Genetics* **79:**397–418 (1975).

*Sherman, F., M. Jackson, S. W. Liebman, A. M. Schweingruber, and J. W. Stewart. "A deletion map of *cyc1* mutants and its correspondence to mutationally altered iso-1-cytochromes c of yeast," *Genetics* **81:**51–73 (1975).

Parasexual Cycles

*Käfer, E. "An 8-chromosome map of *Aspergillus nidulans,*" *Adv. Genetics* **9:**105–145 (1958).

Mosses, D., K. L. Williams, and P. C. Newell. "The use of mitotic crossing-over for genetic analysis in *Dictyostelium discoideum: Mapping of linkage group II,*" *J. Gen. Microbiol.* **90**:247–259 (1975).

Pontecorvo, G., and E. Käfer. "Genetic analysis based on mitotic recombination," *Adv. Genetics* **9**:71–104 (1958).

Roper, J. A. "The parasexual cycle." In *The Fungi*, Vol. 2 (G. C. Ainsworth and A. S. Sussman, Eds.). New York: Academic, 1968, p. 589–617.

Somatic Cell Genetics

*Chaleff, R. S., and P. S. Carlson. "Somatic cell genetics of higher plants," *Ann. Rev. Genet.* **8**:267–278 (1974).

Chu, E. H. Y., and S. S. Powell. "Selective systems in somatic cell genetics," *Adv. Hum. Genet.* **7**:189–258 (1976).

Creagan, R. P., S. Chen, and F. H. Ruddle. "Genetic analysis of the cell surface: Association of human chromosome 5 with sensitivity to diphtheria toxin in mouse-human somatic cell hybrids," *Proc. Nat. Acad. Sci. U.S.* **72**:2237–2241 (1975).

Ephrussi, B., and M. C. Weiss. "Hybrid somatic cells," *Sci. Am.* **220**:26–35 (1969).

Goss, S. J., and H. Harris. "New method for mapping genes in human chromosomes," *Nature* **258**:680–684 (1975).

Kao, F., C. Jones, and T. T. Puck. "Genetics of somatic mammalian cells: Genetic, immunologic, and biochemical analysis with Chinese hamster cell hybrids containing selected human chromosomes," *Proc. Nat. Acad. Sci. U.S.* **73**:193–197 (1976).

Ledoux, L., R. Huart, and M. Jacobs. "DNA-mediated genetic correction of thiamineless *Arabidopsis thaliana*," *Nature* **249**:17–21 (1974).

*Nabholz, M., V. Miggiano, and W. Bodmer. "Genetic analysis with human-mouse somatic cell hybrids." *Nature* **223**:358–363 (1969).

Ruddle, F. H., and R. P. Creagan. "Parasexual approaches to the genetics of man," *Ann. Rev. Genet.* **9**:407–486 (1975).

Siminovitch, L. "On the nature of hereditable variation in cultured somatic cells," *Cell* **7**:1–11 (1976).

*Willecke, K., and F. H. Ruddle. "Transfer of the human gene for hypoxanthine-guanine phosphoribosyl transferase via isolated human metaphase chromosomes into mouse L-cells," *Proc. Nat. Acad. Sci. U.S.* **72**:1792–1796 (1975).

(a) (b)

(a) Burke Judd (University of Texas) studies gene-chromomere relationships in the *zeste-white* region of *Drosophila*. (b) Fred Sherman (University of Rochester) has analyzed the fine structure of the cytochrome-*c* gene in yeast.

Questions and Problems

1. A *Neurospora* zygote was constructed with the genotype *AaBb*. From ordered tetrad data, how would you tell
   (a) if *A* and *B* are linked
   (b) the genotype of the parental strains
2. For the cross

   list all the different kinds of *Neurospora* tetrads that can be obtained from a *single* crossover event. Recall that crossovers can occur between *a* and the centromere, between *a* and *b*, and below *b*. They might also involve different pairs of strands.
3. The following genes are known to be on the same autosome: *d, e, i, k, l, n.* On the basis of the following recombination frequencies, what is the gene order? (If you are right, you will know it).

1-d	25%
1-i	10%
d-e	4%
d-k	7%
n-e	6%
i-n	5%
k-l	18%
l-n	15%

4. (a) In a cross between an $a+c/+b+$ male and an *abc/abc* female, list the classes of offspring which you would expect to find in the F_1 in the order of decreasing frequency, assuming the order of the genes is as written. The $a - b$ distance is 16 and the $b - c$ distance is 6. (b) If you knew that the classes listed in the order of decreasing frequency were:

   (1) $a+c/abc$
       $+b+/abc$
   (2) $+bc/abc$
       $a++/abc$
   (3) abc/abc
       $+++/abc$
   (4) $ab+/abc$
       $++c/abc$

   What would be the order of the genes?

5. In *Drosophila* three linked genes, *a*, *b* and *c* gave the following results when *ac* females were crossed to *b* males and the F_1 inbred:

F_1: the phenotypes were + female and *ac* males

F_2:

Phenotypes	Number of Females	Number of Males
a	49	2
b	0	428
c	49	48
+	451	23
ab	0	47
ac	451	428
bc	0	1
abc	0	23
	1000	1000

(a) Are the three genes sex-linked or autosomal? (b) What are the parental and F_1 genotypes? (c) What is the percent recombination between the three genes? (d) What is the sequence of the three genes in their chromosome? (e) What is the coefficient of coincidence and is there any interference?

Hint: There are "hidden" recombinants whose phenotypes are different from their genotypes. Their frequency must be inferred from the data.

6. Two male plants are scored for the phenotypes of the pollen (review Sections 3.13 and 4.4) they produce. The possibilities are: *brown (br)*, *crinkly (cr)*, *glossy (gl)*, green *(gr)* and *small (s)*, and the results are:

Plant #1		Plant #2	
Pollen Type	**Number**	**Pollen Type**	**Number**
br cr gl	7	*gl s* +	129
+ + *cr*	330	*s* + +	4
+ *gl cr*	123	*gr* + +	138
+ *cr br*	670	+ *gl* +	873
+ *gl br*	366	+ + +	337
br + +	99	*s* + *gr*	846
+ + +	4	*gl* + *gr*	2
+ + *gl*	721	*gr gl s*	361
Total	2320	Total	2690

You learn from the published literature that

$$R_{gr,br} \cong 0.04$$

(a) Give the genotypes of plant #1 and plant #2 (b) Calculate recombination values. (c) Map the genes. (d) What aspect of the data (or the results derived from them) lacks internal consistency?

7. In the cross

$$\begin{array}{cc} ♀ & ♂ \\ \end{array}$$

$$Q \parallel \begin{array}{cc} q & q \\ \times \end{array} \parallel$$

$$k \parallel + \quad k \parallel$$

where:

Q = dominant mutation, quiet voice
q = normal allele, loud voice
k = recessive mutation, krinkly whiskers
$+$ = normal allele, straight whiskers

The F_1 progeny types are:
(1) Quiet, krinkly
(2) Loud, straight
(3) Loud, krinkly
(4) Quiet, straight
(a) What is the significance of these progeny types resulting from this cross?
(b) Design a cross to demonstrate that these genes are sex-linked. (c) How could the cross $\dfrac{Qk}{q+} \times \dfrac{Q+}{\longleftarrow}$ be used to obtain valid data for mapping studies?

8. Given that four chromatids can cross over in 2, 3, and 4 strand double crossovers and your general knowledge of genetics, which of the following statements are true? Explain your answers. (a) The more strands involved in crossing over, the fewer parental type chromatids and the more recombinant type chromatids. (b) The same two chromatids will very likely exchange genetic material via crossovers several times. (c) If two chromosomes (4 chromatids) work it right, when they get finished exchanging information no resultant chromatid will carry the same information as either parent. Diagram.

9. Suppose mapping studies showed that the gene for sickle cell anemia was near the end of a linkage group, while *in situ* hybridization of labelled hemoglobin mRNA from reticulocytes showed a major band of grains one-half way between the telomere and the centromere of a single chromosome type. Are these data compatible? Why or why not?

10. Diagram a series of not more than five crosses that will produce homozygous *white miniature* females starting with Morgan's two mutant male *Drosophila,* one *white* and one *miniature,* and a stock of wild-type females (recall that *w* and *m* are sex-linked genes).

11. Diagram the meiotic events occurring in the cross shown in Figure 12-1.

12. If a genetic marker is present in the X and Y chromosomes in the region of their homology, would it behave in crosses as a sex-linked trait? If not, how would its presence be detected in the X and Y chromosomes (assume that crossing over occurs equally in both sexes).

13. Duchenne-type muscular dystrophy is an inherited human disease, and one type of color blindness is sex-linked. Two healthy parents with normal color vision have fully normal daughters, half their sons are healthy but color blind, and the other half has normal color vision but Duchenne's disease. (a)

Is Duchenne's disease an autosomal or a sex-linked trait? (b) Diagram the probable genotypes of the parents. (c) Could such parents ever have a normal son? If so, under what circumstances?

14. In *Drosophila*, the markers *a*, *b*, and *c* are crossed in all three possible pairwise combinations to give the following recombination frequencies:

$$R_{a,b} = 0.06$$
$$R_{a,c} = 0.10$$
$$R_{b,c} = 0.06$$

What is the expected frequency of *a* + *c* eggs produced by a female with an $\dfrac{a\ \ b\ \ c}{+\ +\ +}$ genotype?

15. The recessive genes *an* (*anther* ear), *br* (*brachytic*), and *f* (*fine* stripe) all lie in chromosome 1 of maize. When a plant heterozygous for these markers is test-crossed with a homozygous recessive plant, the following results are obtained (data from R. A. Emerson):

Test-cross Progeny	Numbers
wild type	88
fine	21
brachytic	2
brachytic fine	339
anther	355
anther fine	2
anther brachytic	17
anther brachytic fine	55
	879

Determine the sequence of the three genes and their map distances.

16. Calculate the coefficient of coincidence for the phage cross given in Section 10.21.

17. In *Drosophila* the sex-linked genes *cut* (*ct*), *lozenge eyes* (*lz*), and *forked bristles* (*f*) are the following map distances apart: *cut* to *lozenge* 7.7 units, and *lozenge* to *forked* 29.0 units. Assuming that there is no interference, what are the expected numbers of genotypes out of 1000 flies recovered from the cross $\dfrac{ct\ \ lz\ \ f}{+\ +\ +} \times \dfrac{ct\ lz\ f}{\longrightarrow}$? What effect will a coefficient of coincidence of 0.5 have on your answer?

18. (a) In the absence of interference, what is the theoretical maximum recombination frequency (in percent) between two markers? (b) Can the total genetic map length of a chromosome ever exceed this value? Explain. (c) If a certain pair of homologues exhibits an average of 2.5 chiasmata per meiosis, what is the expected genetic map length of the entire chromosome?

19. Female corn flowers were made heterozygous for markers *c* (*colorless* aleurone), *wx* (*waxy* endosperm) and *sh* (*shrunken* endosperm) and were pollinated by male flowers homozygous for these markers. Derive a linkage map for these three loci from the following data, where traits are wild-type unless denoted as mutant.

Progeny Phenotype	Number
wild type	17,959
colorless	524
waxy	4455
shrunken	20
colorless, waxy	12
colorless, shrunken	4654
waxy, shrunken	509
colorless, waxy, shrunken	17,699

20. An analysis of tetrads in a cross of *Chlamydomonas* involving the loci *pf* (*paralyzed flagella*) and *nic* (requirement for the vitamin nicotinamide) gave the following results: $PD = 70$, $NPD = 2$, and $T = 28$. Calculate the linkage relationship between *pf* and *nic*.

21. In *Chlamydomonas* the genes designated here as *a*, *b*, and *c* are linked. The following are the tetrads recovered from a cross involving these three loci.

	(1)	(2)	(3)	(4)	(5)
	a *b* *c*	*a* *b* *c*	*a* *b* *c*	*a* *b* *c*	*a* + +
	a + +	*a* + *c*	*a* *b* *c*	*a* + +	*a* *b* +
	+ *b* +	+ *b* +	+ + +	+ *b* *c*	+ + *c*
	+ + *c*	+ + +	+ + +	+ + +	+ *b* *c*
Number of tetrads	10	10	440	160	10

	(6)	(7)
	a *b* +	*a* *b* *c*
	a + *c*	*a* *b* +
	+ *b* *c*	+ + *c*
	+ + +	+ + +
	10	360

(a) What are the genotypes of the parents? (b) Determine the map distances between *a*, *b*, and *c*. (c) Determine the simplest origin of each of the seven classes of tetrads. (d) What is the coefficient of coincidence?

22. Analysis of ordered tetrads in *Neurospora* from a cross *a b c* $\times$ + + + gave the following tetrad types:

	(1)	(2)	(3)	(4)	(5)	(6)
	a *b* *c*	*a* *b* +	*a* *b* *c*	*a* *b* +	*a* *b* *c*	*a* *b* +
	a *b* *c*	*a* *b* +	+ *b* *c*	+ *b* +	*a* + *c*	*a* + +
	+ + +	· + + *c*	*a* + +	*a* + *c*	+ *b* +	+ *b* *c*
	+ + +	+ + *c*	+ + +	+ + *c*	+ + +	+ + *c*
Number of tetrads	300	300	100	100	100	100

(a) Are any of the genes linked? If so, which ones? (b) If any of the genes are linked, how far apart are they? (c) Calculate the distance between each gene and its centromere. (d) Construct a genetic map from the analysis of the tetrad data.

23. Two recessive mutations, *a* and *b*, are located in the same chromosome in a diploid yeast strain and the homologous chromosome is wild type.

Consider first that *a* and *b* lie on the same side of the centromere and answer the following questions. (Assume that *a* is the more distal marker.) (a) If a single crossover occurs between *a* and *b* at meiosis, what will be the genotypes of the meiotic products? (b) If a single crossover occurs between *a* and *b* at mitosis, what will be the genotypes of the products? (c) If the centromere is between *a* and *b*, would your answer to (a) remain valid? If not, give the correct answer. (d) If the centromere is between *a* and *b*, would your answer to (b) remain valid? If not, give the correct answer.

24. In *Neurospora* the second-division segregation frequencies of *ser*, *thr*, and *arg* are found from the study of ordered tetrads to be 10 percent, 10 percent, and 16 percent, respectively.

From the cross *a* + + + + × *A ser thr arg ad* the following recombination frequencies are found:

$$R_{ad,ser} = R_{ad,arg} = R_{ad,thr} = 50 \text{ percent}$$
$$R_{ser,thr} = 9.8 \text{ percent}$$
$$R_{thr,arg} = 12.6 \text{ percent}$$
$$R_{a,ser} = R_{a,thr} = R_{a,arg} = R_{a,ad} = 50 \text{ percent}$$

Draw the indicated linkage group(s). Indicate the position of all markers and any centromeres called for by the data.

25. Diagram the origin of tetrads 1–9 in Table 12-3, showing the crossover events, if any, that would produce the nine tetrad genotypes.

26. Individuals from a stock of *Chlamydomonas* containing genes *a*, *b*, *c* were mated to those from a wild-type stock. The following *unordered* tetrads were formed.

1) + + +	2) *a* + +	3) *a* *b* +	4) *a* + +
+ + +	*a* + *c*	*a* *b* +	+ + +
a *b* *c*	+ *b* +	+ + *c*	+ *b* *c*
a *b* *c*	+ *b* *c*	+ + *c*	*a* *b* *c*
268	12	20	180

5) *a* + +	6) + + *c*	7) + + +	8) *a* + +	9) + *b* *c*
a + +	+ + *c*	*a* + *c*	+ + +	*a* *b* +
+ *b* *c*	*a* *b* +	+ *b* +	+ *b* *c*	*a* + *c*
+ *b* *c*	*a* *b* +	*a* *b* *c*	*a* *b* *c*	+ + +
18	12	142	18	20

(a) Are the genes linked? If so, determine the order and linkage distance. (b) Is it more correct to calculate the intermediate distance between a group of genes and then add them or calculate the distance between the two outermost genes in one step? Why?

27. The following *Neurospora* cross was made,

$$+ sp\ inos \times ala + +$$

where *inos* and *ala* are auxotrophic markers for inosine and alanine requirements, and *sp* is a mutant allele resulting in stunted growth. The data below were obtained from "random spores". The spores were grown in minimal media or with supplements as indicated:

Type of growth (% total)	Medium I: Alanine and Inositol	Medium II: Alanine	Medium III: Inositol	Medium IV: Minimal
wild	50.6	43.4	3.7	0.2
stunted	49.4	6.5	46.1	3.4

(a) Determine the gene order (b) Determine the distances between the genes. Hint—Predict the genotypes expected to grow on each of the 4 types of media. (c) Can the centromere be located using this data? If not, what additional types of data would be required?

CHAPTER
13
Recombination Mechanisms

INTRODUCTION

In the three previous chapters, as strategies for mapping chromosomes and genes were outlined, it was assumed at the outset that markers would recombine when physical exchanges occurred between chromosomes. Cytological evidence for the occurrence of such exchanges during meiosis was presented in Section 12.3. In this chapter we ask how such chromosomal exchanges come about at a molecular level, using data derived from biochemical, structural, and genetic studies.

The chapter begins with a model for **general recombination** (the process that mediates most forms of chromosomal exchange) known as the **Holliday model.** The Holliday model may not be correct in detail, since the actual mechanism of general recombination is by no means firmly established, and alternate recombination models have also been proposed. The Holliday model suffices, however, to organize and offer a conceptual framework for the molecular and genetic observations presented in the ensuing sections of the chapter. At the conclusion of the chapter we examine briefly two additional kinds of recombinational events that have been recognized in the biological kingdom, namely, **illegitimate** and **site-specific** recombination.

GENERAL RECOMBINATION: THE HOLLIDAY MODEL

In 1964, R. Holliday proposed a model to explain general recombination. It has since been subjected to a number of refinements by others, including Holliday himself (see References at the end of this chapter). Therefore, a more accurate (but more cumbersome) designation of the scheme we consider would be a "Modified Holliday" model.

13.1 DESCRIPTION OF THE HOLLIDAY MODEL

According to the Holliday Model, the first event in recombination is a recognition and alignment of homologous nucleotide sequences in two duplex chromosomes (Figure 13-1a). Breaks are next made at comparable positions in two of the polynucleotide chains having the same polarity (Figure 13-1b). Each broken chain then "invades" the opposite helix (Figure 13-1c), establishing base pairs with complementary nucleotides (Figure 13-1d). Ligase enzymes (Figure 5-9) then seal the discontinuities (Figure 13-1e) to produce a molecule with the shape of the Greek letter chi (χ); this structure is generally referred to as a **Holliday intermediate** (other terms are **chi form** or **half-chromatid chiasma.**)

Once created, a Holliday intermediate need not be static. The cross-connection is presumably free to move to the right (as in Figure 13-1f) or to the left as both duplex chromosomes rotate in the same direction. Such shifts are called **bridge migrations.**

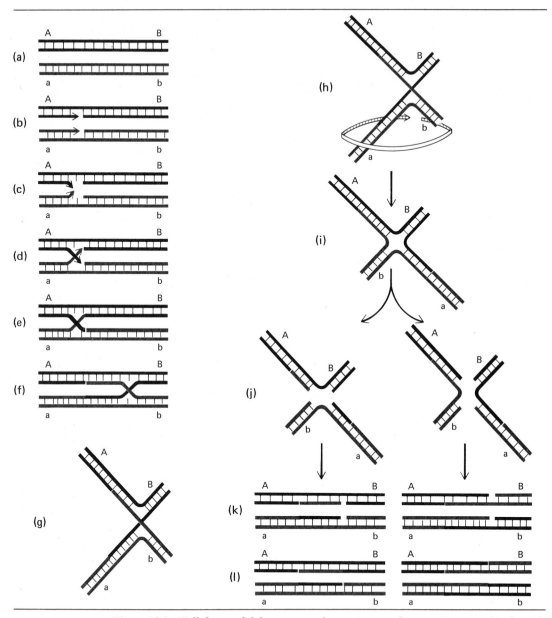

Figure 13-1 Holliday model for reciprocal genetic recombination (see text for details). (From H. Potter and D. Dressler, *Proc. Nat. Acad. Sci. U.S.* **73**:3000, 1976, after R. Holliday, *Genetics* **78**:273, 1974 and previous publications cited therein.)

The Holliday intermediate can be drawn as in Figure 13-1e and f. It can also be drawn in its planar dimensions, as in Figure 13-1g, where the four ends are simply pulled apart. Finally, two arms of the intermediate can be rotated with respect to the two others (Figure 13-1h) to generate the structure shown in Figure 13-1i. While these are all equivalent molecular configurations, the final features of the model are easiest to visualize when the Holliday intermediate is drawn as in Figure 13-1i.

The next event predicted by the model is that the Holliday intermediate is cut in the bridge region to produce two independent duplexes. As drawn in Figure 13-1j, such cuts can occur either along an "east-west" axis or along a "north-south" axis. Cuts along the east-west axis generate two duplexes with the outside markers *AB* and *ab* remaining in the parental configuration; north-south cuts generate *Ab* and *aB* recombinant duplexes (Figure 13-1k). In both cases, the DNA between the flanking *A* and *B* loci contains in part a **hybrid overlap,** that is, the polynucleotide chains derive in part from two different parents (in the diagram, chains depicted in red are shown as being base-paired with chains depicted in black). Each molecule in Figure 13-1k contains a discontinuity or gap in one of its two polynucleotide chains. These gaps can be filled in by DNA polymerase and ligase enzymes to yield intact DNA duplexes (Figure 13-1l).

13.2 MODIFICATION OF THE MODEL FOR NONRECIPROCAL EXCHANGES

Figure 13-1 describes a **reciprocal** exchange: chromosomes "trade" lengths of DNA duplex in such a way that all four polynucleotide chains participating in the exchange event are conserved. Reciprocal exchanges are thought to accompany most general recombination events, but there are important cases—transformation and transduction, for example—in which exchange appears to be **nonreciprocal.** In such nonreciprocal events, a piece of donor DNA is inserted into a recipient chromosome and any "extra" or "dangling" DNA is degraded by exonucleases. Figure 13-2 diagrams a model for nonreciprocal recombination as visualized for bacterial transformation. A comparison of Figures 13-1 and 13-2 reveals that the two models are variations on the same general "Holliday concept" of how recombination comes about.

13.3 THEORETICAL CONSIDERATIONS OF THE HOLLIDAY MODEL

Most features of the Holliday model have been subjected to close theoretical scrutiny. The recognition-alignment phase (Figure 13-1a), for example, is usually thought of in terms of Watson-Crick base pairing between parental duplexes. Many investigators suspect that both prokaryotic and eukaryotic chromosomes can fold or "kink" in such a way that their internal bases

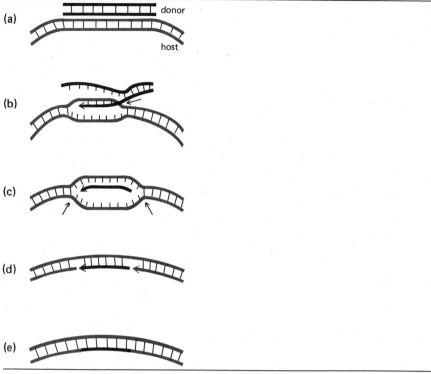

Figure 13-2 Model for nonreciprocal genetic recombination during bacterial transformation of host by donor DNA. (a) Recognition and alignment. (b) Denaturation of host and donor DNA and pairing of host and donor strains. Endonuclease nicks donor DNA at arrow. (c) Endonucleases nick host DNA at arrows. (d) Inserted donor DNA with gaps. (e) Gaps repaired by polymerase and ligase. (After M. Fox, *J. Gen. Physiol.* **49**[suppl.]: 183, 1966.)

are locally exposed. Such kinking could, for instance, be the consequence of the superhelical twists (Figure 5-2) adopted by prokaryotic chromosomes or the nucleosome structure (Figure 2-15) adopted by eukaryotic chromosomes. The resultant DNA would be effectively single-stranded and, therefore, available for recognizing and aligning with exposed bases in a "kinked" region of a homologous chromosome.

The kinked conformation of chromosomal DNA is also postulated to play a role in the ensuing "strand-switching" phase of recombination (Figure 13-1c). The energy stored in the folded chromosome would presumably be released once the DNA was nicked (Figure 13-1b), and this energy might well be utilized in the reannealing phase of the model.

The Holliday model's next prediction, the formation of a Holliday intermediate (Figure 13-2e), has been given theoretical consideration by a number of geneticists. N. Sigal and B. Alberts, for example, have constructed space-filling molecular models of Holliday intermediates and concluded that such macromolecules would be very stable since they would contain no unpaired bases, even at the points of exchange (the cross bridges in Figure 13-1e). M. Meselson has calculated, moreover, that only about 20 seconds would be required for such a bridge to "migrate" (Figure 13-1f) through 1000 base pairs of DNA under physiological conditions. In theory, therefore, extensive regions of hybrid-overlap DNA could be created in a matter of seconds between the time a Holliday intermediate first forms and the time it is cut in the final "maturation phase" (Figure 13-1i) of the model.

EXPERIMENTS SUPPORTING THE HOLLIDAY MODEL OF GENERAL RECOMBINATION

The Holliday model makes the following molecular predictions about general recombination: 1) enzymes of DNA metabolism (nucleases, ligases, and so on) should be active participants in the process, and mutations affecting the activity of these enzymes should affect recombination; 2) a Holliday intermediate should form during recombination and; 3) recombinant DNA should be physically hybrid. In the next few sections we present experimental support for these predictions.

13.4 GENE LOCI AND ENZYMES CONCERNED WITH GENERAL RECOMBINATION

Genes that mediate recombination in *E. coli* are designated *rec*. Of these, the gene marked by the *recA* mutation appears to be the most specifically involved in general recombination. Cells that are *recA⁻* are virtually incapable of recombining during conjugation, transduction, or transformation. It is not yet clear how the *recA* gene product functions; as noted in Section 6.10, its function during postreplication repair is equally unclear. Some speculate that *recA* may specify a regulatory protein which influences the synthesis or effectiveness of several proteins required for DNA breakage and rejoining. Others speculate that it may somehow interact directly with chromosomes to mediate recognition and alignment. Since the *recA* gene has recently been placed in a λ*precA* transducing phage, these speculations should soon be replaced by experimental observations.

Mutations in the *recB* and *recC* genes reduce conjugal and transductional recombination rates to 0.3–20 percent of wild-type rates, in contrast to the virtual abolition of recombination in *recA* cells. The *recBC* genes are known to code for the two subunits of the ATP-dependent exonuclease V of *E. coli*, and since this protein also exhibits endonuclease and DNA unwinding activity, it represents the type of enzyme one would expect to participate in the breakage and rejoining of chromosomes. It is therefore widely believed that the *recBC* gene products are normally utilized for recombination, and probably for other cell activities as well, but that other enzymes can substitute for these operations, albeit inefficiently, in *recBC⁻* cells, thus explaining their leaky phenotype.

In addition to *rec* gene products, general recombination can be shown to require the normal functioning of proteins that also participate in DNA replication and repair (Section 5.7). For example, a T4 strain carrying an *amber* mutation in gene 43 fails to form recombination intermediates when it infects a nonsuppressor host. Since gene 43 codes for a phage-specific DNA unwinding protein, this and other similar observations suggest that unwinding proteins participate in general recombination, perhaps by holding single strands open and protecting them from intracellular nucleases during the recognition-alignment phase (Figure 3-1a). A second example of participation of DNA metabolism genes in recombination is provided by ligase-defective gene 30 *amber* mutants of T4: in this case, recombination intermediates form in nonsuppressor hosts but cannot "mature," and instead accumulate as gap-containing **joint molecules** (Figure 13-3). If these joint molecules are isolated and presented with DNA ligase and DNA polymerase *in vitro,* they are repaired to form covalently joined recombinant chromosomes (Figure 13-3).

In eukaryotes, the evidence for enzymatic involvement in recombination comes mainly from the studies of Y. Hotta and H. Stern on meiosis in the lily. These investigators have reported the existence of a pachytene-specific endonuclease which, they propose, nicks bivalent DNA to allow for matching and alignment of homologues during synapsis. Hotta and Stern have also reported that a DNA unwinding protein is specifically synthesized by early prophase I cells.

13.5   VISUAL EVIDENCE FOR HOLLIDAY INTERMEDIATES

The second specific prediction of the Holliday model is that reciprocal recombination should result in the formation of the chi-shaped Holliday intermediates depicted in Figure 13-1e–i. H. Potter and D. Dressler have

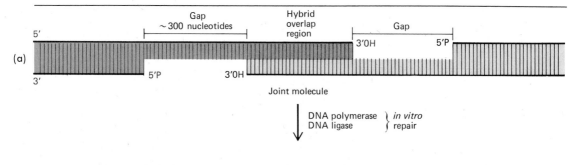

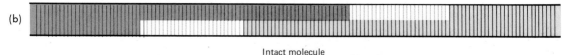

Figure 13-3   (a) Structure of the joint molecule formed during the T4 infection cycle, with red and gray indicating DNA from two different phages. This can be repaired *in vitro* to form an intact molecule (b), with light gray indicating newly synthesized DNA.

recently visualized just such intermediates. The investigators grew *E. coli* cells under conditions that caused them to fill up with large numbers of Col E1 plasmids (Section 5.1; see also Chapter 14). The cells were then lysed and the plasmid fraction of the lysate was examined with an electron microscope. In addition to the usual circular plasmid molecules, Potter and Dressler frequently observed double-sized DNA molecules, each in the shape of a figure-8 (Figure 13-4a). To determine whether these structures

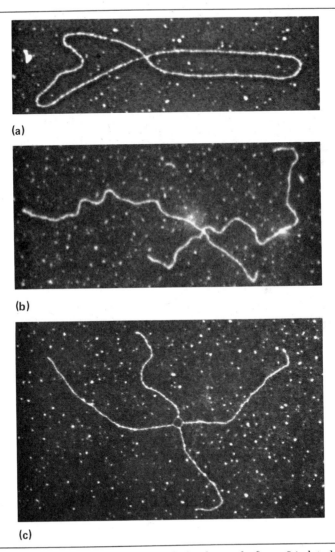

(a)

(b)

(c)

Figure 13-4 (a) A double-sized plasmid with the shape of a figure 8 isolated during the Potter and Dressler experiment. (b) A chi-shaped DNA molecule isolated after *Eco*R1 treatment of the plasmid preparation. (c) A chi-shaped DNA molecule slowing single strands in the crossover region. (From H. Potter and D. Dressler, *Proc. Nat. Acad. Sci. U.S.* **74**:4168, 1977.)

represented two plasmids in the act of recombination, they subjected the preparation to the restriction enzyme *Eco*RI (Section 7.1) which is known to cleave each plasmid circle at a single unique position. The result of *Eco*RI digestion was a collection of the expected rod-shaped chromosomes plus a sizable number of chi-shaped molecules (Figure 13-4b). These inevitably possessed two pairs of equal-length arms, with the lengths of two pairs varying from molecule to molecule. This is precisely the outcome predicted if the two loops of the figure-8's were being held together at diverse homologous regions and were undergoing reciprocal exchanges, as diagrammed in Figure 13-5. Potter and Dressler also observed that the "bridge

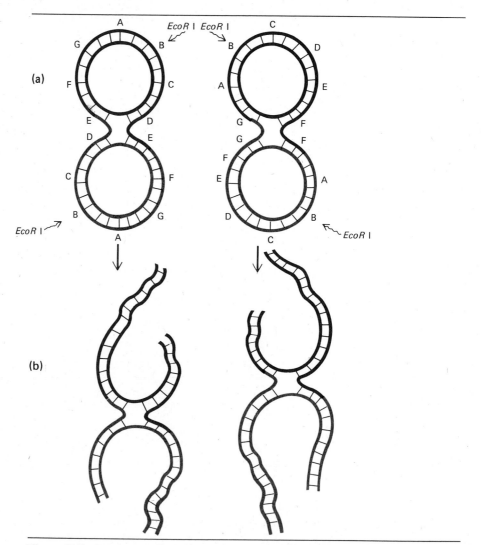

Figure 13-5 Conversion of two figure-8 molecules into chi-shaped molecules with two pairs of equal length arms, assuming the figure 8's represent plasmid molecules undergoing reciprocal recombination. Each Col E1 plasmid has one *Eco*RI site.

region" of the chi-shaped cross often contained spans of single-stranded DNA (Figure 13-4c), just as predicted from the structure drawn in Figure 13-2i. To further strengthen their argument that the structures represented intermediates in recombination, Potter and Dressler demonstrated that while $recBC^-$ cells retained their ability to generate figure-8's, the $recA$ mutation completely abolished the formation of such structures. Finally, Potter and Dressler noted that their Holliday-intermediate-shaped molecules were stable only at 0°; when warmed to 37°C, the molecules "rolled apart" within a matter of minutes into two separate rods. This bridge migration is, of course, a key feature of the Holliday hypothesis (Figure 13-1e and f). Thus the Potter/Dressler experiments provide strong support for several aspects of the Holliday model.

13.6 HYBRID DNA FORMATION DURING RECOMBINATION

The third prediction of the Holliday model is that recombinant DNA molecules should be physically hybrid; that is, they should possess polynucleotide chains derived from both parents, as shown in Figure 13-1k and l. Experiments demonstrating physically hybrid recombinant chromosomes have been performed with a number of phages; here we describe the experiments of T. Gurney Jr. and M. Fox, summarized in Figure 13-6, which show that physical hybrids form during the nonreciprocal act of bacterial transformation (Section 11.8).

Gurney and Fox first grew *D. pneumoniae* (pneumococcus) cells in a medium containing ^{15}N and ^2H (deuterium) so that the DNA of the cells was heavy, much as in the Meselson-Stahl experiment described in Section 5.4. This "heavy" DNA strain carried a genetic marker, ery^r, for resistance to the antibiotic erythromycin, and a marker str^s for sensitivity to the antibiotic streptomycin. The genetic constitution of the "heavy" strain can thus be written as $\frac{ery^r\ str^s}{ery^r\ str^s}$, where the red lines symbolize the heavy isotope label (Figure 13-6a).

The DNA from cells of this strain was isolated, and during the isolation process the DNA became fragmented into pieces that were the appropriate size for transformation. Included in this collection were fragments of the $\frac{ery^r}{ery^r}$ and $\frac{str^s}{str^s}$ variety; fragments carrying both ery^r and str^s were not present, however, since the two genes are far enough apart in the chromosome not to be included together in a piece of DNA in the 0.3 to 8×10^6 dalton range. The fragments were then presented to competent cells having light (^{14}N, ^1H) chromosomes with the genetic constitution $\frac{ery^s\ str^r}{ery^s\ str^r}$; in other words, the recipients were erythromycin-sensitive (ery^s) but streptomycin-resistant (str^r). (Figure 13-6b).

After the donor fragments had been taken up and about 25 minutes was

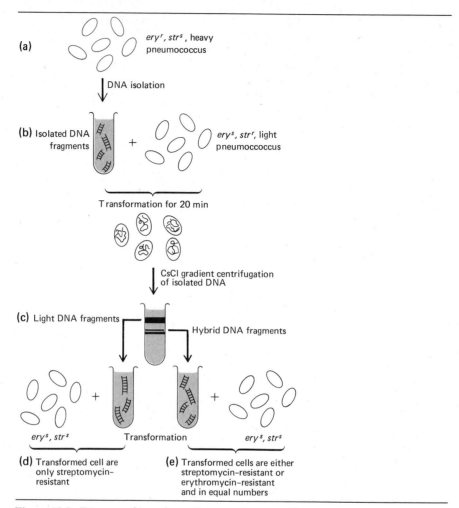

Figure 13-6 Diagram of transformation experiments with pneumococcus performed by T. Gurney and M. Fox. Heavy (^{15}N, ^{2}H-labeled) DNA is depicted in red and light (^{14}N, H-labeled) DNA is depicted in black.

allowed for transformation, the recipient cells were themselves broken and their fragmented DNA was subjected to centrifugation in CsCl (Box 5.1). These fragments distributed themselves in the gradient according to their density and two peaks were found: a large peak with the density of a fully light DNA duplex and a smaller peak with the density of a **hybrid** DNA—a duplex composed of one heavy and one light polynucleotide chain (Figure 13-6c). Almost no DNA was found in that region of the gradient where fully heavy DNA would be expected to equilibrate, suggesting that the heavy donor DNA fragments had somehow been converted into hybrid DNA after being taken up by the cells.

The fragments from the two peaks were then isolated and each fraction was tested for its ability to transform a *third* strain of pneumococcus, a strain sensitive to *both* streptomycin and erythromycin and thus having the constitution $\frac{ery^s\ str^s}{ery^s\ str^s}$. It was found that when the light fragments were used as donor DNA only one class of transformed cells could be recovered, all of which were streptomycin-resistant (Figure 13-6d). In other words—as expected—the light DNA fragments were derived solely from the chromosomes of the original recipients, namely, the streptomycin-resistant cells. In contrast, when the hybrid fragments were used as donor DNA, *two* classes of transformed cells were found with equal frequency: one class was resistant to streptomycin and the other to erythromycin (Figure 13-6e). The physically hybrid DNA fraction thus clearly carried information from both the original donor (*ery^r*) and the original recipient (*str^r*); it can thus be predicted that the DNA in this fraction had the composition $\frac{str^s}{str^r}$ and $\frac{ery^r}{ery^s}$, being hybrid in both density and in the genes it carried on its two strands. Hybrid DNA, in short, appeared to be the result of a successful transformation event.

It may occur to you to wonder why the cells that emerged in these experiments were resistant to only one or the other of the two antibiotics but not to both. Presumably, this results from the fact that a successful transformation at one chromosomal region is a rare event. The chance that *two* such rare events will occur in the *same* chromosome is therefore so unlikely that doubly resistant cells would emerge only very infrequently in the experiments we have been describing.

HETERODUPLEX DNA FORMATION AND SEGREGATION DURING GENERAL RECOMBINATION

13.7 PROPERTIES OF HETERODUPLEX DNA

The models for both reciprocal (Figure 13-1) and nonreciprocal (Figure 13-2) general recombination predict the formation of "black-red" hybrid-overlap DNA, in the reciprocal case as a consequence of bridge migration, in the nonreciprocal case as a consequence of strand insertion. Figure 13-7 redraws these models with two closely linked markers, m_1 and m_2, occupying the DNA between the outside markers *a* and *b*. Looking at the reciprocal model (Figure 13–7a), it is clear that a genetic exchange to the left of the m_1 locus would, if followed by branch migration, yield a physically hybrid piece of DNA that is a "genetic hybrid" $\frac{m_1\ +}{+\ m_2}$ as well. Genetic hybrids are similarly present in the nonreciprocal exchange (Figure 13-7b). If we assume the simplest case, namely that m_1 and m_2 differ from their + counterparts by a single base, then if the m_1 sense strand (Section 8.4) is 5' . . . ACAAT . . . 3'

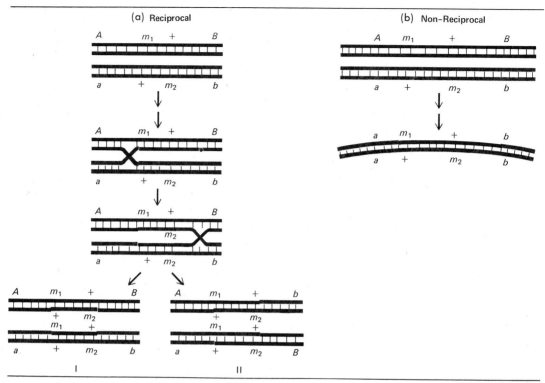

Figure 13-7 Creation of heteroduplex DNA during reciprocal (a) and nonreciprocal (b) recombination.

and its $+$ counterpart reads $5' \ldots$ ACA**GT** $\ldots 3'$ (so that its antisense strand is $3' \ldots$ TGTCA $\ldots 5'$), then the $m_1/+$ hybrid region will have the structure

$$5' \ldots \text{ACAAT} \ldots 3'$$
$$3' \ldots \text{TGTCA} \ldots 5'$$

This hybrid DNA carries a **mismatched** A-C base pair. It is therefore said to contain **heterologous** base sequences and is called a **heteroduplex** (the term **heterozygous DNA** is also used). At least one site of mismatch will also be present at the $+/m_2$ site.

Figure 13-8 diagrams two of the possible "fates" that can befall a sector of such heteroduplex DNA once it has formed. In Figure 13-8a, the heteroduplex region is simply replicated, the result being two daughter **homoduplexes**; in such cases, the heterologous sequences are said to **segregate** from one another. Figure 13-9b illustrates an alternate possibility, known as **mismatch repair** (recall Figure 6-12), during which one strand of the heteroduplex is removed and replaced by a new stretch of homologous DNA.

The following sections describe experiments that demonstrate both the

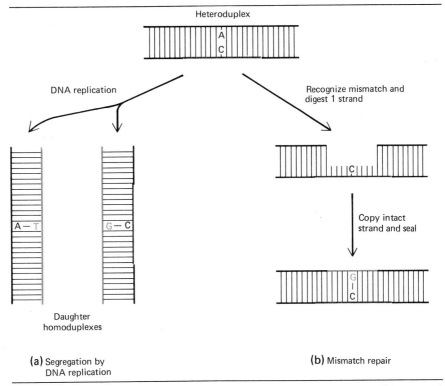

Figure 13-8 Resolution of DNA heteroduplexes by (a) DNA replication or (b) mismatch repair.

generation of heteroduplex DNA during general recombination and its "resolution" by segregation. We then consider evidence for heteroduplex "resolution" by mismatch repair.

13.8 HETERODUPLEX FORMATION AND SEGREGATION IN TRANSFORMATION

The first example of heteroduplex DNA is one we have already encountered, namely, that generated in the Gurney-Fox experiment (Figure 13-6). It should now be apparent that the physically hybrid chromosomes resulting from transformation will also be "genetically hybrid," that is, heteroduplex. Additional experiments by Gurney and Fox, diagrammed in Figure 13-9, showed this to be the case. Cells that were sensitive to both streptomycin and erythromycin, $\frac{ery^s\ str^s}{ery^s\ str^s}$ were transformed by the hybrid donor fragments, just as in the foregoing experiment. This time, however, *single* transformed cells were isolated and each isolate was allowed to undergo repeated cell divisions. When the progeny of any single cell were tested for their drug sensitivity, two distinct clones of cells were found: one clone, for example,

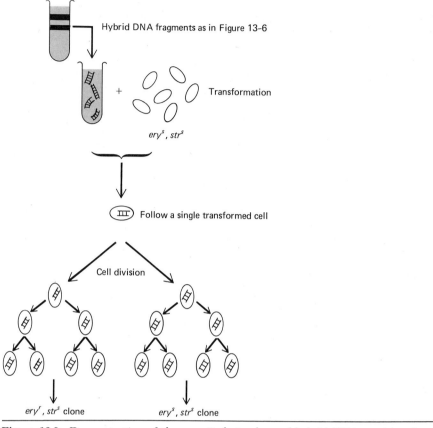

Hybrid DNA fragments as in Figure 13-6

\+   Transformation

ery^s, str^s

Ⅲ Follow a single transformed cell

Cell division

ery^r, str^s clone ery^s, str^s clone

Figure 13-9 Demonstration of the genetic heterology of hybrid DNA resulting from transformation in the Gurney/Fox experiment (see also Figure 13-6).

might be erythromycin resistant (and streptomycin-sensitive), the other, erythromycin-sensitive (and streptomycin-sensitive). As drawn in Figure 13-9, these results can best be explained by assuming that, as a result of transformation, the original cell isolate came to have a chromosome of the $\dfrac{ery^r\ str^s}{ery^s\ str^s}$ type, whose ery^r sequence was inserted into it by recombination. When this chromosome underwent a round of semiconservative DNA replication, one daughter came to have the chromosome structure $\dfrac{ery^r\ str^s}{ery^r\ str^s}$ while the other daughter had the structure $\dfrac{ery^s\ str^s}{ery^s\ str^s}$. The two daughters, genetically distinct with respect to erythromycin sensitivity, then each went on to generate two distinct clones. Thus the ery^r and ery^s sequences, once together in the same cell, subsequently segregated and found themselves in two different cells.

13.9   HETERODUPLEX FORMATION AND SEGREGATION
       IN MIXED PHAGE INFECTIONS

As a second example of heteroduplex DNA and its segregation we can examine recombination in phage T4. When mixed infections are carried out between r^+ and r (rapid-lysis) T4 phages and the progeny phages are allowed to form plaques on a bacterial lawn, most of the plaques are either small and fuzzy-edged (wild-type) or large and clear (r mutant)—as is to be expected for a one-factor cross. Occasional plaques, however, are found to have a "mottled" appearance: roughly half the plaque is fuzzy-edged and half is clear (see Figure 6-4). Since wild-type phage particles can be picked from the fuzzy side of the plaque and r phages from the clear side, such plaques are interpreted to have been formed by phages that were **partial heterozygotes** (or **partial hets**) with heteroduplex $\frac{r}{r^+}$ chromosomes.

It has been postulated that a partial het represents a phage that participated in recombination, acquired heterologous sequences in the r locus, and then became packaged into a protein coat before it had time to replicate its DNA. When such a phage lands on a bacterial lawn and replicates its DNA during the course of its infection cycle, it will generate two types of phage progeny, and will hence produce a mottled plaque.

Support for the notion that T4 partial hets arise during recombination comes from experiments such as the following: Two parental phages are each constructed to carry outside markers, so that a three-factor cross, $A\ r^+\ B \times a\ r\ b$ is performed. Progeny phages are then plated, r^+ and r phages are picked from either side of mottled plaques (as before), and the genetic constitution of these phages is assessed. It is found that a sizable proportion of the progeny are indeed recombinant for the outside markers.

HETERODUPLEX FORMATION AND MISMATCH REPAIR DURING GENERAL RECOMBINATION

That heteroduplex DNA occasionally forms, and segregates, during general recombination seems well established. The alternate proposed fate of heteroduplex DNA, its resolution by mismatch repair (Figure 13-8b), has been inferred by a variety of observations, some of which we consider in the next few sections. We first describe *in vitro* experiments with phage λ showing that heteroduplex DNA can, indeed, be repaired essentially as diagrammed in Figure 13-8b. We then describe genetic phenomena in both prokaryotes and eukaryotes that are best explained by assuming that heteroduplexes in fact form during recombination and are resolved by mismatch repair.

13.10   MISMATCH REPAIR OF HETERODUPLEXES CREATED *IN VITRO*

R. Wagner, J. Wildenberg, and M. Meselson have recently demonstrated that heteroduplex chromosomes are efficiently repaired by *E. coli* cells. They first created, by the procedures outlined in the upper portion of Figure 13-10, heteroduplex λ chromosomes with the general structure $\dfrac{am1\ +}{+\ am2}$, where *am* stands for *amber*. These chromosomes were presented to *E. coli* cells under *in vitro* conditions that allow the uptake of naked DNA, and the resultant progeny were analyzed for the presence of wild-type (am^+) phages. Since Wagner and colleagues were interested in learning whether such am^+ phages could arise by mismatch repair alone, they used *recA E. coli* cells (Section 13.7) and λ strains defective in phage-recombination genes to prevent any am^+ phages from arising as the consequence of recombination. The results of these experiments were clear: λam^+ phages were indeed generated from heteroduplex DNA under nonrecombination conditions. Several observations, moreover, were consistent with the notion that the am^+ chromosomes had been generated by mismatch repair processes of the sort illustrated in the lower portion of Figure 13-10.

One set of observations involved the use of heteroduplexes that were not only mismatched at *am1* and *am2* but also at a third locus, *cl*. These heteroduplexes were used to generate am^+ progeny as before, and the progeny were subsequently tested for their ability to produce *clear* (*cl*) or *turbid* (cl^+) plaques. It was found that if the original heteroduplex had the structure $\dfrac{am1\ \ cl\ \ \ +}{+\ \ \ cl^+\ am2}$ and if *am1* mapped close to the *cl* gene relative to *am2*, then the am^+ progeny were far more frequently cl^+ than *cl*. If, on the other hand, *am1* mapped relatively far from *cl* compared to *am2*, then the am^+ progeny were more often *cl* than cl^+. These results can be explained by assuming that, in the first instance, a **repair tract** digesting the *am1* DNA will frequently digest the neighboring *cl* DNA as well, leaving only cl^+ information, whereas the opposite will be true in the second instance. By creating numerous triply marked heteroduplexes of this sort and calculating the "co-correction frequencies" of various markers, it was possible to estimate that once a typical repair tract initiates at a given mismatch site, it proceeds in a preferred 5′ ⟶ 3′ direction for an average distance of about 3000 nucleotide pairs.

The Wagner data also made it possible to calculate the frequency with which individual base mismatches are repaired. Some mismatches proved to be repaired far more often than others, with values ranging from 2.2 percent for *am53*/ + heteroduplexes to 20.7 percent for *am2*/ + heteroduplexes. Such differences in rate are best explained by the hypothesis of **provoked mismatch repair** which holds that certain mismatches are far more likely than others to be recognized as "wrong" by the repair-enzyme apparatus, and that these mismatches are more likely to instigate a repair tract.

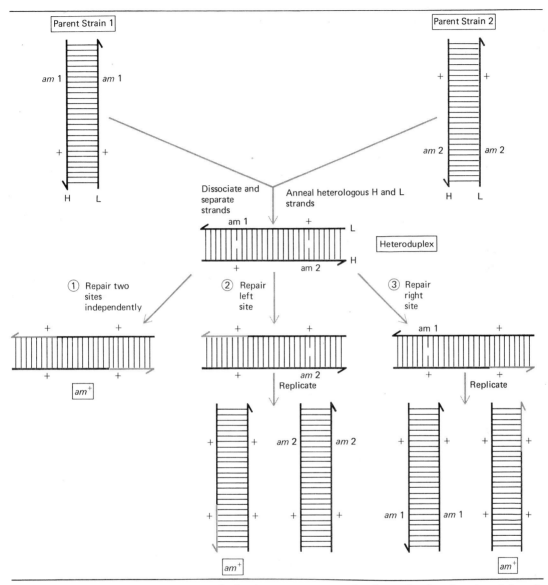

Figure 13-10 Heteroduplex formation *in vitro* using the *am1* and *am2* strains of phage λ, as performed by Wagner, Wildenberg, and Meselson, and the three avenues by which such a heteroduplex can be repaired to produce wild-type (*am+*) DNA. Avenues ② and ③ require DNA replication as well as mismatch repair. DNA produced by repair synthesis is shown in gray.

The experiments of Wagner and coworkers, in short, not only demonstrate that heteroduplexes can be repaired, but also predict two features of the repair process in *E. coli*: 1) different mismatches are likely to be repaired at different rates; and 2) closely spaced mismatches (within 3000 base pairs of one another) are more often "co-repaired" than distantly spaced mismatches.

With these experiments in mind we can turn to consider three phenomena that are encountered when closely linked markers are followed in genetic crosses, namely, **intragenic marker effects, high negative interference,** and **gene conversion.** All these phenomena are readily explained by proposing that short stretches of heteroduplex DNA are often created during the course of recombination and are often resolved by a mismatch-repair mechanism.

13.11 INTRAGENIC MARKER EFFECTS AND MISMATCH REPAIR

When very closely linked markers are being followed in a prokaryotic or eukaryotic cross, recombination frequencies between various markers do not necessarily reflect the physical distances between them. This phenomenon is well illustrated by the data of L. Norkin on *lac* operon mutations in *E. coli*. Norkin found that individual *lac* nonsense mutations could be ordered with respect to one another by two techniques that do not depend on recombination frequencies, namely, deletion mapping (Section 10.19) using partially transducing *lac* phages (Section 11.18), and gene-product mapping (Section 12.15) using the lengths of several "nonsense-fragment" polypeptides produced by the mutation alleles in his collection. These two approaches proved to yield congruent physical maps of the *lac* mutations, as shown in Figure 13-11a. The various mutations were then crossed against one another in *Hfr* X *F⁻* matings, and the frequency of *lac⁺* recombinants from these matings was scored for each cross. Figure 13-11b shows a typical set of results; in this case, recombination frequencies are plotted for crosses involving mutant 118 and other markers in the collection. From the figure, it is clear that mutant 118 recombines far more often with its neighboring marker 545 than with the more distant 498; that it recombines most often with the still more distant marker 698; and so on. Most striking was Norkin's demonstration that the substitution of one nucleotide pair for another at a mutant site could change recombination frequencies by several orders of magnitude.

Such **intragenic marker effects** on recombination frequencies are readily explained by the mismatch repair hypothesis. As with the similar observations made by Wagner, Wildenberg and Meselson, it is only necessary to postulate that certain mismatches provoke repair more readily than others and that such mismatch effects can overshadow or obscure true recombinational exchanges between closely linked loci.

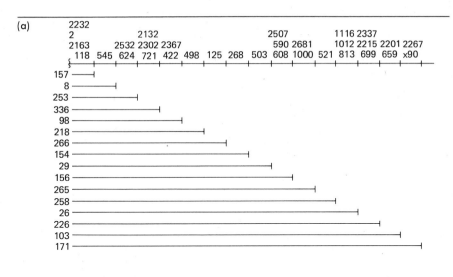

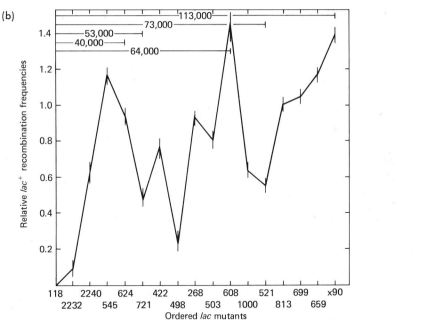

Figure 13-11 (a) Deletion map of the *lacZ* gene of *E. coli* showing point mutations above the line and deletions below. (b) Relative *lac*⁺ recombination frequencies from crosses of mutation 118 with the *lac* mutations shown on the abcissa. The molecular weights of the nonsense-fragment polypeptides produced by some of these mutant strains are given at the top of the graph to provide some indication of real distances. (From L. C. Norkin, *J. Mol. Biol.* **51**:633, 1970. Copyright by Academic Press Inc. (London) Ltd.)

13.12 HIGH NEGATIVE INTERFERENCE AND MISMATCH REPAIR

In Section 12.6 we considered a phenomenon called **interference,** which relates to the inhibitory effect exerted by one recombinational event on the probability of a second recombinational event. We further defined a parameter called the **coefficient of coincidence(s),** which measures the amount of interference occurring in a particular three-factor cross, and we noted that if the number of double exchanges in the cross is less than predicted, s will be a fractional number.

For bacteriophages s typically has a value greater than 1; in phage λ, for example, $s = 5$ for most sets of three markers. The generally high coefficient of coincidence values in phage crosses are thought to be the result of the following mechanical constraints. Recombination will occur only when two genetically distinct chromosomes participate in an exchange; furthermore, such exchanges will occur only if phages of one genotype find themselves undergoing replication and recombination in the same geographic portion of a bacterium as phages of a second genotype. Once two such chromosomes achieve sufficient proximity to undergo a recombinational exchange at one locus, it is probable that they will remain in the same vicinity long enough to undergo a second (and perhaps a third or fourth) recombinational exchange, these additional exchanges occurring at nearby loci. This phenomenon has come to have the cumbersome name of **low negative interference,** a term that originated as follows: interference was first defined, as noted above, as the *inhibitory* effect that one crossover exerted on the establishment of a second; when it was subsequently learned that these are cases in which one genetic exchange seems to *enhance* the probability of a second, the original phenomenon was renamed **positive interference** and the new phenomenon was called **negative interference,** with the double negation in fact connoting an enhancement of recombination.

In addition to demonstrating low negative interference, bacteriophages (and bacteria and eukaryotes as well) exhibit a distinct phenomenon called **localized high negative interference.** This is seen only if the markers followed in a cross lie within a short interval on the genetic map—that is, when the markers are highly localized. In such cases the value of s becomes dramatically high.

A concrete example is given by P. Amati and M. Meselson's analysis of closely linked *sus* markers in the right arm of the λ chromosome. The map order of the markers involved, shown in Figure 13-12, was first deduced by two-factor crosses, and values for R_1 and R_2 were determined for the recombinational distances separating individual pairs of markers, as shown in the first two columns of Table 13-1. Three-factor crosses were then performed, and values of $R_{1,2}$ were obtained for various sets of three markers. Using these values, Amati and Meselson calculated coefficients of coincidence for a number of the map intervals being studied. The resultant data are shown in the third column of Table 13-1. Amati and Meselson also

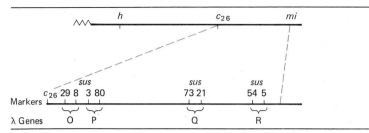

Figure 13-12 Portion of the genetic map of phage λ showing the *sus* markers used by Amati and Meselson in their experiments on high negative interference.

Table 13-1 Crosses Involving c_{26} and Various Pairs of *sus* Markers in Phage λ

sus Markers	R_1, %	R_2, %	$s = \dfrac{R_{1,2}}{R_1 R_2}$
29,8	0.48	0.30	72
29,3	0.48	0.51	41
29,73	0.48	1.4	18
29,21	0.48	1.3	17
29,54	0.48	3.3	11
8,3	0.64	0.32	28
8,80	0.64	0.40	21
8,54	0.64	3.5	8
3,80	1.1	0.25	35
3,73	1.1	1.35	15
3,21	1.1	1.5	13
3,54	1.1	3.1	8
80,73	1.1	1.5	15
80,21	1.1	1.1	15
80,54	1.1	2.1	11
80,5	1.1	3.7	10
73,21	2.6	0.08	12
73,54	2.6	1.0	8
73,5	2.6	1.7	5.5
21,54	2.7	1.0	8
21,5	2.7	1.6	6
54,5	3.0	0.35	17

From P. Amati and M. Meselson, *Genetics* **51**:369 (1965).

collected data from crosses for which more widely separated λ markers were followed, and they plotted all of the data in the graph shown in Figure 13-13. The graph illustrates, first, that even widely separated markers have an *s* value of about 5, an example of low negative interference. In addition, it is clear that when two markers are separated by a distance equivalent to one percent recombination or less, the apparent number of double exchanges is many times higher than expected.

Two explanations have been offered for localized high negative interference; both may well prove to be applicable to particular situations. The first proposes that physical exchanges indeed occur in **clusters,** the idea being that once chromosomes become "effectively paired" for one break-and-rejoin event, the attendant enzymes may well proceed to mediate numerous such events in one highly localized region. The alternative explanation holds that high negative interference is simply an exaggerated view of the marker effects that result from provoked mismatch repair. Supporting this argument are observations by Norkin on the *lac* operon mutations in *E. coli* (Section 13.14); he found that those *lac* markers that gave the highest apparent recombination frequencies in two-factor crosses (Figure 13-11b) also gave the highest values of high negative interference in three-factor crosses (see Problem 13 at the end of this chapter).

13.13   GENE CONVERSION AND MISMATCH REPAIR

Gene conversion was first observed in fungi that can be subjected to tetrad analysis (Sections 12.12 and 12.13). The phenomenon is best illustrated by recalling first what is expected when the markers *a* and *b* are linked and an

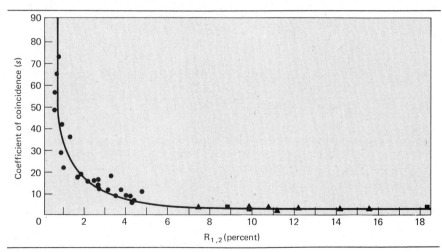

Figure 13-13   The relationship between *s* (the coefficient of coincidence) and $R_{1,2}$ (a measure of the simultaneous occurrence of exchanges in both the *a-b* and the *b-c* intervals in *a b c* X + + + three-factor crosses). Symbols ■ and ▲ represent data from crosses using widely separated markers; the symbol ● represents data from the closely linked markers used by Amati and Meselson (Figure 13-12). (From P. Amati and M. Meselson, *Genetics* **51**:369, 1965.)

$a+ \times +b$ cross is performed: $a+$ $a+$ $+b$ $+b$ parental (P) tetrads will be numerous, while $++$ $++$ ab ab nonparental ditype (NPD) tetrads will be rare, and $a+$ $++$ ab $+b$ tetratype (T) tetrads will be intermediate in frequency. If the same cross is performed but a and b are very closely linked—if for example, a and b lie in the same gene—then, in addition to P, T, and very rare NPD tetrads, four new tetrad classes are found:

$a +$	$a +$	$a +$	$a +$
$+ +$	$a\ \boldsymbol{b}$	$a +$	$a +$
$+ b$	$+ b$	$\boldsymbol{a}\ b$	$+ +$
$+ b$	$+ b$	$+ +$	$+ b$
A	B	C	D

These tetrads violate the Mendelian Law of a reciprocal 2:2 segregation of alleles; in each case, the marker shown in boldface is represented three times and its allele only once. Looking at tetrad A, for example, it is as though one of the $+$ genes had non-reciprocally "converted" one of its a alleles to adopt a wild-type nucleotide sequence. These new tetrad classes are therefore often described as arising from **gene conversion.**

To give a specific example of gene conversion, S. Fogel and D. Hurst performed a careful analysis of the *histidine-1* (*his1*) gene in yeast, crossing various *his1* auxotrophs against one another. In a cross that can be designated as $his^a + \times + his^b$, Fogel and Hurst isolated and analyzed 1081 asci in which at least one member of the tetrad was a $++$ prototroph. They found that 101 of these asci were classical tetratypes, 847 were convertants for the proximal his^a allele (convertant-tetrad class A above), and 133 were convertants for the distal his^b allele (convertant-tetrad class D above). In other words, in crosses involving auxotrophic markers within the *his1* gene, prototrophic cells arise by conversion far more often than by "classical" patterns.

Evidence that Gene Conversion Results from Mismatch Repair   Figure 13-14 diagrams the origin of a convertant tetrad assuming mismatch repair. The parent homologues are shown synapsed in Figure 13-14a. Figure 13-14b shows an exchange comparable to the Type I event illustrated in Figure 13-7a: the outside markers *EF* and *ef* retain their parental configurations, and mismatch regions have formed in two of the four chromatids. By drawing the repair tracts as in Figure 13-14c, a convertant tetrad of the Class A type is created wherein the proximal locus exhibits a $3+:1a$ ratio (Figure 13-14d). If different sets of repair tracts are drawn, convertant tetrads of the B, C, and D classes can readily be generated; you should convince yourself of this with diagrams.

A mismatch-repair explanation for conversion, although not yet proven, is supported by several kinds of observations.

First, it is clear from the *his1* data that the his^a mutation is subjected to conversion almost 7 times as often as his^b, an asymmetry that immediately recalls the marker effects described in Section 13.14 for the *lac* mutations. A "mismatch-repair explanation" in this case would state that a heteroduplex

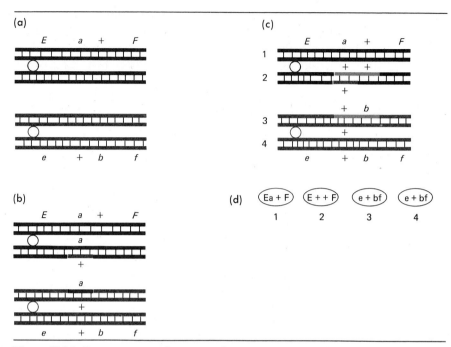

Figure 13-14   Gene conversion by mismatch repair at two sites. (a) Parent homologues in meiotic prophase. (b) A recombination event occurs between the two inner chromatids, generating two mismatches between *a* and + sequences (see Figure 13-7a). (c) Excision and DNA synthesis (gray color) repairs both mismatches. (d) The two meiotic divisions generate a tetrad with a 3:1 conversion for the + allele of *a*.

containing *his*^{a} and its + allele "provokes" repair 7 times more frequently than a *his*^{b}/+ heteroduplex.

Second, conversion is often, but not necessarily, accompanied by the recombination of outside markers. The conversion tetrad diagrammed in Figure 13-14 happens to retain parental EF linkages, but similar events can be diagrammed starting with recombinant chromatids of the Type II class in Figure 13-7a. In an extensive analysis of 907 conversion events in six genes in three yeast chromosomes, Fogel and colleagues found that 49.1 percent of these events were in fact associated with reciprocal recombination of outside markers while 50.9 percent retained the parental disposition of outside markers. It should at once be pointed out that such a tidy 50-50 proportion of recombinant-to-nonrecombinant outside markers is not invariably found in convertant tetrads; the Meselson-Radding paper cited at the end of this chapter summarizes the data that have been published for various crosses and suggests a slight modification of the nicking-and-invading phase of the Holliday model (Figure 13-1b–d) which can account for such imbalances.

Third, a number of crosses demonstrate that two markers very close together in the same gene will often **coconvert,** meaning that both exhibit a

1:3 (or 3:1) segregation in a tetrad whereas outside markers are segregating in a 2:2 ratio. Markers judged to be relatively far apart, on the other hand, are usually converted at one or the other site within a single tetrad, "double" events being rare in this case. The modal length of a coconverted segment has been estimated to be about 1000 nucleotide pairs, which is of the same magnitude as the estimated length of a "corepaired" excision tract in the artificially created heteroduplex DNA of phage λ described in Section 13.10.

Fourth, at least two mutations have been shown to affect both repair activities and conversion rates simultaneously. In the smut fungus *Ustilago,* Holliday and colleagues isolated a strain defective in gene conversion and in DNase I, an enzyme which specifically recognizes and excises heteroduplex DNA *in vitro.* In yeast, W. Boram and H. Roman describe a mutation which prevents repair of radiation-induced single-strand chromosome breaks and enhances gene conversion; they propose that the unrepaired single-strand ends in the mutant are prone to invade homologous chromosomes (Figure 13-1c) and create heteroduplex DNA, which is then subjected to mismatch repair.

Finally, we have noted (Figure 13-8) that heteroduplex DNA has two possible "fates"—either repair or segregation by replication—and the same possibilities obtain for meiotic tetrads. As illustrated in Figure 13-15, a tetrad containing one repaired and one unrepaired heteroduplex is expected to

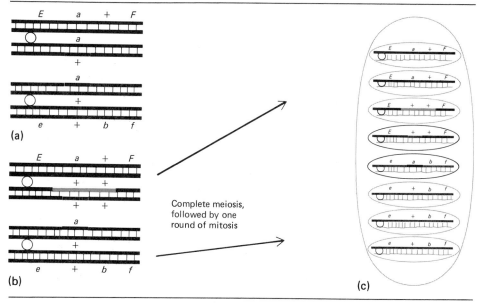

Figure 13-15 Gene conversion by mismatch repair at one site and segregation at another. (a) Meiotic prophase chromatids with two heteroduplexes, generated as in 13-14b. (b) Repair of only one mismatch. (c) Ascus of the 8 products resulting from one round of post-meiotic mitosis. Newly replicated DNA daughter strands are drawn as wavy gray lines. The markers at the proximal locus are present in a 5+ :3*a* ratio.

undergo **postmeiotic mitotic segregation** and yield a 5:3 ratio of +:a marker. Such ratios are in fact sometimes observed for closely linked markers in ordered tetrads of fungi (Figure 4-8), where, as illustrated in Figure 13-15, the 8 postmeiotic mitotic products are held together in a single ascus.

ILLEGITIMATE RECOMBINATION

All models of general recombination include a stage in which a precise recognition and matching-up of homologous DNA sequences occurs (for example, Figure 13-1a). Such precision assures that the resultant chromosomes contain the same kinds and numbers of genes, albeit in different combinations, and most genetic analyses reveal that such precision indeed operates during recombinational events. In recent years several apparent exceptions have emerged to this rule, and the exceptional cases have been given the catchall terms **illegitimate** or **nonhomologous recombination.** Two examples of this phenomenon are considered briefly here.

13.14   ILLEGITIMATE RECOMBINATION BY PHAGE MU-1

The first example of illegitimate recombination is given by the temperate bacteriophage known as **mu-1.** In contrast to phages such as λ which have specific bacterial integration sites (Section 11.13), *mu-1* is able to integrate efficiently at multiple sites, if not at random, in the *E. coli* chromosome. Whenever *mu-1* happens to insert somewhere in the middle of a gene, the gene is no longer functional; therefore, *mu-1* acts as a powerful mutagen so that, after lysogenization by *mu-1*, perhaps 1-3 percent of the resultant lysogenic bacteria carry new auxotrophic mutations that are located throughout the genome. That the phage chromosome becomes physically inserted within bacterial genes can be demonstrated either by P1 transduction or by mating experiments: in both cases there is 100 percent linkage between a given *mu*-induced mutation and the *mu-1* prophage itself. Genetic analysis of 75 independent *mu*-induced mutations in the *z* gene of the *lac* operon, moreover, has revealed that most if not all *mu-1* insertion events occur at different sites within the gene. Thus *mu-1* integration (and presumably its excision as well) clearly occurs independently of any extensive sequence homology. Both events occur efficiently in *recA⁻ E. coli* cells, moreover, which further distinguishes *mu-1* recombination from general recombination.

13.15   ILLEGITIMATE RECOMBINATION DURING TUMOR VIRUS INTEGRATION

The second example we shall consider involves tumor virus integration. Tumor-producing animal viruses are of two major classes. The **DNA tumor viruses** are exemplified by SV40 which, as described in Section 5.12, integrates into the genome of its host during a nonproductive (and often transforming) infection. The **RNA tumor viruses** contain several single-stranded RNA chromosomes within the virion. The virion also contains a virus-coded **RNA-dependent DNA polymerase** (also called **reverse transcriptase**) which, upon infection, copies each RNA chromosome into a duplex DNA chromosome known as a **provirus.** The provirus then proceeds to integrate into the host genome both during productive and nonproductive infections, and viral genes are transcribed. The integrated provirus is also replicated along with the host chromosome and is transmitted from parent to offspring of the host in a Mendelian fashion.

Both the DNA and RNA tumor viruses, therefore, insert double-stranded DNA versions of their genomes into host chromosomes. In the case of SV40, M. Botchan, W. Topp, and J. Sambrook have demonstrated that the virus integrates at multiple sites within the rodent genome; moreover, restriction-enzyme and hybridization analysis of the host DNA adjacent to integrated viral DNA reveals that the two DNAs bear no obvious sequence relationships in common. Thus it appears quite improbable that the integration of SV40 involves site-specific recombination enzymes coded by the virus, and Botchan and coworkers suggest that SV40 sequences may become inserted in cellular DNA purely as a consequence of a chance encounter.

SITE-SPECIFIC RECOMBINATION

The final mode of recombination that has been analyzed is known as **site-specific recombination,** which, like *mu-1* integration, proceeds in *E. coli* cells carrying *recA* mutations. We consider two examples of this phenomenon that are particularly pertinent to genetic analyses.

13.16 PHAGE λ INTEGRATION

The best-studied example of site-specific recombination occurs when a circularized λ chromosome inserts into the *E. coli* chromosome. We have already considered the general features of this event in Sections 5.11 and 11.13: a region of the λ chromosome we can call *attP* interacts and then recombines with the *attB* region of the host, and lysogeny results.

Mutations affecting the ability of λ to lysogenize are of two types: those that can be complemented by co-infecting wild-type phages (also known as **helpers**) and those that cannot. In the former class are mutations that map to the *int* and *xis* loci (Figure 10-8); *int⁻* phages are unable to integrate without helpers, while both *int⁻* and *xis⁻* prophages are unable to excise without helpers. In other words, integration requires a functional *int* gene product while excision requires both the *int* and *xis* proteins.

The second class of mutations—those that cannot be complemented by wild-type gene products—map to the *attP* region or its vicinity (Figure 10-8). Some of these are deletions of the *attP* region. Other mutant strains, however, possess a normal amount of *attP* DNA but remain incapable of integration.

Analysis of various *attP* deletions by R. Davis and J. Parkinson indicated that *attP* is no more than 20 to 50 base pairs in length. Their data suggested that perhaps as few as 12 of these base pairs, located internally, participate in the actual recombination event, with the bases on either side being less directly involved. These genetic predictions have recently received striking confirmation: A. Landy and W. Ross have sequenced *attP* and *attB* DNA (Box 7.2) and have found they carry in common a sequence 15 nucleotide pairs in length, as illustrated in Figure 13-16. This same 15-nucleotide-pair stretch, moreover, is found at both the leftward and rightward ends of the integrated prophage (Figure 13-16). Therefore, the actual site of crossover for both integration and excision must take place within, or at the boundaries of, the 15-nucleotide sectors (convince yourself of this with diagrams).

A 15-nucleotide stretch of homologous DNA is believed to be too short to mediate recognition and alignment by nucleotide pairing alone, and other mechanisms for the interaction between *attB* and *attP* have therefore been postulated. An attractive proposal of M. Shulman and M. Gottesman suggests that the *int* protein behaves as a sequence-specific restriction enzyme (Section 7.1), making staggered nicks in the *att* regions so as to create cohesive single-stranded ends in both the phage and bacterial genomes; these would proceed to inter-anneal and then be sealed. According to this proposal, the nondeleted *att* mutations would represent alterations in the nucleotide sequences of *att* regions such that *int*-specified endonuclease would not recognize them. Such proposals can clearly now be put to experimental test by sequencing mutant *att* regions.

If wild-type λ phages infect mutant *E. coli* cells in which the *gal-bio* region (and therefore the bacterial *att*) has been deleted, λ insertion occurs, with 200-fold lower frequency, into other sites in the genome. In contrast to the situation with *mu* phages, this "illegitimate" insertion is not random: a hierarchy of preferred **secondary sites** exists. Some of these secondary insertion sites are within genes and the resultant lysogens are mutant; the fact that prophage excision from such mutants restores wild-type gene function attests to the remarkable precision of the excision process. Since secondary insertion requires *int* and excision requires *int* and *xis,* the restriction-enzyme model predicts that the secondary sites possess sufficiently similar sequences to the bacterial *attB* sequence (Fig. 13-16) that they are acted upon by the *int* protein.

13.17 INSERTION ELEMENTS

A recently discovered kind of site-specific recombination is carried out by **insertion elements.** These are short stretches of DNA (ranging from 700 to 1400 base pairs) that have the remarkable ability to move from place to place in the *E. coli* chromosome; they also move in and out of plasmids and phage DNA. They are unlike *mu* in that they are not viruses and are not, for

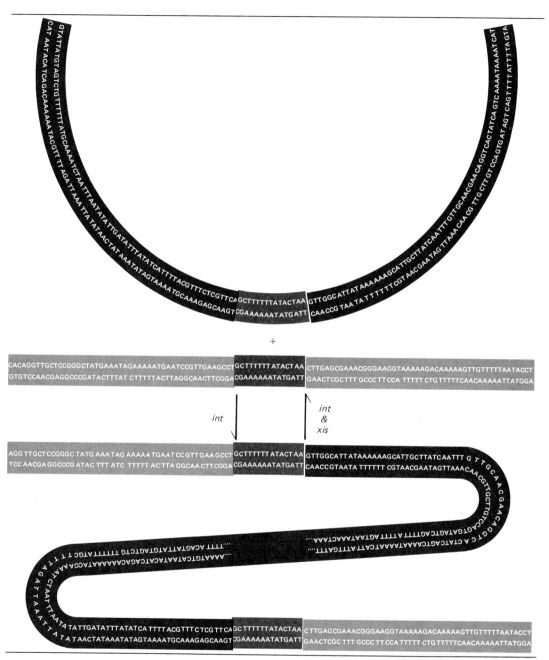

Figure 13-16 Site-specific recombination between the circular λ chromosome (top) and the *E. coli* chromosome to yield an integrated λ prophage (bottom). Integration is mediated by the *int* protein and excision by *int* and *xis*. The λ chromosome is shown in black and the *E. coli* chromosome in gray except for the 15-nucleotide sequence that is common to all 4 recombination sites. (From A. Landy and W. Ross, *Science* **197**:1147, September 16, 1977. Copyright 1977 by the American Association for the Advancement of Science.)

example, ever packaged into protein coats, nor is their integration into the chromosome random. Instead, much as with the secondary sites exploited by phage λ, a given insertion element exhibits a "hierarchy" of preferred sites in the *E. coli* chromosome, many of these sites occurring within genes. Indeed, a prominent hypothesis holds that an insertion element is in fact a gene coding for an *int*-like protein which recognizes particular sequences and mediates its own integration.

The role of insertion elements in *F* element integration and in the inheritance of genes conferring antibiotic resistance is described in Chapter 14.

References

Reviews on Recombination

Fogel, S., and R. K. Mortimer. "Recombination in yeast," *Ann. Rev. Genetics* **5**:219–236 (1971).
Grell, R., Ed. *Mechanisms in Recombination.* New York: Plenum, 1974.
Hastings, P. J. "Some aspects of recombination in eukaryotic organisms," *Ann. Rev. Genet.* **9**:129–145 (1975).

Recombination Models

Boon, I., and N. D. Zinder. "Genotypes produced by individual recombination events involving bacteriophage fl," *J. Mol. Biol.* **58**:133–151 (1971).
*Fox, M. S. "On the mechanism of integration of transforming deoxyribonucleate," *J. Gen. Physiol.* **49** (Pt 2):183–196 (1966).
*Holliday, R. "A mechanism for gene conversion in fungi," *Genet. Res. Camb.* **5**:282–304 (1964).
*Meselson, M. S., and C. M. Radding. "A general model for genetic recombination," *Proc. Nat. Acad. Sci. U.S.* **72**:358–361 (1975).
*Sigal, N., and B. Alberts. "Genetic recombination. The nature of a crossed strand-exchange between two homologous DNA molecules," *J. Mol. Biol.* **71**:789–793 (1972).
Thompson, B. J., M. N. Camien, and R. C. Warner. "Kinetics of branch migration in double-stranded DNA," *Proc. Nat. Acad. Sci. U.S.* **73**:2299–2303 (1976).
Wagner, R. E., Jr., and M. Radman. "A mechanism for initiation of genetic recombination," *Proc. Nat. Acad. Sci. U.S.* **72**:3619–3622 (1975).

Recombination in Viruses and Bacteria

Cassuto, E., T. Lash, K. S. Sriprokash, and C. M. Radding. "Role of exonuclease and β protein of phage λ in genetic recombination. V. Recombination of λ DNA *in vitro,*" *Proc. Natl. Acad. Sci. U.S.* **68**:1639–1643 (1971).
*Gurney, T. Jr., and M. S. Fox. "Physical and genetic hybrids formed in bacterial transformation," *J. Mol. Biol.* **32**:83–100 (1971).
Henderson, D., and J. Weil. "Recombination-deficient deletions in bacteriophage λ and their interaction with *chi* mutations," *Genetics* **79**:143–174 (1975).
*Hosada, J. "Role of genes 46 and 47 in bacteriophage T4 reproduction. III. Formation of joint molecules in biparental recombination," *J. Mol. Biol.* **106**:277–284 (1976).
McEntee, K., J. E. Hesse, and W. Epstein. "Identification and radiochemical purification of the *recA* protein of *Escherichia coli* K-12," *Proc. Nat. Acad. Sci. U.S.* **73**:3979–3983 (1976).
*Norkin, L. C. "Marker-specific effects in genetic recombination," *J. Mol. Biol.* **51**:633–655 (1970).

*Denotes articles described specifically in the Chapter.

Paul, A. V., and M. Riley. "Joint molecule formation following conjugation in wild type and mutant *Escherichia coli* recipients," *J. Mol. Biol.* **82**:35–56 (1974).

*Potter, H., and D. Dressler. "On the mechanism of genetic recombination: electron microscopic observation of recombination intermediates," *Proc. Nat. Acad. Sci.* **73**:3000–3004 (1976), and *Proc. Nat. Acad. Sci. U.S.* **74**:4168–4172 (1977).

Ronen, A., and Y. Salts. "Genetic distances separating adjacent base pairs in bacteriophage T4," *Virology* **45**:496–502 (1971).

Sakaki, Y., A. E. Karu, S. Linn, and H. Echols. "Purification and properties of the γ-protein specified by bacteriophage λ: An inhibitor of the host RecBC recombination enzyme," *Proc. Nat. Acad. Sci. U.S.* **70**:2215–2219 (1973).

Stadler, D., and B. Kariya. "Marker effects in the genetic transduction of tryptophan mutants of *E. coli*," *Genetics* **75**:423–439 (1973).

Stahl, F. W., J. M. Crasemann, and M. M. Stahl. "*Rec*-mediated recombinational hot spot activity in bacteriophage lambda. III. *Chi* Mutations are site-mutations stimulating *Rec*-mediated recombination," *J. Mol. Biol.* **94**:203–212 (1975).

*Tomizawa, J. I. "Molecular mechanisms of genetic recombination in bacteriophage: joint molecules and their conversion to recombinant molecules," *J. Cellular Physiol.* **70**(Suppl. 1):201–214 (1967).

*Wagner, R., and M. Meselson. "Repair tracts in mismatched DNA heteroduplexes," *Proc. Nat. Acad. Sci. U.S.* **73**:4135–4139 (1976).

White, R. L., and M. S. Fox. "On the molecular basis of high negative interference," *Proc. Nat. Acad. Sci. U.S.* **71**:1544–1548 (1974).

White, R. L., and M. S. Fox. "Genetic heterozygosity in unreplicated bacteriophage λ recombinants," *Genetics* **81**:33–50 (1975).

*Wildenberg, J., and M. Meselson. "Mismatch repair in heteroduplex DNA," *Proc. Nat. Acad. Sci. U.S.* **72**:2202–2206 (1975).

Recombination in Eukaryotes

Ahmad, A., W. K. Holloman, and R. Holliday. "Nuclease that preferentially inactivates DNA containing mismatched bases," *Nature* **258**:54–55 (1975).

Ballantyne, G. H., and A. Chovnick. "Gene conversion in higher organisms: non-reciprocal recombination events at the *rosy* cistron in *Drosophila melanogaster*," *Genet. Res.* **17**:139–149 (1971).

*Boram, W. R., and H. Roman. "Recombination in *Saccharomyces cerevisiae*: A DNA repair mutation associated with elevated mitotic gene conversion," *Proc. Nat. Acad. Sci. U.S.* **73**:2828–2832 (1976).

Campbell, D. A. "The induction of mitotic gene conversion by X-irradiation of haploid *Saccharomyces cerevisiae*," *Genetics* **74**:243–258 (1973).

*Gelbart, W., M. McCarron, and A. Chovnick. "Extension of the limits of the XDH structural element in *Drosophila melanogaster*," *Genetics* **84**:211–232 (1976).

*Hotta, Y., and H. Stern. "Persistent discontinuities in late replicating DNA during meiosis in *Lilium*," *Chromosoma* **55**:171–182 (1976).

*Hurst, D. D., S. Fogel, and R. K. Mortimer. "Conversion-associated recombination in yeast," *Proc. Natl. Acad. Sci. U.S.* **69**:101–105 (1972).

Jacobson, G. K., R. Piñon, R. E. Esposito, and M. S. Esposito. "Single-strand scissions of chromosomal DNA during commitment to recombination at meiosis," *Proc. Nat. Acad. Sci. U.S.* **72**:1887–1891 (1975).

Moore, P. D., and R. Holliday. "Evidence for the formation of hybrid DNA during mitotic recombination in Chinese hamster cells," *Cell* **8**:573–579 (1976).

Sochacka, J. H. M., and R. C. Woodruff. "Induction of male recombination in *Drosophila melanogaster* by injection of extracts of flies showing male recombination," *Nature* **262**:287–289 (1976).

Stadler, D. R., A. M. Towe, and J. Rossignol. "Intragenic recombination of ascospore color mutants in *Ascobolus* and its relationship to the segregation of outside markers," *Genetics* **66**:429–447 (1970).

Wildenberg, J. "The relation of mitotic recombination to DNA replication in yeast pedigrees," *Genetics* **66**:291–304 (1970).

Illegitimate and Site-Specific Recombination

*Botchan, M., W. Topp, and J. Sambrook. "The arrangement of simian virus 40 sequences in the DNA of transformed cells." *Cell* **9**:269–287 (1976).

Bukhari, A., J. Shapiro, and S. Adhya, Eds. *DNA Insertion Elements, Plasmids, and Episomes.* Cold Spring Harbor, N.Y.: Cold Spring Harbor Press, 1977.

*Campbell, A. M. "Episomes," *Adv. Genetics* **11**:101–145 (1962).

Couturier, M. "The integration and excision of the bacteriophage Mu-1," *Cell* **7**:155–163 (1976).

*Davis, R. W., and J. S. Parkinson. "Deletion mutants of bacteriophage lambda: III. Physical structure of *att*φ," *J. Mol. Biol.* **56**:403–423 (1971).

Gottesman, M. E. "The integration and excision of bacteriophage lambda," *Cell* **1**:69–72 (1974).

*Landy, A., and W. Ross. "Viral integration and excision: Structure of the lambda *att* sites," *Science* **197**:1147–1160 (1977).

*Weil, J., and E. R. Signer. "Recombination in bacteriophage λ. II. Site-specific recombination promoted by the integration system," *J. Mol. Biol.* **34**:273–279 (1968). [Reprinted in G. L. Zubay, *Papers in Biochemical Genetics.* New York: Holt, Rinehart and Winston, 1968.]

(a)

(b)

(a) Matthew Meselson (left) of Harvard University and Frank and Mary Stahl (right and front) of the University of Oregon collaborated in the classical demonstration of semiconservative DNA replication (Chapter 5); each has since made important contributions in analyzing the molecular basis of phage recombination. (b) Maurice Fox (Massachusetts Institute of Technology) has worked out many molecular features of recombination during bacterial transformation.

Questions and Problems

1. In the Gurney/Fox experiment outlined in Figure 13-6, hybrid DNA was isolated immediately after transformation had occured. If the DNA had instead been allowed to undergo a round of replication in light medium and hybrid DNA had then been isolated, how would its transformation properties differ when presented to the doubly sensitive third strain as in Figure 13-6? Explain using diagrams.

2. Three mutations of unknown map order lie within the sex-linked *vermilion* eye gene of *Drosophila*. They are designated as *a*, *b*, and *c*. Females of the constitution

$$\frac{w\ (ab\ +)\ m}{+\ (+\ +\ c)\ +}$$

where *w* and *m* are outside markers give rise to a few males with wild-type eye color. These are nearly all wild-type with respect to the two outside markers. What can you say about the map order of the mutations *a*, *b*, and *c* if you assume gene conversion does not occur in *Drosophila*? How does your answer change if you assume gene conversion can occur?

3. Is there any unexpected feature of the base compositions in the regions depicted in red in Figure 13-16? Do you expect that these regions would be denatured readily or with difficulty? What base composition might you expect to find in secondary *attB* sites in *E. coli*?

4. Would you postulate that the production of a specialized transducing phage (Figure 11-11) entails generalized, illegitimate, or site-specific recombination? Explain.

5. A stock of female *Drosophila* have attached X ($\widehat{XX}$) chromosomes (Figure 4-19) and $\widehat{XX}Y$ karyotypes. The genes on one X chromosome are $t\ r^a\ +\ f$; the allelic genes on its homologue are $+\ +\ r^b\ +$. All marked loci are distal to the centromere and all mutations are recessive. The r^a and r^b mutations lie within the same gene. The attachment between the homologues prevents the homologues from separating but does not prevent centromere division and chromatid separation.

   (a) Diagram a meiosis-without-crossing-over in such females.
   (b) What are the genotypes and karyotypes of the surviving progeny if the resultant eggs are test-crossed to males carrying recessive mutations at all 4 loci (refer to Figure 4-19 to answer this question).
   (c) Diagram two reciprocal exchange events that could produce surviving progeny that are wild-type with respect to gene *r* function.
   (d) Diagram two conversion events that would have the same result as (c).
   (e) Explain why genetic analysis performed with $\widehat{XX}$ chromosomes is called **half-tetrad analysis.**

6. Mutations in the *CYC1* gene of yeast (Section 12.15) can be mapped by **X-ray induced mitotic recombination:** diploid cells with a non-complementing $\frac{cyc1\ +}{+\ cyc2}$ genotype, for example, are subjected to X rays to induce rare recombination events that generate $\frac{+\ +}{cyc1\ cyc2}$ cells; such recombinants are detected by their ability to use lactate as a carbon source. The results obtained for various pairwise combinations are shown on the next page, where the numbers are given in X-ray mapping milliunits which represent recombinants/10^{11} survivors/roentgen of X ray administered. A dash means no data; a 0 means any value less than 1 milliunit.

	1	2	3	4	5	6	7	8	9	10	11	12	13	14	15	16
1	0															
2	0	2														
3	0	87	0													
4	0	73	70	0												
5	0	78	120	0	0											
6	0	90	165	0	0	0										
7	a	11	—	—	—	105	—									
8	0	63	0	—	—	183	—	—								
9	a	112	—	—	—	67	—	134	—							
10	0	81	67	7	8	6	73	—	42	0						
11	0	2200	a	a	a	b	—	—	—	—	—					
12	0	14	—	—	—	103	1	—	—	—	—	—				
13	0	350	—	—	—	176	—	—	10	169	—	—	—			
14	0	35	—	—	—	52	—	—	—	16	—	—	—	—		
15	0	134	—	—	—	115	39	80	—	200	—	—	390	—	0	
16	0	48	—	—	—	11	43	—	—	17	—	—	170	25	1	3

[a] A cross for which no clear result was obtained due to high residual growth or high spontaneous recombination frequency.

[b] It is clear that the cross of $cy_{1\text{-}6}/cy_{1\text{-}11}$ gives a value of 2000 or greater but accurate values have not been obtained because of the extremely high spontaneous recombination frequency.

(a) What is unusual about mutant 1? How could you explain its properties?
(b) What explanations can you give for the 2 × 2 and 16 × 16 results?
(c) Which mutations appear to affect identical positions in the *cyc1* gene?
(d) Order markers 2, 3, 6, 9, and 13 with respect to one another. How can you explain any nonadditive distances?
(e) Which value utilizing marker 15 is unexpected? How might it be explained?

7. Gene-product mapping (Section 12.15) of three *cyc1* mutants indicates that each is 13 base-pairs apart:

cyc1 13   cyc1 239   cyc1 179

13   13

C. Moore and F. Sherman made diploids carrying pairwise combinations of these mutations, and prototrophs were recovered that arose by X-ray-induced, sunlamp-induced, uv-induced, or spontaneous mitotic recombination. The diploids were also allowed to sporulate, and meiotic prototrophic recombinants were scored. The resulting data are summarized on the next page.
(a) Diagram the relative distances between these markers indicated by the 5 methods.
(b) What genetic principle(s) are illustrated by this study? Explain.

	Mean Prototrophic Frequencies				
Cross	X ray	Sunlamp	UV	Spontaneous	Meiotic
cycl 13 × *cycl* 239	2200	2175	1345	6.1	78
cycl 239 × *cycl* 179	84	297	143	1.7	40
cycl 13 × *cycl* 179	340	1309	641	3.7	21

8. Two pairs of markers are used in two *Drosophila* matings. With the first pair of markers (*a,b*) the interference (observed from the progeny phenotypes) is 0.85. When a second pair of markers (*P,Q*) is used an interference of 0.15 is observed.
(a) Which of these two matings possesses the highest coefficient of coincidence?
(b) Which mating will possess the fewest numbers of double crossovers in the F_1?
(c) Is it necessary to have a third marker between *a* and *b* and between *P* and *Q* in these matings? Explain.

9. The *rosy* (*ry*) locus in chromosome 3 of *D. melanogaster* contains the structural gene for xanthine dehydrogenase (XDH); strains homozygous for *ry* mutations have a *rosy* eye color and are killed by high levels of exogenous purine. When flies of the genotype $\dfrac{kar\ ry^x\ l}{+\ ry^y\ +}$ are test-crossed with *kar ry l*, where *kar* and *l* represent flanking outside markers, and the progeny are reared in high-purine medium, only *ry*⁺ recombinants survive. These are found to fall into two crossover classes with nonparental combinations of outside markers, plus two conversion classes. Write out the genotypes of all 4 classes and diagram how they arise.

10. In a cross involving $\dfrac{kar\ ry^{41}\ l}{+\ ry^{502}\ +}$ females, 1.48×10^6 zygotes were sampled and 40*ry*⁺ progeny were recovered, of which 24 were *kar ry*⁺, 7 were *ry*⁵⁰² convertants, and 9 were *ry*⁴¹ convertants. What are the relative positions of the two *ry* markers with respect to the flanking markers? Do the convertants yield any information in this regard? Explain.

11. Two additional markers lie between *ry*⁵⁰² and *ry*⁴¹, these are concerned with the electrophoretic mobility (*e*) of XDH (Box 7.1). The combination S1 S2 leads to a *slow*-migrating XDH; the combination F1 F2 leads to a *fast*-migrating protein. The females in the above cross can be rewritten as.

$$\frac{kar\ \ +\ \ \ S1\ \ S2\ \ ry^{41}\ \ l}{+\ \ \ ry^{502}\ \ F1\ \ F2\ \ +\ \ \ +}$$

The electrophoretic mobilities of the XDH produced by the 40*ry*⁺ progeny were examined and scored as S1 S2, F1 F2, S1 F2, or F1 S2, the last two recombinant types being electrophoretically distinct from one another.
(a) Of the 24 *kar ry*⁺ crossover progeny, 23 prove to be S1 F2 while one is

S1 S2. What does this result indicate about the relative distances between the 4 sites?

(b)   Would you expect S2 and ry^{41} to coconvert? Explain.

(c)   Six of the 7 ry^{502} convertants were coconvertants for the F1 site. What electrophoretic class of XDH do these 6 progeny synthesize? What does this result indicate about the relative distances between the 4 sites?

12.   A. Ronen and Y. Salts found in 1971 that recombination frequencies between adjacent nucleotides in the *rII* genes of T4 vary 1000-fold when 12 different sites are compared. How does this observation relate to the Benzer map of these genes published 10 years earlier?

13.   Norkin performed *E. coli* crosses of the following type: *Hfr* 721 $\times$ F$^-$ 624 lac_x, where 721, 624, and lac_x represent 3 mutations in the *lacZ* gene, and scored for lac^+ recombinants. The results for various lac_x markers were as follows:

lac_x	lac^+ Observed	lac^+ Expected
608	0.08	0.0012
1012	0.003	0.00033
200	0.007	0.00030
90	0.02	0.00063
707	0.08	0.0012

(a)   Calculate the coefficient of coincidence for each cross.

(b)   At which site(s) in the gene would you expect to find the most pronounced marker effects? The least pronounced? Explain.

14.   Conversion ratios of 6 : 2 are sometimes encountered in analyzing fungal asci. Show with diagrams similar to Figure 13-15 how these could arise.

15.   The structure of the *attP* region of phage λ is usually written as *POP'*, where O represents the 15-nucleotide region of homology; the *attB* region of the *E. coli* chromosome is written *BOB'*. Following λ integration, these sites are found in the following order: *BOP'* (λ prophage) *POB'*. Moreover, the sequence of λ genes in the prophage is found to be *P' N R A J P*, whereas it is *A J POP' N R* in the linear chromosome (Figure 10-8). Recalling Figure 5-15, draw a model for the integration and excision of the λ chromosome which accounts for these facts.

CHAPTER
14

Extranuclear Genetic Systems

INTRODUCTION

This chapter considers a collection of genetic systems that have, at various times and in various contexts, been individually described as **non-Mendelian, nonchromosomal, extrachromosomal, extranuclear, cytoplasmic, uniparental,** and **maternal.** None of these terms applies closely to all the systems we shall consider—although extranuclear perhaps comes closest—for they share few common properties. Thus bacterial cells such as *E. coli* possess a single main chromosome and often have extra DNA elements called plasmids; eukaryotic cells possess a main complement of chromosomes plus extra DNA in their mitochondria and chloroplasts. Qualifying also as extrachromosomal elements are certain viruses, bacteria, and algae which take up residence within other cells and often develop a permanent and mutually dependent relationship with their hosts. Finally, properties of certain eukaryotic cells suggest that their surfaces may possess genetic information that can be modified by the environment and inherited independently of their main chromosomes.

The various extranuclear genetic systems are considered together in this chapter partly because of convenience: they do not readily fit elsewhere in our treatment of genetics. They also have certain qualitative similarities; their mode of inheritance is often, but not always, distinct from the inheritance of the main chromosome(s), and often, their genetic material is replicated at a different time, or by a different set of enzymes, than the genetic material in the main chromosome. The genetic information carried by the elements involved in these systems is often unnecessary for the survival of the organism, although this is not true in several important cases. The elements usually have a physical location outside the nucleus, but even the term "extranuclear" is not uniformly applicable, for a genetic element that resides within the nucleus but does not follow the hereditary patterns of the main chromosomes would also qualify as the kind of genetic system we are considering.

A compelling reason for considering all of these phenomena in one chapter is to convey the actual genetic complexity of cells and organisms. There is no question that a major chromosome or chromosome set carries most of an organism's genes and that the patterns of transmission of these chromosomes dominate observed patterns of heredity. At the same time, these major chromosomes usually reside in cells in which extranuclear genes are also making their modest but often critical contribution to the phenotype of the organism. Once we have described these additional contributors we shall be in a position to determine how the full phenotype is constructed and regulated, which will be the subject of the next four chapters.

In this chapter major scrutiny is given to plasmids and to mitochondrial DNA. Since plasmids represent the smallest bacterial genomes and mitochondrial DNAs the smallest eukaryotic genomes known, their complete genetic analysis is a goal attainable within the next decade. The approaches now being taken to carry out such analyses make use of many of the principles and techniques presented in previous chapters; the reader will, therefore, hopefully derive from this chapter a more integrated view of the manner in which molecular genetics is practiced. In addition, plasmids are rapidly becoming important elements in genetic research (and public health as well); therefore, a detailed understanding of their properties can be argued to be as important to basic genetic training as a detailed understanding of meiosis or recombination mechanisms.

PLASMIDS

14.1 GENERAL PROPERTIES OF PLASMIDS

Plasmids are autonomous genetic elements found in almost every known form of bacterial cell. Plasmids can enjoy an independent, self-replicating existence quite separate from the main chromosome of the cell; additionally, certain (but not all) plasmids can integrate into or out of the main chromosome, such plasmids being designated **episomes.** All known bacterial plasmids exist within the cell as covalently closed duplex DNA rings (Section 13.5). Small circular molecules of DNA, which are possibly plasmids, have also recently been reported in yeast and in *Drosophila* but since little is yet known about these, we focus here on bacterial plasmids.

There are several ways in which bacterial plasmids can be classified. One classification is based on the kind of specialized genetic information a plasmid carries. By this criterion there exist plasmids carrying **antibiotic resistance** genes and known as **R plasmids,** plasmids which specify proteins known as **colicins** (to be described shortly), and which are called **col** plasmids, and so forth. Table 14-1 lists some of the proteins encoded by naturally occurring plasmids.

A second way in which to think of plasmids is that some are **conjugal** (also called **transmissible**) and others are **nonconjugal (nontransmissible).** The best known example of a conjugal plasmid is the *F* element, already described in Section 11.1, the genes of which dictate the formation of surface pili and the conjugal transfer of *F*-element DNA from F^+ to F^- cells. It turns out that similar if not identical genes are present in certain resistance and colicin plasmids as well. Nonconjugal plasmids lack functional "conjugation genes" and are therefore ordinarily transmitted from parent to daughter cells in an "asexual" manner. Nonconjugal plasmids may, however, be **co-transferred** or **mobilized** along with conjugal plasmids if both of these elements are present in the same cell.

14.2 DETECTING THE PRESENCE OF A PLASMID

How can one determine genetically that a bacterial population carries plasmids? When the plasmid has infectious properties, its presence is readily detected by the rapid transmission of a certain trait or traits from cell

Table 14-1 Some Properties Coded by Naturally Occurring Plasmids

Property	Exemplified By
Fertility—ability to transfer genetic material by conjugation	F, R1, Col1
Production of bacteriocins	CloDF13 (*Enterobacterium cloacae*) ColE1
Antibiotic production	SCPI plasmid of *Streptomyces coelicolor*
Antibiotic resistance	R1, R6, RP4
Heavy metal resistance (Cd^{2+}, Hg^{2+})	p1258 (*S. aureus*), R6
Ultraviolet resistance	Col1b, R46
Enterotoxin	Ent
Virulence factors, hemolysin K88 antigen	ColV, Hly
Metabolism of camphor, octane, and so on	Cam, Oct (*Pseudomonas*)
Tumorigenicity in plants	T1-plasmid of *Agrobacterium tumefaciens*
Restriction/modification	Production of *Eco*R1 endonuclease and methylase by plasmid of RY13

Plasmids listed are indigenous to *E. coli* unless otherwise indicated.
Reprinted from S. N. Cohen, *Nature* **263**:731, 1976.

to cell. These traits, moreover, will exhibit no linkage (in transduction experiments, for example) to any markers located in the main chromosome. Finally, in a clone of cells carrying plasmids, segregants that have lost their plasmids will often arise.

The loss of a plasmid can be spontaneous or induced. In some cases a copy of the plasmid may fail to be transmitted to a daughter cell. Alternatively, such treatments as exposure to acridines, starvation for thymidine, or X-irradiation have been found to selectively abolish the replication of certain plasmids without affecting the replication of the main chromosome, a fact which underscores the replicative independence of the two. In either case, the segregants will prove to be viable but will lack those traits determined by plasmid genes. That the segregants have not arisen by mutation is most simply attested to by the frequency with which they arise and by the fact that many traits are lost in a single event.

In the following sections we consider F elements, R plasmids, and Col plasmids, focusing first on the traits these plasmids confer on *E. coli* and then on the genetic specification of these traits. We conclude with a description of how plasmids are used as vehicles for "genetic engineering."

14.3 *F* ELEMENT (*F* SEX FACTOR)

E. coli cells harboring an *F* or *F'* element are endowed with a number of phenotypic traits (Section 11.1):1) They possess pili and can transfer their plasmid DNA to recipient (F) cells. 2) They are sensitive to infection by single-stranded RNA phages and certain single-stranded DNA phages (the so-called **male-specific phages,** since F^+ cells are sometimes called "male" *E. coli*). 3) They resist the growth of such **female-specific phages** as T3 and T7. 4) They exclude the acquisition of additional F elements, first by a **surface-exclusion** (*Sfx*) of F-element uptake, and second by an **immunity** mechanism which prevents the retention of any additional F elements that manage to enter. 5) They can be converted into *Hfr* cells when the F element expresses its episomal properties and inserts into the main bacterial chromosome.

This rich array of F-controlled traits has permitted the isolation of mutant F and F' elements; some, for example, carry mutations conferring male-phage resistance while others carry sfx^- mutations. Complementation analysis between various point mutations has been performed by forcing the formation of mutant$_1$ F/mutant$_2$ F diploids and looking for a restoration of conjugal function before immunity eliminates one of the F elements. Deletion mapping has been performed by investigating whether point mutants can recombine with, and thereby restore conjugal function to, $F'lac$ deletion plasmids formed as a consequence of faulty excision (Section 11.18). Such analyses, initiated by M. Achtman, N. Willetts, and A. J. Clark and pursued by others, has led to the F-element map shown in Figure 14-1.

The dense cluster of genes in the right one-third of the F-element map constitutes the *tra* (DNA *tra*nsfer) operon and its controlling elements. Some of the *tra* genes specify F-pilus formation while others are more specifically involved in the transfer of DNA. The *traS* and *ilzA* genes, also within the operon, mediate the *Sfx* and immunity properties of F^+ cells. If bacteriophage *mu-1* (Section 13.14) inserts into one of the proximal *tra* genes (for example, *traA*), the expression of all distal *tra* genes is abolished. Since *mu-1* is known to carry a nonsense codon, such polar effects (Section 9.16) suggest that the *tra* cluster is an operon which, being 30,000 base-pairs long, represents the largest bacterial operon yet described. Whether this operon is transcribed into a single RNA molecule of comparable length has yet to be established.

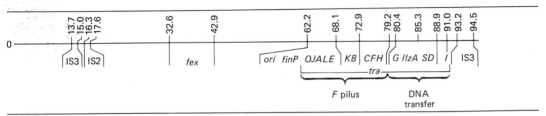

Figure 14-1 Map of the F element of *E. coli*. The total length of the DNA is 94,500 nucleotide pairs or 94.5 **kilobases.** Numbers, in kilobases, refer to physical distances between genes as determined by cloning *EcoR1* restriction fragments of the F DNA. Abbreviations: *ori*, a sequence required for initiation of transfer; *fin*, fertility inhibition. Other abbreviations are explained in the text. (Redrawn from Skurray et al. *Proc. Nat. Acad. Sci. U.S.* 73:64–68, 1976.)

The remaining two-thirds of the *F*-element map includes the *fex* gene(s) mediating the *female*-specific phage *exclusion* phenotype, plus long un-mapped expanses that can be presumed to be involved with the initiation and control of plasmid replication. Included also in these long expanses are the insertion sequences IS2 and IS3. Since insertion sequences (Section 13.17) are capable of moving about from one chromosomal location to the next, N. Davidson and colleagues have proposed that the episome phenotype of an F element is mediated by its IS regions: a plasmid IS, for example, might somehow recognize a homologous IS region in the main chromosome of *E. coli* and insert at that position. The various locations and orientations of *F*-insertion sites in various *E. coli* strains (Figure 11-6) are explained by this hypothesis, since it is only necessary to suppose that main-chromosome IS regions will vary in location and orientation from strain to strain.

14.4   R PLASMIDS

Conjugal R plasmids can be thought of as possessing two distinct types of information: antibiotic-resistance genes and genes for conjugal transfer. We shall consider these two groups of genes separately and then consider how they interact with one another.

Antibiotic-Resistance Genes   The drug-resistance genes in R plasmids are symbolized differently than the comparable genes in the *E. coli* chromo-some. Thus the chromosomal streptomycin-resistance gene is denoted *str-r* while the plasmid-borne gene is written Sm; other common plasmid genes include the gene for resistance to ampicillin (Ap), chloramphenicol (Cm), kanamycin (Km), sulfonamides (Su), and tetracycline (Tc). The chromosomal and plasmid resistance genes generally express their effects in different ways; for chromosomal genes, resistance is typically the result of an altera-tion in a ribosomal protein (Section 9.8), whereas the plasmid genes typi-cally dictate the synthesis of enzymes that inactivate the antibiotics as they enter the cell. For instance the Cm gene directs the synthesis of chloram-phenicol acetyl transferase, an enzyme that removes a critical acetyl group required for the activity of the chloramphenicol molecule.

Some R plasmids carry only one resistance gene while others carry two or more. These plasmids can be mapped by transduction with phage P22 so that, for example, the R222 plasmid has been found to have the gene order Su-Sm-Cm-Tc.

Conjugal-Transfer Genes   Conjugal R plasmids contain genes that dictate both the synthesis of pili and the transfer of DNA. Therefore, for the purpose of discussion, we can say that a conjugal R (and Col) plasmid possesses a *tra* operon, although its component genes appear to differ somewhat from the *tra* genes of the *F* element. A major distinction between the *F* and R plasmids is that the *tra* genes in most conjugal R (and Col) plasmids are highly repressed, so that naturally occurring R^+ strains transfer their plasmids only at very low frequency. This repressed phenotype, known as **fertility inhibition** or fi^+, can be mutated to an fi^- "de-repressed" state such that the R plasmid is infectiously transmitted at high frequency.

Conjugal R plasmids in the fi^+ and fi^- states are clearly distinct from nonconjugal R plasmids, which are unable to transfer their DNA at all. Loss of the ability to transfer may arise by point mutations or deletions in the *tra* genes of an R element, but it can also occur "naturally": a given conjugal R plasmid may separate into two smaller plasmids, one containing the antibiotic-resistance genes (the **R-determinant component**) and the other containing the conjugal-transfer genes (this piece is usually referred to as the **resistance transfer factor** or **RTF,** a somewhat misleading but seemingly entrenched term). Both the R-determinant component and the RTF retain the ability to replicate independently. They also retain the ability to recombine to form a conjugal R plasmid once again.

Interactions Between R-Determinants and RTFs The RTFs that split off from various plasmids appear to be similar or identical to one another, as judged by the high degree of homology seen when their DNA is dissociated and their component strands form hybrid duplexes (Section 7.5). Various R determinants, on the other hand, carry different antibiotic-resistance genes. The mechanism for the reversible associations between these elements was therefore obscure until it was discovered that each possesses IS elements. Thus, as diagrammed in Figure 14-2a, a conjugal plasmid containing a large RTF component and a small R-determinant is believed to generate independent RTFs and nonconjugal R plasmids by a recombination event involving IS regions.

A related and important phenomenon is that a given R plasmid with, say, Km resistance can "pick up" a Tc gene very efficiently from another plasmid. A Tc gene is also capable of becoming incorporated into phage DNA, into bacterial DNA, and into Col plasmids. The "hopping" of such resistance-determinants from plasmid to phage to chromosome is once again believed to be mediated by insertion sequences (Figure 14-2b) or, in some cases, by analogous but distinct classes of sequences that we do not consider further here.

R Plasmids and Public Health The mobility of the R-determinant components and the ability of a given R plasmid to "pile on" large numbers of resistance genes has major implications for public health, since R plasmids can be transmitted not only from cell to cell within a species but also across species lines. Thus, for example, *E. coli, Proteus,* and other non-pathogenic members of the intestinal flora often harbor R plasmids that have permitted the bacteria to survive high doses of antibiotics ingested by their hosts. These R plasmids can subsequently be transferred to infecting pathogenic bacteria such as *Salmonella, Shigella* (which causes dysentery), or *Haemophilus influenzae* type b (which can cause meningitis) so that these organisms also become resistant to the same spectrum of antibiotics.

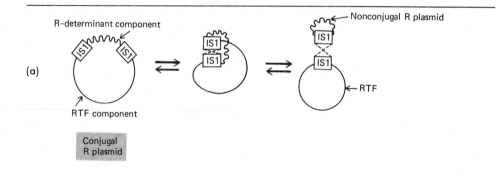

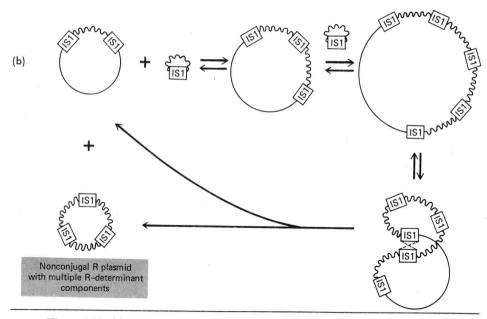

Figure 14-2 (a) Proposed mechanism for reversible dissociation of a conjugal R plasmid into an RTF and a nonconjugal R plasmid at sites of IS1 insertions. (b) Formation of multiple R-determinant elements by the same mechanism. (From K. Ptashne and S. N. Cohen. *J. Bacteriol.* **122**:776, 1975.

Even more disturbing is the finding that such transfers take place freely in sewers and in polluted rivers. In Japan, where resistance statistics have been gathered, drug-resistant *Shigella* rose from about 0.2 percent in 1953 to 58 percent in 1965, with most apparently harboring R agents. In the same survey 84 percent of *E. coli* and 90 percent of *Proteus* collected from hospital patients were similarly resistant to antibiotics. There is thus little question that the large-scale and indiscriminate application of antibiotics has created a pool of drug-resistant enterobacteria that can transmit their resistance genes to infecting pathogens and to other bacteria in polluted environments. Obviously, medical efforts to arrest infection are greatly impeded in such cases, and drug-resistant pathogens may create serious epidemics in such vulnerable locales as hospitals or crowded living areas. The situation should be reversible if drug therapy is limited in the future to persons with acute cases of infectious disease; in many countries of the world, however, such limits are unfortunately not in force at present.

14.5 COL PLASMIDS

Bacterial cells that harbor **Col plasmids** synthesize proteins known as **colicins** which act specifically to kill *E. coli* cells. Such bacteria also synthesize **immunity proteins** which render the Col-harboring cells insensitive to the bactericidal effects of the colicins they produce; this explains how *E. coli* cells are able to carry Col plasmids in the first place. Plasmids specifying other bacteriocins are also known; for example, a **cloacin** encoded by the CloDF3 plasmid is toxic to *Enterobacterium cloacae;* **vibriocins** are directed against *Vibrio cholerae,* and so forth. Within the Col family, moreover, a large spectrum of colicin species is known—colicin Ib, colicin E1, and so on—each specified by a distinct Col plasmid.

Of the colicins, E2 and E3 have been most extensively studied. Both are physically similar proteins and compete for the same receptor on their target bacteria; however, their mode of killing is quite different. E2 proves to be an endonuclease which, when it enters a nonimmune target cell, degrades the cell's DNA. E3, on the other hand, brings about the cleavage of a fragment from the 16S rRNA of *E. coli* ribosomes, thereby destroying the ability of the target cell to synthesize protein.

The sector of a conjugal Col plasmid coding for the colicin and immunity proteins is separable from its conjugal-transfer component in much the same way as we saw for R plasmids (Figure 14-2a). This sector can, moreover, pick up antibiotic-resistance genes, as diagrammed in Figure 14-2b.

CLONING DNA

In 1973, A. Chang and S. Cohen reported that they had been able to isolate pieces of main-chromosome DNA from the bacterium *Staphylococcus* and splice individual pieces into nonconjugal plasmids; when these **recombinant plasmids** were introduced into *E. coli;* the *Staphylococcus* genes had been correctly expressed in and transmitted by their *E. coli* hosts. Chang and Cohen realized that any one clone of *E. coli* host cells would probably carry only one recombinant plasmid and therefore only one piece of *Staphylococcus* DNA; therefore, by growing large numbers of a particular *E. coli* clone, large quantities of a **particular** segment of the *Staphylococcal* genome could be obtained. The Chang-Cohen technique has therefore come to be called **DNA cloning,** and its many uses for genetic research are considered shortly. First, however, the technique must be described in more detail.

14.6   CONSTRUCTING RECOMBINANT DNA

Figure 14-3 outlines a typical protocol for the construction of recombinant DNA. A plasmid (or phage) **cloning vehicle** or **vector** is selected which can be cut by a restriction enzyme such as EcoR1 (Section 7.1) without losing its ability to self-replicate. Popular cloning vehicles include Col El (which can replicate many times during an *E. coli* cell cycle), and pSC101 (which carries a Tc gene). The "linearized" plasmid (Figure 14-3a) is next treated with an exonuclease derived from phage λ which attacks and digests the 5' ends of duplex DNA to create single-stranded "tails" at the 3' ends (Figure 14-3b). Tracts of poly-A are then added to these tails in a reaction mediated by the enzyme terminal transferase (Figure 14-3c). Meanwhile, the DNA of interest has been similarly treated with the exonuclease (Figure 14-3d) and then with terminal transferase (Figure 14-3e). In this case, however, the terminal transferases are provided only with TTP and therefore only add tracts of poly-T to the 3' ends of the DNA (Figure 14-3e). Therefore, when the two preparations are mixed together, the poly-A tracts anneal with the poly-T tracts to form circular structures (Figure 14-3f) which, in the final step (Figure 14-3g), are covalently sealed with ligase.

Uptake of the recombinant plasmids by *E. coli* cells is then induced. Plasmid-bearing *E. coli* clones are subsequently selected, for example, by their colicin-resistance (for Col El vectors) or tetracyline resistance (for pSC101 vectors).

14.7   KINDS OF CLONES AND SELECTION PROCEDURES

Cloning experiments can be designed in one of two general ways. In the first, the DNA introduced into the plasmid is of known composition and function. For example, it is possible to purify hemoglobin mRNA from reticulocytes, make a duplex DNA copy with reverse transcriptase (Section 13.15) and DNA polymerase, and then splice this cDNA into a plasmid. Such cDNA would be expected to contain leader mRNA sequences but would not be expected to include promoters or other untranscribed controlling elements (Section 8.2). This technique is clearly limited to situations in which it is possible to isolate specific RNA species.

The alternative approach to plasmid cloning entails cloning DNA whose composition and function is initially largely unknown. Thus one might clone the restriction fragments obtained from a particular virus, or one might shear the total DNA of an organism into pieces of uniform size and

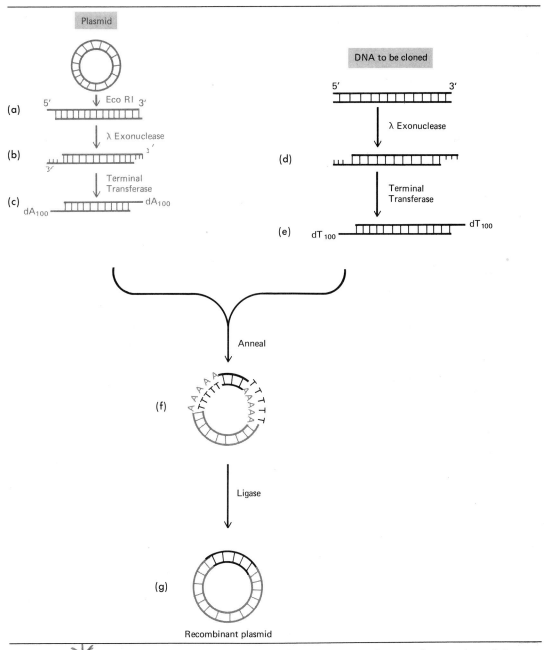

Figure 14-3 Diagram illustrating the construction of a recombinant plasmid. See text for details.

clone random sectors of the genome. Such experiments must be coupled with some selection procedure to obtain clone(s) of interest, and many such procedures have been devised.

One such selection procedure involves looking for complementation between the DNA in a recombinant plasmid and the genome of the host. Thus K. Struhl, J. Cameron, and R. Davis introduced cloning vehicles containing random fragments of the yeast genome into a nonreverting histidine auxotroph of *E. coli* and were able to isolate histidine prototrophs containing, presumably, intact and functional *his* genes from yeast. This experiment was among the first to demonstrate that eukaryotic genes can indeed be transcribed and translated correctly in *E. coli*.

A more indirect but widely applicable selection procedure is illustrated in Figure 14-4. Clones of bacteria containing unidentified sectors of, for example, yeast DNA are plated to nutrient agar in a grid-like fashion. The colonies are then replica-plated (Figure 4-2) to a nitrocellulose filter, as diagrammed in Figure 14-4a. The picked-up cells are lysed *in situ* and their DNA is dissociated into single strands (Figure 14-4b). The filter is now placed in contact with a solution containing, for example, a purified yeast mRNA species that carries a radioactive label (Figure 14-4c). DNA/RNA hybridization ensues (Section 8.4), after which the filter is washed to remove all unreacted RNA. Finally, the filter is subjected to autoradiography (Figure 14-4d). Any radioactively labeled spot on the filter marks the position of an *E. coli* clone which carries yeast DNA homologous to the yeast mRNA, and the *E. coli* clone is then recovered by matching the spot to the original grid of *E. coli* colonies.

14.8   POTENTIAL HAZARDS OF THE CLONING TECHNOLOGY

The technology of genetic engineering carries with it potential hazards. Bacteria harboring recombinant plasmids could conceivably pose risks to health or the environment, with the nature and extent of these risks being theoretically very great (as an example, an extremely virulent strain of *E. coli* could be created carrying tumor-causing genes that are transmitted infectiously). In attempts to minimize such potential hazards, molecular geneticists in the United States have formulated guidelines that must be followed in carrying out experiments with recombinant DNA; the guidelines specify that for many types of eukaryote-prokaryote combinations, recombinant DNA experiments must be performed in so-called "containment laboratories" and the *E. coli* strains must be of an **EK-2** variety, having a genotype that makes it extremely unlikely that the bacteria could colonize the human intestine. The strain $\chi1776$ for example, which was developed by R. Curtiss, carries a constellation of mutations that render it sensitive to the bile salts present in the intestine. The $\chi1776$ cells also require thymine and diaminopimelic acid (DAP), a cell-wall constituent, so that unless these constituents are provided externally, the *E. coli* fail to replicate their DNA or die *via*

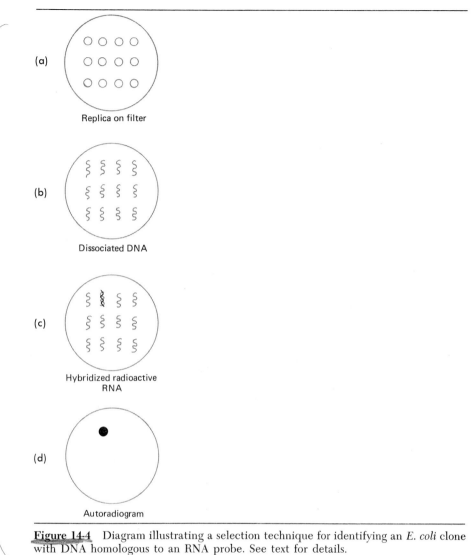

(a)

Replica on filter

(b)

Dissociated DNA

(c)

Hybridized radioactive
RNA

(d)

Autoradiogram

Figure 14-4 Diagram illustrating a selection technique for identifying an *E. coli* clone with DNA homologous to an RNA probe. See text for details.

osmotic lysis. As a result, χ1776 cells fail to survive passage through the intestinal tracts of experimental animals, even when ingested in large numbers.

14.9 APPLICATIONS OF CLONING TECHNOLOGY

The potential usefulness of cloned DNA for genetic research can hardly be overemphasized. We list here a few of the many kinds of experiments that can be performed with such DNA.

uses of

Cloning

First, cloning is useful for obtaining large amounts of a particular gene or sector of the genome. D. Brown and colleagues, for example, isolated the redundant rRNA genes (Section 8.15) from a single toad, cut the DNA into pieces with restriction enzymes, and cloned the fragments so that they had an abundant supply of distinct sectors of the redundant rRNA gene cluster. A comparison of these clones revealed a lack of sequence homology between spacers derived from an individual animal.

Second, cloning is useful for obtaining small sectors of DNA containing particular genes. For example, L. Clarke and J. Carbon have sheared wild-type *E. coli* DNA into random fragments, spliced individual fragments into Col El vectors, and introduced the plasmids into F^+ strains of *E. coli*. A "bank" of 2000 colicin-resistant *E. coli* strains was in this way created. To determine which clone carried, for example, the wild-type allele of a temperature-sensitive *dna* mutation, each strain was mated to an F^- *dna* mutant by mass-mating techniques (Section 11.6). As are noted earlier, F^+ cells often transmit nonconjugal plasmids along with their F element to F^- cells; in this case, therefore, it was simply necessary to identify that clone which transmitted thermoresistance to the *dna* mutant. The plasmid carrying this dna^+ gene could then be isolated and treated very much as a transducing phage; the dna^+ gene product could in theory be transcribed and translated *in vitro;* contiguous genes could be identified to generate a physical map of the region; controlling elements could be isolated by restriction enzymes and sequenced; and so forth.

Third, the cloning of individual eukaryotic genes with their contiguous controlling elements opens up the possibility of understanding how gene expression is regulated in eukaryotes, an understanding that has proved very difficult to obtain as long as the genes remain enmeshed in the gigantic eukaryotic genome with its bewildering complexity. One might, for example, be able to study the effect of a particular hormone on the transcription of a cloned eukaryotic gene. The insights into eukaryotic gene structure provided by the hemoglobin gene clone have already been described in Section 8.19.

Finally, cloning may prove to be of major medical and agricultural importance. Insulin genes have, for example, already been cloned into bacteria, and considerable progress has been made toward introducing nitrogen-fixing genes into the genomes of crop plants. Whether these genes will function in their new hosts to create, for example, prokaryotic "insulin factories" remains to be demonstrated at the time this text is being written.

MOLECULAR STUDIES OF MITOCHONDRIAL GENETIC FUNCTIONS

14.10   THE MITOCHONDRIAL GENETIC APPARATUS

Mitochondria are cytoplasmic organelles found in all eukaryotes (Figure 2-1). They contain the enzymes of the Krebs cycle, carry out oxidative phosphorylation, and participate in fatty-acid biosynthesis. Mitochondria also possess their own endowment of DNA and ribosomes, both of which are distinctive from their counterparts in the nucleus and cytoplasm of the cell. In this and the following sections we consider the mitochondrial genetic apparatus and its functions in some detail; we briefly consider the analogous system in chloroplasts; and we conclude with speculations as to the evolutionary origins of organelle genomes.

Table 14-2 summarizes some of the physical properties of representative mitochondrial DNAs (mt DNAs) and ribosomes. Probably most mitochondrial genomes take the form of a "naked" closed circular duplex, without any evidence of histones or other features of nuclear chromosomes. For animals these circular duplexes are very small ($\sim$10 $\times$ 10^6 daltons; for reference, the Col El plasmid is 4.2 $\times$ 10^6 daltons and the *F* element is 35 $\times$ 10^6 daltons), whereas they are larger in fungi ($\sim$50 $\times$ 10^6) and higher plants ($\sim$70 $\times$ 10^6). In all cases, mitochondrial DNA replication has been shown to be carried out by mitochondrial DNA polymerases that function independently of their nuclear counterparts.

In addition to varying in the size of their mitochondrial DNAs, eukaryotes also vary in the size of their mitochondrial ribosomes and rRNAs (mt rRNAs) (Table 14-2); animal ribosomes are small (55S, with 16S and 12S rRNA) whereas *Neurospora* ribosomes are large (80S, with 22S and 15S rRNA). Regardless of ribosomal size, protein synthesis on mitochondrial ribosomes is generally sensitive to the same spectrum of antibiotics (streptomycin, chloramphenicol, and so forth) as is protein synthesis on bacterial ribosomes; this is clearly an additional reason why the indiscriminate use of antibiotics is potentially harmful.

Mitochondria from all eukaryotes contain RNA polymerases which are unique in that they possess but a single polypeptide (recall from Section 8.3 that the RNA polymerases of both bacteria and the eukaryotic nucleus are high- molecular-weight complexes). Some mitochrondrial RNA polymerases are sensitive to the antibiotic rifampicin (Section 8.3) whereas others are not.

Table 14-2 Physical Properties of Mitochondrial (mito) Compared with Cytoplasmic (cyto) Components in Three Organisms.

	Human (Hela)		Yeast		*Neurospora*	
	Mito	Cyto	Mito	Cyto	Mito	Cyto
Length of genome	5 μm	—	25 μm	—	20 μm	—
Ribosomes	60S	74S	75S	80S	73S	77S
Large subunit	45S	—	53S	60S	50S	60S
Small subunit	35S	—	35S	40S	37S	37S
Ribosomal RNA						
Large subunit	16S	28S	21S	26S	25S	28S
Small subunit	12S	18S	14S	18S	19S	18S

Adapted from P. L. Altman and D. D. Katz, Eds. *Cell Biology*. Bethesda: Fed. Am. Soc. Exp. Biol., 1976, Cell Biology Federation of American Societies for Experimental Biology, Bethesda, Md.

14.11 NUCLEAR VERSUS MITOCHONDRIAL SPECIFICATION OF MITOCHONDRIAL COMPONENTS

In the mid–1960s, as the molecular properties of the mitochondrial genetic apparatus were analyzed, it became quite clear that the numerous enzymes, membrane components, ribosomal proteins, and other components found in mitochondria could not possibly all be encoded within only $5–25\mu$ of mt DNA (an animal mitochondrial chromosome, for example, has only enough DNA for perhaps 15-18 structural genes). Therefore, a number of laboratories investigated which mitochondrial components were specified by the nuclear/cytoplasmic genetic system and which were specified by the mitochondria themselves.

One approach taken was to poison the organelle system with such antibiotics as rifampicin or chloramphenicol and then to test which components continued to be synthesized and which were inhibited. An alternate approach was to isolate RNA molecules found within mitochondria and to test whether they annealed to mt DNA in DNA-RNA hybridization studies (Section 8.4). A third approach was to study the mitochondrial biosynthetic capacities of certain mutants of yeast (described in more detail below) which contain grossly altered mt DNA or none at all. The net results of such studies are summarized in Tables 14-3 and 14-4. Most mitochondrial components are indeed encoded by the nucleus and synthesized in the cytoplasm. The mitochondrial genome is devoted to the synthesis of certain RNA molecules and to certain subunits of three mitochondrial proteins. Genetic evidence for these statements is given shortly.

Table 14-3 Mitochondrial Components in Yeast Mutants with Grossly Altered mt DNA.

Absent
Functional respiratory chain; functional cytochrome bc_1 and cytochrome oxidase
Functional energy-transfer system
Functional protein-synthesizing system; ribosomes

Present
Outer membrane
Inner membrane (altered)
Parts of respiratory chain (cytochromes c and c_1, some subunits of cytochrome aa_3)
Parts of the energy-transfer system: F_1-ATPase
Transport systems
Krebs cycle
DNA and RNA polymerases, ribosomal proteins, elongation factors

From P. Borst, in *International Cell Biology*, B. R. Brinkley and K. R. Porter, Eds. New York: Rockefeller University Press, 1977.

Table 14-4 Biosynthesis of Major Mitochondrial Enzyme Complexes in Yeast

| | Number of Polypeptide Subunits | | |
Enzyme Complex	Total	Made in Cytoplasm	Made in Mitochondria
Cytochrome oxidase (cytochrome aa_3)	7	4	3
Cytochrome bc_1 complex	7	6	1
ATPase (oligomycin-sensitive)	9	5	4
Large ribosomal subunits	30	30	0
Small ribosomal subunits	22	21	1

From P. Borst, in *International Cell Biology,* B. R. Brinkley and K. R. Porter, Eds. New York: Rockefeller University Press, 1977.

14.12 PHYSICAL MAPPING OF THE MITOCHONDRIAL GENOME

Because of its small size, mt DNA has proved eminently suitable for physical mapping studies. Denaturation mapping (Section 8.19), for example, permits long runs of AT-rich DNA to be visualized as "denaturation bubbles," and these serve as landmarks which identify one sector of the chromosome versus the next. The AT-rich regions also permit a sensitive estimate of the lengths of particular deletions. Finally, denaturation mapping has served to demonstrate that the large fungal mt DNAs have extensive AT-rich sectors that are absent from animal genomes.

A second mapping procedure has been to take RNA molecules that are known to hybridize with mt DNA, attach **ferritin** molecules to them, perform a ferritin-RNA/DNA hybridization using separated H and L strands of mt DNA (Section 8.4), and look for the position of the large and electron-dense ferritin molecules along the chromosome with the electron microscope. When this technique is applied, for example, to the HeLa mitochondrial genome (Figure 14-5), the large (16S) and small (12S) species of mt-rRNA map adjacent to one another in the H strand; at least 12 4S (presumably tRNA) sites are widely spaced over the entire H strand, and an estimated 7 4S sites localize to the L strand. Similar approaches using yeast mt-DNA reveal that the large (21S) and small (14S) rRNA genes map to opposite sides of the circle (Figure 14-6)—the only established exception to the rule that these genes are adjacent to one another (or are separated only by a tRNA gene) throughout the biological kingdom (Section 8.14 and 8.15). Many of the yeast tRNA genes have been specifically identified by hybridization; as shown in Figure 14-6, most tRNA genes lie within three discrete sectors of the yeast mt genome. Thus physical mapping has revealed that various mitochondrial genomes are by no means identical in their genetic organization, even though all appear to code for the same classes of gene products.

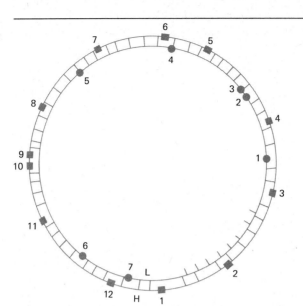

Figure 14-5 Map of mitochondrial genome from cultured human cells (HeLa), showing 4s RNA positions (presumably tRNA genes) on the H and L strands as squares and circles. The rRNA gene region is denoted by the heavy gray sectors. (From A. Angerer, et al., *Cell* **9**:81–90, 1976, © MIT.)

A number of mt DNAs are presently being subjected to restriction mapping (Section 7.1), and the restriction fragments in many cases have been cloned in *E. coli* (see preceding sections of this chapter). Such analyses have already confirmed gene localizations made by electron microscopy, and cloned fragments of mt DNAs will undoubtedly be used to drive coupled transcription-translation systems to assess the full coding potential of a particular mitochondrial genome. Meanwhile, information on the structural genes present in mt DNA has relied on the analysis of mitochondrial gene mutations.

GENETIC ANALYSIS OF THE YEAST MITOCHONDRIAL CHROMOSOME

Mutations that are known or presumed to reside in mitochondrial DNA have been reported for a number of eukaryotes. In *Neurospora*, for example, the *poky* (*mi-1*) mutant strain, first isolated by M. and H. Mitchell, grows slowly, respires poorly, and has aberrant mitochondria; if *poky* mitochondria are isolated and injected into wild-type hyphae, the trait is transmitted. Extensive analysis of *poky* mitochondria by D. Luck and his associates has shown that these mitochondria are deficient in small ribosomal subunits, possibly because they lack a mitochondrial-coded ribosomal protein. Other

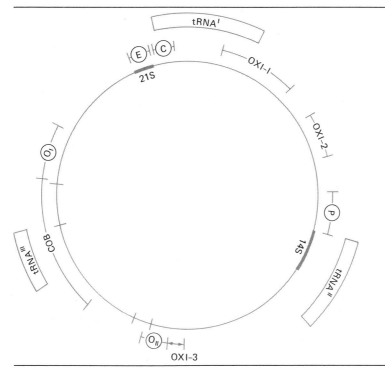

Figure 14-6 Map of mitochondrial genome from *Saccharomyces cerevisiae*, the total length of the circle being 76,000 base pairs (76 kilobases or kb). The three tRNA gene regions (I–III) are indicated. The 14S and 21S rRNA genes are denoted by heavy gray sectors. E, C, P, and O are loci for erythromycin, chloramphenicol, paromomycin, and oligomycin resistance, respectively. Loci OXI-1, OXI-2, and OXI-3 code for the three cytochrome oxidase subunits specified by the mitochondrion; COB codes for the 1 mitochondrion-specified subunit of the cytochrome bc_1 complex. (From A. Lewin, *et al.* In: *Mitochondria*, W. Bandlow, R. J. Schweyen, K. Wolf, and F. Kaudewitz, eds. Berlin; Walter de Gruyter, 1977.)

examples include a mutation to chloramphenicol-resistance in cultured human cells which has been attributed to mitochondrial DNA, and the *cmsT* mutation in corn, which renders crops susceptible to several leaf-blight diseases.

Interesting as these individual mitochondrial mutations may be, a large collection of different mutations is required by the geneticist if a genome is to be mapped and, to date, the yeast *Saccharomyces cerevisiae* is the one organism to meet this criterion. The major advantage in studying yeast is that it is a **facultative anaerobe** and can grow by fermenting glucose in the complete absence of mitochondrial respiratory functions. Mutations to respiratory deficiency are therefore automatically conditional mutations in yeast: they do not prevent normal growth by fermentation but place limitations on growth when yeast cells are forced to utilize a carbon source (for example, ethanol) which they must respire.

14.13 *PETITE* MUTATIONS

In the late 1940s, B. Ephrussi and his colleagues in France observed that when yeast cells are plated under conditions in which either respiratory or fermentative metabolism can take place, an occasional **petite** colony appears among normal **grande** colonies. When cells of the *petite* class were tested for their ability to respire, it was found to be nil or negligible. In other words, *petite* yeast cells can derive energy only by the relatively inefficient process of fermentation, so that their growth is expectedly slow, particularly on an agar plate where glucose is limiting. They are not, however, debilitated in the same way as an obligate aerobe when deprived of its normal mitochondrial function.

Perhaps the most surprising and still unexplained property of the *petite* state is the frequency with which it arises. Spontaneously, one in 100 to 1000 cells in a yeast population exhibits the *petite* phenotype, and if a populaton is treated with such acridine mutagens (Section 6.13) as ethidium bromide, 100 percent of the cells can be converted to the *petite* state.

Some *petite* yeast cells prove to carry mutations in their nuclear genes (see Figure 12-16; petite markers denoted as *p*) and are called **nuclear** or **segregational petites.** These exhibit classical Mendelian inheritance: in a cross between *p* and + cells, the diploids are +/*p* wild-type and the meiotic products of sporulation are 2+:2*p* (the yeast life cycle is diagrammed in Figure 3-10). Most petite yeast, however, prove to carry mutations that exhibit **non-Mendelian inheritance.** These mutations were designated **rho⁻** (also symbolized ρ^-) in contrast to *rho⁺* (ρ^+) wild-type cells.

Two classes of *rho⁻* petites—*neutral* and *suppressive*—are known, and their properties are described separately below.

Neutral *Petites* The non-Mendelian inheritance of neutral *petites* is apparent when a *rho⁺* X *rho⁻* cross is analyzed. The diploid progeny of the cross prove to be *rho⁺* in phenotype, as with the recessive nuclear *petites,* but when these diploid progeny undergo sporulation, all 4 meiotic products are *rho⁺*. If nuclear gene markers (for example, mating-type alleles) are followed in the same cross, they segregate in the expected 2:2 ratio. Moreover, if the *rho⁺* progeny are back-crossed to the original neutral *petite* strain, the emergent meiotic progeny are again *rho⁺*. Ephrussi interpreted these studies to mean that the 4:0 pattern of inheritance of the *rho⁻* mutation was the consequence of its localization in some genetic element that resided in the cytoplasm rather than the nucleus. The loss of *rho⁻* in the cross could then be attributed to a "takeover" by *rho⁺* elements and the concomitant dilution of *rho⁻* in successive cell divisions.

That mt DNA represents the cytoplasmic genetic element affected by neutral *petite rho⁻* mutations seemed likely in view of the fact that *rho⁻* cells are respiratory-deficient. A direct demonstration came when it was shown that most neutral *petites* lack mt DNA altogether (the so-called *rho°*

strains). In other words, it is clear in the case of neutral *petite* yeast that heritable alterations in mitochondrial DNA lead to heritable alterations in mitochondrial phenotype.

Suppressive *Petites* The majority of *rho⁻* mutations convert yeast cells into suppressive *petites*. Suppressive strains are also disabled in mitochondrial protein synthesis but differ from neutral *petites* in that they are expressed in diploid cells: a fully suppressive *petite* will convert all diploids arising from a *rho⁺* × *rho⁻* cross to the *rho⁻* phenotype, whereas partially suppressive strains affect only a portion of the diploids. Similar effects arise when such diploids are sporulated; diploids created with highly suppressive *petites* yield 0 wild-type: 4 *petite* ascospore ratios. While a number of theories on the mechanism underlying suppressiveness have been offered, the actual mechanism is not as yet understood.

Suppressive-*petite* mitochondria usually possess about as much DNA as wild-type mitochondria, but the buoyant density of their DNA is typically shifted towards a lower GC content. Further analysis of the DNA reveals that the suppressive *rho⁻* mutations generally originate as deletions. Since the sequences that are not deleted proceed to reiterate and sometimes rearrange themselves, the result can be a highly scrambled version of the original chromosome. As might be expected, neither suppressive nor neutral *petite* mutations can undergo reversion.

Since genes coding for essential elements of the mitochondrial protein-synthesizing apparatus are distributed throughout mt DNA (Figure 14-6), it is hardly surprising that most if not all deletions and rearrangements of the mitochondrial genome result in defective mitochondrial protein synthesis; the synthesis of all proteins translated in the mitochondrion is, in turn, disrupted. This explains why virtually all suppressive and neutral *petites* exhibit the same phenotype regardless of the composition of their surviving mt DNA: all are primarily affected in their ability to carry out a normal mitochondrial protein synthesis.

14.14 STRUCTURAL GENE MUTATIONS IN YEAST MITOCHONDRIAL DNA

The *rho°* strains of yeast (and *rho⁻* strains as well) are unable to synthesize functional cytochrome oxidase, cytochrome bc_1, and oligomycin-sensitive ATPase. Closer examination revealed that only certain polypeptides of these multimeric enzymes are missing when mitochondrial protein synthesis is blocked (Table 14-4). While this fact is consistent with the notion that the structural genes for the missing polypeptides reside in the mitochondrial genome, it is also conceivable that these structural genes might reside in the nucleus but that their expression requires an intact mitochondrial protein-synthesizing apparatus, a commodity lacking in *petite* cells. Therefore, the task of identifying mitochondrial structural genes by genetic techniques has required isolating cytoplasmically inherited mutant strains that possess a functional translation system but are nonetheless mutant in mitochondrial function.

Figure 14-7 outlines the procedures used for obtaining such mutants. It is seen that the mutants fall into two classes. The *mit⁻* mutants are affected in phosphorylative respiration, and several have been shown to carry alterations in polypeptides of the cytochrome oxidase, cytochrome bc_1, and ATPase enzymes. The *syn⁻* mutants, on the other hand, are affected in their protein-synthetic ability, and mitochondrially-coded rRNAs and tRNAs appear to be affected. The cytoplasmic nature of each mutation can be established by its non-Mendelian segregation ratios during meiosis.

Since mitochondria from two different parents have the capacity to fuse and recombine their DNAs in diploid zygotes, A. Tzagaloff, P. Slonimski, and many others have proceeded to map the various *mit⁻* and *syn⁻* mutations obtained by such selection procedures. A variety of mapping approaches

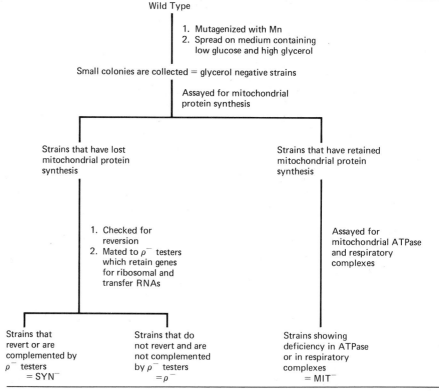

Figure 14-7 Selection for *mit⁻* (no respiration) and *syn⁻* (defective in mitochondrial protein synthesis) mutants of yeast. Mn has been shown to induce point mutations and deletions primarily in mitochondrial DNA. Cells lacking adequate mitochondrial function are unable to grow on glycerol. (From A. Tzagaloff, in *The Genetic Function of Mitochondrial DNA*, C. Saccone and A. M. Kroon, Eds. Amsterdam: North-Holland, 1976.)

has been used, including deletion mapping with various *rho⁻* tester strains and direct two-factor and three-factor crosses. No one approach is reliable in itself (for example, the frequent occurrence of rearrangements in *rho⁻* DNA prevent one from assuming that a given *rho⁻* tester carries a simple deletion), but coherent results using several methods have allowed the structural genes in yeast mt DNA to be ordered as shown in Figure 14-6. With such a relatively clear picture of the informational content of mt DNA, the way is now open to learn how these genomes are regulated and to ask how nuclear and mitochondrial genes cooperate and coordinate themselves to produce a functional organelle.

14.15 MOLECULAR STUDIES OF CHLOROPLAST GENETIC FUNCTIONS

The DNA of higher plant chloroplasts can be isolated as covalently closed duplex circles with molecular weights in the $85\text{-}95 \times 10^6$ dalton range. The potential informational content of these chromosomes is thus somewhat greater than that of mitochondrial DNA. Like mitochondria, chloroplasts have distinct transcriptional and translational components. Chloroplast ribosomes are small ($\sim$67S) and show sensitivity to a variety of bacterial protein-synthesis inhibitors. Chloroplast tRNAs and RNA polymerases are also different from their nuclear/cytoplasmic counterparts.

Molecular analyses of chloroplast genetic functions have taken the same form as those described for mitochondria: antibiotic studies have distinguished chloroplast and cytoplasmic translational activities, for example, and DNA/RNA hybridization studies have identified chloroplast-encoded RNA species. Chloroplast DNA clearly contains genes for rRNAs, for a number of tRNAs, and probably for certain ribosomal proteins. Genes are also present for the three large subunit polypeptides of the multimeric Fraction I protein, a protein possessing ribulose diphosphate carboxylase and oxygenase activities important in photosynthesis. The small-subunit polypeptides of Fraction I protein, on the other hand, are known to be encoded in nuclear DNA and translated on cytoplasmic ribosomes. Finally, chloroplast DNA appears to specify one or more chloroplast membrane polypeptides which are essential to photosynthetic electron transport.

Restriction mapping and cloning of chloroplast DNA promises to yield much additional information on the nature of its genes. J. Bedbrook, P. Kolodner, and L. Bogorad, for example, have been able to generate the map of corn (*Zea mays*) chloroplast DNA shown in Figure 14-8. This genome is seen to carry two copies of the three rRNA sequences (16S, 23S, and 5S), one copy on each strand. The position of the ribulose diphosphate carboxylase gene has also been localized (Figure 14-8).

14.16 MATERNAL INHERITANCE OF CHLOROPLAST MUTATIONS

A cytoplasmically inherited mutation in yeast does not show a strict 2:2 segregation from its wild-type allele at meiosis; instead, the ratios differ

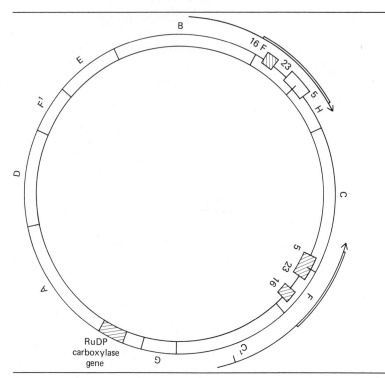

Figure 14-8 Map of chloroplast genome from corn, showing the two rRNA sequences, one on each strand. The sectors beneath the two arrows are inverted repeats of one another. The lettered sectors represent restriction fragments obtained with the enzyme *Sal* I. The RuDP carboxylase sequence is contained within the A fragment in a sector (shaded) cleaved by the enzyme *Bam*. (Courtesy of J.R. Bedbrook, R. Kolodner, D.M. Coen and L. Bogorad.)

from cross to cross depending, for example, on whether a *rho⁻* strain is neutral or suppressive. In most higher animals and plants, on the other hand, cytoplasmic genetic elements usually exhibit a very strict, albeit non-Mendelian, pattern of inheritance known as **maternal inheritance.** In this phenomenon, the cytoplasmic marker is almost invariably transmitted to progeny via the female parent, regardless of nuclear or cytoplasmic genotype. In other words, if a cytoplasmic trait is carried solely by the female in a cross, all of the offspring (male and female) inherit the trait; if the trait is carried solely by the male, none of the offspring inherit it. The basis for this inheritance pattern appears to reside largely in gametic anatomy: eggs are rich in cytoplasm, whereas sperm and pollen usually contribute only a nucleus at fertilization. Therefore, egg cells alone are capable of transmitting cytoplasmic genes to a zygote.

Numerous examples are known of chloroplast maternal inheritance, dating back to 1909 when C. Correns first described this phenomenon for the plant *Mirabilis jalapa,* also known as the Four O'Clock. The distribution of green pigment in a given *Mirabilis* plant varies from leaf to leaf: the leaves on certain branches may be fully green; other branches will have patchy leaves in which green tissue is interspersed with pale-green-to-white tissue in a **variegated** pattern; still others will bear fully pale leaves. Microscopic examination of the green leaves and the green areas of the variegated leaves reveals that the cells contain normal chloroplasts and chlorophyll pigment, whereas the pale leaves and the pale patches lack normal chloroplasts and pigment. These cells cannot, of course, carry out photosynthesis, but they are fed by other portions of the plant.

When the inheritance of these pigment traits is examined, it is found that the progeny inherit the phenotype of the female parent. For example, if ovules derive from fully green portions of the plant, then, regardless of the source of the pollen, only fully green plants will result and the variegated character will not reappear in subsequent generations. Similarly, when ovules derive from a wholly white branch, sickly white plants emerge, even when the pollen comes from a green-branch flower. These results are summarized in Table 14-5.

When the ovule instead derives from variegated branches, three types of seed are produced in variable numbers, again regardless of the male parent: some give rise to pure green, some to pure white, and the majority to variegated offspring (Table 14-5). The classical explanation for this outcome is to assume that the variegated ovule contains a mixture of normal and abnormal plastids. Should the embryo sac happen to receive only normal plastids from this collection, a green plant will result; if it receives only abnormal plastids, a white plant will result. When the embryo sac receives a mixture of both plastids, which should happen most often, the plant will be variegated, the patchiness and the occasional pure-colored branches arising as the normal and abnormal plastids **sort out** during vegetative cell divisions.

Another example from higher plants illustrates the interplay between

Table 14-5   Chloroplast Inheritance in Variegated Four-O'Clocks

Branch of Origin of the Male Parent	Branch of Origin of the Female Parent	Progeny
Green	Green	Green
	Pale	Pale
	Variegated	Green, pale, variegated
Pale	Green	Green
	Pale	Pale
	Variegated	Green, pale, variegated
Variegated	Green	Green
	Pale	Pale
	Variegated	Green, pale, variegated

nuclear and cytoplasmic genomes. A gene in corn called *iojap* (*ij*) has been mapped by M. Rhoades to nuclear linkage group VII (see Figure 12-13). Plants homozygous for *ij* are either inviable white seedlings or variegated with a characteristic white striping, the phenotype being known as *striped*. When the variegated plants serve as females in a cross, they give rise to green, white, and striped progeny, regardless of the nuclear genotype of the paternal parent. Thus, if the pollen derives from a normal green *Ij/Ij* plant as in Figure 14-9a, the resulting progeny will be *Ij/ij* heterozygotes, but many will exhibit abnormal plastid pigmentation: the presence of the "normal" *Ij* gene has no curative effect. In the reciprocal *Ij/Ij* ♀ × *ij/ij* ♂ cross, on the other hand, the *Ij/ij* progeny are all normally pigmented (Figure 14-9b).

The *iojap* trait therefore exhibits classical maternal inheritance once it has become established in an *ij/ij* plant. Moreover, once established, it becomes independent of the *ij* gene, as can be demonstrated by crossing F_1 *Ij/ij* variegated females to *Ij/Ij* normal males. As shown in Figure 14-9c, a mixture of green, striped, and white progeny again results, even though some of the striped and white plants now have an *Ij/Ij* genotype. Thus the *iojap* trait, once established, is permanent.

While variegated-plant phenomena are usually thought of in terms of chloroplast DNA mutations, either spontaneous or, in the case of *iojap*, induced in *ij/ij* homozygotes, it is important to bear in mind that other interpretations exist. A variegated phenotype might well also arise if some stable cytoplasmic genome other than chloroplast DNA—a virus, for example, or an episome of some kind—were present in the mutant organism and affected plastid viability in some way. Such elements might well be inherited through the egg cytoplasm and mimic plastid mutations. In cases

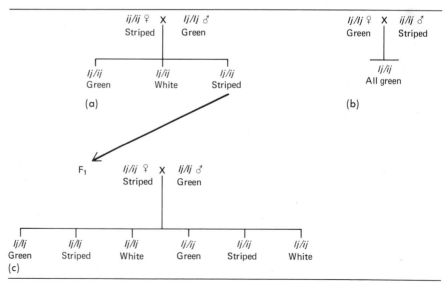

Figure 14-9 (a) Cross between variegated (*iojap*) and green (normal) corn plants. (b) Reciprocal of cross (a). (c) Cross of F_1 variegated females to normal males.

such as *iojap,* a particular nuclear genotype might be required for such elements to take up active residence in the plant cytoplasm. While it seems likely that most maternally inherited plastid disabilities will in fact prove to be chloroplast gene mutations, such a proof requires more than the demonstration that chloroplast aberrations are inherited via the maternal parent. Ultimately, chloroplast-DNA alterations (analogous to *petite* mutations) or specific mutant gene products must be demonstrated.

14.17 THE ORIGIN OF ORGANELLE GENETIC SYSTEMS

Two major theories have been advanced for the evolutionary origin of DNA-containing organelles in eukaryotes; presented here are very abbreviated versions of each. Interested readers are referred to the references cited at the end of the Chapter. 1) The **endosymbiosis theory,** first suggested by R. Altmann in 1890, postulates that organelles are modern descendants of bacterialike organisms that took up residence within primitive, perhaps even then eukaryotic, cells. The presence of these endosymbionts was presumably advantageous to the host cells, and with time the endosymbionts evolved into present-day organelles. 2) The **plasmid theory,** originated by H. Mahler and R. Raff, postulates that in bacterialike protoeukaryotes, certain plasmids first "picked up" genes for tRNA, rRNA, and selected respiratory or photosynthetic proteins from the main chromosome; these plasmids then became attached to, and surrounded by, one or more membranes and, with time, evolved into present-day organelle genomes. The remainder of the protoeukaryotic genome and cytoplasm, meanwhile, went on to evolve in eukaryotic directions.

Both theories seek to explain why organelle genetic systems have prokaryotic features. Both theories also attempt to explain why eukaryotes "bother" to retain independent organelle genomes at all when most organelle-specific information is nuclear-coded: they propose that the proteins synthesized within organelles are, for various reasons, much more efficiently utilized if they are made *in situ.* While it may be impossible to prove or disprove either theory, since the postulated primal events must have occurred about a billion years ago, both theories have been—by the criterion of having stimulated interesting research—highly successful.

14.18 ENDOSYMBIOSIS

In preceding sections of this chapter it was noted that apparent organelle mutations might in some cases be caused by viruses or other infectious agents; the "invasion" of protoeukaryotes by bacteria was also described as a prominent theory for the origin of organelles. Such speculations are firmly rooted in the findings of modern biology: a number of present-day eukaryotic cells or organisms are known to maintain populations of viruses or microorganisms which are transmitted via the cytoplasm. For example,

Paramecium bursaria harbors a collection of small eukaryotic algae which cannot be maintained in culture once they are removed from the *Paramecium* cytoplasm. Since the algae clearly derive some essential component(s) from *Paramecium* and since they are known to provide their hosts with photosynthetic products, the interaction is regarded as mutually beneficial and is called **endosymbiosis.** In other cases the "mutant benefits" are not as apparent. For example, a maternally inherited infectious virus known as **sigma** is found in the cytoplasm of CO_2-*sensitive* strains of *D. melanogaster;* such flies become permanently paralyzed when exposed to concentrations of CO_2 that do not affect normal flies.

The most extensively studied case of endosymbiosis is given by *Paramecium aurelia,* which can harbor some 10 different types of Gram-negative bacterial species in its cytoplasm (Figure 14-10), the best known being called **kappa** particles. Maintenance of each bacterial species requires particular nuclear genes of the host *Paramecium.* The bacteria may be transfered from one cell to the next during conjugation (Figure 3-11), and often elaborate toxins.

14.19 INHERITANCE OF PREFORMED STRUCTURES

The transmisson of endosymbionts and organelles can be attributed to extranuclear DNA found either in infectious particles or in organelles. There are, however, hereditary traits whose origin is not easily ascribed either to nuclear or extranuclear DNA. The best studied of these characteristics concern the cortex of ciliated protozoa such as *Paramecium.* The mouth and contractile vacuole are two prominent features of the cortex of *Paramecium,* and T. Sonneborn and his co-workers have shown that these

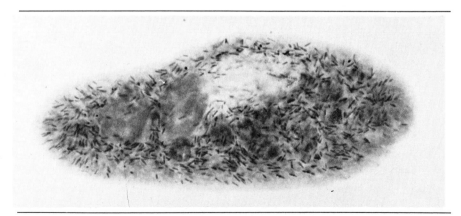

Figure 14-10 *Paramecium tetraurelia* containing endosymbiotic bacteria known as lambda. (Courtesy L. B. Preer, from *Bact. Rev.* **38**, 113, 1975.)

existing, or **preformed,** structures can be transmitted from one cell generation to the next, independent of the transmission of nuclear genes and cytoplasmic genetic factors.

During normal sexual reproduction in *Paramecium* (Figure 3-11) two cells conjugate and then separate after they have exchanged nuclei. On occasion, however, the conjugants fuse and form a double animal (a **doublet**) with two sets of cortical structures. When such doublets reproduce asexually by mitosis, they give rise to doublets, showing that the double nature of the cortical structures has a genetic basis. Significantly, when doublet *Paramecia* are mated with normal (singlet) *Paramecia,* the progeny of the doublet exconjugant are doublets and the progeny of the singlet exconjugant are singlets. Furthermore, when the progeny go through autogamy (Figure 3-12) they maintain their doublet or their singlet properties. Nuclear gene markers introduced into the crosses between doublets and singlets are transmitted and segregate according to the pattern expected for nuclear genes.

The mode of inheritance of the duplicated structures therefore appears to be independent of the mode of inheritance of nuclear genes. It also appears to be unaffected by exchange of cytoplasm, since doublet and singlet cells retain their identities and their ability to reproduce true to type even after they are allowed to conjugate under conditions in which cytoplasm as well as nuclei are exchanged.

An additional observation supporting the idea that the cortical structures of *Paramecium* are genetically autonomous concerns the natural grafting of a piece of cortex from one conjugant to the other. This grafting happens only rarely, but when it does an animal that has duplex cortical structures is recovered. When this animal reproduces, the new structures reproduce autonomously.

Clearly, the preformed cortical structures are maintained by cell division; they appear to be essential for their own reproduction, and their inheritance is not under the control of nuclear genes or cytoplasmic genetic factors. Fifteen years ago, Sonneborn suggested that different parts of the cortex might serve as sites for the "specific absorption and orientation of molecules derived from the milieu and genetic action" and that preexisting cortical structures could act by "determining where some gene products go in the cell, how these combine and orient, and what they do." While such phrases are carefully stated generalizations, it is not yet possible to explain the *Paramecium* phenomena in more specific terms. For other eukaryotes, however, membrane-associated cytoplasmic DNA is regularly reported, and in some cases this DNA is found to have molecular properties distinct from nuclear or organelle DNA. In addition, RNA has been shown to be intimately associated with basal bodies (structures, akin to centrioles, that give rise to cilia and flagella), and basal bodies are known to have certain autonomous properties. Therefore, there is some reason to believe that such phenomena as cell symmetry and form may, at least in some cases, prove to depend on strategically localized species of extranuclear nucleic acids.

References

Plasmids

Betlach, M., V. Hershfield, L. Chow, W. Brown, H. M. Goodman, and H. W. Boyer. "A restriction endonuclease analysis of the bacterial plasmid controlling the *Eco*RI restriction and modification of DNA," *Fed. Proc.* **35:**2037–2043 (1976).

*Bowman, C. M., J. E. Dahlberg, T. Ikemura, J. Konisky, and M. Nomura. "Specific inactivation of 16S ribosomal RNA induced by colicin E3 *in vivo*," *Proc. Natl. Acad. Sci. U.S.* **68:**964–968 (1971).

*Schaller, K., and M. Nomura. "Colicin E2 is a DNA endonuclease," *Proc. Natl. Acad. Sci. U.S.* **73:**3989–3993 (1976).

*Skurray, R. A., H. Nagaishi, and A. J. Clark. "Molecular cloning of DNA from *F* sex factor of *Escherichia coli* K-12," *Proc. Natl. Acad. Sci. U.S.* **73:**64–68 (1976).

Stanfield, S., and D. R. Helinski. "Small circular DNA in *Drosophila melanogaster*," *Cell* **9:**333–345 (1976).

Cloning DNA in Plasmid and Phage Vectors

Abelson, J. "Recombinant DNA: Examples of present-day research," *Science* 196:159–161 (1977) (and many other articles in this issue of *Science*).

Bukhari, A. I., J. Shapiro, and S. Adhya, Eds. *DNA Insertion Elements, Plasmids, and Episomes.* Cold Spring Harbor Laboratory, Cold Spring Harbor, N. Y., 1977.

Campbell, A. M. *Episomes.* New York: Harper and Row, 1969.

*Clarke, L., and J. Carbon. "A colony bank containing synthetic ColEI hybrid plasmids representative of the entire *E. coli* genome," *Cell* **9:**91–99 (1976).

*Cohen, S. N. "Structural evolution of bacterial plasmids: Role of translocating genetic elements and DNA sequence insertions," *Fed. Proc.* **35:**2031–2036 (1976).

*Cohen, S. N. "Transposable genetic elements and plasmid evolution," *Nature* **263:**731–738 (1976).

*Curtiss, R. III. "Genetic manipulation of microorganisms: Potential benefits and biohazards," *Ann. Rev. Microbiol.* **30:**507–533 (1976).

Glover, D. M., R. L. White, D. J. Finnegan, and D. S. Hogness. "Characterization of six cloned DNAs from *Drosophila melanogaster,* including one that contains the genes for RNA," *Cell* **5:**149–157 (1975).

Guerineau, M., C. Grandchamp, and P. P. Slonimski. "Circular DNA of a yeast episome with two inverted repeats: Structural analysis by a restriction enzyme and electron microscopy," *Proc. Natl. Acad. Sci. U.S.* **73:**3030–3034 (1976).

Kramer, R. A., J. R. Cameron, and R. W. Davis. "Isolation of bacteriophage λ containing yeast ribosomal RNA genes: Screening by in situ RNA hybridization to plaques," *Cell* **8:**227–232 (1976).

Maniatis, T., S. G. Koe, A. Efstratiadis, and F. C. Kafatos. "Amplification and characterization of a β-globin gene synthesized *in vitro*," *Cell* **8:**163–182 (1976).

Meynell, G. G. *Bacterial Plasmids.* Cambridge, MIT Press, 1972.

Novick, R. P., R. C. Clowes, S. N. Cohen, R. Curtis III, N. Datta, and S. Falkow. "Uniform nomenclature for bacterial plasmids: A proposal," *Bact. Rev.* **40:**168–189 (1976).

*Southern, E. M. Detection of specific sequences among DNA fragments separated by gel electrophoresis," *J. Mol. Biol.* **98:**503–517 (1975).

Struhl, K. J. R. Cameron, and R. W. Davis. "Functional genetic expression of eukaryotic DNA in *Escherichia coli*," *Proc. Natl. Acad. Sci. U.S.* **73:**1471–1475 (1976).

*Denotes articles described specifically in the chapter.

Organelle Genomes and Heredity

*Angerer, L., N. Davidson, W. Murphy, D. Lynch, and G. Attardi. "An electron microscope study of the relative positions of the 4S and ribosomal RNA genes in HeLa cell mitochondria," *Cell* **9**:81–90 (1976).

*Bedbrook, J. R., and L. Bogorad. "Endonuclease recognition sites mapped on *Zea mays* chloroplast DNA," *Proc. Natl. Acad. Sci. U.S.* **73**:4309–4313 (1976).

Bücher, Th., W. Neupert, W. Sebald, and S. Werner, Eds. *Genetics and Biogenesis of Chloroplasts and Mitochondria.* Amsterdam: North-Holland, 1976.

Ephrussi, B. *Nucleo-cytoplasmic Relations in Microorganisms.* New York: Oxford University Press, 1953.

Gillham, N. W. "Genetic analysis of the chloroplast and mitochondrial genomes," *Ann. Rev. Genet.* **8**:347–392 (1974).

Gillham, N. W. *Organelle Heredity.* New York: Raven, 1978.

Harris, E. H., J. E. Boynton, N. W. Gillham, C. L. Tingle, and S. B. Fox. "Mapping of chloroplast genes involved in chloroplast ribosome biogenesis in *Chlamydomonas reinhardi*," *Molec. Gen. Genetics* (in press) (1978).

Kirk, J. T. O. "Chloroplast structure and biogenesis," *Ann. Rev. Biochem.* **40**:161–196 (1971).

*Lambowitz, A. M., and D. J. L. Luck. "Studies on the *poky* mutant of *Neurospora crassa*," *J. Biol. Chem.* **251**:3081–3095 (1976).

Levings, C. S. III, and D. R. Pring. "Restriction endonuclease analysis of mitochondrial DNA from normal and Texas cytoplasmic male-sterile maize," *Science* **193**:158–160 (1976).

*Mahler, H. R., and R. A. Raff. "The evolutionary origin of the mitochondrion: A nonsymbiotic model," *Int. Rev. Cytol.* **43**:1–124 (1975).

*Rifkin, M. R., and D. J. L. Luck. "Defective production of mitochondrial ribosomes in the *poky* mutant of *Neurospora crassa*," *Proc. Natl. Acad. Sci. U.S.* **68**:287–290 (1971).

Saccone, C., and A. M. Kroon, Eds. *The Genetic Function of Mitochondrial DNA.* Amsterdam, North-Holland, 1976.

Sager, R. "Mendelian and non-Mendelian inheritance of streptomycin resistance in *Chlamydomonas*," *Proc. Natl. Acad. Sci. U.S.* **40**:356–363 (1954).

Sager, R. *Cytoplasmic Genes and Organelles.* New York, Academic Press, 1972.

Sager, R. "Genetic analysis of chloroplast DNA in *Chlamydomonas*," *Adv. Genetics* **19**:287–340 (1977).

*Schweyen, R. J., U. Steyrer, F. Kaudewitz, B. Dujon, and P. P. Slonimski. "Mapping of mitochondrial genes in *Saccharomyces cerevisiae*," *Molec. Gen. Genet.* **146**:117–132 (1976).

*Slonimski, P., and A. Tzagoloff. "Localization on yeast mitochondrial DNA of mutations expressed in a deficiency of cytochrome oxidase and/or cytochrome QH_2—cytochrome c reductase," *Eur. J. Biochem.* **61**:27–41 (1976).

*Tzagoloff, A., F. Foury, and A. Akai. "Assembly of the mitochondrial membrane system. XVIII. Genetic loci on mitochondrial DNA involved in cytochrome *b* biosynthesis," *Molec. Gen. Genet.* **149**:33–42 (1976).

Endosymbiosis

Beale, G. H., A. Jurand, and J. R. Preer. "The classes of endosymbiont of *Paramecium aurelia*," *J. Cell Sci.* **5**:69–91 (1969).

d'Héritier, P. "The hereditary virus of *Drosophila*," *Adv. Virus Res.* **5**:195–245 (1958).

Oishi, K., and D. F. Poulson. "A virus associated with SR-spirochaetes of *Drosophila nabulosa*," *Proc. Natl. Acad. Sci. U.S.* **67**:1565–1572 (1970).

Poulson, D. F., and B. Sakaguchi. "Nature of "sex-ratio" agent in *Drosophila*," *Science* **133**:1489–1490 (1961).

*Preer, J. R., L. B. Preer, and A. Jurand. "Kappa and other endosymbionts in *Paramecium*," *Bact. Rev.* **38**:113–163 (1974).

Sonneborn, T. M. "Kappa and related particles in *Paramecium*," *Adv. Virus Res.* **6:**229–356 (1959).

Inheritance of Preformed Structures

Beisson, J., and T. M. Sonneborn. "Cytoplasmic inheritance of the organization of the cell cortex in *Paramecium aurelia*," *Proc. Natl. Acad. Sci. U.S.* **53:**275–282 (1965).

Meinke, W., and D. A. Goldstein. "Reassociation and dissociation of cytoplasmic membrane-associated DNA," *J. Mol. Biol.* **86:**757–773 (1974).

Nanney, D. L. "Ciliate genetics: Patterns and programs of gene action," *Ann. Rev. Genetics* **2:**121–140 (1968).

Sonneborn, T. M. "Does preformed structure play an essential role in cell heredity?" In *The Nature of Biological Diversity*, J. M. Allen, Ed. New York: McGraw-Hill, 1963, p. 165–221.

Sonneborn, T. M. "Gene action in development," *Proc. Roy. Soc. London* **176:**347–366 (1970).

(a)   (b)

(a) Tracey Sonneborn (Indiana University) has made many contributions toward an understanding of the extranuclear genetics of *Paramecium*. (b) Ruth Sager (Sidney Farber Cancer Center) discovered extranuclear inheritance in *Chlamydomonas reinhardi* and now studies its molecular basis.

Questions and Problems

1. You suspect that a certain trait in *E. coli* is conferred by a plasmid gene. How could you demonstrate this in the case of a conjugal plasmid? A non-conjugal plasmid?

2. The trait *yellow embryo* in a mammal does not affect the viability of either sex. When *yellow embryo* females are crossed to normal males, the F_1 are all *yellow embryo*. When the F_1 is inbred, 3/4 of the progeny are *yellow embryo* and 1/4 are normal. Is the trait maternally inherited? Explain.

3. To what extent are plasmids and temperate bacteriophages like endosymbionts and to what extent are they different?

4. How are plasmids similar to temperate phages and how are they different?

5. When females of a certain mutant strain of *D. melanogaster* are crossed to wild-type males (or to males of any strain), all of the viable offspring are

females. This result could be the consequence of either a sex-linked lethal mutation or a maternally inherited factor that is lethal to males. What kinds of crosses would you perform in order to distinguish between these alternatives?

6. Certain plants from the cross (c) given in Figure 14-9 are subjected to further crosses. What progeny genotypes and phenotypes do you expect from the following:
   (a) *Ij/Ij* white females X *Ij/Ij* green males
   (b) *Ij/Ij* white females X *Ij/Ij* green males
   (c) *Ij/Ij* green females X *Ij/Ij* striped males
   (d) *Ij/Ij* striped females X *ij/ij* striped males

7. You suspect that certain species of mRNA from the nucleus are translated exclusively in organelles. Devise hypotheses to explain how it is that such organelle-destined mRNA is not translated in the cytoplasm while en route to the organelle by postulating unique organelle tRNA species.

8. It has been proposed that cortical patterns in *Paramecium* are determined by DNA located within the *Paramecium* cell membrane. Discuss the experiments concerning the inheritance of preformed structure in light of this hypothesis.

9. The endosymbiotic bacterium *kappa* is maintained only by *Paramecium* cells that carry at least one dominant nuclear gene called *K*; in the absence of *K*, the bacteria are lost from cells after several rounds of mitosis. A *K/k* strain of *Paramecium* harboring *kappa* is crossed with a *k/k* strain. Diagram all possible outcomes of this cross (a) when conjugation involves nuclear exchange only; and (b) when conjugation proceeds under conditions in which cytoplasm (and therefore endosymbionts) as well as nuclei are exchanged.

10. *Paramecium* cells that harbor *kappa* are called "killers" because a toxin released by the endosymbionts is lethal to "sensitive" cells that lack *kappa*, although it is not harmful to cells that possess *kappa*.
    (a) Killer strain *A* and sensitive strain *B* are allowed to conjugate without cytoplasmic exchange. The two exconjugant clones *C* and *D* are cultured separately; each is then induced to undergo autogamy (Figure 3-12). When the exautogamous cells deriving from clone *C* are examined several generations later, half are found to be killers and half sensitives, whereas those deriving from clone *D* prove to be all sensitives. Give the nuclear genotypes of the various strains and clones and explain the results.
    (b) How would the outcome differ, if at all, if the same crosses were performed except that cytoplasmic exchange is allowed during the initial conjugation?

11. When a *Neurospora* cross is made such that the protoperithecium derives from a *poky* strain and the conidia are from a wild-type strain (see Figure 3-9 for the *Neurospora* life cycle) what mitochondrial genotype do you expect of the eight ascospores assuming maternal inheritance of the trait?

12. Mitochondria from the respiratory-deficient *abn-1* strain of *Neurospora* (Figure 6-2) were isolated and physically injected into wild-type *Neurospora* hyphae. What is the expected outcome if the *abn-1* phenotype is controlled

by a nuclear mutation? A recessive mitochondrial mutation? A suppressive mitochondrial mutation?

13. When an erythromycin-resistant (*ery-r*) strain of *Chlamydomonas* is crossed with an erythromycin-sensitive strain, the following data are obtained:

$$ery\text{-}r\ mt^+ \times ery\text{-}s\ mt^- \qquad ery\text{-}s\ mt^+ \times ery\text{-}r\ mt^-$$
$$\downarrow \qquad\qquad\qquad\qquad \downarrow$$

All progeny are *ery-r*   All progeny are *ery-s*

(a) Devise a hypothesis to explain these results.

(b) Both mating types of *Chlamydomonas* contribute apparently equal amounts of cytoplasm to the zygote. How does this affect your answer to (a)?

(c) Explain these results according to a model which proposes that the *mt* loci encode restriction endonuclease and modification enzymes (Section 5.9).

14. A collection of *mit⁻* mutations of yeast (Fig. 14-7) called *A-F*, are all deficient in cytochrome *b*. Pairwise *a* × *α* crosses (Figure 3-10) are made between various *mit⁻* strains, and the resultant diploids are tested for their ability to grow on a nonfermentable medium. The following results are obtained, where + = growth and − = nongrowth.

	a	A	B	C	D	E	F
α A		−	+	−	+	−	−
B			−	+	−	+	+
C				−	+	−	−
D					−	+	+
E						−	−
F							−

(a) How many genes involved with cytochrome *b* synthesis are marked by the strains? Which strains mark the same genes?

(b) Describe how you would determine the linkage relationships between these cytochrome-*b* genes using sporulation techniques.

(c) Describe how you would determine their linkage relationships using a collection of *ρ⁻* tester strains.

15. The offspring of male horses and female donkeys are known as hinnies; the offspring of reciprocal crosses are known as mules. You discover that the restriction endonuclease patterns (Section 7.1) from horse mitochondrial DNA are very different from donkey patterns. Outline experiments utilizing this information which would determine whether or not mitochondria are maternally inherited in ungulate mammals.

16. Explain why an *fi⁺* plasmid is a far less desirable vector for cloning recombinant DNA than is an *fi⁻* plasmid.

17. To avoid contamination of plates carrying EK-2 bacteria with normal bacteria, EK-2 bacteria are usually made to carry one or more genes to antibiotic resistance, and antibiotic-containing growth media are used. Would you expect these resistance genes to be carried by the main chromosome or by plasmids? Explain.

CHAPTER
15

Related Genes: Alleles, Isoloci,
and Gene Families

INTRODUCTION

Trait is a useful term in genetics: it indicates an aspect of the phenotype, sometimes a specific and sometimes a rather generalized one. Thus the ability to utilize lactose, the possession of short fingers, and a sensitivity to loud noise can all be considered traits. In this and the next four chapters we consider how single genes and combinations of genes produce particular phenotypic traits. We have, of course, been considering such material throughout the text—whenever we have stated, for example, that gene *A* gives rise to a particular polypeptide or that gene *A* is dominant to its allele *a*. Now, however, we explore the construction of the phenotype more carefully, using examples from human genetics whenever feasible, since the human phenotype has an obvious inherent interest.

This chapter is particularly concerned with traits specified by genes that are similar to one another. We first consider the simplest case of such traits in diploid organisms—namely, a trait dictated by one locus having two or more alleles. In such a case, the phenotype depends on which two alleles are present in the diploid and on their relative degrees of dominance in a heterozygote. We then consider traits that are determined by two or more distinct, but related, gene loci. Since each such locus will be represented twice in a diploid, and since dominant or recessive alleles may be found at each locus, considerable variability is possible in such cases.

TRAITS CONTROLLED BY A SINGLE GENE LOCUS

In the sections that follow, human traits are considered that are controlled by a single gene locus that may have two, three, or multiple alleles. For any two alleles in a diploid, several relationships are possible. On the one hand, one allele may be **fully recessive,** so that its presence exerts no detectable influence on the phenotype of a heterozygote (the allele may, for example, produce a nonfunctional enzyme or a nonsense fragment of a structural polypeptide). On the other hand, an allele may be **fully dominant** so that its presence inevitably defines the phenotype, regardless of which other allele is present. Finally, an allele may be **partially** or **incompletely dominant** so that its presence in a heterozygote invariably influences the phenotype but need not necessarily define the phenotype, depending on the nature of the allele at the paired locus.

With the above definitions as guidelines, we can proceed to consider the rich phenotypic detail of specific human traits.

15.1   INBORN ERRORS OF METABOLISM

The term "inborn error of metabolism," first coined by A. Garrod in 1902, has come to refer to a pathological human trait caused by an inherited defect in a single enzyme. Usually, although not invariably, the affected individual is homozygous for a recessive gene which codes for the enzyme, as in the following four examples.

Alkaptonuria This condition was the first to be described by Garrod. Approximately one person in every million is homozygous for the alkaptonuria gene, known as a, and thereby suffers from the disorder. Affected individuals may develop arthritic ailments in later life, but otherwise appear to be quite healthy except for the color of their urine, which becomes black shortly after exposure to air. A single biochemical reaction has been shown to be absent in alkaptonuric individuals. Thus, normal individuals catabolize the amino acids phenylalanine and tyrosine via a substance called homogentisic acid, which in turn is oxidized to fumaric and acetoacetic acid (Figure 15-1); alkaptonurics lack an active homogentisic acid oxidase enzyme and therefore excrete homogentisic acid into their urine. It is this homogentisic acid which turns black upon oxidation by air.

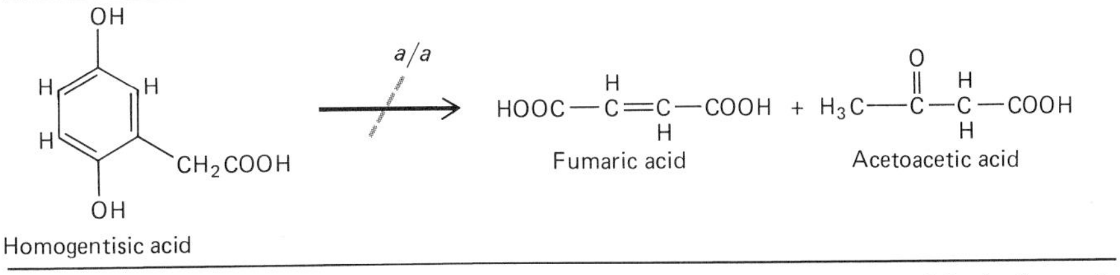

Figure 15-1 The normal metabolism of homogentisic acid, a reaction blocked in *a/a* persons suffering from alkaptonuria.

Phenylketonuria (PKU) The most common form of phenylketonuria is produced by a defect in the enzyme phenylalanine hydroxylase. This enzyme normally converts phenylalanine to tyrosine (Figure 15-2); in phenylketonuric individuals, however, phenylalanine accumulates and is converted to other phenyl derivatives (Figure 15-2). One of these, phenylpyruvic acid, can be detected in the urine. If left undiagnosed and untreated in infancy, the accumulated phenyl compounds of phenylketonuria produce irreversible and severe mental retardation. Therefore, the urine of newborn infants is routinely tested for the presence of phenylpyruvic acid, and affected infants are placed on a diet low in phenylalanine, a treatment which allows normal brain development. Since, in Caucasian populations, about one in 15,000 newborns is a *p/p* homozygote, the PKU screening and therapy programs represent major contributions of genetics to human welfare.

Lesch-Nyhan Syndrome A recessive human gene in the X chromosome causes afflicted males to develop a complex spectrum of traits. Such individuals are typically subnormal in intelligence, spastic, destructive, and

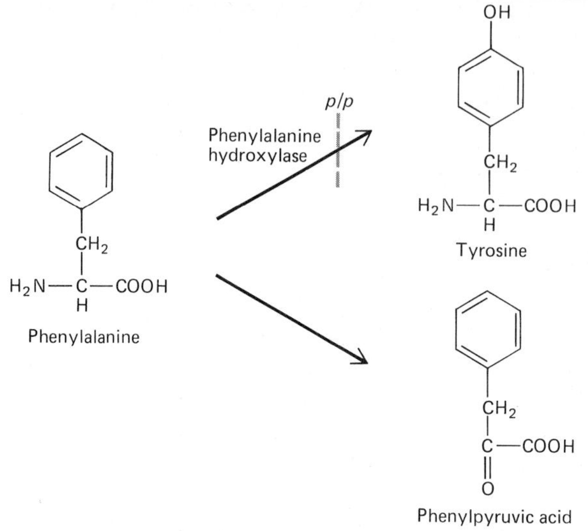

Figure 15-2 The normal breakdown of phenylalanine into tyrosine and into phenylpyruvic acid by p/p phenylketonuric individuals.

self-mutilating, with a particular tendency to bite their fingers and lips. These children prove to lack activity of the enzyme hypoxanthine guanine phosphoribosyl transferase (HGPRT), an enzyme normally involved in purine metabolism (Figure 15-3) (see also Section 12.21).

While it is not yet known why HGPRT inactivity produces such a bizarre pattern of behavior, the enzyme deficiency clearly exerts a **pleiotropic** effect, meaning that many facets of the phenotype are altered as a consequence of a single mutational lesion. Most inborn errors of metabolism are, in fact, highly pleiotropic, reflecting the multiple instances during fetal and infant development in which the absence of an enzyme (or the excess of a metabolite) can cause damage. Thus phenylketonuria not only produces brain defects but also a reduced pigmentation in hair and skin and a defective metabolism of the amino acid tryptophan. Pleiotropy often proves to be a major obstacle in determining the primary defect in an inherited clinical syndrome and, therefore, in devising therapy for the syndrome.

Tay-Sachs Disease Homozygosity for a number of mutations in the human gene pool produces a variety of diseases known collectively as **lysosomal storage diseases.** The individual affected by such diseases lacks a specific lysosomal enzyme that normally acts to break down some type of complex macromolecule (a polysaccharide, lipid, protein, or nucleic acid), with the result that such macromolecules accumulate in the individual's tissues.

Tay-Sachs disease is perhaps the best known of such storage diseases. Infants homozygous for the recessive gene hex A, which is located in chromosome 15 (Figure 12-22), are defective in N-acetylhexosaminidase A.

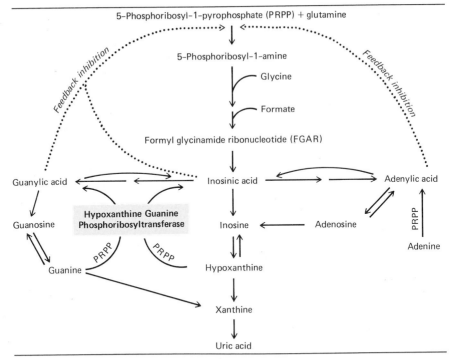

Figure 15-3 Normal pathways in purine biosynthesis, showing the essential role of HGPRT, the enzyme missing in Lesch-Nyhan individuals.

This enzyme normally cleaves the terminal hexosamine from a brain ganglioside lipid known as GM_2 (Figure 15-4). In Tay-Sachs infants, the unmetabolized GM_2 ganglioside accumulates within the brain cells, leading to cerebral degeneration and death by three years of age.

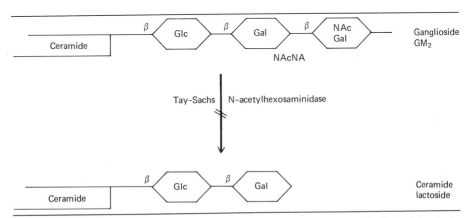

Figure 15-4 The normal breakdown of ganglioside GM_2, a reaction blocked in Tay-Sachs individuals. Abbreviations: NAcGal, N-acetylgalactosamine; NAcNA, N-acetylneuraminic acid; Gal, galactose; Glc, glucose. Ceramide is a long-chain amino alcohol linked to a long-chain fatty acid.

Therapy for storage diseases clearly requires that some means be found to supply target tissues (for example, brain cells) with the missing enzymes. A promising approach, which has already been used successfully to treat a related lysosomal disorder known as **Gaucher's disease,** involves packaging needed enzymes inside membrane vesicles and injecting the vesicles into the patient's blood. The vesicles proceed to fuse with the target-cell membranes, introducing their enzyme contents into the cytoplasm where they can act on accumulated macromolecules.

Detecting Heterozygous Carriers of Inborn Errors of Metabolism   Although most inborn errors of metabolism produce clinical symptoms only in homozygotes, it is possible in many instances to ascertain whether a person is a heterozygous carrier of a recessive mutant allele. For example, parents of phenylketonuric infants exhibit a slight, but significant, increase in blood phenylalanine levels which has no effect on their health; persons heterozygous for the Tay-Sachs allele have reduced serum levels of N-acetyl-hexosaminidase A; and Lesch-Nyhan heterozygotes can be detected by reduced HGPRT activity in their hair follicle cells. In many cases a heterozygote will exhibit only 50 percent of the level of normal activity for a particular enzyme. This means that fully normal persons have, in fact, twice as high a level of that enzyme as needed for unimpaired health.

Detecting Homozygotes for Inborn Errors of Metabolism in Utero   A sample of amniotic fluid can be withdrawn from a pregnant uterus **(amniocentesis)** and its constituent fetal cells can be subjected to karyotyping or biochemical assays. Fetuses with abnormal chromosome constitutions (Section 6.17) and a number of metabolic diseases (Table 15-1) can in this way be detected. Such prenatal diagnosis, coupled with therapeutic abortion, has permitted Tay-Sachs heterozygotes, for example, to escape the anguish of bearing children who die in infancy.

With the increasing understanding of human genetic diseases and the availability of heterozygote screening and prenatal diagnostic methods, **genetic counseling** is an increasingly important service provided by hospitals and other centers. Genetic counselors are trained not only to provide potential parents with an accurate estimate of the probabilities that they will have a child with a genetic disorder, but also to help parents adjust to the birth of such a child or to reach a decision about prenatal diagnosis and possible abortion.

15.2   THE ABO BLOOD-GROUP LOCUS

The ABO locus, which maps to chromosome 9 in the human genome (Figure 12-22), has three common alleles, two of which (I^A and I^B) are codominant and the third of which (I^O) is recessive. Since the locus controls an important surface property of human erythrocytes, it is considered in some detail.

Table 15-1 Inborn Errors of Metabolism Diagnosable before Birth

Lipidoses

Cholesterol-ester storage disease*
Fabry disease†
Farber disease*
Gaucher disease†
Generalized gangliosidosis (GM$_1$ gangliosidosis Type 1)†
Juvenile GM$_1$ gangliosidosis (GM$_1$ gangliosidosis Type 2)†
Tay-Sachs disease (GM$_2$ gangliosidosis Type 1)†
Sandhoff's disease (GM$_2$ gangliosidosis Type 2)†

Juvenile GM$_2$ gangliosidosis (GM$_2$ gangliosidosis Type 3)*
GM$_3$ sphingolipodystrophy*
Krabbe's disease (globoid-cell leukodystrophy)†
Metachromatic leukodystrophy
Niemann-Pick disease Type A
Niemann-Pick disease Type B*
Niemann-Pick disease Type C*
Refsum disease*
Wolman disease†

Mucopolysaccharidoses

MPS I—Hurler†
MPS I—Scheie*
MPS—Hurler/Scheie*
MPS II A—Hunter†
MPS II B—Hunter†
MPS III—Sanfilippo A†
   Sanfilippo B*

MPS IV—Morquio's syndrome
MPS VI A—Maroteaux-Lamy syndrome*
MPS VI B—Maroteaux-Lamy syndrome*
MPS VII—β-glucuronidase deficiency*

Amino Acid and Related Disorders

Argininosuccinic aciduria†
Aspartylglucosaminuria*
Citrullinemia†
Congenital hyperammonemia*
Histidinemia*
Hypervalinemia*
Iminoglycinuria*
Isoleucine catabolism disorder*
Isovaleric acidemia*

Cystathionine synthase deficit (homocystinuria)*
Cystathioninuria*
Cystinuria†
Hartnup disease*
Methylenetetrahydrofolate reductase deficiency*
Ornithine-α-ketoacid transaminase deficiency*

Maple-syrup-urine disease:
   Severe infantile†
   Intermittent*
Methylmalonic aciduria
   Unresponsive to vitamin B$_{12}$*
   Responsive to vitamin B$_{12}$†

Disorders of Carbohydrate Metabolism

Fucosidosis*
Galactokinase deficiency*
Galactosemia†
Glucose-6-phosphate dehydrogenase deficiency*
Glycogen-storage disease (Type II)†
Glycogen-storage disease (Type III)*

Propionyl CoA carboxylase deficiency (ketotic hyperglycinemia)*
Succinyl-CoA: 3 ketoacid CoA-transferase deficiency*
Vitamin B$_{12}$ metabolic defect*

Glycogen-storage disease (Type IV)†
Mannosidosis*
Phosphohexose isomerase deficiency*
Pyruvate decarboxylase deficiency*
Pyruvate dehydrogenase deficiency*

Miscellaneous Hereditary Disorders

Acatalasemia*
Acute intermittent porphyria*
Adenosine deaminase deficiency†
Chediak-Higashi syndrome*
Congenital erythropoietic porphyria*
Congenital nephrosis†
Lysosomal acid phosphatase deficiency†
Lysyl-protocollagen hydroxylase deficiency*
Myotonic muscular dystrophy*
Nail-patella syndrome*
Orotic aciduria*

Cystinosis†
Familial hypercholesterolemia*
Glutathionuria*
Hypophosphatasia†
I-cell disease†
Leigh's encephalopathy*
Lesch-Nyhan syndrome†
Protoporphyria*
Saccharopinuria*
Sickle cell anemia*
Testicular feminization*
Thalassemia†
Xeroderma pigmentosum*

* Prenatal diagnosis potentially possible.
† Prenatal diagnosis made
From A. Milunsky, *New Eng. J. Med.* **295**:377, 1976.

The A, B, and H Glycolipids   The ABO locus is concerned with the production of **glycosyltransferases**—enzymes that mediate the synthesis of polysaccharides. We are concerned here in particular with polysaccharides that attach to lipids and associate with cell membranes, forming **cell-surface glycolipids.** By influencing which glycosyltransferases are present in a cell, therefore, the ABO locus influences the types of surface glycolipids a cell can produce.

Glycosyltransferases act in a very precise fashion. If we denote a polysaccharide as R-glucose-glucose-galactose, where the R represents the end of the chain that will attach to the lipid and the galactose represents the free end of the chain, then one type of glycosyltransferase may add a fucose molecule to the chain to produce R-glucose-glucose-galactose-fucose. A second type of glycosyltransferase may recognize the same polysaccharide chain but will add a mannose, instead of a fucose, to it, producing an R-glucose-glucose-galactose-mannose chain.

The I^A allele of the ABO-locus codes for a glycosyltransferase known as α-N-acetylgalactosamyl transferase. This enzyme recognizes the polysaccharide chain drawn in Figure 15-5a and abbreviated as **H.** Recognition is followed by the addition of the sugar α-N-acetylgalactosamine and, with this addition, the H polysaccharide is converted to a polysaccharide known as **A** (Figure 15-5b). The I^B allele of the ABO locus, on the other hand, codes for an α-D-galactosyltransferase. This enzyme also recognizes H but, instead of adding an α-N-acetylgalactosamine, it adds instead a galactose residue, producing the **B** polysaccharide (Figure 15-5c). In both cases, some H chains fail to be acted upon by either enzyme and retain their H structure.

Expression of the I^A, I^B, and I^O Alleles   The H, A, and B polysaccharides are found as components of glycolipids on the erythrocyte surface, and their distribution depends strictly on the ABO locus genotype. Thus, an I^A/I^A homozygote produces erythrocytes that carry H and A chains but never B chains, while an I^B/I^B homozygote produces erythrocytes with H and B, but never with A chains. The I^A/I^B heterozygote possesses both enzymes and therefore carries H, A, and B polysaccharides on his or her red-cell surfaces. The I^A and I^B alleles are therefore said to be **codominant,** since each allele makes a comparable contribution to the phenotype of the heterozygote.

The third I^O allele of the ABO locus codes for neither the A nor B type glycosyltransferase enzyme; therefore, an I^O/I^O homozygote is unable to produce either A or B polysaccharides and places only "unmodified" H glycolipids on the erythrocyte surface. As might be predicted, the I^O allele is fully recessive to I^A and to I^B, so that I^A/I^O heterozygotes produce H and A chains and I^B/I^O heterozygotes produce H and B chains.

Detecting the ABO Blood Type   Polysaccharides A, B, and H can act as **antigens,** meaning that they are capable of eliciting specific anti-A, anti-B, and anti-H antibodies when injected into experimental animals (Box 9.2). The antisera from such animals can then be used to carry out **blood typing**

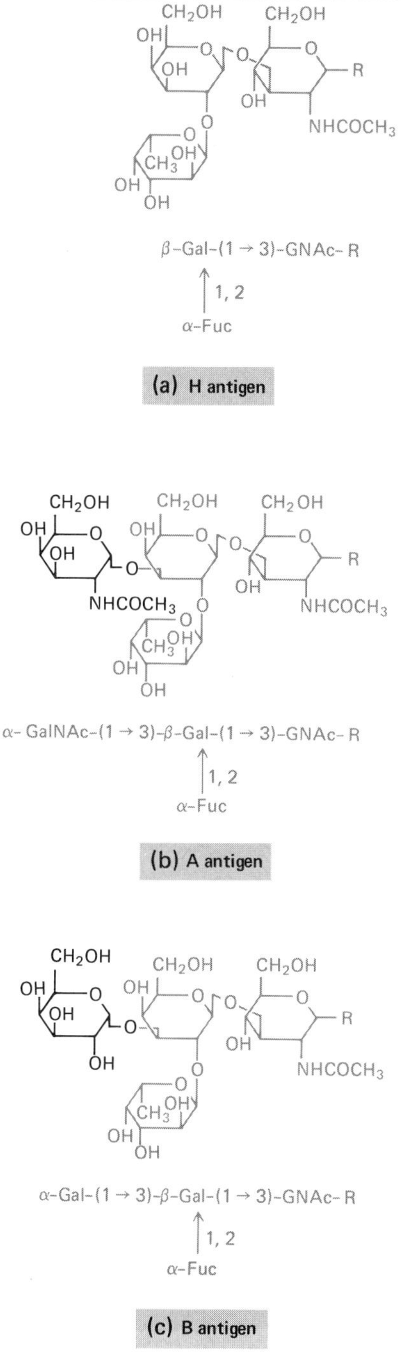

β-Gal-(1 → 3)-GNAc- R

↑ 1, 2

α-Fuc

(a) H antigen

α- GalNAc-(1 → 3)-β-Gal-(1 → 3)-GNAc- R

↑ 1, 2

α-Fuc

(b) A antigen

α-Gal-(1 → 3)-β-Gal-(1 → 3)-GNAc- R

↑ 1, 2

α-Fuc

(c) B antigen

Figure 15-5 Terminal sugar sequences in polysaccharide chains of glycolipids which confer A, B and H specificity. Gal: D-galactose; Fuc: L-fucose; GNAc: N-acetyl-D-glucosamine; GalNAc: N-acetyl-D-galactosamine.

of human beings. If a sample of red blood cells from a person is found to **agglutinate** (stick together) only in the presence of the anti-H serum, then the cells clearly possess only the H polysaccharide, and the person being tested must be an I^O/I^O homozygote. The person is then said to be of **Blood Type 0.** If a sample of cells from a second individual agglutinates with both anti-H and anti-A antisera, then the cells possess H and A and the individual is said to be of blood **Type A;** correspondingly, cells that agglutinate with anti-H and anti-B sera derive from **Type B** persons; and cells that agglutinate in the presence of all 3 antisera are of **Type AB.**

The cell-surface polysaccharides A, B, and H are not only antigenic to experimental animals. They are also potential antigens in humans—with one important proviso: **a normal individual will not produce antibodies against him- or herself.** Thus the serum of a person with Type A blood will contain neither anti-A nor anti-H antibodies; however, since the B polysaccharide is not an element of that person's "self," anti-B antibody production is not "forbidden," and proceeds apace. Similarly, the blood sera of Type B persons contains anti-A antibodies, the sera of Type 0 persons contains both anti-A and anti-B antibodies, and the sera of Type AB persons contains no antibodies against this group of cell-surface polysaccharides. These relationships are summarized in Table 15-2.

The clinical importance of the blood type can now be appreciated, for when a person is in need of transfused erythrocytes, the donor blood type must be carefully checked. The injection of Type A cells into Type 0 or Type B persons, for example, will result in potentially fatal agglutination of the transfused erythrocytes by circulating anti-A antibodies in the sera of the recipient.

15.3   PEPTIDASE A AND OTHER DIMERIC ENZYMES

The codominance of the I^A and I^B blood type alleles in a heterozygote results from the fact that each glycosyltransferase gene product operates independently. A slightly more complex kind of relationship between pairs

Table 15-2   The Agglutination Reactions Observed with the A, B, AB, and O Blood Groups

Blood Group	Blood Type-Specific Sugar	Type(s) of Antibody Produced	Types of Red Blood Cells Agglutinated	Transfusions Accepted from
A	Galactosamine	Anti-B	B, AB	Type A, Type O
B	Galactose	Anti-A	A, AB	Type B, Type O
AB	Galactosamine + galactose	None	None	Any donor
O	None	Anti-A and Anti-B	A, B, and AB	Type O only

of alleles is illustrated by the human locus coding for **peptidase A,** an enzyme found in erythrocytes which acts to hydrolyze dipeptides into their individual amino acids.

When erythrocyte extracts from various persons are subjected to gel electrophoresis (Box 7.1) and the gel is stained to reveal peptidase A activity, three types of enzyme patterns are found (Figure 15-6): some individuals possess only the peptidase A enzyme (called 1) which migrates relatively slowly toward the cathode; some individuals possess a more rapidly migrating peptidase A species (called 2); and some individuals possess three types of peptidase A enzymes—namely, enzyme 1, enzyme 2, and an enzyme denoted 2-1 with intermediate electrophoretic mobility (Figure 15-6). Such patterns are explained in the following way:

1. Peptidase A is a dimeric protein (Section 9.3) having the composition $\alpha\alpha$.
2. The locus coding for the α polypeptide of peptidase A exists in two allelic forms, PEP A^1 and PEP A^2.
3. The gene product of PEP A^1, called α^1, has a slower electrophoretic mobility than the α^2 product of the PEP A^2 allele.
4. PEP A^1/PEP A^1 homozygotes produce $\alpha^1\alpha^1$ **homodimer** enzymes equivalent to enzyme 1; PEP A^2/PEP A^2 homozygotes produce $\alpha^2\alpha^2$ homodimers equivalent to enzyme 2; and PEP A^1/PEP A^2 heterozygotes produce, in addition to the two types of homodimers, a **heterodimer** enzyme, $\alpha^1\alpha^2$, with an electrophoretic mobility equivalent to enzyme 2-1.

The validity of this series of postulates for peptidase A is demonstrated by the following experiment. Enzymes 1 and 2 were isolated and dissociated into their component polypeptides by incubating in mercaptoethanol and urea to reduce S-S bridges (Section 9.3). The polypeptides derived from dissociation of each enzyme were then mixed together, allowed to reasso-

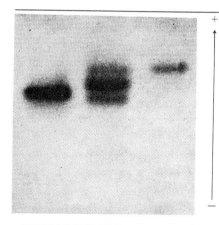

+

Figure 15-6 Electrophoretic patterns of peptidase A types 1, 2-1, and 2. (From H. Harris, *The Principles of Human Biochemical Genetics,* Amsterdam: North-Holland, 1975.)

Origin

| 1 | 2-1 | 2 |

Peptidase A types

ciate, and subjected to gel electrophoresis. The gel electrophoresis patterns then revealed not only enzymes 1 and 2 but also enzyme 2-1 which, in this case, had formed *in vitro* from enzyme 1 and enzyme 2 subunits.

In humans, numerous cases have been described that are qualitatively similar to the peptidase A situation in which a single locus, having two or more alleles, codes for a dimer-forming polypeptide, with pure homodimers present in homozygotes and both homo- and heterodimers present in heterozygotes. What are the effects of such heterozygosity on the phenotype? In many cases, homozygotes can be distinguished from heterozygotes only by electrophoretic analysis of their component enzymes: the heterodimer enzymes appear to function quite as well as the homodimer enzymes, and heterozygosity has no other measurable effect on the phenotype. In some cases, however, the heterodimer may be more active, or more stable, than either homodimer. In still other cases, one type of homodimer may be particularly unstable. For example, individuals heterozygous for the PEP A^8 allele possess no active $\alpha^8\alpha^8$ homodimers in their red blood cells and thus appear to have only two electrophoretic variants of the peptidase A enzyme.

15.4 DOMINANT MUTATIONS: PENETRANCE AND EXPRESSIVITY

Dominant mutant genes are relatively uncommon as compared to "recessives" or "partial dominants." In theory, dominant mutations should be of particular value for genetic analysis since they should exert an immediate effect on the phenotype of diploid organisms. In fact, however, many dominant mutant genes prove to be rather variable determinants of an individual's phenotype, particularly in humans. This variability is due to two phenomena known as penetrance and expressivity.

Penetrance is an all-or-nothing concept: a gene is either completely penetrant or incompletely penetrant. An incompletely penetrant gene finds expression in some individuals but not others. Figure 15-7, for example, shows two identical or **monozygotic** twins who derived from the same fertilized egg and can therefore be assumed to have identical genotypes. The genetically based developmental abnormality leading to the production of a harelip has clearly "penetrated" the twin on the left but not the twin on the right.

Incomplete penetrance is a property of many dominant genes. The human dominant gene *D*, for example, causes a bent and **stiff little finger.** Pedigrees are known, however, in which a woman exhibits this trait and her son does not, while two of his three children do. Clearly, the son was at least heterozygous for the gene, since he was able to pass it on to his children, but the gene did not affect his own finger development at all. Another example of incomplete penetrance is given by some forms of the psychotic disease **schizophrenia:** present evidence indicates that in some cases the illness may be controlled by two incompletely penetrant dominant genes.

Once a gene is penetrant (finds expression) it will frequently exhibit a

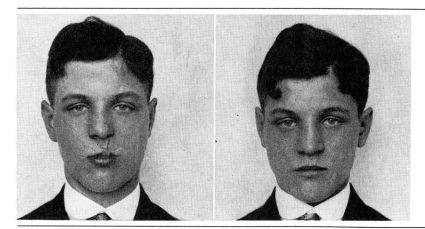

Figure 15-7 A pair of identical twins: *left:* penetrance of harelip; *right:* lack of penetrance. (From F. Claussen, *Zeitschr. Abstgs. und Vererbgsl.* **76**:30, 1939.)

variable expressivity, meaning that the same gene will produce a range of phenotypes in various "penetrated" individuals. There are many examples of genes with variable expressivity in *Drosophila,* higher plants, and animals. Again using a human example, we can cite **polydactyly,** a condition leading to an excessive number of fingers and toes. It is clear from the pedigree shown in Figure 15-8 that this trait is determined by a dominant gene and that the expressivity of this gene varies widely: some persons have normal

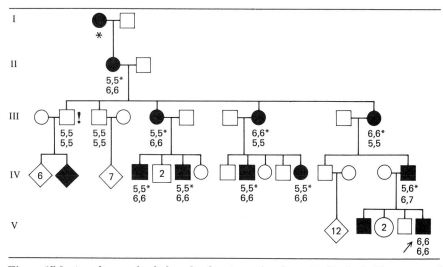

Figure 15-8 A pedigree of polydactyly, showing only selected individuals. The groups of four numbers below affected individuals represent the number of fingers (above) and toes (below) on the left and the right. An asterisk indicates that the nature of the polydactyly was not reported unequivocally. Arrow points to the propositus (affected individual who attracts the geneticist's attention). (From *Principles of Human Genetics,* Third Ed.; by Curt Stern. W. H. Freeman and Co. Copyright © 1973, after Lucas, *Guy's Hosp. Rpts.,* 3rd Ser. **25**, 1881.)

hands but six-toed feet, others have six-fingered hands and normal feet; one individual has more digits on his right extremities than on his left, and one has six digits on all four extremities. The pedigree for polydactyly also illustrates the incomplete penetrance of the gene: the male marked (!) does not exhibit the trait and yet passes it on to a child. As a general rule, genes that are incompletely penetrant are found to show variable expressivity as well, and indeed "no penetrance" may often represent one extreme in the observed range of variation, with "highly abnormal" perhaps representing the other.

Factors Influencing Penetrance and Expressivity The variation in the degree of expression attained by the genes we are now discussing is thought to be due to two factors: **genetic background** and **environmental influence.** **Genetic background** refers to the fact that an individual inherits a genotype and not a particular gene. A particular gene product will therefore normally be operating in the presence of countless different combinations of other gene products. When a particular gene product is especially sensitive to the nature of other gene products in a cell, its expression may well be variable from one individual to the next and, in some cases, inhibited altogether.

The effect of genetic background can be reduced or eliminated by studying highly inbred strains or monozygotic twins, and it is here that the effects of **environmental influence** can best be appreciated. A straightforward example is given by the dominant trait *Curly* of *D. melanogaster.* The *Curly* (*Cy*) gene gives rise to flies with curled-up wings when pupae are maintained at 25°C; when the pupae are maintained at 19°C, however, many individuals have apparently normal wings (incomplete penetrance), while the rest show a range of abnormalities (variable expressivity). It appears that the *Curly* gene produces wings that are unusually sensitive to heat at the time the adult fly emerges from the moist pupal case; at 19°C the wings dry more or less normally, whereas at 25° the upper and lower portions of the wings dry at different rates and a *Curly* wing results.

The environmental influence in the case of *Curly* wing is easy to analyze, but this is not always the case. The intrauterine environment of mammals, even for two monozygotic twins, is believed to be quite variable—twins will occupy different positions in the uterus, placental connections may differ, and so forth, and the array of effects that may influence gene expression after birth is so vast that it would be quite impossible to specify them all. In other words, when a gene exhibits incomplete penetrance and variable expressivity, particularly in humans, it is usually difficult to determine which environmental factors are involved.

ISOZYMES SPECIFIED BY ISOLOCI

The remainder of this chapter describes traits that are specified by two or more distinct, but related, gene products. We first consider traits that are controlled by **isoloci** and carried out by **isozymes,** terms that are best defined by examples.

15.5 HUMAN PHOSPHOGLUCOMUTASE ISOZYMES

Suppose the phenotypic trait being followed is the ability to transfer phosphate groups from glucose-1-P to glucose-6-P during carbohydrate metabolism, a reaction catalyzed by an enzyme activity called **phosphoglucomutase.** Phosphoglucomutase enzymes can be detected in cell extracts subjected to gel electrophoresis, much as with peptidase A (Figure 15-6), and these enzymes are known to function as monomeric proteins. Therefore, if human phosphoglucomutase activity were dictated by a single gene locus, as with peptidase A, one would expect either one band (for homozygotes) or two bands (for heterozygotes) when erythrocytes from a particular individual are subjected to electrophoresis and stained for the enzyme. Instead, H. Harris and coworkers have found multiple phosphoglucomutase bands in every human extract tested. Pedigree and somatic cell hybridization studies (Sections 4.3 and 12.21) reveal that one group of these bands is specified by the PGM_1 locus, located in human chromosome 1. The PGM_2 locus, in chromosome 4, is found to specify a second group of bands; and the PGM_3 locus in chromosome 6 specifies a third group. The various phosphoglucomutase monomers present in the electrophoresis gels are therefore collectively known as **isozymes**—enzymes that have similar substrates but distinctive electrophoretic or biochemical properties—and the PGM_1, PGM_2, and PGM_3 loci can be considered as a set of **isoloci.**

Since each PGM locus is found in a different human chromosome, alleles at any one locus would be expected to segregate at anaphase I of meiosis, whereas markers at unlinked loci would be expected to assort independently of one another (Chapter 4). This is indeed found to be the case. Thus, two alleles at the PGM_1 locus, known as PGM_1-1 and PGM_1-2, give rise to two distinct phosphoglucomutase isozymes. A PGM_1-1/PGM_1-2 heterozygote will transmit either PGM_1-1 or PGM_1-2 to individual children, never both. This same person may also be heterozygous at the PGM_2 locus, with a PGM_2-1/PGM_2-3 genotype. Again, only one or the other PGM_2 allele will be transmitted to individual offspring. Each child will, however, receive one PGM_1 allele and one PGM_2 allele from each parent, the identity of each being the chance consequence of independent assortment.

15.6 MOUSE ESTERASE ISOZYMES

As a second example of isozymes we can consider the **esterases.** Esterase enzymes resemble phosphoglucomutases in that they function as monomers and exist as **electrophoretic variants** detectable by subjecting electrophoresis gels to specific staining reactions. Several mouse esterases, however, prove to be specified by isoloci that are closely linked, in contrast to the unlinked isoloci responsible for the phosphoglucomutases. Specifically, F. Ruddle and colleagues have determined that the mouse esterase loci *Es-1,* *Es-2,* and *Es-5* are all found within about 10 map units of one another in linkage group XVIII of the mouse. At each of the three loci, at least two allelic genes are known: the *Es-la* and *Es-lb* alleles, for example, give rise, respectively, to *fast* and *slow* electrophoretic variants of the Es-1 esterase. In this case, of course, the genes at the various isoloci will not assort independently; instead, a chromosome carrying, for example, genes *Es-1a, Es-2b,* and *Es-5a* will usually be inherited as a unit, with occasional recombinational events breaking up the general linkage.

The esterase isozymes specified by *Es-1, Es-2,* and *Es-5* are in many functional respects similar to one another: all are found in the serum and all are resistant to the inhibitor eserine. A fourth esterase isolocus in the mouse, *Es-3,* is not linked to the others and encodes a distinct kind of esterase: it is largely restricted to kidney cells, and it is sensitive to eserine. Two general principles are illustrated by this example. First, isozymes that catalyze the same enzymatic reaction are not necessarily structurally similar (indeed, the three classes of human phosphoglucomutase enzymes described above prove to differ in size, in tissue distribution, in substrate specificity, and in stability). Second, clustered isoloci are more likely to be related than are unlinked isoloci (although this rule has its exception), and are indeed likely to represent a **gene family.**

GENE FAMILIES

Most genes in both prokaryotes and eukaryotes are believed to be present in only a single copy per haploid genome, but there are a number of important cases in which each haploid genome contains two or more genes that are either identical to, or very similar to, one another in their nucleotide sequence. Such gene sets are usually denoted as **gene families,** and it is generally believed that the genes in such families originated as duplications of some common ancestral sequence. The subsequent evolutionary history of the duplicated sequences varies considerably from one example to the next: in some cases the reiterated genes remain clustered together; in other cases their linkage has been broken and they may even reside on different chromosomes; finally, there are cases in which some members of a gene family are clustered and others dispersed throughout the genome.

In the following sections we consider four well-studied examples of gene families: the rRNA genes, the hemoglobin genes, the histocompatibility genes, and genes specifying immunoglobulins.

15.7 THE rRNA GENE CLUSTERS

Gene families that we have already considered in detail in Sections 8.14 and 8.15 are the ribosomal RNA gene clusters found in both prokaryotes and eukaryotes. Many studies report these rRNA sequences to be identical but since recent studies reveal some sequence dissimilarity, the term near-identical appears most appropriate to describe them.

The near-identity of the many rRNA genes in any one species is apparent testimony to how critical "correct" rRNA sequences are to proper ribosomal function. As we noted in Section 8.14, if an organism could tolerate much variation in the nucleotide sequences of its rRNA species, then one would expect that over the millions of years of that organism's existence, mutations in its redundant rRNA genes would accumulate, leading modern organisms to produce several or many distinctly different types of rRNA. Since this is not the case, it is proposed that for a eukaryotic organism having, for example, 400 rRNA sequences in its nucleolar organizer, a single defective rRNA gene would result in 1/400 of its ribosomes being faulty. This phenotype, it is argued, would be lethal because many essential mRNAs bound to the defective ribosomes would remain untranslated (or be incorrectly translated), and growth and development of the organism would thereby be impaired.

Since the spacer DNA between eukaryotic ribosomal genes is clearly free to diverge both in its length and in its sequence (Section 8.15), it seems probable that ribosomal genes experience mutagenic events. Some researchers believe that most such mutations are immediately lethal (the **selection theory** for rRNA gene homogeneity), whereas others espouse a **correction theory** which holds that grossly mutant sequences are often repaired before they have a chance to exert lethal effects. The actual mechanism for maintaining sequence homogeneity in large gene clusters, however, remains unknown.

15.8 THE HEMOGLOBIN GENES

Hemoglobin (Figure 9-5) is a tetramer with the general structure x_2y_2, where x and y are two of the 6 known globin chains—α, β, δ, γ, ϵ, and ζ. The most common form of adult hemoglobin has the structure $\alpha_2\beta_2$ and is known as HbA; a minor form of adult hemoglobin, $\alpha_2\delta_2$, is called HbA_2; and the most prominent fetal hemoglobin, HbF, is $\alpha_2\gamma_2$. The ϵ and ζ chains are present only in the first few weeks of embryonic life and little is yet known about them; we therefore focus on the α, β, δ, and γ chains.

Figure 15-9 depicts the proposed evolutionary relationships between the four globin chains of hemoglobin and its close relative, myoglobin (Figure 9-4), which is found in muscle tissues. These two globins are believed to have come from a single ancestral gene, which is thought to have duplicated. One copy of this duplicate went on to evolve into the modern myoglobin gene, while the second copy duplicated again. One copy of this second pair then went on to evolve into the modern α gene, while the second copy of this second pair duplicated again, and so forth. By this scheme, therefore, the β and δ chains of hemoglobin would be expected to be most closely related to one another and most distantly related to myoglobin.

Evidence for these relationships comes from several sources. First, amino-acid analysis of the globin chains reveals that all are of about the

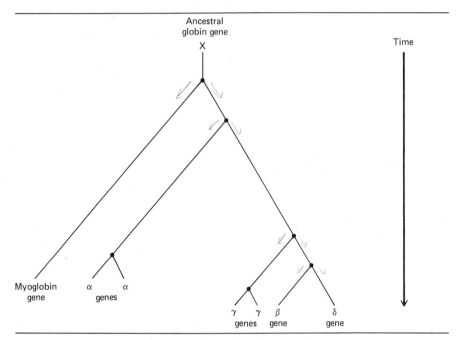

Figure 15-9 Proposed evolutionary relationships between the various globin genes. Gene duplications are represented by filled circles.

same length (153 amino acids for myoglobin, 141 amino acids for the hemoglobin α chain, and 146 amino acids for the hemoglobin β, γ, and δ chains). Myoglobin is quite dissimilar from these others in its amino acid composition; α and β prove to differ by 84 amino acids, β and γ by 30 amino acids, and β and δ by only 10. These results are consistent with the evolutionary sequence proposed.

Another line of evidence for a common genetic origin for the globin chains derives from the linkage relationships between the various hemoglobins. In mouse-human somatic cell hybrids (Section 12.21), one cell line has been identified that retains human α-chain DNA but lacks β; another line has lost α but retains β. Thus the α and β genes appear to reside on separate chromosomes. On the other hand, as described in more detail below, the β, δ, and γ genes all appear to be closely linked to one another. It is believed, therefore, that the β, δ, and γ genes derive from relatively recent duplications that have not yet "had time" to evolve many sequence dissimilarities or separate chromosomal positions, while the α sequence is a far more distant relative in the hemoglobin gene family, deriving from an ancient duplication and presently well separated from the others.

The α Genes In the above discussion it was tacitly assumed that the human haploid genome possesses only one α gene; however, this appears not to be the case. Careful hybridization studies (Section 8.4) reveal the presence of two very similar or identical α gene sequences per haploid genome (Figure 15-9); these are assumed, though not yet proven, to represent the result of a relatively recent tandem duplication event. Thus the normal human diploid cell carries not 2, but 4 α gene copies.

Genetic analysis indicates that a similar α gene duplication exists in mice. One mouse strain, known as SEC, produces two types of α chain. One of these types has a serine at position 68 and the other has a threonine. Since SEC is highly inbred and presumably homozygous at most of its gene loci, the persistence of these two forms of the α chain polypeptide suggests the presence of two copies of the α gene in the mouse genome. That these two copies are closely linked can be demonstrated by crossing SEC mice with C57BL mice, a second inbred strain that produces only one class of α chain, having an asparagine at position 68. When the heterozygous F_1 of the cross are backcrossed to either parent (Table 15-3), the ability to synthesize α 68Ser and α 68Thr is inevitably inherited together: no recombinants were found in 551 progeny examined, indicating a recombination frequency of less than 0.005 for the two loci (Table 15-3). It therefore appears that SEC carries an α gene duplication and that the two genes have come to differ by a single codon. A similar duplication may, of course, be carried by C57BL mice as well, but this fact cannot be detected genetically as long as the two genes are identical to one another in sequence.

The β and δ Genes The close linkage of the human β and δ genes is best demonstrated by the **Lepore hemoglobins,** named after the location where the trait was first detected. Amino-acid analysis of these hemoglobins reveals that their α chains are normal whereas their non-α or Lepore chains are distinctly abnormal: as diagrammed in Figure 15-10, the N-terminal

Table 15-3 Progeny of (C57BL × SEC) F_1 Mice Backcrossed to Either C57BL or SEC Mice

	Matings	
Alpha-Chain Genotype	(C57BL × SEC)F_1 × C57BL	(C57BL × SEC)F_1 × SEC
$\alpha^{68asn}/\alpha^{68asn}$	63	–
$\alpha^{68asn}/\alpha^{68ser,thr}$	73	216
$\alpha^{68ser,thr}/\alpha^{68ser,thr}$	–	199
Possible recombinants	0	0
Totals	136	415

From R. A. Popp, *J. Hered.* **60:**131 (1969), Table VII.

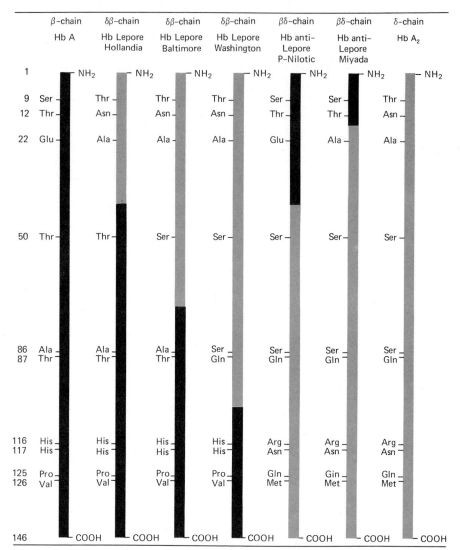

Figure 15-10 Relationships of the amino acid sequences of the non-α chains of various Lepore (δ-β) and anti-Lepore (β-δ) hemoglobins to those of the normal β- and δ-hemoglobin chains. Only the amino acids which differ in the normal β- and δ-polypeptide sequences are shown. (From H. Harris, *The Principles of Human Biochemical Genetics*, Amsterdam: North Holland, 1975.)

amino acids of Lepore chains are identical to those in a δ chain whereas the C-terminal sequence is identical to that of a β chain. In other words, the three types of Lepore hemoglobin chains depicted in Figure 15-10 are all δ-β **hybrid polypeptides,** each differing only in the position at which the δ → β switch occurs.

The formation of hybrid Lepore chains is readily explained by assuming that the δ and β chains are very similar in nucleotide sequence and are arranged in tandem linkage. When homologous chromosomes bearing these sequences undergo synapsis during meiotic prophase, it is supposed that pairing occasionally occurs out of register so that a β sequence pairs with a δ sequence, as illustrated in Figure 15-11a. If a crossover then occurs in the mispaired region, one chromosome will emerge with a δ-β Lepore sequence, (Figure 15-11b), a phenomenon known as **unequal crossing over.**

The unequal crossing over model predicts that for each δ-β Lepore gene produced, a β-δ **antiLepore** gene sequence should also be generated (Figure 15-11c). This prediction has recently been substantiated: Figure 15-10 summarizes amino-acid sequence data for Hb-antiLepore chains, and their β-like N-termini and δ-like C-termini are evident.

The γ Genes Normal human HbF contains two types of γ chains; one of these carries glycine at position 136 ($^G\gamma$), while the other carries an alanine at 136 ($^A\gamma$). Furthermore, both $^A\gamma$ and $^G\gamma$ hemoglobin variants have been described wherein amino acids other than those at position 136 have undergone substitution. When a newborn is heterozygous for one such variant (for example, $^A\gamma/^A\gamma'$), the amino-acid sequence of its $^G\gamma$ chains is unaffected, and the HbF present in that infant contains $^A\gamma$ and $^A\gamma'$ as well as $^G\gamma$. These data indicate that at least two structural genes for γ chains must exist per haploid genome (Figure 15-9), a conclusion confirmed by hybridization experiments.

The close linkage of the γ genes to the δ-β gene cluster has been demonstrated by two kinds of genetic evidence:

1. The human variant **hemoglobin Kenya** has been found to be a "fusion hemoglobin" in the same sense as Lepore and antiLepore, but it contains $^A\gamma$ sequences at its N-terminus and β sequences at its C-terminus.
2. Humans with a certain form of **thalassemia** (a general term for inherited hemoglobin abnormalities) are found to carry a deletion that removes $^A\gamma$ and δ gene sequences but leaves $^G\gamma$ and β genes intact, indicating a $^G\gamma$-$^A\gamma$-δ-β gene order.

While the γ genes therefore appear to lie very close or adjacent to their β and δ relatives, they are distinct both in sequence and in the control of their

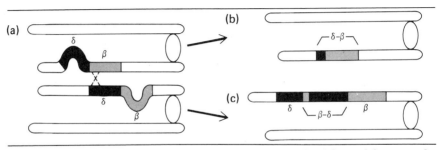

Figure 15-11 Generation of Lepore (δ-β) and anti-Lepore (β-δ) hemoglobin genes by mispairing and unequal crossing over.

synthesis, the γ genes normally being transcribed exclusively during fetal life. Presumably, therefore, the fetal and adult genes possess distinct kinds of controlling elements that are subject to very different regulatory signals.

15.9   THE MAJOR HISTOCOMPATIBILITY LOCUS

In mammals, the fate of a tissue graft is determined by the **histocompatibility system** of the donor and host. Two organisms are said to be histocompatible if they can accept solid tissue transplants from each other and incompatible if they cannot. Compatible organisms have the same spectrum of **histocompatibility antigens** associated with their cell surfaces, and are therefore unable to produce the same spectrum of antibodies (Section 15.2). In addition to defining "self" versus "non-self," the histocompatibility antigens presumably play additional roles in cell-surface physiology that are as yet poorly understood.

In all species that have been studied, an organism possesses a single **major histocompatibility locus** or **complex (MHC)** and multiple minor loci. (The difference between major and minor is operational: the major locus gives rise to antigens that elicit a more vigorous immune response in an incompatible host than the antigens specified by the minor loci.) Thus at least 30 loci have been identified in the mouse, including the major **H-2** locus, and many are present in humans in addition to the major **HLA** locus. Many of these loci are not linked, nor is there any reason to expect them to be linked since the products they specify undoubtedly mediate quite different cellular functions.

Most intensively studied of the MHCs have been the mouse H-2 locus in chromosome 17 and the human HLA locus in chromosome 6. The H-2 locus occupies 0.5 recombination units of DNA and the HLA locus occupies more than 1.0 unit—which represents enough DNA to code for several thousand polypeptides. Whether or not the full coding potentials of these loci have been realized, there is little question that numerous genes map to each region, as shown in Figure 15-12a and b.

Of the H-2 locus genes, those denoted K and D are responsible for

Figure 15-12   (a) MHC locus of mouse and its location in chromosome 17. The K and D regions specify the major histocompatibility antigens found in the cell membrane, and the TL region specifies transplantation antigens on the cell surface. The various genes in the I region specify lymphocyte cell-membrane components known as Ia antigens (for immune response-associated) which appear to play key roles in the various functions carried out by B and T lymphocytes. The S region controls one or more of the serum complement proteins involved in the recognition and lysis of foreign material. The G region specifies an erythrocyte antigen. The *Glo* locus contains the structural gene for the enzyme glyoxalase I. (b) MHC locus of humans and its location in chromosome 6. The B and A regions specify the major histocompatibility antigens found in the cell membrane; C and D represent more recently recognized minor loci. The number of alleles presently known for each of these loci is: A, 18; B, 22; C, 5; and D, 6. Genes specifying complement proteins C2, C4, C8, and properdin (Bf) are present, as are two erythrocyte-antigen genes, Ch and Rg. (From D. C. Shreffler, *Prog. Immunol. III*, Canberra City: Aust. Acad. Sci., 1978, and R. Payne, in *HLA and Disease*, J. Dausset and A. Svejgaard, Eds., Baltimore: Williams and Wilkins, 1977.)

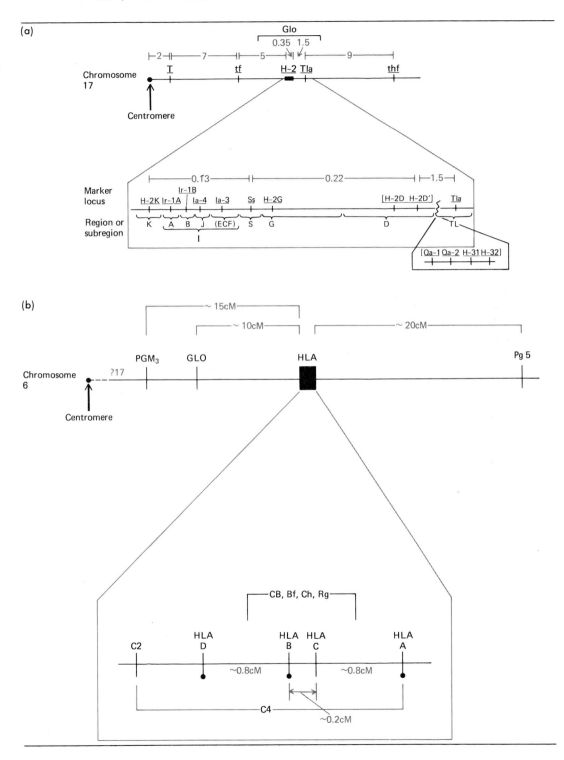

producing the major cell-surface histocompatibility antigens; the corresponding human genes are called HLA-A and HLA-B. Each gene is multiply allelic, there being on the order of 20 known alleles for the HLA-A and HLA-B genes, and each allele is denoted by a number (for example, HLA-A1, HLA-A2, HLA-A3 . . . and HLA-B1, HLA-B2, HLA-B3 . . .).

There appears to be a strong relatedness between genes K and D, and also between genes HLA-A and HLA-B: amino acid analysis of their respective glycopolypeptide gene products has established close sequence homologies. Moreover, the mouse K/D family is clearly related in sequence to the human HLA-A/HLA-B family. In many respects, therefore, the MHC genes are reminiscent of the hemoglobin gene families described in the preceding section.

As is apparent from Figure 15-12, genes K and D are separated by a number of genes, as are HLA-A and HLA-B. The "intervening" genes specify proteins that play a variety of roles in the immune system. As detailed in the legend to Figure 15-12, some of these proteins affect the nature of lymphocyte interactions, others control features of the immune response, and others specify or control a number of the serum complement proteins that function in many immunologic reactions. The genes that code for these proteins do not appear to be related in sequence to the K/D or HLA-A/B families. Their presence in the MHC is therefore presumed to reflect some functional requirement that the genes governing these related activities be inherited as a closely linked unit along with the two major histocompatibility genes.

The closely linked genes in an MHC locus are ordinarily inherited as a block: thus a person with an $\dfrac{A1\ B7}{A3\ B8}$ genotype produces gametes that are usually either A1 B7 or A3 B8; only infrequently do recombinant A1 B8 or A3 B7 gametes arise. The A1 B7 combination of alleles is referred to as a **haplotype** (an abbreviation of a "haploid genotype"), as is the A3 B8 combination. Thus an $\dfrac{A1\ B7}{A3\ B8}$ individual is said to have an A1 B7 haplotype and an A3 B8 haplotype, and this individual normally transmits one or the other haplotype to each of his or her children.

A striking feature of the MHC is that certain haplotypes are found in human populations far more often than would be expected by chance; moreover, certain haplotypes are strikingly correlated with susceptibility to particular human diseases. These aspects of MHC genetics are considered again in Chapter 20.

IMMUNOGENETICS

15.10 PROPERTIES OF THE IMMUNE SYSTEM

The gene families concerned with producing **immunoglobulins** (also called **antibodies**) belong in a category of their own, for they exhibit types of interactions not yet known for any other genes. To appreciate these distinctive interactions it is important to have in mind first the role that antibodies play in the physiology of a mammalian organism.

A mammal can synthesize many thousands of different kinds of antibodies. Each antibody is an oligomeric protein molecule whose structure is described shortly, and each has the ability to complex with and thus remove from the tissues or blood of the body a particular antigen or family of antigens. The immunoglobulins are synthesized by blood cells called **B lymphocytes** and by differentiated forms of these lymphocytes known as **plasma cells.** Each lymphocyte is highly specialized to synthesize only one (or perhaps a few) different kinds of antibody. It must follow, then, that an organism contains thousands of different lymphocyte clones, each committed to the synthesis of a particular antibody or group of antibodies.

Full details of the immune response are beyond the scope of this text; the interested reader is referred to the several excellent books cited at the end of this chapter. For our purposes we can give a vastly simplified account of what occurs, using as an example a human lymphocyte committed to synthesize an antibody against a protein found in the coat of the smallpox virus. We can call this antibody *antipox*. The lymphocyte and its clonal descendants synthesize antipox throughout their lifetimes, *whether or not they ever actually encounter the smallpox protein*. In other words, the synthesis of the antibody does not require the presence of the antigen as some sort of stimulant. When, however, the antigen is introduced, a series of events occurs. In this series, the antigen first becomes complexed with molecules of antipox that are located on the surface of the antipox lymphocytes. This interaction stimulates the lymphocytes to undergo many rounds of mitosis and to differentiate into specialized, antipox-secreting plasma cells. As a result, an organism that has been exposed to the smallpox virus soon acquires high blood-serum titers of the antipox antibody. This is the basis for immunizing persons by giving inoculations of attenuated forms of pathogenic viruses.

Plasma cell cancers (plasmacytomas or multiple myelomas) may lead to the proliferation of a single type of plasma cell. Blood or urine samples from such patients are often used to analyze the chemical properties of the single types of immunoglobulins that are overproduced. From such studies a number of facts are now known about human immunoglobulins. We must explore the properties of these molecules in some detail in order to understand the genes that specify them.

15.11   GENERAL PROPERTIES OF IgG ANTIBODIES

Figure 15-13 shows a model of an immunoglobulin G **(IgG)** antibody, the most prominent of the 5 immunoglobulin classes produced by humans (Table 15-4). In overall topology, an IgG molecule is seen to resemble a two-pronged fork. Four polypeptide chains, interconnected by disulfide bridges, are found in each IgG molecule. Two of these chains are longer than the other two, so that each IgG is said to have two **heavy chains,**

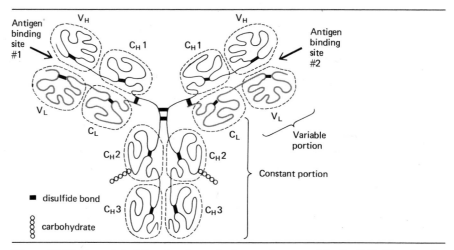

Figure 15-13 A model of the structure of a human immunoglobulin G (IgG) molecule. Light polypeptide chains are shown in gray and heavy in black. The variable regions of heavy and light chains (V_H and V_L), the constant region of the light chain (C_L), and the homology regions in the constant region of heavy chain (C_H1, C_H2, and C_H3) are thought to fold into compact domains (delineated by dotted lines), but the exact conformation of the polypeptide chains has not been determined. (From J. A. Gally and G. M. Edelman, *Ann. Rev. Genetics* **6**:1, 1972.)

(drawn in black in Figure 15-13) and two **light chains** (drawn in gray). In any one molecule, the two heavy chains are identical and the two light chains are identical.

Here we come to a critical concept in immunogenetics, one that must be carefully understood if the rest of this chapter is to be meaningful. Up to

Table 15-4 Classes of Human Immunoglobulins

	IgG	IgA	IgM	IgD	IgE
Heavy chains:					
Class	γ	α	μ	δ	ε
Subclass	$\gamma_1\gamma_2\gamma_3\gamma_4$	$\alpha_1\alpha_2$	–	–	–
Mol wt $\times 10^{-4}$	5.5	~6	6–7	~6	~7.5
Light chains:					
Class	κ, λ	κ, λ	κ, λ	κ, λ	κ, λ
Mol wt $\times 10^{-4}$	2.2–2.3	2.2–2.3	2.2–2.3	2.2–2.3	2.2–2.3
Whole molecule:					
Formula	$\kappa_2\gamma_2$ or $\lambda_2\gamma_2$	$(\kappa_2\alpha_2)_n$ or $(\lambda_2\alpha_2)_n$ $n = 1, 2, 3 \cdots$	$(\kappa_2\mu_2)_5$ or $(\lambda_2\mu_2)_5$	$\kappa_2\delta_2$ or $\lambda_2\delta_2$	$\kappa_2\varepsilon_2$ or $\lambda_2\varepsilon_2$
Mol wt $\times 10^{-5}$	~1.5	1.6	9–10	1.7–1.8	~2
Carbohydrate %	2.9	7.5	7.7–10.7	12	10.7

After H. Metzger, *Ann. Rev. Biochem.* **39**:889 (1970), Table I.

now we have been considering complete genes and their complete poly-peptide products. In considering immunoglobulins, however, the unit of interest is not a complete polypeptide but rather a **portion** of a polypeptide. These polypeptide units are known as **domains,** and they are delineated in Figure 15-13 by dashed lines. Domains labeled V_H, C_H1, V_L, and C_L (the meaning of these abbreviations will become apparent shortly) are located in each "prong" of the fork, while domains C_H2 and C_H3 are found in the "handle" of the fork. Amino-acid analyses have revealed that all six of these domain types are generally homologous with one another in their amino-acid sequence. It is therefore believed that a single block of DNA under-went duplication and subsequent diversification, much as we have de-scribed for hemoglobin or H-2, to generate the six homologous DNA sequences.

What is novel, of course, is that, in contrast to a family of polypeptides being encoded by a family of genes, we are here describing a family of homologous IgG domains as being generated by a **family of DNA se-quences.** In the case of IgG, it turns out that such domain-specifying DNA sequences are the units of information that are inherited, with functional genes specifying complete heavy and light polypeptide chains being cre-ated later, and probably via recombination, in the lymphocyte cell line. Since it is the DNA sequences that are transmitted through the germ line, these are the units that first occupy our attention. At the conclusion of the chapter we consider how these individual DNA sequences are thought to combine with one another in lymphocytes to produce functional genes.

15.12 VARIABLE AND CONSTANT PORTIONS OF IgG MOLECULES

Of the six homologous domains in an IgG molecule, the V_L and V_H domains are seen in Figure 15-13 to reside at the tip of each prong of the fork, while the C_L and the 3 C_H domains make up the central portion of the molecule. Furthermore, since X-ray studies of antigen-antibody interactions reveal that the tip of each prong of an IgG molecule represents an antigen-combining site, an antibody molecule is said to be **bivalent,** each antigen-combining site being formed by a V_L-V_H combination (Figure 15-13). As we stressed in the preceding section, the antigen-binding sites of different types of anti-body molecules must each be different, so as to account for the thousands of different antigens that are specifically recognized by the immune system. It follows that each antigen-combining prong is called the **variable portion** of the IgG molecule (Figure 15-13), with the symbol V_L denoting the **variable (V-) region** of the light chain and V_H the variable region of the heavy chain.

The remaining, non-antigen-combining portion of an IgG molecule is known as the **constant portion** of the molecule (Figure 15-13) although, as we shall see, this term is in some ways a misnomer. The constant portion includes the **constant (C-) region** of the IgG light chains (C_L) plus the three constant regions of the heavy chains (C_H1, C_H2, and C_H3). Carbohydrate is associated with the C_H2 regions (Figure 15-13), and at least a part of the constant portion is involved in binding to the surface of lymphocytes.

15.13   CONSTANT AND VARIABLE REGIONS OF H AND L CHAINS

We are now in a position to explore at a molecular level what is meant by "constant" and "variable." As noted earlier, a given myeloma cell line will produce a single type of IgG molecule with unique antigen-combining sites. A large number of myeloma IgG species have by now been isolated, purified, separated into their component H and L chains, and subjected to amino-acid sequencing (Section 9.2). The various sequences have then been compared with one another.

Such sequencing studies reveal, first, the general, overall homology between the six IgG domains that we have already mentioned in the preceding section. Second, they reveal that the various C_L regions fall into two subclasses, known as κ and λ, while the C_H regions can be categorized as γ_1, γ_2, γ_3, and γ_4. Within each subclass of both heavy and light chains, the amino-acid sequences are very similar to one another. Thus, if 20 types of myeloma IgG sequences are compared, 5 of the 20 might prove to have a C_H-region falling into the γ_1 class, 3 of 20 might have a γ_2 region, 8 of 20 a γ_3 region, and 4 of 20 a γ_4 region. In the same collection, 12 of the 20 might contain C_L-regions of the κ class (also called C_κ) and 8 of the 20 C_L-regions of the λ class (or C_λ). Thus to say that C regions are "constant" is misleading in the sense that they are by no means invariant; they are, however, relatively constant by comparison with the large amount of variation found in the V-regions of the IgG molecule.

The variable regions of light and heavy chains are diagrammed in Figure 15-14. Amino-acid sequencing data reveal 3 sectors of the V_L region and 4

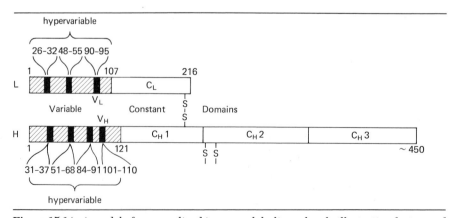

Figure 15-14   A model of a generalized immunoglobulin molecule illustrating features of the amino acid sequences of light (L) and heavy (H) chains that are described in the text. Numbers indicate amino-acid positions in the chains, where 1 is the N-terminal amino acid. (From A. Williamson, *Ann. Rev. Biochem.* **45**:467, 1976.)

sectors of the V_H region that are denoted as **hypervariable** sectors; within these sectors, there is a vast amount of divergence in the amino acids present. This diversity, of course, generates the variability of the V regions and, therefore, the thousands of different antigen-combining sites which an organism's immune system may display.

With this much of an overview of antibody structure, we can proceed to ask how genetic information for antibody molecules is encoded and organized in the genome, focusing primarily on the IgG class.

15.14 GENETIC SPECIFICATION OF C_H REGIONS

The four classes of human C_H-regions—γ_1, γ_2, γ_3, and γ_4— are encoded by four distinct DNA sequences in the human genome (Figure 15-15). Alleles of these four sequences are also known: for example, two alleles give rise to γ_1 sequences with either lysine or arginine at amino-acid position 214. Four distinct γ DNA sequences also exist in the mouse genome, and when crosses are made between differentially marked mouse strains, all 4 sequences are found to be inherited as a unit, with no recombinants found among over 2000 tested progeny.

The astute reader may understandably be astonished by the idea that the IgG molecules of 2000 mice were subjected to amino-acid sequencing in the above-cited cross. In fact, the immunogeneticist usually analyses IgG constant-regions not by amino-acid analysis but with a collection of antisera that distinguish what are called the **allotypes** of these regions. Allotypic antisera are prepared in the following way: 1) a rabbit (or guinea pig) is injected with a particular spectrum of mouse IgG molecules; 2) the rabbit recognizes specific features of the mouse IgG molecules as being foreign and proceeds to produce antibodies against them; 3) the rabbit is then bled, yielding a serum which recognizes, and will immunoprecipitate (Box 9.2), the variety of mouse IgG molecules to which it was exposed. With further refinements, this rabbit serum can be made to recognize, and immunoprecipitate, only a single class of mouse IgG molecules. Thus, as a

		V–sequences	C–sequences										
Light chains	κ	κI κII κIII	κ										
	λ	λI λII λIII λIV λV	λ lys λ arg										
Heavy chains		HI HII	γ_4	γ_2	γ_3	γ_1	α_1	α_2	μ_2	μ_1	δ	ϵ	

Figure 15-15 The proposed minimum number of DNA sequences required for human immunoglobulins. V sequences are found expressed in association with products of C sequences of the same row. The order of sequences is largely arbitrary. (From C. Milstein and A. J. Munro, *Ann. Rev. Microbiol.* **24**, 335, 1970)

hypothetical example, mouse IgG molecules with tryptophan at position 170 of the γ_1 C_H-region might be precipitated by one rabbit antiserum, while IgGs carrying leucine at this position might be ignored by this rabbit serum but precipitated by a second serum. Each serum would then be said to detect a mouse C_H allotype. An allotype, in short, can be defined as a constant-region phenotype, and allotypic variants are animals carrying allelic versions of a particular constant-region amino acid sequence.

Before leaving the C_H-regions we should note that the 5 major classes of immunoglobulins, summarized in Table 15-4, are so classified because of distinctive features of their C_H-regions, with IgG molecules possessing γ regions, IgM molecules possessing μ regions, IgA molecules possessing α regions, and so forth (Figure 15-15). Just as the four copies of the γ sequences are found to be closely linked, so is it believed (but not yet established) that *all* of the C_H loci—those for γ, μ, α, etcetera—may be closely linked and represent a single sequence family in tandem array.

15.15   GENETIC SPECIFICATION OF C_L REGIONS

Considerable genetic evidence indicates that humans possess but one locus specifying the C_κ-domain of the immunoglobulin molecule (Figure 15-15). This locus is known as *Inv*, and it exists in two allelic forms, *Inv (1)* and *Inv (3)*. *Inv 1* specifies a κ region with leucine at position 191 and *Inv 3* specifies a κ with valine at this position. A single locus for κ regions is also present in the mouse genome.

The genetics of C_λ regions is less well studied than that of C_κ regions, but present estimates call for at least 2 (Figure 15-15) and perhaps 3 or 4 distinct λ DNA sequences per haploid genome. These sequences are assumed, but not yet proved, to be clustered.

It has been reported for the rabbit that C_λ allotypic markers assort independently of C_κ and that neither shows linkage to C_H allotypic markers. On the other hand, very recent studies of mouse allotypes indicate that all of the C-region and V-region sequences are linked together in mouse chromosome 17. If the mouse studies are confirmed and prove generally applicable, then the interesting possibility arises that the various homologous IgG sequences are all part of a giant "sequence family."

15.16   GENETIC SPECIFICATION OF V_L- AND V_H-REGIONS

The V_L-region makes up approximately the first 107 amino acids of a light chain, beginning at the N-terminus. As noted earlier, 3 "hypervariable regions" in the light chain (Figure 15-14) carry most of the unique sequences that distinguish one V_L-region from the next, and these unique sequences are believed to be located in that portion of the prong of the immunoglobulin molecule where antigen recognition takes place. The remaining amino-acid sequences in the variable regions of the light chain are more highly conserved, making it possible to recognize patterns within them. In this way the various V_L amino acid sequences have been classified into major subgroups.

The most obvious pattern found in the light chains of immunoglobulins is that certain kinds of amino acid sequences are associated only with C_κ-regions, whereas others are associated only with C_λ. Because of this constancy, it is thus also possible to speak of V_κ and V_λ regions. Moreover, since all light chains will be either $V_\kappa C_\kappa$ or $V_\lambda C_\lambda$, it is possible to speak of κ and λ light chains (Figure 15-15).

For κ light chains three basic kinds of sequences have been recognized in the V_κ-region: κI, κII, and κIII (Figure 15-15). Within each subgroup the V_κ-regions have the same total number of amino acids, certain "diagnostic" amino-acids at specific positions, and certain patterns of amino-acid combinations. The simplest explanation for these phenomena, and one we shall adopt, is that three different classes of nonallelic V_κ genetic sequences exist in humans. Similar kinds of evidence indicate the existence of five subgroups of V_λ (λI-λV) and thus the existence of five different classes of nonallelic V_λ sequences (Figure 15-15).

The V_H regions of heavy chains are similar to the V_L regions of light chains in that they carry hypervariable regions (Figure 15-14) but also retain sufficient relatedness so that they can be classified into subgroups. Specifically, therefore, the human genome appears to carry at least two V_H loci, HI and HII (Figure 15-15).

A major distinction between the V_H and V_L regions is that a given type of V_H-region can associate with more than one class of C_H-region. Thus a V_{H1}-region may be associated with a C_H-region of the γ class, the μ class, or the α class. In other words, the specificity we noted between $V_\kappa C_\kappa$ and between $V_\lambda C_\lambda$ groups is not observed in the heavy chains.

15.17 LINKAGE OF V AND C SEQUENCES

A tenet of immunogenetics holds that V_κ DNA sequences are linked to C_κ sequences, that V_λ sequences are linked to C_λ sequences, and that V_H sequences are linked to C_H sequences. Genetic evidence for this tenet involves the use of **idiotypes** as V-region markers. Just as allotypes represent antigenic groups located in the constant portion of an IgG molecule, so can idiotypes be generally defined as V-region-specific antigens that are recognized by specific antisera. For a given cross, therefore, animals must be made heterozygous for two allotype-idiotype combinations and then backcrossed to determine whether recombination has occurred between them. Despite the fact that idiotypes are difficult to analyze and interpret, such crosses have consistently indicated close V-C linkages. Indeed, as we noted above, recent experiments with mice suggest linkage between all of the V-region and C-region DNA sequences.

15.18   GERM-LINE vs. SOMATIC-DIVERSITY THEORIES OF ANTIBODY DIVERSITY

One of the most striking facts about the variable regions of immunoglobulins, in both their light and heavy chains, is one we have already stressed: their vast diversity. Within each heavy- or light-chain subgroup there are probably thousands of related, but unique, amino acid sequences, the variability being clustered in the hypervariable regions. Two very different hypotheses have been formulated to explain this diversity.

The **germ line theory** (Figure 15-16a) makes the most obvious assumption, namely, that the genome contains thousands of related, but unique DNA sequences, each of which specifies one of the V-regions synthesized by an organism. This theory thus predicts that a large percentage of the mammalian genome is devoted to carrying variable-sequence families and states that each person will inherit and transmit $V_{\lambda Ia}$, $V_{\lambda Ib}$, $V_{\lambda Ic}$, . . . , $V_{\lambda In}$, $V_{\lambda IIa}$, $V_{\lambda IIb}$, . . . , $V_{\lambda IIn}$ and so on through his or her germ line. Each lymphocyte will then selectively express only one of these genes for the V portion of a light chain and only one for the V portion of a heavy chain. Because most lymphocytes will happen to select different combinations of sequences to express, the organism will come to possess its vast diversity of antibody-producing cells.

The alternate **somatic diversity theory** (Figure 15-16b) supposes that only a small number of V sequences are transmitted from parent to offspring, perhaps as few as three V_{κ} sequences, five V_{λ} sequences and a few V_H sequences. These gene sequences then experience thousands of different base-pair changes (Chapter 6) in individual lymphocyte lines, creating the observed diversity. The somatic theory therefore predicts that one lymphocyte clone will come to carry sequences $V_{\lambda Ia}$ and $V_{\lambda IIc}$, another clone $V_{\lambda Ib}$ and $V_{\lambda IIe}$, and so forth. Two of these sequences are then selected for expression in a given lymphocyte clone.

The most critical test of the two theories of antibody diversity has been made by nucleic-acid hybridization studies which investigate whether radioactively labeled mRNA or cDNA probes carrying variable-region sequences find a large or a small number of complementary sequences with which to hybridize in the embryonic genome. To date, such experiments reveal a small number of V sequences, a result that supports a nongerm-line theory for the origin of antibody diversity. There is as yet no consensus, however, on how a few germ-line V sequences might come to acquire such diversity in the various lymphocyte clones. One idea is that somatic mutations are selectively incurred by V-region sequences; other theories hold that somatic crossing over (Section 12.17) accompanied by mispairing and perhaps gene conversion (Section 13.13) may generate the somatic diversity. All such models must ultimately seek to explain how it is possible for a lymphocyte to subject a small sector of its genome to somatic diversity while leaving the rest of its genome intact and invariant.

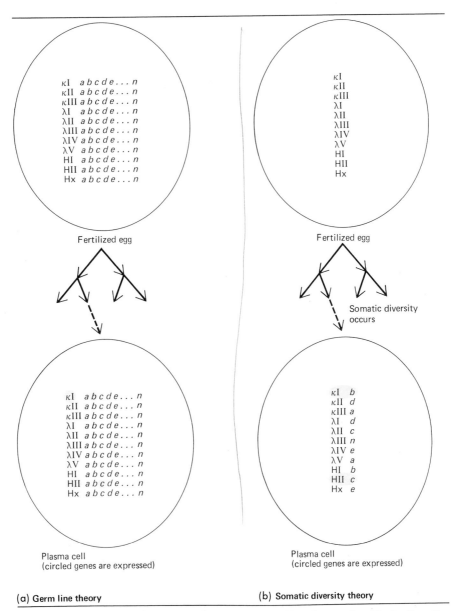

Figure 15-16 The germ line versus the somatic mutation theory of antibody diversity, diagrammed for the V sequences.

15.19 INTEGRATION OF V AND C REGIONS

Whether by direct inheritance through the germ line or by some process of somatic diversity, it is clear that an organism possesses or comes to possess an enormous variety of V DNA sequences, while having only a small and fixed number of C sequences. It was therefore postulated some years ago that the thousands of different VC combinations in an organism's immunologic repertoire are generated in somatic cells by some mechanism that allows V information to combine directly with C information.

In theory, such combinations could occur at any of three levels: separate portions of polypeptides could fuse to create a single light chain or heavy chain; separate mRNA molecules could fuse before being translated; or separate DNA sequences could fuse into single genes before being transcribed. Fusion at the polypeptide level became an unnecessary hypothesis when it was discovered that plasmacytoma cells produce single mRNA species containing V-region information at their 5′ ends and C-region information at their 3′ ends. The choice, therefore, remained between an mRNA-fusion hypothesis and a DNA-fusion hypothesis at the time that N. Hozumi and S. Tonegawa performed the experiments described below and in Figure 15-17.

Hozumi and Tonegawa prepared ^{125}I-labeled mRNA from a mouse plasmacytoma line known as MOPC 321, which produces large quantities of a single type of κ chain. This labeled RNA, they reasoned, could be used as

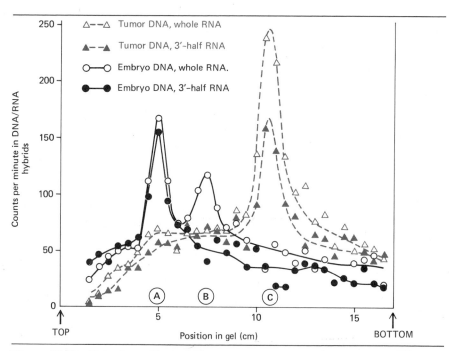

Figure 15-17   Data from Hozumi-Tonegawa experiment, showing DNA/^{125}I-RNA hybridization at positions in the gel labeled A, B, and C. See text for details. (From N. Hozumi and S. Tonegawa, *Proc. Nat. Acad. Sci. U.S.* **73**:3628, 1976.)

a probe, in DNA/RNA hybridization experiments, for DNA sequences specifying both the V_κ and the C_κ regions of the κ chain. They also cleaved this mRNA in two and isolated a "3'-half" mRNA fraction which they similarly labeled with ^{125}I. The 3'-half of an mRNA, you will recall from Chapter 9, specifies the carboxy-terminal half of a polypeptide chain. Since, in a κ polypeptide, the carboxy-terminal half contains the constant region (Figure 15-14), the labeled "3'-half" mRNA should in fact serve as a probe for DNA sequences specifying the C_κ region alone. *adult*

Hozumi and Tonegawa then isolated MOPC 321 tumor DNA, digested it into unique fragments with restriction endonuclease (Section 7.1), fractionated the resulting fragments according to size by gel electrophoresis (Box 7.1), sliced the gel into sections, and determined which gel slices contained DNA sequences that would hybridize with their two ^{125}I-labeled mRNA probes. As shown in Figure 15-17, both the whole mRNA and the 3'-half mRNA hybridized with tumor DNA from the same slice, labeled C, near the bottom of the gel. This result was consistent with the idea that both the V_κ DNA and the C_κ sequences reside in the same small restriction fragment and are, therefore, closely linked or contiguous in the tumor-cell genome.

The most striking part of the Hozumi-Tonegawa experiment can now be appreciated. Using the same RNA probes, they tested where hybridization would occur when restriction fragments were made not from adult DNA but rather from mouse embryo DNA. Here, a very different result was obtained (Figure 15-17): the 3'-half and the whole mRNA found homologous DNA sequences in a fragment located in slice A near the top of the gel. Moreover, the whole mRNA, but not the 3'-half, found homologous DNA in slice B that moved to the middle of the gel. Therefore, in the adult tumor cell line the restriction endonuclease appeared to generate a small DNA fragment with both V_κ and C_κ sequences in it, while in the embryo, V_κ sequences were located in a different fragment from C_κ sequences, and neither of these was cut by the endonuclease in the same manner as the tumor-line sequence. These results are summarized and interpreted in Figure 15-18.

The Hozumi-Tonegawa experiments give experimental evidence for the widely held theory, stated earlier in the chapter, that V-region and C-region DNA sequences are inherited separately and become joined as genes at some stage of lymphocyte differentiation. Figure 15-19 shows various models that have been offered to explain how such joining might occur; details of the models are given in the figure legend. All have clearly been influenced by known examples of chromosomal fusions, the most conspicuous influence being the insertion of an episome into a bacterial chromosome (Figure 5-15). Regardless of which, if any, of the models proves correct, there is little question that basic research into recombination mechanisms (Chapter 13) is now fully convergent with research on the molecular genetics of antibody production.

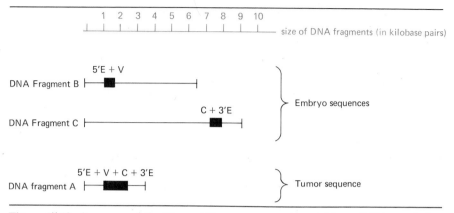

Figure 15-18   Summary of the Hozumi-Tonegawa experiment. Mouse DNA restriction fragments are from gel positions A, B, and C in Figure 15-17. 5'E and 3'E designate the leader and trailer base sequences of the κ-chain mRNA molecule (Section 8.5). V and C designate base sequences corresponding to variable and constant regions. (From N. Hozumi and S. Tonegawa, *Proc. Natl. Acad. Sci. U.S.* **73**:3628, 1976.)

15.20   SEQUENCE HOMOLOGIES BETWEEN IMMUNOGLOBULINS AND HISTOCOMPATIBILITY ANTIGENS

Amino-acid sequence analyses indicate that the histocompatibility antigens, described in Section 15.9, share homologies with the 6 homologous domains found in IgG molecules. Since the mouse H-2 locus maps to chromosome 17, the proposed location of mouse IgG sequences (Section 15.15), it seems likely that all derive from a common ancestral nucleotide sequence. It is fascinating to speculate on the evolutionary forces operating on a single DNA sequence which, on the one hand, could go on to generate genes that define self and therefore remain highly constant within an individual and, on the other hand, could go on to generate sequences that define nonself and are extraordinary in their vast diversity in a given individual.

15.21   DISTINGUISHING ALLELES FROM ISOLOCI AND GENE FAMILIES

We close this chapter by reviewing the subtle but important distinctions between (multiple) alleles at one locus and a set of isoloci (or a gene family) each of which may, in turn, be multiply allelic. The formal distinction can be straightforwardly visualized by recalling our definition of a gene locus as a physical location in a particular chromosome. Thus, if gene locus 3 is defined as that group of nucleotides transcribed from promoter #3 when counting from one end of human chromosome 7, then all extant nucleotide sequences that occupy this particular locus are alleles of one another. The nucleotides following promoter #9 on this same chromosome

(a) Copy-insertion

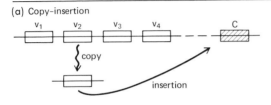

(b) Excision-insertion

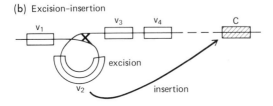

(c) Deletion

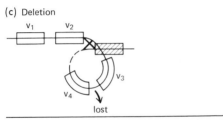

Figure 15-19 Three proposed mechanisms for the integration of V and C DNA sequences. In the copy-insertion model (a), a specific V sequence is duplicated and the copy is inserted at a site adjacent to a C sequence. In the "excision-insertion" model (b), a specific V sequence is excised into an episome-like structure, and this in turn is integrated adjacent to a C sequence. In the "deletion" model (c) DNA in the interval between a particular V sequence and the corresponding C sequence loops out, is excised, and diluted out upon subsequent cell multiplication. (From N. Hozumi and S. Tonegawa, *Proc. Natl. Acad. Sci. U.S.* **73**:3628, 1976.)

may well code for a similar polypeptide, and multiple versions of this gene may also exist in the human "gene pool." Nevertheless, a gene occupying locus 9 is *never* considered allelic to a gene occupying locus 3. Instead, genes 3 and 9 are either considered to occupy isoloci or to belong to a gene family, depending on their degree of relatedness.

The experimental distinction between alleles at one locus and genes marking isoloci is readily made when isoloci are loosely linked or unlinked. In this case, markers for isoloci will recombine freely or assort independently at meiosis, whereas alleles of a single locus will almost always segregate at anaphase I. When isoloci are closely linked, on the other hand, the distinction is more difficult. It is especially important to guard against the erroneous concept that the distinction can be made by recombination frequencies alone. The fact that markers a and *b*, for instance, are "never observed to recombine," whereas both recombine at measurable frequency with marker *c*, **cannot** be used to argue that *a* and *b* are allelic and are not alleles of *c*. We noted in Chapter 13 how marker effects and related phenomena can cause recombination frequencies to be quite unreliable for closely linked markers. Moreover, if two isoloci lie next to one another in a chromosome, markers in adjacent genes may well recombine with one another less often than will widely spaced markers in the same gene. Therefore, a test of independent function [for example, some form of a complementation test (Section 10.10)] must ultimately be performed to show that two genes are not alleles of one another.

References
Single-Locus Traits

Bowman, B., and J. A. Mangos. "Cystic fibrosis," *New Eng. J. Med.* **294:**937–938 (1976).

Desforges, J. F. "Genetic implications of G-6-PD deficiency," *New Eng. J. Med.* **294:**1438–1440 (1976).

Friedman, P. A., D. B. Fisher, E. S. Kang, and S. Kaufman. "Detection of hepatic phenylalanine 4-hydroxylase in classical phenylketonuria," *Proc. Natl. Acad. Sci. U.S.* **70:**552–557 (1973).

Gilbert, F., R. Kucherlapati, R. P. Creagan, M. J. Murnane, G. J. Darlington, and F. H. Ruddle. "Tay-Sachs' and Sandhoff's diseases: The assignment of genes for hexosaminidase A and B to individual human chromosomes," *Proc. Natl. Acad. Sci. U.S.* **72:**263–267 (1975).

Goldstein, J. L., M. K. Sobhani, J. R. Faust and M. S. Brown. "Heterozygous familial hypercholesterolemia: Failure of a normal allele to compensate for mutant allele at a regulated genetic locus," *Cell* **9:**195–203 (1976).

Harris, H. *Prenatal Diagnosis and Selective Abortion.* Cambridge: Harvard University Press, (1974).

Harris, H. *The Principles of Human Biochemical Genetics.* Amsterdam: North-Holland, (1975).

Johnson, L. A., R. B. Gordon, and B. T. Emerson. "Two populations of heterozygote erythrocytes in moderate hypoxanthine guanine phosphoribosyltransferase deficiency," *Nature* **264:**172–174 (1976).

*Karlsson, J. L. "A double dominant genetic mechanism for schizophrenia," *Hereditas* **65:**261–268 (1970).

Karlsson, J. L. "Genetic association of giftedness and creativity with schizophrenia," *Hereditas* **66:**177–182 (1970).

Kolodny, E. H. "Lysosomal storage diseases," *New Eng. J. Med.* **294:**1217–1220 (1976).

Landsteiner, K., and A. S. Wiener. "Studies on an agglutinogen (Rh) in human blood reacting with anti-rhesus serum and with human isoantibodies," *J. Exp. Med.* **74:**309–320 (1941). [Reprinted in S. H. Boyer, *Papers on Human Genetics.* Englewood Cliffs, N.J.: Prentice-Hall, 1963.]

Milunsky, A. "Prenatal diagnosis of genetic disorders," *New Eng. J. Med.* **295:**377–380 (1976).

Pederson, H., and N. Mygind. "Absence of axonemal arms in nasal mucosa cilia in Kartagener's syndrome," *Nature* **262:**494–495 (1976).

Race, R. R., and R. Sanger. *Blood Groups in Man,* 6th ed. Oxford: Blackwell, (1975).

Rosenfield, R. E., F. H. Allen, Jr., and P. Rubinstein. "Genetic model for the Rh Blood-group system," *Proc. Natl. Acad. Sci. U.S.* **70:**1303–1307 (1973).

Sang, J. H. "Penetrance, expressivity, and thresholds," *J. Hered.* **54:**143–151 (1963).

Schenkel-Brunner, H., and H. Tuppy. "Enzymatic conversion of human O into A erythrocytes and of B into AB erythrocytes," *Nature* **223:**1272–1273 (1969).

*Denotes articles described specifically in the chapter.

Watkins, W. M. "Blood-group substances," *Science* **152**:172–181 (1966).

Watkins, W. M., and W. T. J. Morgan. "Possible genetical pathways for the biosynthesis of blood group mucopolysaccharides," *Vox Sang.* **4**:97–119 (1959). [Reprinted in S. H. Boyer, *Papers on Human Genetics.* Englewood Cliffs, N.J.: Prentice-Hall, 1963.]

*Woolf, C. M., and R. M. Woolf. "A genetic study of polydactyly in Utah," *Am. J. Hum. Genetics* **22**:75–88 (1970).

Isozymes

Hopkinson, D. A., and H. Harris. "Recent work on isozymes in man," *Ann. Rev. Genetics* **5**:5–32 (1971).

*Ruddle, F. H., T. B. Shows, and T. H. Roderick. "Esterase genetics in *Mus musculus:* Expression, linkage, and polymorphism of locus *Es-2,*" *Genetics* **62**:393–399 (1969).

Shaw, C. R. "Electrophoretic variation in enzymes," *Science* **149**:936–943 (1965).

Gene Duplications

Bridges, C. B. "The *Bar* "gene" a duplication," *Science* **83**:210–211 (1936). [Reprinted in J. A. Peters, *Classical Papers in Genetics.* Englewood Cliffs, N.J.: Prentice-Hall, 1959.]

Kikuchi, A., and L. Gorini. "Similarity of genes *argF* and *argI,*" *Nature* **256**:621–624 (1975).

Ohno, S. *Evolution by Gene Duplication.* New York: Springer-Verlag, 1970.

Tartof, K. D. "Unequal mitotic sister chromatid exchange as the mechanism of ribosomal RNA gene magnification," *Proc. Natl. Acad. Sci. U.S.* **71**:1272–1276 (1974).

Tartof, K. D. "Redundant genes," *Ann. Rev. Genet.* **9**:355–385 (1975).

Wellauer, P. K., R. H. Reeder, I. B. Dawid, and D. D. Brown. "The arrangement of length heterogeneity in repeating units of amplified and chromosomal ribosomal DNA from *Xenopus laevis,*" *J. Mol. Biol.* **105**:487–505 (1976).

Hemoglobin genes

*Baglioni, C. "The fusion of two peptide chains in hemoglobin lepore and its interpretation as a genetic deletion," *Proc. Natl. Acad. Sci. U.S.* **48**:1880–1886 (1962).

Deisseroth, A., R. Velez, and A. W. Nienhuis. "Hemoglobin synthesis in somatic cell hybrids: Independent segregation of the human alpha- and beta-globin genes," *Science* **191**:1262–1264 (1976).

Friedman, S., and E. Schwartz. "Hereditary persistence of foetal haemoglobin with β-chain synthesis in *cis* position ($^G\gamma$-β^+-HPFH) in a negro family," *Nature* **259**:138–140 (1976).

Hilse, K., and R. A. Popp. "Gene duplication as the basis for amino acid ambiguity in the alpha-chain polypeptides of mouse hemoglobins," *Proc. Natl. Acad. Sci. U.S.* **61**:930–936 (1968).

Ingram, V. M. "Gene evolution and the haemoglobins," *Nature* **189**:704–708 (1961). [Reprinted in S. H. Boyer, *Papers on Human Genetics.* Englewood Cliffs, N.J.: Prentice-Hall, 1963.]

Kabat, D., and R. D. Koler. "The thalassemias: Models for analysis of quantitative gene control," *Adv. Hum. Genet.* **5**:157–222 (1975).

Kendall, A. G., P. J. Ojwang, W. A. Schroeder, and T. H. J. Huisman. "Hemoglobin Kenya, the product of a γ-β fusion gene: studies of the family," *Am. J. Hum. Genet.* **25**:548–563 (1973).

Old, J., J. B. Clegg, D. J. Weatherall, S. Ottolenghi, P. Comi, B. Giglioni, J. Mitchell, P. Tolstoshev, and R. Williamson. "A direct estimate of the number of human γ-globin genes," *Cell* **8**:13–18 (1976).

Steinheider, G., H. Melderis, and W. Ostertag. "Evidence for the Lepore-like hybrid globin β chain genes in mice," *Nature* **257**:712–714 (1975).

Weatherall, D. J. "The genetics of the thalassaemias," *Brit. Med. Bull.* **25**:24–29 (1969). [Reprinted in L. Levine, *Papers on Genetics.* St. Louis: Mosby 1971.]

Weatherall, D. J., and J. B. Clegg. *The Thalassaemia Syndromes,* 2nd ed. Oxford: Blackwell, (1972).

Weatherall, D. J., and J. B. Clegg. "Molecular genetics of human hemoglobin," *Ann Rev. Genet.* **10**:157–178 (1976).

Histocompatibility Genes

Bach, F. H., and J. J. van Rood. "The major histocompatibility complex—genetics and biology," *New Eng. J. Med.* **295:**806–813 (1976).

Blumenthal, M. N., D. B. Amos, H. Noreen, N. R. Mendell, and E. J. Yunis. "Genetic mapping of Ir locus in man: linkage to second locus of HL-A," *Science* **184:**1301–1303 (1974).

Eichmann, K. "Genetic control of antibody specificity in the mouse," *Immunogenetics* **2:**491–506 (1975).

Götze, D., Ed. *The Major Histocompatibility Complex.* New York: Springer-Verlag, 1976.

Klein, J. *Biology of the Mouse Histocompatibility-2 Complex.* New York: Springer-Verlag, 1975.

Lachmann, P. J., D. Grennan, A. Martin, and P. Demant. "Identification of Ss protein as murine C4, "*Nature* **258:**242–243 (1975).

Munroe, A., and S. Bright. "Products of the major histocompatibility complex and their relationship to the immune response," *Nature* **264:**145–152 (1976).

Shreffler, D. C., and C. S. David. "The H-2 major histocompatibility complex and the I immune response region: Genetic variation, function, and organization," *Adv. Immunol.* **20:**125–163 (1975).

Svejgaard, A., M. Hauge, C. Jersild, P. Platz, L. P. Ryder, L. S. Nielsen, and M. Thomsen. *The HLA System.* Monographs in Human Genetics **7.** Karger, Basel, 1975.

Immunogenetics

Gally, J. A., and G. M. Edelman. "The genetic control of immunoglobulin synthesis," *Ann. Rev. Genetics* **6:**1–46 (1972).

Hobart, M. J., and I. McConnell. *The Immune System.* Oxford: Blackwell Scientific, 1975.

Hood, L., J. H. Campbell, and S. C. R. Elgin. "The organization, expression, and evolution of antibody genes and other multigene families," *Ann. Rev. Genet.* **9:**305–353 (1975).

*Hozumi, N., and S. Tonegawa. "Evidence for somatic rearrangement of immunoglobulin genes coding for variable and constant regions," *Proc. Natl. Acad. Sci. U.S.* **73:**3628–3632 (1976).

Kabat, E. A. *Structural Concepts in Immunology and Immunochemistry.* New York: Holt, Rinehart and Winston, 1976.

Milstein, C., and A. J. Munro. "The genetic basis of antibody specificity," *Ann. Rev. Microbiol.* **24:**335–358 (1970).

Nisonoff, A., J. E. Hopper, and S. B. Spring. *The Antibody Molecule.* New York, Academic Press, 1975.

Roitt, I. *Essential Immunobiology.* Oxford: Blackwell, 1974.

Weigert, M., M. Potter, and D. Sachs. "Genetics of the immunoglobulin variable region," *Immunogenetics* **1:**511–523 (1975).

Williamson, A. R. "The biological origin of antibody diversity." *Ann. Rev. Biochem.* **45:**467–500 (1976).

George Snell (The Jackson Laboratory, Bar Harbor) has made fundamental contributions to the immunogenetics of the mouse.

Questions and Problems

1. A schizophrenic marries a nonschizophrenic person. Assuming the disease to be determined by two incompletely penetrant dominant genes, *P* and *Q*, list all possible genotypes of the two parents.

2. How would you determine whether a phenotype that differs in the degree of its expression is determined by a gene with variable expressivity versus a series of multiple alleles?

3. Huntington's chorea is a fatal disease leading to the degeneration of the nervous system. It is found in persons heterozygous for the dominant allele *Ht*, but has variable expressivity in the sense that a person may develop symptoms as early as childhood or as late as 60 years old, with the mean age of onset at about 40 years.

   (a) A man died of Huntington's chorea. The disease is not known in his widow's family. What are the chances that his son carries the gene?

   (b) In certain families known to carry the gene, the disease is found to "skip" several generations before reappearing. How might this be explained?

4. A woman having blood type A had a type O and a type B child. What is the genotype and phenotype of the father? What are the genotypes of the woman and her children? Could the legitimacy of children with blood type AB or blood type A have been questioned?

5. Persons known as *secretors* have the same AB blood group antigens in their saliva as are found on their blood cell surfaces, whereas *nonsecretors* do not have antigen in their saliva. The gene *Se* (*secretor*) is dominant to its allele *se* (*nonsecretor*), and the *secretor* locus is not linked to the ABO locus. In the following matings, what are the expected proportions of A *secretors*, B *secretors*, AB *secretors*, and *nonsecretors* among the progeny?
   (a)   AB *secretor* (*Se/se*) ✕ O (*Se/se*)
   (b)   AB *secretor* (*Se/se*) ✕ O (*se/se*)

6. The antigens M and N are located on human red cells along with antigens A and B. All persons are either M, N, or MN, and the trait is determined by a pair of codominant alleles, L^M and L^N. Persons do not make antibody against M or N antigens, whereas rabbits do. Therefore persons are typed as M, N, or MN by mixing a sample of their blood with anti-M or anti-N serum from rabbits: anti-M will agglutinate M or MN cells, whereas anti-N will agglutinate N or NM cells.
   In an O,M ✕ AB,MN mating, the progeny are A,M, B,M, A,MN, and B,MN in equal numbers.
   (a)   Is the *L* locus linked to the *I* locus?
   (b)   What are the genotypes of the parents in this mating?
   (c)   List all possible genotypes and phenotypes in the progeny of an A,M ✕ B,N mating.

7. A child has the blood type O,N and is a *secretor*. Which of the following matings could have produced the child and which could not? Explain in each case.
   (a)   O,N *secretor* ✕ O,N *secretor*
   (b)   A,N *secretor* ✕ A,N *secretor*
   (c)   A,N *nonsecretor* ✕ A,N *nonsecretor*
   (d)   AB,N *secretor* ✕ O,N *nonsecretor*
   (e)   A,N *nonsecretor* ✕ B,N *secretor*

8. A highly simplified view of the human Rh blood groups divides persons into two categories: those whose blood cells agglutinate with the antiserum anti-Rh_0 (also called anti-D) and are therefore **Rh-positive**, and those whose cells do not react with this serum and are therefore **Rh-negative**. An Rh-positive person can be designated as R/R or R/r, and an Rh-negative person as r/r. Knowing that this locus is not linked to other blood group loci, give all possible genotypes of the parents of persons with the following phenotypes:
   (a)   O,MN,Rh-negative
   (b)   A,M,Rh-negative
   (c)   AB,MN,Rh-positive

9. Discuss the kinds of dominance relationships described in this chapter in terms of the concept of complementation (Section 10.10).

10. In inbred mice the enzyme isocitric dehydrogenase is made up of two identical subunits. The electrophoretic band patterns of this subunit in two inbred strains of mouse (BALB/c and C57/b) and in their F_1 hybrid are as follows:

Pole

+

	_____		_____
			_____
		_____	_____
−	BALB/c	C57/b	BALB × C57

Explain these results. What other pattern might you have expected in the F$_1$ gel?

11. Listed below are the HLA phenotypes of a family.

(a) Identify the genotypes of parents and offspring and write out their haplotypes.

(b) Child #2 requires a kidney transplantation. Which, if any, of the parents and/or sib(s) would be fully histocompatible donors? Explain.

Mother: A2; B7, B8
Father: A1, A3; B7, B9
Child #1: A1, A2; B8, B9
Child #2: A2, A3; B7, B8
Child #3: A1, A2; B7, B8

12 (a) If two alleles at a particular locus are codominant, how might a monosomic or trisomic condition aid in the chromosomal location of the gene locus?

(b) If two alleles are dominant and recessive, which of the following aberrations would aid in locating them to a specific chromosome: duplication, deficiency, monosomy, trisomy? Explain.

13. Four babies were born in a hospital on the same night, and their blood groups were later found to be O, A, B, and AB. The four parents were:

O and O
AB and O
A and B
B and B

Assign the four babies to their correct parents.

14. Are the following statements true or false? Explain in each case.

(a) All immunoglobulin molecules in an individual have a common polypeptide chain (*i.e.*, one with an identical amino acid sequence).

(b) All IgG immunoglobulin molecules in an individual have the same heavy chain and differ from each other only by the light chain type they contain.

(c) Variable light regions (V$_L$) within a given immunoglobulin molecule associate with either a kappa or a lambda C$_L$ subgroup.

(d) An IgG immunoglobulin molecule consists of two polypeptide chains linked by a disulfide bridge.

(e) γ$_2$ and α$_1$ C$_H$ chains can be found in an immunoglobulin molecule.

(f)   A given C_H gamma chain is specified by multiple gene loci at which there are multiple alleles.

15. Explain why model (a) of Figure 15-19 is incompatible with the results of Hozumi and Tonegawa.

16. Diagram how hemoglobin Kenya would be generated assuming a $^G\gamma^A\gamma\delta\beta$ gene order and unequal crossing over.

17. Two *Drosophila* isoloci specify two esterases. Each has two codominant alleles, and all 4 products can be distinguished electrophoretically. Each enzyme is active as a dimer, the products of one locus do not associate with the products of the other locus. Show with diagrams the expected progeny phenotypes and their relative proportions when flies heterozygous at both isoloci are inbred and the isoloci are (a) unlinked; or (b) closely linked.

18. (a)   Persons homozygous for the Lepore hemoglobin gene develop moderate to severe hemolytic anemia. Which hemoglobin(s) would you expect to be abnormal in these persons and which normal? Explain.

(b)   Would you expect persons homozygous for antiLepore hemoglobin to be as afflicted as Lepore homozygotes? Examine Figure 15-11 carefully in making your answer.

19. In a technique known as **peptide mapping**, polypeptides are isolated and digested with an enzyme such as trypsin; the resultant **tryptic peptides** (Section 9.2) are then separated by chromatography to generate a **peptide map.** If you prepared peptide maps of the non-α hemoglobin chains of normal persons and Lepore homozygotes, what would you expect to find?

20. Type O persons are sometimes called universal donors and type AB persons universal recipients. Explain.

21. The sex-linked *Bar* (*B*) mutation results from a duplication in the X chromosome (Figure 4-14). Females homozygous for *Bar* and heterozygous for the closely linked outside markers *f* (*forked* bristles) and *fu* (*fused* wings) were crossed to *forked fused* males, and rare flies recombinant for the outside markers were recovered. Some of these proved to have normal eyes; others exhibited an extreme eye phenotype called **ultrabar.** Cytological examination of polytene chromosomes revealed that the normal progeny carried one dose of the 16A segment (Figure 4-14) while the *ultrabar* progeny carried three. Explain these results, using diagrams.

22. In a Hungarian family, two α-chain variants are known that are called Buda and Pest. One individual in this family was found to have 25% Buda, 25% Pest, and 50% normal hemoglobin A. What does this suggest about the number of genes specifying the α chain? How would you write this person's genotype?

CHAPTER
16

Genes That Cooperatively
Produce a Phenotype

INTRODUCTION

This chapter explores the properties of genes that are apparently not structurally related to one another—that is, are not alleles or members of gene families—but nonetheless cooperate to generate a particular phenotypic trait. In some cases such genes prove to be grouped together in one region of the chromosome, as with **operons** in bacteria, **gene clusters** in fungi, and **complex loci** in *Drosophila*. In other cases the genes are not linked but nonetheless function in a cooperative, regulated way. And finally, there are numerous traits such as egg yield, height, and kernel size that are influenced by an unknown but sizeable number of genes that have not been individually characterized. The inheritance of such traits is the subject of a discipline known as **quantitative genetics,** to which we give a generalized introduction at the end of this chapter.

CLUSTERED GENES SPECIFYING ONE TRAIT

16.1   OPERONS

In bacteria and bacteriophages different genes concerned with the same trait are often (but by no means always) found clustered together in a group known as an **operon.** In the operon, the component genes are transcribed as a unit and are thereby under joint control. We have made reference to operons on several occasions throughout this text. In this section we examine operons from a functional point of view. Regulation of operon transcription is covered in Chapter 17.

The genes in a bacterial operon most frequently control sequential metabolic steps. A straightforward example of an operon of this type is the *gal* operon of *E. coli* (Figure 16-1), where the sequence of genes in the operon directly mirrors the sequence followed by the gene products as they carry out the degradation of galactose into glucose-1-phosphate and UDP-glucose. A slightly more complex example is given by the tryptophan (*trp*) operon of *E. coli* (Figure 16-2), where the sequence of genes again reflects the sequence of reactions of the gene products, but where various gene products form enzyme complexes that act in concert; specifically, the *trpE* and *trpD* gene products form an enzyme complex, as do the *trpB* and *trpA* products. We should note that the organization of the *trp* operon in *Salmonella typhimurium* is identical to that shown for *E. coli*.

Exceptions to these simple types of operons are well known. In the *leu* operon of *S. typhimurium* (Figure 16-3), for example, the *leuA* gene is in phase with the reaction sequence of the gene products, whereas the *leuB* gene is out of order: its gene product, a dehydrogenase, acts *after* the combined gene products of *leuC* and *leuD* catalyze an isomerization reaction. Far greater disparity is found when the gene order of the *his* operon in *S. typhimurium* is compared with the sequence of reactions in the

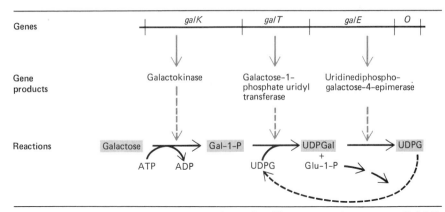

Figure 16-1 The galactose (*gal*) operon of *E. coli*. *Abbreviations:* O, operator; Gal-1-P, galactose-1-phosphate; UDPG, uridine diphosphoglucose; UDPGal, uridine diphospho-galactose; Glu-1-P, glucose-1-phosphate. (After S. Adhya and J. Shapiro, *Genetics* **62**:231, 1969.)

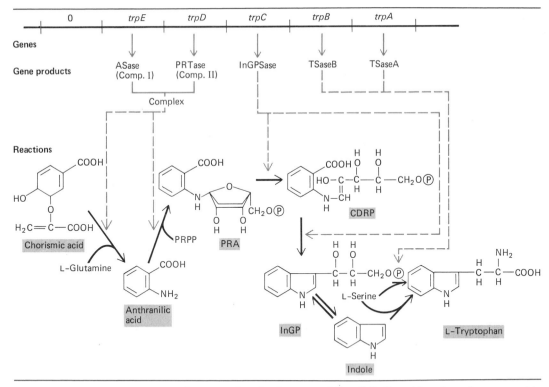

Figure 16-2 The tryptophan (*trp*) operon of *E. coli*. In *S. typhimurium*, the comparable genes are called trpA, B, E, D, and C. *Abbreviations:* O, operator; ASase, anthranilate synthetase; PRTase, phosphoribosyl transferase; InGPSase, indole glycerol phosphate synthetase; TSase, tryptophan synthetase; PRA, phosphoribosyl-anthranilic acid; CDRP, carboxyphenylamino-1-deoxyribulose-5-phosphate; InGP, indole-3-glycerol-phosphate. (From G. Wuesthoff and R. H. Bauerle, *J. Mol. Biol.* **49**:171, 1970.)

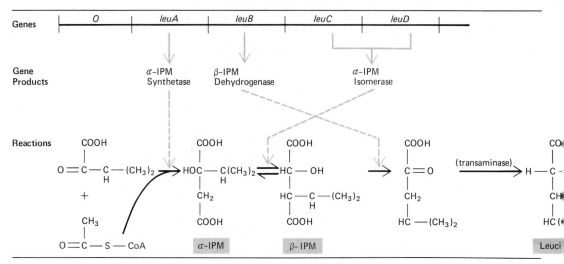

Figure 16-3   The leucine (*leu*) operon of *S. typhimurium. Abbreviations:* O, operator; α-IPM, α-isopropylmalonate; β-IPM, β-isopropylmalonate. (After J. Calvo and H. Worden, *Genetics* **64**:199, 1970.)

pathway of histidine biosynthesis. As seen in Figure 16-4, there is no apparent relation between the sequence of *his* genes in the operon and the order in which the gene products act.

A somewhat tangential, but nonetheless important, point about bacterial operons should be made at this point: the clustering together of certain functionally related genes in a bacterial operon does not preclude there being other related genetic loci elsewhere in the genome. For example, in addition to the *his* operon, at least five other *his* loci that relate to histidine biosynthesis are found in the *S. typhimurium* genome. None of these appears to specify an enzyme involved in the major histidine biosynthetic pathway, but all relate to normal histidine production.

The *lac* operon of *E. coli* (Figure 16-5) represents a very different sort of operon from those we have been describing: among its three structural genes (*lacZ, lacY,* and *lacA,* which are usually referred to as *z, y,* and *a*), only the *z* gene specifies an enzyme known to be directly involved in a metabolic pathway. The *z* gene product is a single, long polypeptide chain (134,000 daltons); this combines with three other, identical chains to form the tetrameric enzyme **β-galactosidase** which catalyzes the hydrolysis of lactose (and other β-galactosides) into galactose and glucose. The *y* gene product, on the other hand, does not specify a catabolic enzyme at all but rather a molecule known as **M protein** or **galactoside permease** which is located in the bacterial cell membrane and acts to facilitate the specific uptake of lactose (and similar molecules) from the external medium. The last gene in the operon, *a*, specifies the enzyme **thiogalactoside transacetylase** which can, *in vitro,* transfer the acetyl group from acetyl coenzyme A to isopro-

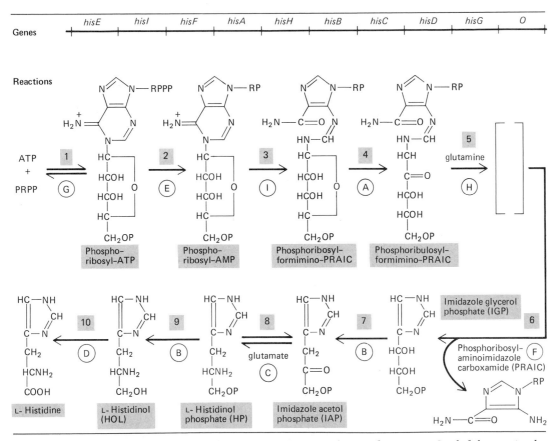

Figure 16-4 The histidine (*his*) operon of *S. typhimurium*. Circled letters in the biosynthetic pathway relate enzymes to the lettered *his* operon genes from which they derive; boxed numbers refer to the 10 sequential steps in the pathway. The structure of the compound in brackets is uncertain. *Abbreviations:* R, ribosyl; P, phosphate; PRAIC, phosphoribosylaminoimidazole carboxamide. (From M. Brenner and B. N. Ames, in *Metabolic Regulation*, H. J. Vogel, Ed. New York: Academic, 1971.)

pyl-D-thiogalactoside **(IPTG),** a synthetic galactoside derivative. The function of the enzyme *in vivo* is unknown, but since strains carrying deletions of the *a* gene behave normally in all respects the *a* gene appears not to specify an enzyme essential for lactose utilization. The *lac* operon, in summary, specifies three proteins that all interact with β-galactoside, but do not do so in a series of related catabolic reactions.

This survey of representative operons has stressed their dissimilarities, but all operons appear to be more or less eminently logical ways of organizing functionally related genes. The evolutionary origin of an operon is, however, difficult to visualize. If we assume that the different genes in ancient organisms arose through more or less random events, it is not readily apparent how several quite different genes concerned with the same trait

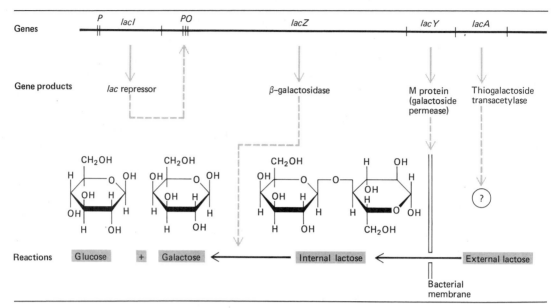

Figure 16-5   The lactose (*lac*) operon of *E. coli. Abbreviations: P,* promoter; *O,* operator; *lacI,* regulatory gene (to be discussed in Chapter 17).

came to find themselves next to one another in the same segment of a chromosome. One can only suppose that when such an arrangement did, perhaps by chance, come into being, it endowed an organism with such superior adaptive value that the arrangement was immediately conserved.

16.2 "GENE CLUSTERS" IN FUNGI: THE *HIS4* LOCUS

Fungi such as yeast and *Neurospora* possess a number of so-called **gene clusters,** each of which contains genetic information for several steps in a biosynthetic or degradative pathway. Table 16-1 summarizes the known gene clusters in yeast; in addition, M. Case and N. Giles have performed extensive genetic analyses of the *arom* gene cluster (encoding 5 enzymatic activities in the biosynthetic pathway of aromatic amino-acids) and the *qa* gene cluster (encoding 3 enzymatic activities in the catabolism of aromatic amino-acids) in *Neurospora*.

As a specific example of a gene cluster we can consider the *HIS4* locus in yeast, studied extensively by G. Fink and his colleagues. This cluster specifies 3 of the 10 enzymatic activities involved in histidine biosynthesis that are summarized in Figure 16-4. A combined genetic and complementation analysis of this locus defines three regions, *HIS4A, B,* and *C* (Figure 16-6) that behave as 3 distinct genes. Thus missense mutations defective in the same biochemical reaction were found by Fink and his coworkers to map to

Table 16-1 Gene Clusters in Yeast

Gene Cluster	Enzyme Activities Encoded
1. *HIS4*	1. Dehydrogenase 2. Cyclohydrolase 3. Pyrophosphorylase
2. *ARO2*	1. Dehydroquinase 2. DHQ synthetase 3. DHS reductase 4. Shikimic acid kinase 5. EPSP synthetase
3. *ADE3*	1. Tetrahydrofolate ligase 2. Methenyl tetrahydro-folate cyclohydrolase 3. Methenyl tetrahydro-folate reductase
4. *TRP5*	1. InGP → Indole 2. Indole + serine → tryptophan
5. *FAS1*	1. Keto reductase 2. Condensing enzyme 3. Acyl carrier protein
6. *FAS2*	1. Dehydratase 2. Enoyl reductase
7. *URA3*	1. Carbamyl phosphate synthetase 2. Aspartate trans-carbamylase

the same region of the locus: mutants defective in hydrolase activity (step 3 in the histidine biosynthetic pathway, Figure 16-4) all map to *HIS4A*; mutants defective in a pyrophosphohydrolase activity (step 2) all map to *HIS4B*; and mutants defective in a dehydrogenase (step 10) all map to *HIS4C*. Moreover, missense mutations in any one region were found to complement missense mutations in either of the other two regions.

Additional experiments in the Fink laboratory revealed that the *HIS4* locus functions as a single unit of transcription. Thus nonsense mutations (Section 9.16) in the *HIS4A* region were found to abolish *HIS4A, B,* and *C* activities; nonsense codons in *HIS4B* affected *HIS4B* and *HIS4C* expression, and so forth. Such polarity of a proximal *HIS4A*) nonsense mutation on the

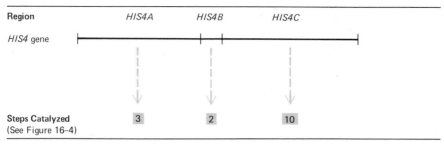

Figure 16-6   The 3 regions of the *HIS4* locus of yeast and the steps of histidine biosynthesis controlled by each region.

expression of distal genes is, of course, observed also for bacterial operons (Section 9.16). In many respects, therefore, *HIS4* resembles an operon.

A major distinction exists between *HIS4* and a bacterial operon, however, for the *HIS4* transcript proves to specify a **single polypeptide chain,** with a molecular weight of 95,000, that carries all three enzymatic activities. In other words, *HIS4* is in fact a single gene, as we have defined a gene in Section 8.2, whereas a bacterial operon, although transcribed into a single mRNA, contains internal AUG "start" codons and is translated into distinct, separate polypeptide chains.

Are the other known "gene clusters" in fungi (Table 16-1) similar to the *HIS4* locus in being single genes specifying multifunctional polypeptides? Although the answer to this question is not yet known, the tendency of the enzymatic activities specified by fungal gene clusters to remain associated through extensive purification procedures argues that this may be the case. Indeed, Fink speculates that a major difference between prokaryotes and eukaryotes may be that eukaryotes are incapable of reinitiating translation in response to the sequence AUG . . . UAAAUG, perhaps because eukaryotic ribosomes actually dissociate from the message when they encounter UAA termination signals, and can "reenter" an mRNA only at the 5' end. Granted such a difference between the structure of bacterial operons and fungal "gene clusters," the fact remains that the fungi have often found it advantageous to encode several related enzymatic activities in a common transcriptional unit. Such an arrangement clearly allows the expression of these activities to be regulated as a unit.

In yeast, three steps in histidine biosynthesis are carried out by enzymatic activities that are specified by *HIS4,* while the remaining seven steps are specified by genes unlinked to *HIS4.* In bacteria, on the other hand, we have seen that all 10 *his*-related enzymes are encoded in a single giant operon under the control of a single operator region (Figure 16-4). We may well ask how and why these two divergent arrangements of similar genetic information have evolved in prokaryotes and eukaryotes.

16.3 COMPLEX LOCI IN *DROSOPHILA:* THE *RUDIMENTARY* LOCUS

The rudimentary (*r*) locus is representative of a number of so-called **complex loci** in *Drosophila* (other examples include *white, Notch, dumpy, miniature* and *lozenge*). These complex loci appear to contain a number of genes concerned with a single trait. Mutations in the *r* locus, for instance, all produce similar phenotypes, namely, truncated wings and female sterility. A number of independent *r* mutants have been shown to fall into seven complementation groups, I-VII, based on their ability to yield normal wings in diploid flies. S. Nørby was able to show that *r*/*r* larvae are in fact auxotrophic for pyrimidines or pyrimidine precursors. Since pyrimidines are normally supplied in the *Drosophila* diet, wing development and female fertility are usually the only traits affected in *r*/*r* homozygotes. When enzyme assays are performed, however, flies homozygous for *r* alleles in complementation group VI are found to lack carbamyl phosphate synthetase (CPSase), the first enzyme in the pyrimidine biosynthetic pathway (Figure 16-7). Enzyme assays also show that flies homozygous for *r* alleles in complementation groups I or III lack the second enzyme in the pathway (aspartate transcarbamylase or ACTase); while complementation class VII mutants lack the third enzyme in the sequence (dihydroorotase or DHOase). A complementation map of the *r* locus region can therefore be drawn as in Figure 16-8, where the boundaries of the regions specifying ACTase and CPSase are seen to be as yet unclear. The fourth enzyme of the pyrimidine biosynthetic pathway, dihydroorotate dehydrogenase (DHOdehase), is unaffected by *r* mutations and presumably maps outside the *r* locus.

The complementation map of the *r* locus is in fact more complex than shown in Figure 16-8 in that a mutation in complementation group I, for example, may restore normal wing morphology in one group-VI mutant, but

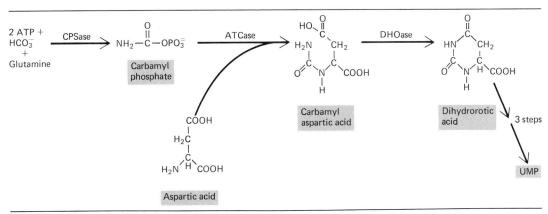

Figure 16-7 The early steps of the pyrimidine biosynthetic pathway. CPSase, Carbamyl phosphate synthetase; ATCase, Aspartate transcarbamylase; DHOase, Dihydroorotase.

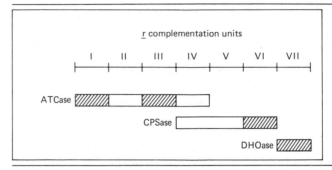

Figure 16-8 Enzyme-specifying regions of the *rudimentary* complementation map of *D. melanogaster*. Complementation units are designated as I to VII. Horizontal bars indicate those units for which control of ATCase, CPSase, or DHOase levels is demonstrated (cross-hatched) or suspected (open). (From J. M. Rawls and J. W. Fristomi, *Nature* **255:**739, 1975.)

does this only poorly with another group-VI mutant. Such patterns allow at least two interpretations. Since in wild-type *Drosophila* CPSase and ACTase activities are found to reside in a large multimeric complex with a molecular weight of 340,000, one possibility is that CPSase activity resides in one class of polypeptide and ACTase in a second class, and that final levels of enzyme activity are dependent, in part, on the way the polypeptides of these two classes associate with one another. A given mutant group-I polypeptide might, in this case, associate in a stable fashion with one mutant form of group-VI polypeptide but poorly with a second mutant form, with overall levels of enzymatic activity fluctuating accordingly. The alternate possibility, of course, is that the product of the *r* locus is a single, multifunctional polypeptide—much like the yeast *HIS4* product—which associates with itself to form multimeric complexes and which exhibits various levels of intragenic complementation (Section 10.11). Studies are clearly needed to determine whether a single class or multiple classes of polypeptides make up the 340,000-dalton enzyme aggregate.

A fine-structure map (Section 10.18) of the *r* locus has been constructed by P. Carlson, who finds the most distant *r* alleles to lie about 0.07 map units from one another. This distance corresponds to an estimated 24,000 base-pairs of DNA, which is much more DNA than would be necessary to encode the three pyrimidine-pathway enzymes actually encoded by the *r* locus. It is not yet known whether the "extra" DNA in the locus specifies additional enzymatic function(s), whether it serves regulatory functions, or whether it contains intervening sequences (Section 8.21).

16.4   FUNCTIONALLY RELATED GENE CLUSTERS IN "HIGHER" EUKARYOTES

Examples of functionally related gene clusters become increasingly meager as one moves "up" the eukaryotic evolutionary ladder. Two notable cases of such clusters in higher animals, however, are the major histocompatibility loci of mice and humans. As described in Chapter 15 (Figure 15-12), each of these loci contains members of a related gene family that presumably arose by gene duplication. In addition, however, the major histocompatibility loci include genes that give no evidence of being structurally "part of a family," and yet all serve in various capacities to mediate the immune responsiveness of the organism. Therefore, it appears that examples of clusters of functionally related genes can be found throughout the biological kingdom.

DISPERSED GENES SPECIFYING ONE TRAIT

The overwhelmingly common finding in the eukaryotes is that genes concerned with the same trait are dispersed throughout the genome. Thus a glance at the linkage map of *C. reinhardi* (Figure 12-15) shows that genes involved with thiamin synthesis (*thi*) map to several different chromosomes, as do the *arg* genes involved with arginine synthesis. The same is found in yeast (Figure 12-16), *Neurospora* (Figure 12-17), *Drosophila* (Figure 12-11) and the mouse (Figure 12-13). Even in *E. coli* related genes are not necessarily clustered. The genes that have been identified for the different aminoacyl-tRNA synthetases in *E. coli,* for example, map to widely dispersed regions of the chromosome, as do the genes for arginine biosynthesis.

The dispersed loci that specify a common trait do so by dictating the structures of polypeptides that act in concert to create a certain phenotype. Various kinds of such interactions have been recognized, depending on whether the trait being monitored is dependent on a single protein, a biochemical or developmental sequence of reactions, or a composite of two, several, or many proteins. These possibilities are considered in the following sections.

16.5 DISPERSED GENES SPECIFYING ONE ENZYME: HUMAN LACTATE DEHYDROGENASE

Because proteins are often oligomeric, dispersed genes will often specify the same trait by virtue of their specifying subunits of the same oligomeric protein. An example is given by the genes specifying human lactate dehydrogenase (LDH). LDH catalyzes the formation of lactic acid from pyruvate. When human cell extracts are subjected to electrophoresis and stained for lactate dehydrogenase activity, 5 isozymes (Section 15.6) are typically found, the relative proportions of each varying from one tissue to the next. Each isozyme proves to be a tetramer (composed of 4 polypeptide chains), and *in vitro* analysis reveals that two different kinds of polypeptides, known as A and B, give rise to the 5 isozymes. The two polypeptides do this in the following manner: LDH 1 is composed of four B chains (B4); LDH 2 contains one A and 3 B chains (AB_3); LDH 3 is A_2B_2; LDH 4 is A_3B_1; and LDH 5 is A_4. Somatic cell hydridization has demonstrated that the A polypeptide chain is specified by the *A* locus in human chromosome 11, while the B chain is coded by the *B* locus in human chromosome 12. Allelic forms of both genes are known, generating more complex isozyme patterns in heterozygotes, but most humans are homozygous at both loci.

16.6   DISPERSED GENES SPECIFYING A MULTIENZYME COMPLEX

Multienzyme complexes in fungi can be specified by a single locus, as we saw for the *HIS4* product in yeast (Figure 16-6). Multienzyme complexes may also be composed of polypeptides from unlinked genes, and the interactions between them may become exceedingly complex, as exemplified by the *trp*-1 and *trp*-2 genes of *Neurospora crassa*. These genes are not linked and each controls the synthesis of a polypeptide chain. Four of the *trp*-1 polypeptides interact with two of the *trp*-2 polypeptides to form a hexameric protein that exercises three enzymatic activities in the pathway of tryptophan biosynthesis, namely, anthranilate synthetase, phosphoribosyl-anthranilic acid (PRA) isomerase, and indole-3-glycerol-phosphate (InGP) synthetase activity. Mutations in the *trp*-2 gene lead to the loss of anthranilate synthetase activity, and mutant *trp*-2 cells produce a tetrameric protein that represents an aggregate of the *trp*-1 product. Mutations in *trp*-1 lead to a variety of phenotypes: various *trp*-1 mutant strains exhibit alterations in any or all of the activities normally catalyzed by the enzyme complex. Such pleiotropy suggests that the *trp*-1 polypeptides are not only essential to PRA isomerase and InGP synthetase activity but that they also create the appropriate macromolecular configuration for anthranilate synthetase activity, presumably by their interaction with the *trp*-2 polypeptides.

Some generalized principles should be stated at this point regarding the genetic analysis of multimeric proteins. Most eukaryotic proteins, and particularly enzymes, function *in vivo* as multimers which can become extremely complex, having many polypeptides, catalyzing several distinct reactions, and encoded by several genes. This is also the case for proteins that perform membrane-associated activities, as exemplified by the components of the mitochondrial electron transport chain (Section 14.10). Analysis of the genes that specify such multimeric-protein polypeptides requires considerable care, for the kinds of pleiotropic effects described above for the *trp*-1 mutations can give erroneous impressions as to the function of the gene product being analyzed. It is essential for such studies that many mutant alleles of the gene in question be obtained and that their phenotypic effects be monitored both alone and in combination with other alleles.

BIOCHEMICAL GENETICS

Mutations in genes that control sequential biological steps, be the genes clustered or dispersed, have proved to be invaluable for "dissecting" the number and order of reactions in a given metabolic pathway. In particular, a field of study known as **biochemical genetics** has combined genetics and biochemistry to elucidate the nature of metabolic pathways, most notably in haploid organisms whose growth requirements are known and whose gene expression is not complicated by allelic interactions.

The approach followed in biochemical genetics is to assemble a collection of mutant strains that cannot synthesize a particular metabolic component and thus require it for their growth. These strains are then subjected to complementation tests to estimate how many separate genes are involved in the synthesis of the component, and the genes are mapped to determine their linkage relationships. Strains in a given complementation group are then tested for their ability to grow when supplied with known metabolic precursors of the final component. The following general rule is then applied: if they **can** grow in the presence of a certain precursor, they must suffer from a genetic lesion affecting a step **before** the synthesis of that precursor; if instead they **cannot** grow when a certain precursor is supplied, the genetic lesion must affect a step that **follows** the synthesis of the precursor.

16.7 BIOCHEMICAL GENETICS OF HAPLOID ORGANISMS

As a first example of a biochemical genetic analysis we can cite mutant strains of *E. coli* that require arginine for growth. Complementation analysis of partial diploids (Section 11.17) places these strains into eight complementation groups, and crosses show that the strains carry mutations in eight genes, *argA* through *argH*. The *argH* strains can grow only if they are given arginine; no other precursors will do. The *argG* strains, on the other hand, will grow either in the presence of arginine or of argininosuccinic acid. The *argF* strains can grow only if supplied with arginine, argininosuccinic acid, or citrulline; and so on. By comparing the growth requirements of the different strains it is possible to determine the sequence of precursors in the pathway leading to arginine biosynthesis. These results are illustrated in Figure 16-9.

Biochemical genetics can also determine whether a biosynthetic pathway is linear (as is the arginine pathway in *E. coli*) or branched. For example, three unlinked complementary genes that we can call *thi*-1, *thi*-2, and *thi*-3 are known to be involved in thiamin biosynthesis in *Neurospora*. The growth requirements of strains carrying these mutations are summarized in Table 16-2. When an attempt is made to order these precursors into a linear pathway, however, the data argue against such a pathway: a thiazole ⟶ pyrimidine ⟶ thiamin sequence suggests that *thi*-1 should grow when supplemented with pyrimidine—which is not observed—and a pyrimidine ⟶ thiazole ⟶ thiamin sequence suggests that *thi*-2 should grow when supplemented with thiazole—which is also not observed. The correct pathway, shown in Figure 16-10, shows that both thiazole and pyrimidine are common precursors of thiamin synthesis.

A final example of the use of biochemical genetics concerns the obser-

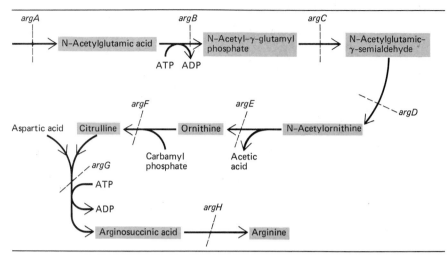

Figure 16-9 Pathway of arginine biosynthesis and a spectrum of mutant strains of *E. coli* blocked at the indicated positions.

vation that single-gene mutations in *Neurospora* block the biosynthesis of two quite distinct amino acids—methionine and threonine. A study of the accumulated precursors of these two amino acids reveals that the mutations prevent the biosynthesis of homoserine from aspartic acid (Figure 16-11, site 1) and that homoserine serves as a precursor for the biosynthesis of both methionine and threonine along independent pathways. Other sites in which mutations are known in these pathways are also indicated in Figure 16-11.

16.8   BIOCHEMICAL GENETICS OF DIPLOID ORGANISMS: MODIFIED DIHYBRID RATIOS AND EPISTASIS

Biochemical genetic analysis in diploid organisms is best described by giving specific examples. We can first consider the genes participating in the synthesis of the brown ommochrome pigments that are prominent in the

Table 16-2   Growth Requirements of *Neurospora* Strains Requiring Thiamin

| Strain | Precursor | | Thiamin |
	Thiazole	Pyrimidine	
thi-1	+	−	+
thi-2	−	+	+
thi-3	−	−	+

A plus sign indicates growth.
A minus sign indicates no growth.

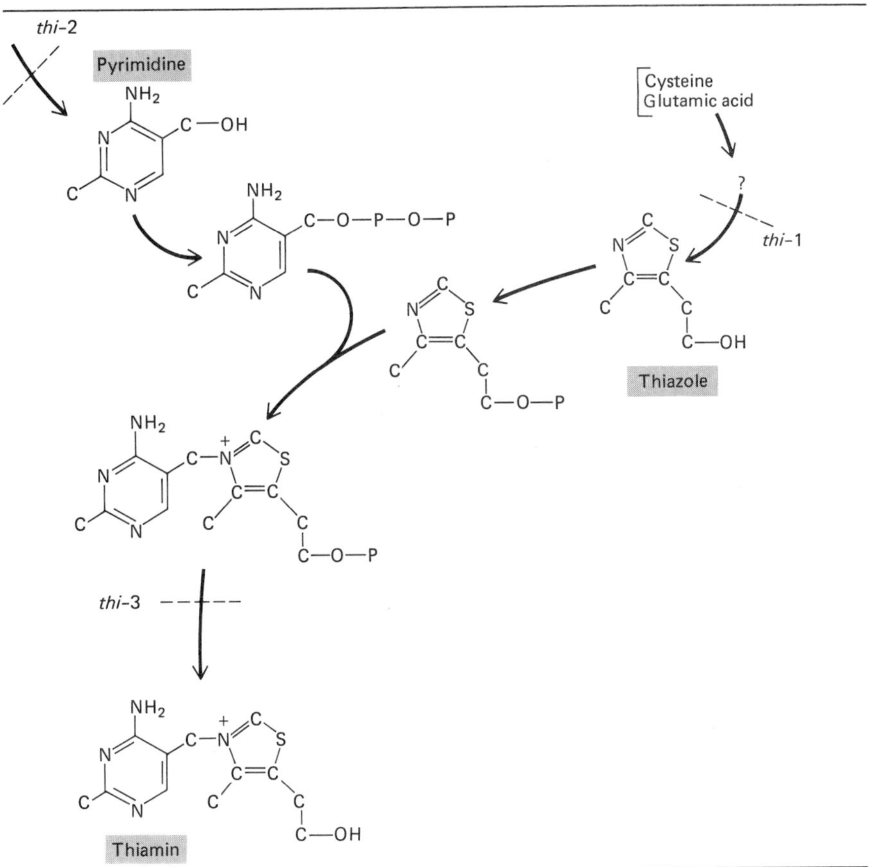

Figure 16-10 Pathway of thiamin synthesis in *Neurospora* showing sites blocked in the mutant strains *thi*-1, *thi*-2, and *thi*-3. (From R. P. Wagner and H. K. Mitchell, *Genetics and Metabolism*, 2nd ed. New York: John Wiley, 1964.)

wild-type eye of *D. melanogaster*. Three loci, one sex-linked (*vermilion, v*), one in chromosome II (*cinnabar, cn*), and one in chromosome III (*scarlet, st*), are implicated as being involved in ommochrome biosynthesis, since flies homozygous or hemizygous for mutations in either *v, cn,* or *st* have red eyes and synthesize only the red pterin eye pigments. When the ommochrome pigment precursors accumulated by the homozygous mutant strains are isolated and analyzed, it becomes possible to construct the pathway shown in Figure 16-12, in which the *v, cn,* and *st* gene products are seen to control sequential steps in the biosynthesis of the final ommochrome pigment.

The fact that the *v, cn,* and *st* loci are involved in specifying a common trait is also evident when the various eye-color mutants are crossed to one another. If, for example, *scarlet* females are mated with *vermilion* males and the F_1 is inbred, then a 9:3:3:1 ratio of phenotypes (Section 4.8) is not

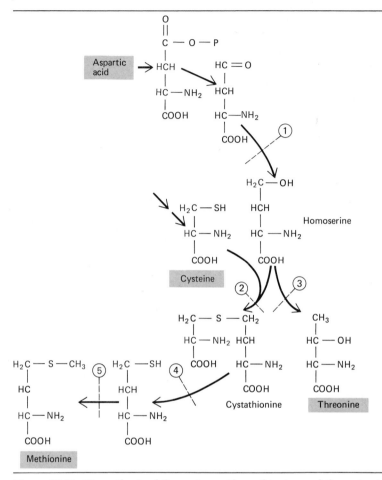

Figure 16-11   Biosynthesis of the amino acids methionine and threonine in *Neurospora* showing reactions blocked in various mutant strains (①︎ to ⑤︎). (From R. P. Wagner and H. K. Mitchell, *Genetics and Metabolism*, 2nd ed. New York: John Wiley, 1964.)

observed among the F_2, even though the alleles can be shown to assort independently. Instead, as diagrammed in Figure 16-13, the F_2 progeny exhibit a 9:7 phenotypic ratio. More generally, if a 9:3:3:1 is considered the "standard" obtained when **dihybrid** $\left(\dfrac{+}{a}\ \dfrac{+}{b}\right)$ organisms are inbred, then variations such as 9:7 are termed **modified dihybrid ratios,** and they signal that the two loci are interacting to produce the trait.

A second and more complex example can be given in which the biochemical basis for the phenotype is not yet fully understood but in which interrelated pathways also appear to be involved. This concerns two independently assorting loci, c and a, that control coat color in mice. Mice homozygous for the recessive gene c cannot synthesize pigment anywhere

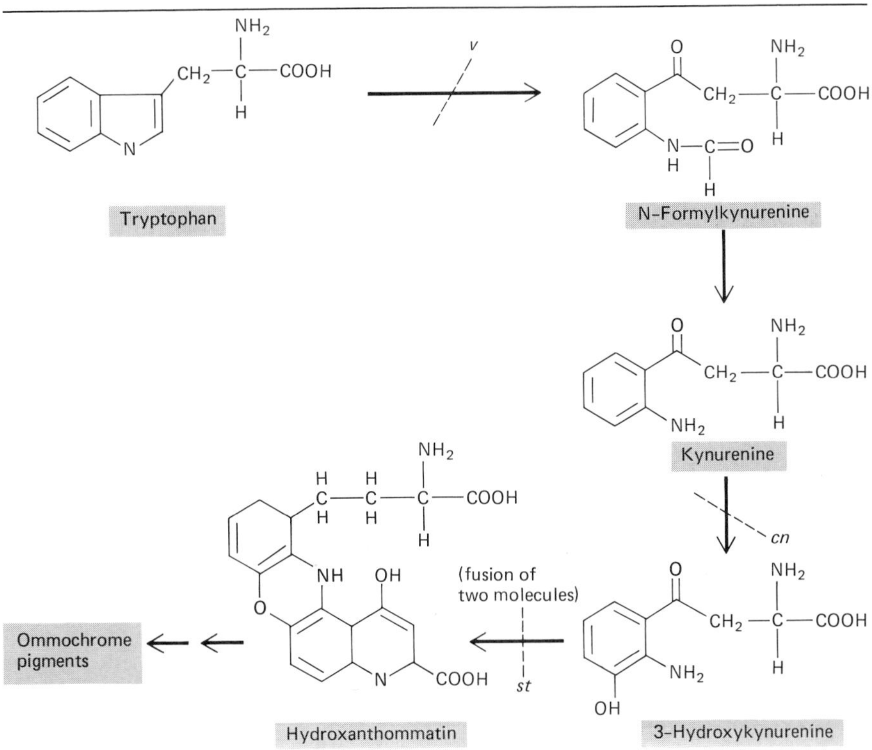

Figure 16-12 Synthesis of the ommochrome pigments in *Drosophila melanogaster* showing sites blocked in mutant strains homozygous for *vermilion* (*v*), *cinnabar* (*cn*), and *scarlet* (*st*).

in their bodies and have white hair (*albino*). Mice that are homozygous for *a* produce completely black hair. When *white* and *black* mice are crossed $\frac{+}{+} \frac{c}{c} \times \frac{a}{a} \frac{+}{+}$, the $\frac{+}{a} \frac{+}{c}$ mice in the F_1 all have the grayish coat known as *agouti*. Individual agouti hairs are black with a yellow band near the tip. When the F_1 mice are inbred, 9/16 of the F_2 are *agouti*, 3/16 are *black*, and 4/16 are *white*. The most reasonable way to explain this result is to propose that the *c* locus constitutes the structural gene for tyrosine oxidase, an enzyme that acts early in the biosynthesis of melanin (Figure 16-14). The *a* locus, we can then imagine, is involved with the placement of the melanin pigment in the hair. The melanin is dispersed throughout the hair when the *a* gene alone is present, whereas its wild-type allele specifies the unusual pigment arrangement of agouti hair. You should work this out for yourself using the steps outlined in Figure 16-13 as a model.

A term that is frequently encountered when such complex phenotypic traits are beng analyzed is **epistasis,** which means literally "standing above." A gene or gene pair at one locus is said to be epistatic to a gene or gene at a second locus when the gene product of the first locus masks the expres-

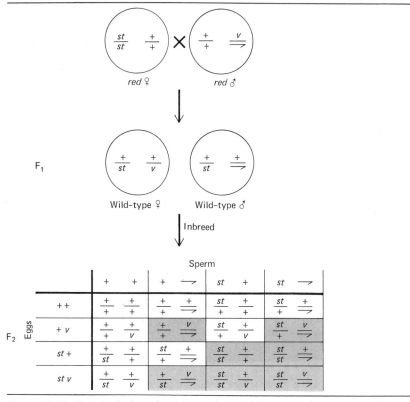

Figure 16-13 Result of inbreeding the F_1 obtained from a *scarlet*(*s*) × *vermilion* (*v*) cross of *D. melanogaster*. The 16 genotypic classes of F_2 progeny are shown in the checkerboard; the seven shaded classes will have a mutant (*red eye*) phenotype and the nine unshaded classes will have a wild-type (purple-brown eye) phenotype.

sion of the second. Thus the homozygous *albino* gene pair in the mouse is said to be epistatic to the *agouti* and *black* genes. Additional examples of epistasis are given in the problems at the end of the chapter.

16.9   GENETIC DISSECTION OF COMPLEX PATHWAYS

The approach of biochemical genetics can be used to dissect very complex metabolic pathways. Perhaps the most elegant of these applications, initiated by R. Edgar and W. Wood, concerns morphogenesis of the phage T4, a process that involves the elaborate and sequential self-assembly of numerous different kinds of proteins.

The analysis begins by isolating conditional lethal strains of T4 that carry *amber* or temperature-sensitive mutations in genes essential to the assembly of mature phages. Each strain is allowed to infect *E. coli* cells under non-

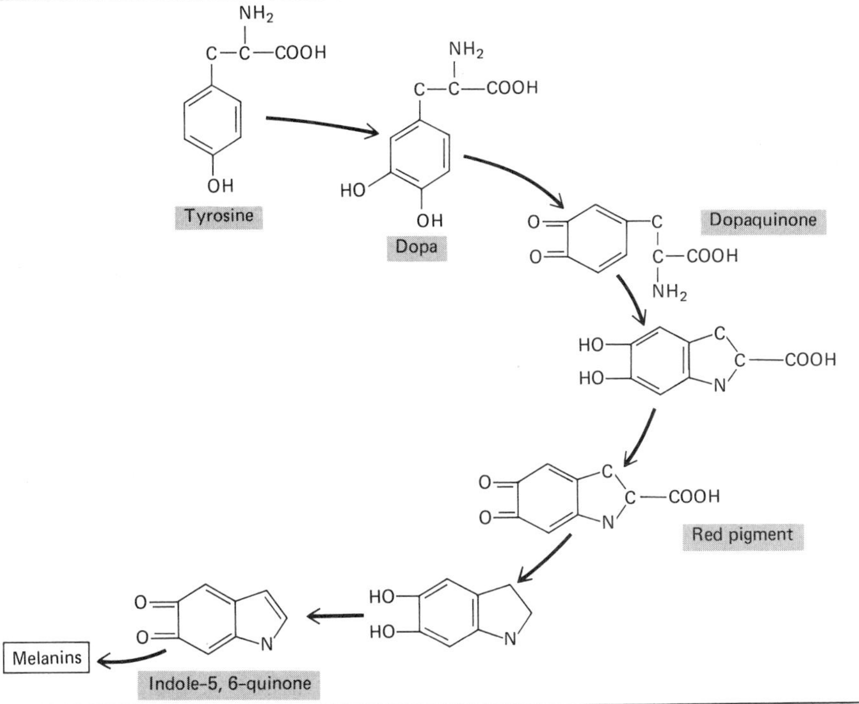

Figure 16-14 Pathway of melanin biosynthesis from tyrosine. (From R. P. Wagner and H. K. Mitchell, *Genetics and Metabolism*, 2nd ed. New York: John Wiley, 1964.)

permissive conditions, whereupon phage assembly proceeds up to the point at which an essential protein is missing or inactive. Depending on the protein that is absent, different precursor phage structures will accumulate in the infected cells. If the cells are now lysed artificially, the precursor structures can often be isolated and observed with the electron microscope; they can also be subjected to *in vitro* complementation tests (Section 10.13) with precursor structures derived from other lysates. The complementation patterns of the various T4 strains can also be studied *in vivo*. In this way, some 50 genes have been shown to be involved in T4 morphogenetic processes, either by directly specifying some structural protein of the phage coat or by specifying a protein required in smaller amounts for some delicate assembly process. The sequence of events and the relevant genes, as they are presently known, are shown in Figure 16-15. The mutant T4 strains have clearly permitted an understanding of phage assembly in exquisitely intricate detail.

In principle the study of any morphogenic or biochemical sequence can be analyzed with the use of mutant strains, and many laboratories are currently isolating mutant strains blocked in pathways leading to the development of the brain or of particular behavior patterns (see references

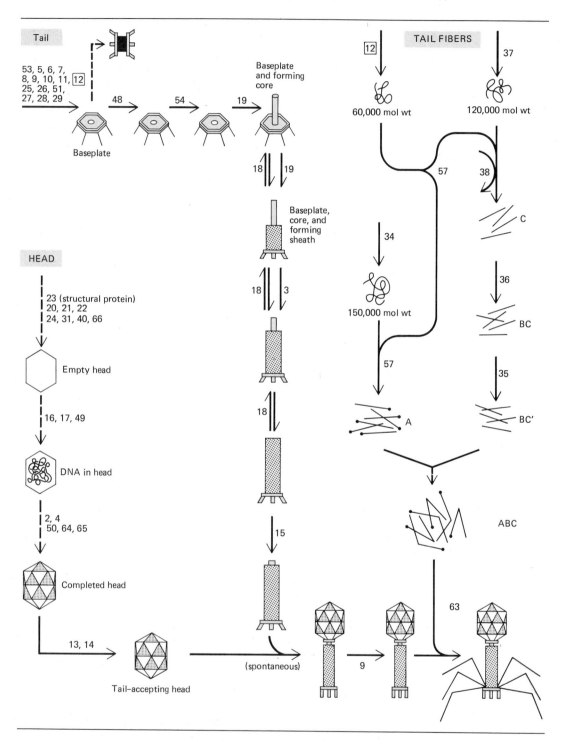

Figure 16-15 Morphogenetic pathway of phage T4 showing individual pathways for head, tail, and tail-fiber assembly. Dashed arrows indicate steps that have not yet been demonstrated *in vitro*. Number(s) associated with each arrow indicates the gene controlling that step or protein in the assembly process. Gene 12 is seen to participate both in the tail and in the tail fiber sequence. In the tail fiber sequence, molecular weights of the intermediates are given, and capital letters denote antigenic compositions. (From M. Levine, *Ann. Rev. Genetics* **3:**323, 1969; F. R. Frankel, H. L. Batcheler, and C. K. Clark, *J. Mol. Biol.* **62:**439, 1971; and J. King, *J. Mol. Biol.* **58:**693, 1971.)

cited at the end of this chapter). In practice, as the trait to be analyzed becomes more complex, it becomes increasingly difficult to recognize and maintain mutant strains that are *directly* affected in the pathway of interest. Many strains that have, for example, abnormal brain development may prove to be suffering from a metabolic lesion that affects the brain only indirectly (for example, phenylketonuria); other mutations exhibit considerable pleiotropy. This does not mean that complex sequences will not eventually be understandable in terms of the genes that control them but only that achievement of this goal appears distant at the present time.

16.10 DISPERSED GENES CONTROLLING COMPOSITE TRAITS

There are many examples in which the phenotype of a diploid eukaryote is best categorized as a **composite trait.** The relevant gene products do not appear to interact to produce a single protein, nor do they appear to be acting in sequence in a biosynthetic pathway. Instead, the gene products seem to interact by masking, modifying, or enhancing the effect of one another.

In some cases the components of a composite trait can be distinguished; in such cases the genetic analysis becomes relatively straightforward. For example, the purple eye color of *D. melanogaster* is a composite of two classes of pigments—the brown ommochromes and the red pterins. We have already analyzed mutations such as *st* and *v* which produce red-eyed flies when homozygous and affect the ommochrome pathway (Figure 16-12). A second class of eye-color mutation in *D. melanogaster,* typified by *brown* (*bw*), gives rise to *brown*-eyed flies in homozygotes. To demonstrate that the synthesis of red pigments and brown pigments proceeds by independent, additive pathways a cross can be made between *bw/bw* and *st/st* flies and the F_1 can be inbred. In the F_2, we find that 9/16 have purple eyes, 3/16 have *scarlet* eyes, 3/16 have *brown* eyes, and 1/16 have *white* eyes. In the last, doubly homozygous recessive organisms, neither pigment pathway is operative and neither trait is present. Otherwise, the phenotypic ratio is that expected for two kinds of gene products that do not interact, a statement you should verify for yourself with diagrams.

Genetic analysis of a composite trait becomes increasingly difficult as the biochemical basis of the trait becomes less well known and as the number of loci controlling the trait becomes greater than two. The loci concerned with the red kernel color in wheat can serve as an example. Three loci have been identified, r_1, r_2, and r_3. A monohybrid cross ($R/r \times R/r$) for any gene pair results in the expected phenotypic ratio of 3 *red* :1 *white* and a genotypic ratio of 1 RR:2Rr:1 rr. Dihybrid crosses (for example, $\frac{R_1}{r_1}\frac{R_2}{r_2} \times \frac{R_1}{r_1}\frac{R_2}{r_2}$) give an overall phenotypic ratio of 15 *red* :1 *white,* with only the doubly homozygous recessive kernels producing no pigment. Among the red progeny, however, a discontinuous gradation of intensity is observed between dark red and light red, with each category corresponding to a particular combination of R and r alleles. Thus the phenotypic ratio becomes 1 dark red:4 medium dark:6 medium:4 light red:1 white (Table 16-3).

When trihybrid crosses are performed $\left(\frac{R_1}{r_1}\frac{R_2}{r_2}\frac{R_3}{r_3} \times \frac{R_1}{r_1}\frac{R_2}{r_2}\frac{R_3}{r_3}\right)$, 64 combinations are possible (Figure 16-16), only one of which is *white.* The 63 *red* classes again show a gradation of intensity and can be grouped into

Table 16-3   Results of $\frac{R_1}{r_1}\frac{R_2}{r_2} \times \frac{R_1}{r_1}\frac{R_2}{r_2}$ Dihybrid Cross in Wheat

		Gametes		
		R_1R_2	r_1R_2	R_1r_2	r_1r_2
G a m e t e s	R_1R_2	Dark $\frac{R_1\ R_2}{R_1\ R_2}$	Medium Dark $\frac{R_1\ R_2}{r_1\ R_2}$	Medium Dark $\frac{R_1\ R_2}{R_1\ r_2}$	Medium $\frac{R_1\ R_2}{r_1\ r_2}$
	r_1R_2	Medium Dark $\frac{R_1\ R_2}{r_1\ R_2}$	Medium $\frac{r_1\ R_2}{r_1\ R_2}$	Medium $\frac{R_1\ R_2}{r_1\ r_2}$	Light $\frac{r_1\ r_2}{r_1\ R_2}$
	R_1r_2	Medium Dark $\frac{R_1\ R_2}{R_1\ r_2}$	Medium $\frac{R_1\ R_2}{r_1\ r_2}$	Medium $\frac{R_1\ r_2}{R_1\ r_2}$	Light $\frac{R_1\ r_2}{r_1\ r_2}$
	r_1r_2	Medium $\frac{R_1\ R_2}{r_1\ r_2}$	Light $\frac{r_1\ R_2}{r_1\ r_2}$	Light $\frac{r_1\ r_2}{R_1\ r_2}$	White $\frac{r_1\ r_2}{r_1\ r_2}$

Figure 16-16   Kernel color in wheat determined by three loci marked by gene pairs R_1/r_1, R_2/r_2, and R_3/r_3. Homozygous parental lines are crossed and the F_1 is self-crossed to give the F_2 distribution of color shown in the histogram. (Reprinted with permission of Macmillan Publishing Co., Inc. from *Genetics* by M. Strickberger. Copyright © 1968 by M. W. Strickberger.)

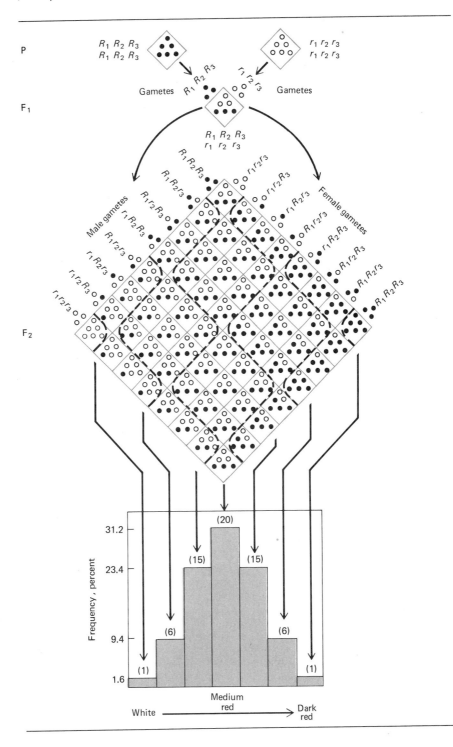

classes with a 1:6:15:20:15:6 ratio of phenotypes. In this case, however, the discontinuities between one class and the next are less distinct. In other words, had one simply inbred two red strains and obtained the progeny shown in Figure 16-16, it would be no mean task to classify them into their phenotypic groupings and conclude that three independently assorting pairs of alleles were involved in determining red color. Instead, it would appear that a rather **continuous variation** in red coloring was present among the progeny.

QUANTITATIVE GENETIC PRINCIPLES

16.11 GENERAL PROPERTIES OF QUANTITATIVE TRAITS

The preceding example concerned with wheat kernel color illustrates the "spread" in the expression of a trait within an interbreeding population when it is controlled by several loci. This is most conspicuously true for such **quantitative traits** as height and weight. An early demonstration of what has come to be known as **quantitative genetics** is given by the experiments of R. Emerson and E. East, published in 1913. These geneticists developed two strains of maize, one giving rise to long (13 to 21 cm) ears and the other to short (5 to 8 cm) ears of corn (Table 16-4 and Figure 16-17). The range in ear length within each class was judged to be due largely to environmental influences, since each parental line was highly inbred. When the two strains were crossed, the F_1 was intermediate in length between the two parents (Table 16-4 and Figure 16-17), a result consistent with the expectation that the F_1 represents a genetically identical heterozygous population that is subject to the same environmental influences as the parents. When the F_1 was inbred, however, the spread of the variation increased significantly even though the environment remained the same: ears were found that were several centimeters shorter and several centimeters longer than any of the F_1 ears (Table 16-4 and Figure 16-17), whereas the mean length for the F_1 and F_2 remained about the same. It should be noted that no ears were longer or shorter than those produced by the original

Table 16-4 Frequency Distributions of Ear Length in Maize

	Length of Ear, cm																				
	5	6	7	8	9	10	11	12	13	14	15	16	17	18	19	20	21	N	$\bar{x}$	σ	S.E.
Parent 60	4	21	24	8														57	6.632	0.816	0.108
Parent 54								3	11	12	15	26	15	10	7	2		101	16.802	1.887	0.188
F_1 (60 × 54)					1	12	12	14	17	9	4							69	12.116	1.519	0.183
F_2				1	10	19	26	47	73	68	68	39	25	15	9	1		401	12.888	2.252	0.112

After Emerson and East, *Nebraska Research Bull.* **2,** 1913.

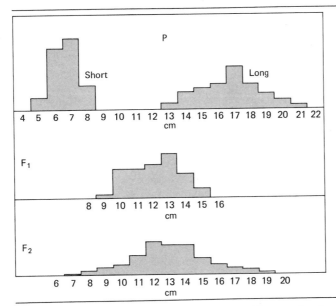

Figure 16-17 The Emerson-East experiment with corn showing the distribution of ear length in the parental lines and in the F_1 and F_2. The histogram bars show percentages of the various populations having particular ear lengths. (From Sturtevant and Beadle, *An Introduction to Genetics*. Philadelphia: W. B. Saunders Co., 1940.)

parents. With regard to the "spread" of a trait, then, as inbreeding is extended to successive generations, the chances increase that all possible combinations of alleles will appear among the progeny and this increases the range of the variability; the new combinations, however, do not create new quantitative extremes.

Perhaps the most familiar example of the spreading of additive traits through interbreeding concerns skin pigmentation in humans. Skin pigmentation is controlled, first, by genes that dictate the synthesis of melanin (Figure 16-14). Because mutations in these genes lead to the *albino* condition, they can readily be identified. In addition, a number of gene loci are concerned either with the distribution of **melanocytes** (pigment-producing cells) beneath the skin or else govern the amount of melanin produced by the melanocytes.

The inheritance of these genes in humans has been studied by G. Harrison and J. Owen. They examined areas of the body that are not normally exposed to sunlight, and measured the reflectance of skin exposed to different wavelengths of light between 420 and 680 nm. When such measurements were made of "true-breeding" European (Caucasoid) and African (Negroid) populations, an expectedly marked difference in average reflectances (hence pigmentation) was found (Figure 16-18, curves *European* and *African*). Measurements made on the F_1 mulatto progeny of African and Caucasian matings gave values intermediate between the two

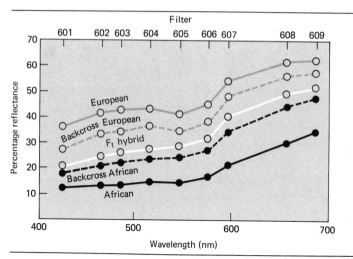

Figure 16-18 Mean reflectance from the skin of Europeans, Africans, and various hybrid groups as measured with a reflectance spectrophotometer. (From G. Harrison and J. Owen, *Ann. Human Genetics* **28**:27, 1964, Cambridge University Press.)

parents (Figure 16-18, curve F_1 *hybrid*), again as one would expect. When a backcross occurs between such mulattoes and either Europeans or Africans, however, the curves obtained from the progeny fall between the intermediate and the extreme curves (Figure 16-18, curves *backcross European* and *backcross African*), a result that indicates at once that the trait is determined by more than two loci (work this out for yourself).

Another, earlier study, made by C. Davenport, follows instead the results of mulatto $\times$ mulatto matings. As seen in Figure 16-19, a wide range of

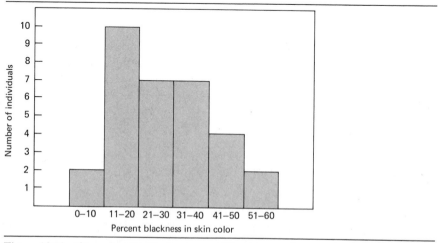

Figure 16-19 Skin color distribution among 32 offspring of F_1 mulattos married to each other. (From P. A. Moody, *Genetics of Man*, 2d ed. New York: Norton, 1975; based on data of C. B. Davenport, 1913.)

pigmentation is observed among the progeny, the only consistent pattern being that a child is usually no darker than his or her darker parent.

It is possible to take the kinds of experiments outlined for wheat, corn, and humans and make certain estimates of the number of gene loci involved in determining a quantitative trait. We saw in the case of red color in wheat that as the number of loci increases from one to three, the proportion of extreme phenotypes among the progeny decreases markedly. Thus in the monohybrid cross 1/4 were dark red and 1/4, white; in the dihybrid cross 1/16 were dark red and 1/16, white; and in the trihybrid cross only 1/64 were dark red and 1/64, white. In other words, when F_1 individuals are heterozygous at n loci, then $(1/4)_n$ of the F_2 offspring will have the phenotype and genotype of the original homozygous parents. As the value of n increases, this number becomes vanishingly small, as does the number of some of the less extreme phenotypes in the F_2. When such lines of reasoning are applied to human skin color pedigrees, it can be estimated that at a minimum perhaps four gene loci determine this trait.

16.12   THE STATISTICAL ANALYSIS OF QUANTITATIVE TRAITS

Quantitative traits are usually analyzed in statistical terms. Since fairly sophisticated statistical concepts must be mastered to perform these operations, and since these lie outside the scope of this text, only an outline of what is involved will be presented here. Texts listed at the conclusion of this chapter should be consulted for a thorough treatment of the subject.

The analysis of quantitative traits usually begins by calculating the quantitative mean and the variance of the trait. The **mean** ($\bar{x}$) value of a trait in a population is obtained by summing all the values observed (Σx) and dividing by the number of values ($\Sigma x/n$). In the Emerson-East experiment (Table 16-4) the mean was 6.6 for the short corn ears, 16.8 for the long ears, 12.1 for the F_1 and 12.9 for the F_2. The **variance** (V or σ^2) describes the way in which these values are dispersed about the mean. One first observes by how much each individual value differs from the mean (d). These differences are squared (d^2) and all the d^2 values are summed (Σd^2). This sum is then divided by $n - 1$ to reflect the number of degrees of freedom in the sample (see Section 4.9). Thus

$$\sigma^2 = \frac{\Sigma d^2}{n - 1}$$

The square root of the variance is σ, the **standard deviation,** which expresses dispersion in the same units of measurement as the mean.

It turns out that the distribution of values for most continuously varying characters in a population assumes a bell-shaped pattern known as a **normal distribution.** Thus the histogram data from the crosses concerned with wheat-kernel color, corn-ear length and skin color (Figures 16-16, 16-17 and 16-18) can all be approximated by such a bell-shaped, normal curve, where the distribution of values is fairly symmetrical about the mean and 95 percent of the values fall within the interval between $\bar{x} - 2\sigma$ and $\bar{x} + 2\sigma$. The variance observed in two different normal distributions may differ widely, however, as is apparent in the figures. In particular, we saw in the Emerson-East experiment that the variance in the F_2 was significantly greater than in the F^1 (Figure 16-17).

16.13 FACTORS CONTRIBUTING TO PHENOTYPIC VARIANCE

We noted in our description of the Emerson-East experiment that there was a considerable range of sizes (in fact, a normal distribution of sizes) in the initial long- and short corn-ear samples, even though each sample derived from highly inbred lines. Most of this variability was attributed to environmental influences, an attribution that can be expressed more quantitatively. One can speak of V_P, the **total phenotypic variance** observed in the population, as being composed of three additive factors: V_E, the **environmental variance;** V_G, the **genetic variance,** and V_{GE}, the variance due to **genetic and environmental interactions.** Thus

$$V_P = V_E + V_G + V_{GE}$$

In estimating the effects of the environment on a trait, highly inbred lines or monozygotic twins are studied, in which cases V_G can be assumed to be negligible or zero. It is also common to ignore the V_{GE} component, since it is difficult to analyze. One can then define a parameter called **heritability** (H) as being equal to V_G/V_P. A trait with a heritability of 1 is said to be expressed without any environmental influence, whereas a trait with a heritability of 0.5 would have half its variability (from individual to individual) determined by environmental factors (for example, diet and incidence of disease) and half by genotypic factors. Table 16-5 summarizes some representative heritability values, where all are expressed as percentages. We should note that ear length in corn has a heritability of only 0.17, so that the variation seen in Table 16-4 can indeed be attributed to environmental factors.

While heritability estimates have proved highly useful to agriculture and animal husbandry, they are, in the final analysis, only approximations. For example, the genetic variability (V_G) is considered to be due only to genes that interact in an additive manner (such as the red kernel genes in wheat). Genes that exhibit dominance or epistatic types of interactions, since they complicate the picture significantly, are generally not included (or are ignored). Second, the heritability estimate ignores all situations covered by the term V_{GE}, where the relative performances of different genotypes vary with the environments in which these genotypes are placed. Thus, if a certain gene product is unable to function at all under a given set of environmental circumstances, but another product is able to do so, such phenomena are not included in the estimate. In other words, the heritability estimate glosses over genes that show incomplete penetrance and variable expressivity (Section 15.4), even though such genes may well contribute to the determination of quantitative traits.

Table 16-5 Estimates of Heritability (in Percent) for Various Characters in Different Varieties of Farm and Laboratory Animals

	Percent Heritability
Cattle	
Birth weight (Angus)	49
Gestation length (Angus)	35
Calving interval (Angus)	4
Milk yield (Ayrshire)	43
Conception rate (Holstein)	3
White spotting (Friesians)	95
Sheep	
Birth weight (Shropshire)	33
Weight of clean fleece (Merino, 22 months)	47
Length of wool fiber (Rambouillet, 14 months)	36
Multiple birth (Shropshire)	4
Chickens	
Body weight (Plymouth Rock, 8 weeks)	31
Shank length (New Hampshire)	50
Egg production (White Leghorn)	21
Egg weight (White Leghorn)	60
Hatchability (composite)	16
Mice	
Tail length (6 weeks)	60
Body weight (6 weeks)	35
Litter size	15
Drosophila melanogaster	
Abdominal bristle number	52
Thorax length	47
Wing length	45
Egg production	18

Reprinted with permission of Macmillan Publishing Co., Inc. from M. Strickberger, *Genetics*. Copyright 1968, Monroe W. Strickberger.

References

Operons

*Adhya, S. L., and J. A. Shapiro. "The galactose operon of *E. coli* K-12. I. Structural and pleiotropic mutations of the operon," *Genetics* **62**:231–247 (1969).

*Ames, B. N., R. F. Goldberger, P. E. Hartman, R. G. Martin, and J. R. Roth. "The histidine operon." In *Regulation of Nucleic Acid and Protein Biosynthesis,* V. V. Konigsberger and L. L. Bosch, Eds. Amsterdam: Elsevier, 1967.

*Denotes articles described specifically in the chapter.

*Beckwith, J. R., and D. Zipser, Eds. *The Lactose Operon.* New York: Cold Spring Harbor Laboratory, 1970.

Brenner, M., and B. N. Ames. "The histidine operon and its regulation," In *Metabolic Regulation,* H. J. Vogel, Ed. New York: Academic, 1971, p. 349–387.

Cordaro, J. C., and E. Balbinder. "Evidence for the separability of the operator from the first structural gene in the tryptophan operon of *Salmonella typhimurium,*" *Genetics* **67:**151–169 (1971).

Imamoto, F., and Y. Kano. "Inhibition of transcription of the tryptophan operon in *Escherichia coli* by a block in initiation of translation." *Nature New Biology* **232:**169–173 (1971).

Gene Clusters

*Bigelis, R., J. Keesey, and G. R. Fink. "The *his* 4 fungal gene cluster is not polycistronic," in: *Molecular Approaches to Eukaryotic Genetic Systems,* G. Wilcox, J. Abelson, and C. F. Fox, Eds., New York: Academic Press, 1977, Vol. 8, pp. 179–187.

Carlson, P. S. "A genetic analysis of the *rudimentary* locus of *Drosophila melanogaster,*" *Genet. Res.* **17,** 53–81 (1971).

Case, M. E., and N. H. Giles. "Revertants and secondary *arom-2* mutants induced in non-complementing mutants in the *arom* gene cluster of *Neurospora crassa,*" *Genetics* **77:**613–626 (1974).

Case, M. E., and N. H. Giles. "Gene order in the *qa* gene cluster of *Neurospora crassa,*" *Molec. Gen. Genet.* **147:**83–89 (1976).

Chalmers, J. H., Jr., and T. W. Seale. "Supersuppressible mutants in *Neurospora:* mutants at the *tryp-1* and *tryp-2* loci affecting the structure of the multienzyme complex in the tryptophan pathway," *Genetics* **67:**353–363 (1971).

Falk, D. R., "Pyrimidine auxotrophy and the complementation map of the *rudimentary* locus of *Drosophila melanogaster,*" *Molec. Gen. Genet.* **148:**1–8 (1976).

Fink, G. R. "A cluster of genes controlling three enzymes in histidine biosynthesis in *Saccharomyces cerevisiae,*" *Genetics* **53:**445–459 (1966).

*Nørby, S. "The biochemical genetics of *rudimentary* mutants of *Drosophila melanogaster.* I. Aspartate carbamoyltransferase level in complementing and non-complementing strains," *Hereditas* **73:**11–16 (1973).

*Rawls, J. M., and J. W. Fristrom. "A complex genetic locus that controls the first three steps of pyrimidine biosynthesis in *Drosophila,*" *Nature* **255:**738–740 (1975).

*Shaffer, B., J. Rytka, and G. R. Fink. "Nonsense mutations affecting the *his*4 enzyme complex of yeast," *Proc. Natl. Acad. Sci. U.S.* **63:**1198–1205 (1969).

Biochemical Genetics

Beadle, G. W., and E. L. Tatum. "Genetic control of biochemical reactions in *Neurospora,*" *Proc. Natl. Acad. Sci. U.S.* **27:**499–506, 1941 [Reprinted in L. Levine, *Papers on Genetics.* St. Louis: Mosby, 1971, p. 338, and in J. A. Peters, Ed., *Classical Papers in Genetics,* Engelwood Cliffs, N.J.: Prentice-Hall, 1959.]

Harris, H. *The Principles of Human Biochemical Genetics,* 2nd ed. Amsterdam: North Holland, 1975.

MacIntyre, R. J., and S. J. O'Brien. "Interacting gene-enzyme systems in *Drosophila,*" *Ann. Rev. Genet.* **10:**281–318 (1976).

Wagner, R. P., and H. K. Mitchell. *Genetics and Metabolism,* 2nd ed. New York: Wiley, 1964.

Wright, S. "Color inheritance in mammals," *J. Hered.* **8:**224–235 (1917). [Reprinted in J. A. Peters, Ed., *Classical Papers in Genetics,* Englewood-Cliffs, N.J.: Prentice-Hall, 1959.]

Ziegler, I. "Genetic aspects of ommochrome and pterin pigments," *Adv. Genetics* **10:**349–403 (1961).

Genetic Dissection of Complex Pathways

Bishop, R. J., and W. B. Wood. "Genetic analysis of T4 tail fiber assembly. I. A gene 37 mutation that allows bypass of gene 38 function," *Virology* **72:**244–254 (1976).

Childs, B., J. M. Finucci, M. S. Preston, and A. E. Pulver. "Human behavioral genetics," *Adv. Hum. Gen.* **7:**57–97 (1976).

Ehrman, L. *The Genetics of Behavior.* Stamford, Conn.: Sinauer, 1976.

Hartwell, L. H., R. K. Mortimer, J. Culotti, and M. Culotti. "Genetic control of the cell division cycle in yeast: V. Genetic analysis of *cdc* mutants," *Genetics* **74:**267–286 (1973).

Hotta, Y., and S. Benzer. "Abnormal electroretinograms in visual mutants of *Drosophila*," *Nature* **222:**354–356 (1969).

*King, J., and U. K. Laemmli. "Polypeptides of the tail fibers of bacteriophage T4," *J. Mol. Biol.* **62:**465–477 (1971).

Levine, M. "Phage morphogenesis," *Ann. Rev. Genetics* **3:**323–342 (1969).

Levine, R. P. "The genetic dissection of photosynthesis," *Science* **162:**768–771 (1968).

McClearn, G. E. "Behavioral genetics," *Ann. Rev. Genetics* **4:**437–468 (1970).

A. R. Kaplan, Ed., *Genetic Factors in "Schizophrenia"* Springfield, Ill.: Charles C Thomas, 1972.

Paulson, J. R., S. Lazaroff, and U. K. Laemmli. "Head length determination in bacteriophage T4: The role of core protein P22," *J. Mol. Biol.* **103:**155–174 (1976).

Quantitative Genetics

Brewbaker, J. L. *Agricultural Genetics.* Englewood-Cliffs, N.J.: Prentice-Hall, 1964.

*Davenport, C. B. "Heredity of skin color in Negro-White crosses," Carnegie Institution, Publication No. 554, Washington, D.C.

East, E. M. "A Mendelian interpretation of variation that is apparently continuous," *Am. Nat.* **44:**65–82 (1910). [Reprinted in L. Levine, *Papers on Genetics.* St. Louis: Mosby, 1971.]

Falconer, D. S. "Introduction to Quantitative Genetics," Edinburgh: Oliver and Boyd, 1960.

Feldman, M. W., and R. C. Lewontin. "The heritability hang-up," *Science* **190:**1163–1168 (1975).

Goldstein, A. *Biostatistics, An Introductory Text.* New York: Macmillan, 1964.

*Harrison, G. A., and J. J. T. Owen. "Studies on the inheritance of human skin color," *Ann. Hum. Genetics* **28:**27–37 (1964). [Reprinted in L. Levine, *Papers on Genetics.* St. Louis: Mosby, 1971.]

Lerner, I. M. *The Genetic Basis of Selection.* New York: Wiley, 1958.

Lerner, I. M. *Heredity, Evolution, and Society.* San Francisco: Freeman, 1968.

Mather, K. "Polygenic inheritance and natural selection," *Biol. Rev.* **18:**32–64 (1943).

Mather, K. *Statistical Analysis in Biology.* London: University Paperbacks, 1965.

Mather, K., and J. L. Jinks. *Biometrical Genetics. The Study of Continuous Variation.* Ithaca, N.Y.: Cornell U. Press, 1971.

Thompson, J. N. "Quantitative variation and gene number," *Nature* **258:**665–668 (1975).

(a) (b)

(a) Bruce Ames (University of California, Berkeley) worked out much of the genetics of the histidine operon in *Salmonella* and developed the "Ames test" for mutagenic agents (Section 6.20). (b) I. Michael Lerner established much of the theoretical and experimental framework for quantitative genetics.

Questions and Problems

1. Spore color in *Neurospora* is autonomous, meaning that it is determined by the genotype of the haploid spore nucleus. The wild-type genotype gives a *black* spore color. Two mutations singly result in *tan* or *gray* spores, but when the two mutations are carried in the same spore it is *colorless*. Tetrad analysis of a cross yields the following results:

	gray	*black*	*black*
	gray	*black*	*gray*
	tan	*colorless*	*tan*
	tan	*colorless*	*colorless*
Percentage	93	3	4

(a)   Classify each type of tetrad as PD, NPD, or T.
(b)   What are the genotypes of the parental strains?
(c)   Are the *gray* and *tan* loci linked? If so, how many map units separate them?

2. As shown in Figure 16-11, mutant strains of *Neurospora* can be blocked at several positions in the pathways leading to methionine and threonine synthesis. Locate the sites where the following strains carry genetic lesions based on their growth requirements (+ means growth, − no growth).

Mutant Strain	Aspartic Acid	Homoserine	Cysteine	Threonine	Cystathionine	Methionine
A	−	−	−	−	+	+
B	−	+	+	+	+	+
C	−	−	−	−	−	+
D	−	−	−	+	−	−

3. Two new arginine-requiring strains of *C. reinhardi*, *arg*-8 and *arg*-9, are crossed. The resulting tetrads are found by replica plating to be of three types: Type 1: all four tetrad products require arginine for growth; Type II: 3/4 of the products require arginine for growth; Type III: 1/2 of the products require arginine for growth. Type I tetrads represent 80 percent of the total, Type II represent 18 percent, and Type III represent 2 percent.
(a)   Are the mutations allelic or in different genes? Explain.
(b)   If in different genes, is there linkage between them? By how many map units?
(c)   Give the genotypes of the various tetrad products.

4. Three *trp* mutant strains of *E. coli* (*trp*-1 through *trp*-3) are isolated, all requiring tryptophan for growth. Each is tested to determine whether it will grow in the presence of tryptophan precursors, with the following results (see Figure 16-2 for a key to abbreviations):

Mutant Strain	Chorismic Acid	Anthranilic Acid	PRA	CDRP	InGP	Tryptophan
trp-1	—	—	—	+	+	+
trp-2	—	—	—	—	—	+
trp-3	—	—	+	+	+	+

(a)   In which gene (or genes) of the *trp* operon (Figure 16-2) could each mutation lie? Explain ambiguous cases.

(b)   Which is the most likely candidate as a polar mutation? Where would such a mutation be most likely to map in the *trp* operon?

5. A *vermilion* female is crossed to a *scarlet* male and the resulting F_1 is inbred. Give the phenotypes and genotypes of the F_1 and F_2 flies, each classified according to sex.

6. A *scarlet* female is crossed to a *vermilion* male and the resulting F_1 is inbred. Give the phenotypes and genotypes of the F_1 and F_2 flies, each classified according to sex.

7. A hen with a *rose*-shaped comb is crossed with a rooster having a *pea*-shaped comb. The F_1 all have *walnut*-shaped combs. When the F_1 is inbred, 9/16 of the F_2 have *walnut* combs, 3/16 have *pea* combs, 3/16 have *rose* combs, ad 1/16 have *single*-shaped combs.

(a)   How would you explain these results?

(b)   What progeny would you expect from a cross *rose* $\times$ *pea*? What would be the phenotypes in the F_2 when this F_1 was inbred?

8. An *agouti* mouse is crossed with a *white* mouse. Their progeny are 3/8 *agouti*, 1/2 *white*, and 1/8 *black*. What are the genotypes of the parents and of the progeny?

9. In corn, the three dominant genes *A*, *C*, and *R* are all necessary to produce a colored seed. Plants with the genotype $\dfrac{A \;\; C \;\; R}{\rule{0.5em}{0.4pt} \;\; \rule{0.5em}{0.4pt} \;\; \rule{0.5em}{0.4pt}}$ have colored seeds; all others are colorless. A colored plant is subjected to crosses with two tester plants with the following results:

(a)   A cross with $\dfrac{a \; c \; R}{a \; c \; R}$ plants yields 50 percent colored seeds.

(b)   A cross with $\dfrac{a \; c \; r}{a \; c \; r}$ plants yields 25 percent colored seeds.

What is the plant's genotype?

10. Two different strains of summer squash produce a *spherical*-shaped fruit, and both breed true. When they are crossed, the F_1 all have a *disc*-shaped fruit. Inbreeding the F_1 yields the following ratio among the F_2: 9/16 *disc*, 6/16 *spherical*, and 1/16 *long*. Explain these results.

11. *Barred* feathers in the chicken are produced in the presence of the dominant sex-linked gene *B* (*Note:* Review the chromosomal basis of sex determination in birds in Section 4.13); a chicken hemizygous or homozygous for its *b* allele has *non-barred* feathers. Chickens also possess the non-sex-linked *C/c* alleles for pigmentation: *c/c* chickens have solid white feathers regardless of the

presence of the B allele whereas birds heterozygous or homozygous for C have colored feathers.

Predict the outcome of the following cross in terms of sex, genotype, and phenotype of progeny chicks:

$$\frac{B\;C}{b\;c}\,\delta \;\times\; \frac{b\;C}{W\;c}\,\female$$

12. The plant *Bursa* (shepherd's purse) can have seed capsules in two shapes: *triangular* and *oval*. The F_1 progeny of a *triangular* $\times$ *oval* cross is all *triangular;* when the F_1 is inbred the F_2 are in the ratio 15 *triangular*: 1 *oval*. How might this result be explained?

13. Four adjacent genes code for enzymes required in the tryptophan biosynthetic pathway in an organism. Complementation tests were made by crossing the haploid *trp⁻* strains 1 through 5 in different pairwise combinations, and the resulting diploid cells were tested for their ability to grow on minimal medium. Results were:

1	0				
2	+	0			
3	+	+	0		
4	+	+	+	0	
5	0	0	0	0	0
	1	2	3	4	5

+ = growth on minimal
0 = no growth on minimal

Another haploid *trp⁻* strain (6) was discovered which could complement strains 1, 3, and 4. When this new auxotroph was crossed with strain 5, a very few cells of the resulting diploid strain were found to have the ability to grow on minimal medium. These rare cells probably arose as a result of one of the following:

(a) complementation.

(b) reversion to wild-type of the strain 6 mutation in the diploid cells.

(c) rare events in which a triple recombination exchange occurred between 3 different loci along the chromosome carrying the mutant genes.

(d) a four-strand double crossover between the mutant genes in strains 5 and 6.

Explain your choice of answer.

14. Four mutant strains of *E. coli* which required serotonin for growth were tested for their growth responses to substances thought to be intermediates in the pathway of serotonin synthesis. The following results were obtained (+ = growth when substance was added to minimal medium; − = no

growth). What is the order of the intermediate substances in the serotonin pathway and the position of the mutant blocks?

Mutant	Serine	Tryptophan	Anthranilic acid	Serotonin
a	−	+	−	+
b	−	−	−	+
c	+	+	−	+
d	−	+	+	+

15. Three *arg* mutant strains of *E. coli* (see Figure 16-9) are streaked close together (but not touching) on a petri dish of minimal medium in the following fashion:

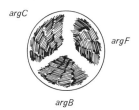

argC

argF

argB

Each mutation completely blocks the activity of its respective enzyme (Figure 16-9). Assuming that cross-feeding can take place, what growth response might you expect to see several days after streaking?
(a)   No growth of any of the strains on the dish.
(b)   Cells of *arg C* nearest those of *arg F* growing; and cells of *arg B* nearest those of *arg C* growing.
(c)   Cells of *arg F* nearest those of *arg C* growing.
(d)   Cells of *arg C* nearest those of *arg B* growing.
Explain your choice of answer(s).

16. An essential multimeric enzyme in rabbits is composed of 2 different protein subunits. Gene locus 1 (allele A dominant to allele a) codes for one subunit, while gene locus 2 codes for the other (allele B dominant to b). The homozygous recessive condition at locus 2 (b/b) results in the inability to produce a functional subunit, and is lethal at a very early embryonic stage. Homozygous recessives for locus 1 (a/a) produce a marginally effective enzyme subunit which leads to metabolic disturbances invariably resulting in dwarfism. All other genotypes are phenotypically normal.

If rabbits heterozygous at both loci are crossed, what is the proportion of *viable* progeny which would have a *normal* phenotype? Explain your answer.

17. A scientist was studying mutants of *E. coli* and isolated a number which required compound G to grow. A number of compounds in the pathway to G were known and each was tested for the ability to support the growth of the

mutant. A+ indicates that growth occured when the gene product was added.

Gene Products

Mutant	A	B	C	D	G	
1	−	−	−	+	+	
2	+	+	−	+	+	
3	−	−	−	−	+	
4	−	+	+	+	+	
5	−	+	−	+	+	(will grow if both A + C are added together)

(a)   What is the pathway?

(b)   At which step is each mutant blocked?

18. Two different true-breeding strains of sweet peas having white flowers were crossed. The flower color of the F_1 was purple. When the F_1 was inbred 9/16 of the progeny had purple flowers and 7/16 had white flowers. Explain these results in terms of epitastic relationships between two gene loci concerned with the biosynthesis of the purple pigment anthocyanin.

CHAPTER
17

Control of Gene Expression in Bacteria and Bacteriophages

INTRODUCTION

In thinking of an animal or a higher plant, it is perhaps not difficult to conceive of its growth and differentiation as an intricate chain of events in which some genes function at all times, others are switched on or off as the cellular or extracellular environment changes, and still others function only at specific places and times. Such a concept may be less obvious for an organism like *E. coli,* whose chromosome appears to be in a continuous state of replication, whose life cycle can take as little as 20 minutes, and whose capacity to differentiate in the eukaryotic sense appears negligible. The concept might also, at first sight, seem inapplicable to a bacteriophage whose genes are limited in number, whose life cycle is very short, and whose phenotypic capacities also appear to be limited.

In fact, however, the genes of bacteria and of their phages are under exquisitely sophisticated kinds of controls, and our deep understanding of these controls stands in sharp contrast to our present ignorance of how gene expression is regulated in eukaryotes (Chapter 18). This chapter focuses on the different kinds of control mechanisms that have evolved to regulate gene expression in bacteria and their phages. We begin with some general definitions so as to give an overview of what is involved in gene expression, and then turn to several specific examples that give experimental evidence for our introductory statements.

GENERAL FEATURES OF GENE REGULATION

17.1 CONSTITUTIVE VERSUS REGULATED GENES

Basic distinctions can be made between genes whose expression is regulated and genes whose expression is not. The latter can be called **constitutive genes.** A constitutive gene is expressed continuously, so that its product (for example, an enzyme) is invariably found in the cell at roughly the same concentration, *regardless of growth conditions and regardless of whether the substrate for that enzyme is ever presented to the cell or not.* The expression of a **regulated gene,** on the other hand, is subject to modification by chemical stimuli that appear in and disappear from the gene's environment.

It should be stressed at once that a constitutive gene by no means experiences unregulated expression in the sense, say, that cancerous growth is unregulated. There are many levels at which control is exerted over the expression of *all* genes, including constitutive genes. Three ways in which such control may be exercised are the following:

1. The structure of a promoter may influence the frequency with which transcription of a gene is initiated by a DNA-dependent RNA polymerase; and this feature can then govern how much of the gene's product is present in the cell.

2. The structure of a gene's mRNA transcript might well determine how rapidly the message is digested by cellular ribonucleases after being created.
3. The base sequence associated with the AUG start signal on a transcript might influence how readily its translation is initiated.

A more precise distinction between constitutive and regulated genes, therefore, might be that the controls governing expression of a constitutive gene are invariant, whereas the controls governing regulated gene expression are subject to modification.

There seems to be some logic as to which enzymes of *E. coli* are synthesized constitutively and which are synthesized "on call." The constitutive enzymes are generally those that are needed at all times. Thus the enzymes involved with glucose metabolism—enzymes of glycolysis and the hexose monophosphate pathway—are constitutive, and one can imagine that the metabolism of glucose is sufficiently central to the physiology of *E. coli* that these pathways should be continuously functional. The metabolism of less common sugars such as lactose, arabinose, and galactose, on the other hand, is effected by regulated enzymes, and it is clearly economical to elaborate such enzymes only on those occasions when their substrates are present. Regulated genes also specify the enzymes of most biosynthetic pathways—pathways for the synthesis of amino acids, purines, pyrimidines, and so on—and it again seems logical that these enzymes would be synthesized only as there was a demand for their biosynthetic products.

17.2 GENERAL PROPERTIES OF GENES THAT ARE REGULATED

Two general mechanisms can be imagined for limiting the expression of regulated genes to specific occasions. Control could be exerted at the **transcriptional level,** meaning that a regulated gene would be transcribed only when conditions are appropriate. Alternatively, control could be exerted at a **translational level** so that an mRNA from a regulated gene would not be translated into protein unless, say, certain ribosomal factors, influenced by some feature of the environment, came to associate with the ribosomes. Much research has focused on whether specialized forms of ribosomes, tRNAs, aminoacyl-tRNA synthetases, or other components of the protein-synthesizing apparatus (Chapter 9) are able to recognize only certain types of gene transcripts and thereby exert translational control. Only a few examples of such controls, however, have been found for prokaryotes. In contrast, a wealth of data support the existence of transcriptional controls. Thus we adopt as our current hypothesis the concept that prokaryotic genes are, in general, regulated at a transcriptional level.

If a gene's transcription is to be regulated, we can imagine that there must be some genetic elements associated with the gene that respond to environmental signals and either allow or disallow mRNA to be synthesized from that gene. Two major classes of such **controlling elements** have been identified in bacteria and in DNA-containing phages. Both are portions of the chromosome and both lie very close to the gene(s) they govern. One is the **promoter** and the other the **operator.**

A promoter, as we saw in Section 8.7, represents a sequence of bases that is recognized by a DNA-dependent RNA polymerase, with promoter-polymerase binding initiating transcription of the neighboring gene. In many senses the promoter of a regulated gene is likely to be similar to the promoter of a constitutive gene. Presumably, the major difference between the two types of promoters is that promoters of constitutive genes have relatively invariant initiation properties, whereas promoters of regulated genes are more less active in promoting transcription of contiguous genes, depending on the presence or absence of specific proteins known as **regulator proteins.** Examples of regulator proteins and their effects on transcription will be presented shortly.

The second main class of controlling element is the operator. An operator also represents a sequence of bases that interacts with a protein. The protein is also called a regulator molecule, and operator-regulator interaction will either prevent or promote transcription of the regulated gene, depending on which operator is being considered. The major distinction to be made between an operator and a promoter element is that the operator is not normally the site of attachment of an RNA polymerase. Instead, it ordinarily lies between the promoter and the beginning of a regulated gene. The operator thus comes to influence whether a polymerase is able to progress from its attachment site at the promoter to the gene itself.

Let us pause here to clarify some terminology. The operator was first postulated by F. Jacob and J. Monod in 1961 as a controlling element for the genes involved in lactose utilization. These genes constitute the *lac* operon, an **operon** (Section 8.2) being a group of related genes that is transcribed into a single molecule of mRNA. Therefore the initial, and very simple, definition of an operator was that it controlled the transcription of an operon. More recently, however, single genes have been found that are regulated, and these possess operator loci in the same way that operons do. These single genes are occasionally called operons, but it seems preferable to conserve the term operon for *groups* of genes that are co-transcribed. Therefore in this text an operator refers to any locus that interacts with a regulator molecule and does not serve as a promoter, with the understanding that an operator locus may lie next to a single regulated gene or next to a regulated operon.

17.3 GENERAL PROPERTIES OF MOLECULES INVOLVED IN REGULATION

The preceding section has focused on a regulated gene. We now describe the features of the diffusible molecules that do the regulating. As already noted, the regulator proteins (also called regulators, affector proteins or affectors) are key molecules in any regulation process. They have the potential to bind specifically to controlling elements, the specificity residing in their three-dimensional configurations.

If gene transcription is to be regulated by the binding and dissociation of regulator proteins to or from controlling elements, then there must obviously exist some additional class of molecules which governs whether such binding in fact takes place. These additional molecules are known as **effector molecules.** An effector molecule is classically a small molecule—a sugar, an amino acid, or a nucleotide—that can bind to a regulator protein and thereby change its ability to interact with a controlling element. Perhaps the most important concept to come from the work of Jacob and Monod in the late 1950s was the idea that a regulator protein could bind to an operator and inhibit operon expression as long as the effector was not present. Once the effector molecule appeared in the intracellular environment, however, it would bind to the regulator and thereby "pull" the regulator off the operator, thus allowing operon expression to proceed. These relationships are diagrammed in Figure 17-1.

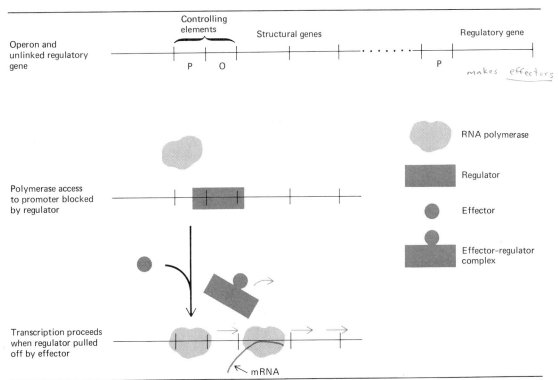

Figure 17-1 Relationships between promoter (p), operator (o), and two regulated structural genes in an operon, showing RNA polymerase attachment to promoter, a regulator protein binding to the operator, and, in the presence of effector, the effector-regulator binding which removes the regulator from the operon. This diagram presents the essence of the 1961 Jacob-Monod model for *lac* operon regulation although some of the molecular details have since been added.

Regulator molecules, being proteins, must themselves be specified by genes; these are called **regulatory genes.** The reader should fix clearly in mind the distinction between a gene that is regulated and a regulatory gene, since this distinction is basic to an understanding of this chapter.

Four key elements thus emerge as essential to genetic regulation in bacteria and phages: (1) the regulated structural gene (or gene cluster) itself; (2) the controlling elements (including the promoter and the operator); (3) the regulator protein (and its structural gene, the regulatory gene); and (4) the effector. All are diagrammed in Figure 17-1. As we go on to consider actual examples of genetic control it will become apparent that highly complex kinds of gene regulation can be understood in terms of the interplay between these four kinds of molecules or regions of DNA.

17.4   INDUCIBLE VS. REPRESSIBLE OPERONS

Before we proceed with specific examples of gene regulation, the final general concept that is important to grasp is the distinction between **inducible** and **repressible** operons.

Operons such as *lac* (*lac*tose), *gal* (*gal*actose) and *ara* (*ara*binose) are said to be inducible: full levels of their transcription do not occur unless their sugar effectors—lactose, galactose, or arabinose—are presented. The effectors are therefore often referred to as **inducers.** Operons such as *trp* (*tryp*tophan), *his* (*his*tidine) and *arg* (*arg*inine), on the other hand, are said to be repressible: full levels of transcription occur until such time as their amino acid effectors—tryptophan, histidine, or arginine—reach a critical concentration, at which point transcription is inhibited. Effectors of repressible operons are often referred to as **corepressors.** In short, then, inducible operons are "off" unless the effector is present, whereas repressible operons are "on" until the effector is present. Both kinds of control can clearly accomplish the same goal, but by reciprocal means.

As a rule of thumb, inducible operons in bacteria typically carry genes for "gratuitous" enzymes (that is, enzymes not required for survival), and the enzymes are usually involved in the catabolism of a substrate not always present in the bacterial medium. Thus it "makes sense" for the enzymes of arabinose utilization to be inducible by arabinose: arabinose may be only infrequently taken up by a bacterial cell, and it is energetically most efficient to produce enzymes for arabinose degradation only "on call." Repressible operons, in contrast, typically carry genes for "necessary" enzymes that participate in biosynthetic processes. Thus histidine is in constant demand by a growing cell, and a steady output of *his*-encoded gene products is generally occurring; such a cell is said to be **derepressed.** Should free histidine levels become too high, however (as, for example, when rates of protein synthesis abate) the expression of these genes is repressed and histidine biosynthesis soon abates as well.

With this much background we can turn to specific descriptions of two representative bacterial operons, the inducible *lac* operon and the repressible *trp* operon of *E. coli,* followed by a description of several phage operons. Additional examples of bacterial and phage operons are included in the problems at the end of the chapter.

REGULATION OF LACTOSE UTILIZATION

17.5 GENERAL FEATURES OF *lac* REGULATION

As described in detail in Section 16.1, lactose utilization in *E. coli* involves the products of the three structural genes that constitute the *lac* operon; these products are **β-galactosidase,** a lactose-splitting enzyme specified by gene *z;* **M protein,** a molecule necessary for lactose uptake and specified by gene *y;* and **transacetylase,** a poorly understood enzyme specified by gene *a.* A map of the *lac* operon is found in Figure 16-5 and should be reviewed at this time.

Analysis of *lac* operon regulation has reached a high degree of sophistication, in part because, as we have seen in Section 11.19, pure *lac* DNA can be obtained from transducing phages. The entire nucleotide sequence of the *lac* controlling region has more recently been determined, and the relevant regulator molecules have been isolated and analyzed. The molecular basis for *lac* regulation can therefore be described in some detail. Briefly stated, *lac* regulation exhibits the following features.

1. The operon is under **negative control,** meaning that when it interacts with a regulator protein known as the *lac* repressor, operon transcription is inhibited. The repressor is specified by the *i* regulatory gene, which maps just outside the operon (Figure 16-5). p 622
2. The operon is also under **positive control,** meaning that when it interacts with a regulator known as **CAP,** operon transcription is enhanced. The regulatory gene for CAP is known as *crp* and maps elsewhere in the *E. coli* genome.
3. The *lac* repressor binds to the *lac* operator, thereby preventing transcription, unless lactose or a lactose analogue is presented to the cell; in the presence of either substance, the repressor leaves the operator, permitting *lac* transcription.
4. The CAP protein binds to a controlling element near the *lac* promoter, thereby enhancing transcription, unless glucose is presented to the cell; the addition of glucose causes CAP to leave its controlling element and *lac* transcription drops to very low levels.

In the sections that follow we present the genetic and molecular evidence for this picture of *lac* regulation.

17.6   ALLOLACTOSE AS THE TRUE EFFECTOR

When wild-type *E. coli* cells are grown on a carbon source other than lactose, the intracellular levels of β-galactosidase, M protein, and transacetylase are only about 1/1000 of the levels these substances attain when lactose is present (it is important, however, to realize that low levels of all three of these proteins are present at all times). Introduction of lactose (Figure 17-2a) can be shown to stimulate the transcription of the *lac* operon almost immediately, and *de novo* synthesis of all three proteins commences within 3–4 minutes. Early observers of such phenomena therefore concluded that lactose was able to induce the *lac* operon, and lactose was termed an inducer.

Recently, however, it has been shown that once lactose enters the cytoplasm, it is acted upon by one of the few molecules of β-galactosidase that manage to be synthesized in the uninduced cell. The enzyme transforms lactose into **allolactose** (Figure 17-2b), and it is this allolactose that acts as the true effector molecule, interacting directly with the repressor protein to allow *lac* transcription.

The fact that allolactose and not lactose is the natural effector of *lac* transcription explains why mutations in the z and y genes of the *lac* operon cause cells to be uninducible by lactose. A z^- cell cannot produce a functional β-galactosidase and therefore cannot transform entering lactose into allolactose, while a y^- cell cannot produce even a few molecules of functional M protein and therefore cannot transport lactose molecules into the cytoplasm for conversion to allolactose. To minimize these effects, it is customary in experimental studies to induce the *lac* operon with the lactose

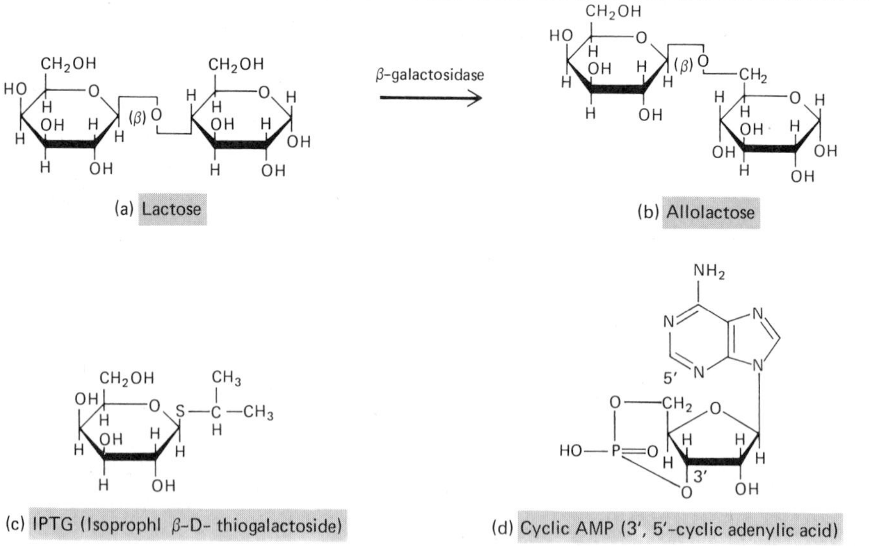

(a) Lactose

β-galactosidase

(b) Allolactose

(c) IPTG (Isoprophl β-D- thiogalactoside)

(d) Cyclic AMP (3', 5'-cyclic adenylic acid)

Figure 17-2   Small molecules active in *lac* operon regulation.

analogue isopropylthiogalactoside (IPTG) (Figure 17-2c). This molecule can be taken up by y^- cells; moreover, it does not require β-galactosidase modifications since it can be shown to interact directly with the *lac* repressor.

17.7 THE *lac* REPRESSOR AND THE *i* GENE

The "target" of either IPTG or allolactose in the *E. coli* cell is the *lac* repressor. The repressor, in its active form, is a tetramer composed of 4 identical polypeptides, each containing 360 amino acids. The amino-acid sequence of the repressor polypeptide is known, as is the complete nucleotide sequence of its structural gene, the *i* gene.

The *i*-gene Promoter   The *i* gene maps close to, but outside of, the *lac* operon proper (Figure 16-5), and its transcription is controlled by an independent promoter. Figure 17-3 shows the nucleotide sequence of this promoter, the mRNA transcribed from it, and the N-terminus of the repressor polypeptide. Shown also is the position at which a CG nucleotide pair is replaced by a TA pair, a change effected by a mutation known as i^Q. The Q stands for quantity, and whereas wild-type cells contain minute concentrations of *lac* repressor (perhaps 10 molecules per cell), i^Q cells produce ten times more. Thus the single i^Q transition clearly results in an **"up" promoter mutation** which allows more frequent initiations of *i*-gene transcription by the RNA polymerase. Some seventy *i*-gene promoter mutants have been isolated, the best of which makes a hundred times more repressor than wild type. Such strains have been invaluable for isolating enough repressor for experimental analysis.

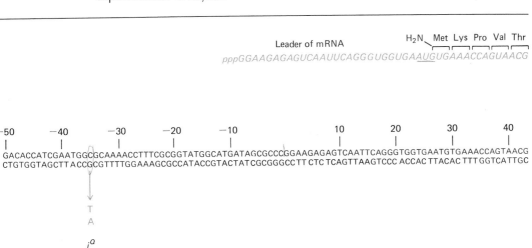

Figure 17-3 Nucleotide sequence of the *i*-gene promoter region. (Courtesy of Dr. M. Calos.)

Mutations in the *i* Gene   Starting with i^Q strains, B. Müller-Hill and J. Miller have proceded to induce large numbers of mutations in the *i* gene proper. The mutant repressors produced by such strains have been isolated and studied *in vitro,* and the mutations have been ordered into a linear sequence by deletion mapping (Section 10.8), producing the fine-structure map shown in Figure 17-4.

Two major classes of *i* gene mutations have emerged from such analyses. First, a large number of *i* gene mutations result in the synthesis of repressor polypeptides that form tetramers and bind IPTG normally *in vitro* but fail to bind to *lac* DNA. Strains carrying such mutations (which belong to a class known as i^{-d}) behave *in vivo* as one would expect if their *lac* repressor fails to bind to the *lac* operator: they synthesize both β-galactosidase and M protein constitutively, whether or not lactose (or IPTG) is present in the medium. The i^{-d} mutations are all found to cluster within the first 58 base pairs of the *i* gene (Figure 17-4), the region encoding the N-terminal end of the repressor. It is concluded, therefore, that the N-terminus of the *lac* repressor is involved in recognizing and binding to the *lac* operator DNA.

The second major class of *i* mutations is denoted i^{-s} (s standing for superrepressed). The i^{-s} strains have the opposite phenotype from i^{-d} mutants since they fail to synthesize either β-galactosidase or M protein unless very high concentrations of inducer are present. When analyzed *in vitro,* i^{-s} repressor polypeptides form tetramers and bind to *lac* DNA, but exhibit very low affinity for IPTG or other effectors. The i^{-s} mutations map to the central portion of the *i* gene (Figure 17-4), and it is therefore concluded

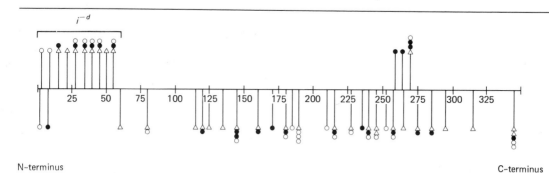

Figure 17-4   Genetic map of the *i* gene, with each vertical line representing missense mutations that fall within a particular deletion that extends into the gene. The circles and triangles represent mutants isolated by different conditions of mutagenesis. All mutants above the line have partial or full IPTG binding activity *in vitro;* mutants below the line have no detectable activity. The IPTG-binding mutants between residues 1 and 53 are all downward, while those between 260 and 270 are not. Distances are given in terms of amino acids of *lac* repressor protein (at the time the map was drawn, this number was thought to be 347). (From J. Miller et. al., in: *Protein-Ligand Interactions,* H. Sund and G. Blante, eds. Berlin: Walter de Gruyter, 1975.)

that the inducer binding site involves "central" amino acids in the *lac* repressor polypeptides.

Behavior of *i*-gene Mutations in Partial Diploids   Interesting and informative results are obtained when various *i*-gene mutations are coupled with the wild-type i^+ gene in partial diploids (Section 11.7). An i^+/i^s partial diploid proves to be fully as non-inducible for *lac* gene expression as is an i^s haploid cell. This i^s dominance is expected: the i^s mutant repressors, being unable to interact with inducer, would presumably come to "monopolize" both the *lac* operator sites in the cell, thereby blocking induction.

Somewhat unexpected is the finding that virtually all mutations of the constitutive i^{-d} variety are also dominant in i^+/i^{-d} partial diploids (hence the appelation i^{-d} for **d**ominant). This effect is explained by proposing that the mutant repressor polypeptides associate with wild-type polypeptides in the partial diploids to form **mixed-hybrid** tetramers which fail to bind properly to the *lac* operator.

17.8   THE *lac* OPERATOR AND OPERATOR MUTATIONS

We can next consider the *lac* operator, the site to which the *lac* repressor binds in such a way as to prohibit transcription of the z gene. The *lac* operator was defined in the original study by Jacob and Monod as the site of o^c (**operator-constitutive**) mutations. These lead to a constitutive synthesis of *lac*-encoded proteins. Unlike the i^{-d} mutations described in the preceding section, however, the o^c mutations were found to map quite far from the *i* gene, and very close to z.

The o^c mutations also differ from i^{-d} mutations in partial diploids, since they are capable of conferring constitutiveness only on the genes located in the same chromosome as themselves. Thus in partial diploids with the genotype $\dfrac{i^+\,o^+\,z^-\,y^+}{F\,i^+\,o^c\,z^+\,y^+}$, both β-galactosidase and M protein are synthesized constitutively; in $\dfrac{i^+\,o^+\,z^-\,y^+}{F\,i^+\,o^c\,z^+\,y^-}$ diploids, inducer is required for M synthesis but not for β-galactosidase production; in $\dfrac{i^+\,o^+\,z^+\,y^-}{F\,i^+\,o^c\,z^-\,y^+}$ cells, inducer is required for β-galactosidase but not for M synthesis; and so forth. In genetic terminology, the *lac* operator is seen to exert its effect on genes in the *cis* position only, a phenomenon known as **cis-dominance.** Such a phenomenon, it should be noted, is not found with i^{-d} mutations. Diploids with the genotype $\dfrac{i^{-d}\,o^+\,z^+\,y^-}{F\,i^+\,o^+\,z^-\,y^+}$, for example, synthesize both β-galactosidase and M protein in a constitutive fashion.

The *cis*-dominance of o^c mutations was correctly interpreted by Jacob and Monod to signify that the operator locus is not a gene specifying a diffusable gene product that can act on all chromosomes of the cell, but instead a controlling element that can only influence the genes directly contiguous to it. This effect is perhaps best appreciated by considering *cis*-dominant o^c mutations that confer only partial constitutiveness: in i^+ $o^c z^+$ cells of this type, intermediate levels of β-galactosidase are synthesized at all times, but full levels are not produced unless inducer is added. Such o^c operators are visualized as binding repressor protein, but less effectively than wild-type operators, hence allowing RNA polymerase molecules an "intermediate" access to the adjacent structural genes in the absence of inducer. Such mutant operators would not be expected to allow an increased access to structural genes that are located in a different chromosome.

Studies *in vitro* have strongly suggested that the o^c mutations mark the *lac* repressor recognition site: for example, *lac* DNA isolated from o^c strains fails to bind repressor tetramers. It was therefore not surprising when nucleotide-sequencing studies localized the o^c mutations within the center of the operator region. Figure 17-5 shows a partial nucleotide sequence of the *lac* operon, and on the left side of the figure is indicated that portion of the *lac* DNA which is protected from DNase digestion by repressor binding. Within the repressor-protected segment, 8 o^c mutations have been identified (Figure 17-5). Therefore, it seems clearly established that the lac repressor recognizes and singles out this sequence of nucleotides (Figure 17-5, bracket) from among the millions present in the *E. coli* cell.

How does the repressor in fact recognize this operator sequence? Since *lac* repressor fails to recognize single-stranded operator DNA, we are left with the following possibilities: 1) repressor interacts with exposed bases in partially or fully denatured operator DNA; 2) repressor interacts with the edges of the bases exposed in the major or the minor grooves of the operator DNA helix (Figure 1-10); or 3) repressor recognizes some feature of the sugar-phosphate backbone of operator DNA that is modified in some fashion as a consequence of the specific base sequence in the interior. Intensive research is presently focused on determining which of these possibilities (or a combination thereof) is correct.

17.9   HOW REPRESSOR BINDING INHIBITS *lac* OPERON TRANSCRIPTION

We noted in the previous section that *lac* repressor appears to recognize a very small group of nucleotide pairs but that, if only by its sheer bulk, it covers a much larger region of the *lac* DNA (Figure 17-5). On the right-hand side of Figure 17-5 is indicated the portion of *lac* DNA that is protected from nuclease digestion by RNA polymerase binding. Considerable overlap is seen to exist between the polymerase binding site and the repressor binding site. Therefore, as might be expected, binding of repressor to the operator and binding of RNA polymerase to the promoter are mutually exclusive events *in vitro* and, one assumes, *in vivo* as well.

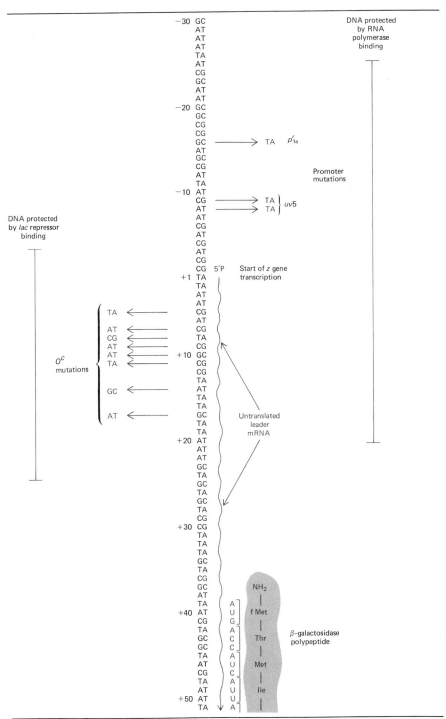

Figure 17-5 Sequence of nucleotides in the *lac* control region and in the initial sector of the *z* gene. (Courtesy W. Gilbert.)

17.10 THE CAP PROTEIN, AND CATABOLITE REPRESSION

For many years, regulation of *lac* operon transcription was thought to reside solely in a competition between repressor and RNA polymerase for access to the same controlling element, much as in our initial general diagram of negative control (Figure 17-1). Recently, however, it has become apparent that the *lac* operon is also under a positive form of control exerted by a dimeric regulator molecule generally known as **CAP** (**C**atabolite **A**ctivator **P**rotein). The term **CRP** (**C**yclic AMP **R**eceptor **P**rotein) is also used. Both names include essential information about this protein's function, as we now describe.

It has been known for some time from the work of B. Magasanik and others that the presence of glucose inhibits the synthesis of enzymes involved with utilization of such sugars as lactose, arabinose, or galactose. This phenomenon is called **catabolite repression** or the **glucose effect** and it occurs in many microorganisms, including *E. coli* and yeast. Recent experiments with *E. coli* have shown that the glucose effect is brought about by two molecular intermediaries, namely, the CAP protein and a small nucleotide known as **cyclic AMP** (**cAMP**) (Figure 17-2d). CAP is sensitive to intracellular levels of cAMP. As long as levels are high, CAP stimulates the expression of such operons as *lac, ara,* and *gal.* Introduction of glucose causes, by some unknown mechanism, a fall in intracellular cAMP levels and, therefore, a reduction in CAP activity. As a result, *lac* (and *ara* and *gal*) transcription drops to low levels. The "purpose" of this control system is presumably to allow the preferential catabolism of glucose whenever it becomes available since, energetically, glucose is the optimum carbon source for microbial growth.

Strains of *E. coli* that carry mutations in *crp* (the structural gene for CAP) and *cya* (the gene specifying adenylate cyclase, the enzyme necessary for cAMP production) are capable of only 2% of the maximal level of *lac* operon expression. Moreover, when isolated *lac* operon DNA is presented with RNA polymerase and ribonucleotides, active transcription will occur only when an extract of crp^+ (but not crp^-) cells is included and cAMP is provided. Such observations indicate that CAP functions as an activator, stimulating *lac* transcription by binding to the *lac* promoter, and that cAMP acts as an effector molecule, interacting with CAP and thereby enhancing its activity.

How does CAP-plus-cAMP enhance the transcription of the *lac* operon? A clue comes from experiments that identify the sector of *lac* DNA protected from nuclease digestion by CAP binding. This sector proves to be far "upstream" from the polymerase-binding sector identified in Figure 17-6, centering instead around the −60 position (Figure 17-6). Several mutations that prevent CAP binding to *lac* DNA *in vitro* are also found to map to the −60 region (Figure 17-6).

Taken together, therefore, the various observations on *lac* transcription suggest that the *lac* promoter itself is a relatively **low-level promoter,**

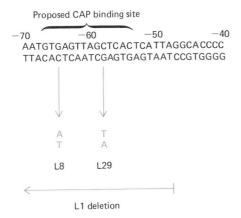

Figure 17-6 Nucleotide sequence in the proposed CAP protein binding site. Numbers extend upstream from those in Figure 17-5. (From W. Gilbert, in *RNA Polymerase*, R. Losick and M. Chamberlin, eds. Cold Spring Harbor, N.Y.: Cold Spring Harbor Laboratories, 1976, p. 199.)

meaning that even when repressor is "pulled off" the *lac* DNA, RNA polymerases initiate transcription only relatively infrequently. When, on the other hand, CAP interacts with *lac* DNA at its contiguous site, initiation frequencies are greatly enhanced. Figure 17-7 suggests one possible way that CAP binding might stimulate CAP transcription: CAP may interact directly with RNA polymerase as well as with DNA, thereby increasing polymerase binding to its promoter.

17.11 OVERLAP BETWEEN THE *z* GENE AND THE OPERATOR

Before leaving the *lac* operon we should consider Figure 17-5 one more time and note an interesting feature of its organization, namely, that *lac* operator DNA serves a double role. Nucleotides 1–21 of the operator interact with repressor in the uninduced cell and then, when the cell is induced and repressor leaves the operator, this same sequence is transcribed into the non-translated "leader sequence" of *lac* operon mRNA. Operator sequences, it should be noted, need not be transcribed, as we shall see when we consider the *trp* operon.

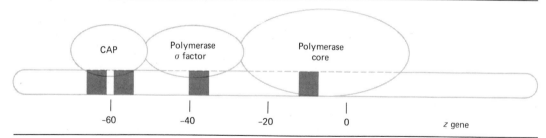

Figure 17-7   A hypothetical arrangement of the CAP factor and the RNA polymerase subunits on the *lac* promoter. If all 3 proteins are laid out along DNA, assuming that each protein has a 2:1 axial ratio, then they can contact each other as well as the DNA recognition regions (gray boxes). (From W. Gilbert, in *RNA Polymerase*, R. Losick and M. Chamberlain, eds. Cold Spring Harbor, N.Y.: Cold Spring Harbor Laboratories, 1976, p. 199.)

REGULATION OF TRYPTOPHAN BIOSYNTHESIS

17.12   GENERAL FEATURES OF TRYPTOPHAN REGULATION

The biosynthesis of each amino acid in a bacterial cell is performed by a series of enzymes. The tryptophan pathway, presented in Figure 16-2, involves five distinct enzymatic activities, all of which are specified by a single operon called *trp*. Regulation of these enzymes in *E. coli* proves to be carried out at three distinct levels; these are described in the following sections. Since similar mechanisms are believed to govern many other biosynthetic pathways, tryptophan regulation in *E. coli* can serve as our model for biosynthetic control in bacteria.

17.13   FEEDBACK INHIBITION

The first level of control for tryptophan biosynthesis, known as **feedback inhibition,** does not involve genes and is therefore not of direct relevance to this text. It is, however, an important form of control for most biosynthetic pathways and must be mentioned for completeness. During feedback regulation a product of a biosynthetic pathway, usually the end product, inhibits the activity of an early enzyme in the pathway. Specifically, in the presence of increasingly high concentrations of tryptophan, the enzyme anthranilate synthetase (Figure 16-2, ASase) is found to increasingly reduce its affinity for its two substrates, glutamine and chorismate. Such a control system can clearly function, at the phenotypic level, to assure relatively constant rates of tryptophan biosynthesis.

17.14   OPERATOR-REPRESSOR REGULATION OF THE *trp* OPERON

Regulation of tryptophan biosynthesis at the gene level occurs at two separate sites in the *trp* operon, the first of these being the trp operator locus o (Figure 16-2). As with the *lac* operon, the sequence of nucleotides in the *trp* operator is recognized by a repressor protein. The repressor is encoded by the *trpR* gene, a gene that lies outside the operon and, in fact, maps far away from the operon (in contrast to the contiguous *i* gene of the *lac* operon). While the detailed properties of the *trp* repressor are not yet established, it appears to be a multimer with a molecular weight of about 60,000.

The *trp* operon is said to be repressible, a term we defined earlier in this chapter: transcription of its structural genes proceeds as long as tryptophan concentrations are low and falls when tryptophan concentrations rise. This effect is due, in part, to the fact that the *trp* repressor alone is incapable of binding to the *trp* operator. Binding can occur only when tryptophan, acting as an effector or "corepressor," associates to form a repressor-tryptophan complex. This complex then proceeds to recognize the operator and inhibit *trp* transcription in wild-type strains. In *cis*-dominant o^c mutants or in *trpR*⁻ mutants, binding fails and transcription becomes constitutive. Studies *in vitro* have shown that the bound repressor-tryptophan complex excludes RNA polymerase from its promoter binding site; it has also been shown that RNA polymerase, once bound to the *trp* operon, cannot be inhibited by subsequent addition of repressor-tryptophan. Therefore, the two proteins appear to compete for binding to the same sector of the genome, much as we saw for *lac*.

The *trp* operon has been transduced to both λ and φ80 phages (Section 11.13), so that it has been possible to isolate and study *trp* DNA and *trp* mRNA *in vitro*. Such studies, most notably those of G. Zubay, C. Yanofsky, and their coworkers, have generated the information summarized in Figure 17-8. The operator is contained in the region to the left of the purine #1 (A) that initiates *trp* mRNA transcription, and it appears not to be transcribed. The hyphens and bars drawn in Figure 17-8 indicate two regions of **nucleotide symmetry** in the *trp* operator: if, for example, one starts at the dot in Figure 17-8 and reads to the right on the upper strand, the sequence encountered is the same as the sequence found on the opposite strand reading to the left of the dot. Sequences bearing two-fold symmetry are also

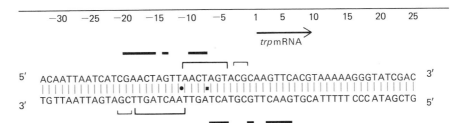

Figure 17-8 Nucleotide sequence in the proposed *trp* operator and mRNA initiation region. The initiation site for *trp* mRNA synthesis is numbered position 1. The centers of the hyphenated symmetries are indicated by a dot and a square between the two DNA strands. The nucleotide pairs involved in each symmetry are denoted by the presence of similar bars above and below the sequence. (From Bennett, C. N., M. E. Schweingruber, K. D. Brown, C. Squires, and C. Yanofsky, *Proc. Natl. Acad. Sci. U.S.* **73**:2351, 1976.)

present in the *lac* operator (Figure 17-5), in the CAP binding site (Figure 17-6), and in a number of other control regions that have been sequenced. It is believed, therefore, that regulator molecules may recognize such symmetrical sequences and, in cases where the regulators are dimers or tetramers, may make symmetric contacts with the controlling-element DNA.

17.15   ATTENUATOR REGULATION

The *trp* operon, and repressible operons generally, do not respond to catabolite repression and appear to have no regulation at the promoter level comparable to that exerted by CAP in the *lac* operon. Transcription of *trp* is, on the other hand, affected at the level of a new class of controlling element known as the **attenuator region.**

The existence of the attenuator was first suspected from the phenotype of *E. coli* strains carrying deletions which covered neither the operator region nor the *trpE* structural gene, yet led to a significant and *cis*-dominant increase in the expression of the *trp* operon. Subsequent analysis of the 5′ end of *trp* mRNA revealed that the sector of the operon covered by the deletions is normally transcribed into a long (160-nucleotide) "leader" sequence (Section 8.6) preceding the AUG codon that initiates translation of the operon; this is shown in Figure 17-9. By comparing the effects of various deletions on *trp* transcription, C. Yanofsky and his collaborators

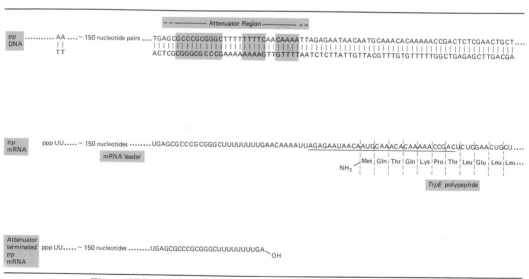

Figure 17-9   Nucleotide sequence of the attenuator region of the *trp* operon. (From Bertrand, K., L. Korn, F. Lee, T. Platt, C. L. Squires, C. Squires, and C. Yanofsky, *Science* **189**:22, 1975. Copyright 1975 by the American Association for the Advancement of Science.)

were able to pinpoint a new controlling-element region in a sequence of 30-60 nucleotides lying close to *trpE* (Figure 17-9). They were further able to show that when wild-type *trp* DNA is transcribed *in vitro,* the RNA polymerase usually "falls off" the template at this site, resulting in a short RNA transcript of the first 130 nucleotides of the leader sequence (Figure 17-9). Finally, they were struck by the fact that the nucleotide sequence of this region contains not only considerable twofold symmetry but also a block of G-C pairs followed by a block of AT pairs (Figure 17-9). Since similar sequences are found at the ends of genes and are believed to signal the termination of transcription (Section 8.8), Yanofsky and his colleagues were led to propose that the attenuator represents a transcription-termination signal that resides at the end of the "leader region" in *trp* operon DNA. In other words, Yanofsky and his coworkers envision that once transcription of the *trp* operon has been initiated at the promoter, it may or may not terminate at the attenuator, depending on what regulator molecules are associated with the attenuator region. Termination at the attenuator would, of course, prevent *trp* operon expression.

What might be accomplished by this sort of control? The Yanofsky group proposes that the repressor-operator region and the attenuator region effectively back each other up: as tryptophan levels drop, a two-stage response is launched during which repression at the operator is first lifted, followed by a relaxation of transcriptional attenuation. This would perhaps enable the cell to regulate the expression of *trp* and production of mRNA over a greater range of tryptophan concentrations than possible by the repression system alone.

17.16 REGULATION OF GENE EXPRESSION DURING PHAGE T7 INFECTION

The bacteriophages have developed numerous strategies for subverting host-directed DNA and protein synthesis into phage-directed DNA and protein synthesis. All of these strategies appear at present to be mediated by the following kinds of macromolecules: 1) novel RNA polymerases; 2) novel RNA polymerase subunits; 3) repressor proteins, and 4) activator proteins. In this section we describe the lytic cycle of T7 which utilizes the first strategy; in subsequent sections we describe the phages SPO1 and λ, which utilize the three subsequent strategies.

The infection cycle of phage T7 was described in Figure 5-12, and a map of the T7 chromosome is found in Figure 17-10. Control over transcription of the T7 genome proves to be elegantly simple. At the extreme "left end" of the T7 chromosome (the "top end" in Figure 17-10) lie at least three sequences recognized as promoter sites by *E. coli* RNA polymerases. When a T7 chromosome enters an *E. coli* host cell, therefore, a host RNA polymerase proceeds to initiate transcription from one of these sites and produces

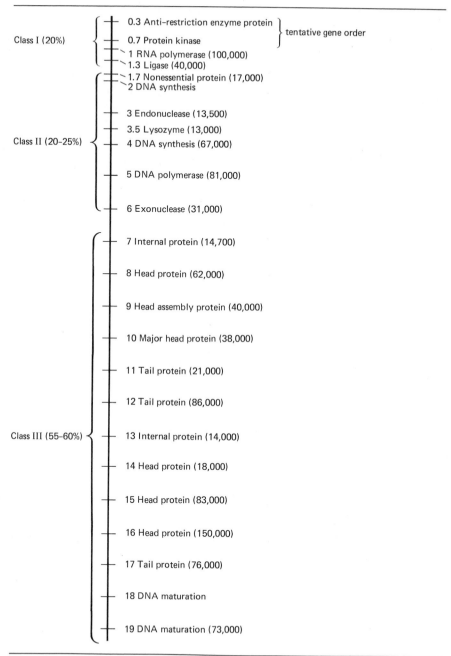

Figure 17-10 Genetic map of T7 showing essential genes 1 to 19 and unessential genes mapping in between. Classes I to III represent the three transcriptional gene classes described in the text, and percentages indicate the proportion of the chromosome occupied by each class. Numbers in parentheses give the molecular weights of the various gene products. (After F. W. Studier, "Bacteriophage T7," *Science* **176**:367–376, 1972.)

a single long transcript covering genes 0.3, 0.7, 1, 1.1, and 1.3, which are known as Class I genes (Figure 17-10). This transcript is next cleaved by host RNase III molecules into 5 mRNA molecules, each encoding one of the five gene products, and the mRNAs are translated on host ribosomes. The gene 0.3 product acts to overcome the restriction/modification system of the *E. coli* host (Section 5.9) and is the first to act. Translation of a functional gene-1 product is, however, the pivotal event, for gene 1 encodes a new T7 RNA polymerase which alone is able to recognize the promoters governing the remaining genes in the T7 chromosome.

The appearance of the T7 polymerase allows the remaining 80 percent of the genome to be expressed; this remainder of the genome encodes the class II gene products, which are involved in phage DNA replication (along with the gene 1.3 product, a DNA ligase), and the class III gene products involved with phage morphogenesis, DNA packaging, and host lysis (Figure 17-10). At the same time, the gene 0.7 product, a protein kinase, acts to shut off expression of the T7 **"early"** genes (Class I genes) as well as host mRNA synthesis, presumably via some effect on host RNA polymerases. The result, therefore, is that nucleotide precursors for RNA synthesis are almost entirely diverted toward transcribing T7 **"late"** genes to assure a rapid and efficient phage production and lysis of the host.

17.17 REGULATION OF GENE EXPRESSION DURING PHAGE SPO1 INFECTION

Phage SPO1 is a large virus that carries out a virulent infection (Section 5.10) of *B. subtilis*. Many features of its lytic cycle are similar to those of T4 (Figure 5-13), the *E. coli* bacteriophage that has been the more intensively studied. But whereas the molecular basis for gene control in T4 remains a matter of controversy and appears to be fairly complex, a description of the analogous control in SPO1 can now be given in some detail.

Gene expression during the SPO1 cycle occurs in a temporally defined sequence summarized schematically in Figure 17-11. **Early genes** are transcribed by *B. subtilis* RNA polymerases almost immediately after infection, exactly as with phage T7. One of the products of early transcription is the

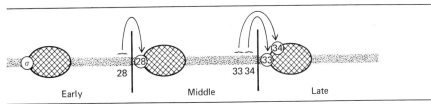

Figure 17-11 Proposed model for the temporal control of phage SPO1 gene expression. The figure illustrates schematically the three major categories of SPO1 gene transcription; the figure is not intended to represent the actual physical arrangement of early, middle and late genes on the phage genome or the number and arrangement of promoter sites. The cross-hatched ellipses represent the host subunits of RNA polymerase other than sigma factor. (From T. D. Fox, *Nature* **262:**748, 1976.)

polypeptide specified by phage gene *28* which is required for subsequent events in the cycle, again as in the case of T7. Gene *28*, however, does not specify a new phage-specific polymerase but rather a polypeptide known as gene *28* product which associates with the host polymerase, apparently in place of the host σ factor (Section 8.3), and directs the polymerase toward the specific transcription of phage **middle genes** (Figure 17-11). Two of these middle genes are genes *33* and *34*, which specify two other polypeptides called the gene *33* and gene *34* products. These polypeptides interact with the host polymerase and direct the expression of **late genes.** In other words, the transition from early to middle and from middle to late in each case involves a modification of the host RNA polymerases rather than a substitution for them. Such a mechanism might eliminate the need for an equivalent of the 0.7 gene product of T7 which shuts off host synthesis: here, host transcription abates as polymerases become modified in phage-specific directions.

Evidence that the SPO1 genome controls its expression as described above comes largely from the work of J. Pero and her colleagues, who have used a variety of molecular approaches. Critical to their argument was the demonstration that *amber* mutations in genes *28, 33,* and *34* not only prevent progression to the next temporal phase of the SPO1 infection cycle—as first shown by D. Fujita and E. P. Geiduschek—but also abolish the synthesis of the polypeptides found to associate with and modify the host RNA polymerase.

REGULATION OF GENE EXPRESSION DURING PHAGE λ INFECTION

More is known about the control of gene expression in phage λ than in any other phage; comparable detail at a molecular level is available only for the *lac* operon. An understanding of λ regulation is therefore immensely rewarding, but it is not easily acquired. A number of different operons, regulatory proteins, and "feedback loops" are involved that cannot be accurately described in any simplified way. The rather detailed account of the lytic and lysogenic cycles of λ presented here will allow a full appreciation of their complexity of design.

As noted at several points in previous chapters, phage λ can follow one of two pathways after it infects an *E. coli* cell: it can direct a lytic cycle that results in the ultimate lysis of the host cell, releasing a burst of about 100 phage progeny, or it can lysogenize the host, during which the infecting phage chromosome inserts into the *E. coli* chromosome (see Figure 5-15) The decision as to whether λ pursues a lytic or lysogenic course depends on the relative levels of a number of λ-coded regulator proteins, some which are lysis-promoting and others of which are lysogeny-promoting. To simplify the presentation, a description is first given of a λ infection that results in lysis, where the lysogeny-promoting genes are ignored. A λ infection is then described that results in lysogeny; this time the lysis-promoting genes are ignored. Finally, consideration is given to the competition between these opposing sets of gene products during an actual λ infection.

17.18   REGULATION DURING THE λ LYTIC CYCLE

The λ lytic cycle occurs in three phases, much as with the SPO1 cycle, but for λ these have come to be called the **immediate-early, delayed-early,** and **late** phases.

The immediate-early phase begins immediately after a λ chromosome enters an *E. coli* cell. Like the SPO1 early phase, it is directed solely by host RNA polymerases. The *E. coli* enzymes initiate transcription at two promoter sites. One λ immediate-early promoter recognized by *E. coli* RNA polymerase is adjacent to a "sense" sequence (Section 8.4) that is transcribed in a rightward direction; this sequence is therefore said to reside in the **R** (right) DNA strand, and this promoter is called **P_R**. The second immediate-early promoter, **P_L**, is located in the left (**L**) DNA strand and is contiguous to genes transcribed in a leftward direction. These relationships are diagrammed in Figure 17-12.

The immediate-early transcription initiated at P_R leads to the expression of a gene called *cro* (for *c*ontrol of *r*epressor and *o*ther things) (Figure 17-12). Since the Cro protein resulting from this transcription is directly involved in the lysis-vs.-lysogeny "decision", we consider it in a later section. The immediate early transcription initiated at P_L leads to the expression of gene *N* (Figure 17-12) and the production of N protein, a protein essential for lysis. N protein is a regulator molecule; specifically, it functions as an activator to stimulate transcription of the delayed-early genes. Its role in the lytic cycle can therefore be analogized to the gene-28 protein of SPO1.

The delayed-early genes transcribed under the influence of N protein (Figure 17-13) include *cII* and *cIII,* which are relevant only to lysogeny; genes *O* and *P,* which are required for the replication of the λ chromosome and therefore qualify as "lysis genes"; and gene *Q.* The product of gene *Q* is required for the expression of the late phage genes in much the same way that the *N* product is required for expression of the delayed-early genes: phages that are Q⁻ express late genes only at a very low level. Since Q functions to activate transcription, its role in the lytic cycle is comparable to the role of the gene-33 and gene-34 proteins of SPO1.

During the late phase of the lytic cycle, phage DNA replication (which began earlier) is in full operation. Under the influence of *Q,* genes necessary for the production of λ heads and tails (genes *A* through *J;* see Figure 10-8) are transcribed, as are genes necessary for bacterial lysis (*S* and *R*), and the cycle is completed.

We can now consider what is known about how the N and Q activator molecules exert their critical roles in the lytic cycle. Much information on the N protein derives from mutant strains. Thus if phages are *N⁻* and

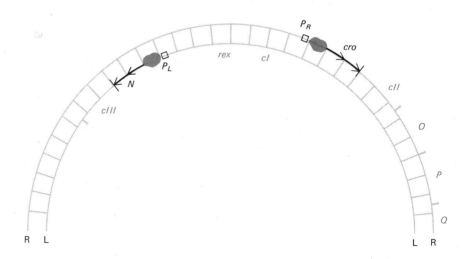

Figure 17-12   Immediate-early transcription of genes *N* and *cro* from the P_L and P_R promoters by *E. coli* RNA polymerase (gray object) in the circular chromosome of phage λ. Strand L is "sense" for leftward transcription; strand R is "sense" for rightward transcription.

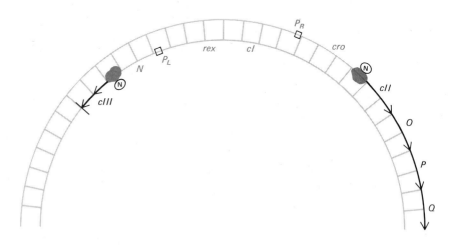

Figure 17-13   Delayed-early transcription of the λ chromosome stimulated by N protein.

produce nonfunctional N protein, an RNA polymerase remains "frozen" at the end of the N gene; if phages produce normal N protein, on the other hand, RNA polymerases are capable of moving past this stopping point and transcribing the following gene, namely, *cIII* (Figure 17-13). The N product therefore appears to function at an attenuator site, much as we saw with the *trp* operon (Figure 17-9), enabling RNA polymerases to read through a termination sequence at the *N-cIII* boundary.

Genetic evidence also indicates that the N protein exerts this effect by combining with *E. coli* RNA polymerases so as to alter their transcriptional properties. C. Georgopoulos, for example, isolated mutant strains of *E. coli*, called *gro*, that were unable to support a lytic cycle after phage λ infection. Among these strains he singled out *gro* strains carrying mutations that map in or near the *rpoA* locus (Figure 8-3), the site of the gene specifying the α subunit of the *E. coli* RNA polymerase. The polymerases isolated from such *gro* strains indeed proved to differ from wild-type polymerases in their *in vitro* transcriptional properties. Georgopoulos then proceeded to select for λ mutants capable of lysing *gro* strains; some of these λ mutants proved to carry mutations mapping to gene N. These results are interpreted to mean that a *gro* mutation in *E. coli* alters the polymerase in such a way that it cannot interact with wild-type N protein, whereas the mutant *E. coli* polymerases can interact with certain mutant forms of λ N protein and proceed to read through the termination signals.

Studies *in vitro* have established that N protein will exert its anti-terminator influence on *E. coli* RNA polymerase only if it combines with the polymerase at a site at or near the P_L or P_R promoters. In other words, once an N-free polymerase transcribes an immediate-early gene and freezes at the termination signal, the addition of N cannot relieve the blockage; N must combine with the polymerase before or at the time transcription is initiated. Presumably, therefore, N protein either modifies the polymerase in some way at the promoter site or else travels with it to the site of N action.

The Q gene product is believed to act similarly, allowing read-through of termination signals at delayed-early/late gene boundaries. A detailed picture of N and Q action awaits isolation of N and Q proteins, both of which prove to be unstable proteins.

17.19   REGULATION DURING THE λ LYSOGENIC CYCLE

A λ lysogenic cycle, like a lytic cycle, is critically dependent on the action of N protein in allowing read-through past termination signals. As noted earlier, such "relief" in the P_L-initiated transcription permits expression of gene *cIII*. Similarly, the P_R-initiated transcript freezes at the end of *cro* unless N protein is present, whereupon gene *cII* is expressed (Figure 17-13).

The *cII* and *cIII* products are essential to lysogeny in that they permit the transcription of gene *cI*, which codes for **λ repressor protein.** Since the λ repressor protein is the critical regulator molecule in setting up a lysogenic state, phages that are cI^- cannot lysogenize *E. coli* and consequently produce only *clear* phages on a bacterial lawn (Figure 6-4). Phages that are cII^- cI^+ or $cIII^-$ cI^+ are similarly incapable of lysogeny and turbid-plaque formation since they are also unable to express their *cI* genes.

The *cII* and *cIII* gene products activate cI transcription, by unknown mechanisms, from a promoter known as $\mathbf{P_{RE}}$ (promoter for **r**epressor **e**stablishment. As diagrammed in Figure 17-14, the P_{RE} promoter lies in the L strand of the λ chromosome and is located about a thousand nucleotides to the right of gene *cI*. Transcription from this promoter proceeds in a leftward direction and covers a portion of the "antisense" strand of the *cro* gene before *cI* and the adjacent *rex* gene are copied (Figure 17-14). The resultant transcript therefore has a long "leader" of RNA (Section 8.6). Translation of this message results in the production of a large amount of λ repressor protein plus the *rex* gene product (which, in some unknown fashion, prevents λ lysogens from being lysed by infecting T4 phages carrying mutation in the *rII* gene).

Once the λ repressor appears in the cell, it recognizes and binds to two operator sites, called O_L and O_R, which overlap the P_L and P_R promoters. A detailed description of these controlling elements is presented in the next section. Here it is sufficient to note that repressor binding prevents further RNA polymerase initiations at the P_L and P_R promoters, much as was diagrammed in Figure 17-1. The transcription of genes N and *cro* is therefore prevented and, since the N and Cro proteins are highly unstable, their intracellular levels fall rapidly. Also blocked is the N-activated read-through transcription of genes O, P, and Q, which is necessary for delayed-early and late events in the lytic cycle. Therefore, if sufficient repressor is made and becomes bound to the O_L and O_R operator sites, the lysogenic stage is set.

The establishment of lysogeny requires not only that lysis be prevented

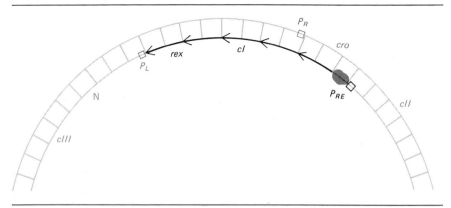

Figure 17-14 Transcription of genes *cI* and *rex* from the P_{RE} promoter during the onset of a lysogenic cycle.

so that the host cell is "saved." It also requires that the infecting λ genome integrate into the host genome as a stable prophage. As detailed in Section 13.16, this integration requires an interaction and recombination between the *att* sequences in λ and in *E. coli;* it also requires products of genes in the "middle" of the λ chromosome, including *int, exo, β*, and γ. Transcription of these genes proves to be initiated at yet another promoter, called P_{INT} which is activated by the *cII* and *cIII* proteins. In other words, *cII* and *cIII* products encourage lysogeny by activating transcription at two distinct promoters: the P_{RE} promoter of the *cl* gene, and the P_{INT} promoter governing the integration genes.

17.20 REPRESSOR-OPERATOR INTERACTIONS IN PHAGE λ

The λ repressor is a stable protein (in contrast to the other regulator molecules specified by λ) and has been purified and partially sequenced by M. Ptashne and collaborators. The monomer has a molecular weight of 26,000 and binds to duplex operators either as a dimer or a tetramer, apparently making contacts primarily within the major groove of an operator-DNA helix.

The two λ operators, O_L and O_R, have also been isolated and sequenced by members of the Ptashne laboratory, and the results reveal a surprising fact: each operator contains not one but three repressor binding sites (O_L1, 2, 3 and O_R1, 2, 3), as diagrammed in Figure 17-15. Each site is 17 base-pairs long and is separated from the next site by short A-T rich "spacers". If low concentrations of repressor are presented to operator DNA, only the O_L1 and O_R1 sites are protected from DNase digestion; however, as more repressor is added, the two additional binding sites of each operator become protected as well. It is concluded, therefore, that the O_L1 and O_R1 sites, which are directly adjacent to genes *N* and *cro*, bind repressor with the highest affinity and are most effective in turning off *N* and *cro* expression.

Genetic studies have yielded considerable information concerning repressor-operator interactions in phage λ. Mutant strains of λ have been isolated that carry the equivalent of *lac o^c* mutations (Section 17.11). These are known as **virulent** strains: they direct the synthesis of normal repressor but cannot respond to it, and are therefore unable to lysogenize. When the operator DNA from such *virulent* strains is isolated, it proves to bind repressor relatively weakly. Sequence analyses, moreover, reveal that the various *virulent* strains carry nucleotide changes within their O_L1, O_L2, O_R1 or O_R2 sequences (Figure 17-15). Promoter mutations affecting P_L and P_R, on the other hand, map to the spacer regions between the operators (Figure 17-15). Since λ repressor will exclude the binding of *E. coli* RNA polymerase to operator DNA *in vitro*, there is clearly considerable sequence overlap between λ operators and promoters, just as we saw for bacterial operons (Figures 17-5 and 17-8).

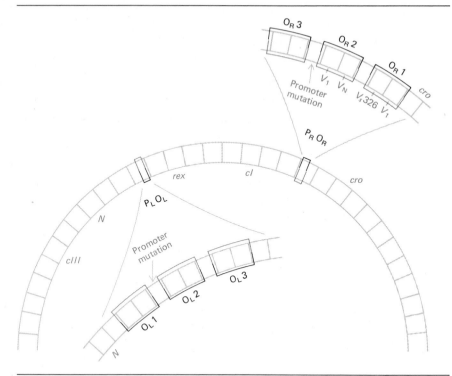

Figure 17-15 Fine structure of the two operator regions of the λ chromosome. Positions of various *virulent* (V) mutations in the O_R are shown, as are two promoter mutations within the AT-rich "spacer" regions.

17-21 THE LYSIS—LYSOGENY "RACE"

We can now consider an actual λ infection and describe the competition between lytic and lysogenic pathways. Although all six regulator proteins of λ are involved in the competition—namely, the *N, Q, cII, cIII, cI* and *cro* gene products—the action of the Cro protein is perhaps the most pivotal.

The *cro* gene, we have noted, is expressed immediately after a λ infection and does not require *N* function (Figure 17-12). The Cro protein therefore comes to be present in the cell somewhat before the *cII-* and *cIII-*products appear and before *cI* expression and repressor synthesis are activated. Present evidence indicates that Cro is also a repressor and that it interacts with some of the same binding sites in the O_L and O_R operators as does the *cI-*specified repressor; its binding, for example, is sensitive to some of the same *virulent* mutations. The Cro protein does not, however, bind nearly as efficiently as the *cI-*repressor, nor is it nearly as stable. Therefore, as it binds to O_L and O_R, it slows down but does not completely inhibit expression of genes *N* and *cIII* in the leftward operon and genes *cro, cII, O, P* and *Q* in the rightward operon.

We now come to the critical point in the overall process. It turns out that the P_{RE} promoter (Figure 17-14), which governs *cI* transcription and is activated by the *cII*- and *cIII* products, is more sensitive to a decline in *cII*- and *cIII*-protein levels than the expression of late phage genes is sensitive to a decline in N and Q levels. Therefore, if enough Cro is produced and the transcription of the *cII* and *cIII* genes is sufficiently turned off, then the *cI* gene is never adequately activated. As a result, no stable inhibition of phage DNA replication and head and tail synthesis is achieved, and a lytic cycle ensues. If, on the other hand, insufficient Cro is produced or its binding to the two operators is not effective enough to abate *cII* and *cIII* expression, then abundant *cI* repressor will be synthesized and lysogeny will ensue. Such factors as temperature, the metabolic state of the host, and the genotypes of the host and the infecting phage all appear to influence the level of Cro production so that, in any given infection, the final balance between Cro-mediated "pseudo-repression" and *cI*-mediated "true" repression may tip one way or the other.

17.22 THE LYSOGENIC STATE

A diagnostic property of bacteria lysogenic for λ [*E. coli*(λ)] is that they are **immune to superinfection** by additional λ phages. In other words, when a culture of *E. coli*(λ) is presented with λ particles, the phage DNA is able to enter the bacterial cells but no lysis of the bacteria occurs. This immunity results from the important fact that the prophage dictates the synthesis of a low level of cI repressor. The repressor serves to block transcription of the P_LO_L and P_RP_R operons in the prophage, thereby sustaining the lysogenic condition; it also represses the P_LO_L and P_RO_R operons of any superinfecting λ chromosomes that may happen to enter the cell. As a result, the infecting chromosomes cannot replicate, nor can they ordinarily insert into the host chromosome, since the λ site of the latter is already occupied. They therefore simply remain in the host cytoplasm and are diluted out by successive host cell divisions.

The low levels of repressor found in an *E. coli*(λ) lysogen prove to be transcribed not from the P_{RE} promoter we have described earlier, but from a second promoter called **P_{RM}** (**p**romoter for **R**epressor **M**aintenance). As diagrammed in Figure 17-16, the P_{RM} is directly adjacent to the *cI* gene, in

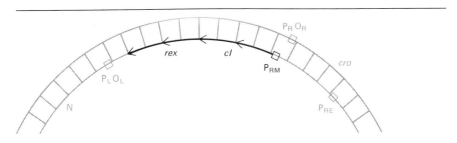

Figure 17-16 Transcription of genes *cI* and *rex* from the P_{RM} promoter once lysogeny is established.

contrast to the more distant P_{RE}. Unlike the P_{RE}, moreover, the P_{RM} does not require cII or $cIII$ activation—indeed, neither cII nor $cIII$ is expressed in a lysogen since both the $P_L O_L$ and $P_R O_R$ operons are repressed. Control over P_{RM} transcription is instead exerted by the cI protein itself in a process known as **autoregulation:** when repressor levels become low, repressor protein has the fascinating property of being able to promote its own synthesis; when repressor levels "overshoot," transcription from P_{RM} is turned off by repressor. How these effects are exerted is not yet clear, but the mechanism is clearly designed to ensure that repressor synthesis continues at a slow, steady rate in the lysogenic bacterium.

17.23   PROPHAGE INDUCTION

The spontaneous loss of a λ prophage from an *E. coli* chromosome occurs only rarely. Three experimental means have therefore been developed to bring about the induction of a lytic cycle in lysogenic cells. All share a common feature: the cI repressor protein is somehow prevented from maintaining repression of the λ prophage operons.

The first approach to lysis induction is known as **zygotic induction.** Matings are induced between lysogenic *Hfr* donor cells and nonlysogenic recipient cells, as described in Section 11.2. When the chromosomal segment containing the prophage enters the recipient cell, it encounters a cytoplasm free of repressor protein. Transcription of the $P_L O_L$ and $P_R O_R$ operons therefore commences, and a lytic cycle frequently ensues. No such events occur when the recipient cells are also lysogenic for λ.

The second approach involves treating a culture of lysogenic cells with such agents as ultraviolet light, mitomycin, nitrogen mustard, and indeed, any carcinogen (c.f. Section 6.20). One theory holds that such treatments physically damage the repressor protein. The alternate theory holds that such treatments generate single-stranded breaks in *E. coli* DNA to which repressor protein temporarily binds, thereby producing a temporary drop in free repressor levels and introducing the possibility of induction.

The third method of prophage induction utilizes a cI mutation known as cI 857. This mutation results in the synthesis of a thermolabile repressor protein, one that is stable at 30°C but becomes inactive at 42°C. Induction in this case simply requires raising the temperature of a lysogenic *E. coli* culture. Since the outcome of this procedure is far more reliable than either ultraviolet, chemical, or zygotic induction, the cI 857 strain is now routinely used in studies of the lysogenic process.

17.24   GENE REGULATION DURING SV40 INFECTION

We conclude our consideration of viral infectious cycles with SV40 which, although very much a virus like λ in its infection pattern, can grow only in mammalian cells and therefore conducts its infection in a eukaryotic environment.

The life cycle of SV40 was presented in Section 5.12 and should be reviewed at this time. Shortly after infection, the small SV40 chromosome, (Figure 7-2), is transcribed by host enzymes from one strand (the **early** or **E.** strand) into a 19S mRNA. This early mRNA is then translated into a single protein, with a molecular weight of about 90,000, known as **T antigen.** The single early gene encoding the T antigen is known as *gene A.* Mutant strains that produce a thermolabile T antigen (*tsA* mutants) cannot direct either a lytic or a "transforming" infection at nonpermissive temperatures. Specifically, such mutant strains cannot initiate SV40 DNA replication or initiate late SV40 transcription, and the thermolabile T antigen they produce fails to bind to the SV40 chromosome, in contrast to its wild-type counterpart. Therefore, the T antigen is formally analogous to essential "early" regulator proteins, such as the λ N protein, in a bacteriophage cycle.

A major difference between T antigen and analogous early phage proteins, however, is that only some of the T-antigen activities are carried out upon the SV40 chromosome itself. T antigen also stimulates chromosome replication and mitosis in the host nucleus, and there is strong evidence that it directly affects both the translational properties of host ribosomes and the surface properties of the host plasma membrane. How this single early gene product plays such a multiplicity of roles during an SV40 infection is at present unknown.

T antigen appears to regulate its own synthesis. When *tsA* mutants are shifted to restrictive temperature shortly after infection, existent thermolabile T antigen undergoes degradation. This is followed first by a 15-fold increase in the rate of early 19S mRNA transcription and then by a similar increase in the synthesis of thermolabile T antigen (which then proceeds to undergo degradation). In other words, T antigen appears to exert autoregulation of its own synthesis in much the same way as the λ repressor. Whether these forms of control by SV40 and λ are any more than coincidentally similar remains to be seen.

Expression of gene A in a permissive monkey cell allows transcription of late genes which reside on the opposite or **late (L)** strand of the SV40 chromosome. Strand-switching, therefore, is a device employed by both phage and animal viruses. The two late transcripts, late 16S and late 19S mRNA, encode the three viral capsid proteins, VP1, VP2, and VP3. Furthermore, most or all of the sequences coding for VP3 are included in those sequences specifying VP2; in other words, VP3 appears to be some sort of a cleavage product of VP2. The lytic cycle is complete when all three proteins assemble into mature capsids surrounding replicated SV40 chromosomes.

In a mouse cell, expression of gene A sets the stage for either an abortive infection or for a transformation event which, as we saw in Section 13.15, involves integration of the SV40 genome into the mouse genome. That expression of *gene A* is vital for transformation can again be demonstrated with *tsA* mutants: at high temperatures, such mutants are unable to initiate a transformation event. Even more dramatically, if cells are transformed by *tsA* mutants at permissive temperature and are then shifted to a restrictive temperature, the transformed state is in most cases aborted, and the cells revert to their normal state. Such events are, of course, formally analogous to λ prophage induction in the *cI* 857 strain that produces temperature-sensitive repressor proteins. It perhaps goes without saying that intensive research is presently focused on determining the molecular basis for SV40 transformation and the role of T antigen in this process.

References

Regulation of the *lac* Operon

Beckwith, J., and D. Zipser, eds. *The Lactose Operon.* New York: Cold Spring Harbor Laboratory, 1970.

Carpenter, G., and B. H. Sells. "Regulation of the lactose operon in *Escherichia coli* by cAMP," *Int. Rev. Cytol.* **41:**29–58 (1975).

*Davies, J., and F. Jacob. "Genetic mapping of the regulator and operator genes of the *lac* operon," *J. Mol. Biol.* **36:**413–417 (1968).

*Dickson, R. C., J. Abelson, W. M. Barnes, and W. S. Reznikoff. "Genetic regulation: The *lac* control region," *Science* **187:**27–35 (1975).

Dickson, R. C., J. Abelson, P. Johnson, W. S. Reznikoff, and W. M. Barnes. "Nucleotide sequence changes produced by mutations in the *lac* promoter of *Escherichia coli*," *J. Mol. Biol.* **111:**65–76 (1977).

*Eron, L., R. Arditti, G. Zubay, S. Connaway, and J. R. Beckwith. "An adenosine 3′:5′-cyclic monophosphate-binding protein that acts on the transcription process," *Proc. Natl. Acad. Sci. U.S.* **68:**215–218 (1971).

*Eron, L., and R. Block. "Mechanism of initiation and repression of *in vitro* transcription of the *lac* operon of *Escherichia coli*," *Proc. Natl. Acad. Sci. U.S.* **68:**1828–1832 (1971).

*Gilbert, W., and B. Müller-Hill. "Isolation of the *lac* repressor," *Proc. Natl. Acad. Sci. U.S.* **56:**1891–1898 (1966).

*Gilbert, W., and B. Müller-Hill. "The *lac* operator is DNA," *Proc. Natl. Acad. Sci. U.S.* **58:**2415–2421 (1967).

*Gilbert, W. "Starting and stopping sequences for the RNA polymerase," in *RNA Polymerase.* R. Losick and M. Chamberlin, eds., New York: Cold Spring Harbor Laboratory. pp. 193–205 1976.

*Gilbert, W., J. Gralla, J. Majors, and A. Maxam. "Lactose operator sequences and the action of *lac* repressor," in *Protein-Ligand Interactions,* H. Sund and G. Blauer, eds. Berlin: Walter de Gruyter & Co. pp. 193–210, 1975.

Hopkins, J. D. "A new class of promoter mutations in the lactose operon of *Escherichia coli*," *J. Mol. Biol.* **87:**715–724 (1974).

*Jacob, F., and J. Monod. "Genetic regulatory mechanisms in the synthesis of proteins," *J. Mol. Biol.* **3:**318–356 (1961). [Reprinted in G. L. Zubay, *Papers in Biochemical Genetics.* New York: Holt, Rinehart and Winston, 1968.]

*Jacob, F., and J. Monod. "Genetic mapping of the elements of the lactose region in *Escherichia coli*," *Biochem. Biophys. Res. Comm.* **18:**693–701 (1965). [Reprinted in G. L. Zubay, *Papers in Biochemical Genetics.* New York: Holt, Rinehart and Winston, 1968.]

*Jobe, A., and S. Bourgeois. "*Lac* repressor-operator interactions. VI. The natural inducer of the *lac* operon," *J. Mol. Biol.* **69:**397–408 (1972).

Maizels, N. "*E. coli* lactose operon ribosome binding site," *Nature* **249:**647–649 (1974).

*Denotes articles described specifically in the chapter.

Müller-Hill, B. "*Lac* repressor and *lac* operator," *Prog. Biophys. Molec. Biol.* **30**:227–252 (1975).

*Pardee, A. B., F. Jacob, and J. Monod. "The genetic control and cytoplasmic expression of inducibility in the synthesis of β-galactosidase by *E. coli*," *J. Mol. Biol.* **1**:165–178 (1959). [Reprinted in E. A. Adelberg, *Papers on Bacterial Genetics,* 2nd ed. Boston: Little, Brown, 1966.]

Richmond, T. J., and T. A. Steitz. "Protein-DNA interaction investigated by binding *Escherichia coli lac* repressor protein to poly [d(A·U − HgX)]". *J. Mol. Biol.* **103**:25–38 (1976).

*Riggs, A. D., S. Bourgeois, and M. Cohn. "The *lac* repressor-operator interaction. III. Kinetic studies," *J. Mol. Biol.* **53**:401–417 (1970).

Regulation of the *trp* Operon

*Bennett, G. N., M. E. Schweingruber, K. D. Brown, C. Squires, and C. Yanofsky. "Nucleotide sequence of region preceding *trp* mRNA initiation site and its role in promoter and operator function," *Proc. Natl. Acad. Sci. U.S.* **73**:2351–2355 (1976).

*Bertrand, K., L. Korn, F. Lee, T. Platt, C. L. Squires, C. Squires, and C. Yanofsky. "New features of the regulation of the tryptophan operon," *Science* **189**:22–26 (1975). [Research articles relating to this review are found in *J. Mol. Biol.* **103**:319–420, 1976.]

Morse, D. E., and A. N. C. Morse. "Dual-control of the tryptophan operon is mediated by both tryptophanyl-tRNA synthetase and the repressor," *J. Mol. Biol.* **103**:209–226 (1976).

Morse, D., and C. Yanofsky. "Amber mutants of the *trpR* regulatory gene," *J. Mol. Biol.* **44**:185–193 (1969).

Control of Gene Expression in Virulent Phages

Daejelen, P., and E. Brody. "Early bacteriophage T4 transcription. A diffusible product controls rIIA and rIIB RNA synthesis," *J. Mol. Biol.* **103**:127–142 (1976).

*Fox, T. D. "Identification of phage SPO1 proteins coded by regulatory genes *33* and *34,*" *Nature* **262**:748–753 (1976).

Pachl, C. A., and E. T. Young. "Detection of polycistronic and overlapping bacteriophage T7 late transcripts by *in vitro* translation," *Proc. Natl. Acad. Sci. U.S.* **73**:312–316 (1976).

*Petrusek, R., J. J. Duffy, and E. P. Geiduschek. "Control of gene action in phage SPO1 development: Phage-specific modifications of RNA polymerase and a mechanism of positive regulation," in *RNA Polymerase*. R. Losick and M. Chamberlin, eds. New York: Cold Spring Harbor Laboratory, pp. 567–585, 1976.

Snyder, L., L. Gold, and E. Kutter, "A gene of bacteriophage T4 whose product prevents true late transcription on cytosine-containing T4 DNA," *Proc. Natl. Acad. Sci. U.S.* **73**:3098–3102 (1976).

*Studier, F. W. "Bacteriophage T7," *Science* **176**:367–376 (1972).

Studier, F. W. "Gene 0.3 of bacteriophage T7 acts to overcome the DNA restriction system of the host," *J. Mol. Biol.* **94**:283–295 (1975).

Thermes, C., P. Daegelen, V. de Franciscis, and E. Brody, "*In vitro* system for induction of delayed early RNA of bacteriophage T4," *Proc. Natl. Acad. Sci. U.S.* **73**:2569–2573 (1976).

*Tjian, R., and J. Pero. "Bacteriophage SP01 regulatory proteins directing late gene transcription *in vitro,*" *Nature* **262**:753–757 (1976).

Control of Gene Expression in Phage λ

Beyreuthen, K., and B. Gronenborn. "N-terminal sequence of phage lambda repressor," *Molec. Gen. Genet.* **147**:115–117 (1976).

*Echols, H., and L. Green. "Establishment and maintenance of repression by bacteriophage lambda: the role of the cI, cII, and cIII proteins," *Proc. Natl. Acad. Sci. U.S.* **68**:2190–2194 (1971).

Echols, H., D. Conit, and L. Green. "On the nature of *cis*-acting regulatory proteins and genetic organization in bacteriophage: the example of gene Q of bacteriophage λ," *Genetics* **83**:5–10 (1976).

*Eisen, H., P. Brachet, L. Pereira da Silva, and F. Jacob. "Regulation of repressor expression in λ," *Proc. Natl. Acad. Sci. U.S.* **66:**855–862 (1970).

Folkmanis, A., Y. Takeda, J. Simuth, G. Gussin, and H. Echols. "Purification and properties of a DNA-binding protein with characteristics expected for the Cro protein of bacteriophage λ, a repressor essential for lytic growth," *Proc. Natl. Acad. Sci. U.S.* **73:**2249–2253 (1976).

*Georgopoulos, C. P. "Bacterial mutants in which the gene N function of bacteriophage lambda is blocked have an altered RNA polymerase," *Proc. Natl. Acad. Sci. U.S.* **68:**2977–2981 (1971).

Hershey, A. D., Ed. *The Bacteriophage Lambda.* New York: Cold Spring Harbor Laboratory, 1971.

Herskowitz, I. "Control of gene expression in bacteriophage lambda," *Ann. Rev. Genet.* **7:**289–324 (1973).

Honigman, A., S.-L. Hu, R. Chase, and W. Szybalski: "4S *oop* RNA is a leader sequence for the immunity-establishment transcription in coliphage λ," *Nature* **262:**112–116 (1976).

Lederberg, E. M., and J. Lederberg. "Genetic studies of lysogenicity in *Escherichia coli,*" *Genetics* **38:**51–64 (1953).

Lieb, M. "Mapping missense and nonsense mutations in gene cl of bacteriophage lambda: Marker effects," *Molec. Gen. Genet.* **146:**285–290 (1976) [and two subsequent articles].

*Maniatis, T., M. Ptashne, K. Backman, D. Kleid, S. Flashman, A. Jeffrey, and R. Maurer. "Recognition sequences of repressor and polymerase in the operators of bacteriophage lambda," *Cell* **5:**109–114 (1975).

Oppenheim, A. B. "Internal inversion used to study regulation of the *int* gene in bacteriophage λ," *Nature* **261:**615–617 (1976).

*Ptashne, M., K. Bachman, M. Z. Humayun, A. Jeffrey, R. Maurer, B. Meyer, and R. T. Sauer; Autoregulation and function of a repressor in bacteriophage lambda," *Science* **194:**136–161 (1976).

*Ptashne, M. "Isolation of the λ phage repressor," *Proc. Natl. Acad. Sci. U.S.* **57:**306–313 (1967).

*Ptashne, M. "Specific binding of the λ phage repressor to λ DNA," *Nature* **214:**232–234 (1967). [Reprinted in G. L. Zubay, *Papers in Biochemical Genetics.* New York: Holt, Rinehart and Winston, 1968.]

*Reichardt, L., and A. D. Kaiser. "Control of λ repressor synthesis," *Proc. Natl. Acad. Sci. U.S.* **68:**2185–2189 (1971).

*Reichardt, L. F. "Control of bacteriophage lambda repressor synthesis after phage infection. The role of the *N, cII, cIII* and *cro* products," *J. Mol. Biol.* **93:**267–288 (1975) [and following article].

Sussman, R., and H. Ben Zeev. "Proposed mechanism of bacteriophage lambda induction: Acquisition of binding sites for lambda repressor by DNA of the host, *Proc. Natl. Acad. Sci. U.S.* **72:**1973–1976 (1975).

Wulff, D. L. "Lambda *cin-1*, a new mutation which enhances lysogenization by bacteriophage lambda, and the genetic structure of the lambda cy region," *Genetics* **82:**401–416 (1976) [and subsequent article].

Control of Gene Expression in SV40

Ahmad-Zadeh, C., B. Allet, J. Greenblatt, and R. Weil. "Two forms of simian-virus-40-specific T-antigen in abortive anal lytic infection," *Proc. Natl. Acad. Sci. U.S.* **73:**1097–1101 (1976).

*Goldberger, R. F. "Autogenous regulation of gene expression," *Science* **183:**810–816 (1976).

Grodzicker, T., J. B. Lewis, and C. W. Anderson. "Conditional lethal mutants of adenovirus type 2-simian virus 40 hybrids. 11. Ad2⁺ ND1 host-range mutants that synthesize fragments of the ad 2⁺ ND1 30K protein," *J. Virol.* **19:**559–571 (1976).

Levine, A. J. "SV40 and adenovirus early functions involved in DNA replication and transformation," *Biochim. Biophys. Acta* **458:**213–241 (1976).

*Martin, R. G., and K. Chou. "Simian virus 40 functions required for the establishment and maintenance of malignant transformation," *J. Virol.* **15:**599–612 (1975) [and accompanying articles].

*Reed, S. I., G. R. Stark, and J. C. Alwine. "Autoregulation of simian virus 40 gene A by T antigen," *Proc. Natl. Acad. Sci. U.S.* **73**:3083–3087 (1976).

*Tegtmeyer, P., M. Schwartz, J. K. Collins, and K. Rundell. "Regulation of tumor antigen synthesis by simian virus 40 gene A," *J. Virol.* **16**:168–178 (1975).

(a)　　　　　　　　(b)　　　　　　　　(c)　　　　　　　　(d)

(a) Suzanne Bourgeois (Salk Institute), (b) Benno Müller-Hill (University of Cologne, Germany), and (c) Walter Gilbert (Harvard University) are all students of repressor-operator interactions in the *lac* operon of *E. coli*. (d) Mark Ptashne (Harvard University) studies the mode of action of the repressor protein specified by phage λ.

Questions and Problems

1. The synthesis of enzyme E is inducible and dependent on three loci: a regulatory gene (R), a structural gene (S), and an operator (o). Haploid bacterial strains are found with the following properties:
   - (a) Enzyme synthesis occurs only when inducer is present.
   - (b) Enzyme synthesis never occurs.
   - (c) Enzyme synthesis occurs in the presence or absence of inducer.
   - (1) List all possible genotypes producing each of the phenotypes listed above.
   - (2) When the F' element $F\,R^+\,o^+\,S^-$ is introduced into these haploid cells, which of the resultant partial diploids can synthesize E and which cannot?
   - (3) Of those partial diploids that can synthesize E, which require the presence of an inducer?

2. Give the phenotypes of *E. coli* cells with the following *lac*-related genotypes:
   - (a) $i^s\,o^c\,z^+\,y^+$
   - (b) $i^+\,o^+\,z^-\,y^+$ where z carries a promoter-proximal nonsense mutation
   - (c) $i^s\,o^+\,z^+\,y^-$
   - (d) $i^{-d}\,o^c\,z^-\,y^+$
   - (e) $\dfrac{L1\,o^c\,z^+\,y^+}{F\,i^+\,o^+\,z^-\,y^-}$ (see Figure 17-6 for location of $L1$)

3. In what respect is a repressor protein similar to an aminoacyl-tRNA synthetase molecule?

4. Four haploid strains of *E. coli* denoted A, B, C, and D carry mutations affecting the expression of the *lac* operon (none of the mutations lie in genes *y* or *a* or in promoter regions). The following observations are made:

    (a) Strains B and C do not synthesize β-galactosidase, even in the presence of lactose.

    (b) Strains A and D continuously synthesize β-galactosidase regardless of the presence of lactose.

    (c) When the *F'* element *F $i^+ z^-$* is introduced into strain A, the resultant partial diploid synthesizes β-galactosidase when lactose is present but not when it is absent. When the same *F'* element is introduced into strain D, no effect on strain D's phenotype is observed.

    (d) Strain C synthesizes a material that is immunologically cross-reactive with β-galactosidase in a precipitin test (see Box 9.2). Strain B makes no cross-reactive material, but it will do so if an *amber* suppressor gene is introduced into the cell via an *F'* element.

    (e) When the *lac* operon of strain A is transposed to an *F'* element and this element is introduced into strain B, the resulting partial diploid is inducible with lactose.

    What are the genotypes of the four strains with regard to their *i*, *o*, and *z* loci? Give your reasoning in each case. Indicate any cases where uncertainty remains.

5. Partial diploids of *E. coli* that are i^s/i^+ are noninducible, whereas i^s/i^Q partial diploids can be induced to 25 percent of the normal level. Discuss these results in terms of the fact that the *lac* repressor is a tetrameric protein.

6. Three mutations, *a*, *b*, and *c*, affect the expression of a repressible operon of *E. coli* where *abc* is the correct map order as determined by transduction. Strains carrying various combinations of these markers are tested to see whether they form the regulated enzyme in the presence or absence of corepressor. The following results are obtained (P means the enzyme is present in large amount; A means it is absent or present in only trace amounts):

	Corepressor Present	Corepressor Absent
a + +	A	A
+ *b* +	P	P
+ + *c*	P	P
+ + +/*F a b c*	A	P
+ + *c*/*F a b* +	A	P
+ *b* +/*F a* + *c*	P	P
a + +/*F* + *b c*	P	P

Which mutation lies in the regulatory gene for the operon? Which lies in the structural gene for the regulated enzyme? Which lies in the operator locus of the operon? Give your reasoning in each case.

7. The feedback inhibition of enzyme activity occurs immediately, whereas the loss of enzyme activity by genetic repression requires several generations during which existent enzymes are diluted out. The following table summarizes the effects that two products of biosynthesis, A and B, exert on the activities of three biosynthetic enzymes (Immed. = immediate effect; Delay = effect after several generations; + means enzymatic activity; − means no enzymatic activity)

	Presence of A		Presence of B		Presence of A + B	
	Immed.	Delay	Immed.	Delay	Immed.	Delay
Enzyme 1	+	+	+	+	−	−
Enzyme 2	+	−	+	+	+	−
Enzyme 3	+	+	+	−	+	−

Which enzymes are inhibited and which are repressed by the biosynthetic products? Based on these results, suggest a reasonable metabolic pathway leading to the production of A and B.

8. What phenotypes would you expect of phage λ carrying the following mutations? Specifically, can they carry out lysogeny or a lytic cycle? What proteins can they synthesize?
   (a) A deletion of *cI*
   (b) A deletion of Q
   (c) A deletion of N
   (d) A deletion of *cro*
   (e) A deletion of P_{RE}
   (f) A deletion of P_R
   (g) A deletion of P_L
   (h) A deletion of P_{RM}

9. The enzyme tryptophanyl-tRNA synthetase (Figure 8-13) can exist in two states: "charged" (carrying tryptophan) and "uncharged" (free of tryptophan). A model for regulation at the *trp* attenuator proposes that control is exerted by the relative levels of these two forms of the enzyme, with one form of the enzyme preventing RNA polymerases from "falling off" at the attenuator and the other form lacking this ability. Which form of the enzyme would you expect to prevent attenuation in such a model? Explain in the context of *trp* regulation.

10. The **immunity region** of a temperate-phage chromosome contains its repressor-encoding gene and the operators controlled by the repressor. The temperate phage 434 is sufficiently similar to λ that the two chromosome types will undergo recombination in a mixed infection, yet phage 434 will readily superinfect and lyse an *E. coli* (λ) lysogen.
   (a) Explain why this information suggests that the immunity regions of λ and 434 are different.
   (b) A hybrid phage is constructed by recombination between λ and 434. Describe two ways you might test whether the hybrid contains a λ or a 434 immunity region.

11. Hybrid λ phages are constructed in which the *lac z* gene is fused next to the λP_{RM} promoter. Levels of β-galactosidase expression were then observed in *E. coli* cells lysogenic for the hybrid phages. When the hybrid phages were cI^+, 2600 units of enzyme activity were detected per cell. Would you expect, in the following two cases, enzyme levels to be higher, lower, or the same as for the cI^+ hybrid? Explain.
    (a) The hybrid is cI^-.
    (b) The hybrid is cI^+ and the *E. coli* lysogens also contain a recombinant plasmid carrying a *cI* gene that greatly overproduces *cI* protein.

12. Would you expect *amber* o^c mutations? Explain.

13. Cells carrying a deletion covering the *lac* repressor binding site and the *lac z* gene continue to express the *lac a* gene.
    (a) Would you expect this expression to be regulated or constitutive? Explain.
    (b) Would you expect this mutation to be *cis*-dominant or also *trans*-dominant? Explain.
    (c) What does this mutation reveal about the relative locations of the *lac* promoter and *lac* operator?

14. How would you distinguish between $trpR^-$ and $trpO^c$ mutations in *E coli*?

15. A mutation at the *hisT* locus affects the expression of the repressible *his* operon of *E. coli* (Figure 16-4). It is found that the mutation changes one nucleotide in the tRNA^{His} anticodon such that the tRNA^{His} fails to recognize the *his* codon in mRNA. Would you expect cells carrying this mutation to have high or low levels of *his* operon expression? Explain your answer.

16. A phage T7 strain carries an *amber* mutation in gene 1. Compare the course of an infection by this strain of permissive and nonpermissive *E. coli* (Section 9.19).

17. The mRNA transcribed from P_{RE} (Figure 17-14) has a much longer leader (Section 8.6) than the mRNA initiated at P_{RM}. Explain how this observation might help account for the different levels of λ repressor in a lysogen compared with levels found after the first few minutes of a λ infection.

18. A mutation in the *cI* gene of λ, called ind^-, renders the repressor insensitive to uv induction; a second mutation, cI^s, renders the repressor unusually sensitive to uv induction. How would you explain these mutations by the two theories for uv-induction mechanisms offered in Section 17.23?

19. When *lac* repressor proteins are briefly exposed to trypsin, 50–60 amino acids are clipped off one end; the resultant "core" tetramer continues to bind IPTG normally but no longer binds to operator DNA. Does trypsin cleave at the N-terminus or the C-terminus? Explain.

20. The behavior of i^{-d} mutations in partial diploids has been termed **negative complementation**. Explain the meaning of this term.

CHAPTER
18

Control of Gene Expression in Eukaryotes

INTRODUCTION

Two different classes of regulatory phenomena can be recognized in eukaryotes. The first, which we can call **short-term** or **reversible regulation,** corresponds to the kind of regulation we have studied in bacteria (Chapter 17). It represents a cell's response to fluctuations in the environment; specifically, short-term regulation involves changes in activities or concentrations of enzymes as particular substrates or hormone levels rise and fall and as the cell cycle (Section 2.1) is traversed. The second kind of eukaryotic regulation can be termed **long-term** or (usually) **irreversible** ("usually" in that there are occasions, such as cancerous growth, when the process appears to reverse itself). Long-term regulation includes the phenomena associated with **determination, differentiation,** or, more generally, **development:** it is involved in the numerous steps by which a fertilized egg becomes an organism of perhaps trillions of cells which have diverse and, ultimately, quite permanent roles to play in the maintenance of the whole.

We should stress at the outset that short-term regulation will be a feature of both developing and fully differentiated eukaryotic cells; in other words, even as a cell is undergoing differentiation it will also be responding to its environment and experiencing the fluctuations of its cell cycle. Moreover, during differentiation, certain features of a short-term regulatory process may change dramatically: in the early embryo, for example, DNA synthesis occurs rapidly and the cell cycle exhibits very short, if any, G_1 and G_2 periods; as differentiation proceeds these periods lengthen and ultimately assume cell-specific proportions.

In this chapter we first consider examples of short-term regulation in fungi, in insects, and in mammals and review prominent theories as to how such regulatory phenomena might be related to known features of nuclear and chromatin organization. We then examine a number of long-term differentiation phenomena that occur during eukaryotic embryology, focusing on studies that use genetic approaches to elucidate these phenomena.

SHORT-TERM REGULATION IN FUNGI

Many studies have been made of short-term control systems in fungi, most being concerned either with biosynthetic or catabolic pathways. The three examples presented here illustrate relatively straightforward, but representative, modes of control. Emphasized are similarities and differences between these and analogous bacterial control systems.

18.1   THE *qa* GENE CLUSTER IN *NEUROSPORA*

A *Neurospora* cell contains all the enzymes required for the synthesis of unsaturated ring structures known as **aromatic molecules,** of which the amino acids tyrosine and phenylalanine (Figure 9-1) are familiar examples. When levels of such aromatic compounds in the cell become too high, several "scavenging enzymes" are induced which break down the compounds into nonaromatic forms. The induction of three of these enzymes is accomplished by elevated levels of an aromatic metabolite known as quinic acid, and the three induced enzymes—a dehydroquinase, a dehydrogenase, and a dehydrase—are encoded by the linked genes *qa-2*, *qa-3*, and *qa-4*. (Figure 18-1). Once quinic acid is catabolized, enzyme synthesis is turned off. The *qa-2*, *qa-3*, and *qa-4* genes in the cluster can, therefore, be classified as inducible genes, much as the inducible catabolism-related genes in *E. coli* (Section 17.4).

We saw in Chapter 17 that the induction of gene expression in prokaryotes proceeds under the influence of positive or negative control mechanisms: in a positive-control system such as that exerted by CAP or N-protein, gene transcription is stimulated directly; in a negative-control system such as that involving the *lac* or λ repressors, induction involves the lifting of transcriptional repression. For the *qa* genes of *Neurospora*, a positive form of control is found: the inducer (quinic acid) appears to combine with a regulator (the product of a closely linked gene known as *qa-1*) to turn on *qa* expression. Evidence for this model of *qa* gene regulation comes from the properties of *qa-1* mutations. A number of *qa-1⁻* mutant strains have been isolated by M. Chase and N. Giles. All of these produce a **pleiotropic-negative** phenotype, meaning that quinic acid cannot induce any of the *qa*-encoded enzymes. Several of these strains, moreover, prove to carry deletions in the *qa-1* gene. Since deletions in a repressor-protein gene would be expected to lead to constitutive enzyme synthesis, whereas deletions in a positive-regulator gene would be expected to generate a pleiotropic-negative phenotype, it has been concluded that the *qa-1* gene product somehow acts in a positive manner.

Since *qa-1* maps close to the regulated genes, an alternate possibility is that it might be a control element analogous to a bacterial or phage operator. Several observations, however, indicate that *qa-1* is a gene and not a control element. First, heterokaryons (Section 12.16) which carry two (non-deletion) *qa-1⁻* alleles can be constructed, and certain of these are found to complement one another; that is, they exhibit intragenic comple-

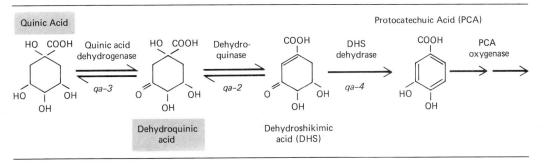

Figure 18-1 Catabolic quinate-shikimate pathway in *Neurospora crassa*.

mentation (Section 10.11). Such results suggest that the *qa-1* gene product is a multimeric protein. Moreover, certain *qa-1⁻* mutations are temperature-sensitive. Since almost all temperature-sensitive gene products are polypeptides, it again appears that *qa-1* encodes a protein having regulatory properties.

The *qa-1* gene can also mutate to a constitutive form. In such *qa-1ᶜ* strains, all three *qa* enzymes are synthesized even when quinic acid is absent. The *qa-1* protein produced by such strains has, therefore, apparently lost its requirement for inducer binding as a prerequisite for exerting positive effects. As would be expected from studies of prokaryotes, some such *qa-1ᶜ* mutations prove to be dominant to their wild-type allele in heterokaryons.

The *qa* gene cluster, in short, appears to be regulated by a closely-linked gene, *qa-1*, with genetic properties familiar to readers of Chapter 17. Lacking thus far are any *cis*-dominant mutations that mark controlling elements adjacent to the 3 structural *qa* genes. It is not known, therefore, with what the *qa-1* protein/inducer complex interacts in order to stimulate expression of the catabolic enzymes.

18.2   THE *GAL* GENE CLUSTER IN YEAST

Galactose fermentation in yeast is carried out by three enzymes—a kinase, a transferase, and an epimerase—in the same fashion as for *E. coli* (Figure 16-1). These enzymes are specified by three closely-linked loci known as *GAL1*, *GAL7*, and *GAL10*. Whether these *GAL* loci are in fact 3 genes or encode 3 portions of a multifunctional polypeptide (see Section 16.2) is not known; for present purposes we consider them as 3 genes. The three enzyme activities are coordinately induced by galactose. In many respects, therefore, the system is analogous to the *qa* cluster in *Neurospora* described above.

The yeast *GAL* cluster differs from *qa*, however, in that the *GAL* cluster is controlled by the product of an unlinked gene. Moreover, since most mutations in this gene lead to a constitutive synthesis, and not an inhibition, of the galactose catabolizing enzymes, this unlinked gene appears to encode a repressor-like protein. Additionally, mutations in the gene prove to be recessive to wild-type in diploids and thus behave very much like certain *i⁻* mutations in the *lac* operon. For this reason, the mutant genes were denoted as *i⁻* by H. Douglas, D. Hawthorne and their colleagues, and the gene regulating the *GAL* gene cluster has been called *i*.

Further study has shown that the analogy between the *i* genes of yeast and *E. coli* goes only so far. It turns out that the *i* repressor of yeast acts not on the *GAL* gene cluster itself but on yet another unlinked gene, *GAL4*. Mutations in the *GAL4* gene (denoted as *gal4*) generate a pleiotropic negative phenotype for the kinase, transferase, and epimerase enzymes. Moreover, *gal4 i⁺* and *gal4 i⁻* mutants are both noninducible by galactose.

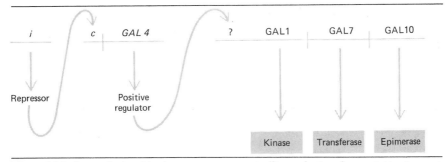

Figure 18-2 Interrelationships between genes controlling galactose fermentation in yeast.

Therefore, the *GAL4* locus appears to encode a positive regulator protein, much like the *qa-1* protein, which directly stimulates transcription in the *GAL* gene cluster; this *GAL4* locus, in turn, is negatively controlled by the *i* gene product. These relationships are diagrammed in Figure 18-2.

Associated with *GAL4* is a region called *c* which behaves formally like an operator locus in that it can undergo mutations to an o^c-like condition. Such mutations, known as *C*, map right next to *GAL4* and can confer constitutive enzyme synthesis on the *GAL* gene cluster only if they are *cis* to a wild-type *GAL4* allele. In this condition, moreover, they are dominant to the wild-type allele, so that a $\dfrac{CGAL4}{cGAL4}$ diploid produces enzymes constitutively. Presumably, therefore, the *c* locus is the usual site for *i*-repressor binding, (Figure 18-2), a property that can be lost by mutation.

Finally, the *i* locus can experience i^s mutations that are formally analogous to i^s mutations in *E. coli*: such mutations produce a repressor that is no longer susceptible to inducer action, and the resultant cell is noninducible by galactose. As one would expect, such i^s mutations are dominant to *i* in diploids but are not expressed in haploid strains carrying *C* operator mutations.

Lacking in this and all other descriptions of eukaryotic regulatory mechanisms to date is any notion of what a controlling element such as *c* might look like at a molecular level and, consequently, what the molecular nature of the *i* protein and its interaction with *c* might be. It can be expected that when genes such as *GAL4* are cloned in recombinant plasmids (Section 14.6), sufficient *c* DNA will be made available that its sequence and modes of repressor binding can be analyzed.

18.3 REGULATION OF NITROGEN METABOLISM IN *ASPERGILLUS*

The *qa* and *GAL* control systems described above are analogous to *lac* and *trp* in *E. coli* in that regulator molecules are apparently specialized to control a single gene or gene cluster. We have also encountered, in Chapter 17, a second class of regulator molecule which acts not at one but at several unlinked sites in the genome. A prominent example of this second class of regulator molecule is the CAP protein which responds to fluctuations in glucose/cAMP levels by altering the expression of several discrete operons in *E. coli,* all of which are related to sugar catabolism.

In eukaryotes, regulatory mechanisms of this kind must be very common since, as noted in Chapter 16, functionally related genes in eukaryotes are usually not linked in operon-like clusters, yet are often co-induced by the same effector molecules or environmental stimuli. One can therefore predict that regulatory genes like *crp,* which specifies the CAP protein in *E. coli,* should exist in eukaryotes and that mutations in these genes should disallow the induction of a variety of related, but unlinked, genes.

The *intA* (for *int*egrator) locus in linkage group II of *Aspergillus nidulans* appears to contain such a gene. Mutations in *intA* affect the expression of three inducible structural genes involved in nitrogen metabolism: *amdS* (in linkage group III) which specifies acetamidase, and *gatA* (linkage group VII) and *gabB* (linkage group VI) which specify a transaminase and a permease involved in ω-amino acid metabolism. Specifically, *intA^c* mutations lead to high constitutive levels of all three activities, while *intA$^-$* mutations reduce the levels of the 3 enzymes.

The *intA* product responds to at least two effector molecules, namely, the amino acid β-alanine and GABA (γ-amino-n-butyric acid), and all three structural genes are sensitive to both effectors. With molecular details still unavailable, readers of Chapter 17 are again presented with familiar forms of control circuits, and it is tempting to speculate that since the fungi lie at the "primitive" end of the eukaryotic spectrum, they will prove to have control mechanisms similar to those of bacteria.

SHORT-TERM REGULATION IN HIGHER EUKARYOTES

18.4 HIGHER EUKARYOTES AND THE HORMONAL RESPONSE

A fundamental feature of the higher eukaryotes is their exploitation of cell specialization, wherein a given cell type becomes able to perform particular functions and concomitantly unable to perform other functions. Although cell specialization can and does occur in some simple eukaryotes, it becomes a predominant feature of complex eukaryotes as seen, for example, in Table 18-1.

The extreme forms of cell specialization found in mature complex animals are possible because of the remarkable properties of blood (or blood-like fluid such as hemolymph). The composition of the blood is so stabilized that each cell receives the proper balance of ions, sugars, amino acids, and other essential molecules. Consequently, most cells in a complex eukaryote never experience the drastic fluctuations in the quality of the environment that simple eukaryotes must be prepared to face, and the cells of complex eukaryotes are thereby free to focus on particular metabolic activities.

Table 18-1 Distribution of Various Enzymes in Tissues of the Rat

Enzyme	Liver	Kidney	Spleen	Heart	Skeletal Muscle	Small Intestine	Pancreas	Brain
3-Hydroxybutyrate dehydrogenase	+		0	0	0		0	0
Malate dehydrogenase		0		+				
Pyrroline-2-carboxylate reductase		+	0	0				+
Catalase	+	+			0		0	0
Peroxidase	0	0	+	0	0	+		0
Homogentisate oxidase	+	+	0	0	0	0		0
Catechol methyltransferase	+		0	0	0	0		0
Dimethythetin-homocysteine methyltransferase	+		0					0
Ribonuclease	0	0		0	0		+	
Carboxylesterase		0	0		0		+	
Phospholipase	+		+	0	0	+		0
Acetylcholinesterase	0	0			0			+
Cholinesterases		0		+	0	+		
Alkaline phosphatase	0	+	0		0	+	0	0
Acid phosphatase			+	0	0			
Glucose-6-phosphatase	+		0	0	0			0
β-Mannosidase	+	+		0	0		+	0
β-Acetylamino deoxyglucosidase	+	+		0	0		0	0
β-Glucuronidase	+		+		0		0	
Arginase	+		0	0	0	0	0	0
Guanine deaminase	+	+	+		0		+	0
Aldolase	0	0	0	0	+			0
Citrate synthase	0	0		+	0			
Aconitate hydratase		+		+				0
Glutamine synthetase	+	0	0		0		0	+

+ means enzyme is present in amount 40 to 100 percent of that in the tissue where it is most plentiful; 0 means enzyme is essentially absent, that is, less than 8 percent of level of the tissue with the highest activity. If the level falls between 8 and 40 percent, or if data are lacking, space is left blank.
From R. J. Britten and E. H. Davidson, Science **165**:349 (1969), Table 2.

The constancy of the blood is effected by a number of tissues. Focusing on a vertebrate organism, we can cite such tissues as the intestine, liver, and kidney, which exert a direct effect on the blood's composition by adding the appropriate concentrations of ions and substrates and removing any excesses or waste products. Other tissues respond to changes in blood composition more indirectly: they secrete hormones that ultimately act to restore the equilibrium condition. As an example of a hormone-secreting organ we can cite the parathyroid gland: certain parathyroid cells secrete parathyroid hormone when blood calcium levels are below the optimum, and other cells secrete calcitonin when blood calcium levels are above the optimum. Parathyroid hormone ultimately promotes calcium resorption from the bone, increases calcium permeability in the intestine, and reduces calcium excretion in the kidney. Calcitonin has opposite effects. In other words, kidney, liver, and bone cells do not respond to blood calcium levels per se but rather to hormonal signals that dictate what the levels of calcium should be.

In our oversimplified view of a complex eukaryote, therefore, we can think of cell types that act on blood composition directly and of those that act on it more indirectly. A third group of cells, also intimately involved in overall regulation, secretes hormones that perform insulated regulatory functions—"insulated" in the sense that they operate independently of most external influences. A familiar example is the delicate balance of secretions by the pituitary and the ovary which collectively produce all the complex changes in a woman's uterus during a menstrual cycle. The course of the menstrual cycle may not, of course, be completely free of dietary or other external influences, but it is relatively free of such influences when compared with the parathyroid's response to a low-calcium diet.

From this brief discussion, then, we can come away with a picture of short-term regulation in complex eukaryotes as being largely mediated by hormones. The direct-acting organs—the kidney, liver, intestine, and others—are all responsive to a variety of hormones that dictate how the blood should be processed. Cells of other organs and tissues, not participating in blood processing at all, rely heavily on hormonal signals that dictate whether they will produce a given protein. It is hardly surprising, then, that most research on short-term regulation in complex eukaryotes has focused on hormones.

A given hormone may have one or several **target cells.** Parathyroid hormone, for example, has target cells in the kidney, intestine, and bone, and each cell type ultimately responds quite differently to the same hormonal species. In their various target cells, hormones are known to elicit new biosynthetic or metabolic processes, and we are interested here in how this is accomplished.

Many hormones, most notably those classified as **polypeptide hormones,** act at the cytoplasmic level to stimulate or repress levels of specific enzymatic activities. While these events are extremely interesting from the viewpoint of cell physiology, we focus here on hormones that appear to have a direct effect on gene expression.

18.5   ECDYSONE STIMULATION OF GENE TRANSCRIPTION IN DIPTERA

The first report of a specific hormonal effect at the chromosomal level in any organism was made in 1960 by U. Clever and his associates for the Dipteran *Chironomus tentans.* Clever first carefully studied the patterns of polytene chromosome puffing (Figure 8-23) at various stages of *Chironomus* larval development and established that these patterns followed a fixed sequence, with certain bands puffing at one stage and regressing at a later stage. He then took intermolt larvae, in which puffing of band I-19A was known to be particularly active, and injected the animals with **ecdysone,** an insect steroid hormone that triggers molting. He found that within 10 to 15 minutes puff I-19A regressed, band I-18C began to puff, and band IV-2B followed suit. This was exactly the puffing sequence known to take place just before molting, a time at which the insect spontaneously releases ecdysone into the hemolymph. On the reasonably certain assumption that puffing corresponds to the transcription of one or several genes (Section 8.20), Clever's experiments are best interpreted to mean that ecdysone somehow turns on the transcription of particular genes.

More recent studies by M. Ashburner have been made on third-instar *Drosophila* salivary glands cultured *in vitro* and presented with ecdysone at particular intervals. These experiments have established that approximately 6 bands respond very rapidly to ecdysone—bands 23E and 74E of the larval chromosomes, for example, show puffing within 5 minutes of ecdysone presentation—and all reach a maximum size and then regress within the first four hours. These so-called **early puffs** are unaffected by such inhibitors of protein synthesis as cycloheximide, and they regress prematurely if ecdysone is washed from the culture medium during the four-hour period. Between three and 10 hours after exposure to hormone, a group of about 100 **late puffs** can be visualized. Since late puffing does not occur in the presence of cycloheximide, it is concluded that the late puffs are triggered, at least in part, by some ecdysone-stimulated early-puff gene product(s) rather than by ecdysone alone. In other words, the rapid, and transitory, early puffing qualifies as the **primary response** to the hormone, with the more delayed, and massive, late puffing being the **secondary response.**

18.6 STEROID-HORMONE STIMULATION OF GENE TRANSCRIPTION IN MAMMALS

While it is clear from the above experiments that ecdysone stimulates specific patterns of gene transcription very rapidly, and probably without the need for a newly synthesized protein intermediary, it is as yet unclear how such stimulation is brought about. Does ecdysone act at the cell membrane to stimulate the uptake of certain effector molecules which, in turn, stimulate transcription, or does ecdysone itself act as an effector molecule, entering the cell, migrating to the nucleus and there stimulating transcription of specific genes in specific tissues? While the second possibility appears at present to be the more likely, definitive answers to these questions are still lacking for ecdysone.

A great deal more, on the other hand, is known about the patterns of activity for such medically relevant steroid hormones as the **estrogens, androgens,** and **glucocorticoids.** Mammalian and avian steroid hormones have been shown in numerous studies to enter their target cells and rapidly accumulate in the nucleus. Immature female rats, for example, have been exposed to ^{3}H-estradiol and, at various intervals, their uterine epithelium has been subjected either to autoradiography or to a cell fractionation in which nuclei are separated from cytoplasm. In both kinds of experiments, a sizable portion of the label rapidly localizes within the uterine-cell nuclei.

Transport of the hormone to the nucleus has been shown to be brought about by specific **receptor proteins** that reside in the cytoplasm of an uninduced cell. These proteins (perhaps 10,000 per cell) bind particular steroids with very high affinity. This binding causes an allosteric change in the shape of the receptor and is quickly followed by nuclear migration and a primary response to the hormone. Thus, in prokaryotic terminology (Chapter 17), steroids act as effectors and receptor proteins act as regulator molecules, becoming active only when bound to the effector.

Genetic evidence for the importance of steroid/receptor binding and nuclear migration in launching a primary steroid-hormone response has been provided by G. Tomkins, K. Yamamoto, C. Sidley, and associates. These investigators began with the mouse lymphoma cell line S49, which is rapidly killed by physiological concentrations of glucocorticoids, and proceeded to isolate mutant cell lines resistant to such cytotoxic effects of the steroid. These resistant mutants, they reasoned, might be expected to be blocked at various stages of steroid utilization. Analysis of several mutant lines has confirmed the investigators' prediction. Most of the mutant lines are found to be receptor-deficient (r^-), meaning that labeled steroid enters the cells but is not bound to any cytoplasmic or nuclear protein. Certain mutant lines prove to transport steroid to the nucleus with reduced efficiency (nt^-), while others show an excessively large proportion of steroid binding by the nuclear fraction and are called nt^i (increased nuclear transfer). When receptor proteins are isolated from nt^- and nt^i mutants, they exhibit altered molecular properties. Therefore, it is possible that at least some nt^- and nt^i mutations may mark the structural gene(s) for steroid receptor protein(s).

Far less well established is what the steroid/receptor complexes do once they enter a nucleus. That they can stimulate the expression of specific genes has been claimed in numerous experiments. It is presently unclear for many of these demonstrations whether or not the stimulated transcription represents a primary response to the hormone (that is, a very rapid response that is independent of protein synthesis, as with the early puffs in *Drosophila*) or a secondary response: the case is very difficult to prove. Nonetheless, most investigators assume that steroid/receptor complexes do, in fact, stimulate the expression of (probably) a very limited number of genes in target nuclei, the products of which proceed to elicit a (more massive) secondary response as the cell prepares to enlarge, divide and synthesize particular proteins, or whatever the physiological response may happen to be for that tissue.

We arrive, therefore, at the heart of the matter: How do steroid/receptor complexes stimulate specific gene transcription? The answer is unknown, and the available experimental data, while extensive, in fact tell us rather little because the experiments themselves are open to multiple lines of interpretation. We must therefore move to the realm of theories and models.

18.7 THEORIES OF SHORT-TERM REGULATION IN EUKARYOTES

Contemporary theories of eukaryotic gene control are necessarily based on our current understanding of chromatin organization and transcription. As detailed in Chapters 2, 7, and 8, eukaryotic chromosomes are made up of DNA-histone "beads" known as nucleosomes. The DNA itself is organized in blocks of "unique" DNA interspersed with blocks of "moderately repetitive" sequences, and the histones appear to associate with the DNA in stoichiometric proportions. A diverse group of proteins known as the nonhistones also associates with chromatin. Nuclear transcription generates very long pieces of heterogeneous (hn) RNA, at least some of which are "tailored" to create mRNA. Major candidates for regulatory molecules or mechanisms in eukaryotes, therefore, include the nonhistones, the histones, the physical conformation of chromatin, and the moderately-repetitive DNA sequences. These are considered in turn in the following sections.

The Nonhistones   The nonhistone proteins are excellent candidates as regulator molecules: they associate with chromatin; there are many different kinds (several hundred are detected by two-dimensional gel electrophoresis) so that they could potentially carry out numerous different functions; and they reportedly vary greatly in their tissue distribution and their relative abundance at different stages of the cell cycle and cell differentiation. In the steroid field, for example, one report claims that a particular nonhistone protein is synthesized when cortisone, a steroid produced by the adrenal gland, is given to adrenalectomized rats. Other reports claim that particular nonhistone proteins associate with steroid/receptor complexes once these enter the nucleus. The problem is that there is as yet no way to demonstrate cause-and-effect relationships in such studies: a given phenomenon could be the byproduct of hormone action rather than its necessary intermediate. Similarly, the different tissue distributions of the nonhistone proteins could be the consequence, and not the cause, of cell differentiation. Mutant strains are unquestionably essential if one is to prove, as in prokaryotes, that a particular nonhistone protein is in fact acting as a regulator molecule.

The Histones   Current models for the involvement of histones in regulation begin with the generally accepted "dogma" that histones do not vary in overall composition or proportion from tissue to tissue or even from one region of chromatin to the next. It is also widely believed, although not proven, that a "tight" DNA-histone complex is relatively inert to transcription. A fundamental role of histones in gene regulation is therefore perceived to lie with their ability to "turn off" gene expression.

It is also known that histones can be modified by phosphorylation, methylation, and acetylation; the last is particularly pertinent since acetylated histones are known to be relatively short-lived. Some of these modifications might, therefore, (transiently) alter DNA/histone interactions and

Figure 18-3 Electron micrographs of (a) normal and (b) leukemic human lymphocytes. *Abbreviations:* C, centriole; G, Golgi region; nu, nucleolus, ER, endoplasmic reticulum. (From D. R. Anderson, *J. Ultrastruc. Res.* **9, Suppl:1,** 1966.)

allow certain segments of the genome to be transcribed, as described in detail below.

Physical Conformation of Chromatin It was noted in Section 7.13 that highly repetitive eukaryotic DNA is usually found to localize in permanently condensed, deeply staining portions of chromosomes known as constitutive heterochromatin. These repetitive sequences are not transcribed *in vivo*, suggesting that DNA maintained in such a tightly condensed state may be inaccessible to transcription enzymes (although not to replication enzymes). This concept is supported by a second phenomenon, known as facultative heterochromatization (to be described in detail in Section 18.17), wherein the turn-off of expression of a particular chromosome or set of chromosomes during development is accompanied by their becoming dense and deeply staining.

In addition to these two specialized types of heterochromatin, numerous observations demonstrate that highly differentiated cells contain much more condensed chromatin than do relatively undifferentiated cells. This is highlighted in Figure 18-3. Figure 18-3a illustrates a normal white blood cell whose nucleus contains large masses of electron-dense chromatin. Figure 18-3b shows a leukemic white blood cell, a cell engaged in rapid rates of protein synthesis and proliferation, and it is seen to contain virtually no condensed chromatin.

That condensed chromatin is not transcribed *in vivo* can be demonstrated by autoradiography (Box 2.2): when ^{3}H-uridine is administered to cells and autoradiographs are made of specimens prepared for electron microscopy, the observed traces of radioactive emissions tend to appear over euchromatin and not over areas of condensed chromatin. Moreover, when nuclei are subjected to various fractionation procedures that separate actively transcribed chromatin (carrying nascent ^{3}H-uridine-labeled RNA chains) from inactive chromatin, the active fraction is found to have distinctive physical properties that are consistent with the idea that it is in a more open, accessible conformation. For example, chromatin being transcribed is found to be particularly sensitive to DNase digestion.

These various lines of evidence suggest, therefore, that gene transcription may be selectively activated by a decondensation of previously condensed chromatin. In a specific model proposed by K. Yamamoto and B. Alberts, such decondensation events are visualized as occurring in small patches, each containing perhaps 6000 nucleotides, with controlling elements contiguous to the patches. The model envisions that when steroid/receptor complexes enter the nucleus of a target cell and interact with particular target controlling elements, "waves" of chromosomal protein changes—for

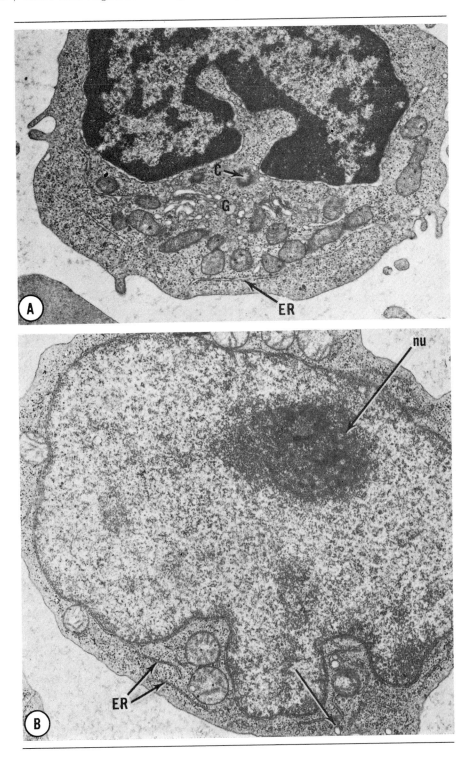

example, histone acetylation—might spread along the adjacent chromatin, opening it up for transcription.

Interspersed, Moderately-Repetitive DNA The Yamamoto-Alberts model for steroid-stimulated gene activation supposes that steroid/receptor complexes, like the CAP/cAMP complexes of *E. coli*, recognize the DNA sequences of controlling elements contiguous to target genes. R. Britten and E. Davidson were the first to point out, in a series of highly influential reviews, that since the moderately repetitive DNA of the eukaryotic genome is interspersed among unique nucleotide sequences (Section 7.14) and since the unique sequences are assumed to contain structural genes, eukaryotic controlling elements might prove to be synonymous with, or reside within, moderately-repetitive DNA. Yamamoto and Alberts incorporate this concept into their model and propose that the binding of a steroid/receptor complex to a moderately repetitive sequence next to a target gene initiates the aforementioned "wave" of conformational change in the adjacent chromatin, and its consequent expression.

Is there any evidence that steroid/receptor complexes bind to chromatin? The answer is yes, and the affirmative is considerably strengthened by the demonstration that glucocorticoid/receptor proteins from *nt⁻* steroid-resistant variants of the lymphoma line S49 have a lower binding affinity for chromatin than do controls. The answer must at once be qualified, however, for it appears that both normal and mutant steroid/receptor complexes have a low affinity for a very large number of sites in the eukaryotic genome, many more than they presumably activate. Therefore, in most binding studies, massive nonspecific binding creates a "noise level" above which any specific binding is very difficult to detect. Similar "noise," we should quickly note, is present in prokaryotic studies as well: CAP protein binds nonspecifically to many sites in the *E. coli* chromosome, and its specific affinity for the *lac* promoter was only demonstrated once purified *lac* DNA became available.

Purified "target DNA" for steroid/receptor complexes will presumably soon become available by the technique of genetic cloning (Section 14.6). Meanwhile, many (although not all) workers in the field assume that there exists in a eukaryotic genome a very small number of controlling elements to which particular steroid/receptor complexes bind with high affinity, these controlling elements being adjacent to "target genes." Yamamoto and Alberts include in their model the concept that a "target controlling element" will have multiply repeated copies of high-affinity binding sequences, reminiscent of the λ operator regions (Figure 17-15). A "wave" of chromatin activation would, in this case, be achieved not when a single steroid/receptor complex binds randomly, and transiently, with a low-affinity site but only when a number of complexes associate, side by side, along the reiterated high-affinity sites next to the target genes.

Similar kinds of models can be proposed, of course, for the activity of integrator-type proteins, repressors, and activators in fungal cells. The

Yamamoto-Alberts model is also immediately applicable to the ecdysone-stimulated puffing of specific polytene bands in the Diptera. Indeed, Yamamoto and Alberts conclude their model by pointing out that each "patch" of chromatin and its associated controlling element (repetitive sequence) may well prove to be synonymous with a chromomere (Section 2.12).

Posttranscriptional Mechanisms   Once the "decision" has been made to transcribe a gene, a number of intracellular mechanisms can govern the extent to which the transcript is ultimately allowed phenotypic expression. We saw in Section 17.22 that the *cl* gene of phage λ is transcribed into an mRNA having either a short or a long leader sequence; the former is translated very infrequently, the latter very frequently, with the result that intracellular λ repressor levels differ markedly in the two cases. Such **"translational control"** also operates to determine the order of expression of the three proteins encoded by such RNA phages as Qβ and R17: the AUG "start" signal for translation of the coat-protein sequence is the only one available to ribosomes when a phage enters an *E. coli* cell; translation of the coat sequence brings about a major conformational change in the structure of the phage RNA, however, with the result that the AUG codons for the replicase and A-protein sequences are unmasked.

For eukaryotes, posttranscriptional control theories focus either on **intranuclear** events or on **cytoplasmic** events. Intranuclear control would most likely occur at the level of "processing" the primary gene transcripts (Section 8.17). If, as seems increasingly likely, primary transcripts prove to be synonymous with the giant molecules called heterogeneous nuclear RNA (hnRNA; Section 8.20), then these transcripts clearly undergo considerable "tailoring" before acquiring their final mRNA form: extra nucleotides are excised, perhaps from their interiors (Section 8.20); poly-A tracts are added to their 3' ends; and so forth. One might then postulate that the relative levels of intranuclear tailoring enzymes could influence which primary transcripts are capable of being translated into protein. Specifically, the cytoplasmic transport, lifetime, and ribosome-binding efficiency of a particular transcript might be variously affected by such enzymes.

Posttranscriptional control at the cytoplasmic level could be exerted by different classes of ribosomes and/or initiation factors that enhance or depress rates of expression of particular mRNAs. Moreover, **post-translational** modifications of the polypeptide product (acetylation, glycosylation, and so on) could affect phenotypic expression if they were to occur at different rates. While any one of these mechanisms may well operate to influence the final expression of a particular gene, mRNAs from one cell type are generally found to be translated very efficiently and accurately by cells or cell extracts from other cell types. J. Gurdon, for example, injects purified mRNAs into *Xenopus* oocytes and follows their translation into radioactively labeled proteins. The list of mRNAs that have been successfully translated in such experiments includes those for calf lens protein, mouse immunoglobulins, and duck globin. The rate of synthesis of globin mRNA by the *Xenopus* oocytes, moreover, is comparable to that carried out by the highly specialized reticulocyte cell. Therefore, most students of eukaryotic regulation assume that short-term control of gene expression will probably, in the main, operate at the nuclear level.

LONG-TERM DIFFERENTIATION: SOME GENERAL FEATURES

Our consideration of long-term differentiation in higher eukaryotes can begin with a very general picture. Differentiation commences with a fertilized egg, itself a highly specialized cell that performs certain functions that no mature cells ever perform. The zygote undergoes a series of mitotic divisions. At various times during this mitotic period individual cells become **determined**—a process that no one understands—such that they and their clones become committed to forming particular cell types—liver, heart, kidney, or whatever. Following the determination event—and often many cell generations later—individual cells undergo **differentiation** into their specialized forms. Differentiation may involve mitosis, cell fusion, cell migration, or intercellular interactions. These events occur in an elaborate, self-reinforcing chain: self-reinforcing in the sense that event A must be completed before event B can possibly begin, that the onset of event B triggers a biosynthetic event or cellular migration necessary for event C, and so forth.

The process of differentiation, strictly speaking, lies outside the scope of genetics, since it represents the acting out of particular developmental programs. The geneticist is more directly concerned with the nature of the developmental programs themselves and how different programs are imposed on particular cells at particular times during embryogenesis.

18.8 THE CONSTANCY OF NUCLEAR DNA

A simple way to visualize a determination event is to propose that a cell destined to give rise to, say, a liver cell simply degrades all of its DNA except the DNA relevant to liver-specific and general, cellular "housekeeping" activities; a cell determined to give rise to a bone cell would similarly discard all but its bone-making and housekeeping information; and so forth. Alternatively, one could propose that a series of directed mutational events might change the informational content of liver DNA so that it would proceed in one direction while bone DNA would proceed in another. The experiments described below indicate that neither of these proposals is a correct description of eukaryotic development, and suggest instead that the nuclei of differentiated cells continue to carry most of the genetic information present in the zygote nucleus.

The first line of evidence for the retention of a complete genome during development comes from nuclear transplantation experiments with *X. laevis* conducted in J. Gurdon's laboratory. Nuclei were taken from differentiated tissues such as brain, liver, and intestine, and individual nuclei were injected into fertilized *Xenopus* eggs whose nuclei had been surgically removed. Most of these hybrids did not survive, presumably because of the trauma incurred during their preparation, but the few that did went on to develop into perfectly normal toads. As proof that the survivors did indeed inherit their genomes from the transplanted nuclei, Gurdon used +/O-nu heterozygotes as his donor strain and +/+ toads as his recipient strain. As noted in Section 8.15, +/O-nu nuclei contain only one nucleolus, whereas +/+ nuclei contain two nucleoli. Gurdon showed that in numerous tissue samples from various survivors only uninucleolate cells were present. Clearly, nuclei from the liver and brain of a toad, then, must continue to carry all the DNA necessary to construct an entire frog.

Somatic cell hybridization experiments offer the second type of evidence for the concept that differentiation is not accompanied by a permanent loss of genetic information. The technique of cell fusion via Sendai viruses was described in Section 12.19 in the context of somatic-cell genetics. The study of such fused cells has also yielded important information about the nature of differentiation. Relevant here are studies from H. Harris's laboratory of the fusion of cultured mouse fibroblasts with chicken erythrocytes. Several features of the erythrocytes must be described.

1. The chicken erythrocyte nuclei are highly condensed and transcriptionally inactive.
2. Before being turned off completely, the erythrocyte nuclei have been almost totally devoted to transcribing information for hemoglobin synthesis.
3. The chick erythrocyte contributes little if any cytoplasm to the heterokaryon formed after fusion with a mouse fibroblast has occurred, meaning that the newly formed heterokaryon contains mouse fibroblast cytoplasm, a mouse fibroblast nucleus, and a chick erythrocyte nucleus.

Shortly after such heterokaryons form, the chick chromatin loses its condensed appearance and begins to synthesize RNA (as judged by autoradiography). Moreover, it does not revert to synthesizing globin messages, as might be predicted considering its previous "incarnation." Instead, it appears to transcribe general housekeeping genes. One of these genes contains information for the synthesis of inosinic acid pyrophosphorylase (IAP), an enzyme involved in ribonucleotide synthesis. When chick erythrocytes are fused with mutant mouse cells deficient in IAP activity, the heterokaryons are found to exhibit IAP activity several hours after fusion, and the enzyme has the biochemical properties of the *chick* enzyme and not of the mouse enzyme. Even the extreme form of nuclear differentiation exhibited by avian erythrocytes, therefore, does not bring about a permanent alteration in the genetic potential of the nucleus.

It should quickly be pointed out that these classical studies with nuclear transplants and cell hybrids in fact demonstrate only that differentiation *need not be* accompanied by a permanent loss of genetic information. They leave open two very important possibilities. First, they cannot prove that permanent modifications of the genome do not accompany some (or even many) differentiation events. Indeed, the experiments of Hozima and Tonegawa (Section 15.20) reveal that rearrangements of the immunoglobulin genes in fact appear to take place during the differentiation of lymphoid cells, and similar events may affect small portions of the genome in other differentiated cells. Second, such experiments are by definition unable to detect "semi-permanent" changes in the genome. If, for example, the arrangement of DNA is altered during differentiation by selected recombination events, (as diagrammed for the immunoglobulin genes in Figure 15-20), then it is possible that these rearrangements might be capable of reversing themselves if the nucleus is transplanted to a different cytoplasm.

18.9 LIMITED DNA TRANSCRIPTION

Since the wholesale elimination or alteration of DNA does not at present appear to accompany most determination or differentiation events, it is logical to next propose that differentiated cells transcribe only a small fraction of their genome and that cells with different specializations transcribe different portions (transcribing in common, perhaps, certain house-keeping genes). Both predictions can be tested by DNA-RNA hybridization, a technique described in Section 8.4. Without presenting the details of these studies, which prove to involve quite sophisticated molecular analyses when performed correctly, we can summarize their conclusions.

As noted in Section 7.10, C_0t-plot data reveal that about 70 percent of the DNA in a eukaryotic nucleus is of the "single copy" class which is assumed to contain structural genes. There is enough single-copy DNA per nucleus for about a million genes, yet several kinds of analyses indicate that a mammalian genome contains a maximum of only about 50,000 genes. Therefore, only about 5 percent of eukaryotic single-copy DNA is believed to code for protein; the function of the remaining 95 percent is unknown.

If we take 50,000 genes as a working number, we can next ask what fraction of these genes is expressed during development and what fraction is expressed by fully differentiated cell types. A recent analysis of this sort, made by G. Galau and colleagues with sea urchins, is shown in Figure 18-4. The largest number of genes, an estimated 20,000–30,000, find expression in the oocyte. As will be described in Section 18-11, intensive transcription during oocyte differentiation generates abundant "maternal mRNAs" that are stored until fertilization and used during subsequent stages of embryon. development which, for the sea urchin, are denoted as the blastula, gastrula and pluteus stages. Therefore, the data for these stages in Figure 18-4 probably represent, at least in large part, oocyte-generated gene tran-

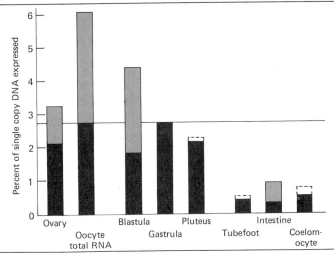

Figure 18-4 Sets of structural genes active in sea urchin embryos and adult tissues. The black portion of each bar indicates the amount of single-copy sequence shared between gastrula mRNA and the RNA preparations listed along the abscissa. The gray portions show the amount of single-copy sequence present in the various RNAs studied but absent from gastrula mRNA. Dashed lines indicate the maximum amount of reaction which could have been present and escaped detection. (From G. A. Galau et al., *Cell* **7**:487, 1976. Copyright © MIT.

scripts. The transcriptional activity during embryogenesis is seen in Figure 18-4 to be markedly greater than that found in such differentiated adult tissues as tubefoot or intestine. In other words, it would appear that most of the genes in the sea-urchin genome encode information required for early events in embryology.

To what extent do the gene transcripts in various differentiated cell types differ from one another? A comparison by R. Axel and colleagues of the RNA sequences present in chicken liver and chicken oviduct has revealed that each tissue type is expressing on the order of 15,000 genes. Of these, 85 percent appear to be "shared," that is, about 13,000 of the single-copy genes transcribed in oviduct nuclei are also transcribed in liver nuclei. These presumably encode proteins that serve housekeeping functions common to both, and perhaps all, tissues in the chicken. The remaining 15 percent of the transcripts appear to dictate the tissue-specific proteins that determine the very different properties of oviduct and liver cells.

The studies and the numbers cited above derive from very recent publications and, in view of the complexity of these experiments, somewhat different numbers will likely be generated in the next few years. The order-of-magnitude concepts, however, are probably valid: a small proportion of the total eukaryotic DNA contains protein-encoding genes; a large percentage of these genes is expressed early in development; a much smaller percentage of these genes is expressed by differentiated cell types;

and only a very small number of genes experience unique (or at least major) expression in specific differentiated cell types. Ultimately, then, an understanding of the differentiation "program" of an organism must include an explanation of how these patterns of gene expression are brought about.

18.10   REGULATORY MOLECULES IN THE CYTOPLASM

A differentiated nucleus clearly experiences constraints on its transcriptional activities. These constraints may involve histones, nonhistones, or other agents. The cytoplasm, however, also plays a major role in long-term differentiation. Such early experimental embryologists as T. Boveri and E. B. Wilson provided numerous examples of cytoplasmic influences over development (many are well described in E. Davidson's excellent book, *Gene Activity in Early Development*). That the cytoplasm can induce nuclear changes is also obvious in the chick-mouse cell fusion experiments and in the toad nuclear transplant experiments described in Section 18.8. The influence of the cytoplasm becomes particularly dramatic in some of the toad experiments: Gurdon has found that when a small, nondividing brain nucleus is injected into an enucleate fertilized egg, the nucleus swells to the size characteristic of a fertilized egg nucleus. It also modulates its rate of RNA synthesis and increases its rate of DNA synthesis in an apparent effort to mimic the biosynthetic properties of a rapidly dividing fertilized egg nucleus. When a brain nucleus is instead injected into an egg undergoing meiosis, its chromosomes condense and abortive spindle formation occurs.

A cornerstone of developmental genetics is therefore the concept that the cytoplasm is of critical importance in maintaining the differentiated state. It is imagined that when certain determined genes turn on, certain proteins are synthesized; some act in the nucleus to sustain the transcription of particular genes and others act in the cytoplasm. The cytoplasm of a differentiated cell is thought to acquire in this manner a collection of regulatory molecules that reinforce the differentiated condition. When a foreign nucleus is introduced into this cytoplasm by transplantation or cell fusion, the nucleus is expected to come rapidly under the regulatory influence of the host cytoplasm.

PERSPECTIVE

The experiments described in the preceding sections enable us to put forward three axioms relating to long-term differentiation: (1) differentiation does not involve massive, permanent changes in nuclear DNA; (2) differentiation involves limited DNA transcription; and (3) differentiation involves self-reinforcing changes in the cytoplasm.

In the sections that follow, individual case histories of well-documented and interesting developmental phenomena are presented that illustrate the above 3 axioms more specifically.

EARLY EVENTS IN LONG-TERM DIFFERENTIATION

18.11 REGULATION IN THE EGG

The differentiation of an oocyte cell into a mature egg has been studied most extensively in the echinoderms and amphibia. As noted in Section 18.9, these oocytes engage in the synthesis of enormous amounts of RNA. Some of this RNA proves to be informational and the rest ribosomal; we consider each in turn.

Informational "Lampbrush RNA" Synthesis We noted in Section 8.20 that the diplotene oocyte chromosomes of an amphibian such as *Xenopus* take on a "lampbrush" morphology (Figure 8-21). The abundant RNA produced during the "lampbrush stage" of amphibian oogenesis is presumably functionally analogous to the RNA produced by sea urchin oocytes (Figure 18-4): once the egg is fertilized and cleavage begins, "lampbrush RNA" appears to direct most of early embryogenesis. Thus, inhibitors of gene transcription have little if any effect on protein synthesis or normal development until the blastula stage (Figure 18-5), at which time the embryo consists of from 6000 to 15,000 cells; moreover, embryonic nuclei seem virtually inactive in RNA synthesis during the early-cleavage stage. The toad egg therefore appears to accumulate well in advance the informational RNA it will need to take it through all the early stages of development.

The storage of maternal informational RNA for use in the zygote is most conspicuous in amphibia, echinoderms, and insects. Relatively little of this RNA appears to be present in the mammalian egg, in which the zygote nucleus commences mRNA synthesis by the 2- to 4-cell stage. It is not known why mammals are distinctive in this feature of development.

Ribosomal RNA Synthesis The second kind of RNA synthesized in the amphibian oocyte and stored for use by the embryo is ribosomal RNA. Its synthesis occurs in the pachytene stage of prophase I, the stage preceding the lampbrush stage, and it occurs by a most extraordinary process. As we saw in Section 8.15, the nucleolar organizer DNA of *Xenopus* is a long stretch of tandemly repeated copies of rRNA genes interspersed with spacer DNA. During oogenesis this DNA undergoes repeated rounds of replication by the same type of rolling-circle mechanism that generates copies of *E. coli* plasmids (Figure 11-3). Eventually, 1000 to 1500 additional copies of nucleolar organizer DNA are produced, so that, at diplonema, the nucleus is found to contain 1000 to 1500 tiny independent nucleoli, each actively engaged in the rapid synthesis of rRNA. The nucleoli function until the first meiotic division, and the rRNA that is synthesized is stored in the egg in the form of **maternal ribosomes.** At meiosis I the extra nucleolar DNA is discarded into the cytoplasm. Nothing is yet known about how this gene-amplification process is restricted to a particular period in amphibian development.

The storage of maternal ribosomes for use by the young zygote is a conspicuous activity of some egg cells and not others, and only fish and amphibia appear to undertake amplification of their ribosomal DNA in order to increase their synthetic potential for rRNA. More generally, a comparative study of oogenesis leads one to conclude that quite different avenues have evolved among different species for the differentiation of an egg cell. Even at its inception, in other words, the zygote of one species may differ dramatically from the zygote of another.

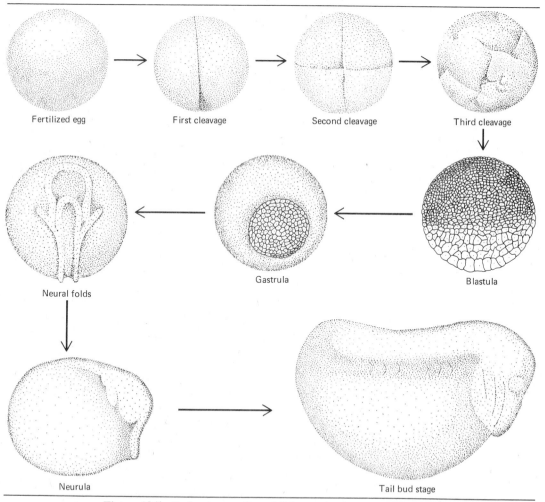

Fertilized egg   First cleavage   Second cleavage   Third cleavage

Gastrula   Blastula

Neural folds

Neurula   Tail bud stage

Figure 18-5   Early stages of amphibian embryogenesis. (From R. Rugh, *A Guide to Vertebrate Development*, Burgess, Minneapolis, 1971, rev. 1977. Used by permission.)

18.12 MATERNAL INFLUENCE IN EARLY DEVELOPMENT

In most embryos, the first two or four cleavage-stage cells (Figure 18-5) can give rise, when dissected apart, to small but complete embryos. This capacity is progressively lost, however, as cleavage proceeds. Instead, each cell undergoes a **primary determination** such that it becomes committed to form only particular portions of the embryo. Primary determination events are relatively crude; they seem to direct a cell along general rather than specific lines of development. Once a cell has become primarily determined as an endodermal cell, however, its descendants experience the more finely tuned determinative events that dictate whether the cell will become liver, gut, or pancreas. In other words, the primary events appear to set the stage for the rest of the developmental process.

In many organisms primary determination events appear to be dictated by **maternal influences.** It seems that during differentiation of the egg the maternal genome directs a specific kind of patterning within the egg cytoplasm so that "determinative molecules"—possibly proteins, possibly RNAs—are differentially placed in different parts of the egg. For example, the eggs of many lower vertebrates appear to contain "endodermal determinants" in one region, "ectodermal determinants" in another, and so on. Then, as the egg cleaves, individual daughter cells arise that contain, let us say, only the endodermal factors, and these cells experience primary determination in the endodermal direction.

Shell Coiling in the Snail   As a first example of maternal influences we can consider the inheritance of the direction of shell coiling in the snail *Limnaea paregra*. The shell may coil to the left or to the right, and the internal organs of the snail will adopt a corresponding handedness. A single genetic locus determines the trait, with the *dextral* (D) allele dominant to the *sinistral* (d) allele. The expression of this locus is unusual, however, in that it occurs during oogenesis. As shown in Figure 18-6, a *d/d sinistral* mother will give rise only to *sinistral* offspring, even when the father provides *D*-bearing sperm and the offspring are all *D/d* heterozygotes. Cytological studies indicate that when eggs derive from *sinistral* mothers the plane of the first zygotic cleavage is at right angles to the plane adopted when eggs derive from *dextral* mothers. Therefore, at some time before the second meiotic division of snail oogenesis, the *D* and *d* alleles in the maternal genome specify some property of the cytoplasm that ultimately determines the orientation of the first cleavage furrow of the zygote.

The Germ Plasm in Amphibia   Certainly the best studied examples of cytoplasmic determinants are those that localize in one sector of amphibian and insect eggs to form what embryologists have labeled the **germ plasm.** This cytoplasm is readily identified because it typically contains conspicuous granules. When the fate of these granules is followed visually during

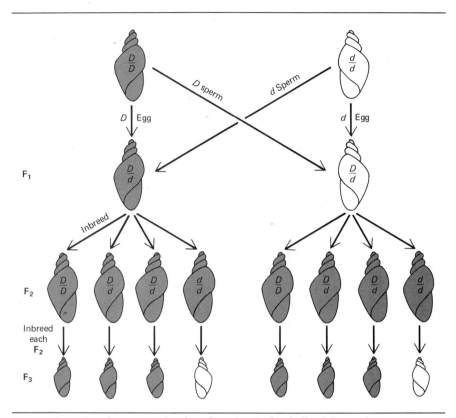

Figure 18-6   The inheritance of coiling direction in the shells of the snail *Limnaea*. Gray snails are dextral, white snails are sinistral. (After E. Sinnot, L. C. Dunn, and T. Dobzhansky, *Principles of Genetics,* 5th ed. McGraw-Hill Book Company. Used with permission of McGraw-Hill Book Company. © 1958)

Xenopus embryogenesis, they are found at gastrulation to localize in only a few cells. If these cells are removed at time of gastrulation, the resultant toad is sterile; therefore, these cells are identified as constituting the primary cells of the germ line. If the germ plasm is instead surgically removed from a fertilized egg, a sterile toad also results. It appears, therefore, that the maternal genome must be capable of both producing and patterning a specific sector of the egg cytoplasm that will, later in embryogenesis, exert a determinative influence.

Polar Plasm in Drosophila   The germ plasm in *Drosophila* is localized at the posterior end of the oval-shaped *Drosophila* egg and is therefore known as the **polar plasm,** with its granular inclusions being known as **polar granules.** At the conclusion of the early cleavage stage of the zygote the polar granules become localized in about 10 **pole cells** which divide several times and then form a ridge of tissue from which are generated the gametes of an adult fly.

Surgical experiments of the type described above for *Xenopus* have indicated that the polar plasm contains the germ-line determinants in *Drosophila*. Genetic evidence also supports this notion: in *D. subobscura* a recessive autosomal mutation known as *grandchildless* (*gs*) affects the formation of normal pole cells with the result that the eggs produced by *gs/gs* homozygotes give rise, when fertilized, to completely sterile male and female progeny that have no gametes in their gonads.

K. Illmensee and A. P. Mahowald have recently demonstrated the determinative properties of *D. melanogaster* polar plasm experimentally. As diagrammed in Figure 18-7, Illmensee and Mahowald took polar-granule-

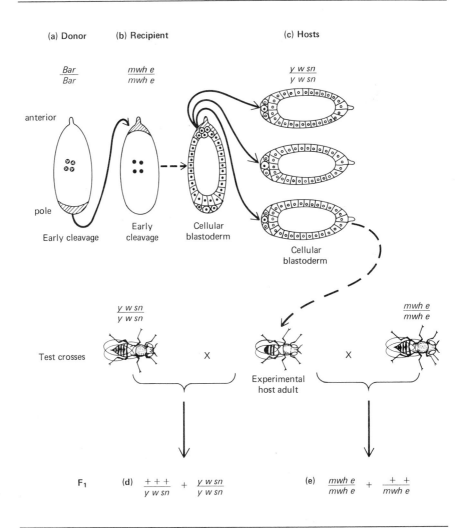

Figure 18-7 Diagram of the Illmensee-Mahowald experiments. See text for details. (From K. Illmensee and A. P. Mahowald, *Proc. Natl. Acad. Sci. U.S.* **71**:1016, 1974.)

containing cytoplasm from **donor** zygotes, injected it into the anterior ends of young **recipient** zygotes, allowed development to proceed for about two hours, and tested whether the anterior cells exposed to the polar plasm had been influenced to become presumptive germ cells. To answer this question it was necessary to transplant the anterior cells to the posterior region of **host** embryos where they would be assured of being included in adult gonadal tissue. It was also necessary to use genetic markers so that donor, recipient and host cell types could be distinguished, as described in detail below. The result was unambiguous: the anterior region of the recipient zygote, which normally differentiates into somatic structures, acquired the ability to differentiate into germ-line cells when exposed to the polar-granule-containing cytoplasm.

The specifics of the Illmensee-Mahowald experiments, diagrammed in Figure 18-7, are as follows. Three sets of genetically marked flies were used. Donor cytoplasm was derived from young zygotes carrying the *Bar* mutation (Figure 18-7a) so that if any nuclei were inadvertently included in the extract, dominant *Bar* genes would be detectable later in the experiment. Recipient embryos were homozygous for the chromosome-3 markers *mwh* (*multiple wing hairs*) and e (*ebony* body) (Figure 18-7b). Host embryos were wild-type for both of these chromosome-3 loci but were homozygous or hemizygous for the X-chromosome markers *y* (*yellow* body), *w* (*white* eyes) and *sn* (*singed* bristles) (Figure 18-7c). Recipient embryos were separated into two groups, the "experimentals" and the "controls." Experimentals received anterior injections of the donor cytoplasm whereas controls received mock injections. Host embryos were similarly divided into two groups, one receiving posterior-region implantations from experimental embryos (Figure 18-7c) and the other receiving implantations from control embryos.

Experimental and control host embryos were then allowed to mature into adult flies, and each was mated with *y w sn* homozygotes (Figure 18-7d). The progeny from the control matings were, as expected, all *yellow white singed.* Four of the experimental flies, however, were found to produce wild-type as well as *yellow white singed* offspring (Figure 18-7d). In other words, these four flies produced gametes carrying the + alleles of *y*, *w*, and *sn.* If these gametes derived from implanted cells, reasoned Illmensee and Mahowald, then their full haploid genotype should be + + + *mwh* e,

meaning that the wild-type progeny flies should be $\dfrac{+\quad +}{mwh\quad e}$ heterozygotes.

To prove this, the wild-type flies were test-crossed to *mwh* e homozygotes (Figure 18-7e), and about half of the test progeny indeed proved to express both *multiple wing hairs* and *ebony* body color.

The Illmensee-Mahowald experiments demonstrate, therefore, that the anterior region of a fertilized *Drosophila* egg, which normally gives rise to various somatic tissues, can be converted to germ-line cells simply by exposure to a polar cytoplasm extract. This cytoplasm, therefore, can cause a cell type to differentiate in a particular direction.

General Considerations If, as seems clearly the case, cytoplasmic factors serve to influence different portions of the early embryo to differentiate in particular directions, then many of the important questions to be asked in developmental genetics are simply pushed back to the oocyte-maturation stage. What kinds of genetic programs "tell" an egg cell to put certain molecules in one sector of the cytoplasm and other molecules in another? Is this all accomplished by gene products of the lampbrush stage (when the egg still carries genes inherited from both parents) or are stages before and after the lampbrush stage also important? And how do the cytoplasmic factors work? None of these questions can presently be answered.

IMAGINAL DISCS AND *DROSOPHILA* DEVELOPMENT

18.13   IMAGINAL-DISC CELLS IN *DROSOPHILA*

In describing the Illmensee-Mahowald experiments we glossed over the process of embryogenesis in *Drosophila,* but since this process begins quite differently from vertebrate embryogenesis and is of intense interest to developmental geneticists, we now examine it more closely.

In *Drosophila,* after fertilization and fusion of the pronuclei, the zygote nucleus undergoes a very rapid and synchronous series of mitoses until about 500 nuclei fill the central portion of the zygote. These nuclei do not mix freely but instead preserve their orientation with respect to one another. Each nucleus then migrates to a position at the periphery of the zygote and most of these divide four more times so that about 6500 nuclei are present. Finally, cytoplasmic membranes form around each nucleus, whereupon the multinucleate **syncytial blastoderm** is transformed into a multicellular structure, one-cell thick, known as the **cellular blastoderm.** (Figure 18-7).

The blastoderm cells proceed to differentiate in one of two general directions: some give rise to body parts of the larva, whereas others will eventually form the body parts of the adult fly. It turns out that throughout the three larval instars of the *Drosophila* life cycle (Figure 3-6), certain cells are set aside as discrete packages of essentially undifferentiated cells, each region being known as an **imaginal disc.** As diagrammed in Figure 18-8, there is a disc for each leg, for each wing, for the genital structures, and so forth. The disc cells retain their undifferentiated morphology until such time as they are exposed to the hormone ecdysone (Section 18.5) whereupon they undergo a transformation into the expected adult tissues.

We noted in Section 18.5 that ecdysone stimulates a very specific puffing sequence in larval polytene tissues. J. J. Bonner and M. L. Pardue designed experiments to ascertain whether the same group of genes is stimulated by ecdysone in the third-instar imaginal discs. Since imaginal-disc nuclei do not undergo polytenization, Bonner and Pardue had to make their inquiry somewhat indirectly. They excised imaginal wing discs, cultured them *in vitro,* and exposed them to ecdysone and ^{3}H-uridine for various lengths of

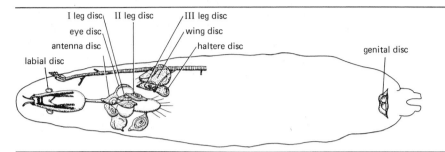

Figure 18-8 Imaginal discs in a mature larva of *D. melanogaster*. A ventral view of the larva is shown and the brain and pharynx are depicted. Each disc except the genital disc is represented twice, producing a total of 17. (From D. Bodenstein, in *Biology of Drosophila*, M. Demerec, Ed. New York: John Wiley, 1950.)

time. The cultured cells were then broken open, their ^{3}H-labeled RNA was extracted, and the poly A-containing mRNA fraction of this RNA was prepared. Finally, the ^{3}H-mRNA was exposed to salivary-gland chromosome preparations under *in situ* hybridization conditions (Section 7.13). The resultant tritium grains indeed localized over specific polytene bands, but none of these proved to correspond to the six bands that undergo early puffing in larval tissues.

These results reveal cell determination in *Drosophila* to be a highly sophisticated process: in larval salivary-gland cells stimulated by ecdysone to synthesize and secrete salivary proteins, ecdysone activates one set of genes, whereas in imaginal wing discs which will form adult structures during pupation, ecdysone activates a second set of genes. If we assume for the moment that ecdysone, like mammalian steroids, complexes with a receptor and migrates to the nucleus, the receptor/ecdysone complex in larval tissues has very different "target genes" than does the complex in disc tissues. Are multiple, tissue-specific receptors involved? Or does one type of receptor/ecdysone complex form and then become modified differentially in different tissues? Or does each determined cell type so modify its target controlling elements that a different subset of genes responds to the identical receptor/hormone complex? These are, of course, central questions concerning the nature of long-term differentiation in all organisms.

18.14 MOSAIC ANALYSIS OF IMAGINAL DISC DETERMINATION IN *DROSOPHILA*

We can now focus on the imaginal discs themselves, and can raise the following specific question. (1) When, during the early stages of *Drosophila* embryology, are imaginal disc cells determined? (2) How many cells are initially determined for each disc? (3) Where in the cellular blastoderm do such **presumptive imaginal disc** cells reside? (4) Is the determination of a presumptive imaginal disc cell a reversible event?

Answers to such questions have come from the analysis of flies known as **mosaics.** Mosaics possess a mixture of cells, some of one genotype and some of another, and they can be induced by two general techniques. In the first technique, the early mitotic divisions of embryonic development are so manipulated that nuclei of two different genotypes come to reside in a single zygote. As a result, roughly half the nuclei of the cellular blastoderm are of one type and half are of the other type. Since sex-chromosome loss usually generates such mosaicism, the resultant flies are known as **sexual mosaics** or **gynandromorphs.** In the second technique, nuclei of two different genotypes are not created until a later stage in embryonic development. Most nuclei of the adult fly are therefore of the original type and relatively few are of a new type. Such flies can be called **recombinational mosaics** since the mosaicism is usually generated by mitotic recombination.

Sexual Mosaics and Fate Maps Of various methods devised for creating sexual mosaics, the most widely employed has involved prompting XX zygotic nuclei to lose one of their X chromosomes during an early mitotic division so that two cell lines are created, one XX (female) and the other XO (male). This elimination of an X chromosome is greatly stimulated in *D. melanogaster* when one of the parents contributes an unstable ring-shaped X chromosome to the zygote. The paternal parent in such crosses is constructed to carry an X chromosome bearing recessive mutations affecting adult cuticular structures, namely, *white* eye, *forked* bristles, and *yellow* body color. The maternal parent carries two X chromosomes bearing their wild-type alleles, one of which is the ring chromosome. Elimination of the ring chromosome, which almost always occurs during the first mitotic division of the zygote, then creates a clone of XO cells that express the *white, forked,* and *yellow* traits in adult tissues, plus a clone of XX cells that are wild in phenotype. These events are summarized in Figure 18-9.

The cleaving nuclei in the early *Drosophila* zygote tend to "stay put": they do not migrate about in the young zygote (Figure 18-9). As a result, the clonal descendants of the XX nucleus generally localize in one half of the cellular blastoderm and the XO descendants in the other half, and the adult fly is usually half female (with wild-type body parts) and half male (with mutant body parts) (Figure 18-9g). In some cases such half-and-half flies are bilaterally symmetrical, but in other mosaics the male-female boundaries follow all sorts of irregular planes (Figure 18-9). Such irregular patterns appear to be the consequence of the fact that the spindle of the first zygotic mitosis can orient at any angle, the result being that the first two daughter nuclei may find themselves in any of a number of orientations with respect to the zygote axis. Since these positions affect those of all subsequent daughter nuclei in the embryonic syncytium, all sorts of planes of mosaicism will result, including a bilateral pattern.

Particularly valuable for developmental studies are instances in which the boundary line between XX and XO nuclei happens to pass between cells in a presumptive imaginal disc. In the resultant flies, the derivative of that

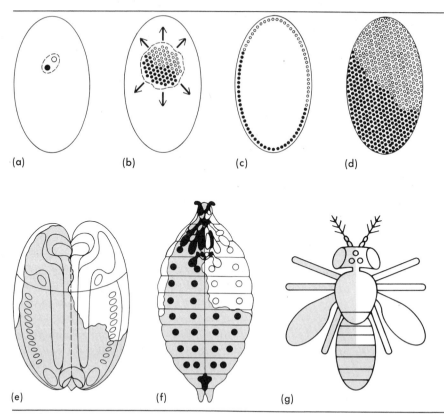

Figure 18-9 Diagram of the formation of a mosaic fly by loss of one X chromosome. (a) Fertilized female egg having undergone one nuclear division. Solid circle represents nucleus containing two X chromosomes, one a ring chromosome and the other carrying recessive genes. Open circle represents nucleus from which the ring X chromosome has been lost during mitosis; its composition is therefore XO and the recessive genes on the remaining X can be expressed. (b) Division of the nuclei. Arrows indicate direction of nuclear migration. (c) Nuclei beneath the egg cortex. (d) Surface view of the multinucleate syncytial blastoderm showing the mosaic boundary separating the two areas. (e) Embryo map. (f) Larva map. (g) Adult fly. (From Y. Hotta and S. Benzer, *Nature* **240:**527, 1972.)

disc—a leg, a wing, or an antenna—will be mosaic for male and female tissue. Mosaics of this sort allow one to estimate how many nuclei, among the thousands in the syncytial blastoderm, actually serve as progenitors for particular adult structures. The approach taken, notably by A. H. Sturtevant, A. Garcia-Bellido, and J. Merriam, has been to examine, for example, a mosaic scutal region (the dorsal part of the thorax) and to estimate the relative size of the smallest **minority area**—a minority area being a patch of male tissue in a predominantly female scutum or the reverse. For the scutum the smallest minority area is found to be about one-eighth of the total area of the adult structure. This result is interpreted to mean that one in every eight nuclei in the scutum derives from a different nucleus than the other seven, so that at least eight nuclei had to have been determined as scutal disc precursors. Similar estimates for other structures yield values of 10 nuclei for a leg disc, six for a wing disc, and very few (perhaps only one) for other discs.

Mosaic flies can be used not only to determine the number of progenitor nuclei for a given body part but also to map the position of these nuclei on the blastoderm surface. The principle of the mapping process is simple: it is assumed that the farther apart two nuclei are in the syncytial blastoderm (Figure 18-9d), the less likely it will be that the mature body parts to which they give rise will happen to lie within the same mosaic region. For example, leg I and leg II are found to be of different types (one normal and one mutant) in 10 percent of a collection of adult mosaic flies, whereas leg I and antenna are of different types in 25 percent of the mosaics, indicating that the distance separating the leg I and antenna primordia on the blastoderm surface is greater than the distance separating the two leg regions. These percentages are converted to map units known as **sturts** (after A. H. Sturtevant): legs I and II are separated by 10 sturts and leg I and antenna by 25 sturts. The resulting **fate map** is shown in Figure 18-10a, and a sketch of the blastoderm surface based on the fate map is shown in Figure 18-10b.

Recombinational Mosaics and the Sequence of Determination of Imaginal Discs E. Hadorn and others have performed numerous experiments in which portions of mature imaginal discs have been transplanted from nonmetamorphosing larvae into the abdomens of metamorphosing larvae. The tissues present in the abdomen of the adult fly are then examined. Such experiments reveal that not only is an imaginal disc specifically determined to form a wing or a leg, but certain regions of a disc normally form only certain parts of adult structures. In this way, "surgical fate maps" of individual discs can be constructed such that even the position of individual bristles can be localized to particular disc regions. Learning when and how such **determinative information** is acquired by larval disc cells is of major interest. Once the original progenitor cells for a wing disc have been determined, for example, when and how do certain cells "know" to become precursors for the top cells of the wing, others for the bottom cells, and so on?

To answer such questions, several laboratories have constructed what we earlier termed **recombinational mosaics.** As with sexual mosaics, the making of recombinational mosaics requires that fertilized eggs or larvae be made heterozygous for such recessive mutations affecting cuticular structures as *y* (*yellow*) or *f* (*forked*). The recombinational mosaics do not, however, carry an unstable ring chromosome and therefore do not lose a chromosome during an early mitosis. Instead, they are irradiated with X-rays at particular times during their development to induce mitotic crossing over. As described in Section 12.17, mitotic crossing over can create cells that are homozygous for particular genes (see Figure 12-19). An X-ray dose is chosen such that induced crossing-over occurs very infrequently; therefore, it can

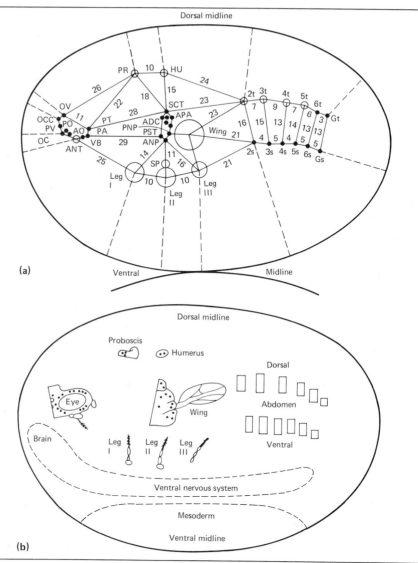

Figure 18-10 (a) Fate map of *D. melanogaster* blastoderm constructed from mosaic mapping showing sites of cells that will eventually develop into the indicated external body parts of the adult fly. Distances are given in sturts. Dotted lines indicate distances to nearest midline. The size of the circle used to represent a site is proportional to the frequency with which the corresponding structure is itself split by the mosaic boundary. *Abbreviations:* ADC, anterior dorsocentral bristle; AO, anterior orbital bristle; ANP, anterior notopleural bristle; ANT, antenna, APA, anterior postalar bristle; HU, humeral bristles; OC, ocellar bristle, OCC, occiput; OV, outer vertical bristle; PA, palp; PNP, posterior notopleural bristle; PO, posterior orbital bristle; PR, proboscis; PST, presutural bristle; PT, postorbital bristles; PV, post-vertical bristle; SP, sternopleural bristles; VB, vibrissae; 1t, first abdominal tergite; 1s, first abdominal sternite. (b) External parts of adult depicted on blastoderm surface. Dotted lines indicate areas which give rise to the nervous system and the mesoderm according to embryological studies. (From Y. Hotta and S. Benzer, *Nature* **240:**527, 1972.)

be assumed that, at most, one cell per presumptive disc region will become homozygous for a marker allele, say *y*, as a consequence of X-irradiation. That homozygous cell, and its clonal descendants, proceed to create a small patch of *yellow* tissue in an otherwise wild-type fly. The power of recombinational-mosaic analysis rests with the fact that irradiation can be performed at selected stages throughout *Drosophila* development, allowing one to ask whether a clone of cells generated at an early phase has a different developmental repertoire from a clone generated at a later stage.

As a specific example of recombinational-mosaic analysis we can consider the determinative sequence leading to the development of the adult mesothorax region, a region that includes the wings and the second pair of legs. The first step in this sequence (Figure 18-11) brings about a change in presumptive mesothorax disc cells such that the marked cells of recombinational mosaics are no longer "totipotent," that is, potentially found in all places and tissues of the mesothorax. Instead, the cells become restricted in their developmental potential so that they will form either anterior (A) or posterior (P) portions of the legs or wings. This determinative event, which occurs during the blastoderm stage, is said to divide the presumptive mesothorax region into two **compartments** separated by an antero-posterior **compartment boundary.** If a clone of genetically marked cells is created by

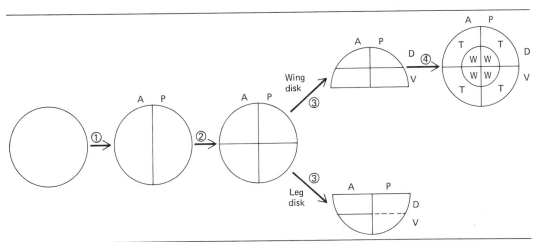

Figure 18-11 Diagram of compartmentalization of the mesothoracic cuticular structures. After each subdivision (represented by a straight line) no clone can include territories on both sides of that line. The first step (1) is the segregation of anterior (A) and posterior (P) polyclones, and this occurs, probably, at the blastoderm stage. The second (2) consists of the separation of the dorsal part of the segment (wing disk) and the ventral part (leg disk). This event takes place before 10 hours of development. These two disks later become physically separated. During the larval period the wing disk becomes subdivided (3) into wing dorsal (D) and wing ventral (V) and about the same time (4) into thorax (T) and wing blade (W) polyclones. For the leg disk there is some evidence of the subdivision of the anterior compartment into dorsal and ventral. (From G. Morata and P. A. Lawrence, *Nature* **265:**211, 1977.)

X-irradiation after this initial step has occurred, the compartment boundary is never crossed: when, for example, marked cells are found in the anterior wing and leg tissues of a mosaic fly, they are never found in its posterior wing or leg tissues. All the cells in the anterior compartment, both marked and unmarked, are therefore termed an anterior **polyclone** since all have experienced a determinative event ("anteriorness") in common. Similarly, the posterior-compartment cells are said to make up a posterior polyclone. It should be emphasized at the outset that the cells of a polyclone are not necessarily a clone at all; regardless of their mitotic ancestry, they experience a common determinative event solely by virtue of their common position within the disc.

Step two in mesothorax disc determination (Figure 18-11) creates a second compartment boundary, this boundary separating those cells that are determined to form dorsal (wing) structures from other cells determined to form ventral (leg) structures. This second determinative event occurs before 10 hours of development and is again defined by the creation of a boundary across which a given clone of genetically marked cells will not pass. As shown in Figure 18-11, the two groups of cells proceed to separate physically to form an imaginal wing disc and leg disc. Focusing on the wing disc, two further subdivisions (steps three and four of Figure 18-11) occur during the larval period; in these steps, cells compartmentalize into wing-dorsal (D) and wing-ventral (V) polyclones and, at about the same time, into wing-blade (W) and thorax (T) polyclones.

It would appear, therefore, that in *Drosophila* a compartment represents a unit of determination. Each compartment is populated by a polyclone, and each founding polyclone represents from 5–50 cells. Determination events, in other words, affect groups of cells during *Drosophila* development rather than simply affecting single cells. Garcia-Bellido has postulated that for each determination event there is activated a **selector gene** which selects a particular developmental pathway for a particular group of cells. More information on the nature of these selector genes is provided in Section 18.22. Granted the existence of selector genes, however, the fundamental question remains: How is it decided which selector gene becomes activated in which group of cells at a particular developmental stage? This question is usually answered by proposing that "positional information" and "polarity information," present in the egg, serve to activate various primary selector genes in different sectors of the early blastoderm; such activation leads to the determination of initial founder polyclones; each founder polyclone becomes so differentiated that new selector genes are activated; and so forth. We are, therefore, thrown back once again to patterned information in the egg as the instigator of *Drosophila* development.

SEX DETERMINATION AND DIFFERENTIATION DURING MAMMALIAN DEVELOPMENT

18.15   GONADAL SEX DETERMINATION IN MAMMALS

The expression of a mammal's **gonadal sex**—whether it will form testes or ovaries—necessitates the expression of the zygote genome and occurs relatively late in development. The gonad primordium region in the human embryo, for example, remains in an undifferentiated "intersex" state for about five weeks, during which time most of the other organ primordia have already begun to give evidence of performing differentiated functions. During this period, moreover, the embryo develops both **Müllerian ducts** (the precursors of oviducts) and **Wolffian ducts** (the precursors of sperm ducts), as if it were indeed uncommitted to the sex it is to become. Then, during the sixth week, Y chromosome-bearing embryos develop seminiferous tubules in their gonadal primordia while the gonads of XX embryos retain their undifferentiated gonadal state, and it is only during the following week that changes indicative of follicle development begin to appear in the embryonic ovary. The development of the testis before the ovary suggests that at a critical point in development, the Y chromosome directly triggers testis determination in the primordial gonad cells. This concept is supported by the genetic data summarized in Section 4.11 and 4.12, which indicate that the presence or absence of a Y chromosome is the critical sex-determining factor in mammals.

A Y-linked gene that is believed by some investigators to be directly involved in testis determination gives rise to a cell-surface protein known as the **H-Y antigen.** First described in mice but since found in other mammals, including humans, the H-Y antigen is expressed in various male cell types but is absent in females. S. Ohno and others have proposed that when primordial germ cells bearing the H-Y antigen migrate to the H-Y antigen-bearing cells of the gonadal ridge, cell-cell interactions between the two like cell surfaces induces seminiferous-tubule formation. Such interactions would, of course, fail to occur between XX cells, which lack H-Y antigen. Whether the H-Y antigen plays such a critical role in development, or whether other Y-linked gene products (also) participate in testicular determination remains to be proved. Meanwhile, it is generally agreed that any sex-determining Y-linked genes function only once during a mammalian male's lifetime, at this crucial interval in his embryological development.

18.16   GENITAL-DUCT SEX DETERMINATION IN MAMMALS

The **genital-duct sex** of a mammal normally includes a penis and sperm duct in the male and a vagina, uterus, and oviducts in the famale. The Wolffian and Müllerian primordia for these organs exist in the embryo before its gonadal sex is determined (Figure 18-12a). After the gonadal sex is deter-

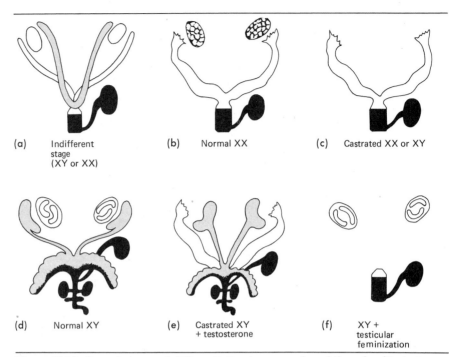

Figure 18-12 Secondary sexual differentiation in mammals and the role of testosterone and the effect of the *Tfm* mutation in the embryonic development of the Wolffian duct (gray), Müllerian duct (outlined), and urogenital sinus (black) which includes bladder and urethra. The pair of bean-shaped bodies at the top of each square signify gonads: indifferent gonad (outlined), ovary (filled with small circles), and testes (containing two tubules). (From S. Ohno, *Nature* **234:**134, 1971.)

mined two events must occur: the "appropriate" primordia must go on to develop into fully formed structures and the "inappropriate" primordia must degenerate.

The normal female genital-duct organs are diagrammed in Figure 18-12b. These organs develop in female embryos whose ovaries have been surgically removed (Figure 18-12c), suggesting that Müllerian ducts will go on to develop into oviducts and a uterus in the absence of any signal from the gonads.

Genital-duct differentiation is more complex in the male embryo, for the normal complement of male organs (Figure 18-12d) will develop only if testes are present. If male rabbit embryos are castrated after testis differentiation takes place but before genital-duct sex determination begins and are then returned to the uterus to complete development, they are found at birth to contain oviducts, a uterus, and degenerate Wolffian ducts (Figure 18-12c). In other words, the male genital-duct sex is apparently induced by some activity of the testis; when the testis is removed, the female course is followed instead.

At least two secretions from the testis appear necessary to induce male genital duct development. The first is the hormone **testosterone.** If testosterone is administered to either male or female castrated rabbit embryos during the time they are *in utero,* the rabbits are born with a normal set of male genital ducts (Figure 18-12e). They are also born with oviducts (Figure 18-12e), indicating that a second, unidentified secretion is also required from the testis to ensure degeneration of the Müllerian ducts.

The requirement for activation by testosterone in male genital-duct determination is dramatized by the **testicular feminization syndrome,** a syndrome that has been described in humans, mice, cattle, and rats. Affected organisms show no evidence of any male genital-duct development, even though they are XY and possess normal testosterone-secreting testes (Figure 18-12f). Ohno has shown that the syndrome is produced in male mice when a recessive mutant gene, called *Tfm,* is present in their X chromosomes. Mice that are *Tfm*/Y are unable to respond to testosterone or testosterone derivatives. This can be demonstrated *in vitro* by taking testosterone target tissues, such as kidney slices, from normal and mutant mouse embryos and presenting them with a radioactively labeled androgen. The normal tissues readily bind the male hormone whereas the mutant tissues do not, and the difference can be shown to be caused by a deficiency in levels of androgen-binding receptor protein in the *Tfm* males. Thus the *Tfm* mutation is formally analogous to i^s mutations in the *lac* repressor gene (Section 17.7) and to the glucocorticoid-resistant mutations described in Section 18.6 for the mouse lymphoma line S49.

It is useful at this point to compare the effects of steroids on short-term and long-term regulation. Specific receptor proteins and differential gene transcription appear to be involved in both cases. The short-term responses are, however, reversible—androgen administration causes hypertrophy while androgen withdrawal causes atrophy to adult male target tissues— whereas the long-term responses are essentially unidirectional; once a Wolffian duct differentiates into a sperm duct, castration will cause the duct to atrophy but not to return to its embryonic state.

18.17 FACULTATIVE HETEROCHROMATIZATION

A phenomenon more indirectly related to sexual differentiation is known as **facultative heterochromatization.** In this phenomenon a chromosome or a set of chromosomes becomes heterochromatic (turned off) in the cells of one sex while remaining euchromatic (turned on) in the cells of the opposite sex.

Three widely divergent cases are known in which facultative heterochromatization of this sort occurs. The first is found in such coccid insects as the mealybug: both maternal and paternal chromosome sets are active during all stages of development in the female, but the entire paternal set is selectively inactivated during the early development of the male embryo so that only the female set is active in the adult male somatic tissues and only the female set is included in spermatozoa. The second case occurs in the kangaroo and other marsupials: the paternally derived X chromosome is selectively turned off in female cells. The third case, and the one on which we shall focus, occurs in female placental mammals: either the maternally derived or the paternally derived X chromosome is turned off. This process is commonly known as **Lyonization,** after M. Lyon, an early and still active investigator of the phenomenon as it occurs in mice, and the turned-off heterochromatic X chromosome is called a **Barr body,** after M. Barr.

The term facultative heterochromatization derives from the fact that the turned-off chromosome is heterochromatic under the microscope, inactive in transcription by autoradiography, and late-replicating like constitutive heterochromatin (as described in Section 2.13). Unlike constitutive hetero-chromatin, however, it is not enriched in repetitive DNA sequences (Section 7.13) and the heterochromatic state is adopted (facultative) rather than permanent (constitutive). Thus at the onset of an XX embryo's development both X chromosomes are euchromatic and both are believed to be active in RNA synthesis; it is not until probably the blastula stage that one of these chromosomes becomes heterochromatic and dysfunctional. Moreover, both X chromosomes remain euchromatic in the female germ line so that the change is not obligatory.

Numerous experiments have demonstrated that at a particular time in mammalian embryological development, each XX cell will turn off either the paternal or the maternal X chromosome. The choice is entirely random—50 percent of the time the maternal X is turned off, and 50 percent of the time the paternal X is turned off. Once the decision has been made by a given cell, however, the same choice is made by all the cell's descendants. This can be shown conclusively with cultured fibroblast cells derived from human females who are heterozygous for sex-linked genes. For example, when single cells are cloned from a woman heterozygous for a mutant allele of the glucose-6-phosphate dehydrogenase (G6PD) gene, each clone is found to produce either the mutant or the wild-type form of the enzyme but never a mixture; in a random sampling of cells, moreover, about 50 percent produce wild-type and 50 percent mutant clones.

The Lyonization phenomenon means that all female mammals, including women, are mosaics. Dramatic visual examples are tortoise-shell and calico cats, which are heterozygous for *black* and *orange* alleles of an X-linked coat color gene. Tortoise-shell cats have a finely mottled black and orange coat while calico cats, which also carry a "spotting" gene, exhibit an irregular patchwork of black and orange sectors all over their bodies, each sector representing a clone of cells derived from a Lyonized hair cell precursor.

Calico cats are almost always female, but an occasional male calico has been reported. These prove to be rare XXY cats and are invariably sterile, as would be expected of human males with the Klinefelter syndrome de-scribed in Section 4.11. The fact that a Y chromosome does not interfere with Lyonization argues that the process is not triggered by sex hormones.

It has been proposed that the function of Lyonization is to reduce the effective X chromosome dosage of a cell. The strongest argument for this hypothesis comes from studies of mutant fetuses carrying more than two X chromosomes. Fetuses that are XXX exhibit two Barr bodies in every cell; fetuses that are XXXX exhibit three Barr bodies; and so on. In other words, the mammalian cell seems to be programmed so that only one X chromosome per cell is allowed to remain euchromatic and all others are inactivated. It is argued that since male mammalian cells function with only one X chromosome, Lyonization has evolved to effect the same genic balance in the female. This argument is obviously teleological and the true reason for the evolution of Lyonization is not known, nor is there any idea how it is initiated, effected, or maintained.

In *Drosophila* there is no clear counterpart of facultative heterochromatization of the X chromosome. Large portions of the X chromosome remain euchromatic in both males and females, and the remaining portions seem to be a mixture of constitutive and perhaps facultative heterochromatin. A physiological effect analogous to facultative heterochromatization however, occurs in *Drosophila,* and is known as **dosage compensation.** Male cells exhibit the same activities of sex-linked enzymes as female cells, and this is apparently accomplished by a stimulated rate of gene transcription from the single male X chromosome.

Before leaving the subject of facultative heterochromatization we should note the existence of a phenomenon known as **allelic exclusion** which occurs in antibody-forming lymphocyte and plasma cells. We learned in Sections 15.15 and 15.16 that the constant region of a heavy or light chain in an antibody molecule is dictated by a single genetic locus, and that two or more alleles frequently exist at such loci. When clones of single plasma cells are studied, however, they are invariably found to synthesize only one version of a kappa chain, only one type of gamma chain, and so on. In other words, even though a plasma cell may often be heterozygous at a given antibody-forming locus, only one of its two alleles is ever expressed in the mature cell. The antibody genes are almost certainly not X-linked. Furthermore, it cannot be postulated that a complete chromosomal set is inactivated in these cells (as in the mealybug), because the cells continue to synthesize both paternally and maternally derived histocompatibility antigens. Whether allelic exclusion will prove to operate by a limited heterochromatization or by some other mechanism remains to be seen.

MUTATIONS AFFECTING LONG-TERM REGULATION

In Chapters 16 and 17 we saw how gene mutations have been put to elegant use in elucidating both the sequences of biochemical pathways and the life cycles of bacteriophages. In theory, mutations should also be of value in determining how long-term regulation is established and maintained in eukaryotes.

The final sections of this chapter present some of the insights into the nature of long-term differentiation that have been acquired by studies of mutant organisms.

18.18   CHROMOSOMAL REARRANGEMENTS AND POSITION EFFECTS

As noted in Section 18.7, the model of R. Britten and E. Davidson for gene organization in eukaryotes proposes that a structural gene is contiguous to its (moderately repetitive) controlling element(s). It follows, therefore, that if a gene is separated from its controlling element by some sort of chromosomal rearrangement (a translocation, inversion, or deletion), control over the expression of that gene might be modified. This is, in fact, clearly the case for prokaryotes: *lac* structural genes have been transposed via transducing phases to sites adjacent to *trp* or *purE* operators, and in *E. coli* cells carrying such "fused" operons, the synthesis of β-galactosidase is sensitive to intracellular concentrations of tryptophan or purines and indifferent to concentrations of lactose. Analogous phenomena in eukaryotes, known as **position effects,** are described below.

The *Bar* Position Effect   When the genes in the 16A segment of the *D. melanogaster* X chromosome are present in a single copy, the chromosome is considered wild type ($+$) whereas a duplication of the 16A segment, known as the *Bar* mutation, creates an X chromosome we can denote as *B* (Figure 4-14). Flies with a $+/+$ or a $+/\longrightarrow$ genotype have normally sized eyes; the eye size of *B*/$+$ females is somewhat reduced, and the eye size of *B*/*B* females and *B*/$\longrightarrow$ males is extremely reduced. Clearly, eye size is *not* affected by the total number of 16A segments present in the genome—two copies are present in both $+/+$ and *B*/$\longrightarrow$ flies, for example. Rather, eye size is affected by the chromosomal position of the 16A segments. More formally, the two 16A segments must be present in *cis* to exert their deleterious effect on ommatidia formation.

Unequal crossing over (Section 15.9) in *B*/*B* flies will occasionally generate *ultrabar* (*BB*) chromosomes that carry three copies of the 16A segment. Such chromosomes prove to have an even more drastic effect on eye size than a *B* chromosome. Whereas a *B*/*B* fly will have an average of 68 ommatidia per eye, a *BB*/$+$ fly will have an average of only 45 ommatidia per eye, even though the two flies carry the same total number of 16A segments.

The 16A segment of the polytene X chromosome lies between segments 15F and 16B and exhibits six bands (Figure 4-14). In the nonpolytene state this portion of the X chromosome is euchromatic. Therefore, for the purpose of our discussion, we can consider that the segment carries six active genes (1 to 6) and we can represent the *Bar* region of the chromosome as 15F-123456123456-16B. From this information we can put forward a logical proposal for the origin of the *Bar* phenomenon: we can assume that a deleterious effect on ommatidia development is generated every time gene

6 finds itself next to a copy of gene 1 instead of its usual neighbor, the first gene in the 16B segment. Two such 6-1 combinations, as in an *ultrabar* chromosome, would create even more damage than one.

How might the presence of new neighboring chromatin influence gene expression? One simple idea is the proposal that whenever the DNA in band 6 makes contact with the DNA in Band 1 it acquires the properties of a controlling element which depresses the transcription of the band 1 gene. The controlling element might, of course, correspond to band 1 and the gene to band 6, and the situation may be far more complex in detail than outlined here. The point is that the *Bar* position effect suggests that the relative location of DNA segments in a chromosome is an important factor in eukaryotic regulation.

V-Type Position Effects in *Drosophila* A reduction in eye size invariably occurs when *Drosophila* flies possess *Bar* chromosomes. The *Bar* phenomenon is therefore known as a **stable** or **S-type position effect,** in contrast to the more prevalent **variegated** or **V-type position effects** known in *Drosophila*. A V-type position effect is characterized by its instability: certain cells (and their clonal descendants) display a mutant phenotype, whereas others do not, with the result that a variegated (mosaic) fly emerges.

V-type position effects are generated by a specific pattern of mutation: a break first occurs within heterochromatin, usually within centric constitutive heterochromatin (Section 2.13) or within the constitutive heterochromatin of the Y chromosome. The broken heterochromatin is then brought into contact with euchromatin, either by a translocation or by an inversion (Figure 6-1). In its new position the heterochromatin exerts its variable effect: in some nuclei it suppresses the expression of genes in the adjacent euchromatin; in others it does not. When appropriate genotypes are chosen, the resultant mosaicism can be visualized by inspection. Thus, when a rearrangement (R) brings heterochromatin next to the wild-type (+) allele of the *white* eye color locus in the X chromosome of *D. melanogaster,* the chromosome can be designated R(+). When this chromosome is coupled in females with an X chromosome that bears the *white* (*w*) allele of the gene, flies emerge with wild-type (red) and *white* eye patches: the red patches are generated by those R(+)/*w* cells whose + gene remains "on," while the *white* patches are produced by the R(+)/*w* cells whose + gene has been turned "off."

The inhibitory effect of broken heterochromatin can extend as much as six map units (50 polytene bands) into adjacent euchromatin, with proximal genes being inhibited before distal genes. One inversion in the X chromosome, for example, brings heterochromatin into the vicinity of the *rst* (*roughest*) locus, a locus concerned with the arrangement of ommatidia in the eye. Several map units away from *rst* lies the *w* locus described above. In R(+ +)/*rst w* flies the white ommatidia invariably exhibit the roughest phenotype, whereas red ommatidia may be either rough or smooth. One therefore speaks of a **spreading effect** associated with V-type phenomena;

the sphere of influence of broken heterochromatin may reach quite distant genes in one cell but only nearby genes in another.

Numerous factors are known to influence the extent to which position-effect variegation will occur in a given organism. If a fly possesses more than one Y chromosome, for example, or additional Y chromosome fragments, variegation is far less pronounced than in normal XY or XX flies. Certain gene loci are also known to suppress the occurrence of variegation, and such factors as the temperature at which the embryo is incubated are also important in determining whether many or few nuclei in the early zygote experience the influence of the broken heterochromatin. These kinds of fluctuations are common in developmentally abnormal phenotypes: such variables as gene dosage, "modifier genes," and temperature combine to influence the expression of the abnormal condition. We are also taken back to our discussion of incomplete penetrance and variable expressivity in Section 15.4. More generally, a complex series of events undoubtedly attends the establishment of many, if not most, phenotypes.

Position Effects in the Mouse   The far-reaching inhibitory properties of heterochromatin are not peculiar to *Drosophila*. It has been shown in the mouse that when translocations occur between an autosome and the X chromosome the translocated segment of the X persists in undergoing Lyonization. As the segment becomes heterochromatic, the turning-off process apparently extends into the adjacent autosomal chromatin, with the result that the mouse becomes variegated for certain autosomal genes as well as for X-linked genes.

As in *Drosophila*, the effect of the mouse heterochromatin on adjacent euchromatin occurs in a spreading manner, with genes distal to the X material being affected only if proximal genes are also affected. Also like *Drosophila*, the heterochromatic influence can affect the expression of genes located long distances away from the broken chromatin. The fact that both facultative and constitutive heterochromatin can exert such similar effects is striking indeed. Even though the two types of heterochromatin are dissimilar in many respects, they evidently share similar regulatory properties once they have undergone breakage. How these broken properties might relate to the regulatory properties of intact heterochromatin, however, is not yet known.

Dissociator-Activator   The *dissociator* (*Ds*) locus represents a class of genetic elements discovered in maize and studied intensively by B. McClintock. Its first remarkable property is that it can move about from one location in the genome to the next; in other words, it may map to one chromosomal position and then, several generations later and in the same strain of maize, map to a completely different position, often in a different chromosome. Its entrance into a new chromosomal position may be accompanied by an increase in chromosomal breakage at that locus, but such mutational events do not necessarily occur and they are not thought to underlie the regulatory effects of the *Ds* element.

Once in position, the *Ds* element brings about an unstable suppression of contiguous genes, with this effect spreading perhaps five map units to either side of the location of the element. As a result, embryonic and adult tissues of the maize plant exhibit a variegated phenotype, much the same as the mosaicism accompanying V-type position effects in *Drosophila* and mice. Indeed, the phenomena are superficially so similar that it is tempting—in the absence of much understanding of either—to equate the two and to propose that *Ds* represents a segment of heterochromatin with a particular propensity to undergo insertion into diverse chromosomal locations.

The *Ds* material is unable to exert its suppressive effect on contiguous euchromatin unless the *activator* (*Ac*) element is also present somewhere in the genome. Like *Ds*, *Ac* can move from one chromosomal position to another, but it appears to have no effect on its neighboring genes. Its only known effect is to permit *Ds* to operate, and such a facilitation can occur even when the two elements are in different chromosomes. The *Ac* locus is of particular interest in that it can have a *temporal* effect on *Ds* suppression. As maize strains are bred to carry more and more copies of *Ac*, *Ds*-induced variegation is initiated at progressively later times in development.

Because we are ignorant of the nature—and even the size—of *Ac* and *Ds*, all comments on their significance constitute speculation. Nonetheless, a number of geneticists have entertained the hypothesis that *Ds* represents a controlling element—analogous to an operator, perhaps—that has "got out of hand" and moves from locus to locus. In other words, it is argued that *Ds* is a mutant version of a normal element that once regulated the expression of a gene or genes with which it made stable contact. These same arguments suggest that the regulatory influence of *Ac* over *Ds* reflects a once normal regulator-operator type of interaction.

18.19 LETHAL MUTATIONS AFFECTING ZYGOTE METABOLISM

Recessive gene mutations that cause lethality during the early development of homozygous zygotes fall into two general classes. Most affect genes required for basic biochemical or metabolic processes, and their deleterious effects are first seen at that stage of development when the zygote either runs out of the required metabolic product or requires the product at high levels. The remaining mutations affect more fundamentally "developmental genes" that are required if the organism is to progress from one stage to the next.

Mutations that fall into the first class include the *rudimentary* (*r*) mutations of *Drosophila*, which cause truncated wings in homozygotes. Female flies homozygous for *r* mutations produce *r* eggs that cannot undergo normal embryological development if fertilized by *r*-bearing sperm, but which develop normally when fertilized by + sperm. Eggs produced by +/*r* females can, on the other hand, develop normally when fertilized by *r* sperm even if, at the final meiotic division, they come to carry only the female *r* allele. In other words, at least one + allele of the *rudimentary* gene must be present either during oogenesis or during embryogenesis; if the wild-type allele is absent from both developmental periods, then at most about 1 percent of the expected number of progeny manage to survive. Such a pattern of inheritance would certainly be expected of a slightly leaky "developmental gene," but it turns out, as outlined in Section 16.3, that the *rudimentary* locus in fact codes for several enzymatic activities involved in pyrimidine biosynthesis. Indeed, eggs from *r/r* females can be "rescued" simply by injecting them with 0.01 μg of pyrimidine nucleosides.

18.20   MATERNAL-EFFECT MUTATIONS AFFECTING EGG PATTERNING

Eggs carrying biochemical-lethal mutations such as the *rudimentary* series in *Drosophila* are typically **rescuable** if the missing factor is not required by the haploid or homozygous egg and if a wild-type allele is provided by the sperm. In other cases, however, development is found to be **nonrescuable** by wild-type sperm, and mutations causing such effects may mark genes that function specifically during oogenesis. This seems to be the case for three **maternal-effect** (*mat*) mutants of *D. melanogaster* isolated by T. B. Price and A. Garen. Eggs produced by females homozygous for these mutations fail to undergo a normal zygotic development even if their nuclear genotype, following fertilization, is +/*mat*, while eggs produced by +/*mat* heterozygous females develop into normal flies even if their nuclear genotype, following fertilization, is *mat/mat*. In other words, the + alleles of the *mat* loci are apparently essential for producing an egg that can give rise to a normal zygote.

What happens to eggs that have differentiated without the benefit of normal *mat* function? Following fertilization, all undergo apparently normal mitoses during the first two hours of development to reach the multinucleate-syncytium stage (Figure 18-9d), and proceed to form polar-granule-containing pole cells (Figure 18-7c). Fertilized eggs developing from *mat(3)1/mat(3)1* females, however, fail to transform the syncytium into a cellular blastoderm at all; the zygotes deriving from *mat(3)3/mat(3)3* females fail to form cells on the posterior-dorsal surface of the blastoderm; and the zygotes from *mat(3)6/mat(3)6* females form cells only at the anterior and posterior ends, leaving a noncellular region in the middle. These selective effects strongly suggest that the *mat(3)3* and *mat(3)6* loci are involved in distributing materials essential for blastoderm formation to particular sectors of the egg, much as the *grandchildless* mutation of *D. subobscura* appears to be involved in polar-granule positioning (Section 18.12). A careful comparison of mutant and wild-type oogenesis may well allow these maternally-produced substances to be identified and characterized biochemically.

18.21 NOTCH MUTATIONS IN *DROSOPHILA*

If the course of development indeed involves a finely tuned differential expression of the zygotic genome in space and in time, then one might predict that many mutations affecting true "developmental genes" might commonly be dominant or at least partially dominant: development might well be disrupted at least somewhat if even one copy of such a critical gene fails to function properly at the correct time and place. The *Notch* locus of *D. melanogaster* can serve to represent such a situation. The locus corresponds to band 3C7 of the X chromosome, and point mutations and deficiencies that map to this band are all partially dominant: females heterozygous for *Notch* mutations exhibit such ectodermal abnormalities as notches and thickened veins in the wings and alterations in body bristle morphology. Embryos hemizygous or homozygous for *Notch* mutations, on the other hand, experience gross developmental abnormalities: a large portion of the ectodermal cells, which would normally give rise to such larval structures as skin, salivary gland, or pharynx, instead give rise to neural cells. As a result the central nervous system of the embryo becomes about three times its normal size, an enormous structure that destroys the orientation of most other developing organs, and death ultimately results.

The problem with mutations such as *Notch* is that it is very difficult to discover what embryological process(es) the locus controls simply by examining the mutant phenotypes. The various syndromes produced by *Notch* mutations are consistent with the notion that the normal gene product(s) of the *Notch* locus function to prevent neurogenesis in ectodermal cells that are not induced to turn on their "neurogenesis program." Other speculative theories can, however, explain the observations equally well, and we are left with a "developmental gene" candidate in search of a function.

18.22 TRANSDETERMINATION AND HOMEOTIC MUTATIONS

In Section 18.14 we considered imaginal-disc determination in *Drosophila* in some detail and noted that at particular stages in this process, selector genes were thought to choose a particular developmental pathway for a particular polyclone.

Evidence for the existence of such selector genes comes from two sources. First, E. Hadorn discovered that when explanted imaginal discs were subcultured for long periods of time in adult hosts and were then placed into metamorphosing larvae, the discs were occasionally found to have undergone a change which Hadorn called **transdetermination.** In this change, a genital disc, for example, might have acquired the capacity to differentiate into wing-disc derivatives. Such occurrences can be interpreted to mean that in the unnatural environment of a fly's abdomen, the selector gene required to maintain "genitalness" in the disc cells fails to function, the determined state is lost in at least a few cells, and these cells go on to activate a different selector gene which sets them up to differentiate into wing and thoracic tissues.

The second line of evidence for the existence of selector genes comes from studies of **homeotic mutations,** which effectively create flies that appear to have undergone transdetermination. The *Antennapedia* mutation, for example, alters the determination of the antennal imaginal disc such that it gives rise to a leg structure, the bizarre result being a fly with a leg emerging from its head (Figure 18-13). Other homeotic mutations cause the development of a wing from an eye, and so forth. Each of these mutations behaves genetically as a single lesion, yet each is clearly causing the redirection of an entire developmental sequence. The inference, therefore, is that at least some homeotic mutations mark specific selector genes.

The *bithorax* locus is perhaps the most intensively studied of such putative selector genes. Homeotic mutations in the *bithorax* locus occur quite frequently. E. B. Lewis has shown that different *bx* alleles bring about different homeotic phenotypes, one "extreme" being found in bx^3/bx^3 homozygotes or in *bx* deficiency strains, where a short appendage known as the anterior haltere is transformed into an anterior wing and the anterior III

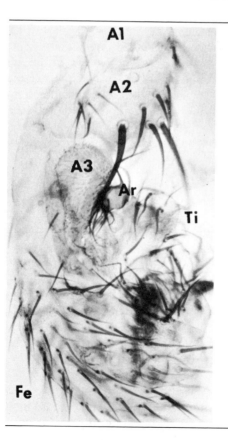

Figure 18-13 Photograph of an *Antennapedia* antennal appendage showing three antennal segments (A1, A2, and A3) and the arista (Ar) continuous with leg segments (Ti, tibia; Fe, femur). (From J. H. Postlethwaite and H. A. Schneiderman, *Proc. Natl. Acad. Sci. U.S.* **64**:180, 1969.)

leg is transformed into an anterior II leg. Such a phenotype suggests that the bx^+ gene is required for an early event in anterior haltere and third-leg determination, and that when bx fails to function, wing and second-leg pathways are followed instead. Support for this concept comes from examining the haltere imaginal disc region in bx/bx first-instar larvae: the disc is found to be much larger than a normal haltere disc but similar in size to a wing disc, as though the "wing pathway" were already being followed by the bx cells at this early stage of development.

The bx^+ gene appears to be required not only for haltere disc determination but also for maintaining that determined state throughout larval development, as seen by the following experiment. When $+/bx$ flies are irradiated so as to produce bx/bx homozygous clones, the clones produce wing structures in the adult even if they are induced in the middle of the third-instar larval period. Transdetermination phenomena predict, or course, that selector genes should indeed be required throughout development.

The genes marked by homeotic mutations are excellent candidates as "master" selector genes (Section 8.14). Their activity in many cases appears restricted to particular polyclones (as defined by recombinational mosaic studies), and their mutation can result in a determinative switch affecting an entire compartment. Whether such genes exist in mammals is not known; their detection may prove very difficult since a mutation in such a gene would very likely be lethal in an organism without a dual larval/imaginal disc pattern of development.

18.23 THE *T* LOCUS IN THE MOUSE

We conclude our consideration of mutations affecting development by describing a fascinating locus in the mouse, known as *T*, which resides in chromosome XVII, quite close to the major histocompatibility locus *H-2* (Section 15.10 and Figure 15-12). The T-locus has been studied extensively by L. C. Dunn and D. Bennett and, more recently, by K. Arntz and F. Jacob.

The normal ($+$) version of the *T* locus appears to specify several molecular species that associate with the surface of preimplantation embryonic cells; these same molecules also associate with cells of the male germ line. Since these molecules are defined by their antigenticity they are called **cell surface antigens,** and at least one of them, called **F9,** is found on human spermatozoa as well as mouse germ cells. Particularly important is the fact that while F9 is not expressed by somatic tissues (brain, kidney, or white blood cells), it is expressed by neoplastic mouse **teratoma** cells which have the capacity to differentiate into many tissue types during malignant growth. There is considerable speculation, therefore, that cell-surface molecules such as F9 play fundamental roles in early development, perhaps acting to mediate crucial cell-cell interactions that lead to determinative events. Such a role would be analogous to the proposed activity of the H-Y antigen in mediating testis differentiation (Section 18.15).

Support for the concept that *T*-locus gene products function in early embryology comes from the properties of mutant strains. The wild-type (+) locus can be readily induced in the laboratory to undergo mutations to dominant forms known as *T* alleles. Mice heterozygous for such mutations (+/*T*) have short tails while homozygous mice (*T*/*T*) inevitably die *in utero*. The remaining alleles at the *T*-locus are known as *t*. These are fully recessive to wild-type, but affect the phenotype in *T*/*t* heterozygotes, where the mouse is normal but tail-less. This phenomenon has allowed recognition of mice carrying recessive *t* mutations, and a large number of *t* alleles have in this way been identified and studied.

Relevant to the subject of this chapter are the phenotypes of mice homozygous for the various *t* alleles. While most *t*/*t* homozygotes appear to develop normally, certain lethal *t* alleles, when homozygous, prove to disallow normal embryogenesis. Moreover, as illustrated in Figure 18-14, t^{12}

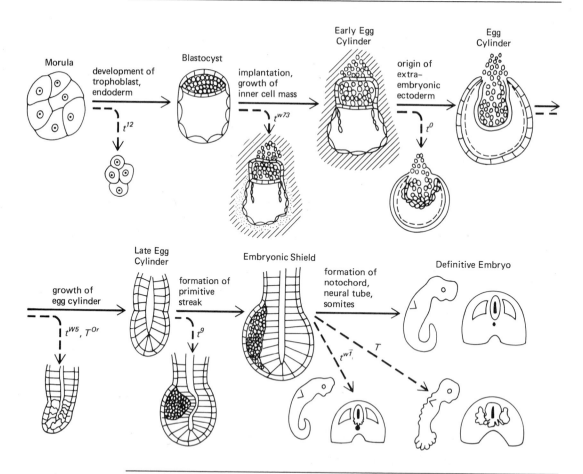

Figure 18-14   A diagrammatic representation of early development in the mouse and the defects seen in embryos homozygous for *T*-locus mutations (From D. Bennett, *Cell* **6**:441, 1975. Copyright © MIT.)

homozygotes fail to form blastocysts, t^{w37} homozygotes fail to implant properly, and so on, as if each lethal t allele affects a distinct embryonic stage. Consistent with this notion is the demonstration that when the sperm cells of $+/t^{12}$ mice are analyzed, roughly half the sperm in the sample are found to carry cell-surface antigens that are not detected on the other half. Extrapolating from this result, one can imagine that the cells of t^{12}/t^{12} embryos might have cell-surface properties that disallow development beyond the blastocyst stage, and so forth.

The various lethal and semilethal t alleles fall into six general complementation groups, and while highly complex complementation patterns are found when the data are examined closely, the T-locus is generally viewed as containing a small number of related genes. Because the lethal t alleles almost totally supress recombination over an extensive portion of chromosome 17, it is not possible to map the region nor to ascertain whether individual t alleles represent chromosome rearrangements, multiple mutations, or whatever.

The "t syndromes" outlined in Figure 18-14 are not mere laboratory curiosities. It turns out that although recessive t alleles have never been generated by mutagenesis in the laboratory, they are widely carried by natural mouse populations. Moreover, they are often transmitted preferentially to offspring simply because, for unknown reasons, sperm bearing lethal t alleles are longer-lived than others and therefore are more likely to reach and fertilize an egg. The "reason" why natural mouse populations maintain such high frequencies of deleterious t loci has not yet been discovered.

A facile interpretation of the "t syndromes" of Figure 18-14 can be made along the lines of the selector-gene theory for *Drosophila*. One can propose that the wild-type product of the t^{12} gene permits cells to differentiate from morula to blastula, that the wild-type product of t^{w73} is expressed at implantation time and permits the following stages of development, and so on. In fact, however, it appears that certain of the "late-acting" t alleles specify cell-surface antigens that are present on cell surfaces during the morula stage. The gene products of the T-locus do not, therefore, seem to control development by making sequential appearances on the cell surfaces of embryonic cells, and their mode of action is unknown. It probably goes without saying that their role in development will be a major subject for research in the coming decade.

References

Regulation in Lower Eukaryotes

*Arst, H. N. "Integrator gene in *Aspergillus nidulans*," *Nature* **262**:231–234 (1976).

*Arst, H. N., and D. W. MacDonald. "A gene cluster in *Aspergillus nidulans* with an internally located *cis*-acting regulatory region," *Nature* **254**:26–31 (1975) (and following article).

*Denotes articles described specifically in the chapter.

*Case, M. E., and N. H. Giles. "Genetic evidence on the organization and action of the *qa-1* gene product: A protein regulating the induction of three enzymes in quinate catabolism in *Neurospora crassa, Proc. Natl. Acad. Sci. U.S.* **72:**553–557 (1975).

*Case, M. E., and N. H. Giles. "Gene order in the *qa* gene cluster of *Neurospora crassa," Molec. Gen. Genet.* **147:**83–89 (1976).

Chovnick, A., W. Gelbart, M. McCarron, B. Osmond, E.P.M. Cardido, and D. L. Baillie. "Organization of the rosy locus in *Drosophila melanogaster:* Evidence for a control element adjacent to the xanthine dehydrogenase structural element," *Genetics* **84:**233–255 (1976).

*Douglas, H. C., and D. C. Hawthorne. "Uninducible mutants in the *gal i* locus of *Saccharomyces cerevisiae," J. Bacteriol.* **109:**1139–1143 (1972).

Greer, H., and G. R. Fink. "Isolation of regulatory mutants in *Saccharomyces cerevisiae,"* in: *Methods in Cell Biology XI,* D. M. Prescott, Ed. New York: Academic Press, pp. 247–272 (1975).

Metzenberg, R. L. "Genetic regulatory systems in *Neurospora," Am. Rev. Genetics* **6:**111–132 (1972).

Thuriaux, P., F. Ramos, A. Piérard, M. Grenson, and J. M. Wiame. "Regulation of the carbamoylphosphate synthetase belonging to the arginine biosynthetic pathway of *Saccharomyces cerevisiae," J. Mol. Biol.* **67:**277–287 (1972).

Insect Hormones and Chromosome Puffing

Ashburner, M. "Function and structure of polytene chromosomes during insect development," *Adv. Insect Physiol.* **7:**1–95 (1970).

*Ashburner, M., C. Chihara, P. Meltzer, and G. Richards. "Temporal control of puffing activity in polytene chromosomes," *Cold Spring Harbor Symp. Quant. Biol.* **38:**655–662 (1973).

Beerman, W., Ed. *Developmental Studies on Giant Chromosomes.* New York: Springer-Verlag, 1972.

Clever, U. "Regulation of chromosome function," *Ann. Rev. Genetics* **2:**11–30 (1968).

Kroeger, H., and M. Lezzi. "Regulation of gene action in insect development," *Ann. Rev. Entomology* **11:**1–22 (1966).

Laufer, H., and T. K. H. Wolt. "Juvenile hormone effects on chromosomal puffing and development in *Chironomus thummi," J. Exp. Zool.* **173:**341–352 (1970).

Steroid Hormone Action

Palmiter, R. D., P. B. Moore, E. R. Mulvihill, and S. Entage. "A significant lag in the induction of ovalbumin messenger RNA by steroid hormones: a receptor translocation hypothesis," *Cell* **8:**557–572 (1976).

Ringold, G. M., K. R. Yamamoto, G. M. Tomkins, J. M. Bishop, and H. E. Varmus. "Dexamethasone-mediated induction of mouse mammary tumor virus RNA: A system for studying glucocorticoid action," *Cell* **6:**299–305 (1975).

*Yamamoto, K. R., and B. M. Alberts. "Steroid receptors: Elements for modulation of eukaryotic transcription," *Ann. Rev. Biochem.* **45:**721–746 (1976).

Yamamoto, K. R., M. R. Stampfer, and G. M. Tomkins. "Receptors from glucocorticoid-sensitive lymphoma cells and two classes of insensitive clones: Physical and DNA-binding properties, *Proc. Natl. Acad. Sci. U.S.* **71:**3901–3905 (1974).

*Yamamoto, K. R., U. Gehring, M. R. Stampfer, and C. H. Sibley. "Genetic approaches to steroid hormone action," *Recent Prog. Horm. Res.* **32:**3–32. (1976).

Models for Eukaryotic Regulation

*Britten, R. J., and E. H. Davidson. "Gene regulation for higher cells: A theory," *Science* **165:**349–357 (1969).

Brown, S. W. "Heterochromatin," *Science* **151:**417–425 (1966).

Dehmus, M. E. "Stimulation of ascites tumor RNA polymerase II by protein kinase," *Biochemistry* **15:**1821–1829 (1976).

*Davidson, E. H., and R. J. Britten. "Organization, transcription, and regulation in the animal genome," *Quart. Rev. Biol.* **48**:565–613 (1973).

Garel, A., and R. Axel. "Selective digestion of transcriptionally active ovalbumin genes from oviduct nuclei," *Proc. Natl. Acad. Sci. U.S.* **73**:3966–3970 (1976).

Krieger, D. E., R. Levine, R. B. Merrifield, G. Vidali, and V. G. Allfrey. "Chemical studies of histone acetylation. Substrate specificity of a histone deacetylase from calf thymus nuclei," *J. Biol. Chem.* **249**:332–334 (1974).

Lewin, B. "Units of transcription and translation: Sequence components of heterogeneous nuclear RNA and messenger RNA," *Cell* **4**:77–93 (1975).

Lodish, H. F. "Translational control of protein synthesis," *Ann. Rev. Biochem.* **45**:39–72 (1976).

McCarthy, B. J., J. T. Nishiura, D. Doenecke, D. S. Nasser, and C. B. Johnson. "Transcription and chromatin structure," *Cold Spring Harbor Symp. Quant. Biol.* **38**:763–771 (1973).

Stein, G. S., T. C. Spelsberg, and L. J. Kleinsmith. "Nonhistone chromosomal proteins and gene regulation," *Science* **183**:817–824 (1974).

General Features of Long-Term Differentiation

Axel, R., P. Feigelson, G. Schutz. "Analysis of the complexity and diversity of mRNA from chicken liver and oviduct," *Cell* **7**:247–254 (1976).

Bernhard, H. P. "The control of gene expression in somatic cell hybrids," *Int. Rev. Cytol.* **47**:289–325 (1976).

Davis, F. M., and E. A. Adelberg. "Use of somatic cell hybrids for analysis of the differentiated state," *Bact. Rev.* **37**:197–214 (1973).

*Galau, G. A., W. H. Klein, M. M. Davis, B. J. Wold, R. J. Britten, and E. H. Davidson. "Structural gene sets active in embryos and adult tissues of the sea urchin," *Cell* **7**:487–505 (1976).

Gurdon, J. B. *The Control of Gene Expression in Animal Development.* Cambridge, Mass.: Harvard University Press, 1974.

*Gurdon, J. B., H. R. Woodland, and J. Lingrel. "The translation of mammalian globin mRNA injected into fertilized eggs of *Xenopus laevis*. I. Message stability in development," *Dev. Biol.* **39**:125–133 (1974).

*Mintz, B., and K. Illmensee. "Normal genetically mosaic mice produced from malignant teratocarcinoma cells," *Proc. Natl. Acad. Sci. U.S.* **72**:3585–3589 (1975).

Rosbash, M., M. S. Campo, and K. S. Gummerson. "Conservation of cytoplasmic poly(A)-containing RNA in mouse and rat," *Nature* **258**:682–686 (1975).

Regulation in the Egg

*Boycott, A. E., C. Diver, S. L. Garstang, and F. M. Turner. "The inheritance of sinistrality in *Limnaea peregra* (Mollusca, Pulmonato)," *Phil. Trans. Royal Soc. London Series B* **219**:51–131 (1930).

Cline, T. W. "A sex-specific, temperature-sensitive maternal effect of the *daughterless* mutation of *Drosophila melanogaster*," *Genetics* **84**:723–742 (1976).

*Fielding, C. J. "Developmental genetics of the mutant *grandchildless* of *Drosophila subobscura*," *J. Embryol. Exp. Morph.* **17**:375–385 (1967).

*Gall, J. G. "Differential synthesis of the genes for ribosomal RNA during amphibian oogenesis," *Proc. Natl. Acad. Sci. U.S.* **60**:553–560 (1968).

*Hourcade, D., D. Dressler, and J. Wolfson. "The amplification of ribosomal RNA genes involves a rolling circle intermediate," *Proc. Natl. Acad. Sci. U.S.* **70**:2926–2930 (1973).

*Illmensee, K., and A. P. Mahowald. "The autonomous function of germ plasm in a somatic region of the *Drosophila* egg," *Exp. Cell Res.* **97**:127–140 (1976).

Rice, T. B., and A. Garen. "Localized defects of blastoderm formation in maternal effect mutants of *Drosophila*," *Dev. Biol.* **43**:277–286 (1975).

Sommerville, J., and D. B. Malcolm. "Transcription of genetic information in amphibian oocytes," *Chromosoma* **55**:183–208 (1976).

Developmental Genetics of *Drosophila*

*Bonner, J. J., and M. L. Pardue. "Ecdysone-stimulated RNA synthesis in imaginal discs of *Drosophila melanogaster*," *Chromsoma* **58:**87–99 (1976).

Crick, F. H. C., and P. A. Lawrence. "Compartments and polyclones in insect development," *Science* **189:**340–347 (1975).

*Garcia-Bellido, A., and J. R. Merriam. "Cell lineage of the imaginal discs in *Drosophila* gynandromorphs," *J. Exp. Zool.* **170:**61–75 (1969).

Gehring, W. J. "Developmental genetics of *Drosophila*," *Ann. Rev. Genet.* **10:**209–252 (1976).

*Hadorn, E. "Transdetermination in cells," *Sci. Am.* **219:**110–120 (1968).

Hall, J. C., W. M. Gelbart, and D. R. Kankel. "Mosaic systems," in: *The Genetics and Biology of Drosophila*. M. Ashburner and E. Novitski, Eds. New York; Academic Press (1976).

Hotta, Y., and S. Benzer. "Courtship in *Drosophila* mosaics: Sex-specific foci for sequential action patterns," *Proc. Natl. Acad. Sci. U.S.* **73:**4154–4158 (1976).

Hotta, Y., and S. Benzer. "Mapping of behavior in *Drosophila* mosaics," *Nature* **240:**527–535 (1972).

Kankel, D. R., and J. C. Hall. "Fate mapping of nervous system and other internal tissues in genetic mosaics of *Drosophila melanogaster*," *Dev. Biol.* **48:**1–24 (1976).

Morata, G., and P. A. Lawrence. "Control of compartment development by the *engrailed* gene in *Drosophila*," *Nature* **255:**614–617 (1975).

*Morata, G., and P. A. Lawrence. "Homeotic genes, compartments, and cell determination in *Drosophila*," *Nature* **265:**211–216 (1977).

*Postlethwaite, J. H., and H. A. Schneiderman. Pattern formation and determination in the antenna of the homeotic mutant *Antennapedia* of *Drosophila melanogaster*," *Dev. Biol.* **25:**606–640 (1971).

Ripoll, P., and A. Garcia-Bellido. "Cell autonomous lethals in *Drosophila melanogaster*," *Nature New Biol.* **241:**15–16 (1973).

Schneiderman, H. A. "New ways to probe pattern formation and determination in insects," in: *Insect Development*, P. A. Lawrence, Ed. Oxford: Blackwell (1976).

Schubiger, G., and G. D. Alpert. "Regeneration and duplication in a temperature-sensitive homeotic mutant of *Drosophila melanogaster*," *Dev. Biol.* **42:**292–304 (1975).

Stern, C. *Genetic Mosaics and Other Essays.* Boston: Harvard University Press, 1968.

*Sturtevant, A. H. "The *claret* mutant type of *Drosophila simulans*, a study of chromosome elimination and cell-lineage," *Z. wiss. Zool.* **135:**323–326 (1929).

Wieschaus, E., and W. Gehring. "Clonal analysis of primordial disc cells in the early embryo of *Drosophila melanogaster*," *Dev. Biol.* **50:**249–263 (1976).

Sex Determination and Lyonization

*Barr, M. L. "Sex chromatin and phenotype in man," *Science* **130:**679–685 (1959).

Cattanach, B. M. "Control of chromosome inactivation," *Ann. Rev. Genet.* **9:**1–19 (1975).

Chandra, H. S. "Inactivation of whole chromosomes in mammals and coccids: some comparisons," *Genet. Res.* **18:**265–276 (1971).

*Davidson, R. G., H. M. Nitowsky, and B. Childs. "Demonstration of two populations of cells in the human female heterozygous for glucose-6-phosphate dehydrogenase variants," *Proc. Natl. Acad. Sci. U.S.* **50:**481–485 (1963).

Lucchesi, J. C., J. M. Rawls, and G. Maroni. "Gene dosage compensation in metafemales (3X; 2A) of *Drosophila*," *Nature* **248:**564–567 (1974).

*Lyon, M. F. "Sex chromatin and gene action in the mammalian X-chromosome," *Am. J. Hum. Gen.* **14:**135–148 (1962). [Reprinted in S. H. Boyer, *Papers on Human Genetics*. Englewood Cliffs, N. J.: Prentice-Hall, 1963.]

Lyon, M. F. "Possible mechanisms of X-chromosome inactivation," *Nature New Biol.* **232:**229–232 (1971).

Maclean, N., and V. A. Hilder. "Mechanisms of chromatin activation and repression," *Int. Rev. Cytol.* **48:**1–54 (1977).

Meyer, W. J., B. R. Migeon, and C. J. Migeon. "Locus on human X chromosome for dihydrotestosterone receptor and androgen insensitivity," *Proc. Natl. Acad. Sci. U.S.* **72:**1469–1472 (1975).

Mittwoch, U. *Genetics of Sexual Differentiation.* New York: Academic Press, 1973.

*Ohno, S. "Simplicity of mammalian regulatory systems inferred by single gene determination of sex phenotypes." *Nature* **234:**134–137 (1971).

Ohno, S., U. Tettenborn, and R. Dofuku. "Molecular biology of sex differentiation," *Hereditas* **69:**107–124 (1971).

*Ohno, S. "Major regulatory genes for mammalian sexual development," *Cell* **7:**315–321 (1976).

*Seecof, R. L., W. D. Kaplan, and D. G. Futch. "Dosage compensation for enzyme activities in *Drosophila melanogaster*," *Proc. Natl. Acad. Sci. U.S.* **62:**528–535 (1969).

Rearrangements and Position Effects

Baker, W. K. "Position-effect variegation," *Adv. Genetics* **14:**133–169 (1968).

Baker, W. K. "Evidence for position-effect suppression of the ribosomal RNA cistrons in *Drosophila melanogaster*," *Proc. Natl. Acad. Sci. U.S.* **68:**2472–2476 (1971).

Eicher, E. M. "X-autosome translocations in the mouse: total inactivation versus partial inactivation of the X chromosome," *Adv. Genetics,* **15:**176–259 (1970).

McClintock, B. "The origin and behavior of mutable loci in maize," *Proc. Natl. Acad. Sci. U.S.* **36:**344–355 (1950). [Reprinted in J. A. Peters, Ed., *Classical Papers in Genetics.* Englewood Cliffs, N.J.: Prentice-Hall, 1959.]

*McClintock, B. "The control of gene action in maize," *Brookhaven Symposia in Biology* **18:**162–184 (1965).

Spofford, J. B. "Position-effect variegation in *Drosophila*," in: *The Genetics and Biology of Drosophila,* M. Ashburner and E. Novitski, Eds. New York: Academic Press (1976).

The T Locus

*Bennett, D. "The T-locus of the mouse," *Cell* **6:**441–454 (1975).

Bennett, D., L. C. Dunn, and K. Artzt. "Genetic change in mutations at the *T/t*-locus in the mouse," *Genetics* **83:**361–372 (1976).

Dunn, L. C. "Abnormalities associated with a chromosome region in the mouse. I. Transmission and population genetics of the *t*-region," *Science* **144:**260–263 (1964).

*Kemler, R., C. Babinet, H. Condamine, G. Gachelin, J. L. Guenet, and F. Jacob. "Embryonal carcinoma antigen and the *T/t* locus of the mouse, " *Proc. Natl. Acad. Sci. U.S.* **73:**4080–4084 (1976).

*Yanagisawa, K., D. Bennett, E. A. Boyse, L. C. Dunn, and A. Dimes. "Serological identification of sperm antigens specified by lethal *t*-alleles in the mouse, "*Immunogenetics* **1:**57–67 (1974).

(a) (b)

(a) Mary F. Lyon (MRC Radiobiology Unit, England) studies the mechanism and effects of facultative heterochromatization of the X chromosome. (b) Barbara McClintock (Cold Spring Harbor Laboratory) has conducted extensive genetic analyses of controlling elements in maize.

Questions and Problems

1. Explain why a straightforward interpretation of the chick-mouse cell hybridization experiments (described in the text) is dependent on the absence of chick cytoplasm from the hybrid.

2. Human XO zygotes do not develop normal ovaries and are born sterile. What does this fact suggest about the facultative heterochromatization process in presumptive ovaries?

3. What can you conclude about human sexual development from the fact that XXY males are sterile and XYY males are fertile?

4. It has been reported that a burst of cell division occurs during the fifth week of human embryological development in the presumptive testes but not in the presumptive ovaries. How might this relate to other observations regarding gonadal sex determination in humans?

5. What would be the effect, if any, on the sex and fertility of humans if there were:
   (a) Mitotic nondisjunction during the first cleavage of a fertilized XY egg to yield one daughter XXY cell and one daughter Y cell, with subsequent separation of the two cells to form twins;
   (b) Mitotic nondisjunction during the first cleavage of a fertilized XY egg to yield one daughter XYY and one daughter X cell, with subsequent separation of the two cells to form twins;
   (c) Injection of testosterone during and after the sixth week of life in an XX embryo;
   (d) Transplantation of a testis to an XX embryo during the sixth week of life;
   (e) Castration (removal of testes) at birth of an XY individual.

6. A 25-year-old woman has never menstruated. An analysis of her karyotype reveals that she is XY. What do you think is wrong? Would hormone therapy be helpful?

7. Two genes in the human X chromosome determine the enzymes A and B, respectively. One gene exists in the allelic forms A_1 and A_2; the other in the forms B_1 and B_2. A women's father was A_1B_1 and her mother was homozygous for A_2B_2. Individual cells from this woman were tested for their enzyme phenotypes. Which of the following, mutually exclusive results would you expect from this study according to the Lyon Hypothesis? Explain.
   (a) There are two types of cells in roughly equal numbers: those that produce only enzymes A_1 and B_1 and those that produce only enzymes A_2 and B_2.
   (b) There are four types of cells in roughly equal numbers: those producing enzymes A_1 and B_1; those producing A_1 and B_2; those producing A_2 and B_1; and those producing A_2 and B_2.

8. Why do you think human females are not known to exhibit a patchy hair coloring in the manner of a calico cat?

9. An enzyme, BVDase, is expressed in human erythrocytes and is dictated by a sex-linked gene. Two alleles of the gene are known, *B* and *b*, where the *b* gene product is inactive. It is found that of women who are known from pedigree studies to be heterozygotes at this locus, about 3 percent exhibit no BVDase activity in any of their erythrocytes whereas most have a mixed population of erythrocytes, some with and some without BVDase activity.

   (a) Explain this result in terms of the Lyon Hypothesis.

   (b) Do you predict that there are many or few erythrocyte precursors in a human embryo at the time X inactivation takes place?

   (c) Predict the frequency of heterozygous women in whom all erythrocytes contain BVDase.

10. Mosaic *Drosophila* can arise in two ways.

    (a) Describe each avenue.

    (b) How would you determine which avenue had been followed in a given sample of mosaic flies?

11. An inversion in chromosome III of *D. melanogaster* produces V-type position effects. When this R $(+ + +)$ chromosome is coupled with a chromosome III carrying the markers *se* (*sepia* eyes), *ca* (*claret* eyes), and *r* (*rough* eyes), the variegated flies have the following phenotypes: eyes are always *rough*, may or may not be *claret*, and are never *sepia*. Relate the positions of these three markers to the position of the inversion.

12. **Sex-limited genes** produce traits that are expressed in only one of the sexes, presumably because some hormonal attribute of femaleness or maleness must be present if the genes are transcribed and/or the gene products expressed.

    The sex-limited human trait known as pattern baldness is expressed as follows: men heterozygous or homozygous for the autosomal gene H^B will lose the hair on top of their heads as they age; women lose their hair only if they are H^B/H^B homozygotes. A man exhibits pattern baldness but his parents do not. His wife is not balding and baldness is not found in her lineage. Give the genotypes of the husband, wife, and parents. What proportion, if any, of the man's sons would you expect to become bald? His daughters? Would the phenotypes of the offspring differ if the H^B gene were a sex-linked dominant gene but not sex-limited? Would they differ if it were a sex-linked and sex-limited dominant gene?

13. A man who considers himself "undersexed" is given testosterone treatment, whereupon he begins losing his hair. Offer a genetic explanation for this event.

14. If a nucleus had been inadvertently included in the cytoplasmic extracts utilized in an Illmensee-Mahowald experiment (Figure 18-7), how would the outcome of the experiment have been different?

15. Malignant cells are sometimes found to synthesize surface molecules known as **tumor-specific antigens,** and in some cases these are reported to be immunologically similar or identical to antigens synthesized by fetal cells. Devise a theory for the origin of such cancers using one or more of the principles presented in this chapter.

16. The *O-nu* mutation of *Xenopus* can be defined as a rescuable maternal-effect mutation. Explain this statement, using the experiments summarized in Figure 8-16 to support your argument.

17. A fetus is diagnosed by amniocentesis as trisomic for chromosome 21 but the parents decide against an abortion. The child proves to have the facial characteristics of a Down's syndrome child but has normal intelligence. How might you explain this?

18. Does it make any difference in terms of progeny phenotype whether a *grandchildless* (*gs/gs*) female fly is mated with a +/+, +/*gs*, or *gs/gs* male? Explain.

19. In *Chlamydomonas reinhardi*, the temperature-sensitive mutation *gam-1* prevents normal sexual agglutination (Figure 3-8) at restrictive temperature. The mutation is not linked to the mating-type (*mt*) locus but is expressed only in *mt⁻* cells; *mt⁺ gam-1* cells agglutinate normally and transmit the trait to their *mt⁻* progeny.

   (a) Give the genotypes, phenotypes, and expected proportions for the tetrads (Section 4.1) emerging from an *mt⁻ gam-1* × *mt⁺* + cross, where the mating is performed at permissive temperature and the phenotypes are assessed at restrictive temperature.

   (b) Discuss how a "sex-limited" mutation such as *gam-1* might exist in a single-celled haploid organism like *C. reinhardi*.

20. B. Mintz and K. Illmensee have been able to inject genetically marked cells from a malignant mouse tumor cell line into blastocysts (young embryos) of a mouse embryo, implant these blastocysts into the uteri of female mice, and examine the phenotypes and genotypes of the resultant mouse progeny. A number of the progeny mice prove to be mosaics, exhibiting, for example, tumor-derived coat-pattern genes and, in some cases, forming gonads from tumor-derived cells and transmitting tumor-derived genes to their offspring. Discuss these results (a) in terms of the totipotency of differentiated cells and (b) in terms of the concept that at least some malignancies derive from developmental aberrations of gene expression rather than mutational alterations in gene structure.

CHAPTER
19

Population Genetics I: Mendelian Populations and Evolutionary Agents

19.1 AN OVERVIEW OF POPULATION GENETICS

This and the next chapter consider genes as they occur in populations and as they have changed in evolution. A treatment of these subjects in two chapters is, by definition, most incomplete, and the reader is urged to explore the field further in the several texts cited in the bibliographies.

In moving from the molecular and chromosomal genetics of preceding chapters to a consideration of population genetics, three major shifts in orientation must be made. First, attention must turn from the individual—the individual base pair, gene, locus, or strain—to a large collection of individuals. Second, attention must focus on **variation.** Whereas the molecular or chromosomal geneticist seeks to minimize variation by using *pure* lines, *stable* genetic markers, *uniform* enzyme assay systems, and so on, the population geneticist is primarily interested in variability, in the amount of variability within a population, and in the way genes change during evolution. Third, one's time scale must change, since some of the phenomena of interest to population geneticists may take thousands or millions of years to occur.

Three major areas of research in population genetics are covered in this and the next chapter: the study of genetic polymorphism; the study of evolution; and the development of a theory of population dynamics. We outline each of these three areas briefly here so that the material developed in subsequent sections can be put into context.

THE STUDY OF POLYMORPHISM

Perhaps the most salient feature of any natural population is its diversity. This diversity is obvious when we consider the human species, for we are attuned to sensing differences in human appearance, personality, sexuality, and so on. It is less obvious, perhaps, when we think of a population of flies or sweet peas, but it exists nonetheless. In genetic terminology natural populations are said to be **polymorphic.**

Polymorphism is most apparent when it affects a visible or behavioral phenotype, but it is not at all restricted to such traits. In 1966 R. Lewontin and J. Hubby undertook the first extensive analysis of protein polymorphisms in natural populations of *Drosophila pseudoobscura.* They did this by subjecting extracts of individual flies to gel electrophoresis, a technique described in Box 7.1, and observing the rates of migration of various proteins which represented 18 genetic loci. They found, quite unexpectedly, that many of the proteins existed in the population in the form of **isoelectric variants,** meaning that for a given type of protein some individuals possessed a *fast*-migrating species and others a *slow*-migrating species, much as the isozymes described in Section 15.7. Numerous subsequent studies of such diverse species as the human, mouse, horseshoe crab, wild oat, and barley have all produced the same result: an abundance of **protein poly-**

morphism is found wherever it is sought. Protein polymorphism signals the existence of **allelism,** and it has been estimated that from 20 to 50 percent of all the structural gene loci in a given species exist in two or more allelic forms in any given population. The discovery of such diversity at a gene and protein level has, of course, greatly enhanced interest in what causes and maintains polymorphism in a population.

Genetic polymorphism is considered in detail in Chapter 20, at which time we shall consider the prevailing theories concerning its origin and maintenance.

THE STUDY OF EVOLUTION

Evolution has been described by Lewontin as a process that converts variation *within* a population into variation *between* populations, both in space (race formation and speciation) and in time (the evolution of phyla). The existence of polymorphism and the occurrence of evolution are therefore closely related subjects.

The most prominent theory regarding the driving force behind evolution is, of course, the Darwin-Wallace theory of natural selection proposed in 1858 by C. Darwin and A. Wallace. The theory holds that as genetic variants arise within a population the fittest will be at a selective advantage and will be more likely to produce offspring than the rest. As the fit continue to enjoy greater survival and reproductivity, new species will eventually evolve.

Driving forces other than natural selection are thought to underlie certain evolutionary phenomena, and these are described in this chapter. Major attention is given to natural selection, however, since it has been the subject of the most extensive research.

POPULATION GENETICS THEORY

A third major research area in population genetics is the development of a quantitative theory to define the laws that govern the genetic structures of natural populations. Were such a theory developed, it might well be applied to existing populations in order to predict, for example, their evolutionary courses or their responses to particular changes in the environment.

The existing theory derives from many sources, most fundamentally perhaps from the works of R. A. Fisher, J. B. S. Haldane, and S. Wright. It is still far from the stage at which it can be used to make detailed predictions for complex populations, but it serves an important function even as it is being developed: once new theoretical ideas are generated they suggest new experimental approaches to the study of natural populations, and the new data emerging from such studies in turn suggest modifications in the theory. In other words, theory and experiment are intimately associated in population genetics, often far more conspicuously than in molecular and chromosomal genetics.

The sections that follow include some elementary theory concerning basic population dynamics. Theory and observation are then discussed together in terms of agents that can bring about evolutionary change.

THE HARDY-WEINBERG EQUILIBRIUM

19.2 EQUILIBRIUM POPULATIONS AND THE HARDY-WEINBERG LAW

Population genetics theory can be said to have its roots in the 1908 publications of G. Hardy and W. Weinberg, who considered the behavior of genes in **idealized populations.** An idealized population has two properties that are only rarely encountered in natural populations: individuals in the population are assumed to mate randomly, and the population as a whole is assumed to be genetically static. In a population in which mating is totally random any mature female, provided she is physiologically competent, is equally likely to mate with any physiologically competent male in the population. The act of mating, in other words, is blind to both genotype and phenotype. In a genetically static population gene frequencies are not influenced by any evolutionary pressures: a given phenotype is no more fit for the environment than any other and no genes in the population are undergoing mutational change.

We can now consider a single gene locus in this idealized population and we can define the **frequency of alleles** at that locus in the population. For the simple case in which there are only two alleles, A and a, we can let p equal the frequency of A and q the frequency of a, and we can state, by definition, that $p + q = 1$. Specifically, if 48 percent, or 0.48, of all the pertinent loci in the population contain the A allele, then $p = 0.48$, and it follows that $q = 0.52$.

By assuming that breeding in the idealized population occurs at random, all of the gametes in the population can be considered to comprise a single pool and the various allelic frequencies can be thought of as probabilities. Thus p indicates the probability that, in a random sampling, an A-bearing gamete will be "drawn" from the pool and q indicates the probability that an a-bearing gamete will be drawn. When we now consider the probability that two particular gametes will be drawn simultaneously, that is, the probability that two particular gametes will unite to form a zygote, we simply multiply the probabilities of these independent events, much as we did in Section 10.7 when we calculated the expected frequency of double crossovers. Thus if p is the probability of drawing an A gamete, the probability that a particular fertilization will produce an A/A homozygote is simply $p \times p = p^2$. Similarly, the probability of producing an a/a homozygote is q^2. Finally, the probability of producing an A/a heterozygote is equal to $pq + pq = 2\,pq$. These various relationships are diagrammed in Figure

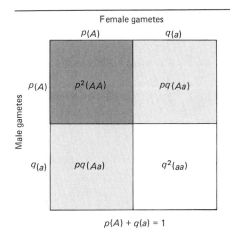

Female gametes

$p(A)$ $q(a)$

Male gametes

$p(A)$ | $p^2(AA)$ | $pq(Aa)$
$q(a)$ | $pq(Aa)$ | $q^2(aa)$

$p(A) + q(a) = 1$

$p^2(A) + 2pq(Aa) + q^2(aa) = 1$

Figure 19-1 The elements of a Hardy-Weinberg equilibrium. Gamete frequencies ($p_{(A)}$ and $q_{(a)}$) are multiplied together to predict genotypic frequencies in the next generation, the sum of which must equal 1.

19-1. Summarizing, therefore, the numbers of homozygous and heterozygous individuals in an idealized population:

$$p^2 = A/A \text{ homozygotes}$$
$$2pq = A/a \text{ heterozygotes}$$
$$q^2 = a/a \text{ homozygotes}$$

We noted earlier that since p and q are frequencies, their sum is equal to 1. Multiplying these frequencies together in various combinations does not alter the fact that their total value must add up to unity. Therefore, we can write the following sum:

$$p^2 + 2pq + q^2 = 1$$

The above sum can now be factored and expressed as:

$$(p + q)^2 = 1$$

Indeed, we now recognize that if $p + q = 1$, then multiplying these frequencies by themselves is simply the same operation as $(p + q)^2$ which must also equal 1 ($1^2 = 1$).

You may have recognized the equation $p^2 + 2pq + q^2 = 1$ as the expression for a binomial distribution, and its application to population genetics has come to be known as the **Hardy-Weinberg Law** or the **Hardy-Weinberg Equilibrium.** The expression is important because it allows one to calculate either the frequency of alleles (defined above as p and q) or the frequency of genotypes (defined above as $p^2 + 2pq + q^2$) for idealized populations.

Suppose, for example, we learn that in a static, randomly breeding human population 36 percent of the persons cannot taste phenylthiocarbamide (PTC), a substance that tastes very bitter to the rest of the population. Knowing that the "tasting" phenotype is controlled by a single locus and that the *nontaster* allele (*a*) is recessive to the *taster* allele (*A*), we can conclude the following:

1. The frequency of *a/a* homozygotes, and therefore the frequency of the *a/a* genotype, is 36 percent = 0.36.
2. This value corresponds to q^2, so that q is simply the square root of $0.36 = 0.6$.
3. The value of p becomes $1 - q = 0.4$.
4. The population must therefore contain $p^2 = 0.16 = 16$ percent homozygous tasters, $2pq = 0.48 = 48$ percent heterozygous tasters, and, as we learned, 36 percent nontasters.

A key feature of an idealized population can now be appreciated: *regardless of the proportions of particular diploid genotypes that establish a population, the genotypes assume a Hardy-Weinberg distribution in the following generation and in all succeeding generations provided that the sexes are equally represented.* If the starting generation, for example, consists of 7000 *A/A* individuals and 3000 *a/a* individuals, the starting genotypic frequencies will be *A/A* = 0.7 and *a/a* = 0.3 and the starting allele frequencies will be, similarly, $p = 0.7$ and $q = 0.3$. At the next generation, when random breeding occurs and a common gamete pool is established, p will still equal 0.7 and q will equal 0.3, but the Hardy-Weinberg formulation will now apply. Therefore we can expect that

$$p^2 = 0.49 = \text{frequency of } A/A$$
$$q^2 = 0.09 = \text{frequency of } a/a$$

and

$$2pq = 2 \times 0.7 \times 0.3 = 0.42 = \text{frequency of } A/a$$

Since half the gametes produced by the *A/a* heterozygotes will carry *A* and since all the gametes produced by *A/A* homozygotes will carry *A*, it follows that the frequency of *A* in the gamete pool will be

$$1/2(0.42) + 0.49 = 0.70$$

which was our original starting frequency of *A*. By similar calculations, the value of q is found again to be 0.3. In other words, as long as the population is genetically static and randomly breeding and as long as the sampling of the gamete pool is done in a random manner, allelic and genotypic frequencies should neither rise nor fall from one generation to the next.

19.3   DEVIATIONS FROM HARDY-WEINBERG EQUILIBRIUM

The idealized population we have been considering is often termed a **Mendelian population** because the behavior of its alleles can be predicted by the laws of probability in much the same way that the F_1 and F_2 ratios were predicted in the Mendelian crosses described in Chapter 4. In Figure

19-1, for example, the genotypic ratios are estimated in the same manner as when a heterozygous F_1 population is inbred (see, for example, Figure 4-3).

A Mendelian population does not experience fluctuations in its allelic or genotypic frequencies from one generation to the next, as we have just seen. In contrast, such fluctuations are usually observed when natural populations are studied over a series of generations. To determine whether sampling errors are responsible, the data can be subjected to a chi-square (χ^2) test by following the procedures outlined in Section 4.9. When a χ^2 analysis indicates that the fluctuations deviate significantly from the predictions of the Hardy-Weinberg theorem, it is inferred that sampling errors are not involved and that the theorem itself is inapplicable to the population in question.

Significant deviations from the ideal can arise for three reasons: the population may not be randomly breeding; it may not be static; or it may be neither randomly breeding nor static. Since the great majority of natural populations are found to deviate significantly from the ideal, it becomes the task of the quantitative population geneticist to modify the Hardy-Weinberg theory so that it will become applicable to nonstatic populations in which mating is said to be **assortative** rather than random. In the process of making such modifications the theory has become mathematically more and more sophisticated to the point at which it is no longer meaningful to most students of genetics and even, it can be said, to many geneticists themselves.

Why, then, master simple concepts like the Hardy-Weinberg Law when the present mathematical descriptions of actual population dynamics bear little relationship to a binomial distribution? Perhaps the best answer is that the Hardy-Weinberg concept, and many of the other simple algebraic expressions we present in the following sections, are useful. They help to define some of the important variables in a population and demonstrate how the population geneticist goes about trying to define the variables more precisely.

Having considered a model Mendelian population that is *not* undergoing evolution, we can now examine **evolutionary agents**—factors that are thought to bring evolution about. Four major "factors of evolution" are usually listed: mutation, selection, migration, and random drift, and these are considered in the remaining sections of this chapter.

MUTATIONS IN EVOLUTION

One can visualize several ways in which mutation might bring about evolutionary change. Mutation might be highly **directed** at a particular locus such that allele a_1 is selectively driven to the a_2 form. Alternatively, the mutation process might be random but, with time, the a_2 form would come to predominate over a_1. Finally, mutation might simply provide a population with new alleles on which any and all evolutionary agents (natural selection, for example) could act.

19.4   DIRECTED MUTATIONS

The concept that there might be highly directional influences on particular alleles to mutate from one form to another has often been advanced; it is, in fact, one way to state the theory, championed by J. B. Lamarck, that acquired traits are inherited, and it is also one way to state the concept that divine forces may play a role in evolutionary change.

From a genetic point of view the best evidence for directed mutation comes from the study of **mutator genes,** genes that influence the mutation process at other loci. A classical example of a mutator gene is the Dt element in maize. This gene acts specifically on an unlinked locus called a which is involved in the synthesis of the purple pigment anthocyanin. Specifically, Dt causes recessive a_1 alleles to "mutate" repeatedly to A_1 forms such that the synthesis of anthocyanin can occur. As a result, $\dfrac{Dt\ a_1}{Dt\ a_1}$ plants exhibit purple dots on the surfaces of otherwise colorless corn kernels, each dot presumably representing a single clone of anthocyanin-producing cells. The mutant corn plants also have purple streaks on their stems and leaves. The Dt element acts specifically on a_1 alleles, and its mode of action is totally obscure.

Mutator genes such as Dt may well have contributed to particular evolutionary events. In general, however, they are not believed to represent major evolutionary agents, and mutations are thought to enter the gene pool in a random fashion.

19.5   MUTATION PRESSURE AS AN EVOLUTIONARY AGENT

Granted that mutation from $a_1 \longrightarrow a_2$ occurs in most cases by a random process, we can next inquire whether this process will disrupt equilibrium conditions in a Mendelian population and so potentially bring about evolutionary change. Stated more formally, if values of p and q no longer remain constant from one generation to the next, will such **mutation pressure** bring about significant evolutionary changes in the absence of other evolutionary agents?

To answer this question we must first define a term that is important to population genetics, namely, the **mutation rate,** $\mu.$ If p is the frequency of allele a_1 in a population and a_1 is mutating to a_2 at a constant mutation rate μ per generation, the change in p per number of generations (t) becomes

$$\frac{dp}{dt} = -\mu p \qquad (1)$$

This equation is solved as follows:

$$\frac{dp}{p} = -\mu dt$$

$$\int \frac{dp}{p} = -\int \mu dt$$

$$\ln p = -\mu t + \ln c$$

$$p = ce^{-\mu t}$$

If we consider the initial frequency of a_1 to be p_o at a time when $t = 0$, then c becomes equal to p_o and

$$p = p_o e^{-\mu t} \tag{2}$$

where $e = 2.72$.

We can now turn to actual mutation rates which have been estimated for various loci in various organisms. These values, summarized in Table 19-1, indicate a standard spontaneous mutation rate of about 10^{-5} for eukaryotes, with 10^{-4} being high and 10^{-6} or less being low. The values are seen to be lower for prokaryotes.

Taking a value of 10^{-5} for μ, what does Equation 2 tell us about the likelihood that evolution can come about via mutation alone? Looking at the equation, it becomes clear that values of t must approach the reciprocal of μ if any significant change is to occur in the value of p. Specifically, if the value of p is to be reduced by about one-third of its original value ($p_o/2.72$) then, when $\mu = 10^{-5}$, t must equal 10^5 or 100,000 generations. This means 100,000 years in an organism with a one-year generation time, and perhaps 20 times as long for humans if we assume a human generation time of 20 years. These time estimates are almost certainly much too small, for Equation 2 ignores the occurrence of $a_2 \longrightarrow a_1$ reversions. Obviously, if the reversion rate (ν) is comparable to the forward mutation rate (μ)—and there is no a priori reason to assume otherwise for point mutations—the value of p will change slowly indeed. It must therefore be postulated that for some reason μ is significantly greater than ν.

These considerations make it seem unlikely that marked changes in the allelic frequencies of a population can be effected by mutation pressure alone. Rather, in the absence of other evolutionary agents, it is expected that some sort of **mutational equilibrium** will be established at each locus. The theoretical expression for this equilibrium condition is as follows. We let p be the frequency of a_1 and q the frequency of a_2 in a population. The change in p per generation (Δp) will then be the gain in $a_2 \longrightarrow a_1$ reversions (νq) minus the loss of $a_1 \longrightarrow a_2$ forward mutations (μp) and Δq will be the gain in forward mutations (μp) minus the loss in reversions (νq).

These relationships are written:

$$\Delta p = \nu q - \mu p = \nu(1 - p) - \mu p$$
$$\Delta q = \mu p - \nu q = \mu(1 - q) - \nu q \tag{3}$$

Table 19-1 Spontaneous Mutation Rates in Different Organisms

Organism	Character	Rate	Units
Bacteriophage	Lysis inhibition, $r \longrightarrow r^+$	1×10^{-8}	Per gene[a]
T$_2$	Host range, $h^+ \longrightarrow h$	3×10^{-9}	per replication
Bacteria	Lactose fermentation, $lac^- \longrightarrow lac^+$	2×10^{-7}	
Escherichia	Phage T$_1$ sensitivity, T$_1$-s $\longrightarrow$ T$_1$-r	2×10^{-8}	
coli	Histidine requirement, $his^- \longrightarrow his^+$	4×10^{-8}	
	$his^+ \longrightarrow his^-$	2×10^{-6}	Per cell
	Streptomycin sensitivity,		per division
	str-s $\longrightarrow$ str-d	1×10^{-9}	
	str-d $\longrightarrow$ str-s	1×10^{-8}	
Algae	Streptomycin sensitivity,		
Chlamydomonas	str-s $\longrightarrow$ str-r	1×10^{-6}	
reinhardi			
Fungi	Inositol requirement, $inos^- \longrightarrow inos^+$	8×10^{-8}	Mutation
Neurospora	Adenine requirement, $ade^- \longrightarrow ade^+$	4×10^{-8}	frequency
crassa			among
			asexual
			spores
Corn	Shrunken seeds, $Sh \longrightarrow sh$	1×10^{-5}	
Zea mays	Purple, $P \longrightarrow p$	1×10^{-6}	
Fruit fly	Yellow body, $Y \longrightarrow y$, in males	1×10^{-4}	Mutation
Drosophila	$Y \longrightarrow y$, in females	1×10^{-5}	frequency
melanogaster	White eye, $W \longrightarrow w$	4×10^{-5}	per
	Brown eye, $Bw \longrightarrow bw$	3×10^{-5}	gamete
Mouse	Piebald coat color, $S \longrightarrow s$	3×10^{-5}	per
Mus musculus	Dilute coat color, $D \longrightarrow d$	3×10^{-5}	sexual
Man	Normal $\longrightarrow$ hemophilic	3×10^{-5}	generation
Homo sapiens	Normal $\longrightarrow$ albino	3×10^{-5}	
Human bone	Normal $\longrightarrow$ 8-azoguanine resistant	7×10^{-4}	
marrow cells	Normal $\longrightarrow$ 8-azoguanosine resistant	1×10^{-6}	Per cell
in tissue			per division
culture			

[a] Correction of the other mutation rates in this table to a per-gene basis would not change their order of magnitude.

From R. Sager and F. J. Ryan, *Cell Heredity*. New York: Wiley, 1961.

We now let $\widehat{p}$ and $\widehat{q}$ signify the **equilibrium allelic frequencies** when $\Delta p = 0$ and $\Delta q = 0$, since at equilibrium there will be no net change in either value. We then get

$$\widehat{p} = \frac{\nu}{\mu + \nu}$$

$$\widehat{q} = \frac{\mu}{\mu + \nu}$$

and

$$\frac{\widehat{p}}{\widehat{q}} = \frac{\nu}{\mu} \tag{4}$$

Mutational equilibrium is expected in a population that is subject to mutational but to no other evolutionary pressures. If μ and ν remain steady, then $\widehat{p}$ and $\widehat{q}$ are to all purposes equivalent to the p and q utilized in the Hardy-Weinberg expression for a nonevolving population.

19.6 MUTATIONAL "CURRENCY"

We are left with the third, and most obvious, avenue by which mutation might contribute to evolution: by providing new alleles (new "currency") on which evolutionary agents such as natural selection can act.

The ways in which mutation can change the informational content of structural and regulatory genes and controlling elements have been described throughout this text. From an evolutionary perspective we can begin by listing mutations under three theoretical categories. (1) Mutations can produce genes which prove, in the evolutionary arena, to be **selectively advantageous** compared to their ancestral alleles. (2) By the same criterion, mutations can be **selectively disadvantageous.** (3) Finally, **neutral mutations** may occur which lead to the production of protein products that are functionally identical to their antecedents. In forthcoming sections we shall be considering how evolutionary agents might act on these three different kinds of mutations.

In addition to point mutations, chromosomal mutations are also believed to have played a major role in evolution. These are considered in Section 20.13.

19.7 MUTATIONAL "LOAD"

In emphasizing the beneficial effects of mutation in evolution, the potentially deleterious effects of mutation must be stressed as well, for while a higher mutation rate in a population heightens the population's potential for evolutionary change, a high mutation rate also increases the potential number of slightly or severely deleterious genes per gamete. Obviously, if evolution is to proceed, an accumulation of deleterious mutations—often referred to as a **mutational "load"**—must not continuously overshadow the progressive effects of allelic diversity.

Some interesting problems associated with the concept of mutational load become evident on studying Table 19-1. Such species as the mouse and human are seen to exhibit at particular loci mutation rates that are, if anything, significantly greater than the mutation rates found at particular loci in *E. coli.* When this observation is coupled with the fact that the mouse and human possess about 1000 times more nucleotide pairs per genome than *E. coli* (3×10^9 versus 4.5×10^6 pairs), a theoretical dilemma arises. Must it be concluded that humans and mice carry a thousand times greater mutational load per generation than *E. coli*? In other words, as the total size of a genome increases, is there a concomitant increase in the total number of potentially deleterious mutational events experienced per organism?

Although the instinctive answer is "no," theoretically plausible arguments to support a negative answer have only recently been offered, most notably by T. Ohta and M. Kimura. They assume, as do many geneticists, that the thousandfold increase in genome size that has accompanied eukaryotic evolution has been brought about via numerous DNA duplications. Within each duplicated segment, as we have just noted, some new genes may be created by mutation and may become important in evolution. Granted the random nature of mutation, however, and the presumed "selective immunity" of duplicated DNA, Ohta and Kimura propose that most of this duplicated DNA progressively accumulates mutations to the point at which, in mammals, it is **mutationally inert.** The dilemma of the mammalian mutational load can then be resolved by proposing that the mutation rate of, say, 10^{-5} mutations per locus per generation will create deleterious mutations in mammals *only* at loci that carry *bona fide* genes or controlling elements. The remaining inert DNA may well experience nucleotide changes at the same rate, but no major additional load will be imposed on individual organisms or on particular species.

On the basis of their theories Ohta and Kimura have estimated the proportion of the human genome that represents inert DNA. The most accurate estimate available of the spontaneous human mutation rate at a molecular level is 8.3×10^{-9} mutation per codon per year, a rate based on analyses of amino acid substitutions in various proteins (which we consider in the next chapter). The human haploid genome contains about 3×10^9 nucleotide pairs and therefore about 10^9 potential codons. If all the DNA in the human genome represents genes or controlling elements, there should then be $8.3 \times 10^{-9} \times 10^9 \cong 8$ mutations per haploid genome per year. If this mutation rate applies to germ-line cells, sperm and eggs would each be expected to have accumulated, during the 20 years they require to reach sexual maturity, $20 \times 8 = 160$ mutations, meaning that a zygote would carry $160 + 160 = 240$ mutations. Various population geneticists have estimated that the human species can in fact support only about one deleterious mutation per gamete per generation. Therefore, a maximum of about $1/160 = 0.6$ percent of the genome should contain genes and controlling elements, the rest being inert DNA that can mutate relatively freely.

The Ohta-Kimura theory is attractive in that it leads to the same kinds of conclusions about the nature of the mammalian genome as conclusions reached by very different lines of reasoning. As already noted in previous chapters, a mammal gives no evidence of synthesizing 1000 times as many different kinds of proteins as a bacterium. Secondly, as noted in Section 18.9, DNA-RNA hybridization experiments indicate that only a small fraction of a eukaryote's genome appears to be transcribed into RNA. Finally, comparative DNA-DNA hybridization studies performed by C. Laird and colleagues indicate that DNA molecules change their nucleotide composition far more rapidly than proteins change their amino acid composition. Thus, in one example, when similar species of artiodactyls (hooved mammals) are compared (Figure 19-2), the extent of nonhomology between their DNAs is found to be three times greater than the extent of difference between the amino acid sequences of their hemoglobin. Even greater disparities are found among rodents.

A particularly dramatic example of this kind of observation has recently come from D. Brown and coworkers. They studied the ribosomal DNA (see Figure 8-18) from *X. laevis* and *X. mulleri,* two closely related species of toad

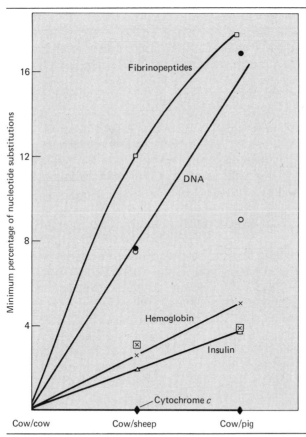

Figure 19-2 Comparison of nucleotide sequence divergence in DNA and amino-acid sequence divergence in proteins among artiodactyls, with sequence differences expressed relative to bovine proteins or DNA. Estimates of gene divergence are shown as circles. Inferred percentages of nucleotide substitutions for the various proteins are based on comparative amino-acid sequence data. *Symbols:* □, fibrinopeptides; ⊠, hemoglobin α chain; X, hemoglobin β chain; △, insulin; ◆, cytochrome *c*. (From C. Laird, B. McConaughy, and B. McCarthy, *Nature* **224**:149–154, 1969.)

that can be induced to form (usually) sterile hybrids in the laboratory. They found that although the informational sequences for rRNA are remarkably similar for the two species, the nontranscribed spacer DNA is exceedingly dissimilar. Apparently, therefore, few if any mutations were tolerated in the informational segments of the ribosomal DNA of *X. laevis* and *X. mulleri* as the two species followed their separate evolutionary paths, whereas numerous mutations have accumulated in those DNA spacer segments regarded as being representative of inert DNA.

Granted that both theoretical considerations and experimental observations indicate that much DNA in eukaryotes can be formally designated as inert, two possibilities at once arise. On the one hand, inert DNA might simply be relictual "junk" in the genome with no function whatever. Alternatively, inert DNA might perform vital functions that do not depend on specific nucleotide sequences. If, for example, chromosomal folding or synapsis depended on the presence of middle-repetitive DNA (Section 7.14) at regular intervals along a chromosome, then a base-pair change in this DNA would by no means have the same potential consequences as a base-pair change in the active-site region of the cytochrome c gene. It would not follow, however, that this DNA was functionless. It may, of course, turn out that some inert DNA is indeed "junk" while the remainder performs important but sequence-independent functions. A major challenge to molecular geneticists in the coming decade will be to elucidate the role of this DNA in the eukaryotic genome.

Before leaving the subject of mutational load we should consider some statistics. Estimates of the percentage of human conceptuses that undergo spontaneous abortion before attaining 22 weeks of age or 500 g in weight are as high as 45 percent. Furthermore, of the embryos and fetuses recovered for study, many are visibly abnormal. Although these observations could be used to argue that the human mutational load is very great, two counterarguments can be made. It can first be pointed out that an undetermined proportion of these abortions may be environmentally induced by, for example, malnutrition or illness in the mother or perhaps by the effects of increased uterine age. It can also be pointed out that an estimated one-third of all spontaneous human abortions involve a fetus with an abnormal number of chromosomes. The tendency to produce chromosomal abnormalities may well be inherited in some cases, but by and large these abnormalities appear to derive from errors occurring during meiosis in germ-line cells. Since these errors are eliminated by abortions, they never enter the human gene pool and thus never have a chance to contribute to the human mutational load.

NATURAL SELECTION: GENERAL PRINCIPLES

Just as mutation is believed to be the primary source of new alleles in a population, so is natural selection generally believed to be the prominent agent for determining the relative frequency of alleles in a population. Natural selection differentiates between phenotypes in a population with respect to their ability to produce offspring. One phenotype may better survive endemic onslaughts of parasites or predators than another; one may penetrate new habitats more effectively than another; one may mate more efficiently than another; one may even prey on the other. The important point is that some natural situation or environmental feature selectively allows one organism to develop and to propagate more efficiently than another, and the genotype of this organism is therefore afforded greater representation in the overall gene pool. If this selective process continues over many generations, allelic frequencies will alter significantly and the potential will arise for evolutionary change.

19.8 RELATIVE FITNESS AND SELECTION COEFFICIENTS

In the theory of natural selection, there arise three terms that are particularly important: **survival rate, relative fitness, and selection coefficient.** To calculate values for these parameters—and thus best understand their meaning—we begin with data from a particular population. For simplicity we shall consider in this population only a single locus defined by the A and a alleles; we shall also assume incomplete dominance so that A/a heterozygotes can be distinguished phenotypically from A/A and a/a homozygotes. We then count the number of A/A, A/a, and a/a individuals in a given generation immediately *before* and immediately *after* some selective event—the introduction of a parasite, a change in temperature—has occurred. Representative numbers are given in Table 19-2. From these we can calculate a **survival rate** for each genotype, as shown in the table.

The genotype with the greatest survival rate is defined as the fittest, and is used as the standard for the **relative fitness (W)** of all other genotypes in that generation. Specifically, in our example, A/A has a survival rate of 0.95 and is thus more fit than A/a or a/a (Table 19-2). In determining values for W, therefore, all survival rates are divided by 0.95 so that the relative fitness of A/A becomes $0.95/0.95 = 1$, the relative fitness of A/a becomes $0.80/0.95 = 0.84$, and the relative fitness of a/a becomes $0.55/0.95 = 0.58$.

The **selection coefficient (s)** is simply $1 - W$. Just as W reflects the chances of an organism's reproductive success, so does s reflect the chances of its reproductive failure due to selection. W and s are two sides of the same coin.

The example in Table 19-2 relates to data taken within the same generation of a population. It is probably more common to encounter data in which phenotypic frequencies are enumerated for two successive generations of a population that has experienced selection. In this case relative fitness can be estimated by comparing the observed second-generation phenotypic frequencies with those expected on the basis of a Hardy-Weinberg equilibrium, much as in the first stage of a χ^2 test. The resulting

Table 19-2   Relative Fitness and Selection Coefficients Calculated from Data Taken in the Same Generation of a Population

Data: Number of Individuals in the Population According to Genotype

	A/A	A/a	a/a
Before selection	4100	5000	2200
After selection (same generation)	3900	4000	1200

Calculations:

(a) Survival rate

$$A/A = \frac{3900}{4100} = 0.95$$

$$A/a = \frac{4000}{5000} = 0.80$$

$$a/a = \frac{1200}{2200} = 0.55$$

(b) Relative fitness W (compared with A/A's maximum survival rate)

$$W_{AA} = \frac{0.95}{0.95} = 1.00$$

$$W_{Aa} = \frac{0.80}{0.95} = 0.84$$

$$W_{aa} = \frac{0.55}{0.95} = 0.58$$

(c) Selection coefficient s

$$s_{AA} = 1 - W_{AA} = 0$$
$$s_{Aa} = 1 - W_{Aa} = 0.16$$
$$s_{aa} = 1 - W_{aa} = 0.42$$

ratios can be considered as survival rates and relative fitness and selection coefficients can be calculated as before. These procedures are illustrated in Table 19-3.

With this much background in the quantitative evaluation of natural selection we can turn to more specific examples of this phenomenon.

19.9   DIRECTIONAL SELECTION: QUANTITATIVE THEORY

A principal pattern followed by selection in natural populations is that of **directional selection.** The effect of directional selection is to eliminate or to reduce the reproductive potential of particular phenotypes in a population. Specifically, if the variance in a particular trait is normally distributed in a population before directional selection begins to operate, the variance might become markedly skewed in one direction afterward (Figure 19-3).

Table 19-3 Relative Fitness and Selection Coefficients Calculated from Data Taken in One Generation before Selection and in a Second Generation after Selection

Data: Number of Mating Individuals in the Population According to Genotype

	A/A	A/a	a/a	Total
Before selection (first generation)	3100	4000	1800	8900
After selection (next generation)	3600	4600	1600	9800

Calculations:

(a) Allelic frequencies in first generation	Frequency of $A = p = \dfrac{2(3100) + 4000}{2(8900)}$
	$= \dfrac{10,200}{17,800}$
	$= 0.57$
	Frequency of $a = q = 1 - p = 0.43$
(b) Expected genotypic frequencies in second generation	Expected $A/A = p^2 = (0.57)(0.57) = 0.32$
	$A/a = 2pq = 2(0.57)(0.43) = 0.49$
	$a/a = q^2 + (0.43)(0.43) = 0.19$
(c) Observed genotypic frequencies in second generation	Observed $A/A = \dfrac{3600}{9800} = 0.37$
	$A/a = \dfrac{4600}{9800} = 0.47$
	$a/a = \dfrac{1600}{9800} = 0.16$
(d) Ratio of observed to expected genotypic frequencies in second generation	Ratio $A/A = \dfrac{0.37}{0.32} = 1.16$
	$A/a = \dfrac{0.47}{0.49} = 0.96$
	$a/a = \dfrac{0.16}{0.19} = 0.84$
(e) Relative fitness W (compared with A/A's maximal ratio)	$W_{AA} = \dfrac{1.16}{1.16} = 1.00$
	$W_{Aa} = \dfrac{0.96}{1.16} = 0.83$
	$W_{aa} = \dfrac{0.84}{1.16} = 0.72$
(f) Selection coefficient s	$s_{AA} = 1 - W_{AA} = 0$
	$s_{Aa} = 1 - W_{Aa} = 0.17$
	$s_{aa} = 1 - W_{aa} = 0.28$

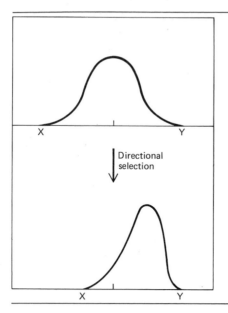

Figure 19-3 The effects of directional selection on a population whose phenotypes are normally distributed between two extremes, X and Y. Directional selection acting against X and in favor of Y changes the distribution of the phenotypic variation.

19.10   COMPLETE ELIMINATION OF RECESSIVES

An extreme form of directional selection occurs when recessive homozygotes (a/a) are systematically eliminated from the population by selection. In such cases the relative fitness of a/a is set at 0 and the selection coefficient for this set of alleles therefore becomes $1 - 0 = 1$. If we also assume, for simplicity, that A/a and A/A are identically fit, and that no other alleles are involved at this locus, then by definition A/A and A/a each have fitnesses of 1 and selection coefficients of 0.

We next define q_o as the starting frequency of a, and we assume that the ideal population we are dealing with is in a Hardy-Weinberg equilibrium before selection. This means that the frequency of a/a individuals will be q_o^2, the frequency of A/A individuals will be p_o^2, the frequency of A/a individuals will be $2p_oq_o$; the total population can then be represented as the sum

$$p_o^2 + 2p_oq_o + q_o^2 = 1$$

Once selective pressure has been imposed, each genotypic frequency is multiplied by its relative fitness. The values of p_o^2 and $2p_oq_o$ remain unchanged, since each is multiplied by 1, but the value of q_o^2 is reduced to zero, since it is multiplied by zero. The total population can therefore be expressed after selection as the sum $p_o^2 + 2p_oq_o$.

When this population engages in reproduction, all the a alleles in the gamete pool will be contributed by the A/a heterozygotes, which constitute $2p_oq_o$ of the total. Since only half the gametes produced by the A/a organisms will carry the *a* allele, the frequency of *a* in the next generation (expressed as q_1) will be

$$q_1 = \frac{1/2(2p_o q_o)}{p_o^2 + 2p_o q_o}$$

$$= \frac{p_o q_o}{p_o(p_o + 2q_o)} \qquad p_o = 1 - q_o$$

$$= \frac{q_o}{1 - q_o + 2q_o}$$

$$= \frac{q_o}{1 + q_o} \tag{5}$$

By performing the same operations for successive generations we get

$$q_2 = \frac{q_1}{1 + q_1} \qquad q_3 = \frac{q_2}{1 + q_2} \qquad q_4 = \frac{q_3}{1 + q_3}$$

and, in general,

$$q_n = \frac{q_o}{1 + nq_o} \tag{6}$$

where q_n is the frequency of *a* in the population after *n* generations.

The elimination of "undesirable" homozygotes from a breeding population by directional selection is sometimes suggested as a way to reduce the frequency of "undesirable" alleles in the gene pool of the population. It can be seen from Equation 6, however, that the frequency of the "undesirable" allele would fall slowly indeed and that the allele would still be present after many thousands of years. Let us suppose, for example, that a **eugenics** program prohibits all albino (*a/a*) humans from having children on the grounds that albino homozygotes can experience certain health difficulties. If the population contains one albino among every 10,000 persons (approximately the *a/a* frequency in human populations), how many generations would elapse before the *a/a* frequency became one in every million persons? The desired genotype reduction, in this case, is from 10^{-4} (q_o^2) to 10^{-6} (q_n^2). By taking the square roots of these values we obtain the corresponding reduction in gene frequency, namely, 10^{-2} (q_o) to 10^{-3} (q_n). We can rewrite Equation 6 solving for *n* as

$$n = \frac{q_o - q_n}{q_o q_n}$$

$$= \frac{1}{q_n} - \frac{1}{q_o}$$

The number of generations required to go from q_o to q_n then becomes

$$n = \frac{1}{10^{-3}} - \frac{1}{10^{-2}} = 1000 - 100 = 900 \text{ generations}$$

or about 18,000 years. Considering that the recorded history of humans began only 5000 years ago and that mutation is constantly providing the human gene pool with new a alleles, such a eugenics program seems most ineffective.

19.11   PARTIAL SELECTION AGAINST RECESSIVES

We can next consider a less complete form of directional selection against homozygous recessive individuals. Here the selection coefficient s is less than 1 and the relative fitness W of the homozygous recessive individuals is $1 - s$, a number greater than zero. Let us once again begin with a Mendelian population and, following selection, multiply each original genotypic frequency by its relative fitness. The proportion of A/A then remains p_o^2, the proportion of A/a remains $2p_oq_o$, and the proportion of a/a becomes $q_o^2(1 - s)$. The *total* population after selection can be expressed as $1 - sq_o^2$, where sq_o^2 represents the a/a alleles that are eliminated by selection.

In the next generation a-bearing gametes derive from two sources: the A/a heterozygotes as before, whose contribution is expressed as $1/2(2p_oq_o)$, and the surviving a/a homozygotes, whose contribution is $q_o^2(1 - s)$. The frequency of a alleles in this next generation is therefore

$$q_1 = \frac{p_oq_o + q_o^2(1 - s)}{1 - sq_o^2}$$
$$= \frac{q_o(1 - sq_o)}{1 - sq_o^2} \tag{7}$$

and, in general,

$$\boxed{q_n = \frac{q_{n-1}(1 - sq_{n-1})}{1 - sq_{n-1}^2}} \tag{8}$$

The only case in which this formula can be directly solved occurs when $s = 1$, in which case we have Equation 6 again. When s does not equal 1, the best that can be done is to express Δq, the change in q in one generation, as

$$\Delta q = q_1 - q_o$$
$$= \frac{q_o(1 - sq_o)}{1 - sq_o^2} - q_o$$
$$= \frac{-sq_o^2(1 - q_o)}{1 - sq_o^2} \tag{9}$$

At this point we can drop the subscript $(_o)$ since we are actually interested in a general value for Δq and not simply the value after the first generation. We can also simplify Equation 9 by making the assumption that s has a small value such that $1 - sq_o^2$ is approximately equal to 1. Equation 9 thereby becomes

$$\Delta q \simeq -sq^2(1 - q) \tag{10}$$

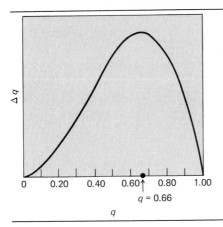

Figure 19-4 Graphical representation of $\Delta q \cong -sq^2 (1 - q)$ where s is a constant. (From C. C. Li, *Population Genetics*, University of Chicago Press, Chicago, 1955, Figure 64.) (Reprinted from *Population Genetics* by C. C. Li by permission of The University of Chicago Press. © 1955 by The University of Chicago. All rights reserved.)

When this equation is plotted for values of q between 0 and 1 (Figure 19-4), it is clear that the rate of change in q becomes very small as q approaches 0 or 1 and that the rate attains a maximum when $q = 0.66$. This fact is of importance, for it means that when an allele is prevalent in a population (q near 1) and is placed under mild selective pressure it will initially be eliminated very slowly. Similarly, if a new and selectively favorable allele arises, its initial rate of increase in the population will also be very slow.

19.12 THE APPLICATION OF DIRECTIONAL SELECTION THEORY TO LABORATORY POPULATIONS

It sometimes becomes possible to determine a selection coefficient for a particular genotype under laboratory conditions and thus to test the applicability of relationships such as Equations 9 and 10. One study, for example, focused on the sex-linked *ras* (*raspberry*-eye) allele of *D. melanogaster* affecting eye color (see Figure 16-12). It was first established that under laboratory conditions *ras/ras* and *ras/→* flies had about the same viability as +/+ or +/*ras* flies but that *ras/→* males mated with only half the efficiency of their +/→ counterparts. This led to the prediction that the *ras/→* genotype should have a relative fitness of 0.5. A modified version of Equation 9, which takes sex linkage into account, was then developed and a predicted rate of disappearance of the *ras* allele from a *Drosophila* population was calculated. This rate is plotted in Figure 19-5 (gray curve). Against this rate is plotted the observed rate of decline in the *ras* allele frequency (black curve) with data derived from experiments in which +/*ras* flies were introduced into population cages (Figure 19-6) and allowed to propagate freely over a period of about 18 generations. The similarity between the predicted and observed curves argues strongly for the relevance of the mathematical equations to such situations.

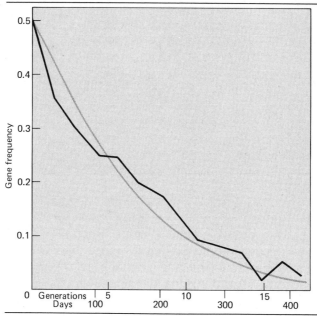

Figure 19-5 Natural selection under laboratory conditions showing decline in frequency of the *ras* allele of *D. melanogaster* (black curve) compared with the predicted (gray) curve using a modified form of Equation 9. (After D. J. Merrell, *Evolution* **7**:287, 1953; data plotted in D. S. Falconer, *Introduction to Quantitative Genetics*, London: Longman, © 1960.)

19.13 INDUSTRIAL MELANISM

We can now turn to several actual examples of directional selection. Perhaps the most often cited example from natural populations is given by **industrial melanism.** Most moth species in England, and specifically the common Peppered Moth (*Biston betularia*), rest on tree trunks when they are not in flight. In rural parts of England—and in all of England before the Industrial Revolution—the trunks of many trees are covered with light gray lichens, and the moths are similarly light in color. As industry became prominent during the middle of the last century, however, the lichens in industrial areas were killed by falling soot and the tree trunks were blackened. At about this time (*ca.* 1850), a collector in Manchester captured an apparently new variety of Peppered Moth with dark (melanic) body and wings. As time passed, this new, dark variety of moth became increasingly more common, and it now predominates in such industrialized regions as Manchester and Birmingham.

The inference from such observations, therefore, is that some sort of selective agents act against light moths in industrial areas and against black moths in rural areas. H. Kettlewell demonstrated the nature of these agents by observing, first, that Peppered Moths are preyed on by various bird

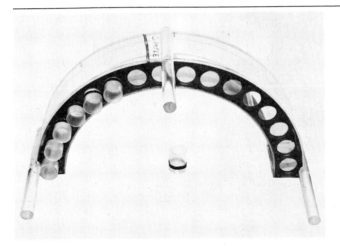

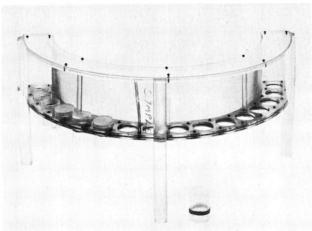

Figure 19-6 Photograph of a population cage. Parent flies are introduced into the cage and the holes are filled with vials that are either empty or contain a medium used both as food and as a site for egg deposition. Food vials can be replenished at appropriate intervals, and adult flies can be sampled by examining those flies that happen to rest in the empty vial marked "sample." (Courtesy of B. Wallace.)

species. He then released known numbers of dark moths into rural woods and light moths into industrial woods and noted their likelihood of becoming prey as compared with the native variety of Peppered Moth. The results were clear: the soot-laden trees afforded excellent camouflage for the melanic moths relative to the light moths and the survival of the melanic moths was accordingly much higher. The reverse was true with the lichen-covered trees (Figure 19-7).

The selective agents in this case turned out to be bird predators and industrial soot. This combination has produced dramatic results in a rela-

Figure 19-7   Photograph of *Biston betularia*, the Peppered Moth, and its black form, *carbonaria*. *Left:* Light and dark forms on a soot-covered oak tree near Birmingham, England. *Right:* The same two forms on a lichen-covered tree in a soot-free region. (Courtesy of H. B. D. Kettlewell.)

tively short period of time, in part because the agents of selection are themselves strong and in part for a genetic reason: dark moths are produced when a dominant allele (*B*)—which is lacking in light (*b/b*) moths—is present in the genome. Because both *B/b* and *B/B* individuals are thus protected in industrial areas, the *B* allele has been able to spread much more rapidly than a recessive gene under similar pressures.

The melanic variety of the black Peppered Moth is far darker today than specimens collected 100 years ago, which suggests that other pigment-producing alleles, in addition to *B*, are now in the process of being preserved by natural selection. The fact remains that the original selective events were most probably directed at a single genetic locus.

19.14   ARTIFICIAL SELECTION

Artificial selection is simply a selection program devised by humans rather than brought about by a natural situation. However, the process of enriching a population for a given trait normally occurs far more rapidly under the artificial situation, since all but the desired phenotypes can be prevented from reproducing.

Countless experiments have shown that it is possible to select for just about any trait present in a population, a major limitation being the heritability of the trait (Section 16.12). Strains with particular shapes, sizes, behavior patterns, temperature optima, sexual preferences, and other traits have been selected. S. Spiegelman and colleagues have even subjected isolated RNA molecules to artificial selection; we describe these experiments to illustrate the design of an artificial selection program.

The goal of Spiegelman's experiments was to select a chromosome fragment derived from the RNA phage Qβ. This chromosome fragment was to have two properties: it was to be the smallest portion of the chromosome still capable of being replicated, and its replication was to proceed at a maximal rate. Isolated Qβ chromosomes were first incubated with Qβ replicase in a medium containing radioactive UTP. At the end of 20 minutes a small sample of RNA was counted for radioactivity to determine how much new RNA had been synthesized during the 20-minute interval; a second small RNA sample was used to inoculate a second tube, also containing radioactive UTP. This process was repeated for a total of 75 tube transfers, and the intervals between the transfers were gradually reduced to five minutes. Samples were then subjected to sucrose gradient centrifugation (Box 8.1) at various times to determine their size; they were simultaneously tested for their ability to direct a productive infection of *E. coli*.

The results of this program were dramatic. After only four transfers the RNA lost its infectious properties (Figure 19-8, inset), presumably because there existed *in vitro* no selective pressures that prohibited the loss of information for coat protein, internal protein, or replicase synthesis. Moreover, after eight transfers the rate of RNA synthesis rose dramatically (Figure 19-8, main graph), as though much smaller RNA molecules were being replicated much more efficiently. Indeed, by the 38th transfer the RNA showed a single 15S peak in contrast to the 28S RNA synthesized after the first transfer. By the 75th transfer the RNA was reduced in size to about 550 nucleotides, a reduction of 83 percent from the size of the original Qβ chromosome. Presumably the selection program preserved only those nucleotides that dictated how the Qβ replicase would attach to the RNA and initiate its replication.

The Spiegelman experiments make it particularly obvious that a rigorous selection for one trait can lead to the drastic exclusion of many other traits. This fact is of no concern for many experimental purposes but becomes of critical concern to plant and animal breeders. Their goal must be the selection of strains that are better fit in the overall sense, and this requires a great deal of skill and patience. Far more often than not, as artificial selection focuses on one trait (the production of sweet corn kernels, for example), undesirable traits are unknowingly selected for as well (susceptibility to fungal infection, perhaps, or poor tolerance of dry soil conditions). The result may therefore be a strain of corn that produces exceptionally fine

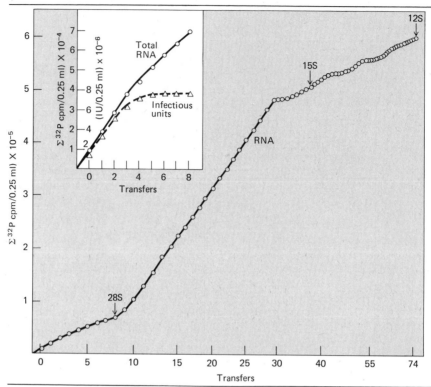

Figure 19-8 Serial transfer of Qβ RNA, with RNA synthesized in one reaction used to prime the second reaction (first transfer), and so on. The first 13 reactions utilized 20-minute incubation periods; transfers 14 to 29 used 15-minute periods; transfers 30 to 38 used 10-minute periods; transfers 39 to 52 used 7-minute periods, and transfers 53 to 74 used 5-minute periods. The inset shows both infectious and total RNA during the early period. (After S. Spiegelman et al., *Cold Spring Harbor Symp. Quant. Biol.* **33**:101, 1968.)

crops under highly controlled (and expensive) conditions, a strain that is of little interest to most farmers. Thus the product of any well-designed selection program, if it is to be useful to agriculture or animal husbandry, must be carefully tested under field conditions.

THE FITNESS OF THE GENOME

The statements made above about animal and plant breeding take us to a central feature of the selection process, be it natural or artificial, namely, that the selection process ultimately acts on entire genomes and not simply on a single allele or locus. In cases such as industrial melanism where the selection pressures are intense and the phenotypes are black *versus* white, it is perhaps valid to focus on a single locus, but most selection processes are less intense and operate on more complex phenotypes. In this section we consider some of the more prominent kinds of gene-gene interactions that appear to affect the selection process.

19.15 GENETIC BACKGROUND

The contribution of genetic background to the selection process is well illustrated by an example involving the *sepia*-eye allele in *D. melanogaster*. This allele is found in very low frequencies in natural populations, which suggests that it is constantly selected against and maintained chiefly by mutation pressure. When, however, native *sepia*-bearing flies are introduced into a population cage with flies from a highly inbred "wild-type" laboratory strain, the frequency of the allele tends to increase and to eventually stabilize at a higher level than that found under natural conditions. This outcome, in itself, is most readily explained by assuming that selective pressures in the laboratory are different from those in the wild. When, however, the experiment is repeated under identical conditions with two different "wild-type" laboratory strains, the *sepia* gene stabilizes at two levels quite different from that of the first experiment. These results suggest that the three "foreign" genetic backgrounds in the laboratory strains of *D. melanogaster* contain **fitness modifier genes** which differentially alter the selection coefficient against *sepia* under laboratory conditions. The reverse situation can arise as well, namely, that the relative fitness of an allele such as *ras*, when estimated in the presence of a laboratory strain's genetic background, will bear little relevance to its fitness in the presence of fitness modifier genes that exist in the "native" genetic background.

19.16 LINKAGE DISEQUILIBRIUM

Linkage disequilibrium (D) describes the tendency of two linked alleles to be inherited together more often than expected. Although difficult to prove, the presence of linkage disequilibrium in a large Mendelian population usually suggests that selection is acting to favor retention in the gene pool of chromosomes carrying both alleles.

Let us at once consider a specific example. The major histocompatibility locus (MHC) of humans (Figure 15-12) contains two loci, *A* and *B*, which recombine at a rate of about 0.8%. As noted in Section 15.10, each locus is multiply allelic, with certain *A* and *B* alleles being more common in particular populations than others. In addition, certain haplotypes (*A, B* combinations) are unexpectedly common in particular human groups. Thus the frequency of the *A1, B8* haplotype in North-European populations is about 9 percent, while the frequency of *A1* alone is 17 percent and the frequency of *B8* alone is 11 percent. If the two loci were independent, then the expected frequency of the *A1, B8* haplotype would be $0.17 \times 0.11 = 0.0187 \cong 2$ percent. The extent of linkage disequilibrium (D) for this combination of alleles is therefore $0.09 - 0.02 = 0.07$.

Is natural selection the agent responsible for linkage disequilibrium at the MHC? Are *A1, B8* Caucasians more resistant to disease, for example, than Caucasians with other haplotypes? Most investigators suspect that selection is responsible, but proof awaits an understanding of the role of the MHC gene products in mammalian physiology. Meanwhile, the example of the MHC reminds us that selection probably can and does act on groups of genes, and some investigators refer to such gene constellations as **super genes.**

19.17   SELECTIVE SUPPRESSION OF RECOMBINATION

If selection indeed acts on groups of genes so that, for example, it can be shown that organisms inheriting an a_1b_1 chromosome will be more fit than those inheriting a_1 alone or b_1 alone, then it follows that the existence of recombination might well be detrimental in such cases. In other words, it would appear advantageous to the species if an "adaptively superior" a_1b_1 chromosome could be transmitted intact from generation to generation instead of being broken up continuously by recombination. Have organisms evolved mechanisms which selectively prevent recombination within "valuable" gene blocks?

Again, unequivocal answers cannot be given, but several suggestive examples exist. One concerns the ability of chromosomal inversions to suppress meiotic recombination. It was noted in Section 6.15 that when a chromosome carrying an inversion attempts meiotic synapsis with its homologue, pairing entails an awkward looping configuration and crossing over within the inverted segments generates inviable chromosomes. Therefore, a block of genes contained within an inversion will usually be transmitted intact as long as the organism is an inversion heterozygote.

The pioneering work of T. Dobzhansky established that natural populations of *Drosophila* possess chromosomal inversions that can be readily detected in polytene chromosome preparations. Moreover, populations of the same species collected in different habitats (the desert *versus* the mountains of California, for example) were found to carry inversions involving different blocks of genes. These observations led Dobzhansky to propose that selected groups of fit alleles were being protected from recombination by inversion. Supporting this notion have been numerous studies, both in the field and in the laboratory, which show that flies heterozygous for inversions enjoy a selective advantage over homozygotes under particular environmental conditions, being able, for example, to produce more progeny or to compete successfully with other species for food.

MIGRATION PRESSURE

19.18   DYNAMICS OF MIGRATION PRESSURE

In addition to mutation and selection, **migration pressure** represents an "evolutionary force" that can influence allelic frequencies in a particular direction. As its name implies, migration pressure describes the effect of introducing new individuals with new genotypes into a Mendelian population. Figure 19-9 diagrams the process. Illustrated are two populations of moths, X and Y, population Y having a far higher frequency of the black allele (*B*) than population X. A random sampling of moths from population X now migrates to, and mixes with, population Y. The immigrants thus come to represent some fraction *m* of the total number of individuals present in the now expanded population of Y. If we let q_X and q_Y represent the original frequencies of the *b* (white) allele in the original populations, then a new frequency, q'_Y, is established in the new population. The value of q'_Y will be

Population X with a frequency q_X of the white allele (*b*)

Migration

Population Y with a frequency q_Y of the white allele (*b*)

Figure 19-9   Migration pressure as an evolutionary agent. The Y population of moths receives some immigrants from the X population, an event that changes its frequency of the white allele. (After E. O. Wilson and W. H. Bossert, *A Primer of Population Biology.* Stamford, Conn.: Sinauer Associates, 1971.)

equal to the contribution made by the immigrants (q_Xm) plus the contribution made by the original Y population [$q_Y(1 - m)$], where $1 - m$ represents the proportion of nonmigrants. Therefore

$$q'_Y = q_Xm + q_Y(1 - m)$$

and the change in q after one generation in such a population becomes

$$\Delta q = q'_Y - q_Y$$
$$= q_Y - mq_Y + mq_X - q_Y \qquad (11)$$
$$= -m(q_Y - q_X)$$

When numerical values are substituted into this equation, it becomes clear that significant changes in the value of q_Y can result in one generation even if the two populations differ only slightly in the frequency of a given allele and if a moderate degree of migration occurs between them.

That migrations have been effective in spreading alleles from one population to the next is suggested by data on the relative frequencies of the three ABO blood-group alleles, L^A, L^B, and L^O (Section 15.2). As summarized in Figure 19-10, particular human populations or races possess these alleles at quite different frequencies, with the L^B allele being particularly common in Asia. When the frequency of the B allele in European countries is assessed, a gradient known as a **cline** is observed (Figure 19-11), with B

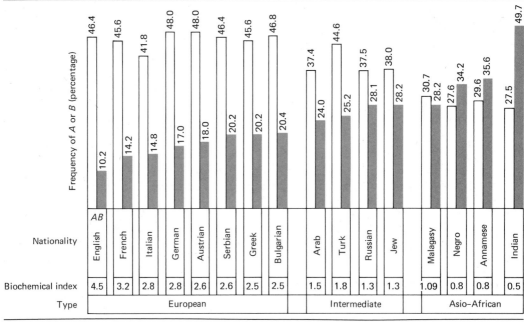

Figure 19-10 Ethnic differences in ABO gene frequencies. The figures given are percentages of positive reactions with anti-A and anti-B reagents. The "Biochemical Index" is the ratio of A to B. (From L. Hirschfeld and H. Hirschfeld, *Anthropologie*, vol. 29, pp. 505–537, 1919, and W. F. Bodmer and L. L. Cavalli-Sforza, *Genetics, Evolution, and Man*, San Francisco, W. H. Freeman and Co., 1976.)

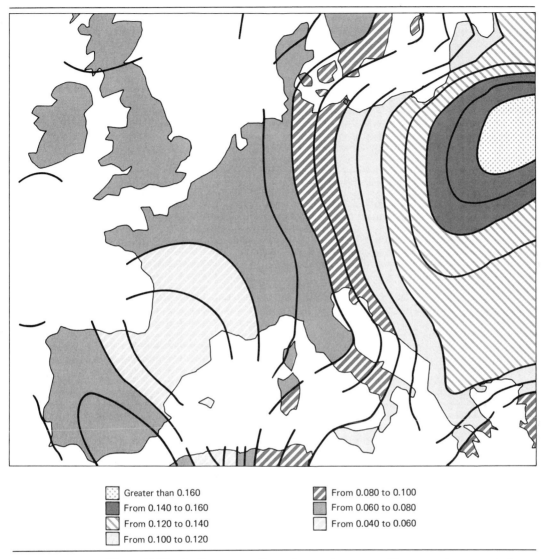

⬚ Greater than 0.160	▨ From 0.080 to 0.100
■ From 0.140 to 0.160	▨ From 0.060 to 0.080
▨ From 0.120 to 0.140	☐ From 0.040 to 0.060
☐ From 0.100 to 0.120	

Figure 19-11 A cline, or gradient of gene frequency. The computer-generated map shows the frequency of the *B* allele of the ABO blood-group system in Europe. Note the gradual change from high frequencies near central Asia toward low frequencies in western Europe. The average "slope" of the cline is about 1 percent per 400 kilometers. The cline is probably a remnant of early population migrations and may be gradually disappearing, but it might also be maintained by selective pressures. (Courtesy of D. E. Schreiber, IBM Research Laboratory, San Jose, and of R. Matessi; taken from W. F. Bodmer and L. L. Cavalli-Sforza, *Genetics, Evolution, and Man*, San Francisco, W. H. Freeman and Co., 1976.)

alleles apparently radiating from a very "high-B" focus in western Asia. While such clines can be interpreted to result from Mongolian invasions in the twelfth century, such an argument is understandably difficult to prove.

19.19 GENE FLOW

A relatively large influx of individuals from one population into a second is considered a migration. Probably more common is a related, but more limited, phenomenon known as **gene flow.** Gene flow occurs when two populations with distinct allelic frequencies become contiguous and individuals "cross the boundary" and interbreed. If individuals from population X contribute alleles to population Y more often than the reverse, then genes are said to flow from X to Y. Eventually, of course, the allelic frequencies in the two populations will become equivalent, but the process may take a long time.

A particularly dramatic example of gene flow in human populations resulted when black Africans and white Americans were brought together several centuries ago as a consequence of the slave trade. Since the interactions between the two groups were such that the "boundary" was usually crossed by white males in extramarital encounters, alleles from the white population tended to flow into the black population far more often than the reverse. Table 19-4 lists the consequences for several loci, and it is clear that, during the 10 or so generations that such gene flow has occurred, the allelic frequencies in the American black populations have come to differ consid-

Table 19-4 Allele Frequency Estimates for African Blacks and for Blacks and Whites Living in Georgia and California

Allele	Black Africa	Black California	Black Georgia	White California	White Georgia
R_0 $\left.\begin{array}{c}\\\\\\\\\end{array}\right\}$ Rh alleles	.617	.486	.533	.023	.022
R_1	.066	.161	.109	.413	.429
R_2	.061	.071	.109	.140	.137
$r + R_0{}^n$	.245	.253	.230	.405	.374
A $\left.\right\}$ ABO alleles	.156	.175	.145	.247	.241
B	.136	.125	.113	.068	.038
Fy^a (Duffy blood group)	.000	.094	.045	.429	.422
Hb^s (Sickle cell hemoglobin)	.092	no estimate	.043	no estimate	.000
G6PD (Glucose-6-P-dehydrogenase) deficiency	.176	no estimate	.118	no estimate	.000

From J. Adams and R. H. Ward, *Science* **180:**1137, 1973. Copyright 1973 by the American Association for the Advancement of Science.

erably from the African and to resemble far more closely the American whites.

"SPLINTER" POPULATIONS

A Mendelian population in a Hardy-Weinberg equilibrium is considered "infinite" in the sense that its gamete pool is visualized as being very large and mixing randomly at each generation. In the remaining sections of this chapter we consider the evolutionary factors that come into play when a small group of individuals leaves a Mendelian population and starts a new population.

19.20   THE FOUNDER EFFECT

The first feature of such small "splinter" populations, known as the **founder effect,** can be appreciated by returning to Figure 19-9. Suppose that the five moths diagrammed as leaving the Mendelian population X do not fly to join population Y, but instead fly to a small island where no other moths of that species reside. These five founders of a new, isolated "splinter" colony carry only a sampling of the total array of alleles present in population X. Since they can only mate with one another, their genotypes will have a major influence on the allelic frequencies that come to characterize the new population. If, for example, one of the five moths happens to be heterozygous for a rare recessive allele *w* which affects wing shape, then even though this allele may be extremely rare in population X as a whole, it will be present in one out of every 10 gametes in the "gamete pool" generated by the 5 founders. It is this founder effect that is believed responsible, in part, for the unusual allelic frequencies (and phenotypes) of many isolated colonies compared to the large populations from which they once derived.

19.21   RANDOM DRIFT

The second phenomenon that influences the allelic frequencies in a small population, known as **random drift,** is most easily visualized in the context of a coin-flipping analogy. If a coin is flipped 1000 times, the number of "heads" is expected to be roughly 500 and the number of "tails" to be roughly 500. We would not be surprised if the numbers came out 507 and 493, but we would suspect the balance of the coin if the result were, for example, 700 and 300. If, on the other hand, a coin were flipped ten times and it came out 7 "heads" and 3 "tails," we would not be startled; in such a small sampling of coin flips, we would say, such a wide fluctuation from the expected 50:50 outcome is not unexpected.

The same principle applies to what can be called gametic sampling. A male W/w moth in our hypothetical founder colony will produce sperm, roughly half of which (by segregation) carry W and half w. Assuming that 10 of his gametes manage to fertilize eggs, it is not unexpected that, say, 7 of these fertilizing sperm will happen to carry w and only 3 will happen to carry W. In a large Mendelian population, this sampling effect would presumably be canceled by a second W/w moth in which 7 W and 3 w sperm were involved in fertilizations. In our example, however, there is no second W/w moth. Therefore, this chance sampling bias has the effect of causing the frequency of W in the second generation to be significantly less than it was among the five original founders.

When gametes are now selected at random from this second generation of moths, two effects come into play: the relative frequency of W-bearing gametes in the total "gamete pool" is reduced; moreover, the total gamete pool is still relatively small so that sampling errors may again occur. Therefore, once a gene such as W begins to experience attrition in a small population, it is likely to move rapidly toward **extinction** ($p = 0$) while its allele

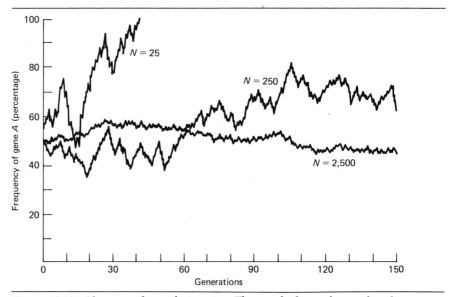

Figure 19-12 Chance and population size. The graph shows the results of computer experiments to simulate chance effects in three populations of different size, each starting with a gene frequency of 50 percent. The smallest population ($N = 25$ individuals) shows fixation of allele A after 42 generations. The medium-size population ($N = 250$) shows less important fluctuations, and has not reached fixation of either allele after 150 generations. Note that the frequency of allele A did become greater than 80 percent shortly before generation 110, but that chance events in succeeding generations happened to carry the frequency back toward 50 percent. In the largest population ($N = 2500$), fluctuations are quite small, and fixation is very unlikely in any particular generation. However, fixation will eventually occur if the experiment is continued long enough. (From *Genetics, Evolution, and Man* by W. F. Bodmer and L. L. Cavalli-Sforza. W. H. Freeman and Company. Copyright © 1976.)

(w) moves towards **fixation** ($q = 1$). A computer simulation of these events is shown in Figure 19-12.

An often cited case of founder effect and drift in human populations concerns the inhabitants of Pingelap, a small cluster of islands in the Pacific. In the late 18th century, all but about 30 Pingelapese were killed by either typhoon or famine. Among the present population of 1600 that descended from these 30 founders, about 5% are homozygous for a rare allele (which we can call a) that causes a form of partial blindness known as achromatopsia. Converting this information into allelic frequences we find for the present population that

$$a/a = q^2 = 0.05$$
$$q = \sqrt{0.05} \cong 0.23$$

Among the 30 founders, on the other hand, it is likely that only one individual was heterozygous for the rare allele a, meaning that $q = \dfrac{1}{60} \cong 0.014$ in the original group. One is confronted, in other words, with an increase in the frequency of the a allele from an estimated 1.4 percent to the present 23 percent, with no evidence of selection for the trait. The most plausible explanation, therefore, is to invoke random drift.

19.22 GENOME SELECTION IN FINITE POPULATIONS

We noted in Section 19.16 that selection may in many cases act on a region of a chromosome or even a whole chromosome rather than on a single gene locus. This effect is particularly dramatic in small populations, as seen in the experiments described below that were performed by J. Powell and R. Richmond.

Powell and Richmond focused attention on the sex-linked T_0 locus of *Drosophila paulistorum,* which codes for the enzyme tetrazolium oxidase. They established that wild populations of *D. paulistorum* in the Brazilian Andes carried alleles for fast (*F*) and slow (*S*) electrophoretic variants (Section 15.7), and they found that neither allele was under noticeable selection pressure in a population-cage environment. Specifically, they demonstrated that if they "founded" a population cage with *F/F* homozygotes whose ancestors included a large number of female flies independently isolated in the Brazilian wild, then the frequency of the *F* allele after 45 generations (900 days) did not change appreciably (Figure 19-13, curve A).

They then examined what the effect would be if they "founded" a population cage with the same number of *F/F* homozygotes but used flies that had descended from only two independently isolated progenitors. As is evident in Figure 19-13, curve B, the frequency of *F* is now seen to fluctuate considerably from one generation to the next, and the allele appears to be on the path to elimination.

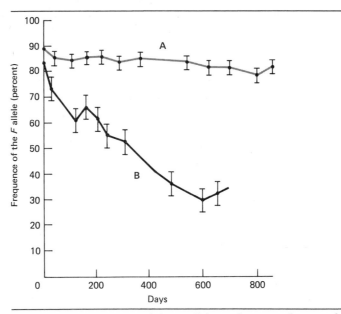

Figure 19-13 Frequency of the *F* allele at the tetrazolium oxidase (T_0) locus in two laboratory populations of *D. paulistorum*, the "diverse" population (A) founded with over 100 X-chromosomes carrying *F*, the "limited" population (B) founded with only about 6 X-chromosomes carrying *F*. (From J. R. Powell and R. C. Richmond. *Proc. Natl. Acad. Sci. U.S.* **71**:1663–1665, 1974.)

To best appreciate these results, visualize a collection of *F*-bearing X-chromosomes from various *D. paulistorum* flies isolated in the wild. One X-chromosome might carry alleles $a_1\ b_2\ c_1\ F\ d_2$ . . . , another might carry $a_2\ b_2\ c_1\ F\ d_1$ . . . and, in general, most of the chromosomes would be different from one another. If we now imagine that some of the gene loci closely linked to T_0 carry alleles with positive or negative effects on survival in population cages, the Powell-Richardson experiments are readily explained. In the case of cage A, where many different X-chromosomes "founded" the population, any deleterious combination of linked alleles would presumably be present in a relatively small proportion of the total chromosomes; moreover, X-chromosomes having a positive effect on survival might also be present so that a canceling-out effect would take place. The net outcome, therefore, would be the maintenance of the *F* allele at its introduced frequency (Figure 19-13). In the case of cage B, on the other hand, where only a few X-chromosomes are chosen to "found" the colony, one of these is likely to carry a combination of linked genes that affects population-cage fitness. Because this chromosome represents a very large proportion of the total, a marked effect on the frequency of *F* will accompany selection at the linked loci. Fitness modifier genes (Section 19.15) in this and other chromosomes may also be disproportionately represented.

In summary, therefore, we can say that random drift and selection pressure allow a "splinter" population to acquire very different allelic frequencies from its parent Mendelian population within a very few generations.

Island species are in fact frequently known to have singularly bizarre aspects to their phenotypes, as though their ancestors were indeed forced to come up with makeshift but viable combinations of traits within short periods of time. This feature of island populations is of major historical importance, for it was the singularity of the species of animals (and particularly finches) on the Galapagos Islands that attracted the attention of Charles Darwin.

References

General Texts and Reviews on Population Genetics

Bodmer, W. F., and L. L. Cavalli-Sforza. *Genetics, Evolution, and Man.* San Francisco: Freeman, 1976.

Cannings, C., and L. Cavalli-Sforza. "Human population structures," *Adv. Human Genetics* **4:**105–172 (1973).

Cavalli-Sforza, L. L., and W. F. Bodmer. *The Genetics of Human Populations.* San Francisco: Freeman, 1971.

Crow, J. F., and M. Kimura. *An Introduction to Population Genetics Theory.* New York: Harper and Row, 1970.

Felsenstein, J. "The theoretical population genetics of variable selection and migration," *Ann. Rev. Genet.* **10:**253–280 (1976).

Fisher, R. A. *The Genetical Theory of Natural Selection.* Oxford: Clarendon, 1930.

Fisher, R. A. *The Theory of Inbreeding,* 2nd ed. Edinburgh: Oliver and Boyd, 1965.

Kimura, M., and T. Ohta. *Theoretical Aspects of Population Genetics.* Princeton, N.J.: Princeton University Press, 1971.

Lerner, I. M. *The Genetical Basis of Selection.* New York: Wiley, 1958.

Lewontin, R. C. *The Genetic Basis of Evolutionary Change.* New York: Columbia University Press, 1974.

Milkman, R. "The genetic basis of natural variation in *Drosophila melanogaster,*" *Adv. Genetics* **15:**55–114 (1970).

Murray, J. *Genetic Diversity and Natural Selection.* Edinburgh: Oliver and Boyd, 1972.

Robertson, A. "Animal breeding," *Ann. Rev. Genetics* **1:**295–312 (1967).

Spiess, E. B., Ed. *Papers on Animal Population Genetics.* Boston: Little Brown, 1962.

Spiess, E. B. "Experimental population genetics," *Ann. Rev. Genetics* **2:**165–208 (1968).

Sprague, G. F. "Plant breeding," *Ann. Rev. Genetics* **1:**269–294 (1967).

Wallace, B. *Topics in Population Genetics.* New York: Norton, 1968.

Wilson, E. O., and W. H. Bossert. *A Primer of Population Biology.* Stamford, Conn.: Sinauer, 1971.

Wright, S. *Evolution and the Genetics of Populations* (2 vols). Chicago: University of Chicago Press, 1969.

Mutation as an Evolutionary Agent

Britten, R. J., and D. E. Kohne. "Repetition of nucleotide sequences in chromosomal DNA." *Handbook of Molecular Cytology* (A. Lima-de-Faria, Ed.). Amsterdam: North Holland, 1969.

Britten, R. J., and D. E. Kohne. "Implications of repeated nucleotide sequences." In *Handbook of Molecular Cytology* (A. Lima-de-Faria, Ed.). Amsterdam: North Holland, 1969.

*Brown, D. D., P. C. Wensink, and E. Jordan. "A comparison of the ribosomal DNA's of *Xenopus laevis* and *Xenopus mulleri:* the evolution of tandem genes," *J. Mol. Biol.* **63:**57–73 (1972).

*Kimura, M., and T. Ohta. "The average number of generations until fixation of a mutant gene in a finite population," *Genetics* **61:**763–771 (1969).

Laird, C. D., B. L. McConaughy, and B. J. McCarthy. "Rate of fixation of nucleotide substitutions in evolution," *Nature* **224:**149–154 (1969).

McCarthy, B. J. "The evolution of base sequences in nucleic acids." In *Handbook of Molecular Cytology* (A. Lima-de-Faria, Ed.). Amsterdam: North Holland, 1969.

Ohta, T. "Mutational pressure as the main cause of molecular evolution and polymorphism," *Nature* **252:**351–354 (1974).

*Ohta, T., and M. Kimura. "Functional organization of genetic material as a product of molecular evolution," *Nature* **233:**118–119 (1971).

Wallace, B. *Genetic Load: Its Biological and Conceptual Aspects.* Englewood-Cliffs, N.J.: Prentice-Hall, 1970.

Wills, C. "Genetic load," *Sci. Am.* **222:**98–107 (1970).

*Yanofsky, C., E. C. Cox, and V. Horn. "The unusual mutagenic specificity of an *E. coli* mutator gene," *Proc. Natl. Acad. Sci. U.S.* **55:**274–281 (1966).

Zamenhof, P. J. "A genetic locus responsible for generalized high mutability in *Escherichia coli,*" *Proc. Natl. Acad. Sci. U.S.* **56:**845–852 (1966).

Directional Selection

*Adams, J., and R. H. Ward. "Admixture studies and the detection of selection," *Science* **180:**1137–1143 (1973).

Clarke, C. A., and P. M. Sheppard. "A local survey of the distribution of industrial melanic forms in the moth *Biston betularia* and estimates of the selective values of these in an industrial environment," *Proc. Royal Soc.* (*Series B*) **165:**424–439 (1966).

Cook, L. M., R. R. Askew, and J. A. Bishop. "Increasing frequency of the typical form of the peppered moth in Manchester," *Nature* **227:**1155 (1970).

*Dobzhansky, T. "Adaptive changes induced by natural selection in wild populations of *Drosophila,*" *Evolution* **1:**1–16 (1947).

Dobzhansky, T., and B. Spassky. "Artificial and natural selection for two behavioral traits in *Drosophila pseudoobscura,*" *Proc. Natl. Acad. Sci. U.S.* **62:**75–80 (1969).

Eriksson, K. "Genetic selection for voluntary alcohol consumption in the albino rat," *Science* **159:**739–741 (1968).

*Franklin, I., and R. C. Lewontin. "Is the gene the unit of selection?," *Genetics* **65:**707–734 (1970).

John, B., and K. R. Lewis. "Chromosomal variability and geographic distribution in insects," *Science* **152:**711–721 (1966).

*Kettlewell, H. B. D. "The phenomenon of industrial melanism in Lepidoptera," *Ann. Rev. Entomology* **6:**245–262 (1961).

*Merrell, D. J. "Selective mating as a cause of gene frequency changes in laboratory populations of *Drosophila melanogaster,*" *Evolution* **7:**287–298 (1953).

*Mills, D., R. L. Peterson, and S. Spiegelman. "An extracellular Darwinian experiment with a self-duplicating nucleic acid molecule," *Proc. Natl. Acad. Sci. U.S.* **58:**217–224 (1967).

Linkage Disequilibrium

Cavalli-Sforza, L. L., and M. W. Feldman. "Evolution of continuous variation: Direct approach through joint distribution of genotypes and phenotypes," *Proc. Natl. Acad. Sci. U.S.* **73:**1689–1692 (1976).

Felsenstein, J. "The evolutionary advantage of recombination," *Genetics* **78:**737–756 (1974).

*Denotes articles described specifically in the chapter.

Hammerberg, C., and J. Klein. "Linkage disequilibrium between *H-2* and *t* complexes in chromosome 17 of the mouse," *Nature* **258**:296–299 (1975).

*Powell, J. R., and R. C. Richmond. "Founder effects and linkage disequilibrium in experimental populations," *Proc. Natl. Acad. Sci. U.S.* **71**:1663–1665 (1974).

Wills, C., and C. Miller. "A computer model allowing maintenance of large amounts of genetic variability in Mendelian populations. II. The balance of forces between linkage and random assortment," *Genetics* **82**:377–399 (1976).

(a)   (b)

(a) Motoo Kimura (National Institute of Genetics, Misima, Japan) has made important contributions to modern population genetics theory. (b) J. B. S. Haldane made fundamental contributions to population genetics theory as it was being developed.

Questions and Problems

1. The MN blood type is determined by a pair of codominant alleles, L^M and L^N. A population of 600 persons was blood-typed with the following results: 300 were type M, 240 were MN, and 60 were N. Does this represent a Mendelian population in a Hardy-Weinberg equilibrium? If not, what should the numbers of persons with each genotype have been?

2. The population described in Question 1 reproduced until it reached a size of 5000 persons. Assuming that this mating was random and that evolutionary pressures were nonexistent, how many M, MN, and N persons do you expect in the expanded population?

3. In a Mendelian population of 500 persons, five are found to be of blood type N. This group now intermarries at random with a second Mendelian population of 500 persons in which 20 are of blood group N.

(a) What is the frequency of the L^N gene in the combined population?

(b) What proportion of the offspring of the combined population will be blood type N? Is this a lower or higher incidence than in the two populations considered separately?

4. In a Mendelian population, the expected number of $A/A \times a/a$ matings is $2(p^2 \times q^2)$, where the expression is multiplied by 2 to account for both $A/A\, ♀ \times a/a\, ♂$ matings and the reciprocal. Similarly, the number of $A/a \times a/a$ matings is expected to be $2(2pq \times q^2)$.

(a) If A shows complete dominance to a, then what are the proportions of the offspring expected to exhibit the dominant and recessive phenotypes in each of these matings?

(b) You now wish to determine the proportion of recessive offspring produced by all the $A/A \times a/a$ and $A/a \times a/a$ matings in the population. Show how this proportion is most simply expressed as $\dfrac{q}{1+q}$ (recall that $p = 1 - q$).

(c) One person in 10,000 expresses a particular recessive trait. If these persons all marry persons who do not exhibit the trait, what percentage of their offspring will be expected to exhibit the trait?

5. The ABO blood groups are controlled by three allelic genes, I^A, I^B, and I^O (see Table 15-2). The Hardy-Weinberg formula can in this case be expanded by one term so that r = the frequency of I^O and $p + q + r = 1$. The zygotes formed in this case will be
$$(p + q + r)^2 = p^2 + q^2 + r^2 + 2pq + 2pr + 2qr = 1.$$

(a) In a certain population, the frequency of gene I^A was estimated to be 0.20; I^B was estimated at 0.08; and I^O at 0.72. What are the expected proportions of blood-group phenotypes in this population, assuming random mating?

(b) When the blood-group phenotypes of the persons in this population were actually tested, 1340 were found to be of type O, 895 of type A, 305 of type B, and 70 of type AB. Do the gene frequencies agree with those expected using the chi-square test? (*Note:* There is one degree of freedom in this situation.) Is mating in fact random?

6. One person out of 2500 in a Mendelian population is sterile because she or he is homozygous for the recessive gene b. What will be the proportion of such sterile persons in the next generation?

7. Utilizing the information in Table 17-2, predict the equilibrium frequencies of the *str-s* and *str-d* genes in a population of haploid *E. coli*.

8. A population of mice is "founded" by two mice, one D/D (normal coat color) and the other d/d (*dilute* coat color). Their offspring and succeeding offspring are allowed to mate randomly for a number of generations until a large population of mice is established. Complete selection against *dilute* mice is now imposed. What will be the change in gene frequencies after ten generations under the selection program?

9. Explain why, in Table 19-3a, certain values are multiplied by 2.
10. Explain the following statement: A gene with incomplete penetrance (Section 15.4) is less sensitive to selective pressures than is a completely penetrant gene.
11. Because of antipollution laws, trees in certain industrial areas of England have lost their sooty coating and black moths are now experiencing predation. Would you expect black moths to disappear at the same rate, more rapidly, or less rapidly than did the white moths in the nineteenth century (assume that birds can recognize black moths against a white background as readily as white moths against a black background)? Explain.
12. A population of 5000 persons contains 1500 that are of blood type M, 3000 of blood type MN, and 500 of blood type N. A representative sample of 30 persons from this population moves to join a second population whose size is 70 persons prior to the migration and where the frequency of $L^M = 0.8$ and $L^N = 0.2$ prior to the migration. What is the change in frequency of the two alleles in the expanded population after one generation?

CHAPTER
20

Population Genetics II: Genetic Polymorphism, Species Formation, and Molecular Evolution

INTRODUCTION

Our consideration of evolutionary agents now takes a shift in point of view. In Chapter 19 we focused on *idealized* populations, and we inquired into the manner by which known or postulated evolutionary agents (mutation pressure, directional selection, migration pressure, and random drift) might disrupt equilibrium conditions. We now consider an established fact about *existing* natural populations, namely, their genetic polymorphism, and we seek those agents that might operate to sustain this polymorphism. We then consider the process of species formation, and conclude by describing some general features of molecular evolution.

GENETIC POLYMORPHISMS

We noted the abundance of protein polymorphisms in natural populations at the beginning of Chapter 19. Underlying this polymorphism is extensive genetic variability. Such variability is commonly assessed in the following manner: numerous individuals in a population are first analyzed for the presence of alleles at a large number of loci, the presence of alleles most commonly being inferred from the presence of isoelectric variants of particular protein gene products. The extent of heterozygosity in an *average* individual in the population is then calculated; such calculations typically reveal heterozygosity at from 20 to 50 percent of the loci in an average individual in a typical population. In the case of the T_0 locus of *D. paulistorum* described in Section 19.20, some 80 percent of wild-type flies are heterozygous for the *F/S* alleles.

What maintains such large percentages of heterozygosity in a population? Should not all the fittest alleles have by this point in time replaced all of the less fit? Such questions become particularly pressing when it is found that a genus such as *Limulus* (the horseshoe crab) possesses extensive heterozygosity, even though the genus is not believed to have undergone major evolutionary change since ancient times.

Intensive study has revealed three basic paths through which polymorphism may arise and be maintained in a population. These paths produce genetic patterns that are called **transient polymorphism, balanced polymorphism,** and **random fixation of neutral mutations.** Few population geneticists would challenge the statement that all three paths have contributed to evolutionary events at one time or another. There has been, however, considerable controversy regarding the relative contributions of each to evolutionary events, as described in more detail in a later section.

20.1 TRANSIENT POLYMORPHISM

Transient polymorphism represents a byproduct of directional natural selection (Section 19.9). If we imagine that allele a_1 has a selective advantage over a_2, then with time a_1 should proceed toward fixation at $p = 1$, while a_2 should proceed toward elimination at $q = 0$. During the time that this process is occurring both a_1 and a_2 will be present in the gene pool and a_1/a_2 heterozygotes will be present in the population.

As its name implies, transient polymorphism represents a temporary situation; if selection is operating rapidly, the transiency should be apparent. During the course of industrial melanization, for example, both dark and light Peppered Moths would be expected to cohabit the Manchester trees, but the proportion of light moths would be seen to diminish with time as the dark moths gradually predominated.

20.2 BALANCED POLYMORPHISM

Balanced polymorphism represents a relatively permanent kind of equilibrium in which alleles a_1 and a_2 are present in the population at some steady-state frequencies. This stands in obvious contrast to transient polymorphism in which alleles are being driven toward fixation or extinction.

Balanced polymorphism is the expected state of a Mendelian population in which, as we saw in Section 19.2, allelic frequencies do not change from their "input" frequencies once equilibrium is established. Since most natural populations give little evidence of conforming to this ideal, other mechanisms have been sought to explain how heterozygosity can be maintained at some equilibrium level.

Two general sorts of mechanisms can accomplish this balance. In the first, a population may simply be exposed to a number of different kinds of evolutionary agents whose effects cancel one another out. Mutation and migration pressure may, for example, continuously introduce allele a_1 into a population while directional selection is continuously eliminating it in favor of allele a_2. As long as the pressures remain constant and the rate of introduction of a_1 remains relatively greater than the rate of its elimination (or the reverse), heterozygosity at the a locus will be sustained. In fact, the existence of the countervailing evolutionary forces can, for most purposes, be ignored altogether and the population treated as being in a Hardy-Weinberg equilibrium.

The alternative possibility is that *selection alone* is acting to generate a balance of alleles. This pattern of selection is obviously different from the directional selection which leads to the fixation or elimination of alleles, and it is frequently referred to as **balancing selection.** Two types of balancing selection have been studied most extensively—**disruptive selection** and **heterozygote advantage**—and we shall explore each in some detail.

20.3 DISRUPTIVE OR DIVERSIFYING SELECTION

Disruptive (also called **diversifying**) **selection** was first described by J. Thoday. It arises when a population inhabits a nonuniform environment.

Let us imagine that within such a population one phenotype proves to be successful in a certain environmental niche, whereas another phenotype finds itself relatively more successful in a second niche. If the first phenotype is dependent on an a_1/a_2 genotype and the second on an a_2/a_2 genotype, and the population prospers best as a whole when both environmental niches are maximally utilized, then both a_1 and a_2 should be maintained in some balanced state.

Some recent experiments by J. Powell have demonstrated how environmental variability can promote genetic diversity. Powell began with flies from natural populations of *Drosophila willistonii,* an average fly from such populations being heterozygous at an estimated 18 percent of its loci. The flies were divided among four sets of population cages that were maintained under four regimes: (1) the environment was kept as constant as possible; (2) the food supply was varied during the course of the experiment; (3) the food supply plus the growth medium were varied; and (4) the food supply, growth medium, and temperature were all varied. At the end of 45 weeks (15 generations), the flies in the "constant" cages exhibited about 8 percent heterozygosity per individual (again testing the same spectrum of loci), whereas the flies in the "three-variable" cages exhibited 13 percent heterozygosity. On the likely assumption that the natural environment is far more complex than the "three-variable" cages, these experiments certainly support the argument that much of the natural 18 percent heterozygosity exhibited by *D. willistoni* may be effected by diversifying selection.

20.4   HETEROZYGOTE ADVANTAGE

The second well-known type of balancing selection results when a heterozygote (*A/a*) enjoys greater fitness than either *a/a* or *A/A* homozygotes. Such a situation, which obviously promotes the preservation of heterozygosity in the population, is known as **heterozygote advantage.** Heterozygote advantage depends on the coexistence of two counteracting selective forces: one operating to reduce the fitness of *a/a* individuals and the other operating to reduce the fitness of *A/A* individuals, each with respect to *A/a* individuals. Since the phenomenon is best understood by example, we shall consider several cases in humans and in *Drosophila.*

Sickle-Cell Heterozygote Advantage   Certainly the best known example of heterozygote advantage in humans is provided by the sickle-cell allele of the gene that dictates the β chain of human hemoglobin (Section 15.9). The β chain specified by the sickle-cell Hb_β^S allele differs from the normal by one amino acid: whereas the normal chain has the N-terminal sequence Val-His-Leu-Thr-Pro-**Glu**-Glu-Lys . . . , the mutant chain reads Val-His-Leu-Thr-Pro-**Val**-Glu-Lys. . . . When such a mutant β chain is included in an $\alpha_2\beta_2$ hemoglobin molecule (Figure 9-5), the protein (known as S hemoglobin) is less soluble than normal A hemoglobin at low oxygen tensions, presumably because the neutral valine residue has replaced the charged glutamic acid. As a consequence the conformation of the erythrocyte at low oxygen tensions has a sickle shape rather than the normal disc shape (Figure 20-1), and these deformed cells are defective in oxygen transport.

Homozygous recessive (Hb_β^S/Hb_β^S) individuals synthesize only hemoglobin S (and fetal hemoglobin F), and, under primitive living conditions, rarely reach maturity (although in modern society they more often survive). Such individuals are said to suffer from **sickle-cell anemia.** Heterozygotes, on the other hand, are largely asymptomatic, even though their erythrocytes contain almost as much hemoglobin S as hemoglobin A, and such individuals are said to exhibit the **sickle-cell trait.**

Because the relative fitness of the Hb_β^S/Hb_β^S genotype is near zero (so that $s \simeq 1$), directional selection models would predict that the frequency of the Hb_β^S allele in a population should be zero or approaching zero. This prediction is confirmed in a number of human populations in which the allele is virtually absent. In many African and Asian populations, however, the frequency of the Hb_β^S allele is estimated to be as high as 0.1 to 0.2, and 10 percent of Afro-Americans are believed to be heterozygous for the gene. Thus some factor must prevent the extinction of the a allele in particular populations.

The factor, it appears, is malaria. The falciparum malarial parasite of humans abounds in tropical regions of Africa; it kills an estimated 15 to 20 young children per thousand, and its direct toll on the population as a whole is greater still. Moreover, many persons who inhabit these regions and who die of other causes are actually so weakened by malaria that their susceptibility to all diseases is greatly enhanced. When the distribution of the sickle-cell gene in Africa and Asia (Figure 20-2a) is compared with the

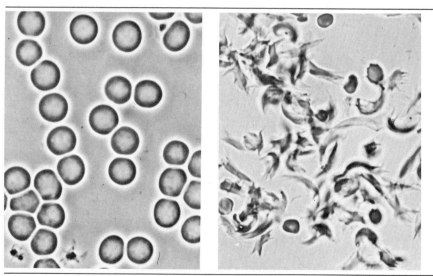

Figure 20-1 Photograph of normal erythrocytes (left) and sickle-shaped erythrocytes (right). (Courtesy of A. C. Allison.)

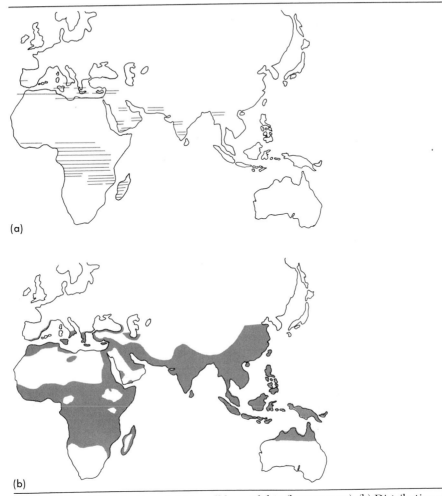

Figure 20-2 (a) Distribution of sickle-cell hemoglobin (bars on map). (b) Distribution of falciparum malaria (shaded areas). (From A. Motulsky, *Human Biology* **32:**43, 45, 1960. By permission of the Wayne University Press, © 1960.)

distribution of high endemic malaria (Figure 20-2b), the correlation is striking, suggesting that malaria serves to balance the deleterious effects of sickle-cell anemia so as to maintain the Hb^S_β gene. How might this be accomplished?

The answer seems to be that $+/Hb^S_\beta$ heterozygotes are significantly more resistant to malaria than $+/+$ homozygotes (the relative resistance of Hb^S_β/Hb^S_β homozygotes is immaterial, since such persons usually die in infancy in any case). The actual basis for heterozygote resistance, although not established, is thought to be as follows: the falciparum parasite develops within human erythrocytes; the erythrocytes of a heterozygote tend to adopt a sickle shape when they enter the capillaries, in which the oxygen tension is particularly low; this sickling process appears to interrupt some stage in parasite multiplication and the infection cycle is broken. Since normal erythrocytes cannot sickle, +/+ individuals do not enjoy this protection from infection and are fully susceptible to malaria.

The mathematical expression of balanced polymorphism by heterozygote advantage is quite simple. We let p^2, $2pq$, and q^2 represent the proportions of A/A, A/a, and a/a individuals before selection occurs, and we let s and t represent, respectively, the selection coefficients against the two homozygotes. The fitness of A/A individuals is then $1 - s$ and the fitness of a/a individuals is $1 - t$, the fitness of A/a individuals being equal to 1. From here we can develop the mathematics much as we did for partial selection against recessives in Section 19.10. Thus, after selection the proportion of A/A individuals is $p^2(1 - s)$, the proportion of A/a individuals remains $2pq$, and the proportion of a/a individuals becomes $q^2(1 - t)$, the total population being equal to $1 - p^2s - q^2t$. The expression can then be written in terms of Δq, as in Equation 9 of Chapter 19, but at this point we can let the whole be equal to 0 since, in balanced polymorphism, there should be no net change in q. When this is done, the equations can be greatly simplified and the expression for $\hat{q}$, the equilibrium frequency of q in balanced heterosis, becomes

$$\hat{q} = \frac{s}{s + t} \tag{1}$$

Try deriving this equation for yourself.

In applying Equation 1 to the sickle-cell example, where A corresponds to + and a to the Hb_β^S allele, we can assume that t, the selection coefficient against Hb_β^S/Hb_β^S homozygotes, is equal to 1. This means that if the selection coefficient s against +/+ homozygotes is equal to 0.25, q will equal 0.2, whereas if s is equal to 0.11, q will equal 0.1. Since the frequency of the sickle-cell gene in African populations is presently estimated at from 0.1 to 0.2, values of s can be assumed to fall in the 0.11 to 0.25 range.

Cystic Fibrosis   Heterozygote advantage may result from resistance to a debilitating disease, as we have just seen. It may also become established if a heterozygote enjoys a reproductive advantage over either homozygote. One study has suggested that this is the case for the allele that produces cystic fibrosis of the pancreas.

In making such a study, an important consideration must be followed. Children homozygous for the lethal cystic fibrosis allele give evidence of illness in infancy, and the heterozygous parents of such children frequently decide against having more children rather than running the risk of bearing another child with the disease. It is therefore not meaningful to analyze the reproductive output of the parents of a child with cystic fibrosis. Instead, the family sizes of the child's grandparents are studied. At least one member of each grandparental pair can be assumed to be heterozygous for the allele, and this heterozygous grandparent can be assumed to be unaware of his/her genetic defect.

The results of these studies indicate that the "cystic fibrosis" grandparents have an average of 4.34 offspring, whereas control couples have an average of 3.43. Such results, which are statistically significant, give us <u>no idea why the "cystic fibrosis" couples tend to have larger families—the explanation could range from a lower incidence of spontaneous abortion to a psychological eagerness to have children</u>—but they suggest that greater productivity may be the basis for the observed polymorphism.

Tay-Sachs Disease The final example we give of a possible heterozygote advantage in humans relates to the autosomal recessive allele in humans that affects the activity of β-galactosaminidase A. Infants homozygous for this allele develop the lethal Tay-Sachs disease, as described in Section 15.1. Table 20-1 summarizes some statistics pertaining to the disease. Although one-third of the cases in the United States now affect children of non-Jewish parents, the allele is particularly prominent among the descendants of the Ashkenazi Jews, who separated from the Sephardic (Mediterranean) and Oriental Jews in 72 A.D. and migrated into Europe. Of relevance here is that when sibships of families known to carry the Tay-Sachs allele are examined the survival rate is reported to be significantly greater than in sibships of families without the allele, suggesting that resistance to an unidentified childhood disease or some similar effect may give heterozygotes an advantage in which the allele is preserved.

Selection Coefficients in Heterozygote Advantage We noted for the case of sickle-cell anemia that the Hb_{β}^{s} gene is maintained at high frequencies (from 0.1 to 0.2) in malaria-ridden populations, which suggests, by Equation 1, a relatively high selection coefficient (perhaps 0.15) against $+/+$ homozygotes. The frequencies of the alleles involved in cystic fibrosis and Tay-Sachs disease, on the other hand, are much lower, the former being about 0.02 in Caucasian populations and the latter about 0.0126 among Ashkenazi Jews and about 0.001 in the non-Jewish population, thus implying that the selection coefficients against the dominant homozygotes must be correspondingly low. To estimate how low these selection coefficients are, we can assume t, the selection coefficient against homozygotes, to be

Table 20-1 Estimates of the Frequency of the Tay-Sachs Disease and Gene in North America

	Total Population	Annual No. of Births	Homozygote Frequency	Allele Frequency	Annual No. of Total Cases	Heterozygote Frequency	Total No. of Heterozygotes
Jewish	6,182,000	107,567	0.00016	0.0126	22	0.026	160,732
Non-Jewish	215,393,326	3,747,844	0.0000017	0.0013	8	0.0029	624,641

Data from J. S. O'Brien, *Adv. Hum. Genetics* (Plenum Publishing Corp.) **3**:39 (1972), and N. C. Myrianthopoulos and S. M. Aronson, *Am. J. Hum. Gen.* (The University of Chicago Press) **18**:313 (1966).

equal to 1 and we can rewrite Equation 1 as

$$s = \frac{\widehat{q}}{1 - \widehat{q}} \qquad (2)$$

Then, for the case of Tay-Sachs disease, we can state

$$s = \frac{0.0126}{1 - 0.0126}$$
$$= 0.0128$$

In other words, a selective advantage of only about 1 1/4 percent on the part of the heterozygote possibly suffices to maintain the Tay-Sachs gene at its low frequency in the Jewish population.

20.5   HETEROSIS

The breeder of a crop plant such as corn seeks to select for strains with more desirable properties than existing strains (Section 19.15). This is usually accomplished by inbreeding: plants are self-fertilized or crossed to close relatives to permit the expression of desirable recessive traits in homozygotes. If two such highly inbred strains are now crossed with one another, the F_1 crops are typically found to be larger and hardier than either parental type (Figure 20-3), a phenomenon known as **heterosis.**

Figure 20-3 Heterosis in corn. Representative plants from the two inbred parental strains are shown at left. A representative of their F_1 hybrid is shown third from left, and representatives of successive inbreedings (F_2 through F_8) are shown to the right of the F_1 plant. (After D. F. Jones, *Genetics* **9**:405, 1924.)

Two theories have been advanced to explain heterosis. The first, known as the **dominance hypothesis,** proposes that in the course of selecting for desirable traits, the breeder has created strains that are also homozygous for somewhat deleterious recessive genes elsewhere in the genome. A hybrid formed between two inbred strains would be expected to acquire normal, dominant alleles at many of these loci, and the resultant plant might well exhibit **"hybrid vigor"** compared to its parents.

The second theory, known as the **overdominance hypothesis,** relates to the phenomenon of heterozygote advantage considered in the previous section of this chapter: it is proposed that the hybrid is more vigorous because it is more heterozygous, regardless of whether the heterozygosity involves dominant alleles. How might the simple fact of heterozygosity provide a more hardy organism? One plausible, although unproved, suggestion points out that since different enzymes are used by different cell types at different developmental stages, the presence of two different versions of many different enzymes might significantly increase the probability that particular stages of an organism's development unfold with particularly optimal rates or yields. In other words, heterozygotes may simply be more flexible, and therefore more likely to come up with favorable sequences of developmental events, than homozygotes. In like fashion, the mature heterozygote may be better equipped to survive fluctuations in the environment than the homozygote.

20.6 HETEROSIS AND BALANCED GENETIC LOAD

Whatever the molecular basis for heterosis, its existence has posed a theoretical dilemma for population geneticists. If it is even only slightly advantageous to be heterozygous at any one locus, and an average organism is heterozygous at only 20 percent of its loci, the cumulative selective disadvantage of being homozygous at the other 80 percent of its loci should result, in theory, in massive levels of unfitness in the population. This postulated burden is often termed the **balanced genetic load,** in contrast to the mutational genetic load discussed in Section 19.7.

Two standard arguments are made to refute the postulated magnitude of balanced genetic load. The first states, in essence, that since natural selection acts on the entire phenotype and not simply on individual loci, it is invalid to sum up thousands of selective processes as though they were individual events; rather, it need only be supposed that heterozygosity at one or a few loci creates individuals that are, in the end, at an advantage under existing selective pressures. Heterozygote advantage at one locus, in other words, can be expected to "pull" a great deal of homozygosity at closely linked loci. This argument affirms the primary importance of selection, and particularly balancing forms of selection, in maintaining polymorphisms.

The alternative argument states that selection is not responsible for much of the polymorphism observed throughout nature. Instead, a large number of the alleles in a population are visualized as being **selectively neutral** so that, for the most part, the presence of two different alleles at one locus contributes no greater selective advantage to an organism than the presence of two identical alleles. This postulate obviously resolves the dilemma of the balanced genetic load, but it forces one to argue that in each of the numerous cases in which heterosis has been demonstrated the greater fitness of the heterozygote can be explained in terms of a relatively small extent of "truly" advantageous heterozygosity, the remaining heterozygous loci being indifferent to selective pressures.

20.7 THE NEUTRAL MUTATION-RANDOM GENETIC DRIFT HYPOTHESIS

So far we have presented two paths that produce polymorphism in a population: directional selection, which leads to transient polymorphism, and balancing selection (frequency-dependent, disruptive, and heterozygote advantage), which leads to balanced polymorphism. A third path producing polymorphism was touched on in the foregoing paragraph when we presented the concept of selectively neutral mutations. The full concept is most aptly called the **Neutral Mutation-Random Genetic Drift hypothesis,** although the term "Non-Darwinian Evolution" has recently come into vogue to emphasize the departure of this theory from a focus on natural selection.

The basic tenets of the Neutral Mutation-Random Genetic Drift hypothesis were set forth by S. Wright, and certain applications have recently been developed by M. Kimura. The hypothesis rests on two major assumptions. The first assumption states that selectively neutral mutations can occur in genes that code for proteins. The idea is that an acidic amino acid might replace another acidic amino acid or a substitution might occur in an "unimportant" region of the protein, with the result that the emergent mutant protein is identical to the original in all functional respects.

The second assumption states that neutral alleles, being neither selectively advantageous nor disadvantageous, simply drift in the gene pool. Thus, if a neutral mutation arises in a woman's germ cell and this germ cell gives rise to a female child, the probability is about 0.5 that the mutant allele will be transmitted to a grandchild and 0.5 that it will not. If it is not, then q becomes equal to zero and the allele is lost. If the allele reaches the third generation (perhaps via two or three individuals if the family is large), the chances are a little better that it will appear somewhere in the fourth generation, but the probability still remains great that its frequency will "backslide" and that q will finally fall to zero. In other words, as long as the allele provides its bearer with no selective advantages or disadvantages whatsoever, its fate in the gene pool is totally indeterminate. It may disappear as soon as it arises (elimination) or it may proceed toward $q = 1$ (fixation) in a seemingly haphazard pattern that is aptly described as a **random walk:** the course taken by the allele depends entirely on how the gamete pool happens to be sampled from one generation to the next. This description is, of course, very similar to that given for the random drift of allelic frequencies in small populations (Section 19.20.)

While a neutral allele is engaging in a random walk, heterozygous individuals will be present in the population and polymorphism will result. This does not represent a balanced polymorphism in that q is constantly changing, but it is clearly different from transient polymorphism in that selection is not involved, no clear directionality is observed, and the overall rate of change is expected to be orders of magnitude smaller than under selective pressures.

The Neutral Mutation-Random Genetic Drift hypothesis makes a number of predictions about the course of events leading to the ultimate fixation of a neutral allele. One is that the rate of its fixation should be approximately equal to the rate (μ) at which the allele arises by mutation—that is, that the average time interval expected to elapse between the origin of a successful neutral mutation and its ultimate fixation is $1/\mu$ generations. With an average mutation rate of 1×10^{-5} at most loci (Table 19-1), this means that neutral mutations should enjoy long intervals of random drift in a population and could therefore, in theory, be strong contributors to heterozygosity.

20.8 EVALUATING THE SELECTION VERSUS DRIFT CONTROVERSY

The Neutral Mutation-Random Genetic Drift hypothesis has attracted both followers and detractors, and many experiments have been performed to test its assumptions and predictions. Regardless of its veracity, the hypothesis has therefore already served a vital function in population genetics. Among the many points that have been debated, two are perhaps most critical.

The Occurrence of Neutral Mutations The first critical question is whether neutral mutations really occur. In other words, are two isoelectric variants of an enzyme completely equivalent from the point of view of natural selection?

Intensive research is devoted to this question. In a typical study electrophoretic variants of an enzyme are isolated and analyzed for their relative activities under various conditions. The results of such a study are summarized in Table 20-2, in which it is seen that in a number of cases polymorphic enzymes turn out to be measurably different from one another. The remaining cases are difficult to evaluate, because the observation of no difference between two variants may simply mean that the two enzymes were not compared at the appropriate pH, temperature, or ionic strength.

The importance of appropriate assay conditions in such studies is well illustrated by cytochrome c. When respiration is assayed *in vitro* by using rat mitochondria, for instance, the rat cytochrome c can be washed out and replaced by numerous other forms of cytochrome c. This result was taken to

Table 20-2   Enzyme Polymorphisms in Man

Polymorphisms Where Quantitative Differences between the Common Phenotypes Have Been Found	Polymorphisms Where as yet No Quantitative Differences Have Been Reported
Glucose-6-phosphate dehydrogenase	Adenosine deaminase
Red-cell acid phosphatase	Peptidase D (prolidase)
Phosphogluconate dehydrogenase	Pancreatic amylase
Adenylate kinase	Pepsinogen
Placental alkaline phosphatase	Glutamate-oxalate transaminase
Peptidase A	Phosphoglucomutase
Peptidase C	Locus PGM_1
Galactose-1-phosphate uridyl transferase	Locus PGM_3
Glutathione reductase	
Liver acetyl transferase	
Red cell NADase	
Salivary amylase	
Serum cholinesterase	
Locus E_1	
Locus E_2	
Alcohol dehydrogenase	
Locus ADH_2	
Locus ADH_3	

From H. Harris, *J. Med. Genetics* **8**:444 (1971).

indicate that amino-acid differences between various cytochrome c's do not affect the function of this protein, suggesting that the amino acid substitutions derive from neutral mutations. It has been shown more recently, however, that the various cytochromes are not interchangeable in supporting mitochondrial ion transport, a process that is probably far more sensitive to protein conformation than the simple flow of electrons to oxygen, nor do they associate with cytochrome oxidases with equivalent affinities. These results have been cited as discounting the neutrality of cytochrome c amino-acid substitutions.

Until very recently it was believed that at least one type of mutation would surely be neutral, namely, a "synonymous" mutation in which one codon is replaced by another codon dictating the same amino acid (Section 9.12). As both prokaryotic and eukaryotic genes have begun to be sequenced, however, it has become clear that synonymous codons are by no means used interchangeably. Thus the ϕX174 chromosome exhibits clear biases for certain codons over others, and the same has been found for insects and vertebrates. One explanation for such imbalances is that certain types of isoaccepting tRNAs (Section 8.10) are more prevalent than others in that species. Thus for an insect gene in which GAA is used to specify glutamic acid far more often than GAG, its expression may occur in cells with high levels of $tRNA_{GAA}^{Glu}$ relative to $tRNA_{GAG}^{Glu}$. For such a gene, the substitution of a GAG for a GAA codon might well modify the rate at which its mRNA product were translated into protein. Such a mutation, in other words, might not be selectively neutral at all. In general, therefore, even advocates of the "neutralist" position agree that it will be very difficult to prove that any one particular mutation is in fact neutral.

Allelic Frequencies in Isolated Populations   The second critical question raised by the Neutral Mutation-Random Genetic Drift hypothesis is whether isolated populations of the same species exhibit similar or dissimilar patterns of allelism. To answer this question, one reasons as follows: if polymorphism arises because various neutral alleles are drifting toward elimination or fixation in a random manner, there should be a randomness in the allelism observed from one population to the next; thus in one population, allele a_1 might be enjoying considerable "luck" and might be present at relatively high frequencies, whereas in another population it might be proceeding toward elimination and be present in low frequencies. If, on the other hand, polymorphic alleles are not neutral and selection is instead maintaining the polymorphism, allelic frequencies should be similar from one population to the next (assuming that all of the populations are experiencing similar kinds of selective pressures).

F. Ayala and his associates tested this question in 10 natural populations of *Drosophila willistonii* taken from diverse locations in the western hemisphere. They looked for isoelectric variants of 28 different kinds of enzymes and defined a locus as polymorphic when the frequency of its most common allele was no greater than 0.95. From this they estimated that about 58 percent of the *D. willistonii* enzyme loci were polymorphic, a result we have by now come to expect. What was of particular interest was the *pattern* of the allelic frequencies for these loci. Ayala found that when an allele was prevalent in one population it was, with high correlation, prevalent in the other nine populations as well; similarly, a rare allele was rare in all 10 populations.

Ayala's results do not support the Neutral Mutation-Random Genetic Drift hypothesis in the short-term sense, but it might be argued that not enough time has elapsed for gene frequencies to change by drift since the time the *Drosophila* populations became geographically separated. Thus the question of the validity of the hypothesis again remains open and will probably continue to be open until many more experiments are performed.

SPECIES FORMATION

Two populations can be said to belong to two species when their genetic divergence is sufficient to produce **reproductive isolation:** individuals from the two populations either fail to mate spontaneously or else fail to produce fertile offspring should a mating occur. A great deal is known about evolutionary patterns and relationships between species, and, although this material extends well beyond the scope of this chapter, it receives excellent coverage in the several books cited at the end of the chapter. We confine our attention here to possible mechanisms by which the polymorphism in a population is translated into the formation of species.

20.9   SPECIATION IN SMALL POPULATIONS

Certainly the most direct way to imagine the creation of a new species is by the invasion of an isolated location by a small number of individuals. As discussed in detail in Section 19.18, a combination of the founder effect and random genetic drift can change the allelic frequencies of such a "splinter population" so that they are very different from the parent population. Moreover, the selection pressures in the new environment (a sandy beach) may differ markedly from the old (a rocky shore). Should genes affecting aspects of reproduction be among those affected by selection or drift, new species may rapidly evolve.

20.10   SPECIATION BY ECOLOGICAL ISOLATION OF POPULATIONS

In Section 20.3 we described diversifying selection as creating two (or more) coexisting groups of phenotypes within a population, each adapted to different niches in the local environment. If the different niches do not serve to isolate the two phenotypes from one another, it is unlikely that this situation will contribute to anything more than a balanced polymorphism in the overall population. If, however, individuals within each locale begin to mate preferentially with one another (a phenomenon known as **assortative mating** or **inbreeding**) so that two different gamete pools are established, the stage is set for speciation.

This mechanism of speciation differs in several respects from the founder mechanism discussed above. First, the two isolated groups would presumably start out by possessing a random sampling of all the alleles in the parental population. Second, genetic drift would not necessarily be an important factor unless one of the isolates were very small (in which case it might more appropriately be considered a founder colony). And finally, the rates of speciation would be expected to be less dramatic, particularly if the two populations occupied similar environments. Indeed, in theory, the two populations might never diverge into two species at all. Required would be the accumulation, by a combination of mutation pressure and various modes of selection, of sufficient differences that the two populations would fail to interbreed if reunited.

20.11   INTRASPECIES VERSUS INTERSPECIES VARIATION

We arrive at a central question in evolutionary genetics theory: what genetic differences actually produce the reproductive and sexual isolation that distinguish one species from a closely related species? Must two evolving species differ at many gene loci, or can changes at a very few "critical" gene loci accomplish the same effect?

The difficulty in answering this question becomes apparent in the following example. We learned in Ayala's study (Section 20.8) that when 10 different populations of the same species of *Drosophila willistonii* were compared, similar allelic patterns were observed for all. In a parallel study Ayala and colleagues instead compared allelic frequencies of four quite distinct species of *Drosophila* and obtained quite different results: for most loci the allelic patterns were found to be significantly different from one species to the next (Table 20-3). From Ayala's research, therefore, one might be drawn to conclude that the process of speciation—at least in *Drosophila*—has involved dramatic fluctuations in allelic frequencies, even though the overall percentage of heterozygosity is about the same from one species to the next. The difficulty with such a conclusion is that the initial event(s) in creating the four distinct *Drosophila* species occurred long before the Ayala study was initiated. Therefore, there is no way to determine whether the present-day differences in allelic frequencies represent the *cause* of species formation or whether they are instead the indirect *consequence* of species formation. One could, for example, postulate that a few loci directly involved with such phenotypes as mating behavior, genital development, or odor formation were instrumental in creating the four species, with most loci "drifting" through the speciation process and stabilizing at quite different levels on the other side.

Needed, therefore, are extensive genetic data on a population before and after a speciation event, plus an understanding of the mutational and selection pressures at work during a speciation event. Such information will be difficult indeed to obtain: the geneticist cannot anticipate "natural"

Table 20-3 Frequencies of Alleles at Twelve "Diagnostic" Loci in Four Species of *Drosophila*. Several Alleles That Occur with Low Frequencies Are Not Included. A Dash Indicates That the Allele Has Not Been Found in the Species

Gene	Alleles	*D. willistoni*	*D. tropicalis*	*D. equinoxialis*	*D. paulistorum*
Lap-5	0.98	0.09	0.02	–	–
	1.00	0.29	0.19	–	–
	1.03	0.50	0.63	0.004	0.004
	1.05	0.09	0.15	0.21	0.08
	1.07	0.007	0.01	0.71	0.86
	1.09	–	–	0.07	0.04
Est-5	0.95	0.03	–	0.03	0.03
	1.00	0.96	–	0.94	0.84
	1.05	0.01	–	0.02	0.13
Est-7	0.96	0.02	0.02	–	–
	0.98	0.16	0.11	–	–
	1.00	0.54	0.62	–	0.002
	1.02	0.23	0.23	–	0.08

Table 20-3 (continued)

Gene	Alleles	*D. willistoni*	*D. tropicalis*	*D. equinoxialis*	*D. paulistorum*
	1.05	0.05	0.03	–	0.78
	1.07	0.003	0.001	–	0.09
Aph-1	0.98	0.02	–	–	–
	1.00	0.84	0.05	0.02	0.01
	1.02	0.08	0.90	0.92	0.93
	1.04	0.06	0.04	0.06	0.03
Acph-1	0.94	0.05	0.95	0.01	–
	1.00	0.92	0.03	0.17	–
	1.04	0.02	0.006	0.81	0.16
	1.06	–	–	–	0.21
	1.08	–	–	0.004	0.62
Mdh-2	0.86	0.001	0.994	0.003	0.001
	0.94	0.02	0.005	0.994	0.993
	1.00	0.97	–	0.004	0.006
Me-1	0.90	–	0.03	–	–
	0.94	–	0.91	–	0.004
	0.98	0.02	0.06	–	0.99
	1.00	0.95	–	0.005	0.005
	1.04	0.02	–	0.99	–
Tpi-2	0.94	0.003	0.01	–	0.02
	1.00	0.98	0.98	0.02	0.98
	1.06	0.01	0.01	0.98	–
Pgm-1	0.96	0.04	–	0.01	0.02
	1.00	0.87	0.01	0.35	0.94
	1.04	0.08	0.98	0.62	0.04
Adk-2	0.96	0.01	0.05	–	–
	0.98	0.05	–	–	–
	1.00	0.88	0.92	0.04	0.98
	1.02	0.05	–	–	–
	1.04	0.004	0.03	0.94	0.02
Hk-1	0.96	0.04	0.02	0.08	–
	1.00	0.95	0.96	0.91	0.01
	1.04	0.006	0.02	0.005	0.97
	1.08	–	0.001	0.002	0.02
Hk-3	1.00	0.98	0.97	0.95	0.07
	1.04	0.006	0.01	0.04	0.92

Symbols for the genes are: *Lap* = leucine aminopeptidase; *Est* = esterase; *Aph* = alkaline phosphatase; *Acph* = acid phosphatase; *Mdh* = malic dehydrogenase; *Me* = malic enzyme; *Tpi* = triose phosphate isomerase; *Pgm* = phosphoglucomutase; *Adk* = adenylate kinase; *Hk* = hexokinase. Allele numbers refer to relative rates of migration in electrophoresis studies.

Data of F. Ayala and J. Powell, *Proc. Natl. Acad. Sci. U. S.* **69:**109 (1972).

speciation events in advance, and if speciation events are instead "induced" in some way, it can be argued that the geneticist is in fact studying artificial selection for reproductive isolation and not the speciation process at all.

20.12 SPECIATION BY INTERSPECIFIC HYBRIDIZATION

The potential for speciation by interspecific hybridization arises when the normal barriers that isolate two species from one another fail to operate. These barriers include differences in habitat, in mating calls, in odor, in morphology of genetalia, in flowering time, and so on. Furthermore, two distinct species will only rarely make a mistake and mate with one another under natural conditions. Equally rare is the likelihood that the hybrid zygote of such a mating will develop into a fertile adult. A hybrid adult, moreover, may not be able to find a compatible mate in the population, although this problem is far less serious for plants (where self-pollination may be possible) than for animals. Should a rare series of events occur, however, in which a fertile interspecific hybrid is formed, a new kind of organism may result, one that could well be the progenitor of a new species.

Speciation by interspecific hybridization has in fact been an important factor in the evolution of plants, particularly the angiosperms. In such a case, as noted in Section 6.18, the formation of a hybrid plant will often be accompanied by a doubling of the ploidy of each parental chromosome set so that proper pairing of homologues can occur in the ensuing production of gametes, a phenomenon known as **allopolyploidy.** A detailed description of evolution in the angiosperms is given in the text of L. Stebbins cited at the end of this chapter.

MOLECULAR EVOLUTION

The branch of genetics known as **molecular evolution** seeks to understand evolutionary relationships between taxonomic groups by comparing the structures of their chromosomes, genes, and gene products. The goal of such studies is to explain how modern genomes evolved from the primitive. Since primitive genomes have left no fossil record, such theories are of necessity based on extrapolations from present-day genomes.

20.13 DUPLICATIONS IN EVOLUTION

It is generally agreed that duplications of informational DNA, be they duplications of genes, of chromosomal segments, or of chromosomes themselves (the last more correctly called changes in ploidy), have been of major importance in evolution. As noted on several occasions in this text, duplicated DNA can be presumed to enjoy a kind of selective "immunity" that vital structural genes cannot experience as long as they are present in only haploid or diploid dosages. As duplicated DNA regions acquire point mutations without affecting the viability of their "hosts," new (and potentially important) genes may evolve from them. Examples of gene-evolution-by-duplication include the rRNA, histocompatibility, and immunoglobulin gene families that are extensively described in Chapter 15 (Sections 15.9 through 15.22), as well as the changes in ploidy that have contributed strongly to the evolution of the higher plants (Section 6.18).

20.14 CHROMOSOMAL REARRANGEMENTS IN EVOLUTION

Chromosomal rearrangements bring about quite a different effect from duplications; they either bring together DNA regions that evolved separately, or else separate once-contiguous sectors of the genome. Again drawing on material presented earlier in the text, we can recall the effects on gene expression brought about by translocations (Section 18.18). We can also recall the postulated role of inversions in keeping together genes important for a particular adaptation (Section 19.17). These inversions may well become the "correct" gene order if the inversion-carrying chromosome comes to stabilize in a newly evolved species.

A pertinent example of chromosomal rearrangement in evolution is found in the primates. Figure 20-4 compares the karyotype of the human with those of the chimpanzee, gorilla, and orangutan. Almost every band present in the human chromosomes is found to have a counterpart in the great-ape chromosomes. At least 8 major inversions distinguish the human and chimpanzee karyotypes, however. In addition, the human chromosome 2 appears to have originated as the result of a fusion between two small acrocentric chromosomes found in the hominoids, resulting in a reduction in haploid chromosome number from 24 to 23. Whether these chromosomal rearrangements played any role in human evolution is not, of course, known.

20.15 PROTEIN EVOLUTION AND THE MOLECULAR CLOCK

Particularly detailed data on patterns of molecular evolution comes from comparing the amino-acid sequences of proteins that are shared by diverse taxonomic groups. Most extensively studied have been cytochrome c (present in all eukaryotes) and hemoglobin (present in all vertebrates and related to more primitive globin polypeptides). A protein such as hemoglobin is typically polymorphic within a species, as we saw for humans (Section 15.9). In the analyses we are about to describe, however, such variations are ignored and the most prominent version of hemoglobin in one species is singled out. Its amino acid sequence is then compared with the amino acid sequence of the hemoglobin in another species, and the

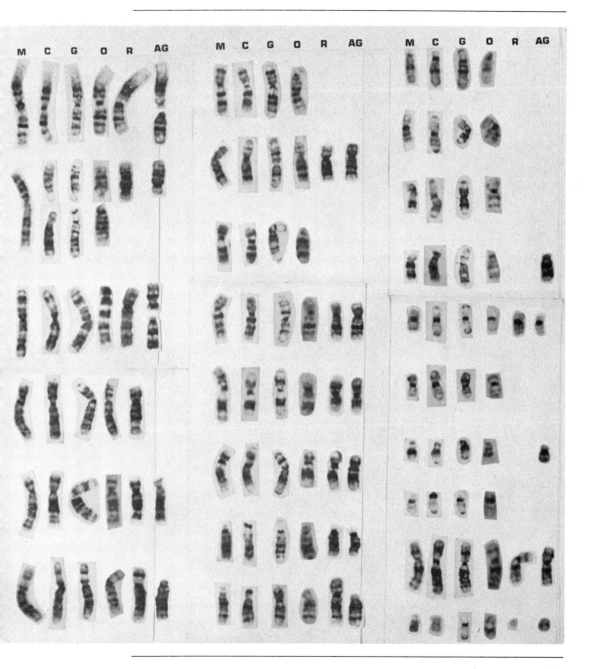

Figure 20-4 Giemsa-banded chromosomes of humans and the great apes arranged to show homology with the human karyotype. The human chromosomes are on the left (M), followed by those of chimpanzee (C), gorilla (G), orangutan (O), Rhesus monkey (R), and African Green monkey (AG), respectively. (Courtesy of Dr. J. J. Garver, Instituut voor Anthropogenetica, Leiden.)

minimal number of nucleotide replacements required to shift from one amino-acid sequence to the other is calculated.

Figure 20-5 shows a resultant "genealogy" of globin polypeptides, where the lengths in each linking line represent the numbers of nucleotide replacements deduced as having occurred between adjacent ancestor and descendant sequences. The line from a primitive globin to the vertebrate β-α hemoglobin ancestor is depicted as having given off, along the way, annelid, insect, mollusc, lamprey, and mammalian myoglobin, and the entire genealogy entails 1629 nucleotide replacements. Shown on the ordinate is a time scale, in millions of years, based on paleontological versions of metazoan evolution, and it is seen that the "protein clock" and the "paleontological clock" agree closely.

From such genealogies it becomes possible to estimate nucleotide replacement rates for hemoglobin genes during various evolutionary periods, and here interesting patterns emerge. A very rapid rate of nucleotide replacement is calculated to have occurred during the period when the

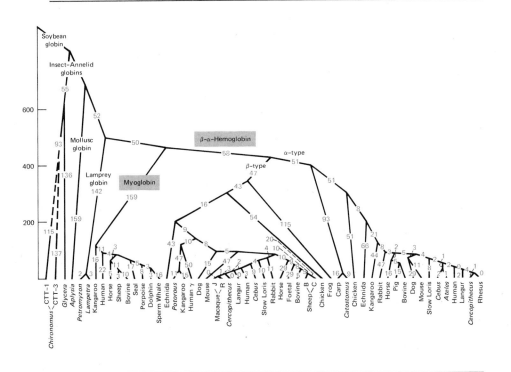

Figure 20-5 Globin genealogical tree. Link lengths are the numbers of nucleotide replacements between adjacent ancestor and descendant sequences. The ordinate is a time scale in Myr based on paleontological views concerning the ancestral separations of the organisms from which the globins came. (From M. Goodman, G. W. Moore, and G. Matsuda. *Nature* **253**:603, 1975).

primitive globin gene, whose polypeptide product functions as a monomer, evolved into a gene whose polypeptides associate as a tetramer capable of cooperative functions. Moreover, the amino acids affected during this "burst" of replacements appear to have been those involved with heme contacts and the interchain cooperativity which facilitates oxygen binding.

A second period of rapid nucleotide replacement is also deduced from the hemoglobin genealogies. This appears to have occurred about 500 million years ago, at the time of the postulated α-β gene duplication, and again a functional interpretation can be given. The β_4 tetramer is adaptively inferior to an $\alpha_2\beta_2$ tetramer, and it is proposed that following the gene duplication, the newly formed α gene accumulated a sufficient number of nucleotide substitutions to generate a novel α polypeptide which could combine with the β chain.

In summary, therefore, the two peak periods for nucleotide substitutions in hemoglobin genes can be correlated with periods in which natural selection was presumably intensively operating to produce optimal versions of α and β chains. During the 300 million years since the α-β divergence, nucleotide replacement rates in the two genes have fallen from a peak of 109 to an average of 15 per 100 codons per 10^8 years, and these more "conservative" replacements are claimed to have created α and β chains that are increasingly better adapted in the transport and release of molecular oxygen. Thus molecular evolutionists at present seem to favor the notion that natural selection has strongly guided protein evolution.

References

Polymorphism

Cobb, G. "Polymorphisms for dimerizing ability at the esterase-5 locus in Drosophila pseudoobscura," Genetics **82:**53–62 (1976).

Dobzhansky, T., and F. J. Ayala. "Temporal frequency changes of enzyme and chromosomal polymorphisms in natural populations of Drosophila," Proc. Natl. Acad. Sci. U.S. **70:**680–683 (1973).

*Hubby, J. L., and R. C. Lewontin. "A molecular approach to the study of genetic heterozygosity in natural populations. I. The number of alleles at different loci in Drosophila pseudoobscura," Genetics **54:**577–594 (1966).

Lewontin, R. C. "An estimate of average heterozygosity in man," Am. J. Hum. Genet. **19:**681–685 (1967).

*Lewontin, R. C., and J. L. Hubby. "A molecular approach to the study of genic heterozygosity in natural populations. II. Amount of variation and degree of heterozygosity in natural populations of Drosophila pseudoobscura," Genetics **54:**595–609 (1966).

Lubbs, H. S., and F. H. Ruddle. "Chromosomal polymorphism in American Negro and white populations," Nature **233:**134–136 (1971).

McDowell, R. E., and S. Prakash "Allelic heterogeneity within allozymes separated by electrophoresis in Drosophila pseudoobscura," Proc. Natl. Acad. Sci. U.S. **73:**4150–4153 (1976).

*Powell, J. R. "Genetic polymorphisms in varied environments," Science **174:**1035–1036 (1971).

*Denotes articles described specifically in the chapter.

Ramot, B. *Genetic Polymorphisms and Diseases in Man.* New York: Academic Press, 1974.

Ranney, H. M. "Clinically important variants of human hemoglobin," *New England J. Med.* **282**:144–152 (1970).

Steinberg, A. G. "Globulin polymorphisms in man," *Ann. Rev. Genetics* **3**:25–52 (1969).

Balancing Selection and Heterosis

*Allison, A. C. "Protection afforded by sickle-cell trait against subterian malarial infection," *Brit. Med. J.* **1**:290–294 (1954). [Reprinted in L. Levine, *Papers on Genetics.* St. Louis: Mosby, 1971.]

Allison, A. C. "Population genetics of abnormal human hemoglobins," *Acta Genetica* **6**:430–434 (1956). [Reprinted in E. B. Spiess, Ed., *Papers on Animal Population Genetics.* Boston: Little, Brown, 1962.]

Gilpin, M., M. Soulé, A. Ondriak, and E. A. Gilpin. "Overdominance and U-shaped gene frequency distributions," *Nature* **263**:497–499. (1976).

Gowen, J. W., Ed. *Heterosis.* Ames, Iowa State College Press, 1952.

*Knudson, A. G., L. Wayne, and W. Y. Hallett. "On the selective advantage of cystic fibrosis heterozygotes," *Am. J. Human Genetics* **19**:388–398 (1967).

Livingston, F. B. "Malaria and human polymorphisms," *Ann. Rev. Genetics* **5**:33–64 (1971).

Morgan, P. "Frequency-dependent selection at two enzyme loci in *Drosophila melanogaster,*" *Nature* **263**:765–766 (1976).

*Myrianthopoulos, N. C., and S. M. Aronson. "Population dynamics of Tay-Sachs disease. I. Reproductive fitness and selection," *Am. J. Human Genetics* **18**:313–327 (1966).

Singh, R. S., J. L. Hubby, and R. C. Lewontin. "Molecular heterosis for heat-sensitive enzyme alleles," *Proc. Natl. Acad. Sci. U.S.* **71**:1808–1810 (1974).

Taylor, C. E. "Genetic variation in heterogeneous environments," *Genetics* **83**:887–894 (1976).

Thoday, J. M., and J. B. Gibson. "The probability of isolation by disruptive selection," *Am. Nat.* **104**:219–230 (1970).

Willis, C., and L. Nichols. "Single-gene heterosis in *Drosophila* revealed by inbreeding,"*Nature* **233**:123–125 (1971).

Zouros, E. "Hybrid molecules and the superiority of the heterozygote," *Nature* **262**:227–230 (1976).

Neutral Mutation-Random Genetic Drift

*Ayala, F. J., J. R. Powell, M. L. Tracey, C. A. Mourão, and S. Pérez-Salas. "Enzyme variability in the *Drosophila willistoni* group. IV. Genic variation in natural populations of *Drosophila willistoni,*" *Genetics* **70**:113–139 (1972).

Bryson, V., and J. J. Vogel, Eds. *Evolving Genes and Proteins.* New York: Academic, 1965.

Cavalli-Sforza, L. L., I. Barrai, and A. W. F. Edwards. "Analysis of human evolution under random genetic drift," *Cold Spring Harbor Symp. Quant. Biol.* **29**:9–20 (1965).

Crow, J. F. "The theory of neutral and weakly selected genes," *Fed. Proc.* **35**:2083–2086 (1976).

DeJong, G., and W. Scharloo. "Environmental determination of selective significance of neutrality of amylase variants in *Drosophila melanogaster,*" *Genetics* **84**:77–94 (1976).

Harris, H. "Molecular evolution: the neutralist-selectionist controversy," *Fed. Proc.* **35**:2079–2082 (1976).

King, J. L. "Progress in the neutral mutation-random genetic drift hypothesis," *Fed. Proc.* **35**:2087–2091 (1976).

*Kimura, M. "Evolutionary rate at the molecular level," *Nature* **217**:624–626 (1968).

*King, J. L., and T. H. Jukes. "Non-Darwinian evolution," *Science* **164**:788–798 (1969).

Laird, C. D., B. L. McConaughy, and B. J. McCarthy. "Rate of fixation of nucleotide substitutions in evolution," *Nature* **224**:149–154 (1969).

Ohta, T., and M. Kimura. "Amino acid composition of proteins as a product of molecular evolution," *Science* **174**:150–152 (1971).

Thoday, J. M. "Non-Darwinian "evolution" and biological progress," *Nature* **255**:675–677 (1975).

Speciation and Human Evolution

*Ayala, F. J., C. A. Mourão, S. Pérez-Salas, R. Richmond, and T. Dobzhansky. "Enzyme variability in the *Drosophila willistoni* group. I. Genetic differentiation among sibling species," *Proc. Natl. Acad. Sci. U.S.* **67**:225–232 (1970).

*Cavalli-Sforza, L. L., and W. F. Bodmer. *The Genetics of Human Populations.* San Francisco: Freeman, 1971.

Carlson, H. L. "Speciation and the founder principle," *Stadler Genet. Symp.* **3**:51–70 (1971).

Dobzhansky, T. *Genetics and the Origin of Species,* 3rd ed., New York: Columbia University Press, 1951.

Dobzhansky, T. *Genetics of the Evolutionary Process.* New York: Columbia University Press, 1970.

Dobzhansky, T., and C. Epling. *Contributions to the Genetics, Taxonomy, and Ecology of Drosophila pseudoobscura and its relatives.* Carnegie Institution of Washington Publication 554, 1944.

Ehrlich, P. R., R. W. Holun, and P. H. Raven, Eds. *Papers on Evolution,* Boston: Little, Brown, 1969.

Jacobs, P. A., W. H. Price, and P. Law, Ed. *Human Population Cytogenetics,* Baltimore: Williams and Wilkins, 1970.

Lerner, I. M. *Heredity, Evolution, and Society.* San Francisco: Freeman, 1968.

Mayr, E. *Population, Species, and Evolution.* Cambridge: Harvard University Press, 1970.

McKusick, V. A. "Human genetics," *Ann. Rev. Genetics* **4**:1–46 (1970).

Schull, W. J., and J. W. MacCluer. "Human genetics: structure of population," *Ann. Rev. Genetics* **2**:279–304 (1968).

Stebbins, G. L. *Variation and Evolution in Plants.* New York: Columbia University Press, 1950.

Stern, C. *Principles of Human Genetics,* 3rd ed. San Francisco: Freeman, 1975.

Tracey, M. L., and S. A. Espinet. "Sex chromosome translocations and speciation," *Nature* **263**:321–323 (1976).

Molecular Evolution

Benveniste, R. E., and G. J. Todaro. "Evolution of type C viral genes: evidence for an Asian origin of man," *Nature* **261**:101–108 (1976).

Blundell, T. L., and S. P. Wood. "Is the evolution of insulin Darwinian or due to selectively neutral mutation?" *Nature* **257**:197–203 (1975).

Dayhoff, M. O. "The origin and evolution of protein superfamilies," *Fed. Proc.* **35**:2132–2138 (1976).

Dayhoff, M. A. "Atlas of protein sequence and structure," Silver Spring, Md., Natl. Biomed. Research Foun., 1969.

*Dayhoff, M. A. "Computer analysis of protein evolution," *Sci. Am.* **221**:87–95 (1969).

*Fitch, W. M., and E. Margoliash. "Construction of phylogenetic trees," *Science* **155**:229–284 (1967).

Fitch, W. M., and C. H. Langley. "Protein evolution and the molecular clock," *Fed. Proc.* **35**:2092–2097 (1976).

Ford, P. J., and R. D. Brown. "Sequences of 5S ribosomal RNA from *Xenopus mulleri* and the evolution of 5S gene-coding sequences," *Cell* **8**:485–493 (1976).

*Goodman, M., G. W. Moore, and G. Matsuda. "Darwinian evolution in the genealogy of hemoglobin," *Nature* **253**:603–608 (1975).

Holmquist, G. "Organisation and evolution of *Drosophila virilis* heterochromatin," *Nature* **257**:503–506 (1975).

Koshland, D. E., Jr. "The evolution of function in enzymes," *Fed. Proc.* **35**:2104–2111 (1976).

Margoliash, E., S. Ferguson-Miller, C. H. Kang, and D. L. Brautigan. "Do evolutionary changes in cytochrome c structures reflect functional adaptations? *Fed. Proc.* **35**:2124–2130 (1976).

(a)

(a) A group of population geneticists at a not-so-recent meeting. *Bottom row* (left to right): Daniel Marien (Queens College); Richard Lewontin (Harvard University); Lee Ehrman (State University of New York, Purchase); Theodosius Dobzhansky (Rockefeller University and University of California, Davis); Howard Levene (Columbia University); and Francisco Ayala (University of California, Davis) *Middle row* (left to right): Sergey Polivanov (Catholic University of America); Bruce Wallace (Cornell University); Louis Levine (City College of New York); Irwin Herzkowitz (Hunter College); Marvin Druger (Syracuse University); and Timothy Prout (University of California, Riverside). *Top:* Richard Levins (left) (Harvard University) and Leigh VanValen (right) (University of Chicago).

Moore, G. W., M. Goodman, C. Callahan, R. Holmquist, and H. Moise. "Stochastic *versus* augmented maximum parsimony method for estimating superimposed mutations in the divergent evolution of protein sequences. Methods tested on cytochrome c amino acid sequences," *J. Mol. Biol.* **105**:15–37 (1976) (and following article).

Ohno, S. "Ancient linkage groups and frozen accidents," *Nature* **244**:259–262 (1973).

Questions and Problems

1. Would the Neutral-Mutation Random Genetic Drift hypothesis expect the discontinuous rates of nucleotide replacements in hemoglobin genes postulated to have occurred during phylogeny? Explain.

2. Histone polypeptides are very highly "conserved," being extremely similar in amino-acid sequence throughout the modern eukaryotes. The fibrinopeptides, on the other hand, which act as small structural elements in blood clots, are extremely diverse in amino-acid sequence even in closely related organisms (Figure 19-2). How would you explain these two statistics?

3. R. Milkman has reported that natural populations of *E. coli* have levels of polymorphism comparable to *Drosophila* and humans. Discuss this finding in terms of the theory that polymorphism is maintained by heterosis.

(b)

(b) A group of population geneticists at a more recent meeting. *Bottom row* (left to right): Sergey Polivanov (Catholic University of America); Theodosius Dobzhansky, Elliot Spiess (University of Illinois, Chicago Circle); Francisco J. Ayala (University of California, Davis); *Middle row* (left to right): Jeffrey R. Powell (Yale University); Wyatt Anderson (University of Georgia); Abd-El Khalek Mourat (University of Cairo, Egypt); Louis Levine (City College of New York); *Top row* (left to right): Kirshna Sankaranarayanan (University of Leiden, Netherlands); Timothy Prout (University of California, Davis); Rollin Richmond (North Carolina State University); George Carmody (Carleton University, Ottawa).

4. Define and distinguish between the following terms; balanced polymorphism; balancing selection; disruptive selection; heterotic selection; generalized heterosis; balancing genetic load.

5. Based on the frequency of the sickle-cell gene in African and Asian populations, what is the proportion of infants who suffer from sickle-cell anemia per generation in these populations? What proportion have sickle-trait?

6. The allele g exhibits a stable frequency of 15 percent in a mouse population, even though the newborn survival of g/g homozygotes is half that of G/g heterozygotes. Assuming the allele to be maintained by balancing selection, calculate the relative fitness of g/g, G/g, and G/G mice. What is the selection coefficient for each genotype?

7. Mice of the H/h genotype have twice the litter size of H/H mice and four times the litter size of h/h mice. What is the predicted equilibrium frequency of the h gene assuming balancing selection.

8. Derive equation 1 from equation 9 of Chapter 19 as suggested in the text.

9. Individuals from two closely related species can be induced to mate but sterility is the rule, and the occasional offspring produced are usually inviable. Explain this result in terms of the hypothesis that chromosome translocations have contributed to the speciation process.

10. Explain the data for the Hb^s gene in Table 19-4.

Author Index

Subject Index

	pages	# pages	‡
chap 12	442 - 490	48	/	Skimmed
chap 14	540 - 567	27	✓
chap 15	p 574 - 610	36	✓

cloning   fig
14.6,7   recombinant DNA   14.3,4

14.13   petite mutations

14.17   endosymbiosis

12.4   2-factor crosses
12.5   3-factor crosses   fig. 12.4
12.6   double crossovers
12.14   Neurospora
12-19   Somatic cell genetics (hybrids)

15.19   Integration of V and C regions
15.13
15.4   penetrance + expressivity
15.2   ABO blood group

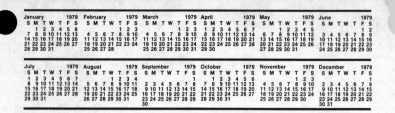

January 1979	February 1979	March 1979	April 1979	May 1979	June 1979
S M T W T F S	S M T W T F S	S M T W T F S	S M T W T F S	S M T W T F S	S M T W T F S
1 2 3 4 5 6	1 2 3	1 2 3	1 2 3 4 5 6 7	1 2 3 4 5	1 2
7 8 9 10 11 12 13	4 5 6 7 8 9 10	4 5 6 7 8 9 10	8 9 10 11 12 13 14	6 7 8 9 10 11 12	3 4 5 6 7 8 9
14 15 16 17 18 19 20	11 12 13 14 15 16 17	11 12 13 14 15 16 17	15 16 17 18 19 20 21	13 14 15 16 17 18 19	10 11 12 13 14 15 16
21 22 23 24 25 26 27	18 19 20 21 22 23 24	18 19 20 21 22 23 24	22 23 24 25 26 27 28	20 21 22 23 24 25 26	17 18 19 20 21 22 23
28 29 30 31	25 26 27 28	25 26 27 28 29 30 31	29 30	27 28 29 30 31	24 25 26 27 28 29 30

July 1979	August 1979	September 1979	October 1979	November 1979	December 1979
S M T W T F S	S M T W T F S	S M T W T F S	S M T W T F S	S M T W T F S	S M T W T F S
1 2 3 4 5 6 7	1 2 3 4	1	1 2 3 4 5 6	1 2 3	1
8 9 10 11 12 13 14	5 6 7 8 9 10 11	2 3 4 5 6 7 8	7 8 9 10 11 12 13	4 5 6 7 8 9 10	2 3 4 5 6 7 8
15 16 17 18 19 20 21	12 13 14 15 16 17 18	9 10 11 12 13 14 15	14 15 16 17 18 19 20	11 12 13 14 15 16 17	9 10 11 12 13 14 15
22 23 24 25 26 27 28	19 20 21 22 23 24 25	16 17 18 19 20 21 22	21 22 23 24 25 26 27	18 19 20 21 22 23 24	16 17 18 19 20 21 22
29 30 31	26 27 28 29 30 31	23 24 25 26 27 28 29	28 29 30 31	25 26 27 28 29 30	23 24 25 26 27 28 29
		30			30 31

8

Thursday
November
1979

8:00	Who — Hooligans
8:30	Loverboy — Loverboy + Get Lucky .... Mark
9:00	Bob Seger — Live Bullet .... Mark
9:30	Kinks — Lola
10:00	Bob Dylan — Greatest Hits .... Donna
10:30	Steely Dan — Greatest .... Zippy
11:00	AC/DC — Dirty Deeds....
11:30	
12:00	
1:00	
1:30	
2:00	
2:30	
3:00	
3:30	
4:00	
4:30	
5:00	
5:30	

October 1979	November 1979	December 1979
S M T W T F S	S M T W T F S	S M T W T F S
1 2 3 4 5 6	1 2 3	1
7 8 9 10 11 12 13	4 5 6 7 8 9 10	2 3 4 5 6 7 8
14 15 16 17 18 19 20	11 12 13 14 15 16 17	9 10 11 12 13 14 15
21 22 23 24 25 26 27	18 19 20 21 22 23 24	16 17 18 19 20 21 22
28 29 30 31	25 26 27 28 29 30	23 24 25 26 27 28 29
		30 31